Homology
Effects

Advances in Genetics

Homology Effects

Edited by

Jay C. Dunlap

Department of Genetics
Dartmouth Medical School
Hanover, New Hampshire

C.-ting Wu

Department of Genetics
Harvard Medical School
Boston, Massachusetts

ACADEMIC PRESS

An Elsevier Science Imprint

San Diego   San Francisco   New York
Boston   London   Sydney   Tokyo

Academic Press
An Elsevier Science Imprint.
525 B Street, Suite 1900, San Diego, California 92101-4495, USA
http://www.academicpress.com

Academic Press
32 Jamestown Road, London NW1 7BY, UK
http://www.academicpress.com

International Standard Book Number: 0-12-017646-7

PRINTED IN THE UNITED STATES OF AMERICA
02   03   04   05   06   07   EB   9   8   7   6   5   4   3   2   1

Contents

v

3 Homologous Chromosome Associations and Nuclear Order in Meiotic and Mitotically Dividing Cells of Budding Yeast   49

Sean M. Burgess

4 The Role of Sequence Homology in the Repair of DNA Double-Strand Breaks in *Drosophila*   91

Gregory B. Gloor

8 **Homology-Dependent Gene Silencing and Host Defense in Plants** 235
Marjori A. Matzke, Werner Aufsatz, Tatsuo Kanno,
M. Florian Mette, and Antonius J. M. Matzke

9 **Quelling in *Neurospora crassa*** 277
Annette S. Pickford, Caterina Catalanotto, Carlo Cogoni,
and Giuseppe Macino

10 **Non-Mendelian Inheritance and Homology-Dependent Effects in Ciliates** 305
Eric Meyer and Olivier Garnier

17 Prions of Yeast as Epigenetic Phenomena: High Protein "Copy Number" Inducing Protein "Silencing"   485
Reed B. Wickner, Herman K. Edskes, B. Tibor Roberts, Michael Pierce, and Ulrich Baxa

Contributors

Numbers in parentheses indicate the pages on which the authors' contributions begin.

Shoshy Altuvia (361) Department of Molecular Genetics and Biotechnology, The Hebrew University—Hadassah Medical School, 91120 Jerusalem, Israel

Werner Aufsatz (235) Institute of Molecular Biology, Austrian Academy of Sciences, A-5020 Salzburg, Austria

Célia Baroux (165) Institute of Plant Biology, University of Zürich, CH-8008 Zürich, Switzerland

Denise P. Barlow (119) ÖAW Institute of Molecular Biology, A5020 Salzberg, Austria

Ulrich Baxa (485) Laboratory of Biochemistry and Genetics, National Institute of Diabetes, Digestive and Kidney Diseases, National Institutes of Health, Bethesda, Maryland 20892

Sean M. Burgess (49) Division of Biological Sciences, Section of Molecular and Cellular Biology, University of California, Davis, California 95616

Caterina Catalanotto (277) Department of Cellular and Hematologic Biotechnology, Section of Molecular Genetics, Università di Roma "La Sapienza," 00161 Rome, Italy

Vicki L. Chandler (215) Department of Plant Sciences, University of Arizona, Tucson, Arizona 85721

Carlo Cogoni (277) Department of Cellular and Hematologic Biotechnology, Section of Molecular Genetics, Università di Roma "La Sapienza," 00161 Rome, Italy

Jay C. Dunlap (xvii) Department of Genetics, Dartmouth Medical School, Hanover, New Hampshire 03755

Herman K. Edskes (485) Laboratory of Biochemistry and Genetics, National Institute of Diabetes, Digestive and Kidney Diseases, National Institutes of Health, Bethesda, Maryland 20892

Olivier Garnier (305) Molecular Genetics Laboratory (CNRS UMR 8541), Ecole Normale Supérieure, 75005 Paris, France

Gregory B. Gloor (91) Department of Biochemistry, The University of Western Ontario, London, Ontario N6A 5C1, Canada

Alla Grishok (339) Program in Molecular Medicine, University of Massachusetts Medical Center, Worcester, Massachusetts 01605

Ueli Grossniklaus (165) Institute of Plant Biology, University of Zürich, CH-8008 Zürich, Switzerland

Tatsuo Kanno (235) Institute of Molecular Biology, Austrian Academy of Sciences, A-5020 Salzburg, Austria

Judith A. Kassis (421) National Institute of Child Health and Human Development, National Institutes of Health, Bethesda, Maryland 20892

James A. Kennison (399) Section on Drosophila Gene Regulation, Laboratory of Molecular Genetics, National Institute of Child Health and Human Development, National Institutes of Health, Bethesda, Maryland 20892

Mitzi I. Kuroda (1) Department of Molecular and Cellular Biology, Howard Hughes Medical Institute, Baylor College of Medicine, Houston, Texas 77030

Jeannie T. Lee (25) Department of Genetics, Howard Hughes Medical Institute, Harvard Medical School, and Department of Molecular Biology, Massachusetts General Hospital, Boston, Massachusetts 02114

Giuseppe Macino (277) Department of Cellular and Hematologic Biotechnology, Section of Molecular Genetics, Università di Roma "La Sapienza," 00161 Rome, Italy

Antonius J. M. Matzke (235) Institute of Molecular Biology, Austrian Academy of Sciences, A-5020 Salzburg, Austria

Marjori A. Matzke (235) Institute of Molecular Biology, Austrian Academy of Sciences, A-5020 Salzburg, Austria

Victoria H. Meller (1) Department of Biology, Tufts University, Medford, Massachusetts 02155

Craig Mello (339) Program in Molecular Medicine, Howard Hughes Medical Institute, University of Massachusetts Medical Center, Worcester, Massachusetts 01605

M. Florian Mette (235) Institute of Molecular Biology, Austrian Academy of Sciences, A-5020 Salzburg, Austria

Eric Meyer (305) Molecular Genetics Laboratory (CNRS UMR 8541), Ecole Normale Supérieure, 75005 Paris, France

Sarah P. Otto (451) Department of Zoology, University of British Columbia, Vancouver, British Columbia V6T 1Z4, Canada

Annette S. Pickford (277) Department of Cellular and Hematologic Biotechnology, Section of Molecular Genetics, Università di Roma "La Sapienza," 00161 Rome, Italy

Michael Pierce (485) Laboratory of Biochemistry and Genetics, National Institute of Diabetes, Digestive and Kidney Diseases, National Institutes of Health, Bethesda, Maryland 20892

B. Tibor Roberts (485) Laboratory of Biochemistry and Genetics, National Institute of Diabetes, Digestive and Kidney Diseases, National Institutes of Health, Bethesda, Maryland 20892

Pascale Romby (361) UPR 9002, CNRS-IBMC, Institut de Biologie Moléculaire et Cellulaire, 67084 Strasbourg Cedex, France

Eric U. Selker (439) Institute of Molecular Biology, University of Oregon, Eugene, Oregon 97403

Lyudmila V. Sidorenko (215) Department of Plant Sciences, University of Arizona, Tucson, Arizona 85721

Frank Sleutels (119) The Netherlands Cancer Institute, 1066 CX Amsterdam, The Netherlands

Jeffrey W. Southworth (399) Section on Drosophila Gene Regulation, Laboratory of Molecular Genetics, National Institute of Child Health and Human Development, National Institutes of Health, Bethesda, Maryland 20892

Charles Spillane (165) Institute of Plant Biology, University of Zürich, CH-8008 Zürich, Switzerland

Maike Stam (215) Department of Plant Sciences, University of Arizona, Tucson, Arizona 85721*

E. Gerhart H. Wagner (361) Institute of Cell and Molecular Biology, Biomedical Center, Uppsala University, 751 24 Uppsala, Sweden

Reed B. Wickner (485) Laboratory of Biochemistry and Genetics, National Institute of Diabetes, Digestive and Kidney Diseases, National Institutes of Health, Bethesda, Maryland 20892

C.-ting Wu (xvii) Department of Genetics, Harvard Medical School, Boston, Massachusetts 02115

Paul Yong (451) Department of Medicine, University of British Columbia, Vancouver, British Columbia V6T 1Z3, Canada

*Department of Developmental Genetics, Vrije Universiteit, 1081 HV Amsterdam, The Netherlands.

Preface

HOMOLOGY EFFECTS: THE DIFFERENCE BETWEEN 1 AND 2

Much of biology revolves around the number 2, or more precisely, the repercussion of >1. The number 1, by itself, establishes existence, whereas 2 and all other numbers >1 suggest persistence. In this way, the first occurrence of replication was a remarkable achievement; it at least doubled the likelihood that an organic element would endure to the next instant. At the same time, that first event of replication, orchestrated or resulting from accidental duplication, introduced another noteworthy situation, the situation of homology. Homology changed the living world in profound ways. Most fundamentally, it introduced the dichotomy of *"same* or *different?,"* a clear step up from the harsh choice of *"exist* or *vanish?"* for an element present only in the singular. That is, once generated, homologs could either remain identical to each other or diverge, and we urge readers to look at the contribution by Sarah Otto and Paul Yong (Chapter 16) in this volume for discussion of this point. Evolution, over the course of billions of years, took the path of *different* from time to time and ultimately produced a magnificence of diversity.

On the other hand, when *different* was not chosen, homologs were perpetuated. At the level of genetic material, the accumulation of homologs led to diploidy, duplications, and more. From here evolved the sexual process, segregation, recombination, and a slew of other phenomena that rest on the presence of homology. Importantly, the ramifications of homology extended well beyond these mechanics of inheritance to the manner in which genetic material could be expressed and, as significantly, silenced. In short, the circumstance of having two or more homologs eventually afforded organisms an enormously powerful yet flexible set of tools not only for maintaining, transmitting, and rearranging genetic material, but also for regulating the function of that material.

This volume of Advances in Genetics addresses the impact, or effects, of homology. It represents the combined efforts of an international panel of experts who have considered homology from multiple angles and who have been generous with their time and ideas (as well as patient during the assembly of this volume). While some might question the causal relationship between homology and a number of the phenomena included, it is clear that each phenomenon reflects, directly or indirectly, a state of homology. All but one of the contributions focus on homology at the level of DNA and/or RNA, and the one exception, a discussion of prion biology by Reed Wickner, Herman Edskes, B. Tibor Roberts, Michael Pierce, and Ulrich Baxa (Chapter 17), highlights the fact that homology can have an impact well beyond the realm of nucleic acids. Finally, this volume underscores the widespread importance of homology effects by including observations of prokaryotes, fungi, ciliates, nematodes, plants, insects, and mammals.

One of the very earliest observations of homology concerned the discovery that many organisms carry two copies, of each chromosome. Another early discovery concerned the observation of a curious "accessory" chromosome that did not always have a homolog. Then, in 1905, Nettie Stevens and Edmund B. Wilson independently reported this "accessory" chromosome, known today as the X chromosome, to be intimately involved in sex determination (Sturtevant, 1965). To acknowledge this history, the volume begins with two discussions, one by Victoria Meller and Mitzi Kuroda (Chapter 1) and the other by Jeannie Lee (Chapter 2), of how *Drosophila* and mammals, respectively, have reconciled disomy of chromosomes in both females and males with the conspicuous exception of monosomy of the X in males. In *Drosophila,* dosage compensation is achieved through upregulation of the male X, while in mammals, one of the two X chromosomes is inactivated in the female. The most salient target of dosage compensation in these organisms is virtually the entire X chromosome, which underscores how extensive the effects of homology can be. Even knowing this, it may surprise readers to learn that, in some organisms, development involves the selective silencing of whole sets of chromosomes and that this silencing can target either the paternal or maternal set. That parent-of-origin can influence gene function brings us to the realm of genomic imprinting. Here, the reader will learn that, in many organisms, some maternally derived genes express themselves differently from their paternally derived homologs and that the differences often reflect the silencing of one homolog. This extraordinary field is represented in the volume by Frank Sleutels and Denise Barlow (Chapter 5) with regard to mammals and by Célia Baroux, Charles Spillane, and Ueli Grossniklaus (Chapter 6) with regard to plants.

X inactivation and genomic imprinting are similar in that they can cause cells and tissues that are otherwise diploid to become functionally hemizygous for the affected genes. In other words, they lead to monoallelic expression in organisms that are otherwise diploid. Monoallelic expression in a diploid organism can also be achieved through loss of heterozygosity, a term used to describe the outcome of events that cause cells, tissues, or organisms that were once heterozygous for two alleles of a gene to become hemizygous or homozygous for just one allele. An extreme example of loss of heterozygosity would be those diploid insects that undergo developmentally regulated loss of the paternal set of chromosomes. Loss of heterozygosity can also be achieved through loss of a single chromosome, chromosome breakage, deletion, mutation, or generation of the homozygous state by mitotic recombination, chromosome nondisjunction, gene conversion, or repair (reviewed by Wijnhoven *et al.*, 2001). Here, readers are directed to the contribution by Vicki Chandler, Maike Stam, and Lyudmila Sidorenko (Chapter 7) describing paramutation in maize. Paramutation is "an interaction between alleles that leads to a mitotically and meiotically heritable change in gene expression." Significantly, in some cases, paramutation transforms one allele into a state indistinguishable from that of its homolog such that tissue homozygous for one allele arises from cells that were originally heterozygous for two different alleles.

Such a plethora of mechanisms by which diploid organisms express but one of two available homologs at the gene, chromosome, or genome level suggests that there are underlying advantages for monoallelism and loss of heterozygosity, advantages in addition to the immediate functions of these phenomena. As explained by Sarah Otto and Paul Yong (Chapter 16), while the advantages of having duplicate copies of a gene have been widely accepted, arguments can also be made for circumstances in which presence of only one copy of a gene is beneficial. For example, although diploid organisms have a greater chance of masking deleterious mutations, haploid organisms more quickly eliminate deleterious mutations from their population and may therefore gain a higher average fitness (Mable and Otto, 1998). In light of this, it may be that monoallelic expression and loss of heterozygosity permit diploid organisms to approximate, and therefore benefit from, haploid advantage by allowing natural selection to act upon allelic variants that would otherwise be masked by heterozygosity (see also Cook *et al.*, 1998). As some mechanisms of monoallelic expression and loss of heterozygosity can be gene-specific, temporally restricted, or spatially constrained to only clonal patches or specific tissue types (or only one gender, as in the case of X-linked genes in *Drosophila* and mammalian males), the forces of natural selection have the potential of being limited in any one instance to just one gene, one tissue type, or one developmental pathway. That is, monoallelic expression and loss of heterozygosity through restricted gene silencing, loss, transformation, or conversion may further afford diploid organisms the opportunity to benefit from a limited approximation of haploid advantage, allowing them to bypass the risks of a prolonged and/or fully haploid developmental stage.

The concept of diploid organisms approximating haploid advantage has been applied to germline tissues. Here, loss of heterozygosity followed by cell-lineage selection can lead to significant changes in the transmission rates of allelic variants (Otto and Hastings, 1998). Monoallelic expression and loss of heterozygosity in strictly somatic tissues may also affect the transmission of allelic variants, even in the absence of cell-lineage selection, in that somatic clones expressing only one variant may affect the fitness of an individual and thereby influence the probability that a variant becomes fixed in the population (see also Sapienza, 1989; Cook *et al.*, 1998). Finally, clonal monoallelism and loss of heterozygosity in the soma may allow natural selection to act on allelic variants that, although recessive lethal or sterile at the organismal level, are cell viable. Such circumstances may influence the divergence of essential genes as well as allow for the positive selection of variants that are advantageous in only one tissue type or are advantageous in one tissue type but deleterious in another. These potential advantages afforded by monoallelic expression and loss of heterozygosity lead one to wonder whether the percentage of any diploid genome that is monoallelically expressed or silenced, either stably or transiently, in a regulated or stochastic fashion at some time during development might be greater than we currently suspect (for further discussion and other arguments for a greater prevalence of monoallelic expression, see Watanabe and Barlow, 1996; Cook *et al.*, 1998; Nutt and Busslinger, 1999; Ohlsson *et al.*, 2001).

The reader will notice that gene silencing is a recurring theme in the volume. The contribution by Marjori Matzke, Werner Aufsatz, Tatsuo Kanno, M. Florian Mette, and Antonius Matzke (Chapter 8) on homology-dependent gene silencing in plants and other organisms reviews and interprets a burgeoning class of observations revealing that silencing, triggered by the presence of homology, can happen at both the transcriptional as well as posttranscriptional steps of gene expression. Excitingly, much progress has been made regarding the mechanistic basis of homology-dependent gene silencing and readers are urged to consult the contributions by Annette Pickford, Caterina Catalanotto, Carl Cogoni, and Giuseppe Macino (Chapter 9) on quelling in *Neurospora* and by Alla Grishok and Craig Mello (Chapter 11) on RNAi with an emphasis on events in *Caenorhabditis elegans*. It appears that important steps of posttranscriptional silencing can be mediated by RNA, especially aberrant and/or double-stranded RNA, and involve rapid degradation of mRNA in a transcript-specific manner. Lest anyone be tempted to believe that RNA-mediated events are all manifest at the level of transcription, however, we encourage consideration of Chapter 10, by Eric Meyer and Olivier Gannier. Here, the reader will learn about the impact of homology on genome rearrangement in ciliates. Finally, this volume acknowledges the tremendously relevant studies of antisense RNA biology in prokaryotes through the contribution by E. Gerhart Wagner, Shoshy Altuvia, and Pascale Romby (Chapter 12).

The "why" of RNA-mediated gene silencing, be it with regard to its origin or its persistence through evolution, has opened fascinating arenas for speculation. Readers will find compelling arguments for its function as a host defense mechanism that can detect and destroy foreign nucleic acids, such as duplicating transposable elements or invading viral entities. RNA-mediated silencing may also provide an efficient means by which populations can be quickly rid of deleterious allelic variants. This is because RNA-mediated silencing, even when triggered by only a single aberrant gene, can inactivate mRNA from other genes homologous to the aberrant gene, regardless of the number and location of the homologs. Such global silencing means that presence of a single, even recessive, mutant allele that produces an aberrant RNA may lead to a null phenotype. If the wild-type function of the affected gene and its homologs is important, the individual bearing the aberrant allele could be rapidly lost from the population. In this way, the consequence of RNA-mediated posttranscriptional gene silencing bears resemblance to that of monoallelic expression and loss of heterozygosity; it can hasten the effect of natural selection by exposing allelic variants that would otherwise be masked by the presence of wild-type homologs. Finally, it is possible that the powerful nature of RNA-mediated posttranscriptional gene silencing has allowed this form of homology effect to contribute to the evolution of exquisite mechanisms that control transcription and minimize opportunities for the generation of aberrant RNA species.

A key question concerning homology effects is how the presence of homology is sensed and telecast. As noted above, many observations have focused attention on RNA-mediated events, with particular attention paid to the potential of an RNA molecule to pair with another RNA molecule or with DNA. That is, RNA, possibly through its ability to form paired structures, may function as a sensor and/or a messenger of homology. If so, might there also exist DNA elements, chromosomal or nonchromosomal, with analogous function? The paired state for any DNA segment would necessarily mean the presence of homology. In fact, studies of recombination, gene conversion, and DNA repair have amply demonstrated that DNA can support and respond to homology searching and pairing. Here, we refer readers to two contributions, one by Greg Gloor (Chapter 4) on the mechanisms of, and the role of DNA homology and pairing in, the repair of double-strand breaks in *Drosophila*, and the second by Sean Burgess (Chapter 3) on double-strand break repair in yeast and the cytogenetic and physical evidence for homologous chromosome interactions in meiotic and mitotically dividing cells.

Can homolog pairing at the level of DNA also affect gene function? This possibility is discussed by a number of the authors mentioned above and is further considered by Eric Selker (Chapter 15) in his contribution concerning the phenomena of RIP (repeat-induced point mutation) and MIP (methylation induced premeiotically) in the haploid fungi *Neurospora* and *Ascobolus*, respectively. RIP and MIP can be triggered by DNA sequence duplications and result in gene silencing through extensive G:C to A:T transition mutations, in the case of RIP, and *de novo* methylation, in the case of MIP, reinforcing the notion that homology effects can act as host defense mechanisms. Intriguingly, both phenomena appear to entail the physical pairing of homologous DNA elements; pairing might be a means of determining whether the genome contains homologous DNA segments, designating a sequence as a repeated element, and/or destroying it. Further support for a role of DNA homolog pairing (be it at the level of DNA or a higher structural level) in gene regulation is provided by Jim Kennison and Jeffrey Southworth in their discussion of transvection (Chapter 13) and by Judy Kassis in her discussion of pairing-sensitive silencing (Chapter 14). Both these topics are reviewed in the context of *Drosophila,* in which homologous chromosomes are paired in somatic cells. In the case of pairing-sensitive silencing, the outcome of pairing is gene repression, whereas consideration of all types of transvection shows that pairing can be associated with both gene activation as well as gene repression. Might these observations suggest that pairing can be a versatile participant in the normal regulation of genes, that degrees of pairing and unpairing can act either as an ON switch or an OFF switch or even a biological rheostat for different levels of gene activity? Readers will find that the authors provide a thoughtful foundation on which to debate this issue.

We come, then, to the end of this preface, hoping that readers will find in the following 17 chapters as much enjoyment and inspiration as we have found. The range of topics is broad, yet we are confident that each chapter will be found as accessible to a newcomer as it is useful for seasoned colleagues. Might there be unifying themes or common issues which run through the chapters? Yes. Among these would be discussions concerning the current-day functions of homology effects, and whether and how these functions relate to the origin of homology effects. Why, for example, are most known homology effects associated with reduced gene function? Does this bias reflect a common origin or selective advantage of homology effects? Or, is the apparent preponderance of silencing among homology effects an artifact of a better ability to detect loss-of-function vs gain-of-function phenotypes? How is homology detected and translated into a consequence, and as the consequences of homology can be profound, how is the process controlled? The answers are emerging in exciting, often unexpected, ways. In fact, the pace of discovery these past few years has led some of us to wonder whether our memories of small meetings, lazy lunches, and articles few and far between are accurate. Thinking back to that more meandering time, we would like to acknowledge the extraordinary support and confidence that Dr. DeLill Nasser gave to the field. Through her efforts at the personal level and at the National Science Foundation, DeLill fostered many of us through financially lean, though always intellectually fantastic, years. We miss her terribly, but take some solace in the fact that before she passed away, she saw her field solidly launched.

Finally, we wish to thank the many authors who dedicated time to this volume. It has been a pleasure to work with them and a privilege to have benefited from their knowledge and insights. To researchers whose work we did not cite above, we extend apologies and an assurance that readers will reach their stories through the chapters that follow. C.-t. Wu would like further to thank Denise Barlow, Vicki Chandler, Jeannie Lee, Marjori Matzke, and Sarah Otto for special consultations, the participants of Genetics 218 for stimulating ideas, George Church, Bill Forrester, and Fred Winston for discussions, and Richard Emmons, Anne Lee, Jill Lokere, Jim Morris, Sharon Ou, and Ben Williams for critical input.

J. C. Dunlap is grateful to Steve Fiering and Jennifer Loros, co-instructors in Genetics 118 (Advanced Topics in Genetics), wherein we and the students realized the need for a good review volume on this topic and thereby started the ball, which ended up as this volume, rolling. C.-t. Wu is supported by the National Institutes of Health (RO1-GM61936) as is J. C. Dunlap (R37-GM 34985, R01-MH44651, and R21-MH62793).

References

Cook, D. L. Gerber, A. N., and Tapscott, S. J. (1998). Modeling stochastic gene expression: Implications for haploinsufficiency. *Proc. Nat. Acad. Sci. USA* **95,** 15641–15646.

Mable, B. K., and Otto, S. P. (1998). The evolution of life cycles with haploid and diploid phases. *BioEssays* **20,** 453–462.

Nutt, S. L., and Busslinger, M. (1999). Monoallelic expression of *Pax5*: A paradigm for the haploinsufficiency of mammalian *Pax* genes? *Biol. Chem.* **380,** 601–611.

Ohlsson, R., Paldi, A., and Marshall Graves, J. A. (2001). Did genomic imprinting and X chromosome inactivation arise from stochastic expression? *Trends Genet.* **17,** 136–141.

Otto, S. P., and Hastings, I. M. (1998). Mutation and selection within the individual. *Genetica* **102/103,** 507–524.

Sapienza, C. (1989). Genome imprinting and dominance modification. *Ann. N.Y. Acad. Sci.* **564,** 24–38.

Sturtevant, A. H. (1965). "A History of Genetics." Cold Spring Harbor Laboratory Press, Cold Spring Harbor, NY.

Watanabe, D., and Barlow, D. (1996). Random and imprinted monoallelic expression. *Genes to Cells* **1,** 795–802.

Wijnhoven, S. W. P., Kool, H. J. M., van Teijlingen, C. M. M., van Zeelan, A. A., and Vrieling, H. (2001). Loss of heterozygosity in somatic cells of the mouse: An important step in cancer initiation? *Mut. Res.* **473,** 23–36.

C.-ting Wu
Boston, Massachusetts
Jay C. Dunlap
Hanover, New Hampshire

1

Sex and the Single Chromosome

Victoria H. Meller

Department of Biology
Tufts University
Medford, Massachusetts 02155

Mitzi I. Kuroda

Department of Molecular and Cellular Biology
Howard Hughes Medical Institute
Baylor College of Medicine
Houston, Texas 77030

Advances in Genetics, Vol. 46

ABSTRACT

Just as homology can trigger a chain of events as described in many of the chapters of this volume, sometimes a lack of homology causes a crisis of a different sort. So it is for the single X chromosome in XY males in many species. Divergent sex chromosome pairs, such as the X and Y chromosomes in mammals and in fruit flies, are thought to have evolved from homologous autosomes. During evolution, the Y chromosome has retained little coding capacity, leaving the male with reduced gene dosage for many functions encoded by the X chromosome. In this chapter we focus on dosage compensation in *Drosophila*, in which most X-linked genes are upregulated by a male-specific ribonucleoprotein complex. This complex is thought to recognize the X chromosome through approximately 35 dispersed chromatin entry sites and then spread *in cis* to dosage compensate most genes on the X chromosome. © 2002, Elsevier Science (USA).

I. THE EVOLUTION OF SEX CHROMOSOMES

Sexual reproduction is overwhelmingly popular among eukaryotic organisms and is universally credited with contributing to species fitness. Although the exact nature of the most significant benefits of sex is still open to debate (Keightley and Eyre-Walker, 2000), one of the important features of sexual reproduction is that it allows the reassortment of genes through recombination. One of the processes that this enables is the elimination of mutations by recombination-mediated exchange with an intact homolog. This ability to salvage damaged chromosomes saves the genome from the slow, irreversible degradation that characterizes asexually reproducing organisms (reviewed in Rice, 1996). However, most eukaryotes carry a sex chromosome that cannot itself recombine and thus remains asexual (or clonal) and is subject to the accumulation of mutations. The gradual loss of information content from this chromosome, ultimately followed by its disappearance altogether, is the consequence of a degradative process driven by the specialized inheritance of the sex-determining chromosomes themselves (Figure 1.1; reviewed in Charlesworth, 1991; Charlesworth and Charlesworth, 2000). This can be observed in both mammals and *Drosophila*, where the small, heterochromatic, and gene-poor Y chromosome has been denied the benefits of sex and is on the road to extinction. The subsequent adjustment that the transcriptional machinery must undertake to accommodate this genetic loss is termed dosage compensation. The origin and mechanism of dosage compensation in *Drosophila* is the subject of this review.

In organisms with a divergent pair of sex chromosomes the heterogametic sex may be either a male, as in the XY males of mammals and *Drosophila*, or

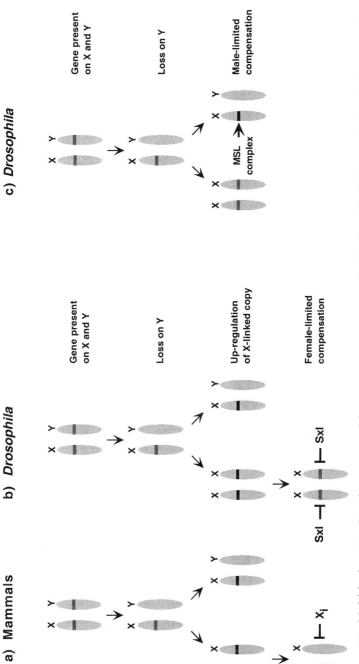

Figure 1.1. Model for the origin of sex chromosomes and dosage compensation in mammals and in *Drosophila*. (a) Proposed pathway for the origin of X inactivation in mammals. (b) Proposed pathway for the origin of downregulation of some X-linked genes in female *Drosophila*. (c) Evolution of male-limited upregulation of most genes on the *Drosophila* male X chromosome by the MSL complex.

female, as in ZW birds. In *Caenorhabditis elegans* the Y chromosome has been lost altogether, and males are designated as XO. For convenience we will limit the following discussion to the *Drosophila* and mammalian karyotype, although the processes driving the evolution of a pair of sex chromosomes can be more widely applied (Bull, 1983). Divergent sex chromosome pairs have evolved from and retain vestiges of their origin as pairs of identical autosomes. The process of sex chromosome differentiation begins with the establishment of a sex-determining locus on one chromosome, and this locus will subsequently be inherited in a strictly sex-limited fashion. Any alleles that are closely linked to this locus will be passed through the heterogametic (male) sex by virtue of their position in the genome, and are thus shielded from selection in the homogametic (female) sex. An artificial evolution experiment using *Drosophila* suggested that alleles displaying a sex-biased benefit are quite common, and the raw material for selection is therefore likely to be already in place when a new sex-determining locus is established (Rice, 1992). The accumulation of strongly male-beneficial genes closely linked to the sex-determining locus can also be seen in fish that carry a Y chromosome that is early in the process of divergence from the X chromosome (Bull, 1983).

The presence of a cluster of genes with sex-biased benefits makes recombination within this region costly to fitness, as it would allow these alleles to pass into the opposite sex. Multiple mechanisms may suppress recombination near the sex-determining locus (see Rice, 1996), but a very simple and effective block to recombination is the inversion of a portion of one chromosome relative to its homolog. Subsequent crossovers that occur between the inversion breakpoints will be lethal because they generate large deletions or duplications. If the inversion includes the sex-determining locus, all of the genes between the break points will be inherited as an undisturbed block in males. This portion of the chromosome will be clonally inherited and deprived of the option to repair mutations through recombination with an intact homolog. The genes within this region are thus consigned to eons of declining function followed by pseudogene status and eventual loss. This process has orchestrated the evolution of the mammalian sex chromosomes, and can be visualized with surprising clarity on the human Y (Lahn and Page, 1999a). The oldest portion of the Y, blocked from recombination with the X by an inversion event that occurred between 240 and 320 million years ago, carries the sex-determining locus, SRY. This region of the Y has suffered the most profound loss of coding potential and harbors only a few genes retaining similarity to ones found on the X. Three subsequent inversions on the Y chromosome have expanded its nonrecombining portion and relegated additional blocks of genes to decay, a process that can be observed in progress on the more recently inverted regions (Lahn and Page, 1999a).

Evolution of the human Y chromosome adheres quite closely to the predicted steps leading to destruction of a nonrecombining sex chromosome:

establishment of a sex-determining locus followed by a block to recombination and subsequent degradation of the nonrecombining chromosome. In broad outline, the origin of the *Drosophila* Y chromosome appears similar. Comparative studies of the genus *Drosophila* support the idea that, as in mammals, degradation of the Y chromosome is a time-dependent result of its recombinational status, which is in turn dictated by sex-limited inheritance of the chromosome. Within the genus the ancestral sex chromosome karyotype, found in *Drosophila melanogaster*, consists of an acrocentric X chromosome (Patterson and Stone, 1952). Translocations in some lineages of the genus have fused autosomal arms to the X or Y chromosome, thus creating neo-X and neo-Y chromosomes that are forced to segregate as sex chromosomes in spite of their autosomal origins. The degree of degradation of the neo-Y correlates with the age of the transposition event that created the arm (Patterson and Stone, 1952; reviewed in Charlesworth, 1991, 1996). For example, the transposition that created a neo-X in *D. americana* is estimated to be only a few hundred thousand years old, and degeneration of its homolog, the neo-Y, is undetectable (Charlesworth *et al.*, 1997). *D. pseudoobscura* has a metacentric X, one arm corresponding to the ancestral X chromosome and a neo-X arm originating from an autosomal fusion estimated to be over 10 million years old. In this species the formerly autosomal neo-Y, which has been forced to segregate opposing the neo-X, has been lost altogether. In a third species, *D. miranda*, the fusion of an autosomal arm to the Y chromosome 2 million years ago once again created neo-X and neo-Y chromosomes, but this time the neo-Y is in an intermediate stage of degradation (Strobel *et al.*, 1978; Steinemann, 1982).

These examples suggest that loss of information from a chromosome is a time-dependent process, but it may be initiated by merely forcing a section of the genome to segregate as a Y chromosome. One point of difference between *Drosophila* and mammals is that the *Drosophila* Y chromosome presently lacks a sex-determining locus, and instead uses the ratio of X chromosomes to autosomes as the primary signal for sexual differentiation. Evolutionary theory suggests that the onset of Y degeneration in the *Drosophila* was triggered by a sex-determining locus on the Y, which was subsequently lost. Primary sex-determination signals evolve with great rapidity, making this a plausible sequence of events (Charlesworth, 1996).

Information loss is not the only process shaping the Y chromosome, as the Y may sometimes acquire new genes. It cannot harbor genes essential for both sexes, but Y-linked genes are frequently required for male fertility. Both *Drosophila* and mammalian Y chromosomes carry genes that appear to have moved from other locations (Kalmykova *et al.*, 1997; Lahn and Page, 1999b). In these cases the identification of highly similar autosomal genes and the lack of introns in the Y-linked copy suggest that retrotransposition moved an autosomal gene to the Y chromosome.

II. SELECTIVE PRESSURE HAS LED TO A VARIETY
OF DOSAGE-COMPENSATION MECHANISMS

The loss of information from the Y chromosome has profound consequences for males: they become haploid for any gene that initially existed on both the X and Y chromosomes but has been mutated on the Y chromosome. The male will suffer a reduction in gene dosage, and the initial reaction to this may be an up-regulation of the X-linked homologue (Figure 1.1). If upregulation occurs in both sexes, rather than just in males, females must respond by downregulating expression of the same gene. Mammalian females deal with this dilemma by silencing most of the genes on one of their X chromosomes through compaction into the heterochromatic Barr body (Lyon, 1961). A comparative evolutionary analysis in mammals revealed that when a functional homologue of an X-linked gene is retained on the Y chromosome, the X-linked gene escapes female silencing. The adoption of female silencing does not occur until the homolog is lost from the Y chromosome (Jegalian and Page, 1998). In this model a step is assumed to occur that may rarely be detected, and that is the upregulation of X-linked genes in both sexes (Charlesworth, 1978; Adler *et al.*, 1997). The subsequent silencing of one of the two female copies suggests that this step has occurred (see Jegalian and Page, 1998).

By contrast, experiments designed to estimate the rate of RNA synthesis from XX and XO nuclei of *Drosophila* gynandromorphs revealed that loss of an X chromosome caused a twofold increase in transcription from the remaining X (Lakhotia and Mukherjee, 1969). Subsequent analyses have shown that *Drosophila* males compensate their haploid, X-linked genes by directing a male-specific regulatory complex to their X chromosome (Figure 1.1c). Staining of polytene chromosomes from various species of *Drosophila* with antibodies that recognize the proteins responsible for upregulation of the male X indicate that the neo-X chromosomes acquire antibody staining as their partner Y chromosomes degenerate. As the Y becomes increasingly eroded, its partner is more and more heavily and uniformly coated with the machinery of compensation (Bone and Kuroda, 1996; Marín *et al.*, 1996; Steinemann *et al.*, 1996), serving to increase transcription of most X-linked genes. Because the protein complex is limited to males, it might be anticipated that female *Drosophila* would have no need to modulate levels of X-linked gene expression. However, a second system of dosage compensation has been proposed to downregulate some X-linked genes in *Drosophila* females (Cline, 1984; Gergen, 1987; Kelley and Kuroda, 1995; Kelley *et al.*, 1995, Figure 1.1b). This mechanism is distinct from both the male branch of compensation and the mechanism of silencing found in mammalian females, as *Drosophila* females are proposed to posttranscriptionally downregulate a subset of X-linked genes. The process of dosage compensation in *Drosophila* can therefore be viewed as a composite of two systems: a male-limited hypertranscription that compensates most of the X chromosome, and a female-limited downregulation of a subset of X-linked genes whose transcription has presumably been enhanced in both sexes. The female-specific *Sex lethal* (*Sxl*) protein regulates both of these pathways. It appears that the mechanism of compensation recruits available molecules that might be easily adapted to the task at hand. *Sxl* does not appear to have sex-specific expression outside of the genus *Drosophila* and thus it is unlikely that its role in dosage compensation is widespread among the Diptera (Meise *et al.*, 1998; Saccone *et al.*, 1998; reviewed in Schutt and Nothiger, 2000).

Dosage compensation in both mammals and *Drosophila* involves the identification and regulation of an entire chromosome. Because linkage to the X is a shared feature of compensated genes, it might be anticipated that this is the deciding factor in whether compensation occurs. Indeed, both mammals and *Drosophila* have X-linked loci that can act as long-range regulators of genes *in cis*. The mammalian X *inactivation center* (*Xic*) is necessary for the silencing of genes on the X (Penny *et al.*, 1996; Marahrens *et al.*, 1997), and the X-linked *roX* genes of *Drosophila* act as sites for entry and spreading of the complexes that mediate up-regulation of the male X chromosome (Kelley *et al.*, 1999). However, in both systems some genes do escape compensation. Mammalian genes that retain active homologs on the Y have no need to undergo compensation (Jegalian and Page, 1998). Similarly, the ribosomal RNA genes of *Drosophila* are present on both the X and Y chromosomes, and are not compensated (Ritossa *et al.*, 1966). In addition, genes with sex-limited expression, and those with autosomal homologs that can blunt the impact of their loss from the Y chromosome, may escape compensation (reviewed in Baker and Belote, 1983; Lucchesi and Manning, 1987). These observations reinforce the notion that compensation occurs on a gene-by-gene basis, as needed to respond to the loss of genetic information. Studies using genes moved between the *Drosophila* X chromosome and the autosomes indicate that (1) location on the X chromosome and (2) sequence information closely linked to genes are both important for accurate genetic compensation (reviewed in Lucchesi and Manning, 1987; Qian and Pirrotta, 1995).

III. DOSAGE COMPENSATION IN *Drosophila* BY THE MSL COMPLEX

The most direct way to rectify the X-chromosome dosage imbalance between the sexes is a male-limited upregulation of X-linked genes in response to their loss from the Y chromosome. This appears to be the predominant mechanism in *Drosophila* (Mukherjee and Beermann, 1965), in which the MSL complex associates with hundreds of sites along the male X chromosome and is thought to upregulate transcription of X-linked genes (Kuroda *et al.*, 1991; but see Bhadra *et al.*, 1999, for an alternative view).

The MSL complex is named for its founding members, the Male Specific Lethal proteins (Kuroda et al., 1991; Palmer et al., 1993; Bashaw and Baker, 1995; Gorman et al., 1995; Kelley et al., 1995; Zhou et al., 1995; Hilfiker et al., 1997). These proteins, in turn, are named for the phenotype that results when the genes that encode them are mutant (Fukunaga et al., 1975; Belote and Lucchesi, 1980a, 1980b; Uchida et al., 1981; Lucchesi et al., 1982; Hilfiker et al., 1997). Mutations in *male-specific lethal one* (*msl1*), *male-specific lethal two* (*msl2*), *male-specific lethal three* (*msl3*), *maleless* (*mle*), and *males absent on the first* (*mof*) all result in failure to localize an active dosage-compensation complex on the X chromosome (Gorman et al., 1993, 1995; Palmer et al., 1994; Lyman et al., 1997; Gu et al., 1998). The complex normally associates with the X chromosome during embryogenesis, and the failure of dosage compensation in mutants appears to lead to a relatively slow decline, with no known specific cause (Fukunaga et al., 1975; Belote and Lucchesi, 1980b; Uenoyama et al., 1982; Belote, 1983; Bachiller and Sanchez, 1989). Presumably, a continuous imbalance of products required for numerous biochemical pathways results in a gradual deterioration in growth and viability. Mutant embryos have no obvious patterning defects, but mutant larvae display retarded development, with eventual death during larval or pupal stages.

The MSL proteins have been identified through positional cloning of the genes that encode them, followed by antibody production against recombinant proteins, and immunolocalization of the endogenous proteins in larval polytene chromosomes (Kuroda et al., 1991; Palmer et al., 1993; Bashaw and Baker, 1995; Gorman et al., 1995; Kelley et al., 1995; Zhou et al., 1995; Hilfiker et al., 1997). In each case, the cognate MSL protein has been found to localize in a reproducible banded pattern along the length of the male polytene X chromosome. The proteins are co-localized on the male X, with only MLE showing, in addition, a more general association with all chromosome arms (Kuroda et al., 1991; Bone et al., 1994).

The MSL proteins do not resemble known DNA-binding factors (Table 1.1). The specificity for binding the X chromosome is dependent on MSL1,

Table 1.1. Molecules Implicated in Dosage Compensation in *Drosophila*

Name	Sequence motifs
MLE	Two double-stranded RNA-binding domains, DExH family of NTP-dependent helicases
MSL1	Novel, acidic
MSL2	RING finger, additional cysteine-rich region
MSL3	Two chromodomains
MOF	Chromodomain, C_2HC-type zinc finger, MYST family of acetyltransferases
JIL-1	Two serine/threonine kinase domains
roX1	Noncoding RNA from chromatin entry site 3F
roX2	Noncoding RNA from chromatin entry site 10C

a novel acidic protein, and MSL2, a RING finger protein (Lyman *et al.*, 1997). Neither factor can stably bind the X chromosome in the absence of the other. It is possible that they form a chromosome-binding configuration only when interacting together, through a segment of MSL1 and the RING finger of MSL2 (Copps *et al.*, 1998; Scott *et al.*, 2000). MSL1 is highly unstable in the absence of MSL2 (Kelley *et al.*, 1995; Chang and Kuroda, 1998). However, if it is overexpressed, MSL1 interacts transiently with all chromosomes, suggesting a general affinity for chromatin. The analogous experiment with overexpressed MSL2 suggests that MSL2 has no general affinity for chromosomes. Therefore, it is possible that MSL1 and MSL2 must work together because MSL1 carries a general affinity for chromatin, while MSL2 provides MSL1 with specificity for the X chromosome. MSL1 also appears to play a central role in complex assembly, with specific affinity for both MOF and MSL3 (Scott *et al.*, 2000).

Three of the MSL proteins are encoded by members of conserved gene families. These are MSL3, a chromodomain protein (Gorman *et al.*, 1995; Koonin *et al.*, 1995; Marín and Baker, 2000), MOF, a histone acetyltransferase (Hilfiker *et al.*, 1997; Pannuti and Lucchesi, 2000), and MLE, a DExH helicase (Kuroda *et al.*, 1991; Lee and Hurwitz, 1993). Neither MSL3, MOF, nor MLE appear to have an essential function in females, suggesting that they are dedicated to dosage compensation. However, in other organisms, members of these gene families clearly play more general roles. MOF and MSL3 may have an ancient relationship, as their closest relatives in yeast function together as members of the NuA4 histone acetylase and chromatin-remodeling complex (Eisen *et al.*, 2001). Several mammalian relatives of MOF and MSL3 have also been reported. Close relatives of MLE are not found among the many helicases in yeast (Marín *et al.*, 2000), but are clearly present in mammals, including RNA helicase A, an abundant essential protein in mouse that has been implicated in a wide variety of cellular processes, from transcription to RNA transport (Lee and Hurwitz, 1993; Lee *et al.*, 1998).

There is substantial evidence that the enzymatic activity of MOF, a histone H4 acetyltransferase, plays a central role in dosage compensation. Recombinant MOF, or partially purified MSL complexes, exhibit specificity for acetylation of Lys16 of histone H4, producing a monoacetylated H4 isoform designated H4Ac16 (Akhtar and Becker, 2000; Smith *et al.*, 2000). The male X is specifically enriched for H4Ac16, while other acetylated isoforms of histone H4 are distributed along all chromosome arms (Turner *et al.*, 1992). This H4Ac16 isoform is dependent on the MSL complex (Bone *et al.*, 1994), and a point mutation in MOF, allowing its incorporation into complexes, but destroying its ability to bind acetyl-coA, results in lack of X-specific H4Ac16 enrichment (Hilfiker *et al.*, 1997). Histone acetylation has emerged as a major strategy to positively regulate gene transcription, by somehow counteracting the negative effect of DNA packaging into chromatin (for review see Kingston and Narlikar, 1999). Therefore, a simple model can be proposed in which the primary or sole function of the MSL complex is to target an active MOF acetyltransferase to the X chromosome, resulting in more accessible chromatin for transcription. Whether the enrichment of H4Ac16 on X-linked genes is sufficient to cause a twofold increase in transcription is not known. Recombinant MOF protein has a much more dramatic effect on transcription of a nonspecific chromatin template *in vitro* (Akhtar and Becker, 2000), suggesting that the achievement of a twofold increase in transcription might even involve an attenuation of MOF activity by the MSL complex.

Recently, evidence that the male X chromosome is also enriched for other posttranslational modifications of histones has emerged through studies of JIL-1, a protein kinase that phosphorylates histone H3 on Serine 10 (Jin *et al.*, 1999, 2000; Wang *et al.*, 2001). JIL-1 is enriched on the X chromosome but is required for viability in both sexes and is found at substantial levels on all chromosomes, suggesting that it has a general function in chromosome organization or transcription. JIL-1 enrichment on the male X chromosome is quite striking in immunostaining experiments, but has been measured as a modest twofold increase by assaying the fluorescence of GFP-JIL-1 expressed from a transgene *in vivo*. While this enrichment is modest, JIL-1 is co-localized with the MSL complex on the polytene X, and can be co-immunoprecipitated with MSL complexes from male tissue culture cells (Jin *et al.*, 2000). Furthermore, the MSL complex regulates JIL-1 enrichment: it is evenly distributed on all chromosomes in *msl* mutants, and becomes enriched on the X chromosomes in females forced to express the MSL complex ectopically. Taken together, the localization of JIL-1 and its histone tail-modification activity suggests that this protein kinase has a general function on all chromosomes, but also participates in dosage compensation on the X chromosome. In mammalian cells, phosphorylation of H3 is coupled to histone acetylation during growth factor-triggered gene activation (Cheung *et al.*, 2000; Clayton *et al.*, 2000), and this relationship may also occur on the male X chromosome (Wang *et al.*, 2001).

More than half of the predicted mass of the MSL complex is attributed to the *RNA on X* (*roX*) RNAs (Akhtar *et al.*, 2000; Meller *et al.*, 2000; Smith *et al.*, 2000). *roX* transcripts coat the male X chromosome, and both *roX* genes are also located on the X chromosome. *roX1* RNA is 3.4 kb, and the major *roX2* RNA species is approximately 500–600 nt (Amrein and Axel, 1997; Meller *et al.*, 1997; Smith *et al.*, 2000). The *roX* RNAs are male-specific, at least in part because they are rapidly degraded if produced in the absence of MSL complexes (Meller *et al.*, 2000). There are conflicting reports with regard to whether *roX* RNAs are detected in females early in embryogenesis (Meller *et al.*, 1997; Franke and Baker, 1999), and it is not known whether unstable *roX* RNAs might be continuously transcribed and degraded in females. Nevertheless, whether through transcriptional repression, a combination of repression and RNA degradation, or solely RNA turnover, the end result is strikingly male-preferential *roX* RNA expression in larvae and adults.

What role(s) do *roX* RNAs play in dosage compensation? Although *roX1* RNA is relatively large and could account for the bulk of RNA in MSL complexes, it appears completely dispensable, as deletion mutant males that lack part of the *roX1* gene are viable, fertile, and have dosage-compensation complexes lacking *roX1* RNA covering their X chromosomes (Meller *et al.*, 1997). A simple *roX2* mutant has not been reported, but X-chromosome deficiencies removing *roX2* and surrounding essential genes have been examined during embryogenesis (Franke and Baker, 1999). Male embryos carrying these deficiencies are destined to die due to the absence of essential gene functions. However, early in embryogenesis these embryos could assemble MSL complexes onto their X chromosomes, suggesting that *roX2* is also dispensable, at least for initial assembly of MSL complexes. In contrast, male embryos carrying a *roX1 Df (roX2)* doubly mutant X chromosome failed to assemble detectable MSL complexes, suggesting that *roX1* and *roX2* are interchangeable, but that at least one of these genes is necessary for MSL complex assembly (Franke and Baker, 1999).

IV. A SERIES OF BINARY SWITCHES REGULATES THE SEX SPECIFICITY OF DOSAGE COMPENSATION IN *Drosophila*

The proper regulation of dosage compensation requires the ability to assess the number of X chromosomes per nucleus and use this information either to assemble an MSL complex on the single X chromosome of males, or to prevent its formation in XX females. The series of binary genetic switches that lead to these outcomes feature regulation at the levels of transcription, RNA processing, translation, and protein complex formation (Figure 1.2). The regulation of dosage compensation is

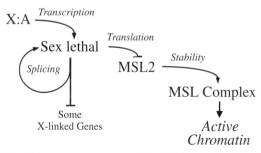

Figure 1.2. Regulation of dosage compensation in *Drosophila* by a series of binary genetic switches. Arrows indicate positive regulation and cross-bars indicate negative control.

initially linked to the well-documented sex-determination hierarchy in *Drosophila* (Cline and Meyer, 1996).

In *Drosophila*, the Y chromosome carries factors required for male fertility, but does not determine sex (Bridges, 1921). Rather, sex is determined by the number of X chromosomes per cell, which is measured in comparison to the number of sets of autosomes (the X:A ratio). One X and two sets of autosomes results in a normal diploid male. Two X chromosomes and two sets of autosomes results in a normal diploid female. However, two X chromosomes and three sets of autosomes results in a triploid intersex, with individual cells expressing either a male or a female phenotype.

The X:A ratio specifies the initial transcriptional state of a key sex deter-mination locus, *Sex lethal (Sxl)* (Cline, 1978, 1984), by the interplay of positive factors encoded by the X and negative regulation by the autosomes. Early in embryogenesis, the *Sex lethal* gene is initially transcribed from an "establishment" promoter, P_e, in embryos with two X chromosomes, but not in embryos with a single X chromosome (Keyes *et al.*, 1992). Extensive genetic analyses have led to the discovery that transcription factors encoded by the X chromosome (*sisA*, *sisB*, *runt*), along with the maternally provided *daughterless* helix-loop-helix transcrip-tion factor, positively regulate transcription of the *Sex lethal* gene (Cline, 1986, 1988; Torres and Sanchez, 1989; Parkhurst *et al.*, 1990; Duffy and Gergen, 1991; Erickson and Cline, 1991, 1993). An additional positive-acting X-chromosome determinant, *sisC* (Sefton *et al.*, 2000), encodes a secreted ligand of the JAK/STAT signal-transduction pathway (Jinks *et al.*, 2000; Sefton *et al.*, 2000). The repres-sive autosomal component of the X:A ratio may be a combination of factors that have weak inhibitory effects on *Sxl* transcription (Younger-Shepherd *et al.*, 1992; Barbash and Cline, 1995).

The X:A ratio provides the key piece of information dictating male or female development, but measuring this delicate balance is required only during a short window of time in early embryogenesis, when the *Sxl* P_e promoter responds to the X-chromosome dosage (Cline, 1984; Keyes *et al.*, 1992). Shortly thereafter, *Sxl* pre-mRNA is transcribed from a "maintenance" promoter, P_m, which does not respond to the X:A ratio and is active in both sexes. However, the stability of the male–female binary decision is assured because *Sxl* protein synthesized in response to the X:A ratio is required for the productive splicing of *Sxl* pre-mRNA transcribed from P_m (Bell *et al.*, 1991). This positive autoregulatory requirement for *Sxl* pre-mRNA splicing allows female development to proceed in embryos with two X chromosomes, but prevents it from occurring in XY males. *Sxl* encodes an RNA-binding protein (Bell *et al.*, 1988), and its function in alternative splicing is required continuously for its own maintenance and for female-specific splicing of *tra* mRNA, which encodes another splicing factor required specifically for female development (Nagoshi *et al.*, 1988). Mutation of *tra* results in XX flies that look and behave male (Sturtevant, 1945). Because mutants lacking *tra* or other female sex-determination functions display a male phenotype but are fully viable, they cannot be necessary for the essential process of blocking inappropriate dosage compensation in females.

V. *Sex lethal* COMMUNICATES THE X:A RATIO TO THE DOSAGE-COMPENSATION PATHWAY THROUGH REPRESSION OF *msl2* TRANSLATION

The regulation of dosage compensation is dependent on *Sex lethal*, but independent of the sex-determination genes that *Sxl* regulates (reviewed in Cline and Meyer, 1996). For this reason, XX *Sxl* mutants are dead. The role of *Sxl* in *msl*-mediated dosage compensation is to prevent MSL complexes from forming in XX females (Gorman *et al.*, 1993). Therefore, the simplest model was that alternative splicing would control one or more *msl* genes. Contrary to this simple prediction, the key role of *Sxl* is to prevent MSL2 translation in females (Bashaw and Baker, 1997; Kelley *et al.*, 1997).

   Sxl regulates the translation of MSL2 protein through binding to multiple sites in the 5′ and 3′ untranslated regions of the *msl2* mRNA (Kelley *et al.*, 1995; Zhou *et al.*, 1995; Bashaw and Baker, 1997). Although *msl2* mRNA undergoes male-specific splicing in its 5′ UTR, this is not the key regulatory step, since males and females expressing unspliceable transgenic mRNAs (the female form) still translate the protein in males and fail to produce the protein in females (Kelley *et al.*, 1997). The key regulatory event is whether SXL protein can bind the *msl2* mRNA in both its 5′ and 3′ UTRs. Although SXL was previously viewed as a nuclear protein dedicated to the regulation of alternative splicing, the translational control of *msl2* mRNA is thought to occur in the cytoplasm (Bashaw and Baker, 1997).

   Sxl repression of MSL2 is sufficient to explain the male-specificity of dosage compensation (Kelley *et al.*, 1995). This is because the other known dosage-compensation proteins are not strictly sex-specific. Each of the *msl* genes is transcribed in both sexes, and all of the proteins except MSL2 are produced in wild-type females. MOF and MLE display comparable levels in males and females (Kuroda *et al.*, 1991; Gu *et al.*, 2000), while MSL1 and MSL3 have lower steady-state levels in females (Palmer *et al.*, 1994; Gorman *et al.*, 1995). Some forms of *msl1* mRNAs also have *Sxl*-binding sites in their 3′ UTRs, and these may contribute to less efficient translation of MSL1 in females (Palmer *et al.*, 1993). However, the principal reason that MSL1 protein levels are low in females is that MSL1 is unstable in the absence of MSL2 (Chang and Kuroda, 1998). When *msl2* mRNA lacking SXL-binding sites is expressed in females, production of MSL2 protein leads to stabilization of MSL1 and assembly of functional MSL complexes on both female X chromosomes (Kelley *et al.*, 1995). Females expressing MSL2 show substantial lethality, presumably due to inappropriate dosage compensation of both X chromosomes. All females die when both MSL1 and MSL2 are expressed ectopically (Chang and Kuroda, 1998).

VI. EVIDENCE FOR THE COEXISTENCE OF A FEMALE-SPECIFIC DOSAGE-COMPENSATION PATHWAY IN *Drosophila*

Although *msl*-mediated dosage compensation is thought to be the principal mechanism, there are several lines of evidence that some X-linked genes in *Drosophila* are dosage-compensated by repression in females rather than by upregulation in males. Females that are mutant for *Sxl* die, at least in part because the *msl* pathway is derepressed. However, *Sxl; msl* double mutants still die, suggesting a second vital function for *Sxl* in females (Skripsky and Lucchesi, 1982; Cline, 1984; Uenoyama, 1984). Evidence that this second function is a separate mode of dosage compensation came from studies of *runt*, an X-linked gene that is dosage-compensated by a mechanism that requires *Sxl* but not the *msl* genes (Gergen, 1987; Bernstein and Cline, 1994). Unexpectedly, it appears that *Sxl* directly downregulates *runt* and a few other X-linked genes in females (Kelley *et al.*, 1995). This hypothesis originated with the discovery that *msl2* and *msl1* mRNAs are downregulated by *Sxl* through binding sites in their 5′ and 3′ UTRs. *msl2* mRNA contains at least five sites distributed between the 5′ and 3′ UTRs, and is completely repressed. *msl1* mRNA contains four sites in its 3′ UTR, and can be partially repressed. Inspection of the *runt* sequence showed that it contains three *Sxl* binding sites in its 3′ UTR, suggesting that its translation or stability could also be partially repressed. A search of the *Drosophila* sequence database in 1995 revealed that out of 22 genes with three or more SXL binding sites in their 3′ UTRs, 20 mapped to the X chromosome. The only autosomal genes were *msl1* and *msl2*, the known targets of *Sxl* repression of *msl*-mediated dosage compensation. Some forms of *Sxl* mRNA also carry many copies of SXL-binding sites, suggesting negative autoregulation, perhaps as a homeostasis mechanism (Kelley *et al.*, 1995; Yanowitz *et al.*, 1999).

The hypothesis that the *Drosophila* X chromosome carries many genes that are upregulated in males, some that are downregulated in females, and a few that are not dosage-compensated can be considered in light of the evolution of sex chromosomes discussed earlier in this chapter. As individual gene function was lost during evolution of the Y chromosome, individual genes came under selective pressure to upregulate in males. Whether this occurred by attraction of a preexisting MSL complex, or through a non-sex-specific mechanism of upregulation, would dictate whether female-specific repression would be coordinately required (Figure 1.1). Now that the entire *Drosophila* genome sequence is available, it will be an interesting evolutionary question to ask whether sequence or gene context can explain why some genes acquired one type of regulation while neighboring genes chose a different mechanism.

VII. MSL-MEDIATED DOSAGE COMPENSATION SPREADS FROM SPECIAL ENTRY SITES *in cis*

As genes came under selective pressure to upregulate, how did they acquire the ability to attract the MSL complex? When MSL proteins were first visualized on the polytene chromosomes, it was apparent that they bound in a reproducible, banded pattern. It was assumed that each band represented a specific DNA-recognition event, and that most X-linked genes would have a sequence element that could attract the MSL complex *in trans*. Although this model appeared straightforward, no X-specific *cis*-acting element emerged from studies of X-linked genes. Furthermore, it was not clear how such a sequence could evolve *de novo* adjacent to most genes (Charlesworth, 1996). These issues have been simplified in a new model for MSL recognition of the X chromosome (Kelley *et al.*, 1999, Figure 1.3). Recent data suggest that the MSL complex initially recognizes a subset of approximately 35 "chromatin entry sites," dispersed along the X chromosome. When a functional complex binds an entry site, it can subsequently spread long distances *in cis*, to regulate flanking genes. This new model postulates that the evolution of only a few chromatin entry sites on the X could be sufficient for the evolution of dosage compensation at many X-linked genes.

Chromatin entry sites are revealed in specific *msl* mutants, where partial complexes are seen at these ~35 sites, rather than in the wild-type pattern of hundreds of sites along the X chromosome (Palmer *et al.*, 1994; Lyman *et al.*,

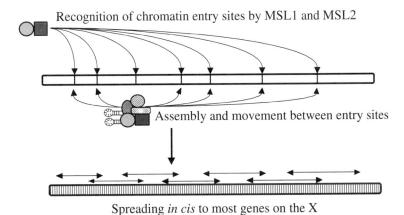

Recognition of chromatin entry sites by MSL1 and MSL2

Assembly and movement between entry sites

Spreading *in cis* to most genes on the X

Figure 1.3. Model for the X-chromosome specificity of the MSL complex. (Top) Initial recognition by MSL1 and MSL2 is thought to occur at a set of 35 dispersed chromatin entry sites, including the *roX1* and *roX2* genes. (Middle) Assembly of additional components, including the *roX2* RNA, is accompanied by the ability of the complex to move between chromatin entry sites. (Bottom) Complete complexes can spread *in cis* to most genes on the X chromosome.

1997; Gu *et al.*, 1998). MSL complexes fail to bind the X chromosome at all in *msl1* or *msl2* mutants, but partial complexes bind the chromatin entry sites in the absence of *msl3*, *mof*, and *mle*. The MSL immunostaining is noticeably weaker in *mle* mutants. The cloning of the *roX1* and *roX2* genes has provided the molecular identity of two of the chromatin entry sites (Amrein and Axel, 1997; Meller *et al.*, 1997; Kelley *et al.*, 1999). Remarkably, these genes are critical nucleation sites for MSL complex spreading, and they also produce RNA components of the complex. Whether this might be a common property of the other chromatin entry sites is not known.

The recent hypothesis that most X-linked genes do not have a specific recognition sequence for the MSL complex comes from experiments in which entry sites have been moved to ectopic locations on autosomes (Kelley *et al.*, 1999). When MSL complexes are attracted to single entry sites on autosomes, they can be seen to spread variable distances into flanking sequences that were never before dosage-compensated. MSL complexes may be observed binding several hundred kilobases from the site of an autosomal *roX* insertion, although the spread into the autosomes is highly variable from nucleus to nucleus and is not uniform along the chromosome. This result suggests that once a complex enters a chromosome, it recognizes sequences or structures that may be common to active genes. However, spreading from single sites on autosomes appears to be less robust than the spreading that is postulated to result in the highly reproducible MSL pattern on the endogenous X chromosome. It is not known whether sequence composition, pairing state, presence of multiple entry sites, or other factors can explain the apparent difference in spreading efficiency on X compared to autosomes.

VIII. THE BEHAVIOR OF MSL COMPLEXES ON PAIRED CHROMOSOMES

The cis-limited spreading of MSL complexes from chromatin entry sites may be a key requirement for the X-chromosome specificity of dosage compensation. Complexes cannot detach from the X chromosome and enter nonhomologous chromosomes unless provided with a transposed entry site. However, some spreading between homologous chromosomes has been detected in transgenic flies (Kelley *et al.*, 1999). The endogenous male X has no homolog, so evidence for the ability to spread *in trans* comes from artificial situations in which *roX* chromatin entry sites have been inserted on paired autosomes. Since the maternal and paternal homolog are tightly synapsed along their entire length, and not easily delineated in polytene chromosomes, the assay for trans spreading also requires a partial disruption of pairing by using multiply inverted balanced chromosomes. Under these unusual conditions, limited spreading of MSL complexes could be observed from a transgenic chromosome (containing an *roX1* entry site) to a nontransgenic homolog. This ability to spread *in trans* has been invoked as a possible explanation for the ability for X-derived transgenes such as *white*, which lack a chromatin entry site, to partially dosage-compensate when moved to autosomes (Kelley *et al.*, 1999).

IX. NONCODING RNAs, EPIGENETIC REGULATION, AND CHROMATIN STRUCTURE

The *roX* RNAs join a growing class of transcripts termed "riboregulators," made up of noncoding RNA molecules performing a broad range of regulatory functions (see Erdmann *et al.*, 2001). This diverse group includes the prokaryotic regulator of RNA polymerase, 6S RNA (Wassarman and Storz, 2000), the *Schizosaccharomyces pombe* meiotic regulator meiRNA (Yamashita *et al.*, 1998), the eukaryotic hormone receptor cofactor SRA (Lanz *et al.*, 1999), *C. elegans* antisense RNAs (Slack *et al.*, 2000), and stress-induced RNAs of prokaryotes and eukaryotes (Lakhotia, 1996; Altuvia *et al.*, 1997; Zhang *et al.*, 1998). The widespread use of RNA molecules as regulators may in part be due to their quick and easy production, as no protein synthesis is required. RNA molecules may also be rapidly destroyed, making them well suited as transient modulators of gene expression. Although a growing number of untranslated RNAs are implicated in regulation of gene expression in eukaryotes, we will limit the following discussion to those that are likely to act through alteration of chromatin structure.

RNA is detected as a component of chromatin, but this has traditionally been attributed to the presence of either nascent transcripts or small RNAs involved in splicing or transcript processing. However, a few studies have broadly implicated RNA in the maintenance and regulation of chromosome structure (Nickerson *et al.*, 1989; Meller and Fisher, 1995), and several recent findings suggest that RNA molecules may have more central roles in chromatin regulation than previously believed. First, dosage compensation in both mammals and *Drosophila* involves large, untranslated RNAs: *Xist* (*X inactive-specific transcript*) in mammalian females (Borsani *et al.*, 1991; Brockdorff *et al.*, 1991; Brown *et al.*, 1992; and see Chapter 2), and the *roX* RNAs in *Drosophila* males (Amrein and Axel, 1997; Meller *et al.*, 1997). Second, the chromodomain, a motif common in proteins that have a role in regulation of gene expression through the formation of complexes that bind chromatin, has been shown to interact with RNA *in vitro* (Akhtar *et al.*, 2000). Although RNA binding has yet to be reported for other chromodomain proteins, the idea that the members of this family might share this property suggests that RNA could be critical in the formation of some of the regulatory complexes that are key to epigenetic gene regulation.

Parallels between *roX* RNAs and *Xist* RNA have been reviewed recently (Kelley and Kuroda, 2000), and *Xist*-mediated X inactivation is the subject of its own chapter in this volume (see Chapter 2). Therefore, we will limit our comparison of RNAs involved in dosage compensation to a few key points. The *roX* RNAs and *Xist* function in processes that arose independently during evolution, yet in each case the sites of synthesis of these noncoding RNAs mark their respective X chromosomes for dosage compensation (Lee and Jaenisch, 1997; Kelley *et al.*, 1999). Furthermore, in each case they are observed to spread *in cis* by unknown mechanisms to cover their target chromosomes. However, although *roX* and *Xist* RNAs are thought to function through alteration of chromatin structure, they must do so with opposite results: upregulation vs repression of an entire chromosome. With special regard to the interaction of homologous sequences (the subject of this volume), the behavior of *roX* RNAs differ notably from *Xist*. Initiation and spreading of *Xist* silencing is limited to the chromosome *in cis*, allowing modification to occur on only one of the two X chromosomes of mammalian females (Clemson *et al.*, 1996; Lee and Jaenisch, 1997). The *roX* RNAs are transcribed from and bind to the single male X chromosome, so *roX* RNAs normally act *in cis*, also. However, *roX* RNA produced from a translocated portion of the X or from an autosomal transgene can bind to X-chromatin targets *in trans* (Meller *et al.*, 1997, 2000; Kelley *et al.*, 1999). In addition, MSL complexes can spread from an autosomal *roX* insertion site into a paired chromosome that does not carry the transgene (Kelley *et al.*, 1999). The molecular basis for these differences in behavior between *Xist* and *roX* is unknown, but their import is clear: the ability to differentiate between two identical X chromosomes is crucial for mammalian dosage compensation, but the single X chromosome of male *Drosophila* need only be distinguished from the autosomes.

Sex chromosomes are frequently unpaired as a consequence of the counterintuitive evolutionary process that dictates the destruction of a chromosome carrying a sex-determination signal. The comparison of two independent solutions to this dilemma, the processes of dosage compensation in mammals and *Drosophila*, reveals that preexisting mechanisms, such as chromatin modification or remodeling activities, may be recruited for the purpose of compensating an entire chromosome. Although independently derived, the unexpectedly similar cis-acting signals, *roX* and *Xist*, direct the regulation of distant regions of the chromosome on which they reside. The fact that both *roX* and *Xist* produce untranslated RNAs that are predicted to function in chromatin remodeling suggests a broader role for RNA in the regulation of chromatin structure. These loci may also be important factors in the rapid evolution of sex chromosomes by allowing compensation to be efficiently directed to genes residing on a chromosome in the process of losing its homolog.

References

Adler, D. A., Rugarli, E. I., Lingenfelter, P. A., Tsuchiya, K., Poslinski, D., Liggitt, H. D., Chapman, V. M., R. W. E., Ballabio, A., and Disteche, C. M. (1997). Evidence of evolutionary up-regulation of the single active X chromosome in mammals based on Clc4 expression levels in *Mus spretus* and *Mus musculus. Proc. Natl. Acad. Sci. USA* **94**, 9244–9248.

Akhtar, A., and Becker, P. B. (2000). Activation of transcription through histone H4 acetylation by MOF, an acetyltransferase essential for dosage compensation in *Drosophila*. *Mol. Cell* **5**, 367–375.

Akhtar, A., Zing, D., and Becker, P. (2000). Chromodomains are protein-RNA interaction modules. *Nature* **407**, 405–409.

Altuvia, S., Weinstein-Fisher, D., Zhang, A., Postow, L., and Storz, G. (1997). A small, stable RNA induced by oxidative stress: role as a pleiotropic regulator and antimutator. *Cell* **90**, 43–53.

Amrein, H., and Axel, R. (1997). Genes expressed in neurons of adult male *Drosophila*. *Cell* **88**, 459–469.

Bachiller, D., and Sanchez, L. (1989). Further analysis on the male-specific lethal mutations that affect dosage compensation in *Drosophila melanogaster*. *Roux's Arch. Dev. Biol.* **198**, 34–38.

Baker, B. S., and Belote, J. M. (1983). Sex determination and dosage compensation in *Drosophila*. *Annu. Rev. Genetics* **17**, 345–393.

Barbash, D. A., and Cline, T. W. (1995). Genetic and molecular analysis of the autosomal component of the primary sex determination signal of *Drosophila melanogaster*. *Genetics* **141**, 1451–1471.

Bashaw, G. J., and Baker, B. S. (1995). The *msl-2* dosage compensation gene of *Drosophila* encodes a putative DNA-binding protein whose expression is sex specifically regulated by *Sex-lethal*. *Development* **121**, 3245–3258.

Bashaw, G. J., and Baker, B. S. (1997). The regulation of the *Drosophila* msl-2 gene reveals a function for Sex lethal in translational control. *Cell* **89**, 789–798.

Bell, L., Maine, E. M., Schedl, P., and Cline, T. W. (1988). *Sex-lethal*, a *Drosophila* sex determination switch gene, exhibits sex-specific RNA splicing and sequence similarity to RNA binding proteins. *Cell* **55**, 1037–1046.

Bell, L. R., Horabin, J. I., Schedl, P., and Cline, T. W. (1991). Positive autoregulation of *Sex-lethal* by alternative splicing maintains the female determined state in *Drosophila*. *Cell* **65**, 229–239.

Belote, J. M. (1983). Male-specific lethal mutations of *Drosophila melanogaster*. II. Parameters of gene action during male development. *Genetics* **105**, 881–896.

Belote, J. M., and Lucchesi, J. C. (1980a). Control of X chromosome transcription by the *maleless* gene in *Drosophila*. *Nature* **285**, 573–575.

Belote, J. M., and Lucchesi, J. C. (1980b). Male-specific lethal mutations of *Drosophila melanogaster*. *Genetics* **96**, 165–186.

Bernstein, M., and Cline, T. W. (1994). Differential effects of *Sex-lethal* mutations on dosage compensation early in *Drosophila* development. *Genetics* **136**, 1051–1061.

Bhadra, U., Pal-Bhadra, M., and Birchler, J. A. (1999). Role of the *male specific lethal (msl)* genes in modifying the effects of sex chromosomal dosage in *Drosophila*. *Genetics* **152**, 249–268.

Bone, J. R., and Kuroda, M. I. (1996). Dosage compensation regulatory proteins and the evolution of sex chromosomes in *Drosophila*. *Genetics* **144**, 705–713.

Bone, J. R., Lavender, J., Richman, R., Palmer, M. J., Turner, B. M., and Kuroda, M. I. (1994). Acetylated histone H4 on the male X chromosome is associated with dosage compensation in *Drosophila*. *Genes Dev.* **8**, 96–104.

Borsani, G., Tonlorenzi, R., Simmler, M. C., Dandolo, L., Arnaud, D., Capra, V., Grompe, M., Pizzuti, A., Muzny, D., Lawrence, C., Willard, H. F., Avner, P., and Ballabio, A. (1991). Characterization of a murine gene expressed from the inactive X chromosome. *Nature* **351**, 325–329.

Bridges, C. B. (1921). Triploid intersexes in *Drosophila melanogaster*. *Science* **54**, 252–254.

Brockdorff, N., Ashworth, A., Kay, G. F., Cooper, P., Smith, S., McCabe, V. M., Norris, D. P., Penny, G. D., Patel, D., and Rastan, S. (1991). Conservation of position and exclusive expression of mouse *Xist* from the inactive X chromosome. *Nature* **351**, 329–331.

Brown, C. J., Hendrich, B. D., Rupert, J. L., Lafreniere, R. G., Xing, Y., Lawrence, J., and Willard, H. F. (1992). The human *XIST* gene: Analysis of a 17 kb inactive-X-specific RNA that contains conserved repeats and is highly localized within the nucleus. *Cell* **71**, 527–542.

Bull, J. J. (1983). "Evolution of Sex Determining Mechanisms." Benjamin/Cummings, Menlo Park, CA.

Chang, K. A., and Kuroda, M. I. (1998). Modulation of MSL1 abundance in female *Drosophila* contributes to the sex specificity of dosage compensation. *Genetics* **150,** 699–709.

Charlesworth, B. (1978). Model for evolution of Y chromosomes and dosage compensation. *Proc. Natl. Acad. Sci. USA* **75,** 5618–5622.

Charlesworth, B. (1991). The evolution of sex chromosomes. *Science* **251,** 1030–1033.

Charlesworth, B. (1996). The evolution of chromosomal sex determination and dosage compensation. *Curr. Biol.* **6,** 149–162.

Charlesworth, B., and Charlesworth, D. (2000). The degeneration of Y chromosomes. *Phil. Trans. Roy. Soc. B* **355,** 1563–2572.

Charlesworth, B., Charlesworth, D., Hnilicka, J., Yu, A., and Guttman, D. S. (1997). Lack of degeneration of loci on the neo-Y chromosome of *Drosophila americana* americana. *Genetics* **145,** 989–1002.

Cheung, P., Tanner, K. G., Cheung, W. L., Sassone-Corsi, P., Denu, J. M., and Allis, C. D. (2000). Synergistic coupling of histone H3 phosphorylation and acetylation in response to epidermal growth factor stimulation. *Mol. Cell* **5,** 905–915.

Clayton, A. L., Rose, S., Barratt, M. J., and Mahadevan, L. C. (2000). Phosphoacetylation of histone H3 on c-*fos*- and c-*jun*-associated nucleosomes upon gene activation. *EMBO J.* **19,** 3714–3726.

Clemson, C. M., McNeil, J. A., Willard, H. F., and Lawrence, J. B. (1996). XIST RNA paints the inactive X chromosome at interphase: Evidence for a novel RNA involved in nuclear/chromosome structure. *J. Cell Biol.* **132,** 259–275.

Cline, T. W. (1978). Two closely linked mutations in *Drosophila melanogaster* that are lethal to opposite sexes and interact with *daughterless*. *Genetics* **90,** 683–698.

Cline, T. W. (1984). Autoregulatory functioning of a *Drosophila* gene product that establishes and maintains the sexually determined state. *Genetics* **107,** 231–277.

Cline, T. W. (1986). A female-specific lethal lesion in an X-linked positive regulator of the *Drosophila* sex determination gene, *Sex lethal*. *Genetics* **113,** 641–663.

Cline, T. W. (1988). Evidence that *sisterless-a* and *sisterless-b* are two of several discrete "numerator elements" of the X/A sex determination signal in *Drosophila* that switch *Sxl* between two alternative stable expression states. *Genetics* **119,** 829–862.

Cline, T. W., and Meyer, B. J. (1996). Vive la difference: Males vs. females in flies vs. worms. *Annu. Rev. Genet.* **30,** 637–702.

Copps, K., Richman, R., Lyman, L. M., Chang, K. A., Rampersad-Ammons, J., and Kuroda, M. I. (1998). Complex formation by the *Drosophila* MSL proteins: Role of the MSL2 RING finger in protein complex assembly. *EMBO J.* **17,** 5409–5417.

Duffy, J. B., and Gergen, J. P. (1991). The *Drosophila* segmentation gene runt acts as a position-specific numerator element necessary for the uniform expression of the sex-determining gene Sex-lethal. *Genes Dev.* **5,** 2176–2187.

Eisen, A., Utley, R. T., Nourani, N., Allard, S., Schmidt, P., Lane, W. S., Lucchesi, J. C., and Cote, J. (2001). The yeast NuA4 and *Drosophila* MSL complexes contain homologous subunits important for transcriptional regulation. *J. Biol. Chem.* **276,** 3484–3491.

Erdmann, V. A., Barchiszewska, M. Z., Szymanski, M., Hochberg, A., de Groot, N., and Barciszewski, J. (2001). The non-coding RNAs as riboregulators. *Nucleic Acids Res.* **29,** 189–193.

Erickson, J. W., and Cline, T. W. (1991). Molecular nature of the *Drosophila* sex determination signal and its link to neurogenesis. *Science* **251,** 1071–1074.

Erickson, J. W., and Cline, T. W. (1993). A bZIP protein, Sisterless-a, collaborates with bHLH transcription factors early in *Drosophila* development to determine sex. *Genes Dev.* **7,** 1688–1702.

Franke, A., and Baker, B. S. (1999). The *roX1* and *roX2* RNAs are essential components of the compensasome, which mediates dosage compensation in *Drosophila*. *Mol. Cell* **4,** 117–122.

Fukunaga, A., Tanaka, A., and Oishi, K. (1975). *Maleless,* a recessive autosomal mutant of *Drosophila melanogaster* that specifically kills male zygotes. *Genetics* **81,** 135–141.

Gergen, J. P. (1987). Dosage compensation in *Drosophila*: Evidence that *daughterless* and *Sex-lethal* control X chromosome activity at the blastoderm stage of embryogenesis. *Genetics* **117**, 477–485.

Gorman, M., Franke, A., and Baker, B. S. (1995). Molecular characterization of the *male-specific lethal-3* gene and investigation of the regulation of dosage compensation in *Drosophila*. *Development* **121**, 463–475.

Gorman, M., Kuroda, M. I., and Baker, B. S. (1993). Regulation of the sex-specific binding of the *maleless* dosage compensation protein to the male X chromosome in *Drosophila*. *Cell* **72**, 39–49.

Gu, W., Szauter, P., and Lucchesi, J. C. (1998). Targeting of MOF, a putative histone acetyl transferase, to the X chromosome of *Drosophila melanogaster*. *Dev. Genet.* **22**, 56–64.

Gu, W., Wei, X., Pannuti, A., and Lucchesi, J. C. (2000). Targeting the chromatin-remodeling MSL complex of *Drosophila* to its sites of action on the X chromosome requires both acetyl transferase and ATPase activities. *EMBO J.* **19**, 5202–5211.

Hilfiker, A., Hilfiker-Kleiner, D., Pannuti, A., and Lucchesi, J. C. (1997). mof, a putative acetyl transferase gene related to the Tip60 and MOZ human genes and to the SAS genes of yeast, is required for dosage compensation in *Drosophila*. *EMBO J.* **16**, 2054–2060.

Jegalian, K., and Page, D. C. (1998). A proposed path by which genes common to mammalian X and Y chromosomes evolve to become X inactivated. *Nature* **394**, 776–780.

Jin, Y., Wang, Y., Johansen, J., and Johansen, K. M. (2000). JIL-1, a chromosomal kinase implicated in regulation of chromatin structure, associates with the Male Specific Lethal (MSL) dosage compensation complex. *J. Cell Biol.* **149**, 1005–1010.

Jin, Y., Wang, Y., Walker, D. L., Dong, H., Conley, C., Johansen, J., and Johansen, K. M. (1999). JIL-1: A novel chromosomal tandem kinase implicated in transcriptional regulation in *Drosophila*. *Mol. Cell* **4**, 129–135.

Jinks, T. M., Polydorides, A. D., Calhoun, G., and Schedl, P. (2000). The JAK/STAT signaling pathway is required for the initial choice of sexual identity in *Drosophila melanogaster*. *Mol. Cell* **5**, 581–587.

Kalmykova, A. I., Shevelyov, Y. Y., Dobritsa, A. A., and Gvozdev, V. A. (1997). Acquisition and amplification of a testis-expressed autosomal gene, *SSL*, by the *Drosophila* Y chromosome. *Proc. Natl. Acad. Sci. USA* **94**, 6297–6302.

Keightley, P. D., and Eyre-Walker, A. (2000). Deleterious mutations and the evolution of sex. *Science* **290**, 331–333.

Kelley, R. L., and Kuroda, M. I. (1995). Equality for X chromosomes. *Science* **270**, 1607–1610.

Kelley, R. L., and Kuroda, M. I. (2000). The role of chromosomal RNAs in marking the X for dosage compensation. *Curr. Opin. Genet. Dev.* **10**, 555–561.

Kelley, R. L., Meller, V. H., Gordadze, P. R., Roman, G., Davis, R. L., and Kuroda, M. I. (1999). Epigenetic spreading of the *Drosophila* dosage compensation complex from *roX* RNA genes into flanking chromatin. *Cell* **98**, 513–522.

Kelley, R. L., Solovyeva, I., Lyman, L. M., Richman, R., Solovyev, V., and Kuroda, M. I. (1995). Expression of Msl-2 causes assembly of dosage compensation regulators on the X chromosomes and female lethality in *Drosophila*. *Cell* **81**, 867–877.

Kelley, R. L., Wang, J., Bell, L., and Kuroda, M. I. (1997). Sex lethal controls dosage compensation in *Drosophila* by a nonsplicing mechanism. *Nature* **387**, 195–199.

Keyes, L. N., Cline, T. W., and Schedl, P. (1992). The primary sex determination signal of *Drosophila* acts at the level of transcription. *Cell* **68**, 933–943.

Kingston, R. E., and Narlikar, G. J. (1999). ATP-dependent remodeling and acetylation as regulators of chromatin fluidity. *Genes Dev.* **13**, 2339–2352.

Koonin, E. V., Zhou, S., and Lucchesi, J. C. (1995). The chromo superfamily—new members, duplication of the chromodomain and possible role in delivering transcriptional regulators to chromatin. *Nucleic Acids Res.* **23**, 4229–4233.

Kuroda, M. I., Kernan, M. J., Kreber, R., Ganetzky, B., and Baker, B. S. (1991). The *maleless* protein associates with the X chromosome to regulate dosage compensation in *Drosophila*. *Cell* **66**, 935–947.

Lahn, B. T., and Page, D. C. (1999a). Four evolutionary strata on the human X chromosome. *Science* **286**, 964–967.

Lahn, B. T., and Page, D. C. (1999b). Retroposition of autosomal mRNA yielded testis-specific gene family on human Y chromosome. *Nature Genet.* **21**, 429–433.

Lakhotia, S. C. (1996). RNA polymerase II dependent genes that do not code for protein. *Indian J. Biochem. Biophys.* **33**, 93–102.

Lakhotia, S. C., and Mukherjee, A. S. (1969). Chromosomal basis of dosage compensation in *Drosophila*. I. Cellular autonomy of hyperactivity of the male X-chromosome in salivary glands and sex differentiation. *Genet. Res.* **14**, 137–150.

Lanz, R. B., McKenna, N. J., Onate, S. A., Albrecht, U., Wong, J., Tsai, S. Y., Tsai, M.-J., and O'Malley, B. W. (1999). A steroid receptor coactivator, SRA, functions as an RNA and is present in an SRC-1 complex. *Cell* **97**, 17–27.

Lee, C.-G., and Hurwitz, J. (1993). Human RNA helicase A is homologous to the maleless protein of *Drosophila*. *J. Biol. Chem.* **268**, 16822–16830.

Lee, C.-G., Soares, V., Newberger, C., Manova, K., Lacy, E., and Hurwitz, J. (1998). RNA helicase A is essential for normal gastrulation. *Proc. Natl. Acad. Sci. USA* **95**, 13709–13713.

Lee, J. T., and Jaenisch, R. (1997). Long-range *cis* effects of ectopic X-inactivation centres on a mouse autosome. *Nature* **386**, 275–279.

Lucchesi, J. C., and Manning, J. E. (1987). Gene dosage compensation in *Drosophila melanogaster*. *Adv. Genet.* **24**, 371–429.

Lucchesi, J. C., Skripsky, T., and Tax, F. E. (1982). A new male-specific lethal mutation in *Drosophila melanogaster*. *Genetics* **100**, s42.

Lyman, L. M., Copps, K., Rastelli, L., Kelley, R. L., and Kuroda, M. I. (1997). *Drosophila* male-specific lethal-2 protein: Structure/function analysis and dependence on MSL-1 for chromosome association. *Genetics* **147**, 1743–1753.

Lyon, M. F. (1961). Gene action in the X-chromosome of the mouse (*Mus musculus* L.). *Nature* **190**, 372–373.

Marahrens, Y., Panning, B., Dausman, J., Strauss, W., and Jaenisch, R. (1997). *Xist*-deficient mice are defective in dosage compensation but not spermatogenesis. *Genes Dev.* **11**, 156–166.

Marín, I., and Baker, B. S. (2000). Origin and evolution of the regulatory gene male-specific lethal-3. *Mol. Biol. Evol.* **17**, 1240–1250.

Marín, I., Franke, A., Bashaw, G. J., and Baker, B. S. (1996). The dosage compensation system of *Drosophila* is co-opted by newly evolved X chromosomes. *Nature* **383**, 160–163.

Marín, I., Siegal, M., and Baker, B. S. (2000). The evolution of dosage compensation mechanisms. *Bioessays* **12**, 1106–1114.

Meise, M., Hilfiker-Kleiner, D., Dubendorfer, A., Brunner, C., Nothiger, R., and Bopp, D. (1998). Sex-lethal, the master sex-determining gene in *Drosophila*, is not sex-specifically regulated in *Musca domestica*. *Development* **125**, 1487–1494.

Meller, V. H., and Fisher, P. A. (1995). Nuclear distribution of *Drosophila* DNA topoisomerase II is sensitive to both RNase and DNase. *J. Cell Sci.* **108**, 1651–1657.

Meller, V. H., Gordadze, P. R., Park, Y., Chu, X., Stuckenholz, C., Kelley, R. L., and Kuroda, M. I. (2000). Ordered assembly of *roX* RNAs into MSL complexes on the dosage compensated X chromosome in *Drosophila*. *Curr. Biol.* **10**, 136–143.

Meller, V. H., Wu, K. H., Roman, G., Kuroda, M. I., and Davis, R. L. (1997). *roX1* RNA paints the X chromosome of male *Drosophila* and is regulated by the dosage compensation system. *Cell* **88**, 445–457.

Mukherjee, A. S., and Beermann, W. (1965). Synthesis of ribonucleic acid by the X-chromosomes of *Drosophila melanogaster* and the problem of dosage compensation. *Nature* **207**, 785–786.

Nagoshi, R. N., McKeown, M., Burtis, K. C., Belote, J. M., and Baker, B. S. (1988). The control of alternative splicing at genes regulating sexual differentiation in *D. melanogaster*. *Cell* **53,** 229–236.

Nickerson, J. A., Krochmalnic, G., Wan, K. M., and Penman, S. (1989). Chromatin architecture and nuclear RNA. *Proc. Natl. Acad. Sci. USA* **86,** 177–181.

Palmer, M. J., Mergner, V. A., Richman, R., Manning, J. E., Kuroda, M. I., and Lucchesi, J. C. (1993). The *male specific lethal-one* gene encodes a novel protein that associates with the male X chromosome in *Drosophila*. *Genetics* **134,** 545–557.

Palmer, M. J., Richman, R., Richter, L., and Kuroda, M. I. (1994). Sex-specific regulation of the *male-specific lethal-1* dosage compensation gene in *Drosophila*. *Genes Dev.* **8,** 698–706.

Pannuti, A., and Lucchesi, J. C. (2000). Recycling to remodel: evolution of dosage-compensation complexes. *Curr. Opin. Genet. Dev.* **10,** 644–650.

Parkhurst, S. M., Bopp, D., and Ish-Horowicz, D. (1990). X:A ratio, the primary sex-determining signal in *Drosophila*, is transduced by helix-loop-helix proteins. *Cell* **63,** 1179–1191.

Patterson, J. T., and Stone, W. S. (1952). "Evolution in the Genus *Drosophila*." Macmillan, New York.

Penny, G. D., Kay, G. F., Sheardown, S. A., Rastan, S., and Brockdorff, N. (1996). Requirement for *Xist* in X chromosome inactivation. *Nature* **379,** 131–137.

Qian, S., and Pirrotta, V. (1995). Dosage compensation of the *Drosophila white* gene requires both the X chromosome environment and multiple intragenic elements. *Genetics* **139,** 733–744.

Rice, W. R. (1992). Sexually antagonistic genes: Experimental evidence. *Science* **256,** 1436–1439.

Rice, W. R. (1996). Evolution of the Y sex chromosome in animals. *Bioscience* **46,** 331–343.

Ritossa, F., Atwood, K. C., and Speigelman, S. (1966). A molecular explanation of the bobbed mutants of *Drosophila melanogaster* as partial deficiencies of "ribosomal" DNA. *Genetics* **54,** 819–834.

Saccone, G., Peluso, I., Artiaco, D., Giordano, E., Bopp, D., and Polito, L. C. (1998). The Ceratitis capitata homologue of the *Drosophila* sex-determining gene Sex-lethal is structurally conserved, but not sex-specifically regulated. *Development* **125,** 1495–1500.

Schutt, C., and Nothiger, R. (2000). Structure, function and evolution of sex-determining systems in Dipteran insects. *Development* **127,** 667–677.

Scott, M. J., Pan, L. L., Cleland, S. B., Knox, A. L., and Heinrich, J. (2000). MSL1 plays a central role in assembly of the MSL complex, essential for dosage compensation in *Drosophila*. *EMBO J.* **19,** 144–155.

Sefton, L., Timmer, J. R., Zhang, Y., Beranger, F., and Cline, T. W. (2000). An extracellular activator of the *Drosophila* JAK/STAT pathway is a sex-determination signal element. *Nature* **405,** 970–973.

Skripsky, T., and Lucchesi, J. C. (1982). Intersexuality resulting from the interaction of sex-specific lethal mutations in *Drosophila melanogaster*. *Dev. Biol.* **94,** 153–162.

Slack, F. J., Basson, M., Liu, Z., Ambros, V., Horvitz, H. R., and Ruvkun, G. (2000). The lin-41 RBCC gene acts in the C. *elegans* heterochronic pathway between the let-7 regulatory RNA and the LIN-29 transcription factor. *Mol. Cell* **5,** 659–669.

Smith, E. R., Pannuti, A., Gu, W., Steurnagel, A., Cook, R. G., Allis, C. D., and Lucchesi, J. C. (2000). The *Drosophila* MSL complex acetylates histone H4 at lysine 16, a chromatin modification linked to dosage compensation. *Mol. Cell Biol.* **20,** 312–318.

Steinemann, M. (1982). Multiple sex chromosomes in *Drosophila* miranda: a system to study the degeneration of a chromosome. *Chromosoma* **86,** 59–76.

Steinemann, M., Steinemann, S., and Turner, B. M. (1996). Evolution of dosage compensation. *Chromosome Res.* **4,** 185–190.

Strobel, E., Pelling, C., and Arnheim, N. (1978). Incomplete dosage compensation in an evolving *Drosophila* sex chromosome. *Proc. Natl. Acad. Sci. USA* **75,** 931–935.

Sturtevant, A. H. (1945). A gene in *Drosophila melanogaster* that transforms females into males. *Genetics* **30,** 297–299.

Torres, M., and Sanchez, L. (1989). The *scute* (T4) gene acts as a numerator element of the X:A signal that determines the state of activity of *Sex-lethal* in *Drosophila*. *EMBO J.* **8,** 3079–3086.

Turner, B. M., Birley, A. J., and Lavender, J. (1992). Histone H4 isoforms acetylated at specific lysine residues define individual chromosomes and chromatin domains in *Drosophila* polytene nuclei. *Cell* **69,** 375–384.

Uchida, S., Uenoyama, T., and Oishi, K. (1981). Studies on the sex-specific lethals of *Drosophila melanogaster*. III. A third chromosome male-specific lethal mutant. *Jpn. J. Genet.* **56,** 523–527.

Uenoyama, T. (1984). Studies on the sex-specific lethals of *Drosophila melanogaster*. VIII. Enhancement and supression of *Sxl*. *Jpn. J. Genet.* **59,** 335–348.

Uenoyama, T., Uchida, S., Fukunaga, A., and Oishi, K. (1982). Studies on the sex-specific lethals of *Drosophila melanogaster*. IV. Gynandromorph analysis of three male-specific lethals, *mle, msl-2,* and *mle(3)132. Genetics* **102,** 223–231.

Wang, Y., Zhang, W., Jin, Y, Johansen, J., and Johansen, K. (2001). The JIL-1 tandem kinase mediates histone H3 phosphorylation and is required for maintenance of chromatin structure in *Drosophila*. *Cell* **105,** 433–443.

Wassarman, K. M., and Storz, G. (2000). 6S RNA regulates *E. coli* RNA polymerase activity. *Cell* **101,** 613–623.

Yamashita, A., Watanabe, Y., Nukina, N., and Yamamoto, M. (1998). RNA-assisted nuclear transport of the meiotic regulator Mei2p in fission yeast. *Cell* **95,** 115–123.

Yanowitz, J. L., Deshpande, G., Calhoun, G., and Schedl, P. D. (1999). An N-terminal truncation uncouples the sex-transforming and dosage compensation functions of Sex-lethal. *Mol. Cell Biol.* **19,** 3018–3028.

Younger-Shepherd, S., Vaessin, H., Bier, E., Jan, L. Y., and Jan, Y. N. (1992). deadpan, an essential pan-neural gene encoding an HLH protein, acts as a denominator element in *Drosophila* sex determination. *Cell* **70,** 911–912.

Zhang, A., Altuvia, S., Tiwari, A., Argaman, L., Hengge-Aronis, R., and Storz, G. (1998). The OxyS regulatory RNA represses rpoS translation and binds the Hfq (HF-1) protein. *EMBO J.* **17,** 6061–6068.

Zhou, S., Yang, Y., Scott, M. J., Pannuti, A., Fehr, K. C., Eisen, A., Koonin, E. V., Fouts, D. L., Wrighteman, R., Manning, J. E., and Lucchesi, J. C. (1995). Male-specific-lethal 2, a dosage compensation gene of *Drosophila* that undergoes sex-specific regulation and encodes a protein with a RING finger and a metallothionein-like cluster. *EMBO J.* **14,** 2884–2895.

2

Is X-Chromosome Inactivation a Homology Effect?

Jeannie T. Lee*

Department of Genetics
Howard Hughes Medical Institute
Harvard Medical School, and
Department of Molecular Biology
Massachusetts General Hospital
Boston, Massachusetts 02114

*To whom correspondence should be addressed: E-mail: lee@frodo.mgh.harvard.edu.

The evolution of unequally sized X/Y sex chromosomes posed a major gene dosage problem for mammals. While females have two X-chromosomes (XX), males have one X and one Y chromosome (XY). Because the Y contains just a tiny fraction of the X chromosome's 2000 genes, this system of sex determination endows the female with nearly twice the number of sex chromosome genes. As first described by Mary Lyon 40 years ago, X inactivation evolved to compensate for this dosage imbalance by silencing one whole X chromosome in XX individuals (Lyon, 1961). Dosage compensation is essential for proper embryonic development, because embryos that fail to achieve it differentiate poorly and die at the time of uterine implantation (Tagaki and Abe, 1990). Molecular switches that control the uniquely mammalian phenomenon of X inactivation have been under intense study for both their clinical relevance as well as their intrinsically curious properties.

Great progress has been made in recent years in understanding the molecular controls governing the tightly regulated steps of X inactivation. To date, only two essential regulatory genes have been identified, and both encode untranslated RNAs which together make a sense and antisense pair (Brown *et al.*, 1991a; Lee *et al.*, 1999a). An understanding of their importance comes at a time when noncoding and antisense RNA genes are generally becoming the focus of related phenomena such as genomic imprinting (see Chapter 5) and RNAi-like processes in posttranscriptional gene silencing (see Chapters 8, 9, and 11). A wealth of intriguing models have been proposed, some with underpinnings in homology-based mechanisms such as those discussed elsewhere in this book.

Is X inactivation a homology effect? Indeed, the initiation of inactivation on one X chromosome depends on the presence of additional X-chromosome homologs. The goal of this chapter is not to provide a comprehensive review, especially since many excellent summaries and historical accounts already exist (Lyon, 1972; Gartler and Riggs, 1983; Heard *et al.*, 1997; Goto and Monk, 1998; Avner and Heard, 2001). Instead, I will concentrate on studies which have helped to bring mechanism into sharper focus, particularly those addressing events at the onset of X inactivation. Here, as in any rapidly evolving field, one will find issues passionately debated. I will discuss the major hypotheses of the field and, in keeping with the theme of this book, pay special attention to current models where homology effects play a role.

I. THE PHENOMENON OF X INACTIVATION

The inactive X is classic heterochromatin in that it is transcriptionally silent (Graves and Gartler, 1986), is condensed and relatively insensitive to DNAse I digestion (Dyer *et al.*, 1985), is hypermethylated within CpG islands at the 5′ ends of X-linked genes (Liskay and Evans, 1980; Mohandas *et al.*, 1981), is late-replicating in S phase (Priest *et al.*, 1967), and is hypoacetylated in the core histones of the nucleosome (Jeppesen and Turner, 1993; Belyaev *et al.*, 1996). These are properties common to many other types of heterochromatin found across the phylogenetic tree. Once established, the X heterochromatin is maintained by mechanisms similar to those at other silent loci. Features which distinguish X inactivation from other types of heterochromatin lie in how silencing is initiated. In this regard, the two forms of X-inactivation—imprinted and random— differ critically in the trigger for silencing (Figure 2.1).

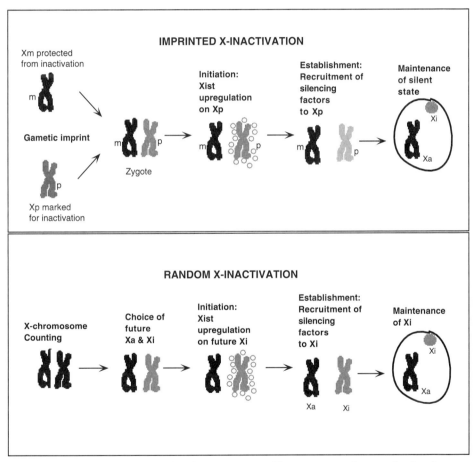

Figure 2.1. Steps of imprinted and random X inactivation. Shown are events of imprinted and random X inactivation in an XX cell. "m" and "p" signify maternal and paternal imprints acquired during oogenesis and spermatogenesis, respectively. Xm, maternal X; Xp, paternal X; Xa, active X; Xi, inactive X. See text for details of events.

A. Imprinted X inactivation

It is believed that X inactivation evolved as an imprinted mechanism more than 130 million years ago in the mammalian subclass Metatheria (Richardson and Czuppon, 1971; Sharman, 1971; Graves, 1996). In this imprinted form, X inactivation occurs only on the X inherited from the father. Imprinted X inactivation is still the primary mode of dosage compensation in extant marsupials such as the kangaroo and opposum, but vestiges can also be found in placental mammals such as the house mouse, *Mus musculus*, in whom paternal inactivation is recapitulated in early ontogeny and specifically in extraembryonic tissues (Takagi and Sasaki, 1975; West *et al.*, 1977).

        The imprint controlling this pattern of nonrandom X inactivation is known to be parentally determined, most likely by signals imparted during gametogenesis. Elegant nuclear transplantation experiments show that reciprocal regulatory signals come from both mother and father (Kay *et al.*, 1994; Latham, 1996; Goto and Takagi, 2000). Embryos carrying one or two paternal X chromosomes shut off all X chromosomes regardless of X-chromosome number. This shows that paternal origin predestines an X chromosome for inactivation. In contrast, embryos carrying two maternal X chromosomes cannot inactivate either X or do so only after a long delay. This implicates the existence of a maternally expressed factor which protects the X chromosome from inactivation. Recent studies suggest that this maternal factor may be the primary signal for imprinting and that paternal inactivation may be a default consequence of maternal resistance to inactivation (Tada *et al.*, 2000). Thus, imprinted X inactivation appears to be a strictly parent-of-origin phenomenon and occurs independently of X-chromosome number.

B. Random X inactivation

In placental mammals (Eutheria), imprinted X inactivation is recapitulated only during early ontogeny (Takagi and Sasaki, 1975; West *et al.*, 1977). The primary mode of inactivation is random, so that either the maternal or paternal X can be silenced. In mice, random X inactivation takes place shortly after implantation in the epiblast lineage which forms the embryo proper (Rastan, 1982). A key difference between random and imprinted inactivation lies in the fact that the random process is not influenced by parent of origin, but depends instead on a zygotic X-chromosome counting mechanism which assesses whether additional X chromosomes are present. The arithmetic aspect of random X-inactivation is therefore similar to that of dosage compensation in *Drosophila melanogaster* and *Caenorhabditis elegans*, where zygotic counting also provides the initial trigger. The counting mechanism measures the relative number of X chromosomes to haploid autosome sets (Kelley and Kuroda, 1995). For X inactivation, studies of variant mouse and human cells show that diploid cells maintain one active X and silence all additional X's (Rastan, 1983; Rastan and Robertson, 1985). Therefore, XY or X0 cells inactive none, XX cells inactivate one X, and XXX cells inactivate two. In contrast, tetraploid cells (4n or having four haploid genomes) can maintain two active X's (Webb *et al.*, 1992), so that XX cells inactivate none, XXX inactivate one, and XXXX cells inactivate two. It is also clear that the Y chromosome does not influence events of X inactivation, because XXY and XXXY cells behave as though they are XX and XXX with respect to inactivation.

Following the counting step, a choice mechanism designates one active and one inactive X chromosome. The ability to select either the maternal or paternal X is another key difference between random and imprinted X inactivation. Choice takes place in the epiblast during early gastrulation and occurs in a cell-autonomous and mitotically stable fashion. Thus, the eutherian XX embryo is a mosaic of cells expressing the X chromosome of mother or father. This means that XX clonal individuals are genotypically identical but phenotypically distinct, a fact that leads to considerable variability of heritable traits in otherwise identical female twins.

II. CONTROL ELEMENTS

A. The X inactivation center

Shortly after Lyon's discovery, the concept of a master control center was introduced by classic geneticists based on the observation that an X-chromosome break frequently resulted in one fragment that could still undergo inactivation and a second that could no longer do as (Russell, 1963; Eicher, 1970; Brown *et al.*, 1991b). Cytogenetic observations pointed to a region in the middle of the mouse and human X chromosome as the critical locus of control. Called the "X-inactivation center" (Xic), this locus must be present in more than one copy for X inactivation to occur and it must be present on the X chromosome designated for inactivation. Positional cloning and chance have enabled identification of several genes lying within or close to the Xic (Figure 2.2), including Xist (Brown *et al.*, 1991a), Tsix (Lee *et al.*, 1999a), Tsx (Simmler *et al.*, 1996), Brx (Simmler *et al.*, 1997), and Cdx4 (Horn and Ashworth, 1995)—two of which (Xist and Tsix) have emerged as essential regulatory genes and will be discussed in detail in ensuing sections.

To delineate minimal sequence requirements at the Xic, studies in the 1990s made use of transgenesis technology in mouse embryonic stem (ES) cells (Figure 2.2)(Heard *et al.*, 1996; Lee *et al.*, 1996; Matsuura *et al.*, 1996; Herzing *et al.*, 1997; Heard *et al.*, 1999a; Lee *et al.*, 1999b). Murine ES cells have provided an excellent *in vitro* model in which to dissect the requirements for X-inactivation,

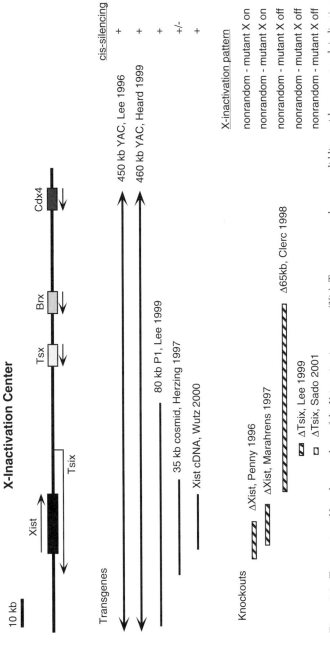

Figure 2.2. Transgenic and knockout analyses of the X-inactivation center (*Xic*). Transgenes are shown as solid lines, with arrows at ends indicating that the transgene extends beyond the region shown. Cis silencing is defined by experimental evidence of *Xic*-induced *Xist* RNA coating or autosomal silencing. +/– indicates that a linked reporter gene was silenced but autosome gene expression was not assessed. Knockouts are shown as hatched bars. The direction of transcription for each *Xic* gene is depicted by arrows.

because they are pluripotent undifferentiated cells that have not committed to X inactivation but can recapitulate its various steps when placed under differentiation conditions in culture. In XY ES cells, introducing a 450-kb fragment from the mouse *Xic* region can reconstitute counting, choice, and inactivation on mouse autosomes (Lee *et al.*, 1996; Lee and Jaenisch, 1997). Despite having only one X chromosome, the XY cells undergo dosage compensation and choose between the X and the transgene-bearing autosome for inactivation. Subsequent work has shown that autosomal transgenes carrying as little as 80 kb of the 450-kb sequence is sufficient to implement silencing of diverse sequences regardless of the chromosome origin, so long as the genes are collinear with the 80-kb sequence (Lee *et al.*, 1999b).

Similar works have been carried out using the human X-inactivation center (XIC) in mouse ES cells (Heard *et al.*, 1999b; Migeon *et al.*, 1999, 2001). A 480-kb fragment from the previously mapped human XIC can operate as an inactivation control center in the heterologous mouse background. Differentiation of the mouse ES cells leads to random choice between the mouse X and transgene-bearing autosome for the subsequent silencing step. Interestingly, cis silencing can be achieved on the mouse autosome even though the inactivation center is of human origin. These results demonstrate close conservation of mechanism between mouse and humans. However, some differences are also observed. The most significant difference is that cis inactivation can been seen in only a subset of cells (Heard *et al.*, 1999b). This difference might be due to some cross-species incompatibilities in the XIC/Xic and the silencing apparatus which lead to a stochastic fluctuation in the extent of cis silencing.

B. *Xist* and gene silencing

Xist expression leads to production of a noncoding RNA discovered 10 years ago in humans by Hunt Willard's group (Brown *et al.*, 1991a; 1992). A moderately conserved homolog was subsequently identified in mice (Brockdorff *et al.*, 1992). *Xist* makes a spliced and polyadenylated transcript that is more than 17 kb in size and has the unique property of being transcribed only from the inactive X chromosome. On the inactive X, *Xist* RNA accumulates along the length of the chromosome in a curious phenomenon that has been described as RNA "painting" or "coating" (Clemson *et al.*, 1996)(Figure 2.3). In human XX cells, *XIST* RNA associates with the inactive X only during interphase, while murine *Xist* RNA remains associated until early metaphase (B. Panning, unpublished observation). The unique ability of the RNA to paint the chromosome has led to the hypothesis that the RNA induces heterochromatin formation by directing silencing factors to cover the entire chromosome.

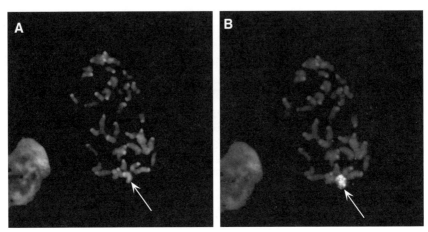

Figure 2.3. *Xist* RNA coats the inactive X-chromosome. (A) All chromosomes of a metaphase spread from a murine female fibroblast are stained by the DNA dye, DAPI. The inactive X is indicated by an arrow. (B) RNA fluorescence *in situ* hybridization (FISH) to a fluorescein-conjugated *Xist* probe shows that the inactive X is uniquely "painted" by *Xist* RNA. The arrow points to the fluorescent speckles representing *Xist* RNA.

Xist expression is developmentally regulated and occurs in three distinct expression states: basal, upregulated, and off (Figure 2.4)(Tai *et al.*, 1994; Beard *et al.*, 1995; Panning *et al.*, 1997; Sheardown *et al.*, 1997a). Prior to the onset of X inactivation and cell differentiation, *Xist* is expressed basally from all X chromosomes in XX and XY cells. At this time, the nascent *Xist* transcripts

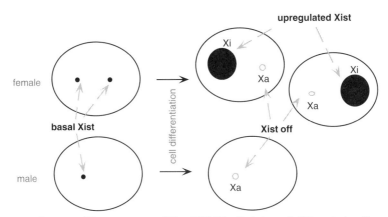

Figure 2.4. Patterns of *Xist* expression in XX and XY ES cells during cell differentiation. Xa, active X; Xi, inactive X.

can be visualized as pinpoint signals by RNA fluorescence *in situ* hybridization (RNA FISH). Cell differentiation is accompanied by changes in *Xist* expression. In XY cells, *Xist* is turned off. In XX cells, *Xist* is also turned off on the active X but is dramatically upregulated on the inactive X. Some estimate a 10- to 20-fold increase in steady-state RNA level (Buzin *et al.*, 1994), while others suggest as much as a 1000-fold increase (Keohane *et al.*, 1996).

RNA FISH analysis of mouse embryos show that upregulation on the future inactive X precedes the downregulation of *Xist* on the future active X (Panning *et al.*, 1997; Sheardown *et al.*, 1997a), implying that the switch from biallelic low-level *Xist* to monoallelic high-level expression occurs in a highly coordinated fashion. These results underscore the idea that some form of trans sensing must occur between the two X chromosomes at the onset of inactivation. The cross-communication ensures that the active and inactive Xs are selected in a mutually exclusive manner to avoid the error of having two active or two inactive Xs per nucleus. Supporting this, XX ES cultures rarely ever demonstrate cells with two high-level *Xist* RNA domains in RNA FISH experiments (J. T. Lee, unpublished observation).

It is widely accepted that *Xist* upregulation is both necessary and sufficient to initiate silencing on the X. Knockout experiments of Penny *et al.* (1996) and Marahrens *et al.* (1997) demonstrate that X chromosomes lacking this locus cannot be inactivated (Figure 2.2). In transgenesis experiments in XY ES cells by Wutz and Jaenisch (2000), high-level expression from a tet-inducible cassette was sufficient to silence autosomal genes in cis (Figure 2.2). Silencing can occur even in undifferentiated cells, cells which had until recently been thought to lack factors for the silencing step of X inactivation.

Timing of *Xist* expression is also critical. *Xist*-induced gene silencing is reversible in undifferentiated cells when *Xist* expression is removed, but becomes irreversible once cells proceed down the pathway of differentiation (Wutz and Jaenisch, 2000). Conversely, *Xist* must be upregulated early in cell differentiation; once cells have differentiated for more than 48 h, high-level *Xist* expression can no longer induce silencing. Thus, there appears to be a narrow time window during which the X-inactivation machinery can be brought to bear. Furthermore, once the inactive state is established, *Xist* and the *Xic* are dispensable and the X heterochromatin appears to become a self-propagating chromatin state (Brown and Willard, 1994; Csankovszki *et al.*, 1999). Nonetheless, *Xist* RNA continues to be synthesized throughout a female's life. Continuous synthesis might ensure proper dosage compensation in those rare instances of somatic X reactivation during female aging (Wareham *et al.*, 1987). Recent work supports this idea (Csankovszki *et al.*, 2001). Loss of *Xist* expresion in somatic female cells appears to destabilize the inactivate state, leading to higher rates of reactivation on the X. Thus, although not absolutely required to maintain X inactivation, continued *Xist* expression seems to be necessary to increase the stability of the inactive state.

While the role played by *Xist* is unequivocal, how it enacts silencing is still unclear. One possibility is that high-level *Xist* transcription alters the chromosome structure and facilitates heterochromatin formation. In light of *Xist* RNA's unique ability to paint the X, a more favored view is that *Xist* makes a functional RNA whose spread along the chromosome recruits silencing proteins to the X (Brockdorff *et al.*, 1992; Brown *et al.*, 1992). Despite intensive investigation, a search for interacting protein partners of *Xist* RNA has not yet been fruitful. Costanzi and Pehrson have proposed that *Xist* RNA might recruit macroH2A, an H2A histone variant that appears to be enriched on the inactive X in somatic XX cells (Costanzi and Pehrson, 1998). Although a direct interaction between the two factors has not yet been demonstrated, a conditional knockout of *Xist* in XX somatic cells shows that loss of *Xist* expression leads to a concomitant loss of macroH2A association (Csankovszki *et al.*, 1999), consistent with the idea that *Xist* RNA recruits macroH2A, although not necessarily directly. MacroH2A as found in the macrochromatin body also exhibits an intriguing pattern of association during the early stages of X inactivation. Prior to the onset of inactivation, a single macroH2A-dense body is found elsewhere in the nuclei of XX and XY cells (Mermoud *et al.*, 1999) in a region now known to co-localize with the centrosome (Rasmussen *et al.*, 2000). MacroH2A then relocates to the inactive X on the seventh day of ES cell differentiation, a time when X inactivation has already been initiated but not completely established (Mermoud *et al.*, 1999). This argues against a role in controlling intiation, but is consistent with a possible role in the establishment or maintenance of the silent state.

The story may be more complex, however, because recent work of Perche *et al.* (2000) suggests that histones in general are concentrated on the inactive X. That is, the higher degree of chromatin compaction on the inactive X might give its associated protein factors the appearance of being enriched on the chromosome. This, therefore, casts doubt on whether macroH2A is functionally any more relevant for X inactivation than other histones. Moreover, losing macroH2A association caused by a conditional deletion of *Xist* does not lead to X reactivation (Csankovszki *et al.*, 1999), indicating that macroH2A is not necessary for the maintenance of the inactive state. Nonetheless, it remains possible that some specific association with the X chromosome is required during the early phases of X inactivation, either in imprinted X inactivation (Costanzi *et al.*, 2000) or in establishment of random inactivation (Mermoud *et al.*, 1999).

Whatever the mechanism of *Xist* RNA binding, transgenic studies show that the ability to spread is context-independent, as ectopically expressed *Xist* RNA can also coat a number of different autosomes (Lee *et al.*, 1996, 1999b; Heard *et al.*, 1999a; Wutz and Jaenisch, 2000). Interestingly, human *Xist* RNA can also coat mouse autosomes when expressed from a transgene *in cis* (Heard *et al.*, 1999b; Migeon *et al.*, 1999). Thus, *Xist/XIST* RNA associates with diverse sequences regardless of chromosome origin so long as *Xist* is synthesized from that chromosome.

What chromosomal elements might mediate *Xist* RNA binding? Lyon (1998) has suggested that L1 repeat elements (LINEs) could serve as the "way stations" originally proposed by Gartler and Riggs (1983) as booster elements for the spread of inactivation from a control center. Indeed, LINEs are present on the X chromosome at higher density than anywhere else in the mouse and human genomes. Furthermore, autosomal silencing in X-to-autosome translocation chromosomes seems to occur more readily when the X fragment is juxtaposed to LINE-rich autosomal regions. Intriguingly, LINEs are most densely clustered around the X-inactivation center, and domains which escape inactivation are often LINE-poor (Bailey *et al.*, 2000).

C. The antisense gene, *Tsix*

We discovered antisense transcription through *Xist* when studying a transgene with an apparent *Xist* promoter truncation but which nonetheless appeared to have ample *Xist* expression (Lee *et al.*, 1999a). Using strand-specific techniques, we found that transcription actually occurs off the opposite DNA strand and that these antisense transcripts are far more abundant than *Xist* RNA in wild-type undifferentiated cells. Named *Tsix*, the antisense gene initiates 15 kb downstream of *Xist* within a prominent CpG island and ends more than 5 kb upstream of *Xist*, encompassing a total of 40 kb. The untranslated RNA is found only in the nucleus in close association with the *Xic* and is expressed only in early development during the time window of X inactivation. While the *Tsix* sequence is conserved at the human *XIC*, it is not known whether the human sequence is also associated with an antisense transcript.

Other groups have suggested that antisense transcription might extend much farther upstream. Debrand *et al.* (1999) showed that antisense transcription could be seen as far as 10 kb beyond the previously mapped *Tsix* start sites. This transcription occurs at very low levels that requires many additional rounds of RT-PCR for detection. Sado *et al.* (2001) later found that at least one spliced form of *Tsix* might originate ~15 kb upstream of the major *Tsix* start sites. This isoform of *Tsix* accounts for only a small fraction of total *Tsix* expression, as knocking out the major *Tsix* promoter abolishes the bulk of antisense transcripts (Lee and Lu, 1999; Sado *et al.*, 2001). Furthermore, knocking out the putative upstream promoter has no obvious consequence for *Tsix* expression (Sado *et al.*, 2001).

Several features immediately suggested *Tsix* as a possible antagonist of *Xist* action. First, *Tsix* expression is dynamically associated with *Xist* expression during cell differentiation (Figure 2.5). Prior to X inactivation, when *Xist* expression is low, *Tsix* expression is found on all X chromosomes in male and female cells. At the onset of X inactivation, *Tsix* RNA disappears on the future inactive X, the chromosome on which *Xist* expression is to be upregulated to high levels. At the same time, *Tsix* remains expressed on the future active X, where *Xist* expression

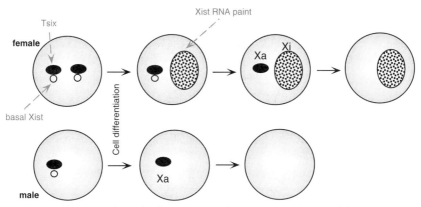

Figure 2.5. Dynamic relationship between *Xist* and *Tsix* expression during cell differentiation.

becomes repressed. Second, *Tsix* lies partly within a 65-kb region whose deletion leads to constitutive *Xist* expression in XX and X0 cells, as shown by Clerc and Avner (1998). The 65-kb knockout argues for a repressive element lying between *Xist* and *Brx* that acts in the counting and choice processes (see Section V). Third, *Tsix* is contained within the 80-kb *Xic* region which recapitulates the regulation of X inactivation in a transgene-based assay (Lee *et al.*, 1999b). Finally, an *Xist* cosmid transgene lacking the 5′ end of *Tsix* shows permissive *Xist* expression (Herzing *et al.*, 1997), consistent with an inhibitory effect of *Tsix* whose deletion enables full expression of *Xist*. These observations led to the proposal that *Tsix* is a repressor of *Xist* expression (Lee *et al.*, 1999), a model that will be discussed in detail in the following section.

III. CURRENT MODELS OF *Xist* REGULATION

With *Xist* as the workhorse of X inactivation, much attention has been centered around factors which control its expression. At least four models have been proposed which fall into distinct but not mutually exclusive categories.

A. Transcriptional regulation

The simplest way of explaining the low, high, and off states of *Xist* is that the gene is transcriptionally regulated. In this model, undifferentiated XX and XY cells would synthesize *Xist* at a low rate so that there is not sufficient RNA accumulation to coat the X chromosome. During differentiation, *Xist* transcription would be increased dramatically to initiate the process of chromosome-wide silencing. Thus, whether X inactivation is intitiated would depend primarily on recruitment of relevant transcriptional factors to silence *Xist* on the designated active X and to induce *Xist* on the designated inactive X. Promoter-bashing experiments have shown that the minimal promoter lies within 400 bp of the major transcriptional start site (Hendrich *et al.*, 1997; Sheardown *et al.*, 1997b). DNAse footprinting, gel retardation assays, and sequence analysis have turned up a number of candidate regulatory elements in the core promoter, including binding sites for Sp1, Tbp, and Yy1, none of which however, is unique to *Xist*. Developmental specificity must therefore lie in other as yet unidentified factors, most likely outside of the core promoter. A transcriptional model is consistent with the transgenic work of Wutz and Jaenisch (2000), which shows that high-level *Xist* transcription alone is sufficient to enable the transcript to accumulate along the X chromosome and effect cis silencing. This implies that the ability to accumulate *Xist* RNA is linked to there being adequate transcription levels.

B. Regulation by RNA stabilization

Arguing against a transcriptional model are two studies indicating that RNA stabilization might be a more relevant mechanism. Panning *et al.* (1997) and Sheardown *et al.* (1997a) hypothesize that *Xist* is controlled by RNA stabilization. Support for this idea comes from nuclear run-on experiments which suggest that the transcriptional rate across *Xist* is equal in undifferentiated cells and in fully differentiated XX cells. Furthermore, when new *Xist* synthesis is blocked by actinomycin D, the steady-state *Xist* RNA level remains high in somatic XX cells whereas *Xist* expression diminishes rapidly in undifferentiated XX ES cells. This argues for developmentally regulated differences in RNA stability as the critical factor rather than stimulation of *Xist* RNA synthesis.

The authors favor a stabilization mechanism involving protection by protein partners. This mechanism is directly analogous to that of dosage compensation in *Drosophila melanogaster,* where the required *roX1* and *roX2* RNAs associate with MSL proteins in ribonucleoprotein complexes decorating the length of the male X chromosome (Amrein and Axel, 1997; Meller *et al.*, 1997; Akhtar *et al.*, 2000). No stabilizing factors for *Xist* RNA have been isolated so far. The authors considered other mechanisms such as alternative splicing or nuclear-to-cytoplasmic RNA transport in altering half-lives. However, alternative splice forms are not evident by RT-PCR, and *Xist* RNA is not obviously found outside of the nucleus.

How can one reconcile the differences between the study by Wutz and Jaenisch and those by Panning *et al.* and Sheardown *et al?* Wutz and Jaenisch suggest that RNA stability might depend on the amount of RNA present, such that low-level expression might lead to quicker RNA turnover. It is also possible that intronic sequences missing in the cDNA transgene contribute to the stability of its transcribed product. Finally, it is possible that both transcriptional and posttranscriptional mechanisms play a role in regulating *Xist* expression.

C. Regulation by alternative promoters

Johnston *et al.* (1998) have proposed that differing *Xist* RNA half-lives result from usage of alternative promoters in *Xist*. By applying RT-PCR and RNA FISH analyses to a region immediately upstream of mouse *Xist*, the authors observe additional transcription extending 6–7 kb beyond the previously mapped promoter. They hypothesize the existence of a novel promoter (P_0) located at -6.6 kb. Transcription from this putative P_0 promoter is seen in undifferentiated XX and XY ES cells and cells of the preimplantation embryo, while transcription from the downstream promoter (P_1/P_2) is seen in somatic XX cells. It is therefore postulated that transcription from the putative upstream promoter produces an unstable RNA and a switch to the downstream promoter mediates the initiation of X inactivation by producing a stable transcript. In this model, the 5′ end of the P_0 transcript contains either a signal for rapid RNA degradation or a transport signal that leads *Xist* RNA away from the X chromosome.

While the P_0 hypothesis raises an interesting mechanism, recent evidence from several groups has not substantiated the existence or function of a P_0 promoter. In a transgenic model, Warshawsky *et al.* (1999) showed that P_0 is not needed for *Xist* expression in undifferentiated cells and that *Xist* transgenes lacking P_0 sequences can nonetheless produce a transcript with a short half-life. Furthermore, strand-specific techniques show that the upstream transcript is oriented in the antisense direction, most likely corresponding to the 3′ end of *Tsix* rather than the 5′ end of *Xist*. Warshawsky *et al.* propose that the expression profile of the P_0 RNA is actually that of *Tsix*. Finally, comparative genomic analysis by Nesterova *et al.* (2001) finds no evidence of sequence conservation around the P_0 promoter among mouse, vole, and human. Further answers regarding the status of P_0 will await knockout analyses currently in progress.

D. Regulation by *Tsix*

We have proposed that *Tsix* might inhibit *Xist* action using one or more of the following mechanisms (Figure 2.6)(Lee *et al.*, 1999a). First, the *Tsix* locus could contain critical DNA elements which silence *Xist* transcription at long range. In this scenario, *Tsix* RNA production would not be critical. Second, movement of an RNA polymerase complex in the antisense orientation could inhibit production of the sense transcript. Third, the RNA itself could be functional. *Tsix* RNA could mask domains within *Xist* RNA and thereby restrict access to silencing proteins. Alternatively, a *Tsix:Xist* RNA duplex could enhance degradation of *Xist* RNA. The last two types of mechanism are reminiscent of posttranscriptional gene silencing (Bass, 2000; and Chapters 8, 9, and 11), a phenomenon found across diverse kingdoms in which gene expression is controlled by double-stranded RNA-mediated degradation of homologous transcripts. Therefore, one possible

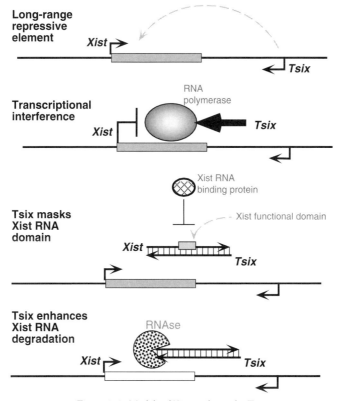

Figure 2.6. Models of Xist regulation by Tsix.

mechanism to consider in the future is an RNAi-like activity for the *Tsix:Xist* RNA duplex.

To test the role of *Tsix* in random X inactivation, we knocked out *Tsix* in ES cells by deleting a 4-kb sequence including the promoter and the CpG-rich 5′ end (Figure 2.2)(Lee and Lu, 1999). This deletion does not affect counting, since *Tsix*+/Y male and *Tsix*+/− female cells still show proper dosage compensation. However, there is a dramatic effect on X-chromosome choice, such that *Xist* expression and X-inactivation are highly highly skewed toward the mutant X in *Tsix*+/− cells. This shows that deleting *Tsix* leads to high-level *Xist* expression on the same X chromosome and renders that chromosome unable to stay active, thereby demonstrating that *Tsix* is a repressor of *Xist* expression in the random form of X inactivation. Additional evidence comes from transgenic analysis by Debrand *et al.* (1999), in which a deletion of the minisatellite, *DXPas34*, located at the 5′ end of *Tsix*, was engineered into an *Xic* YAC transgene. Autosomes carrying the mutated transgene are preferentially inactivated over the X chromosome.

A similar deletion encompassing *DXPas34* was made on the X chromosome by Sado *et al.* (2001) and further demonstrated that *Tsix* is a component of the choice mechanism.

Analysis in mice revealed that *Tsix* shows exclusive maternal expression during preimplantation development (Lee, 2000), an observation that contrasts with *Xist's* exclusive paternal expression during this time (Kay *et al.*, 1993). This raised the hypothesis that *Tsix* might also regulate imprinted X inactivation, a possibility made more intriguing in light of the emerging importance of antisense and noncoding loci in genomic imprinting (reviewed in Tilghman, 1999, and Chapter 5). To test this hypothesis, we transmitted the *Tsix* knockout through the mother or father (Lee, 2000). When inherited from the father, all offspring are normal, presumably because the paternal X normally does not express *Tsix*. When inherited from the mother, there is a sharp decrease in the birthrate of mutation-bearing offspring resulting from postimplantation defects. Mutant embryos show aberrant dosage compensation in extraembryonic tissues, with XY embryos inactivating their only X chromosome and XX embryos inactivating both their X chromosomes. These results argue that *Tsix* might be the putative maternally expressed factor which blocks imprinted X inactivation. Notably, the effects on *Xist* expression differ dramatically between cells that undergo imprinted inactivation (extraembryonic tissues) vs those that undergo random inactivation (epiblast, ES cells). This is presumed to reflect the fact that a counting mechanism operates in epiblast-derived cells and therefore enables inactivation of the proper number of X chromosomes. Similar results have been achieved by transmitting a *DXPas34* deletion through the mouse germline (Sado *et al.*, 2001).

The findings suggest that *Tsix* might harbor an imprinting center on which a mark is placed by the mother and father. The prominent CpG island, including the *DXPas34* marker (Simmler *et al.*, 1993) at the 5′ end of *Tsix*, might serve as such an imprinting center. This would be consistent with emerging evidence for a role of DNA methylation and differentially methylated domains in parental imprinting (Tilghman, 1999). Earlier analysis by Courtier *et al.* (1995) using methylation-sensitive enzymes suggests that *DXPas34* is differentially methylated on the active and inactive X. However, bisulfite sequence analysis of Prissette *et al.* (2001) has not upheld the earlier work. It appears that this CpG-rich region is generally hypomethylated prior to cell differentiation, but then becomes grossly hypermethylated in differentiated ES and somatic cells. This trend is similar in XX vs XY cells and in tissues with imprinted vs random inactivation. Therefore, the role of methylation at the *Tsix* imprinting center is unclear at this time.

Based on genetic analyses at the *Xic*, we have proposed the epistasis model shown in Figure 2.7. In this model, *Tsix* expression represses *Xist* action. On the future active X, continued *Tsix* expression at the onset of cell differentiation

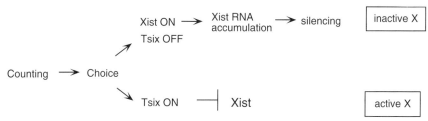

Figure 2.7. An epistasis model for X inactivation.

represses high-level *Xist* induction on that chromosome. On the future inactive X, the loss of *Tsix* expression enables upregulation of *Xist in cis*. Thus, the control of *Tsix* expression at the onset of X inactivation determines whether inactivation will take place on a given X chromosome. We believe that control of antisense expression lies in the mechanisms which specify counting and choice.

IV. EVIDENCE FOR ADDITIONAL CONTROL ELEMENTS IN RANDOM X INACTIVATION

A. Counting

There is general agreement that an X-chromosome counting mechanism determines whether dosage compensation will take place in XX and XY cells. The *Tsix* knockout suggests that counting is genetically separable from choice, because the deletion skews choice without apparently affecting counting. The work of Clerc and Avner (1998) implicates the 65-kb sequence immediately downstream of *Xist* in the role of counting. They engineered an X chromosome bearing a 65-kb deletion that removes 3 kb of sequence at the 3' end of *Xist*, about 15 kb of sequence at the 5' end of *Tsix*, all of the testis-specific gene, *Tsx*, and the 3' end of the brain-enriched gene, *Brx* (Figure 2.2). When the deletion is carried on one X-chromosome in an XX ES line, *Xist* expression is constitutive from the targeted X, leading to nonrandom inactivation of this chromosome, a result that might be explained by the lack of *Tsix* on that chromosome. However, when the deletion is present in a derivative cell line carrying only one intact X chromosome, the single mutant X also undergoes inactivation. This, therefore, points to additional critical elements downstream of *Xist* in the role of X-chromosome counting. Theoretically, this element could include any sequence between the 3' ends of *Xist* and *Brx*, a region encompassing 65 kb. Forthcoming finer deletions of this region will enable pinpointing precise elements.

B. Choice

Following a determination of X-chromosome number, a choice mechanism designates the future active and inactive X's. In the classic model, a single "blocking
factor" is produced by diploid cells and blocks the *Xic* on the future active X; all
remaining X chromosomes, unbound by the blocking factor, undergo inactivation
by default (Ohno, 1969; Eicher, 1970; Lyon, 1971). This would be the simplest
model to explain how diploid cells consistently keep one X on regardless of how
many X chromsomes are present. Choice would involve the selection of only the
X chromosome that will be active, and counting and choosing would essentially be
the same act. However, the *Tsix* knockout phenotype suggests that the mechanism
could be more complex. First, counting and choice are genetically separable, as
choice is skewed in the knockout without affecting counting. Second, XY ES cells
lacking *Tsix* expression still block X inactivation despite a lack of the *Xist* repressor,
suggesting that *Xist* is dually regulated by two mechanisms—one *Tsix*-dependent
and a second of unknown nature (Lee and Lu, 1999). As detailed in Figure 2.8,
it is proposed that the active and inactive Xs must each be purposefully selected.
In this model, the active X is chosen by a blocking factor (repressor) composed
of titrated diffusible X and autosomal factors, while the inactive X is chosen by
a "competence factor"—such as an activator of *Xist*—composed of remaining
untitrated X-linked factors. Such a model would explain why the *Tsix-/Y* male ES
cells did not undergo ectopic inactivation despite lacking *Tsix*: XY cells lack the
production of the competence factor.

   In light of the phenotype in *Tsix* knockout cells during random and
imprinted X inactivation, a role for *Tsix* in choice seems clear. Genetic evidence
indicates that other loci also play a role in this capacity. Cattanach and Papworth

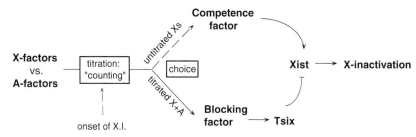

Figure 2.8. A model for dual regulation of *Xist*. In this model, X-chromosome factors (X-factors) and
autosomal factors (A-factors) are produced in limited quantities. The titration of X- and
A-factors (X:A ratio) represents the "counting" step of X inactivation. A complex of X-
and A-factors makes the "blocking factor" that blocks initiation of inactivation on the
future active X via a *Tsix*-dependent mechanism. The remaining X-factors which are not
titrated by A-factors constitute the "competence factor," which then upregulates *Xist* to
initiate inactivation on the future inactive X.

(1981) identified a genetic modifier called the "X-controlling element" (*Xce*), which biases the likelihood of X inactivation. The *Xce* maps within or very near the *Xic* and, in a hybrid background, the X chromosome with a stronger *Xce* haplotype has a higher chance of remaining on. So far, four *Xce* haplotypes have been identified (*Xce*^{a}, *Xce*^{b}, *Xce*^{c}, and *Xce*^{d}, in order of increasing strength). Simmler *et al.* (1993) mapped the *Xce* to a region distal to *Xist* and *Tsix*, perhaps to a region more than 100 kb away. Additionally, Plenge *et al.* (1997) have shown that skewed inactivation patterns can be associated with a point mutation within the human *XIST* promoter. Finally, based on possible phenotypic differences between the knockout of Penny *et al.* (1996) and that of Marahrens *et al.* (1997), Marahrens *et al.* have suggested that a domain between *Xist* exons 1–5 might also determine choice (Marahrens *et al.*, 1998). It would not be surprising if the important decision of choice turns out to involve a complex interplay of many factors.

V. IS X INACTIVATION A HOMOLOGY EFFECT?

Here, I consider this question using the broad definition of homology effect as any effect on gene expression or chromosome behavior relating to the presence of homologous sequences. Using this definition, imprinted X inactivation probably does not qualify as a homology effect, since paternal silencing is strictly the result of gametic imprinting and occurs without regard to total X-chromosome number. In contrast, random X inactivation would indeed qualify as a homology effect. First, X inactivation depends on the presence of more than one X homolog, or more specifically, more than one *Xic* sequence. Second, some form of trans sensing must be operative at the onset of X inactivation, specifically during the process of choice, so that an XX cell does not inadvertently designate one X as both the future active and inactive X. There is presently no knowledge of how this is acomplished.

Finally, although how *Tsix* acts to repress *Xist* expression is still not clear, the fact that *Xist* and *Tsix* produce noncoding homologous RNAs conjures up an analogy to posttranscriptional gene silencing in other organisms (Cognoni and Macino, 2000; and Chapters 8, 9, and 11). In this class of silencing mechanisms, the action of a transcriptionally active gene is nullified by homologous double-stranded RNA molecules produced at ectopic loci in a process referred to as "RNAi," for RNA-inhibition (Fire *et al.*, 1998). The studies of Panning *et al.* (1997) and Sheardown *et al.* (1997a) show that *Xist* transcription does not differ significantly before and after upregulation to high levels. They suggest instead that the gene is posttranscriptionally regulated. In this light, one must seriously consider whether *Tsix* might inhibit *Xist* action by an RNAi-like mechanism in a mammalian form of PTGS.

VI. CONCLUSIONS

How does one sort through the numerous models and arrive at a unifying mechanism for X inactivation? I believe the wealth of models reflects the diverse backgrounds of groups working in X inactivation, a field blessed with human geneticists, mouse geneticists, cytogeneticists, biochemists, and molecular biologists. Work contributed by these many groups show that the X-inactivation center is a very complex region with still many hidden attributes. The vast number of models under consideration need not be mutually exclusive. In fact, I believe that many of the models will eventually be combined once additional data become available. In this review, I have focused on the earliest control steps of X inactivation, specifically on counting and choice. There is as much to learn about the events which occur immediately downstream, particularly with regard to understanding what enables *Xist* RNA to paint the X chromosome and how *Xist* expression induces heterochromatic changes. The field is witnessing rapid and exciting progress, and no doubt the forthcoming answers will be equally complex and satisfying as those which came before. Homology may yet be the key to many of the outstanding questions.

References

Akhtar, A., Zink, D., and Becker, P. B. (2000). A chromodomain-RNA interaction targets MOF to the *Drosophila* X chromosome. *Nature* **407,** 405–409.

Amrein, H., and Axel, R. (1997). Genes expressed in neurons of adult male *Drosophila*. *Cell* **88,** 459–469.

Avner, P., and Heard, E. (2001). X-chromosome inactivation: counting, choice, and initiation. *Nature Rev. Genet.* **2,** 59–67.

Bailey, J. A., Carrel, L., Chakravarti, A., and Eichler, E. E. (2000). Molecular evidence for a relationship between LINE-1 elements and X chromosome inactivation: The Lyon repeat hypothesis. *Proc. Natl. Acad. Sci. USA* **97,** 6634–6639.

Bass, B. L. (2000). Double-stranded RNA as template for gene silencing. *Cell* **101,** 235–238.

Beard, C., Li, E., and Jaenisch, R. (1995). Loss of methylation activates Xist in somatic but not in embryonic cells. *Genes Dev.* **9,** 2325–2334.

Belyaev, N. D., Keohane, A. M., and Turner, B. M. (1996). Differential underacetylation of histones H2A, H3 and H4 on the inactive X chromosome in human female cells. *Hum. Genet.* **97,** 573–578.

Brockdorff, N., Ashworth, A., Kay, G. F., McCabe, V. M., Norris, D. P., Cooper, P. J., Swift, S., and Rastan, S. (1992). The product of the mouse Xist gene is a 15 kb inactive X-specific transcript containing no conserved ORF and located in the nucleus. *Cell* **71,** 515–526.

Brown, C. J., Ballabio, A., Rupert, J. L., Lafreniere, R. G., Grompe, M., Tonlorenzi, R., and Willard, H. (1991a). A gene from the region of the human X inactivation centre is expressed exclusively from the inactive X chromosome. *Nature* **349,** 3844.

Brown, C. J., Hendrich, B. D., Rupert, J. L., Lafreniere, R. G., Xing, Y., Lawrence, J., and Willard, H. F. (1992). The human XIST gene: Analysis of a 17 kb inactive X-specific RNA that contains conserved repeats and is highly localized within the nucleus. *Cell* **71,** 527–542.

Brown, C. J., Lafreniere, R. G., Powers, V. E., Sebastio, G., Ballabio, A., Pettigrew, A. L., Ledbetter, D. H., Levy, E., Craig, I. W., and Willard, H. F. (1991b). Localization of the X inactivation centre on the human X chromosome in Xq13. *Nature* **349,** 82–84.

Brown, C. J., and Willard, H. F. (1994). The human X-inactivation centre is not required for maintenance of X-chromosome inactivation. *Nature* **368,** 154–156.

Buzin, C. H., Mann, J. R., and Singer-Sam, J. (1994). Quantitative RT-PCR assays show Xist RNA levels are low in mouse female adult tissue, embryos and embryoid bodies. *Development* **120,** 3529–3536.

Cattanach, B. M., and Papworth, D. (1981). Controlling elements in the mouse. V. Linkage tests with X-linked genes. *Genet. Res. Camb.* **38,** 57–70.

Clemson, C. M., McNeil, J. A., Willard, H., and Lawrence, J. B. (1996). XIST RNA paints the inactive X chromosome at interphase: Evidence for a novel RNA involved in nuclear/chromosome structure. *J. Cell Biol.* **132,** 259–275.

Clerc, P., and Avner, P. (1998). Role of the region 3′ to Xist in the counting process of X-chromosome inactivation. *Nature Genet.* **19,** 249–253.

Cognoni, C., and Macino, G. (2000). Post-transcriptional gene silencing across kingdoms. *Curr. Opin. Genet. Dev.* **10,** 638–643.

Costanzi, C., and Pehrson, J. R. (1998). MacroH2A1 is concentrated in the inactive X chromosome of female mammals. *Nature* **393,** 599–601.

Costanzi, C., Stein, P., Worrad, D. M., Schultz, R. M., and Pehrson, J. R. (2000). Histone macroH2A1 is concentrated in the inactive X chromosome of female preimplantation mouse embryos. *Development* **127,** 2283–2289.

Courtier, B., Heard, E., and Avner, P. (1995). Xce haplotypes show modified methylation in a region of the active X chromosome lying 3′ to Xist. *Proc. Natl. Acad. Sci.* **92,** 3531–3535.

Csankovszki, G., Nagy, A., and Jaenisch, R. (2001). Synergism of Xist RNA, DNA methylation, and histone hypoacetylation in maintaining X chromosome inactivation. *J. Cell Biol.* **153,** 773–783.

Csankovszki, G., Panning, B., Bates, B., Pehrson, J. R., and Jaenisch, R. (1999). Conditional deletion of Xist disrupts histone macroH2A localization but not maintenance of X inactivation. *Nature Genet.* **22,** 322–323.

Debrand, E., Chureau, C., Arnaud, D., Avner, P., and Heard, E. (1999). Functional analysis of the DXPas34 locus, a 3′ regulator of Xist expression. *Mol. Cell Biol.* **19,** 8513–8525.

Dyer, K. A., Riley, D., and Gartler, S. M. (1985). Analysis of inactive X chromosome structure by in situ nick translation. *Chromosoma* **92,** 209–213.

Eicher, E. M. (1970). X-Autosome translocations in the mouse: Total inactivation versus partial inactivation of the X chromosome. *Adv. Genet.* **15,** 175–259.

Fire, A., Xu, S., Montgomery, M. K., Kostas, S. A., Driver, S. E., and Mello, C. C. (1998). Potent and specific genetic interference by double-stranded RNA in *Caenorhabditis elegans*. *Nature* **391,** 806–811.

Gartler, S. M., and Riggs, A. D. (1983). Mammalian X-chromosome inactivation. *Annu. Rev. Genet.* **17,** 155–190.

Goto, T., and Monk, M. (1998). Regulation of X-chromosome inactivation in development in mice and humans. *Microbiol. Mol. Biol. Rev.* **62,** 362–378.

Goto, Y., and Takagi, N. (2000). Maternally inherited X chromosome is not inactivated in mouse blastocysts due to parental imprinting. *Chromos. Res.* **8,** 101–109.

Graves, J. A. M. (1996). Mammals that break the rules: Genetics of marsupials and monotremes. *Annu. Rev. Genet.* **30,** 233–260.

Graves, J. A. M., and Gartler, S. M. (1986). Mammalian X chromosome inactivation: Testing the hypothesis of transcriptional control. *Somat. Cell Mol. Genet.* **12,** 275–280.

Heard, E., Clerc, P., and Avner, P. (1997). X-chromosome inactivation in mammals. *Annu. Rev. Genet.* **31,** 571–610.

Heard, E., Dress, C., Mongelard, F., Courtier, B., Rougeulle, C., Ashworth, A., Vourch, C., Babinet, C., and Avner, P. (1996). Transgenic mice carrying an Xist-containing YAC. *Hum. Mol. Genet.* **5,** 441–450.

Heard, E., Mongelard, F., Arnaud, D., and Avner, P. (1999a). *Xist* yeast artificial chromosome transgenes function as X-inactivation centers only in multicopy arrays and not as single copies. *Mol. Cell Biol.* **19,** 3156–3166.

Heard, E., Mongelard, F., Arnaud, D., Chureau, C., Vourc'h, C., and Avner, P. (1999b). Human XIST yeast artificial chromosome transgenes show partial X inactivation center function in mouse embryonic stem cells. *Proc. Natl. Acad. Sci. USA* **96,** 6841–6846.

Hendrich, B. D., Plenge, R. M., and Willard, H. F. (1997). Identification and characterization of the human XIST gene promoter: Implications for models of X chromosome inactivation. *Nucleic Acids Res.* **25,** 2661–2671.

Herzing, L. B. K., Romer, J. T., Horn, J. M., and Ashworth, A. (1997). Xist has properties of the X-chromosome inactivation centre. *Nature* **386,** 272–275.

Horn, J. M., and Ashworth, A. (1995). A member of the caudal family of homeobox genes maps to the X-inactivation centre region of the mouse and human X chromosomes. *Hum. Mol. Gene.* **4,** 1041–1047.

Jeppesen, P., and Turner, B. M. (1993). The inactive X chromosome in female mammals is distinguished by a lack of histone H4 acetylation, a cytogenetic marker for gene expression. *Cell* **74,** 281–289.

Johnston, C. M., Nesterova, T. B., Formstone, E. J., Newall, A. E. T., Duthie, S. M., Sheardown, S. A., and Brockdorff, N. (1998). Developmentally regulated Xist promoter switch mediates initiation of X inactivation. *Cell* **94,** 809–817.

Kay, G. F., Barton, S. C., Surani, M. A., and Rastan, S. (1994). Imprinting and X chromosome counting mechanisms determine Xist expression in early mouse development. *Cell* **77,** 639–650.

Kay, G. F., Penny, G. D., Patel, D., Ashworth, A., Brockdorff, N., and Rastan, S. (1993). Expression of Xist during mouse development suggests a role in the initiation of X chromosome inactivation. *Cell* **72,** 171–182.

Kelley, R. L., and Kuroda, M. I. (1995). Equality of X chromosomes. *Science* **270,** 1607–1610.

Keohane, A. M., Belyaev, N. D., Lavender, J. S., O'Neill, L. P., and Turner, B. M. (1996). X-inactivation and histone H4 acetylation in embryonic stem cells. *Dev. Biol.* **180,** 618–630.

Latham, K. E. (1996). X chromosome imprinting and inactivation in the early mammalian embryo. *Trends Genet.* **12,** 134–138.

Lee, J. T. (2000). Disruption of imprinted X inactivation by parent-of-origin effects at Tsix. *Cell* **103,** 17–27.

Lee, J. T., Davidow, L. S., and Warshawsky, D. (1999a). Tsix a gene antisense to Xist at the X-inactivation center. *Nature Genet.* **21,** 400–404.

Lee, J. T., and Jaenisch, R. (1997). Long-range cis effects of ectopic X-inactivation centres on a mouse autosome. *Nature* **386,** 275–279.

Lee, J. T., and Lu, N. (1999). Targeted mutagenesis of *Tsix* leads to nonrandom X-inactivation. *Cell* **99,** 47–57.

Lee, J. T., Lu, N. F., and Han, Y. (1999b). Genetic analysis of the mouse X-inactivation center reveals an 80kb multifunction domain. *Proc. Natl. Acad. Sci. USA* **96,** 3836–3841.

Lee, J. T., Strauss, W. M., Dausman, J. A., and Jaenisch, R. (1996). A 450 kb transgene displays properties of the mammalian X-inactivation center. *Cell* **86,** 83–94.

Liskay, R. M., and Evans, R. (1980). Inactive X chromosome DNA does not function in DNA-mediated cell transformation for the hypoxanthine phosphoribosyltransferase gene. *Proc. Natl. Acad. Sci. USA* **77,** 4895–4898.

Lyon, M. F. (1961). Gene action in the X-chromosome of the mouse (*Mus musculus* L.). *Nature* **190,** 372–373.

Lyon, M. F. (1971). Possible mechanisms of X chromosome inactivation. *Nature New Biol.* **232,** 229–232.

Lyon, M. F. (1972). X-chromosome inactivation and developmental patterns in mammals. *Biol. Rev.* **47**, 1–35.

Lyon, M. F. (1998). X-chromosome inactivation: A repeat hypothesis. *Cytogenet. Cell Genet.* **80**, 133–137.

Marahrens, Y., Loring, J., and Jaenisch, R. (1998). Role of the Xist gene in X chromosome choosing. *Cell* **92**, 657–664.

Marahrens, Y., Panning, B., Dausman, J., Strauss, W., and Jaenisch, R. (1997). Xist-deficient mice are defective in dosage compensation but not spermatogenesis. *Genes Dev.* **11**, 156–166.

Matsuura, S., Episkopou, V., Hamvas, R., and Brown, S. D. M. (1996). Xist expression from an Xist YAC transgene carried on the mouse Y chromosome. *Hum. Mol. Genet.* **5**, 451–459.

Meller, V. H., Wu, K. H., Roman, G., Kuroda, M. I., and Davis, R. L. (1997). roX1 RNA paints the X chromosome of male *Drosophila* and is regulated by the dosage compensation system. *Cell* **88**, 445–457.

Mermoud, J. E., Costanzi, C., Pehrson, J. R., and Brockdorff, N. (1999). Histone macroH2A1.2 relocates to the inactive X chromosome after initiation and propagation of X-inactivation. *J. Cell Biol.* **147**, 1399–1408.

Migeon, B. R., Kazi, E., Haisley-Royster, C., Hu, J., Reeves, R., Call, L., Lawler, A., Moore, C. S., Morrison, H., and Jeppesen, P. (1999). Human X inactivation center induces random X chromsome inactivation in male transgenic mice. *Genomics* **59**, 113–121.

Migeon, B. R., Winter, H., Kazi, E., Chowdhury, A. K., Hughes, A., Haisley-Royster, C., Morrison, H., and Jeppesen, P. (2001). Low-copy-number human transgene is recognized as an X inactivation center in mouse ES cells, but fails to induce cis-inactivation in chimeric mice. *Genomics* **71**, 156–162.

Mohandas, T., Sparkes, R. S., and Shapiro, L. J. (1981). Reactivation of an inactive human X chromosome: Evidence for X-inactivation by DNA methylation. *Science* **211**, 393–396.

Nesterova, T. B., Slobodyanyuk, S. Y., Elisaphenko, E. A., Shevchenko, A. I., Johnston, C., Pavlova, M. E., Rogozin, I. G., Kolesnikov, N. N., Brockdorff, N., and Zakian, S. M. (2001). Characterization of the genomic Xist locus in rodents reveals conservation of overall gene structure and tandem repeats but rapid evolution of unique sequence.

Ohno, S. (1969). Evolution of sex chromosomes in mammals. *Annu. Rev. Genet.* **3**, 495–524.

Panning, B., Dausman, J., and Jaenisch, R. (1997). X chromosome inactivation is mediated by Xist RNA stabilization. *Cell* **90**, 907–916.

Penny, G. D., Kay, G. F., Sheardown, S. A., Rastan, S., and Brockdorff, N. (1996). Requirement for Xist in X chromosome inactivation. *Nature* **379**, 131–137.

Perche, P.-Y., Vourc'h, C., Konecny, L., Souchier, C., Robert-Nicoud, M., Dimitrov, S., and Khochbin, S. (2000). Higher concentrations of histone macroH2A in the Barr body are correlated with higher nucleosome density. *Curr. Biol.* **10**, 1531–1534.

Plenge, R. M., Hendrich, B. D., Schwartz, C., Arena, J. F., Naumova, A., Sapienza, C., Winter, R. M., and Willard, H. F. (1997). A promoter mutation in the XIST gene in two unrelated families with skewed X-chromosome inactivation. *Nature Genet.* **17**, 353–356.

Priest, J. H., Heady, J. E., and Priest, R. E. (1967). Delayed onset of replication of human X chromosomes. *J. Cell Biol.* **35**, 483–487.

Prissette, M., El-Maarri, O., Arnaud, D., Walter, J., and Avner, P. (2001). Methylation profiles of DXPas34 during the onset of X-inactivation. *Hum. Mol. Genet.* **10**, 31–38.

Rasmussen, T. P., Mastrangelo, M.-A., Eden, A., Pehrson, J. R., and Jaenisch, R. (2000). Dynamic relocalization of histone macroH2A1 from centrosomes to inactive X chromosome during X inactivation. *J. Cell Biol.* **150**, 1189–1198.

Rastan, S. (1982). Timing of X-chromosome inactivation in postimplantation mouse embryos. *J. Embryol. Exp. Morphol.* **71**, 11–24.

Rastan, S. (1983). Non-random X-chromosome in mouse X-autosome translocation embryos—Location of the inactivation centre. *J. Embryol. Exp. Morphol.* **78,** 1–22.

Rastan, S., and Robertson, E. J. (1985). X-chromosome deletions in embryo-derived (EK) cell lines associated with lack of X-chromosome inactivation. *J. Embryol. Exp. Morphol.* **90,** 379–388.

Richardson, B. J., and Czuppon, A. B. (1971). Inheritance of glucose-6-phosphate dehydrogenase variation in kangaroos. *Nature New Biol.* **230,** 154–155.

Russell, L. B. (1963). Mammalian X-chromosome action: Inactivation limited in spread and in region of origin. *Science* **140,** 976–978.

Sado, T., Wang, Z., Sasaki, H., and Li, E. (2001). Regulation of imprinted X-chromosome inactivation in mice by Tsix. *Development* **128,** 1275–1286.

Sharman, G. B. (1971). Late DNA replication in the paternally derived X chromosome of female kangaroos. *Nature* **230,** 231–232.

Sheardown, S. A., Duthie, S. M., Johnston, C. M., Newall, A. E. T., Formstone, E. J., Arkell, R. M., Nesterova, T. B., Alghisi, G.-C., Rastan, S., and Brockdorff, N. (1997a). Stabilization of Xist RNA mediate initiation of X chromosome inactivation. *Cell* **91,** 99–107.

Sheardown, S. A., Newall, A. E. T., Norris, D. P., Rastan, S., and Brockdorff, N. (1997b). Regulatory elements in the minimal promoter region of the mouse *Xist* gene. *Gene* **203,** 159–168.

Simmler, M.-C., Cattanach, B. M., Rasberry, C., Rougeulle, C., and Avner, P. (1993). Mapping the murine *Xce* locus with (CA)n repeats. *Mammal. Genome* **4,** 523–530.

Simmler, M.-C., Cunningham, D. B., Clerc, P., Vermat, T., Caudron, B., Cruaud, C., Pawlak, A., Szpirer, C., Weissenbach, J., Claverie, J.-M., and Avner, P. (1996). A 94 kb genomic sequence 3′ to the murine Xist gene reveals an AT rich region containing a new testis specific gene Tsx. *Hum. Mol. Genet.* **5,** 1713–1726.

Simmler, M. C., Heard, E., Rougeulle, C., Cruaud, C., Weissenbach, J., and Avner, P. (1997). Localisation and expression analysis of a novel conserved brain expressed transcript, *Brx/BRX,* lying within the Xic/XIC candidate region. *Mammal. Genome* **8,** 760–766.

Tada, T., Obata, Y., Tada, M., Goto, Y., Nakatsuji, N., Tan, S. S., Kono, T., and Takagi, N. (2000). Imprint switching for non-random X-chromosome inactivation during mouse oocyte growth. *Development* **127,** 3010–3013.

Tagaki, N., and Abe, K. (1990). Detrimental effects of two active X chromosomes on early mouse development. *Development* **109,** 189–201.

Tai, H. H., Gordon, J., and McBurney, M. W. (1994). Xist is expressed in female embryonal carcinoma cells with two active X chromosomes. *Somat. Cell Mol. Genet.* **20,** 171–182.

Takagi, N., and Sasaki, M. (1975). Preferential inactivation of the paternally derived X-chromosome in the extraembryonic membranes of the mouse. *Nature* **256,** 640–642.

Tilghman, S. M. (1999). The sins of the fathers and mothers: genomic imprinting in mammalian development. *Cell* **96,** 185–193.

Wareham, K. A., Lyon, M. F., Glenister, P. H., and Williams, E. D. (1987). Age related reactivation of an X-linked gene. *Nature* **327,** 725–727.

Warshawsky, D., Stavropoulos, N., and Lee, J. T. (1999). Further examination of the Xist promoter-switch hypothesis in X inactivation: Evidence against the existence and function of a Po promoter. *Proc. Natl. Acad. Sci. USA* **96,** 14424–14429.

Webb, S., Vries, T. J. d., and Kaufman, M. H. (1992). The differential staining pattern of the X chromosome in the embryonic and extraembryonic tissues of postimplantation homozygous tetraploid mouse embryos. *Genet. Res.* **59,** 205–214.

West, J. D., Freis, W. I., Chapman, V. M., and Papaioannou, V. E. (1977). Preferential expression of the maternally derived X chromosome in the mouse yolk sac. *Cell* **12,** 873–882.

Wutz, A., and Jaenisch, R. (2000). A shift from reversible to irreversible X inactivation is triggered during ES cell differentiation. *Mol. Cell* **5,** 695–705.

3 Homologous Chromosome Associations and Nuclear Order in Meiotic and Mitotically Dividing Cells of Budding Yeast

Sean M. Burgess*
Division of Biological Sciences
Section of Molecular and Cellular Biology
University of California
Davis, California 95616

*To whom correspondence should be addressed: E-mail: smburgess@ucdavis.edu.

Advances in Genetics, Vol. 46
49

ABSTRACT

This chapter discusses the relationship between nuclear order and the association of homologous DNA sequences in the budding yeast, *Saccharomyces cerevisiae*. Homologous chromosomes functionally interact with one another to repair DNA double-strand breaks (DSBs) introduced either environmentally (e.g., by γ-irradiation) or deliberately by the cell (e. g., during meiosis). DNA homology recognition in these instances often involves the (RecA) homolog *RAD51* and/or the related gene, *DMC1*. Evidence for interactions between homologous chromosomes occurring independent of DSB formation and (RecA) homolog function has also been described in meiotic, premeiotic, and mitotically dividing cells of yeast. These interactions presumably depend upon DNA homology but the molecular details of such associations are poorly understood. Both DSB-dependent and -independent homolog associations may be facilitated by the nonrandom organization of chromosomes in the nucleus, including centromere and telomere clustering, which are also discussed. © 2002, Elsevier Science (USA).

I. INTRODUCTION

Three types of homologous DNA sequences exist naturally in eukaryotic cells. First, diploid cells contain pairs of homologous chromosomes, called homologs, each of which is derived from a different parent. Second, repeated elements including transposons, rDNA repeats, telomeric repeats, as well as multiple clusters of polygenic complexes and repeated genes, are scattered throughout genomes. Third, each chromatid coexists with its exact duplicate or sister following DNA replication. How do homologous DNA sequences, relate (or not relate) to one another during the cell cycle, during meiosis, and during periods of stress and cell damage? This chapter reviews the current status of our understanding of homologous DNA associations in the budding yeast, *Saccharomyces cerevisiae*.

Homologous DNA sequences interact in a number of different states and under a number of different conditions. For each of the examples reviewed here for yeast, these interactions take place within the ordered space of the nucleus and are likely influenced by this order (Section II). DNA double-strand breaks (DSBs) can arise from exposure to ionizing radiation, form during DNA replication of a nicked DNA substrate, or are catalyzed during meiosis. Such breaks are often repaired using homologous DNA sequences, usually a sister chromatid or a homologous chromosome, as a template to replace lost genetic information by homologous recombination. In the case of double-strand break repair (DSBR), DNA homology recognition is mediated, at least in part, by the *recA* homolog, Rad51 (Section III; Figure 3.1A). During meiosis, homologous chromosomes pair along their lengths, undergo meiotic recombination, and ultimately engage in synapsis, where the homologs are held tightly together by the proteinacious synaptonemal complex (SC; Section IV). Homologous chromosomes are also paired in the absence of recombination or the SC in mitotically dividing cells of yeast and during meiosis as seen in situations where recombination and SC are absent or rare (e.g., in mutants where the contributions of other processes is eliminated; Section V). Finally, sister chromatids are held together from the time they emerge from the replication fork until anaphase, when connections between them are severed and the spindle pulls them away from one another. Although sister chromatid cohesion may not be a DNA homology-driven process, the cell's ability to discriminate between homologous chromosomes and sister chromatids is important.

Many epigenetic phenomena in organisms other than yeast appear to occur as a consequence of homologous DNA interactions. These include repeat-associated silencing (Hsieh and Fire, 2000), transvection (Wu and Morris, 1999), paramutation (Hollick *et al.*, 1997), imprinting (LaSalle and Lalande, 1996), and X-chromosome inactivation (Marahrens, 1999). The homology recognition mechanisms used in these diverse processes share features in common with

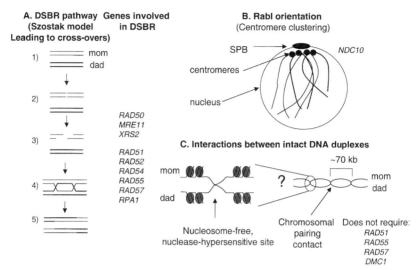

Figure 3.1. Homologous chromosome association/organization in premeiotic and/or vegetative cells. (A) Model for DSB repair by Szostak *et al.* (1983). (1) Intact DNA duplexes; (2) formation of DSB; (3) resection of 5′ ends to reveal 3′ ssDNA tails; (4) dHJ intermediate; (5) Crossed-over DNA products. See Section III.A for details. (B) Organization of chromosomes in the Rabl orientation with centromeres clustered to one side of the nucleus near the SPB. See Section II.B for details. (C) Two types of interactions detected between intact homologous chromosome: (left) Interactions between DNase-hypersensitive regions (i.e., nucleosome-free); (right) pairing contacts between homologous chromosomes detected in spread nuclei by FISH. See text for details.

the homology-related issues discussed here for yeast. The study of naturally occurring homolog interactions in a genetically and cytologically tractable organism such as yeast will enrich our understanding of homology-related processes in these other organisms.

II. THE GLOBAL ARRANGEMENT OF CHROMOSOMES IN THE NUCLEUS

Three prominent features of overall chromosome organization in mitotically dividing cells of yeast are the clustering of centromeres to one side of the nucleus near the spindle pole body (SPB), association of telomeres with the nuclear envelope, and the pairing of homologous chromosomes. During meiosis, three prominent features of nuclear organization include clustering of telomeres near the SPB in a bouquet configuration, homolog pairing, and synapsis. In this review, homolog pairing is defined as the juxtaposition of homologous chromosomes as detected in spread chromosome preparations by fluorescence *in situ* hybridization (FISH).

Pairing is therefore not to be confused with the more specific term of synapsis, where the SC mediates an intimate connection between homologous chromosomes along their lengths, since pairing can occur in either the presence or the absence of the SC (see later). This section reviews the methodologies used to assess chromosome position and pairing in the nucleus and their application to understanding the global arrangement of chromosomes in the yeast nucleus. In Sections III–V, the contribution of chromosome order to homologous chromosome pairing is discussed in detail.

A. Exploring the three-dimensional space of the nucleus

Chromosome organization and homolog pairing has been explored in the yeast nucleus using various methods in both fixed and living cells. In the last decade, the application of FISH analysis to probe specific chromosomal loci in fixed preparations has been developed for both spread and intact nuclei (Guacci *et al.*, 1994; Weiner and Kleckner, 1994; Loidl *et al.*, 1998). Nuclei prepared for FISH analysis can also be processed for immunostaining of specific proteins that bind to known chromosomal elements (Gotta *et al.*, 1996, 1999). Electron microscopy (EM) has also contributed greatly to our understanding of meiotic chromosome structure and chromosome synapsis during meiosis, when chromosomes are condensed and can be visualized by this method (e.g., Byers and Goetsch, 1975; and reviewed in Zickler and Kleckner, 1999).

   In living cells, positions of chromosomal loci can be determined by reconstructing three-dimensional images of GFP-marked chromosomes collected over time. Tagging chromosomes in this way has been carried out by fusing GFP to either the *lacI* repressor or *tet* repressor proteins, which bind to 256 or 336 tandem repeats of the *lacO* or *tetO* elements, respectively, inserted into specific chromosomal loci (Marshall *et al.*, 1997; Straight *et al.*, 1997; Aragon-Alcaide and Strunnikov, 2000). Specific chromosomal regions (e.g., centromeres) can be localized by fusing GFP to known region-specific binding proteins (e.g, kinetochore components; Goshima and Yanagida, 2000; He *et al.*, 2000; see later). The relative spatial proximity of pairs of loci can also be inferred from their rate of collision in living cells as measured using site-specific recombination (Burgess and Kleckner, 1999). Each of these methods has both advantages and disadvantages, which are described here in more detail.

   The yeast nucleus is small compared to nuclei of plant or animal cells, and examination of chromosomes in three-dimensional space offers several challenges. An intact diploid yeast nucleus is about 1.5–2.0 μm in diameter while the diameter of a spread nucleus is usually 6–10 μm across, which is approximately the diameter of a mammalian nucleus. Homolog-pairing interactions in both meiotic and mitotically dividing cells of yeast are most easily detected in nuclear spreads, where homologous interactions can be distinguished from chance colocalization (see Section IV.C). It is thought that physical contacts between homologous chromosomes limit the extent to which homologous chromosomes will separate under the spreading conditions, whereas unconnected, nonhomologous chromosomes will spread randomly (Weiner and Kleckner, 1994). The definition of homolog pairing for both meiotic and vegetatively dividing cells depends critically on the fact that homologous interactions are more frequent than control nonhomologous interactions (see Sections IV.C and V.A). A pairing interaction is indicated by two hybridization signals that touch one another, or nearly touch. Consistent with this notion, distances between the centers of hybridizing foci in nuclear spreads of meiotic pachytene cells plateau at 0.7 μm, which is in agreement with the combined size of chromosome loops and the width of the SC in meiosis (Weiner and Kleckner, 1994).

The points of contact between homologs that are detected in spread nuclei preparations are far apart along the lengths of the chromosomes ($\sim$70 kb for both premeiotic and vegetative cells) and occur at different positions in different cells and possibly at different times (Weiner and Kleckner, 1994; Burgess et al., 1999). In an intact nucleus, then, the volume through which homologous segments are moving, when tethered at such a distance, is not so different from the volume through which nonhomologous DNA segments are moving. Thus, measurements of homolog association in intact cells will not differ substantially compared to random, nonhomolog associations. This notion is consistent with the observation that collisions between allelic loci in vivo are, on average, only twofold greater than nonallelic interactions (Burgess and Kleckner, 1999). In contrast, in spread preparations analyzed by FISH, typically >50% of allelic loci are paired in mitotically dividing cells while only 5–10% of nonhomologous loci are colocalized by random chance (Section V.A; Burgess et al., 1999).

Nuclear order due to centromere and/or telomere clustering is very prominent in yeast (Section II.B), and it also obscures the detection of homolog associations in intact living cells (Burgess and Kleckner, 1999; Aragon-Alcaide and Strunnikov, 2000) and in some fixed spread preparations where centromere clustering is maintained (Jin et al., 2000). This problem can be overcome by analyzing pairing interactions in spread nuclear preparations where centromere clustering is abolished by the forces of spreading (Burgess et al., 1999).

The methods by which data are processed are also likely to be important for interpreting nuclear organization and, in particular, pairing interactions. For example, when measuring the levels of homologous chromosome interactions, data have been processed in binary form (e.g., one focus vs two foci), computed as an average distance between foci, or expressed in a rank-plot form (e.g., arranged in order from the smallest distance to the largest distance). Because it is possible to resolve two foci that are considered to be "paired" (i.e., $d \leq 0.7$ μm apart) in spread nuclei by FISH, the application of binary analysis to either intact or spread nuclei will tend to underestimate pairing contacts. Likewise, average distances are difficult to interpret, especially in intact nuclei, because of the transient nature of the pairing interactions in mitotically dividing cells (see Section V.A). Rank-plot analysis gives an overview of all measurements and has been used to distinguished homolog pairing interactions from chance association of nonhomologous chromosomes in both meiotic and mitotically dividing cells (Weiner and Kleckner, 1994; Burgess et al., 1999).

B. Centromeres in yeast are clustered near the spindle-pole body

In many cell types in many organisms, chromosomes are organized with centromeres clustered to one side of the nucleus and with the arms and telomeres extended in the opposite direction (Figure 3.1.B). This polarized chromosome configuration is referred to as the Rabl orientation, after Carl Rabl's observation of chromosome position in amphibian cells (Rabl, 1885) and has also been observed in cells of *Drosophila*, some plants, and fungi (reviewed in Rabl, 1885; Fussell, 1987). Except during anaphase, centromere clustering is generally rare in mammalian cells (Vourc'h et al., 1993); however, exceptions have been noted in specialized cell types (Sperling and Ludtke, 1981; Leitch, 2000). The evidence for the Rabl orientation in budding yeast is described here.

In budding yeast, clustering of centromeres near the nuclear periphery has been detected using FISH in intact and swollen nuclei, and in some spread nuclei preparations (Guacci et al., 1994, 1997; Jin et al., 1998, 2000). Centromere clustering has also been inferred by subcellular localization of kinetochore components, such as Ndc10, in intact cells (Goh and Kilmartin, 1993). Using pancentromeric DNA probes to detect all 16 centromeres in yeast, Jin and colleagues found that centromeres clustered in both haploid and diploid cells throughout the cell division cycle (Table 3.1A; Jin et al., 1998). In addition, centromeres clustered in G1-arrested cells, after exposure to the mating pheromone, alpha factor (Jin et al., 2000). The dynamic behavior of centromeres during metaphase, however, suggests that centromere clustering may be lost during this period. Fusions of green-fluorescent protein (GFP) to two centromere-binding proteins, Slk19p and Mtw1p, has allowed for the visualization of kinetochores during metaphase in living cells (Goshima and Yanagida, 2000; He et al., 2000). In these studies, tagged kinetochores oscillated between the two spindle axes and separated transiently by distances of up to 0.8 μm near the centromeres, a considerable distance considering that the pole-to-pole length of the short spindle is approximately 2 μm. Loidl and colleagues have suggested that clustering of centromeres may serve to facilitate the attachment of chromosomes to the spindle (Jin et al., 1998). Perhaps centromeres are indeed clustered at the time of microtubule attachment and then become disorganized during metaphase, while they are pulled by the spindle.

Table 3.1. Cell-Cycle Modulation of Chromosome Organization in Mitotically Dividing and in Meiotic Cells

A. Mitotically dividing cells	G1	S	G2	M	Nocodazole arrest	References
Centromeres	Clustered	Clustered	Clustered	Dynamic	Reduced clustering	Jin et al. (1998); He et al. (2000); Goshima and Yanagida (2000)
Telomeres	Multiple clusters	Multiple clusters	Multiple clusters	Multiple clusters	N/A	LaRoche et al. (2000)
Homolog pairing	Paired	Reduced pairing	Paired	Paired (?)	Reduced pairing	Burgess et al. (1999)

B. Meiotic cells	Premeiosis	Premeiotic S	Leptotene	Zygotene	Pachytene	References
Centromeres	Clustered	Centromere/telomere clustering	Dispersed	Dispersed	Dispersed	Trelles-Sticken et al. (1999)
Telomeres	Multiple clusters		Single cluster/bouquet		Dispersed	Trelles-Sticken et al. (1999)
Homolog pairing	Paired	Reduced pairing	Paired	Pairing stabilized by recombination and SC	Pairing stabilized by recombination and SC	Weiner and Kleckner (1994); Loidl et al. (1994); Nag et al. (1995)
Recombination	None	None	DSBs/joint molecules		Crossovers	Reviewed in Zickler and Kleckner (1999)
Synaptonemal complex	None	None	None	Short segments	Full	Reviewed in Zickler and Kleckner (1999)

Two experimental approaches carried out in living cells have also pointed to centromere clustering in yeast. First, Cre/*loxP* site-specific recombination was used to measure the relative probabilities that pairs of loci collide with one another in the nucleus (Burgess and Kleckner, 1999). The Cre-mediated recombination rates between pairs of *loxP* sites integrated at loci on nonhomologous chromosomes varied inversely with the increasing differential between locus-to-centromere distances for the chromosomes tested. These authors suggested that chromosomal interactions are governed by nonspecific centromere clustering. Similar results were obtained using chromosomes tagged with GFP to visualize specific chromosomal loci in living cells, in real time (Aragon-Alcaide and Strunnikov, 2000).

It is commonly thought that the Rabl orientation is a remnant of anaphase, since the pulling forces of the spindle would draw centromeres toward the spindle poles prior to the telomeres. Indeed, yeast chromosomes do exhibit this behavior during anaphase. By using GFP-tagged chromosomal loci to measure centromere and telomere movement in real time in living cells, Straight *et al.* (1997) observed a chromosome-to-pole movement indicative of anaphase A. As expected, the centromeres of the two sister chromatids separated before the telomeres. Interestingly, however, the centromeres were not restricted to the central region of the spindle, but were instead often closer to one of the poles. Thus, centromere oscillations result in their distribution along the length of the spindle axis without a phase of congression typical of metaphase in animal and plant cells (Bajer and Mole-Bajer, 1986; Skibbens *et al.*, 1993). At anaphase the foci moved rapidly in opposite directions along the spindle axis toward opposite spindle poles, with centromeres arriving at the poles first, followed by movement of telomeres toward the pole in a way reminiscent of an elastic contraction (Straight *et al.*, 1997).

At the end of anaphase, centromeres are located near the SPB, and this localization appears to persist throughout the cell cycle (Jin *et al.*, 1998), or at least until the next round of oscillations associated with metaphase (above). While associated with the SPB, the centromeres form a rosette structure with the SPB at its hub (Jin *et al.*, 2000). Evidence that kinetochore function influences centromere clustering comes from the observation that mutation of a kinetochore-specific gene, *NDC10*, resulted in a 90% reduction in centromere clustering (Jin *et al.*, 2000). In addition, cells treated with nocodazole, an inhibitor of microtubule polymerization, exhibited reductions in the levels of centromere clustering (Table 3.1A; Guacci *et al.*, 1997; Marshall *et al.*, 1997; Aragon-Alcaide and Strunnikov, 2000; Jin *et al.*, 2000). Together, these results suggest that centromeres are tethered to the SPB via microtubule attachments to the kinetochore and that this association is the underlying basis for the Rabl orientation.

It may be that the associations with the SPB are sufficient to promote the Rabl orientation, even in the absence of anaphase (Jin *et al.*, 2000). A recent experiment carried out by Loidl and colleagues investigated this question. Normally, when cells enter early meiotic prophase, the Rabl orientation is disrupted and gives way to the bouquet arrangement, where telomeres are clustered and located near the SPB (Section IV.E; Table 3.1B; Trelles-Sticken *et al.*, 1999). When meiotic yeast cells that have lost the Rabl orientation were returned to vegetative growth prior to the first meiotic division, the chromosomes resumed the Rabl orientation without having been subjected to anaphase (Jin *et al.*, 2000). In this case, the Rabl orientation reformed 30 min prior to the formation of a long, bipolar anaphase spindle. Similar results were obtained when a temperature-sensitive *cdc23* mutant (which fails to enter into anaphase) was used.

The Rabl orientation is independent of homolog pairing, and homolog pairing is independent of the Rabl orientation (Section V.A). Formation of the Rabl orientation does not depend on homolog pairing per se, since it occurs in haploid yeast where no homologous chromosomes are present (Jin *et al.*, 1998), and in organisms where somatic homolog pairing apparently does not occur (Bass *et al.*, 2000). Alignment of duplicated regions located on nonhomologous chromosomes, however, could contribute, in part, to the nonrandom organization of chromosomes in the nucleus, especially with respect to the Rabl orientation and telomere clustering (see below). The yeast genome contains several extensive regions of DNA homology that are thought to be relics of a genome duplication event (Wolfe and Shields, 1997). Fifty-five such duplications are on average 55 kb long and together span 50% of the genome. For example, a 30-kb region near the telomere of chromosome VIII is over 90% identical to a region on the right arm of chromosome I (de Steensma *et al.*, 1989; Johnston *et al.*, 1994). The duplications are in the same orientation with respect to the centromere in 50 out of 55 blocks. Conceivably, these duplicated regions could interfere or compete with homologous chromosome interactions in diploid cells.

C. Telomeres associate with one another and with the nuclear Periphery

A diploid yeast nucleus contains 64 telomeres prior to DNA replication. The telomeres in mitotically dividing cells are not scattered randomly within the nucleus, but instead are located near the nuclear periphery in 3–9 clusters (Klein *et al.*, 1992; Gotta *et al.*, 1996). These telomere clusters persist during the course of the mitotic cell cycle (Table 3.1A; Laroche *et al.*, 2000). As cells transit from mitotic growth to meiotic prophase, the association of the centromeres with the SPB is lost while associations of telomeres with the SPB become established. This arrangement of chromosomes is known as the bouquet and has been reviewed extensively by Zickler and Kleckner (1998). The role of telomere clustering with respect to meiotic homolog pairing is described in more detail in Section IV.E.

In mitotically dividing cells the localization of telomeres to the nuclear periphery is dependent on the functions of the *SIR3* and *SIR 4* genes, which are involved in transcriptional silencing, and the *HDF1* and *HDF2* genes, which are involved in nonhomologous end joining (NHEJ) and code for the yeast homologs of Ku70 and Ku80 proteins, respectively (Palladino *et al.*, 1993; Gotta *et al.*, 1996; Laroche *et al.*, 1998). In addition, two myosin-like proteins, *MLP1* and *MLP2*, aid in anchoring the telomeres and the Ku70 and Ku80 subunits to the nuclear membrane (Galy *et al.*, 2000).

It is unknown whether the clusters form randomly or if telomeres of specific chromosomes coexist in defined subgroups. DNA homology present at the proximal domains of telomeres may account for the observed association (Pryde *et al.*, 1997). Several duplicated gene families or polygenic loci exist near the ends of chromosomes. The *MAL* gene family comprises the five polygenic complexes *MAL1*, *MAL2*, *MAL3*, *MAL4*, and *MAL6*, each about 9 kb and all structurally and functionally homologous to one another (Charron *et al.*, 1989). Each *MAL* locus maps to a different telomere, and no chromosome contains more than one complex. In addition, the 10 structural genes encoding alpha-galactosidase, *MEL1-MEL10*, are all located on different chromosomes and map to terminal regions (Vollrath *et al.*, 1988; Naumov *et al.*, 1995). The terminal domains of chromosome III share sequence homology to each other and to the terminal domains of chromosomes V and XI (Johnston *et al.*, 1994; Bussey *et al.*, 1995; Dujon, 1996). Although these repeats cannot account entirely for the observed small numbers of telomeric foci, it would be interesting to examine by FISH whether the DNA homology shared by these sequences correlates with their localization to specific clusters. Ectopic recombination between subtelomeric regions on different chromosomes appears to be influenced, at least in part, by the homology of the smaller composite elements present in different combinations at the ends of most telomeres (e.g., Y'). For example, long Y'-containing elements were found to interact preferentially with telomeres belonging to their own size class, suggestive of localization of this telomere subgroup to specific clusters (Louis and Haber, 1992).

III. HOMOLOGY RECOGNITION DURING DNA DOUBLE-STRAND BREAK REPAIR

The mechanism used by the yeast cells to repair DSBs is the best understood among homology-sensing processes. DSBs are often repaired using a homologous DNA sequence when available (reviewed by Paques and Haber, 1999). Homologous donor sequences acting as a substrate for homologous DSB repair might exist on a homolog in diploid cells, a sister chromatid in G2, or at a homologous DNA segment located at an ectopic position in the genome. Certainly, one potential advantage of homolog pairing would be to bring allelic DNA sequences into close proximity to one another in order to facilitate homologous recombination and to minimize ectopic recombination between nearly homologous sequences. The

Rabl or bouquet arrangements in mitotically dividing cells and during meiosis, respectively, could also serve such a role. If no DNA homology is present in the cells (i.e., in haploid cells in G1), then DSBs can join by the NHEJ pathway (reviewed by Lewis and Resnick, 2000).

A. DNA double-strand breaks are repaired by homologous recombination

Two of several pathways used to repair DSBs by homologous recombination are the DSB repair (DSBR) pathway of Szostak (1983) and the related synthesis-dependent strand annealing (SDSA) pathway (Nassif *et al.*, 1994). Briefly, following DSB formation, the free ends of the break are exonucleolytically processed to reveal 3′ ssDNA tails which participate in strand exchange with a homologous duplex DNA segment (Figure 3.1A; Paques and Haber, 1999). For DSBR the strand exchange event is converted into a double Holliday junction (dHJ), which can be resolved to yield crossed-over DNA segments, as has been demonstrated for meiotic recombination intermediates (see Section IV.B; Schwacha and Kleckner, 1995). For SDSA, the ssDNA pulls out of the duplex following limited DNA synthesis, presumably without the formation of a ligated dHJ intermediate, followed by annealing at regions containing complementary ssDNA.

The Rad51 protein likely plays a prominent role in the homology-sensing mechanism for both of these pathways. Rad51 is homologous to the *Escherichia coli* RecA protein and, like RecA, can carry out DNA strand exchange *in vitro* (Sung, 1994; Sugiyama *et al.*, 1997; Mazin *et al.*, 2000b). During meiosis, Dmc1, another protein with sequence homology to RecA, functions to promote interhomolog-specific interactions (see Section IV.B; Bishop *et al.*, 1992; Rockmill *et al.*, 1995b; Schwacha and Kleckner, 1997). It is still not clear how RecA, Rad51, or Dmc1 mediate homology recognition. Clearly, the ability of RecA and Rad51 proteins to form presynaptic filaments on single-stranded DNA, which are exposed at the sites of breaks following exonucleolytic processing, is important for homology recognition (Kowalczykowski, 1991; Sung and Robberson, 1995).

If the presynaptic filament is key to homology recognition in DSBR, then what exactly is being recognized? Kowalczykowski (2000) has proposed that the presynaptic filament can be viewed as an extended sequence-specific DNA-binding protein that finds homology by a diffusional process, not unlike other sequence-specific DNA-binding proteins. Sequence specificity would thus be conferred by the combination of Rad51 and the ssDNA within the filament and not by Rad51 per se (Kowalczykowski, 2000). The structure of the filament in this respect is intriguing. Like RecA, electron micrographs have shown that Rad51 filaments can extend the helical pitch of ssDNA by 1.5-fold compared to B-form DNA (Sung and Robberson, 1995). Therefore, the homology search involves the comparison

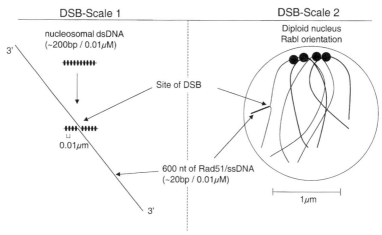

Figure 3.2. The theoretical length of an average Rad51/ssDNA filament following exonucleolytic processing of a DSB relative to chromatin (Scale 1) and to the size of the yeast nucleus (Scale 2).

of two DNA molecules existing on two rather disparate scales (Figure 3.2). Whereas the target of the homology search is chromatin packaged into nucleosomes ($\sim$200 bp/0.01 μm), the equivalent number of nucleotides of Rad51 coated ssDNA would be extended 10-fold this length ($\sim$20 bp/0.01 μm). Certainly, a major obstacle to be overcome during the process of DSBR is the removal of nucleosomes, particularly at the broken ends (Raymond and Kleckner, 1993).

Since recombination involves DNA packaged into chromatin, it is not surprising that additional cellular factors are also important for facilitating homologous DNA pairing and strand exchange. These include Rad52, Rad54, Rad55, Rad57, and Rpa1 (Figure 3.1A; Bianco et al., 1998; Paques and Haber, 1999; Sung et al., 2000). Nearly all of these proteins stimulate Rad51-mediated DNA strand-exchange in vitro (Sung, 1997; New et al., 1998; Petukhova et al., 1998). Rad54 likely aids homology recognition by facilitating local unwinding of the target duplex (Mazin et al., 2000a; Van Komen et al., 2000).

Other mechanisms exist in yeast to repair DSBs if Rad51 function is abrogated, including single-strand annealing (SSA; Fishman-Lobell et al., 1992) and break-induced replication (BIR) (Voelkel-Meiman and Roeder, 1990a, 1990b), both of which require the function of Rad52. For SSA, homology recognition is likely mediated by annealing two complementary ssDNA strands without the intervening strand exchange and synthesis steps of SDSA. For BIR, the basis of homology recognition independent of Rad51 is unknown. Details of these pathways have been extensively reviewed elsewhere (Bianco et al., 1998; Paques and Haber, 1999; Sung et al., 2000; Gloor, 2001).

B. Very short regions of homology are sufficient for homologous recombination

The homology search involved in repairing a DSB by homologous recombination in mitotically dividing cells apparently involves the entire genome, since recombination between two homologous DNA segments can occur at ectopic as well as at allelic positions in the genome, albeit at different efficiencies that depend on their chromosomal location (see Section III.C; Jinks-Robertson and Petes, 1986; Lichten and Haber, 1989). In published studies, the homologous DNA segments under investigation were several kilobase pairs in length or less, demonstrating that very little sequence homology is necessary to identify a donor target correctly. Indeed, very small regions of homology, on the order of 35 bp, are sufficient to target PCR-generated DNA fragments to their homologous sequences in the chromosome following transformation (Baudin et al., 1993; Wach et al., 1994). This feature of homology recognition has been heavily exploited, for example, in the construction of deletion alleles of nearly every gene in the yeast genome (Winzeler et al., 1999). The minimal length of contiguous homology required for successful completion of meiotic recombination was found to be about 150–250 nucleotides (Hayden and Byers, 1992).

C. Homologous recombination is likely influenced by nuclear organization

Although recombination can occur between homologous DNA segments present at various locations in the genome, the spatial organization of chromosomes may influence the efficiency of recombination. For example, when a break occurs, there could be a limited amount of time available to engage in DSBR before NHEJ or some other pathway is used (Haber, 2000). Homolog juxtaposition either by pairing interactions or due to their proximity in the Rabl orientation could make homologous recombination the more likely outcome by presenting a homologous sequence nearby. For example, since chromosomes in yeast are arranged in the Rabl orientation (see Section II.B), one possibility is that the levels of recombination between two loci may vary in accordance with their relative distance from their adjoining centromeres. Allelic loci will always be, by definition, the same distance from their adjoining centromere. Results of analysis of previously collected data from two independent studies comparing ectopic vs allelic recombination suggested an effect of the Rabl on recombination rates (Burgess and Kleckner, 1999). First, recombination between homologous DNA segments located at different distances from their adjoining centromere, and thus located at different latitudes within the nucleus, recombined sevenfold less frequently than the same segments located at allelic positions (Jinks-Robertson and Petes, 1986). In a separate study, homologous DNA segments positioned very close to their adjoining centromeres (25–35 kb away) at either allelic or nonallelic loci recombined at similar rates whether or not the sites were allelic or nonallelic (Lichten and Haber, 1989).

For the spatial proximity of loci to affect recombination rates, the frequency of homologous recombination should be dependent on the concentration of the interacting sites. This condition is met in mitotically dividing cells of yeast, since an increase in the copy number of donor DNA sequences has been shown to increase the frequency of recombination events (Melamed and Kupiec, 1992; Wilson *et al.*, 1994). In addition, the rate of homologous recombination between two DNA segments located on the same chromosome was found to be inversely related to the distance in kilobases that separates the two loci (Lichten and Haber, 1989).

D. Chromosome mobility and the homology search on a nuclear scale

Movement of chromosomes in the nucleus, either by random diffusion or by active processes, could bring nonallelic homologous DNA segments into close proximity by chance and thus contribute to a genome-wide search process. The global organization of chromosomes in the nucleus might constrain the natural movement of chromosomes, however, such that allelic sequences will preferentially interact by virtue of their location in the Rabl orientation. GFP-tagged chromosomal loci have been shown to undergo a constrained diffusion, reflecting true Brownian motion, which occurs even in the absence of active metabolism (Marshall *et al.*, 1997). The observed movement, however, is apparently constrained somewhat by microtubules, since the volume within which the two loci diffused increased after treatment with nocodazole (see Section II.B; Marshall *et al.*, 1997). Such constraint may result from the association of centromeres with the SPB, since the loci under study were closely linked to the centromere, consistent with their association with the SPB.

This model is also consistent with observed levels of homologous recombination, even when the apparently large theoretical size of a Rad51/ssDNA filament in the nucleus is considered. The resection of the 5′ ends of DSBs yields 3′-ssDNA tails that are on average 600 nucleotides long in both meiotic and mitotically dividing cells (Paques and Haber, 1999). The contour length of such tails when coated by Rad51 filament would thus extend 0.3–0.5 μm, assuming a 154% extension of B-form duplex DNA (Sung and Robberson, 1995). Given that the diploid yeast nucleus is about 2 μm, the distance spanned by these filaments is significant, yet does not span the diameter of the entire nucleus (Figure 3.2B). Therefore, allelic sequences and sequences on nonhomologous chromosomes at the same relative latitude within the Rabl orientation should be accessible for repair of DSBs by homologous recombination. For ectopic sequences to interact outside of the Rabl orientation, either the DNA must move within the nucleus or ectopic segments must end up fortuitously co-localized following the dynamics of chromosome movement during anaphase (see Section II.B).

IV. PAIRING AND SYNAPSIS OF HOMOLOGOUS CHROMOSOMES DURING MEIOSIS

Haploid gametes are produced during meiosis via two rounds of chromosome segregation following a single round of DNA replication. Prior to the first round of segregation, homologous chromosomes pair along their lengths and undergo high levels of genetic exchange, then segregate from one another in a reductional division. The second round of segregation is very much like mitosis in that sister chromatids segregate from one another (Roeder, 1997; Zickler and Kleckner, 1999).

Pairing of homologous chromosomes is a hallmark of meiosis. A number of reviews speculate on the nature of homolog association, based on over a century of observations and experiments carried out in a wide variety of organisms (Loidl, 1990; Roeder, 1997; Zickler and Kleckner, 1998, 1999; Walker and Hawley, 2000). This section is intended to provide an overview of meiotic homologous chromosome association in *Saccharomyces cerevisiae*, with an emphasis on the pairing phenotypes exhibited by mutants that are defective for various aspects of meiotic chromosome metabolism.

A. Four mechanisms ensure homologous chromosome alignment during meiosis

Several mechanisms act to initiate, facilitate, or maintain homologous chromosome association or alignment during meiotic prophase. One prominent mechanism is homologous recombination, which is initiated by induced DSB formation and results in crossing over between homologous chromosomes (Figure 3.3A1; reviewed in Keeney, 2001, and Smith and Nicolas 1998; Paques and Haber, 1999). Chromosomes are also organized in the bouquet arrangement with clustered telomeres located near the nuclear envelope (Figure 3.3B1, reviewed in Zickler and Kleckner, 1998). In addition, the synaptonemal complex (SC) acts as a zipper to hold the homologs in very close alignment during the late prophase stages of zygotene and pachytene, also known as synapsis (Figure 3.3A2; reviewed in Roeder, 1997; Zickler and Kleckner, 1999). Finally, homolog pairing contacts, as detected by FISH, occur independent of DSB formation or SC. For example, in situations where the possibility for either recombination or SC formation has been eliminated, homolog pairing is still observed (Loidl *et al.*, 1994; Weiner and Kleckner, 1994; Nag *et al.*, 1995; Cha *et al.*, 2000). Such findings imply the existence of a "DSB-independent" mechanism for homolog recognition and juxtaposition (Figure 3.3B2; Section IV.D). The mechanism underlying these interactions is poorly understood, but closely resembles, the premeiotic and

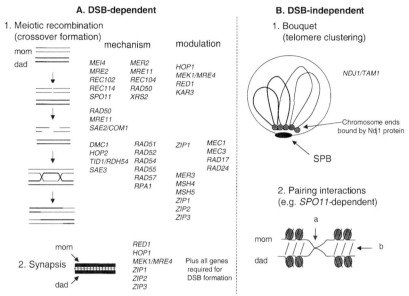

Figure 3.3. Homologous chromosome association in meiotic cells. (A) DSB-dependent interactions. (1) Interhomolog recombination leading to crossover products. Intermediates are the same as shown in Figure 3.1A. Although *REC103/SKI8* and *MER1* are required for DSB formation, it is not clear that they are directly involved in the mechanism of DSB formation and so they are not shown (see text and Keeney, 2001). (2) Side-by-side alignment of homologs connected by the SC. (B) DSB-independent interactions. (1) Bouquet formation; (2) interhomolog pairing interactions detected in the absence of DSB formation. DSB-independent interactions may occur between (a) intact DNA/DNA duplexes via DNA/DNA contacts, or (b) pairing may be driven by the self-association of homolog-specific arrays of DNA-binding proteins.

vegetative pairing interactions involving contacts between presumably intact homologous DNA duplexes (see Sections V.A and V.F).

In yeast, a reasonable model for the temporal progression of these events is DNA replication → pairing ≈ initiation of recombination ≈ bouquet → appearance of SC and crossover formation. There has been considerable debate over the extent to which these processes are mechanistically coupled in yeast and in other organisms where the different mechanisms may assume more or less prominent roles (see Roeder, 1997; Zickler and Kleckner, 1999; Walker and Hawley, 2000). Zickler and Kleckner (1999) have proposed a "bucket brigade" model for interhomolog connectedness, where contacts made early in the meiotic pathway are lost after they have been handed off (and become stabilized) by the next step in the pathway. Such a model may ensure the accuracy of the homolog pairing process, by allowing multiple independent mechanisms to contribute to homolog alignment.

B. Overview of meiotic chromosome dynamics

The S phase preceding the meiotic divisions in yeast takes approximately three times longer than S phase in nonmeiotic cells (Cha *et al.*, 2000). It has been proposed that this additional time is required to establish the foundation on which meiotic chromosome dynamics will be played out. This foundation may include the binding of factors required for homolog pairing, recombination, bouquet arrangement, or SC formation (Zickler and Kleckner, 1999). Immediately prior to meiotic S phase (i.e., in premeiotic G1), homologs are paired along their lengths by multiple, interstitial interactions as detected using FISH (Section V.A; Weiner and Kleckner, 1994; Burgess *et al.*, 1999). During the period of S phase, pairing interactions between homologs are lost and restored (Weiner and Kleckner, 1994; Cha *et al.*, 2000). Cohesion between sister chromatids is likely established during meiotic S phase, as it is established during S phase in nonmeiotic cells (Uhlmann and Nasmyth, 1998).

Following DNA replication, meiotic recombination is initiated by the formation of DSBs catalyzed by the Spo11 protein (Sun *et al.*, 1989; Cao *et al.*, 1990; Padmore *et al.*, 1991; Bergerat *et al.*, 1997; Keeney *et al.*, 1997; Borde *et al.*, 2000). At least 10 genes other than *SPO11* are required for the initiation of DNA double-strand breaks, including *RAD50, MRE11, XRS2, REC102, REC103/SKI8, REC104, REC114, MEI4, MER1, MER2*, and *MRE2*, while mutations in a second class of at least four genes give a 5–10-fold reduction in DSB formation (reviewed by Keeney, 2001; Table 3.2; Figure 3.3A1; Section IV.C). Some members of this second class, including *RED1, MEK1/MRE4*, and *HOP1*, are likely involved in forming higher-order chromosome structure rather than in participating directly in DSB formation (reviewed in Roeder, 1997; Zickler and Kleckner, 1999; Dresser, 2000; and below). Another member of this class includes the motor protein *KAR3*, which may be involved in chromosome movement during meiotic prophase (Bascom-Slack and Dawson, 1997).

Meiotic DSBs, recombination intermediates and products can be detected physically (in real time) during meiosis using gel electrophoresis and Southern blotting techniques. These intermediates include (in temporal order of appearance): 3' ssDNA tails produced by exonucleolytic digest of the 5' ends of breaks, single-end invasions (SEI), double Holliday junctions (dHJ), and mature recombinant DNA products (Cao *et al.*, 1990; Sun *et al.*, 1991; Bishop *et al.*, 1992; Collins and Newlon, 1994; Schwacha and Kleckner, 1994, 1995; Nag *et al.*, 1995; Allers and Lichten, 2000; Hunter and Kleckner, 2001). Mature recombinants arise at the end of pachytene, when full levels of SC are present (Table 3.1B; Roeder, 1997; Zickler and Kleckner, 1999). Many of the genes involved in meiotic recombination and SC formation are listed in Figure 3.3A. The recombined DNA products, in conjunction with cohesion between sister chromatids, hold the homologs on the meiotic spindle and ensure a proper

Table 3.2. Meiotic Mutant Phenotypes

Allele	Pairing levels[a]	Recombination initiation (DSB formation/resection)[b]	SC morphology [axial elements (AE), synaptonemal complex (SC), interaxis connectors (IC)]	References[c] (pairing levels, recombination initiation, SC morphology)
spo11Δ	+/+	ND/NA	AE⁻, SC⁻	[1–3], [4], [3]
spo11-Y135F	++++	ND/NA	ND	[5], [6], [NA]
rad50Δ	++/+	ND/NA	AE⁺/⁻, SC⁻	[1, 3], [4], [7]
rec102Δ	++	ND/NA	AE⁺, SC⁻	[8], [9], [10]
mei4Δ	++	ND/NA	AE⁺, SC⁻	[8], [11, 12], [13]
mer2Δ	++	ND/NA	AE⁺, SC⁻	[2], [2], [2]
hop1Δ	+++/+++	ND to <10% of WT/yes	AE⁻, SC⁻	[1, 3, 8], [14], [3,15]
mer1Δ	+++/++	~10% of WT/yes	AE⁺, SC⁻	[1, 8], [16], [17]
red1Δ	++	ND to 20% of WT/yes	AE⁺, SC⁻	[8], [18], [19]
mek1Δ/mre4Δ	++++	10–20% of WT/yes	AE⁺, SC⁺/⁻	[8], [18], [19]
sae2Δ/com1Δ	+	100%/no		[20], [20, 21], [20]
rad50S	++/++	100%/no	AE⁺, SC⁺/⁻	[1, 3], [7], [7]
mre11S	+	100%/no	AE⁺, SC⁺ (nonhomologous)	[22], [22], [22]
hop2Δ	+ to +++	100%/hyper	AE⁺, SC⁺ (nonhomologous)	[23], [23], [23]
dmc1Δ	+++/+++	100%/hyper	AE⁺, SC⁺ (delayed)	[1, 24], [18, 25, 26], [24, 25]
rad51Δ	+++	100%/hyper	AE⁺, SC⁺ (delayed)	[24], [18, 26, 25], [24]
rad51Δdmc1Δ	+++	100%/hyper	ND	[1, 24], [26], [24]
ndj1Δ/tam1Δ	++++ (delayed)	ND	AE⁺, SC⁺ (delayed)	[27], [ND], [28, 29]
zip1Δ	++++	100%/yes	AE⁺, SC⁻, IC⁺	[8], [18], [24, 30]
zip1Δdmc1Δ	++	ND	AE⁺, SC⁻, IC⁻	[24], [ND], [24]
zip2	++++	ND	AE⁺, SC⁻, IC⁺	[31], [ND], [31]

[a] Pairing levels relative to wild-type as determined by FISH (0–20%, +; 21–40%, ++; 41–80%, +++; 81–100%, ++++). Bold characters show results from experiments where nuclei were scored, regardless of the state of DNA compaction (Weiner et al., 1994; Prinz et al., 1997). All other data were obtained from samples prepared using the spreading method developed by Dresser and Giroux (1988), and where only nuclei with compacted DNA were analyzed. *Sae2/com1* nuclei were prepared by the method of Dresser and Giroux (1988), but all nuclei types were included for analysis.

[b] ND, none detected; NA, not applicable.

[c] [1] Weiner and Kleckner (1994); [2] Rockmill et al. (1995a); [3] Loidl et al. (1994); [4] Cao et al. (1990); [5] Cha et al. (2000); [6] Bergerat et al. (1997); [7] Alani et al. (1990); [8] Nag et al. (1995); [9] Bullard et al. (1996); [10] Bhargava et al. (1992); [11] Keeney et al. (1997); [12] Jiao et al. (1999); [13] Menees et al. (1992); [14] Schwacha and Kleckner (1994); [15] Hollingsworth and Byers (1989); [16] Storlazzi et al. (1995); [17] Engebrecht and Roeder (1990); [18] Xu et al. (1997); [19] Rockmill and Roeder (1991); [20] Prinz et al. (1997); [21] McKee and Kleckner (1997); [22] Nairz and Klein (1997); [23] Leu et al. (1998); [24] Rockmill et al. (1995b); [25] Bishop et al. (1992); [26] Shinohara et al. (1997a); [27] Trelles-Sticken et al. (2000); [28] Chua and Roeder (1997); [29] Conrad et al. (1997); [30] Sym et al. (1993); [31] Chua and Roeder (1998).

reductional division at anaphase I of meiosis (reviewed in Moore and Orr-Weaver, 1998). These connections presumably provide the tension necessary to align chromosomes between the spindle poles (Nicklas 1977).

The conversion of DSBs into recombinant products involves many of the same genes that are required for homologous recombination in mitotically dividing cells (Section III.A; Smith and Nicolas, 1998; Paques and Haber, 1999). The yeast RecA homologs, *RAD51*, and the meiosis-specific *DMC1* gene, are likely involved in homology sensing. Both *rad51*Δ and *dmc1*Δ mutants form meiotic DSBs, yet the breaks become hyperressected and fail to form dHJs with the homologous chromosome (Table 3.2; Bishop *et al.*, 1992; Shinohara *et al.*, 1992; Schwacha and Kleckner, 1995; Xu *et al.*, 1997). While *RAD54* is likely involved in the homology-sensing mechanism for homologous recombination in mitotically dividing cells (see Section V.A), the related gene *TID1/RDH54* likely takes over this function during meiosis. The *tid1*Δ/*rdh4*Δ mutant is severely defective for meiotic recombination, while *RAD54* appears to be dispensable (Shinohara *et al.*, 1997b; Schmuckli-Maurer and Heyer, 2000). The severe recombination defect exhibited by the *rad54*Δ *tid1*Δ/*rdh4*Δ double mutant, however, is suggestive of some overlapping roles for these genes in meiotic cells (Shinohara *et al.*, 1997b). Other genes involved in meiotic recombination include *RAD52*, *RAD55*, *RAD57*, and *RPA1* (Gasior *et al.*, 1998), which are also involved in homologous DSB repair in nonmeiotic cells (see Section III.A).

During meiosis, homologous chromosomes are used as the substrate for recombinational repair of the DSB (Kleckner, 1996; Roeder, 1997; Schwacha and Kleckner, 1997). In contrast, in mitotically dividing cells, DSBs introduced during G2 are repaired using the sister chromatid as a template for repair (Fabre *et al.*, 1984; Kadyk and Hartwell, 1992). *DMC1* and *RED1* likely play important roles in homolog/sister discrimination in meiosis. *RED1* is important for the formation of DSBs that are channeled into an interhomolog-only pathway, and *DMC1* is important for directing such breaks into dHJ formed between homologs (Schwacha and Kleckner, 1997). Defects in homolog/sister bias have also been inferred by observing increased levels of unequal sister chromatid exchange in certain mutant strains. One such mutant includes *mek1*Δ/*mre4*Δ (Thompson and Stahl, 1999). *MEK1/MRE4* has been shown to encode a kinase that phosphorylates Red1 protein (de los Santos and Hollingsworth, 1999; Bailis *et al.*, 2000). Since *mek1*Δ/*mre4*Δ and *red1*Δ mutants exhibit nearly identical meiotic phenotypes, presumably *MEK1/MRE4* is also required for DSB formation along the interhomolog-only pathway (Xu *et al.*, 1997). Another class of factors implied to be involved in interhomolog bias by this genetic criterion includes genes involved in meiotic cell-cycle checkpoint functions, encoded by *RAD17*, *RAD24*, *MEC1*, and *MEC3* (Grushcow *et al.*, 1999; Thompson and Stahl, 1999).

Chromosome organization in the meiotic nucleus also influences the efficiency of recombination, perhaps even more strongly than in nonmeiotic cells. Short homologous DNA segments were found to undergo recombination at an

8- to 17-fold decrease in efficiency when present at ectopic positions on non-homologous chromosomes compared with segments located at allelic positions (Goldman and Lichten, 1996). Furthermore, the efficiency of ectopic recombination was greater for pairs of loci located near the telomeres than for pairs of interstitial loci. One interpretation of these data is that homologous chromosomes are already co-localized before ectopic recombination takes place, with contributions from both pairing interactions and from the bouquet (Goldman and Lichten, 1996). These authors then went on to explore whether homolog associations may limit interactions between homologous DNA segments present at ectopic positions. They found that when interactions between homologous chromosomes were somewhat compromised (e.g., in a $ndj1\Delta/tam1\Delta$ mutant or in the presence of a competing homeologous chromosome from a related species), the levels of ectopic interactions increased (Goldman and Lichten, 2000).

     DSB-independent contacts between homologous chromosomes are also likely to be relevant to DSB formation. For example, the presence of nonhomology on one homolog has been shown to affect the levels of DSB formation at an allelic region on the other homolog in *trans* (Xu and Kleckner, 1995; Bullard *et al.*, 1996; Rocco and Nicolas, 1996). This effect may be similar in nature to the alteration of chromatin structure at an allelic position in *trans* in the presence of heterology, which has been shown to occur in premeiotic cells at a known meiotic DSB hot spot (see Section V.A; Keeney and Kleckner, 1996). Homolog pairing is not absolutely required for DSB formation, since high levels of DSBs were observed when no homolog was present in haploid yeast programmed to enter meiosis (De Massy *et al.*, 1994; Gilbertson and Stahl, 1994). Such DSBs, however, were delayed somewhat in this situation, suggestive of some effect of homolog association. Pairing could thus act to influence the timing of DSB formation or to influence where DSBs occur along the length of a chromosome.

     In summary, meiotic recombination appears to be mechanistically similar to the DSBR pathway that functions in mitotically dividing cells. Meiotic recombination has specialized features which provide programmed formation of DSBs, promote interhomolog-specific interactions, and also promote other features of meiotic chromosome metabolism not discussed here (including crossover control and formation of the SC; see Figure 3.3A). In addition, homolog pairing interactions occurring independent of DSB formation may influence meiotic recombination.

C. Analysis of homolog pairing in mutants defective for meiotic chromosome metabolism

Physical juxtaposition of homologous chromosomes during meiosis in yeast has been assayed primarily using FISH techniques. In general, differentially labeled probes are hybridized to spread chromosome preparations isolated from cells synchronously proceeding through meiosis. The fraction of nuclei exhibiting pairing (i.e., $d \leq 0.7 \mu m$) for any given locus is usually determined by measuring the distances between foci on digitally collected images (Loidl *et al.*, 1994; Weiner and Kleckner, 1994). Pairing levels for allelic loci remain high in spread preparations relative to levels for loci on nonhomologous chromosomes, suggesting that physical contacts holding chromosomes together are maintained during the spreading process. Meiosis-specific homologous chromosome interactions can also be detected *in vivo* by following GFP-tagged loci (Aragon-Alcaide and Strunnikov, 2000) or by *Cre/loxP* site-specific recombination (E. Dean and S. M. Burgess, unpublished). No mutant phenotypes have been reported to date using these assays, however.

Since various processes contribute to the colocalization of homologs (i.e., DSB-independent pairing contacts, recombination, the bouquet, and SC) in wild-type cells, the analysis of meiotic mutants is made difficult. It is also not possible to predict a priori how a mutant defective for DSB-dependent pairing would affect recombination, the bouquet, or the SC. DSB-independent pairing interactions during meiosis have only been detected under two special circumstances: two mutants, *spo11-Y135F* and *hop1*Δ, give no or few DSBs, and no SC yet exhibit high levels of pairing contacts when analyzed by FISH. In contrast, the *spo11*Δ mutant gives no pairing by this assay.

In general, two different methods have been used to prepare and analyze nuclei by FISH analysis. Differences in these methods may contribute to the differences in observed phenotypes for some mutants, which in some cases differ quite dramatically. For example, the *hop1*Δ mutant gives nearly wild-type levels of pairing using one of these methods and very low levels using the other method (Table 3.2, Section V.D). Three factors which could account for the observed differences are discussed below.

First, the criteria used for selecting nuclei to be included in a data set differ between the two methods. The method published by Weiner and Kleckner (1994) considers nuclei from synchronous populations regardless of the condition and state of chromosome compaction, while the other method (e.g., as described in Loidl *et al.*, 1994; Nag *et al.*, 1995) considers only nuclei with compacted DNA. The latter method has the advantage in that if cells condense their chromosomes normally, then the fraction of nuclei analyzed will represent a defined stage of meiosis. On the other hand, if only a small percentage of cells condense their chromosomes properly (e.g., as in the *hop1*Δ mutant—Weiner and Kleckner, 1994; Nag *et al.*, 1995), then the second method will tend to analyze a smaller, and less representative, subset of the population and/or will limit analysis to a different temporal stage of meiosis where initial homolog contacts are possibly abolished, perhaps by the forces of condensation. Another point to consider (although not likely applicable to the *hop1*Δ mutant) is that analysis of pairing in condensed nuclei gives an "all or none" phenotype. That is, any one or few contacts on a chromosome give a positive signal for pairing. In contrast, in nuclei with noncompacted DNA (e.g., early prophase), a pairing signal arises only from a nearby contact.

A second factor contributing to phenotypic variance could come from differences in the degree of spreading using the two methods used to prepare nuclei for FISH. The method described by Loidl *et al.* (1994) and Nag *et al.* (1995) use detergent during the spreading procedure, while the Weiner and Kleckner (1994) method relies on osmotic lysis. Notably, detergent has been reported to disrupt telomere–telomere interactions (Belmont *et al.*, 1989; Klein *et al.*, 1992; Palladino *et al.*, 1993; Hayashi *et al.*, 1998). The presence or absence of detergent, or simply the degree of spreading independent of chemical disruption, could similarly account for the observed differences in pairing levels, especially in cases where pairing interactions have not been stabilized by recombination interactions or SC (see Table 3.2).

Finally, different loci may exhibit different degrees of pairing in a mutant cell. For example, pairing levels measured for the *hop2Δ* mutant were found to range from 15% to 45% of wild type, depending on the locus under consideration (Table 3.2; Leu *et al.*, 1998). Differences in observed pairing levels for interstitial vs telomeric loci have also been noted. For example, telomeric loci remain paired even in a mutant in which pairing at interstitial regions is abolished, as is the case for the *spo11Δ* mutant (Weiner and Kleckner, 1994).

D. Role of meiotic recombination in pairing

Meiotic homolog pairing levels have been assessed using FISH for five mutants defective for DSB formation, including, *spo11Δ*, *rad50Δ*, *rec102Δ*, *mei4Δ*, and *mer2Δ* (Table 3.2). All mutants of this class exhibited less than 40% of wild-type pairing levels, regardless of the spreading method or whether condensed nuclei were scored. The *spo11Δ* mutant was studied in three independent laboratories and exhibited the most severe phenotype, ~10–20% of wild type, pointing to its important role in pairing (Loidl *et al.*, 1994; Weiner and Kleckner, 1994; Rockmill *et al.*, 1995a). Meiotic pairing levels were measured in *rad50Δ*, *rec102Δ*, *mei4Δ*, and *mer2Δ* mutants, and all gave low levels of pairing using both methodologies (Loidl *et al.*, 1994; Weiner and Kleckner, 1994; Nag *et al.*, 1995; Rockmill *et al.*, 1995a).

Spo11 binds covalently to the 5' ends of meiotic DSBs, and based on its homology to a subunit of archaeal type II topoisomerases, is likely involved directly in the catalysis of such breaks (Keeney and Kleckner, 1995; Bergerat *et al.*, 1997; Keeney *et al.*, 1997). A *spo11Δ* mutant goes through meiotic S phase faster compared to wild-type cells and is thus implicated in some mechanism that provides time for events related to the establishment of meiotic chromosome structure to be carried out (Cha *et al.*, 2000). The roles for *SPO11* in prolonging meiotic S phase and in promoting pairing are independent of its catalytic role in DSB formation, however. A mutation that alters the putative catalytic tyrosine to a phenylalanine (*spo11-135F*) completely abolishes DSB formation yet allows for normal S-phase timing and wild-type pairing levels (Bergerat *et al.*, 1997;

Cha *et al.*, 2000). These data provide very strong evidence for DSB-independent pairing.

Consistent with the notion that meiotic pairing interactions occur independent of DSBs, *hop1Δ* and *mek1Δ/mre4Δ* mutants exhibited nearly wild-type pairing levels even though they show 80–90% reductions in DSB levels (Table 3.2; Weiner and Kleckner, 1994; Nag *et al.*, 1995). The *mek1Δ/mre4Δ* result is particularly notable, since this mutant was analyzed under the most stringent conditions for assessing pairing levels, where only those nuclei with condensed DNA were analyzed and where detergent was used during spreading (Nag *et al.*, 1995). It should be considered, however, that full levels of DSBs have been be detected in a *mek1Δ/mre4Δ* mutant in a *rad50S* background, where breaks are not resected (Xu *et al.*, 1997). So, the pairing interactions reported for *mek1Δ/mre4Δ* may not be entirely DSB-independent. Notably, *hop1Δ* mutants exhibited very low levels of pairing using this same methodology. The differences in *hop1Δ* phenotypes assayed using different methods may be accounted for by the loss of pairing contacts following chromosomal condensation (see previous section).

Special separation-of-function mutations, *rad50S* and *mre11S*, and the *sae2Δ/com1Δ* mutation allow for DSB formation but prevent the resection step of meiotic recombination. These mutants gave severe pairing phenotypes, comparable to those exhibited by the DSB-defective mutants described above (Table 3.2; Loidl *et al.*, 1994; Weiner and Kleckner, 1994; Nairz and Klein, 1997).

Pairing defects are generally less severe in mutants blocked at a later step in the meiotic recombination pathway. Mutations in both *HOP2* and *DMC1* cause blocks in the recombination pathway and accumulate hyperrresected DSBs (Bishop *et al.*, 1992; Shinohara *et al.*, 1992; Xu *et al.*, 1997; Leu *et al.*, 1998). Interestingly, *hop2Δ* mutants form SC between nonhomologous chromosomes, whereas *dmc1Δ* mutants tend to form SC between homologous chromosomes, but SC formation is delayed (Rockmill *et al.*, 1995b; Leu *et al.*, 1998). Accordingly, the *hop2Δ* mutant exhibited a slightly more severe defect in homolog pairing than the *dmc1Δ* mutant (Leu *et al.*, 1998). Rad51 does not appear to replace Dmc1 function in promoting pairing because a *rad51Δdmc1Δ* double mutant exhibited high levels of pairing similar to either single mutant alone (Weiner and Kleckner, 1994; Rockmill *et al.*, 1995b).

E. Role of bouquet arrangement in pairing

During meiosis in many organisms the Rabl orientation gives way to the bouquet configuration, in which telomeres are clustered to a region near the nuclear envelope and near the SPB (for reviews, see Dernburg *et al.*, 1995; Zickler and Kleckner, 1998). In contrast, in the Rabl orientation, the centromeres are clustered near the SPB (see Section II.B). The timing of the appearance of the bouquet generally precedes chromosome synapsis and likely follows or is concomitant with presynaptic alignment (Table 3.1B; Zickler and Kleckner, 1998). The bouquet arrangement is seen in meiotic cells of fungi, plants, and animals. In yeast, it appears transiently and has been detected by FISH analysis using telomere-specific probes and by immunostaining of the telomere-specific Ndj1/Tam1 protein (Trelles-Sticken *et al.*, 1999). Meiotic telomere associations have also been inferred from both cytological and genetic studies (Dresser and Giroux, 1988; Klein *et al.*, 1992; Weiner and Kleckner, 1994; Goldman and Lichten, 1996; Chua and Roeder, 1997; Conrad *et al.*, 1997; Rockmill and Roeder, 1998). In premeiotic cells, telomeres are found in multiple clusters, as in mitotically dividing cells (Section II.B). During meiosis, telomeres form a single cluster indicative of the bouquet at the leptotene–zygotene transition and are dispersed by pachytene (Table 3.1B; Trelles-Sticken *et al.*, 1999).

Several independent lines of evidence support the notion that telomere association may facilitate homolog pairing during meiosis, although it is not absolutely required. The only mutant known to disrupt the bouquet in yeast, or in any other organism to date, is *ndj1Δ/tam1Δ*. Analysis of this mutant reveals that *NDJ1/TAM1* is also important for timing of meiotic divisions, chromosome condensation, and SC formation, as well as for proper chromosome segregation and crossover interference (Chua and Roeder, 1997; Conrad *et al.*, 1997; Trelles-Sticken *et al.*, 2000). Homolog pairing is delayed in a *ndj1/tam1* mutant, yet wild-type pairing levels are eventually achieved (Trelles-Sticken *et al.*, 2000). One possibility is that early pairing interactions (e.g., *SPO11*-dependent, DSB-independent) are never established, yet homologs become co-aligned for synapsis due to interactions from DSBR.

Further indication that telomeres may facilitate meiotic homolog pairing in budding yeast comes from a series of observations involving the behavior of a disomic pair of homologs in an otherwise haploid strain (Rockmill and Roeder, 1998). First, two linear homologous chromosomes were found to pair at a slightly higher level than one linear and one circular chromosome that lacked telomeric sequences (98% vs 78%, respectively). Second, haploid cells programmed to enter into meiosis and carrying an extra chromosome (i.e., a disome) exhibited a delay in undergoing the first meiotic division, which, however, was bypassed if the extra chromosome was circular. Third, *NDJ1/TAM1* was found to be required for the disome-induced delay. From these observations, it appears that the presence of a disome generates a signal that reflects the status of homolog association in the cell, which results in a delay of meiosis. Furthermore, this signal is dependent on the interaction of telomeres. Further discussion on the role of the bouquet during meiosis has been presented recently (Zickler and Kleckner, 1998).

F. Role of synaptonemal complex in meiotic pairing

While synapsis of homologous chromosomes is a hallmark of meiosis, with some notable exceptions, SC formation does not depend on DNA homology per se. Instead, synapsis between homologous chromosomes appears to occur because of the prior establishment of pairing interactions and the initiation of meiotic recombination (see above and Section V.B). The genesis of the SC occurs early in meiosis with the formation of axial elements (AE), which are rodlike structures that develop along the lengths of chromosomes around the time of DSB formation (Padmore et al., 1991). Formation of the tripartite SC structure involves the juxtaposition of the two axial elements from each homolog (now called "lateral elements") connected by a central region (reviewed by Zickler and Kleckner, 1999).

ZIP1 encodes a component of the central element of the SC (Sym and Roeder, 1995). A mutant carrying the zip1Δ allele was shown to be defective for SC formation and, though it initiated recombination at nearly wild-type levels, exhibited a twofold reduction in the level of crossover formation (Sym et al., 1993; Xu et al., 1997). Pairing levels in the zip1Δ mutant were at wild-type levels, consistent with the notion that synapsis does not contribute to either the formation or maintenance of homolog pairing interactions (Nag et al., 1995; Rockmill et al., 1995b). A zip2Δ mutation behaved similarly (Chua and Roeder, 1998). In both zip1Δ and zip2Δ mutants, homologous chromosomes were aligned along their lengths and connected by interhomolog bridges, also known as interaxis connectors (IC), when visualized by electron microscopy. These bridges are dependent on meiotic recombination, since they were generally absent in zip1Δdmc1Δ and zip1Δ hop2Δ double mutants (Rockmill et al., 1995b; Leu et al., 1998). For more details on such bridges in yeast and in other organisms, see Zickler and Kleckner (1999).

In yeast, SC can form in the absence of homologous chromosomes in haploid cells that have been genetically programmed to enter into meiosis (Loidl et al., 1991). These interactions were shown to occur between nonhomologous chromosomes and involved several partner switches. This situation is not unlike the diploid hop2Δ or mre11S mutants, where SC can form between nonhomologous chromosomes as described above (Nairz and Klein, 1997; Leu et al., 1998).

V. HOMOLOG INTERACTIONS IN MITOTICALLY DIVIDING CELLS (VEGETATIVE PAIRING)

A. Homolog pairing interactions in premeiotic and in mitotically dividing cells

One of the first indications that homologs communicate with one another in premeiotic cells of yeast came from the observation that at a small nuclease-hypersensitive region, the degree of hypersensitivity seen in diploid strains was greater when homologs had the same DNA sequence as when they had different DNA sequence. These results implied that homologs communicate with one another (i.e., "pair") in such a way as to alter local chromatin structure (Keeney and Kleckner, 1996). In this analysis, DNA heterology was introduced by the addition of 36 bp of DNA containing a BamHI restriction site fragment into a region known to act as a meiotic DSB hot spot. In the same strains, formation of meiotic DSBs at these sites exhibited the same dependence on homology as seen for chromatin structure in premeiotic cells (Xu and Kleckner, 1995). It was suggested from these studies that DSBs may occur preferentially at positions of preDSB pairing (Kleckner, 1996).

Homologous chromosome associations have also been detected by FISH in spread nuclei of premeiotic and in exponentially dividing, or "vegetative" cells of budding yeast (Weiner and Kleckner, 1994; Burgess et al., 1999). In both cases, pairing interactions display the following characteristics: (1) Pairing interactions occur via multiple, independent interstitial interactions between homologous chromosome pairs. (2) Pairing interactions do not require information from both MATα and MATa mating-type loci. In contrast, entry into meiosis is dependent on both loci being expressed (Kupiec et al., 1997). (3) Pairing does not depend on the homology-sensing devices of the Rad51 or Dmc1 proteins, since a strain deleted for the four RecA homologs, rad51Δ, rad55Δ, rad57Δ, dmc1Δ, exhibits wild-type levels of both premeiotic and vegetative pairing. (4) Pairing has been observed in three different strain backgrounds to date (e.g., SK1, S288C, and A364a). (5) Pairing can be observed in spread-nuclear preparations that eliminate the Rabl orientation.

The existence of interhomolog interactions can also be inferred from the greater rates of collision between allelic loci compared to ectopic loci on non-homologous chromosomes as measured by Cre/loxP site-specific recombination in vivo (Burgess and Kleckner, 1999). Overall, the average recombination rates for allelic constructs tested in this analysis were twofold greater than the rates obtained for constructs on nonhomologous chromosomes. Although the Rabl orientation can contribute substantially to chromosome interactions in vivo (See Sections II.B and III.C, homolog-specific interactions occurred at significantly greater levels than expected from the contribution of the Rabl orientation alone for two out of four loci tested. This analysis was done by comparing the observed levels of allelic interactions with the expected levels of nonallelic interactions for loci located at the same relative position in the Rabl orientation (Burgess and Kleckner, 1999).

B. Pairing and the cell cycle

Homolog pairing in vegetative cells of yeast is dynamic and varies over the course of the cell cycle. MATa/MATa diploid cells arrested in G1 by exposure to the mating pheromone alpha factor exhibited high pairing levels, comparable to those measured during meiosis in the spo11-Y135F mutant (see Sections IV.A; Weiner and Kleckner, 1994; Burgess et al., 1999). When cells were released from G1 arrest to allow progression into S phase, a 50% reduction in pairing levels was observed. As cells then traversed into G2, high pairing levels were restored. Reestablishment of pairing occurred during S or in G2 on a locus-by-locus basis (Burgess et al., 1999). Interestingly, centromeric loci unpaired later and unpairing lasted for a shorter period of time compared to other loci (5–8 min, compared to 11–15 min for interstitial probes).

The pattern of loss and reestablishment of pairing interactions in cells traversing S phase was found to be complex and suggestive of three types of determinants involved in pairing and unpairing during this period (Burgess et al., 1999). First, different pairs of loci could behave either independently or nonindependently of one another at the same time point measured during S phase, suggestive of local or regional effects acting on more than one pairing site at a time. (Information describing the independence of pairing at different loci can be obtained by probing two loci at a time and analyzing the distribution of nuclei with pairing at either one locus only, both loci, or neither locus.) Local or regional effects could arise from differences in chromosome sequence, structure/morphology, topology, or spatial position within the nucleus for the different loci under study. Second, an overall reduction in pairing levels was observed for all loci tested during the period of S phase, indicating genome- or nucleus-wide effects. And finally, pairing levels correlated in general with cell-cycle phase transitions, indicating temporal determinants, possibly representing the action of cell-cycle regulatory factors.

In other organisms where somatic pairing has been described, pairing status also varies during the course of the cell cycle. In *Drosophila*, homolog pairing is initiated at embryonic cell cycle 13, which coincides with the addition of G1 and G2 to the cell cycle (Hiraoka et al., 1993; Fung et al., 1998; Gemkow et al., 1998). These result have been interpreted to mean either that S phase or mitosis may not support efficient pairing interactions, or that the length of the interval between mitoses is an important determinant of pairing. In adult tissues of *Drosophila*, homolog pairing is reduced in euchromatic regions in cells inferred to be in S phase by BrdU labeling (Csink and Henikoff, 1998). Finally, while somatic pairing is not a general feature of human chromosomes, pairing at the imprinted Prader-Willi disease locus occurs only in late S of the cell cycle (LaSalle and Lalande, 1996).

C. Pairing and cell-cycle checkpoint functions

It is widely recognized that DNA or spindle damage normally triggers cell-cycle delay or arrest via cell-cycle checkpoint functions. These checkpoints also ensure proper completion of normal cell-cycle events before the initiation of subsequent events, including DNA replication, spindle assembly, and the attainment of cell mass (Elledge, 1996). During cell-cycle delay or arrest, the checkpoint functions signal various changes in the cell that facilitate the repair of DNA damage or allow for the completion of a cellular event or process. Such changes include induced expression of genes involved in DNA damage repair (reviewed in Fried-berg et al., 1995), the suppression of late replication origin firing during S phase, and alteration of DNA replication rates (Paulovich and Hartwell, 1995; Diffley, 1998), relocation of proteins in the nucleus (Mills et al., 1999; Roeder and Bailis, 2000), direct phosphorylation of proteins involved in homologous recombination (Bashkirov et al., 2000), and inhibition of anaphase (Cohen-Fix and Koshland, 1997).

Yeast cells arrested at G2/M of the cell cycle by one of two different treatments exhibit low levels of homolog pairing compared to G1-arrested cells or cells cycling through G2. First, nocodazole treatment triggers regulatory checkpoint arrest due to failure of chromosomes to become attached to a spindle (reviewed in Rudner and Murray, 1996). Average pairing levels dropped to below 50% of wild-type levels in nocodazole-treated cells (Burgess et al., 1999). Similarly, a cdc13-1^{ts} mutation triggers RAD9-dependent DNA damage checkpoint arrest at the non-permissive temperature, due to defective replication and processing of telomeres (Garvik et al., 1995); this arrest was also accompanied by a similar decrease in pairing levels.

Three independent models describe the disruption of homolog pairing in response to cell damage. First, homolog-pairing status could be a downstream target of cell-cycle checkpoint functions in one of two ways. Factors involved in pairing could be specifically modified by cell-cycle regulatory factors or, alternatively, pairing interaction could be disrupted indirectly as a consequence of a change in chromosome organization or structure mediated by checkpoint functions. Second, pairing interactions could be physically disrupted under the conditions that were tested in the experiments described above (e.g., by passage of the DNA replication fork or by microtubule depolymerization). Third, excessive time in the arrested state could cause unpairing. One argument against this final possibility is that cells arrested in G1 by pheromone treatment (and without checkpoint activation) exhibit the highest levels of pairing. So, for the length of arrested time to be an issue, it would have to be specific to the G2 phase of the cell cycle. It is also possible that different cell-cycle stages or different types of damage or checkpoint pathways may behave independently of one another.

The modulation of homolog pairing interactions during the cell cycle and under conditions of checkpoint arrest may influence discrimination between interhomolog and intersister chromatid interactions. That is, disruption of pairing interactions when cell-cycle checkpoints are activated (including during S phase) may favor interactions with the sister chromatid (Burgess et al., 1999). One prediction of this model is that activation of cell-cycle checkpoints during G1 in the presence of DNA damage will not disrupt pairing interactions, since only the homolog would be available for repair under this condition.

D. Heterochromatin-like regions and homologous DNA interactions

Repetitive DNA elements in eukaryotes are often associated with heterochromatin (Henikoff, 2000). As few as three tandem copies of a DNA element in *Drosophila* are sufficient to induce a heterochromatic-like state, as evidenced by the ability to impose position effect variegation, or PEV (Dorer and Henikoff, 1994). PEV occurs when expression of a gene is influenced by its proximity to heterochromatin and these effects can be mediated *in trans* (reviewed in Wakimoto, 1998). Association between heterochromatic sequences has been demonstrated genetically and cytologically in *Drosophila* cells. For example, nonspecific co-localization of heterochromatic sequences can bring homologous DNA sequences together in the same nuclear space (Csink and Henikoff, 1996; Dernburg *et al.*, 1996a). In addition, achiasmate segregation of chromosomes during meiosis in *Drosophila* females is dependent on heterochromatic DNA homology (Hawley *et al.*, 1992; Dernburg *et al.*, 1996b; Karpen *et al.*, 1996).

Over 70% of the yeast genome is thought to be protein coding sequence (Dujon, 1996; Goffeau *et al.*, 1996), a feature generally associated with euchromatin. Repetitive DNA is minimal compared to that found in mammalian chromosomes: ~140 tandem copies of 9-kb rDNA repeated elements are located on chromosome XII. Other repeated chromosomal elements, such as centromeres, telomeric regions, and ARS-consensus sequences, make up 1.2% of the genome. A relatively small fraction of the genome is made up of other types of repetitive DNA, including transposons and solo LTRs. These account for about 2.3% and 1% of the genome, respectively (Dujon, 1996).

The dense staining of heterochromatin as visualized by electron microscopy has not been observed in yeast. Small regions of repetitive DNA, however, do exhibit other qualities typical of heterochromatin. The best-studied regions in this respect are the silent mating type loci, the rDNA repeats, and subtelomeric DNA sequences. Like heterochromatin, these regions are less accessible to DNA-modifying agents, replicate late, are bound by hypoacetylated histone H4, and are able to silence genes introduced nearby (reviewed by Grunstein, 1998). Each of these phenomena is also associated with position effect variegation (PEV) in *Drosophila* (reviewed by Wakimoto, 1998; Henikoff, 2000).

Interactions between heterochromatic-like chromatin in yeast could occur by targeting such sequences to particular regions of the nucleus (e.g., the nuclear periphery) or by a homology-directed mechanism. Two lines of evidence point to both types of mechanism acting in yeast. First, both the silent mating type locus *HML* and telomeric sequences localize to the nuclear periphery in nonmeiotic cells (Laroche *et al.*, 2000). Second, GFP-tagged chromosomal loci associate with one another in a way that is dependent on the homology of the long, repeated operator sequences used for GFP targeting, yet independent of their chromosomal position (Aragon-Alcaide and Strunnikov, 2000). Such interactions are reminiscent of the types of pairing interactions associated with PEV. It would be interesting to know if *lacO* repeats take on properties associated with heterochromatin such as histone hypoacetylation. Also, these regions may form an extensive nucleosome-free region which would make the pairing interactions similar to those observed by Keeney and Kleckner (1996), where a nuclease-hypersensitive site in yeast has been shown to alter chromatin structure *in trans* (see Section IV.A).

E. Possible mechanisms underlying DSB-Independent pairing interactions

The DSB-independent pairing interactions described here include vegetative homolog pairing, meiotic pairing in mutant situations where recombination is blocked, and trans-sensing events that lead to changes of chromatin structure *in trans*. In addition, pairing interactions have been postulated for repeated sequences located at telomeres and elsewhere in the genome. How might such interactions occur? This section speculates on two general mechanisms that could bring intact homologous DNA duplexes together: (1) direct DNA/DNA contacts and (2) patterning of protein arrays along chromosomes. In the first case, homology readout would occur at the nucleotide level and in the latter case the readout would be indirect and based on the binding specificity of the DNA binding proteins involved.

1. Pairing interactions occur between intact DNA duplexes via DNA/DNA interactions

Several lines of evidence point to nuclease hypersensitive regions of chromosomes (e.g., promoters) as being likely sites of pairing interactions (Keeney and Kleckner, 1996; Morris *et al.*, 1999). If such interactions occur generally in nucleosome-free regions, then homolog recognition in these regions may involve direct DNA/DNA interactions (Kleckner and Weiner, 1993; Zickler and Kleckner, 1999). Such interactions may be mediated by local unwinding of DNA (e.g., by a Rad54-like protein) or by stretching of DNA duplex to reveal exposed single-stranded regions (Strick *et al.*, 1998). The ability to pair may be intrinsic to homologous naked DNA sequences. Interestingly, GT-rich regions are associated with loci exhibiting increased genetic exchange and are also preferred binding sequences for *RAD51* protein (Tracy *et al.*, 1997). Subsequently, it has been shown that these sites allow formation of a DNA hairpin on one strand with unstructured DNA on the other strand (S. C. Kowalczykowski, personal communication). Such a structure may be a good candidate for mediating interactions between two DNA duplexes. Perhaps then proteins act to stabilize such interactions.

Any pairing contact would have to be robust enough to maintain a connection under the spreading conditions used to detect homolog pairing. Such contacts may involve plectonemic DNA interactions (i.e., in which DNA duplexes are intertwined). For example, topologically defined domains of chromatin could be physically linked together by a topoisomerase activity. Such a topoisomerase activity could act in concert with a homology-sensing mechanism similar to *RAD51*, but one that could promote duplex–duplex interactions instead of the ssDNA–dsDNA homologous DNA interactions reviewed in Section III.A. Such a model has been proposed to explain achiasmatic pairing in *Drosophila* males (McKee *et al.*, 1992). Interestingly, the RecA protein from *E. coli* has been shown recently to promote interactions between two DNA duplexes. In this study, a RecA-dsDNA nucleoprotein filament was shown to pair with a homologous, supercoiled plasmid and promote strand exchange leading to the formation of a Holliday junction (E. N. Zaitsev and S. C. Kowalczykowski, personal communication). Although neither *RAD51*, *RAD55*, *RAD57*, nor *DMC1* is likely to carry out such a role to promote homolog pairing interactions in yeast, other yeast proteins may be able to carry out this function.

2. Homologs pair via self-association between homolog-specific arrays of DNA binding proteins

While no direct evidence exists for the contribution of self-association between homolog-specific arrays of DNA-binding proteins to pairing, this mechanism may act to stabilize physical contacts between homologs (Cook, 1997). For example, cohesins hold sister chromatids together from the period of S phase through anaphase (Nasmyth, 1999; Koshland and Guacci, 2000; Uhlmann, 2000). It is possible that homolog interactions are also mediated via cohesins, or cohesin-like proteins. Cohesins promote interactions between DNA duplexes and could possibly also mediate connections between homologs, since they would be present at similar sequences due to their sequences-specific binding preferences (Blat and Kleckner, 1999; Laloraya *et al.*, 2000). Apparently, such interactions are robust enough to withstand the forces of spreading, since sister chromatid cohesion is generally not lost during the spreading process (Burgess *et al.*, 1999; B. Weiner and N. Kleckner, personal communication).

In principle, pairing could be stabilized in a similar fashion by self-association of transcription factors *in trans*, i.e., when bound to similar positions along homologous chromosomes (Cook, 1997; Gemkow *et al.*, 1998). If transcription factors were able to link homologous chromosomes together tightly, then it is surprising that somatic/vegetative homolog pairing is not generally observed in more organisms. Also, based on FISH analysis of premeiotic and vegetative pairing interactions in yeast, homologs are connected, on average, once every 60–70 kb, a frequency which is not consistent with self-association of transcription factors (Weiner and Kleckner, 1994; Burgess *et al.*, 1999).

VI. CONCLUDING REMARKS

The fact that several seemingly independent mechanisms have evolved to align homologous chromosomes during meiosis speaks to the importance of homologous chromosome associations for the proper segregation of chromosomes at the reductional division of meiosis. The contributions and roles of DSB-independent homolog interactions in either meiotic or mitotically dividing cells are still not clear. In addition, while only very short sequences of homology are necessary to promote homologous recombination, the extent of DNA homology required to promote the DSB-independent interactions is unknown. The next challenge will be to identify the chromosomal determinants involved in mediating homologous DNA associations and to identify the *trans*-acting factors involved. The molecular, genetic and cytological tools available in yeast make it an attractive organism for addressing these issues.

It is ironic that, although budding yeast is well known for its robust homologous recombination pathways, it exhibits few other homology effects like those documented in other organisms (see other chapters in this volume). One explanation for the lack of observed effects could be that the haploid life-cycle stage of yeast has been more intensively studied than the diploid state, thereby limiting the opportunity to note such naturally occurring events. The further exploration of ploidy-specific phenomenon (e.g., transcription; Galitski *et al.*, 1999) and of the organization of chromosomes in the nucleus may lead to the uncovering of such homology effects.

Acknowledgments

S.M.B. is grateful for support from the Arnold and Mabel Beckman Foundation, the University of California, Cancer Coordinating Committee, and the American Cancer Society (RSG-01-053-01-CCG and IRG-95-125-04). S.M.B. gratefully acknowledges Job Dekker, Joanne Engebrecht, Alastair Goldman, R. Scott Hawley, Wolf-D. Heyer, Scott Keeney, Nancy Kleckner, Steve Kowalczykowski, Joshua Chang Mell, Ted Powers, Beth Weiner, and Ting Wu for many stimulating discussions. The author especially thanks Job Dekker, Joanne Engebrecht, R. Scott Hawley, Scott Keeney, Nancy Kleckner, and Joshua Chang Mell for comments on the manuscript.

References

Alani, E. R., Padmore, R., and Kleckner, N. (1990). Analysis of wild-type and *rad50* mutants of yeast suggests an intimate relationship between meiotic chromosome synapsis and recombination. *Cell.* **61**, 419–436.

Allers, T., and Lichten, M. (2000). A method for preparing genomic DNA that restrains branch migration of Holliday junctions. *Nucleic Acids Res.* **28**, e6.

Aragon-Alcaide, L., and Strunnikov, A. V. (2000). Functional dissection of *in vivo* interchromosome association in *Saccharomyces cerevisiae*. *Nature Cell Biol.* **2**, 812–818.

Bailis, J. M., Smith, A. V., and Roeder, G. S. (2000). Bypass of a meiotic checkpoint by overproduction of meiotic chromosomal proteins. *Mol. Cell Biol.* **20**, 4838–4848.

Bajer, A. S., and Mole-Bajer, J. (1986). Reorganization of microtubules in endosperm cells and cell fragments of the higher plant *Haemanthus* in vivo *J. Cell Biol.* **102**, 263–281.

Bascom-Slack, C. A., and Dawson, D. S. (1997). The yeast motor protein, Kar3p, is essential for meiosis I. *J. Cell Biol.* **139**, 459–467.

Bashkirov, V. I., King, J. S., Bashkirova, E. V., Schmuckli-Maurer, J., and Heyer, W. D. (2000). DNA repair protein Rad55 is a terminal substrate of the DNA damage checkpoints. *Mol. Cell Biol.* **20**, 4393–4404.

Bass, H. W., Riera-Lizarazu, O., Ananiev, E. V., Bordoli, S. J., Rines, H. W., Phillips, R. L., Sedat, J. W., Agard, D. A., and Cande, W. Z. (2000). Evidence for the coincident initiation of homolog pairing and synapsis during the telomere-clustering (bouquet) stage of meiotic prophase. *J. Cell Sci.* **113**, 1033–1042.

Baudin, A., Ozier-Kalogeropoulos, O., Denouel, A., Lacroute, F., and Cullin, C. (1993). A simple and efficient method for direct gene deletion in *Saccharomyces cerevisiae*. *Nucleic Acids Res.* **21**, 3329–3330.

Belmont, A. S., Braunfeld, M. B., Sedat, J. W., and Agard, D. A. (1989). Large-scale chromatin structural domains within mitotic and interphase chromosomes *in vivo* and *in vitro*. *Chromosoma* **98**, 129–143.

Bergerat, A., de Massy, B., Gadelle, D., Varoutas, P. C., Nicolas, A., and Forterre, P. (1997). An atypical topoisomerase II from *Archaea* with implications for meiotic recombination. *Nature* **386**, 414–417.

Bhargava, J., Engebrecht, J., and Roeder, G. S. (1992). The *rec102* mutant of yeast is defective in meiotic recombination and chromosome synapsis. *Genetics* **130**, 59–69.

Bianco, P. R., Tracy, R. B., and Kowalczykowski, S. C. (1998). DNA strand exchange proteins: a biochemical and physical comparison. *Front. Biosci.* **3**, D570–D603.

Bishop, D. K., Park, D., Xu, L., and Kleckner, N. (1992). *DMC1*: a meiosis-specific yeast homolog of *E. coli recA* required for recombination, synaptonemal complex formation, and cell cycle progression. *Cell* **69**, 439–456.

Blat, Y., and Kleckner, N. (1999). Cohesins bind to preferential sites along yeast chromosome III, with differential regulation along arms versus the centric region. *Cell* **98**, 249–259.

Borde, V., Goldman, A. S., and Lichten, M. (2000). Direct coupling between meiotic DNA replication and recombination initiation. *Science* **290**, 806–809.

Bullard, S. A., Kim, S., Galbraith, A. M., and Malone, R. E. (1996). Double strand breaks at the *HIS2* recombination hot spot in *Saccharomyces cerevisiae*. *Proc. Natl. Acad. Sci. USA* **93**, 13054–13059.

Burgess, S. M., and Kleckner, N. (1999). Collisions between yeast chromosomal loci *in vivo* are governed by three layers of organization. *Genes Dev.* **13**, 1871–1883.

Burgess, S. M., Kleckner, N., and Weiner, B. M. (1999). Somatic pairing of homologs in budding yeast: Existence and modulation. *Genes Dev.* **13**, 1627–1641.

Bussey, H., Kaback, D. B., Zhong, W., Vo, D. T., Clark, M. W., Fortin, N., *et al.* (1995). The nucleotide sequence of chromosome I from *Saccharomyces cerevisiae*. *Proc. Natl. Acad. Sci. USA* **92**, 3809–3813.

Byers, B., and Goetsch, L. (1975). Electron microscopic observations on the meiotic karyotype of diploid and tetraploid *Saccharomyces cerevisiae*. *Proc. Natl. Acad. Sci. USA* **72**, 5056–5060.

Cao, L., Alani, E., and Kleckner, N. (1990). A pathway for generation and processing of double-strand breaks during meiotic recombination in *S. cerevisiae*. *Cell* **61**, 1089–1101.

Cha, R. S., Weiner, B. M., Keeney, S., Dekker, J., and Kleckner, N. (2000). Progression of meiotic DNA replication is modulated by interchromosomal interaction proteins, negatively by Spo11p and positively by Rec8p. *Genes Dev.* **14**, 493–503.

Charron, M. J., Read, E., Haut, S. R., and Michels, C. A. (1989). Molecular evolution of the telomere-associated *MAL* loci of *Saccharomyces*. *Genetics* **122**, 307–316.

Chua, P. R., and Roeder, G. S. (1997). Tam1, a telomere-associated meiotic protein, functions in chromosome synapsis and crossover interference. *Genes Dev.* **11**, 1786–1800.

Chua, P. R., and Roeder, G. S. (1998). Zip2, a meiosis-specific protein required for the initiation of chromosome synapsis. *Cell* **93,** 349–359.

Cohen-Fix, O., and Koshland, D. (1997). The anaphase inhibitor of *Saccharomyces cerevisiae* Pds1p is a target of the DNA damage checkpoint pathway. *Proc. Natl. Acad. Sci. USA* **94,** 14361–14366.

Collins, I., and Newlon, C. S. (1994). Meiosis-specific formation of joint DNA molecules containing sequences from homologous chromosomes. *Cell* **76,** 65–75.

Conrad, M. N., Dominguez, A. M., and Dresser, M. E. (1997). Ndj1p, a meiotic telomere protein required for normal chromosome synapsis and segregation in yeast. *Science* **276,** 1252–1255.

Cook, P. R. (1997). The transcriptional basis of chromosome pairing. *J. Cell Sci.* **110,** 1033–1040.

Csink, A. K., and Henikoff, S. (1996). Genetic modification of heterochromatic association and nuclear organization in *Drosophila*. *Nature* **381,** 529–531.

Csink, A. K., and Henikoff, S. (1998). Large-scale chromosomal movements during interphase progression in *Drosophila*. *J. Cell Biol.* **143,** 13–22.

de los Santos, T., and Hollingsworth, N. M. (1999). Red1p, a MEK1-dependent phosphoprotein that physically interacts with Hop1p during meiosis in yeast. *J. Biol. Chem.* **274,** 1783–1790.

De Massy, B., Baudat, F., and Nicolas, A. (1994). Initiation of recombination in *Saccharomyces cerevisiae* haploid meiosis. *Proc. Natl. Acad. Sci. USA* **91,** 11929–11933.

de Steensma, H. Y., de Jonge, P., Kaptein, A., and Kaback, D. B. (1989). Molecular cloning of chromosome I DNA from *Saccharomyces cerevisiae*: localization of a repeated sequence containing an acid phosphatase gene near a telomere of chromosome I and chromosome VIII. *Curr. Genet.* **16,** 131–137.

Dernburg, A. F., Broman, K. W., Fung, J. C., Marshall, W. F., Philips, J., Agard, D. A., and Sedat, J. W. (1996a). Perturbation of nuclear architecture by long-distance chromosome interactions. *Cell* **85,** 745–759.

Dernburg, A. F., Sedat, J. W., Cande, W. Z., and Bass, H. W. (1995). Cytology of telomeres. *In* "Telomeres" (E. H. Blackburn and C. W. Greider, eds.), pp. 295–338. Cold Spring Harbor Laboratory Press, Cold Spring Harbor, NY.

Dernburg, A. F., Sedat, J. W., and Hawley, R. S. (1996b). Direct evidence of a role for heterochromatin in meiotic chromosome segregation. *Cell* **86,** 135–146.

Diffley, J. F. (1998). Replication conrol: choreographing replication origins. *Curr. Biol.* **8,** R771–R773.

Dorer, D. R., and Henikoff, S. (1994). Expansions of transgene repeats cause heterochromatin formation and gene silencing in *Drosophila*. *Cell* **77,** 993–1002.

Dresser, M. E. (2000). Meiotic chromosome behavior in *Saccharomyces cerevisiae* and (mostly) mammals. *Mutat. Res.* **451,** 107–127.

Dresser, M. E., and Giroux, C. N. (1988). Meiotic chromosome behavior in spread preparations of yeast. *J. Cell Biol.* **106,** 567–573.

Dujon, B. (1996). The yeast genome project: What did we learn? *Trends Genet.* **12,** 263–270.

Elledge, S. J. (1996). Cell cycle checkpoints: Preventing an identity crisis. *Science* **274,** 1664–1672.

Engebrecht, J., and Roeder, G. S. (1990). MER1, a yeast gene required for chromosome pairing and genetic recombination, is induced in meiosis. *Mol. Cell Biol.* **10,** 2379–2389.

Fabre, F., Boulet, A., and Roman, H. (1984). Gene conversion at different points in the mitotic cycle of *Saccharomyces cerevisiae*. *Mol. Gen. Genet.* **195,** 139–143.

Fishman-Lobell, J., Rudin, N., and Haber, J. E. (1992). Two alternative pathways of double-strand break repair that are kinetically separable and independently modulated. *Mol. Cell Biol.* **12,** 1292–1303.

Friedberg, E. C., Walker, G. C., and Siede, W. (1995). "DNA Repair and Mutagenesis." ASM Press, Washington, DC.

Fung, J. C., Marshall, W. F., Dernburg, A., Agard, D. A., and Sedat, J. W. (1998). Homologous chromosome pairing in *Drosophila melanogaster* proceeds through multiple independent initiations. *J. Cell Biol.* **141,** 5–20.

Fussell, C. P. (1987). The Rabl orientation: a prelude to synapsis. *In* "Cell Biology: A Series of Monographs: Meiosis" (P. B. Moens, ed.), pp. 275–299. Academic Press, Orlando, FL.

Galitski, T., Saldanha, A. J., Styles, C. A., Lander, E. S., and Fink, G. R. (1999). Ploidy regulation of gene expression. *Science* **285**, 251–254.

Galy, V., Olivo-Marin, J. C., Scherthan, H., Doye, V., Rascalou, N., and Nehrbass, U. (2000). Nuclear pore complexes in the organization of silent telomeric chromatin [see comments]. *Nature* **403**, 108–112.

Garvik, B., Carson, M., and Hartwell, L. (1995). Single-stranded DNA arising at telomeres in *cdc13* mutants may constitute a specific signal for the *RAD9* checkpoint. *Mol. Cell Biol.* **15**, 6128–6138.

Gasior, S. L., Wong, A. K., Kora, Y., Shinohara, A., and Bishop, D. K. (1998). Rad52 associates with RPA and functions with rad55 and rad57 to assemble meiotic recombination complexes. *Genes Dev.* **12**, 2208–2221.

Gemkow, M. J., Verveer, P. J., and Arndt-Jovin, D. J. (1998). Homologous association of the Bithorax-Complex during embryogenesis: consequences for transvection in *Drosophila melanogaster. Development* **125**, 4541–4552.

Gilbertson, L. A., and Stahl, F. W. (1994). Initiation of meiotic recombination is independent of interhomologue interactions. *Proc. Natl. Acad. Sci. USA* **91**, 11934–11937.

Gloor, G. B. (2001). The role of sequence homology in the repair of DNA double-strand breaks in *Drosophila. Adv. Genetics* **46**, 91–117.

Goffeau, A., Barrell, B. G., Bussey, H., Davis, R. W., Dujon, B., Feldmann, H., Galibert, F., Hoheisel, J. D., Jacq, C., Johnston, M., Louis, E. J., Mewes, H. W., Murakami, Y., Philippsen, P., Tettelin, H., and Oliver, S. G. (1996). Life with 6000 genes [see comments]. *Science* **274**, 546, 563–547.

Goh, P. Y., and Kilmartin, J. V. (1993). *NDC10*: A gene involved in chromosome segregation in *Saccharomyces cerevisiae. J. Cell Biol.* **121**, 503–512.

Goldman, A. S., and Lichten, M. (1996). The efficiency of meiotic recombination between dispersed sequences in *Saccharomyces cerevisiae* depends upon their chromosomal location. *Genetics* **144**, 43–55.

Goldman, A. S., and Lichten, M. (2000). Restriction of ectopic recombination by interhomolog interactions during *Saccharomyces cerevisiae* meiosis. *Proc. Natl. Acad. Sci. USA* **97**, 9537–9542.

Goshima, G., and Yanagida, M. (2000). Establishing biorientation occurs with precocious separation of the sister kinetochores, but not the arms, in the early spindle of budding yeast. *Cell* **100**, 619–633.

Gotta, M., Laroche, T., Formenton, A., Maillet, L., Scherthan, H., and Gasser, S. M. (1996). The clustering of telomeres and colocalization with Rap1, Sir3, and Sir4 proteins in wild-type *Saccharomyces cerevisiae. J. Cell Biol.* **134**, 1349–1363.

Gotta, M., Laroche, T., and Gasser, S. M. (1999). Analysis of nuclear organization in *Saccharomyces cerevisiae. Meth. Enzymol.* **304**, 663–672.

Grunstein, M. (1998). Yeast heterochromatin: regulation of its assembly and inheritance by histones. *Cell* **93**, 325–328.

Grushcow, J. M., Holzen, T. M., Park, K. J., Weinert, T., Lichten, M., and Bishop, D. K. (1999). *Saccharomyces cerevisiae* checkpoint genes *MEC1*, *RAD17* and *RAD24* are required for normal meiotic recombination partner choice. *Genetics* **153**, 607–620.

Guacci, V., Hogan, E., and Koshland, D. (1994). Chromosome condensation and sister chromatid pairing in budding yeast. *J. Cell Biol.* **125**, 517–530.

Guacci, V., Hogan, E., and Koshland, D. (1997). Centromere position in budding yeast: Evidence for anaphase A. *Mol. Biol. Cell* **8**, 957–972.

Haber, J. E. (2000). Partners and pathways repairing a double-strand break. *Trends Genet.* **16**, 259–264.

Hawley, R. S., Irick, H., Zitron, A. E., Haddox, D. A., Lohe, A., New, C., *et al.* (1992). There are two mechanisms of achiasmate segregation in *Drosophila* females, one of which requires heterochromatic homology. *Dev. Genet.* **13**, 440–467.

Hayashi, A., Ogawa, H., Kohno, K., Gasser, S. M., and Hiraoka, Y. (1998). Meiotic behaviours of chromosomes and microtubules in budding yeast: Relocalization of centromeres and telomeres during meiotic prophase. *Genes Cells* **3**, 587–601.

Hayden, M. S., and Byers, B. (1992). Minimal extent of homology required for completion of meiotic recombination in *Saccharomyces cerevisiae*. *Dev. Genet.* **13**, 498–514.

He, X., Asthana, S., and Sorger, P. K. (2000). Transient sister chromatid separation and elastic deformation of chromosomes during mitosis in budding yeast. *Cell* **101**, 763–775.

Henikoff, S. (2000). Heterochromatin function in complex genomes. *Biochim. Biophys. Acta* **1470**, O1–O8.

Hiraoka, Y., Dernburg, A. F., Parmelee, S. J., Rykowski, M. C., Agard, D. A., and Sedat, J. W. (1993). The onset of homologous chromosome pairing during *Drosophila melanogaster* embryogenesis. *J. Cell Biol.* **120**, 591–600.

Hollick, J. B., Dorweiler, J. E., and Chandler, V. L. (1997). Paramutation and related allelic interactions. *Trends Genet.* **13**, 302–308.

Hollingsworth, N. M., and Byers, B. (1989). *HOP1*: A yeast meiotic pairing gene. *Genetics* **121**, 445–462.

Hsieh, J., and Fire, A. (2000). Recognition and silencing of repeated DNA. *Annu. Rev. Genet.* **34**, 187–204.

Hunter, N., and Kleckner, N. (2001). The single-end invasion: An asymmetric intermediate at the double-strand break to double-Holliday junction transition of meiotic recombination. *Cell* **106**, 59–70.

Jiao, K., Bullard, S. A., Salem, L., and Malone, R. E. (1999). Coordination of the initiation of recombination and the reductional division in meiosis in *Saccharomyces cerevisiae*. *Genetics* **152**, 117–128.

Jin, Q.-W., Fuchs, J., and Loidl, J. (2000). Centromere clustering is a major determinant of yeast interphase nuclear organization. *J. Cell Sci.* **113**, 1903–1912.

Jin, Q.-W., Trelles-Sticken, E., Scherthan, H., and Loidl, J. (1998). Yeast nuclei display prominent centromere clustering that is reduced in nondividing cells and in meiotic prophase. *J. Cell Biol.* **141**, 21–29.

Jinks-Robertson, S., and Petes, T. D. (1986). Chromosomal translocations generated by high-frequency meiotic recombination between repeated yeast genes. *Genetics* **114**, 731–752.

Johnston, M., Andrews, S., Brinkman, R., Cooper, J., Ding, H., Dover, J., *et al.* (1994). Complete nucleotide sequence of *Saccharomyces cerevisiae* chromosome VIII. *Science* **265**, 2077–2082.

Kadyk, L. C., and Hartwell, L. H. (1992). Sister chromatids are preferred over homologs as substrates for recombinational repair in *Saccharomyces cerevisiae*. *Genetics* **132**, 387–402.

Karpen, G. H., Le, M. H., and Le, H. (1996). Centric heterochromatin and the efficiency of achiasmate disjunction in *Drosophila* female meiosis. *Science* **273**, 118–122.

Keeney, S. (2001). Mechanism and control of meiotic recombination initiation. *Curr. Top. Dev. Biol.* **52**, 1–53.

Keeney, S., and Kleckner, N. (1995). Covalent protein-DNA complexes at the 5′ strand termini of meiosis-specific double-strand breaks in yeast. *Proc. Natl. Acad. Sci. USA* **92**, 11274–11278.

Keeney, S., and Kleckner, N. (1996). Communication between homologous chromosomes: Genetic alterations at a nuclease-hypersensitive site can alter mitotic chromatin structure at that site both in *cis* and in *trans*. *Genes Cells* **1**, 475–489.

Keeney, S., Giroux, C. N., and Kleckner, N. (1997). Meiosis-specific DNA double-strand breaks are catalyzed by Spo11, a member of a widely conserved protein family. *Cell* **88**, 375–384.

Kleckner, N. (1996). Meiosis: how could it work? *Proc. Natl. Acad. Sci. USA* **93**, 8167–8174.

Kleckner, N., and Weiner, B. M. (1993). Potential advantages of unstable interactions for pairing of chromosomes in meiotic, somatic, and premeiotic cells. *Cold Spring Harb. Symp. Quant. Biol.* **58**, 553–565.

Klein, F., Laroche, T., Cardenas, M. E., Hofmann, J. F., Schweizer, D., and Gasser, S. M. (1992). Localization of Rap1 and topoisomerase II in nuclei and meiotic chromosomes of yeast. *J. Cell Biol.* **117,** 935–948.

Koshland, D. E., and Guacci, V. (2000). Sister chromatid cohesion: The beginning of a long and beautiful relationship. *Curr. Opin. Cell Biol.* **12,** 297–301.

Kowalczykowski, S. C. (1991). Biochemistry of genetic recombination: Energetics and mechanism of DNA strand exchange. *Annu. Rev. Biophys. Biophys. Chem.* **20,** 539–575.

Kowalczykowski, S. C. (2000). Initiation of genetic recombination and recombination-dependent replication. *Trends Biochem. Sci.* **25,** 156–165.

Kupiec, M., Byers, B., Esposito, R. E., and Mitchell, A. P. (1997). Meiosis and sporulation in *Saccharomyces cerevisiae*. *In* "Molecular and Cellular Biology of the Yeast *Saccharomyces*" (J. R. Pringle, J. R. Broach and E. W. Jones, eds.), vol. 3, pp. 889–1036. Cold Spring Harbor Laboratory Press, Cold Spring Harbor, NY.

Laloraya, S., Guacci, V., and Koshland, D. (2000). Chromosomal Addresses of the Cohesin Component Mcd1p. *J. Cell Biol.* **151,** 1047–1056.

Laroche, T., Martin, S. G., Gotta, M., Gorham, H. C., Pryde, F. E., Louis, E. J., and Gasser, S. M. (1998). Mutation of yeast Ku genes disrupts the subnuclear organization of telomeres. *Curr. Biol.* **8,** 653–656.

Laroche, T., Martin, S. G., Tsai-Pflugfelder, M., and Gasser, S. M. (2000). The dynamics of yeast telomeres and silencing proteins through the cell cycle. *J. Struct. Biol.* **129,** 159–174.

LaSalle, J. M., and Lalande, M. (1996). Homologous association of oppositely imprinted chromosomal domains. *Science* **272,** 725–728.

Leitch, A. R. (2000). Higher levels of organization in the interphase nucleus of cycling and differentiated cells. *Microbiol. Mol. Biol. Rev.* **64,** 138–152.

Leu, J. Y., Chua, P. R., and Roeder, G. S. (1998). The meiosis-specific Hop2 protein of *S. cerevisiae* ensures synapsis between homologous chromosomes. *Cell* **94,** 375–386.

Lewis, L. K., and Resnick, M. A. (2000). Tying up loose ends: Nonhomologous end-joining in *Saccharomyces cerevisiae*. *Mutat. Res.* **451,** 71–89.

Lichten, M., and Haber, J. E. (1989). Position effects in ectopic and allelic mitotic recombination in *Saccharomyces cerevisiae*. *Genetics* **123,** 261–268.

Loidl, J. (1990). The initiation of meiotic chromosome pairing: the cytological view. *Genome* **33,** 759–778.

Loidl, J., Klein, F., and Engebrecht, J. (1998). Genetic and morphological approaches for the analysis of meiotic chromosomes in yeast. *Meth. Cell Biol.* **53,** 257–285.

Loidl, J., Klein, F., and Scherthan, H. (1994). Homologous pairing is reduced but not abolished in asynaptic mutants of yeast. *J. Cell Biol.* **125,** 1191–1200.

Loidl, J., Nairz, K., and Klein, F. (1991). Meiotic chromosome synapsis in a haploid yeast. *Chromosoma* **100,** 221–228.

Louis, E. J., and Haber, J. E. (1992). The structure and evolution of subtelomeric Y' repeats in *Saccharomyces cerevisiae*. *Genetics* **131,** 559–574.

Marahrens, Y. (1999). X-inactivation by chromosomal pairing events. *Genes Dev.* **13,** 2624–2632.

Marshall, W. F., Straight, A., Marko, J. F., Swedlow, J., Dernburg, A., Belmont, A., Murray, A. W., Agard, D. A., and Sedat, J. W. (1997). Interphase chromosomes undergo constrained diffusional motion in living cells. *Curr. Biol.* **7,** 930–939.

Mazin, A. V., Bornarth, C. J., Solinger, J. A., Heyer, W. D., and Kowalczykowski, S. C. (2000a). Rad54 protein is targeted to pairing loci by the Rad51 nucleoprotein filament. *Mol. Cell* **6,** 583–592.

Mazin, A. V., Zaitseva, E., Sung, P., and Kowalczykowski, S. C. (2000b). Tailed duplex DNA is the preferred substrate for Rad51 protein-mediated homologous pairing. *EMBO J.* **19,** 1148–1156.

McKee, A. H., and Kleckner, N. (1997). A general method for identifying recessive diploid-specific *mutations in Saccharomyces cerevisiae*, its application to the isolation of mutants blocked at intermediate stages of meiotic prophase and characterization of a new gene *SAE2*. *Genetics* **146**, 797–816.

McKee, B. D., Habera, L., and Vrana, J. A. (1992). Evidence that intergenic spacer repeats of *Drosophila melanogaster* rRNA genes function as X-Y pairing sites in male meiosis, and a general model for achiasmatic pairing. *Genetics* **132**, 529–544.

Melamed, C., and Kupiec, M. (1992). Effect of donor copy number on the rate of gene conversion in the yeast *Saccharomyces cerevisiae*. *Mol. Gen. Genet.* **235**, 97–103.

Menees, T. M., Ross-MacDonald, P. B., and Roeder, G. S. (1992). *MEI4*, a meiosis-specific yeast gene required for chromosome synapsis. *Mol. Cell Biol.* **12**, 1340–1351.

Mills, K. D., Sinclair, D. A., and Guarente, L. (1999). *MEC1*-dependent redistribution of the Sir3 silencing protein from telomeres to DNA double-strand breaks. *Cell* **97**, 609–620.

Moore, D. P., and Orr-Weaver, T. L. (1998). Chromosome segregation during meiosis: Building an unambivalent bivalent. *Curr. Top. Dev. Biol.* **37**, 263–299.

Morris, J. R., Geyer, P. K., and Wu, C. T. (1999). Core promoter elements can regulate transcription on a separate chromosome in *trans*. *Genes Dev.* **13**, 253–258.

Nag, D. K., Scherthan, H., Rockmill, B., Bhargava, J., and Roeder, G. S. (1995). Heteroduplex DNA formation and homolog pairing in yeast meiotic mutants. *Genetics* **141**, 75–86.

Nairz, K., and Klein, F. (1997). *mre11S*—A yeast mutation that blocks double-strand-break processing and permits nonhomologous synapsis in meiosis. *Genes Dev.* **11**, 2272–2290.

Nasmyth, K. (1999). Separating sister chromatids. *Trends Biochem. Sci.* **24**, 98–104.

Nassif, N., Penney, J., Pal, S., Engels, W. R., and Gloor, G. B. (1994). Efficient copying of non-homologous sequences from ectopic sites via P-element-induced gap repair. *Mol. Cell Biol.* **14**, 1613–1625.

Naumov, G. I., Naumova, E. S., and Louis, E. J. (1995). Genetic mapping of the alpha-galactosidase *MEL* gene family on right and left telomeres of *Saccharomyces cerevisiae*. *Yeast* **11**, 481–483.

New, J. H., Sugiyama, T., Zaitseva, E., and Kowalczykowski, S. C. (1998). Rad52 protein stimulates DNA strand exchange by Rad51 and replication protein A. *Nature* **391**, 407–410.

Nicklas, R. B. (1977). Chromosome distribution: experiments on cell hybrids and *in vitro*. *Phil. Trans. R. Soc. Lond.* B **277**, 267–276.

Padmore, R., Cao, L., and Kleckner, N. (1991). Temporal comparison of recombination and synaptonemal complex formation during meiosis in *S. cerevisiae*. *Cell* **66**, 1239–1256.

Palladino, F., Laroche, T., Gilson, E., Axelrod, A., Pillus, L., and Gasser, S. M. (1993). *SIR3* and *SIR4* proteins are required for the positioning and integrity of yeast telomeres. *Cell* **75**, 543–555.

Paques, F., and Haber, J. E. (1999). Multiple pathways of recombination induced by double-strand breaks in *Saccharomyces cerevisiae*. *Microbiol. Mol. Biol. Rev.* **63**, 349–404.

Paulovich, A. G., and Hartwell, L. H. (1995). A checkpoint regulates the rate of progression through S phase in *S. cerevisiae* in response to DNA damage. *Cell* **82**, 841–847.

Petukhova, G., Stratton, S., and Sung, P. (1998). Catalysis of homologous DNA pairing by yeast Rad51 and Rad54 proteins. *Nature* **393**, 91–94.

Prinz, S., Amon, A., and Klein, F. (1997). Isolation of *COM1*, a new gene required to complete meiotic double-strand break-induced recombination in *Saccharomyces cerevisiae*. *Genetics* **146**, 781–795.

Pryde, F. E., Gorham, H. C., and Louis, E. J. (1997). Chromosome ends: all the same under their caps. *Curr. Opin. Genet. Dev.* **7**, 822–828.

Rabl, C. (1885). Uber Zelltheilung. *Morphol. Jahrb.* **10**, 214–330.

Raymond, W. E., and Kleckner, N. (1993). *RAD50* protein of *S. cerevisiae* exhibits ATP-dependent DNA binding. *Nucleic Acids Res.* **21**, 3851–3856.

Rocco, V., and Nicolas, A. (1996). Sensing of DNA non-homology lowers the initiation of meiotic recombination in yeast. *Genes Cells* **1**, 645–661.

Rockmill, B., and Roeder, G. S. (1991). A meiosis-specific protein kinase homolog required for chromosome synapsis and recombination. *Genes Dev.* **5**, 2392–2404.

Rockmill, B., and Roeder, G. S. (1998). Telomere-mediated chromosome pairing during meiosis in budding yeast. *Genes Dev.* **12**, 2574–2586.

Rockmill, B., Engebrecht, J. A., Scherthan, H., Loidl, J., and Roeder, G. S. (1995a). The yeast *MER2* gene is required for chromosome synapsis and the initiation of meiotic recombination. *Genetics* **141**, 49–59.

Rockmill, B., Sym, M., Scherthan, H., and Roeder, G. S. (1995b). Roles for two RecA homologs in promoting meiotic chromosome synapsis. *Genes Dev.* **9**, 2684–2695.

Roeder, G. S. (1997). Meiotic chromosomes: It takes two to tango. *Genes Dev.* **11**, 2600–2621.

Roeder, G. S., and Bailis, J. M. (2000). The pachytene checkpoint. *Trends Genet.* **16**, 395–403.

Rudner, A. D., and Murray, A. W. (1996). The spindle assembly checkpoint. *Curr. Opin. Cell Biol.* **8**, 773–780.

Schmuckli-Maurer, J., and Heyer, W. D. (2000). Meiotic recombination in *RAD54* mutants of *Saccharomyces cerevisiae*. *Chromosoma* **109**, 86–93.

Schwacha, A., and Kleckner, N. (1994). Identification of joint molecules that form frequently between homologs but rarely between sister chromatids during yeast meiosis. *Cell* **76**, 51–63.

Schwacha, A., and Kleckner, N. (1995). Identification of double Holliday junctions as intermediates in meiotic recombination. *Cell* **83**, 783–791.

Schwacha, A., and Kleckner, N. (1997). Interhomolog bias during meiotic recombination: Meiotic functions promote a highly differentiated interhomolog-only pathway. *Cell* **90**, 1123–1135.

Shinohara, A., Gasior, S., Ogawa, T., Kleckner, N., and Bishop, D. K. (1997a). *Saccharomyces cerevisiae* recA homologues *RAD51* and *DMC1* have both distinct and overlapping roles in meiotic recombination. *Genes Cells* **2**, 615–629.

Shinohara, A., Ogawa, H., and Ogawa, T. (1992). Rad51 protein involved in repair and recombination in *S. cerevisiae* is a RecA-like protein. *Cell* **69**, 457–470.

Shinohara, M., Shita-Yamaguchi, E., Buerstedde, J. M., Shinagawa, H., Ogawa, H., and Shinohara, A. (1997b). Characterization of the roles of the *Saccharomyces cerevisiae RAD54* gene and a homologue of *RAD54*, *RDH54/TID1*, in mitosis and meiosis. *Genetics* **147**, 1545–1556.

Skibbens, R. V., Skeen, V. P., and Salmon, E. D. (1993). Directional instability of kinetochore motility during chromosome congression and segregation in mitotic newt lung cells: A push-pull mechanism. *J. Cell Biol.* **122**, 859–875.

Smith, K. N., and Nicolas, A. (1998). Recombination at work for meiosis. *Curr. Opin. Genet. Dev.* **8**, 200–211.

Sperling, K., and Ludtke, E. K. (1981). Arrangement of prematurely condensed chromosomes in cultured cells and lymphocytes of the Indian muntjac. *Chromosoma* **83**, 541–553.

Storlazzi, A., Xu, L., Cao, L., and Kleckner, N. (1995). Crossover and noncrossover recombination during meiosis: Timing and pathway relationships. *Proc. Natl. Acad. Sci. USA* **92**, 8512–8516.

Straight, A. F., Marshall, W. F., Sedat, J. W., and Murray, A. W. (1997). Mitosis in living budding yeast: Anaphase A but no metaphase plate. *Science* **277**, 574–578.

Strick, T. R., Croquette, V., and Bensimon, D. (1998). Homologous pairing in stretched supercoiled DNA. *Proc. Natl. Acad. Sci. USA* **95**, 10579–10583.

Sugiyama, T., Zaitseva, E. M., and Kowalczykowski, S. C. (1997). A single-stranded DNA-binding protein is needed for efficient presynaptic complex formation by the *Saccharomyces cerevisiae* Rad51 protein. *J. Biol. Chem.* **272**, 7940–7945.

Sun, H., Treco, D., Schultes, N. P., and Szostak, J. W. (1989). Double-strand breaks at an initiation site for meiotic gene conversion. *Nature* **338**, 87–90.

Sun, H., Treco, D., and Szostak, J. W. (1991). Extensive 3′-overhanging, single-stranded DNA associated with the meiosis-specific double-strand breaks at the *ARG4* recombination initiation site. *Cell* **64**, 1155–1161.

Sung, P. (1994). Catalysis of ATP-dependent homologous DNA pairing and strand exchange by yeast *RAD51* protein. *Science* **265**, 1241–1243.

Sung, P. (1997). Yeast Rad55 and Rad57 proteins form a heterodimer that functions with replication protein A to promote DNA strand exchange by Rad51 recombinase. *Genes Dev.* **11,** 1111–1121.

Sung, P., and Robberson, D. L. (1995). DNA strand exchange mediated by a Rad51-ssDNA nucleoprotein filament with polarity opposite to that of RecA. *Cell* **82,** 453–461.

Sung, P., Trujillo, K. M., and Van Komen, S. (2000). Recombination factors of *Saccharomyces cerevisiae*. *Mutat. Res.* **451,** 257–275.

Sym, M., and Roeder, G. S. (1995). Zip1-induced changes in synaptonemal complex structure and polycomplex assembly. *J. Cell Biol.* **128,** 455–466.

Sym, M., Engebrecht, J. A., and Roeder, G. S. (1993). *ZIP1* is a synaptonemal complex protein required for meiotic chromosome synapsis. *Cell* **72,** 365–378.

Szostak, J. W., Orr-Weaver, T. L., Rothstein, R. J., and Stahl, F. W. (1983). The double-strand-break repair model for recombination. *Cell* **33,** 25–35.

Thompson, D. A., and Stahl, F. W. (1999). Genetic control of recombination partner preference in yeast meiosis. Isolation and characterization of mutants elevated for meiotic unequal sisterchromatid recombination. *Genetics* **153,** 621–641.

Tracy, R. B., Baumohl, J. K., and Kowalczykowski, S. C. (1997). The preference for GT-rich DNA by the yeast Rad51 protein defines a set of universal pairing sequences. *Genes Dev.* **11,** 3423–3431.

Trelles-Sticken, E., Dresser, M. E., and Scherthan, H. (2000). Meiotic telomere protein Ndj1p is required for meiosis-specific telomere distribution, bouquet formation and efficient homologue pairing. *J. Cell Biol.* **151,** 95–106.

Trelles-Sticken, E., Loidl, J., and Scherthan, H. (1999). Bouquet formation in budding yeast: Initiation of recombination is not required for meiotic telomere clustering. *J. Cell Sci.* **112,** 651–658.

Uhlmann, F. (2000). Chromosome cohesion: A polymerase for chromosome bridges. *Curr. Biol.* **10,** R698–R700.

Uhlmann, F., and Nasmyth, K. (1998). Cohesion between sister chromatids must be established during DNA replication. *Curr. Biol.* **8,** 1095–1101.

Van Komen, S., Petukhova, G., Sigurdsson, S., Stratton, S., and Sung, P. (2000). Superhelicity-driven homologous DNA pairing by yeast recombination factors Rad51 and Rad54. *Mol. Cell* **6,** 563–572.

Voelkel-Meiman, K., and Roeder, G. S. (1990a). Gene conversion tracts stimulated by *HOT1*-promoted transcription are long and continuous. *Genetics* **126,** 851–867.

Voelkel-Meiman, K., and Roeder, G. S. (1990b). A chromosome containing *HOT1* preferentially receives information during mitotic interchromosomal gene conversion. *Genetics* **124,** 561–572.

Vollrath, D., Davis, R. W., Connelly, C., and Hieter, P. (1988). Physical mapping of large DNA by chromosome fragmentation. *Proc. Natl. Acad. Sci. USA* **85,** 6027–6031.

Vourc'h, C., Taruscio, D., Boyle, A. L., and Ward, D. C. (1993). Cell cycle-dependent distribution of telomeres, centromeres, and chromosome-specific subsatellite domains in the interphase nucleus of mouse lymphocytes. *Exp. Cell Res.* **205,** 142–151.

Wach, A., Brachat, A., Pohlmann, R., and Philippsen, P. (1994). New heterologous modules for classical or PCR-based gene disruptions in *Saccharomyces cerevisiae*. *Yeast* **10,** 1793–1808.

Wakimoto, B. T. (1998). Beyond the nucleosome: Epigenetic aspects of position-effect variegation in *Drosophila*. *Cell* **93,** 321–324.

Walker, M. Y., and Hawley, R. S. (2000). Hanging on to your homolog: The roles of pairing, synapsis and recombination in the maintenance of homolog adhesion. *Chromosoma* **109,** 3–9.

Weiner, B. M., and Kleckner, N. (1994). Chromosome pairing via multiple interstitial interactions before and during meiosis in yeast. *Cell* **77,** 977–991.

Wilson, J. H., Leung, W. Y., Bosco, G., Dieu, D., and Haber, J. E. (1994). The frequency of gene targeting in yeast depends on the number of target copies. *Proc. Natl. Acad. Sci. USA* **91,** 177–181.

Winzeler, E. A., Shoemaker, D. D., Astromoff, A., Liang, H., Anderson, K., and Andre, B., *et al.* (1999). Functional characterization of the *S. cerevisiae* genome by gene deletion and parallel analysis. *Science* **285,** 901–906.

Wolfe, K. H., and Shields, D. C. (1997). Molecular evidence for an ancient duplication of the entire yeast genome. *Nature* **387,** 708–713.

Wu, C. T., and Morris, J. R. (1999). Transvection and other homology effects. *Curr. Opin. Genet. Dev.* **9,** 237–246.

Xu, L., and Kleckner, N. (1995). Sequence non-specific double-strand breaks and interhomolog interactions prior to double-strand break formation at a meiotic recombination hot spot in yeast. *EMBO J.* **14,** 5115–5128.

Xu, L., Weiner, B. M., and Kleckner, N. (1997). Meiotic cells monitor the status of the interhomolog recombination complex. *Genes Dev.* **11,** 106–118.

Zickler, D., and Kleckner, N. (1998). The leptotene-zygotene transition of meiosis *Annu. Rev. Genet.* **32,** 619–697.

Zickler, D., and Kleckner, N. (1999). Meiotic chromosomes: integrating structure and function. *Annu. Rev. Genet.* **33,** 603–754.

4 The Role of Sequence Homology in the Repair of DNA Double-Strand Breaks in *Drosophila*

Gregory B. Gloor*
Department of Biochemistry
The University of Western Ontario
London, Ontario N6A 5C1, Canada

*To whom correspondence should be addressed: E-mail: ggloor@uwo.ca.

Advances in Genetics, Vol. 46

I. INTRODUCTION

DNA double-strand breaks are among the most serious of lesions that cells have to deal with, and all cells invest significant resources into monitoring and repairing their genomes. Without this investment, the organization and content of the genome would rapidly change, resulting in genome instability. In humans, genome instability is one of the root causes of cancer (Jasin, 2000; Karran, 2000; Kote-Jarai and Eeles, 1999). Several familial cancer syndromes result from mutations in genes that play critical roles in double-strand break repair. Particularly well known examples include the *BRCA1, BRCA2,* and *NBS1* genes (Digweed *et al.,* 1999; Kote-Jarai and Eeles, 1999; Thompson and Schild, 1999; Welcsh *et al.,* 2000). In addition, double-strand break repair plays a central role in recombination and DNA replication. For example, *Spo11*-induced double-strand breaks initiate meiotic recombination in yeast and probably all other eukaryotes (Keeney *et al.,* 1997, 1999; McKim and Hayashi-Hagihara, 1998; Romanienko and Camerini-Otero, 1999; Shannon *et al.,* 1999). Therefore, the study of double-strand break repair helps us to understand aspects of meiotic recombination. In addition, if the DNA replication fork passes through a single-strand nick, it generates a double-strand break (Haber, 1999a; Kuzminov, 1995, 1999; Kuzminov and Stahl, 1999). Repair of such a break reconstitutes the replication fork and allows DNA replication to be completed. Some authors suggest that this is the most important function of double-strand break repair (Haber, 1999a; Kuzminov, 1995). Therefore, the study of double-strand break repair impinges upon many biological processes.

Several recent reviews on DNA double-strand break repair in yeast and mammals discuss this process from different perspectives than will be presented here (Haber, 1999b, 2000a, 2000b; Jasin, 2000; Karran, 2000; Kuzminov, 1999; Michel, 2000; Paques and Haber, 1999). I will focus on double-strand break repair in *Drosophila melanogaster* and will not discuss the role of double-strand break repair in meiotic recombination. In particular, this review will focus on the repair of DNA double-strand breaks made by excision of *P*-transposable elements, although it will mention other systems and phenomena as they relate to the repair of *P*-element-induced double-strand breaks. As described below, *P*-element mobilization provides a convenient means of generating site-specific DNA double-strand breaks in the *Drosophila* genome, the products and frequencies of which can be analyzed quickly and easily.

A. Double-strand break repair pathways

At least three distinct pathways repair DNA double-strand breaks. All three double-strand break repair pathways are active in all eukaryotic cells; however these pathways are not equally active in all cell types or in all organisms. The first pathway is nonhomologous end joining (abbreviated hereafter as NHEJ), and it is diagrammed in Figure 4.1A (Critchlow and Jackson, 1998). In this process joining

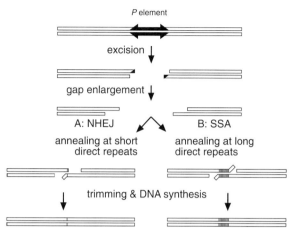

Figure 4.1. Deletion formation by double-strand break repair. *P*-element transposition requires sites on the *P*-element ends and P transposase. *P*-element excision occurs by a double-strand cut in which the two cleavage sites are on opposite DNA strands 17 bp distant from each other. One cut is made at the junction between the *P* element and host sequence, the other is made 17 nt internal to the *P*-element 31-bp inverted terminal repeats. Therefore, each resulting chromosomal double-strand break has 17-nt 3′ extended ends that are composed of the terminal 17 nt of the *P*-element 31-bp inverted terminal repeat. Resection of the resulting ends by exonucleases may occur. The diagram drawn here assumes some degradation of the ends for the sake of completeness. Both the NHEJ and SSA pathways operate without the input of an additional homologous DNA molecule. In the NHEJ pathway, drawn in part A, the resulting single strands may base-pair at very short direct repeat sequences composed of 1 or more nt. This pathway absolutely requires the Ku70 and Ku80 proteins. In the SSA pathway, shown in part B, the single-strand ends base-pair at more extensive direct repeat sequences. Generally, base pairing in this pathway requires at least 50 bp of homology. This pathway occurs in all other eukaryotes and defines the activity of the RAD52 protein in *S. cerevisiae*. In both instances, the non-base-paired termini are removed following base pairing, single-strand gaps are filled by DNA polymerase and sealed by DNA ligase.

of the broken DNA ends occurs with minimal processing or resection, usually at short direct repeats composed of one to a few nucleotides. The product of NHEJ is usually a small deletion. The broken ends in this pathway are promiscuous; two broken ends from different chromosomes or different sites on the same chromosome can join efficiently, leading to a chromosome rearrangement (Haber and Leung, 1996).

The single-strand annealing (SSA) pathway, diagrammed in Figure 4.1B, is defined as a homologous recombination pathway because it requires substantial sequence homology between the interacting partner molecules (Haber, 1995; Paques and Haber, 1999). Resection of the 5′ end at the break site by an exonuclease generates a substantial 3′ extended single-stranded tail on each side of the break. If the break occurs between closely spaced direct repeat sequences, then resection of the ends exposes complementary sequences. A single-strand annealing product results from base pairing between these complementary sequences, removal of any unpaired sequences, and gap filling by DNA polymerase.

On the surface, the NHEJ and SSA pathways appear similar, but they have very different protein requirements. The characteristic protein requirement of the NHEJ pathway is the Ku70/Ku80 heterodimer, which binds to the broken DNA ends (Critchlow and Jackson, 1998). Double-strand break repair by SSA in *Saccharomyces cerevisiae* requires the RAD52 protein, which also binds broken DNA ends. SSA also occurs in all other eukaryotes, and it is assumed that a RAD52 homolog is also required in these organisms (Haber, 1995; Paques and Haber, 1999).

Alternatively, the homologous recombination pathway shown in Figure 4.2 may repair the double-strand break by gene conversion (Haber, 1995, 2000a; Paques and Haber, 1999). In gene conversion, resection of the ends of the double-strand break results in a long single-stranded DNA tail being formed on each side of the break site. Binding of the single-stranded ends by the RAD52 protein enhances binding of the ends by the RAD51 protein, resulting in the production of a RAD51/single-stranded DNA complex (Haber, 1999b; Hiom, 1999; Van Dyck *et al.*, 1999). This complex conducts a homology search, resulting in the recognition of a sequence homologous to the single-strand DNA (Gupta *et al.*, 1998). The single-strand DNA sequence binds to its complement, displacing the other strand. DNA replication follows, and the invading 3' end is extended by copying the donor molecule. Processing of the newly replicated DNA results in a gene conversion with or without associated crossing over (as shown in Figure 4.2). I will differentiate between the two interacting DNA molecules. The broken chromosome that is repaired by accepting information will be referred to as the recipient chromosome or sequence. The template DNA molecule that provides the information for copying will be called the donor chromosome or sequence.

The factors that determine which pathway repairs a given double-strand break are currently unknown. One intriguing model proposes that the KU70/80 heterodimer and the RAD52 protein compete for binding to the broken end and recruit protein factors required for either NHEJ or homologous recombination. Therefore, the relative concentrations or activities of the KU70/80 heterodimer and of the RAD52 protein determine the pathway that repairs a given DNA break (Haber, 1999b; Hiom, 1999; Van Dyck *et al.*, 1999).

B. Monitoring double-strand break repair after *P*-element excision

Most of the experiments in the following discussion monitor the repair of a site-specific double-strand break made in the *Drosophila* genome by the excision of a

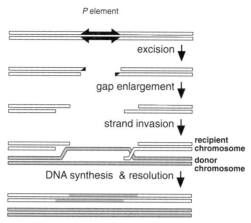

Figure 4.2. Double-strand break repair by gene conversion. P-element excision and processing of the ends occurs as described in the legend to Figure 4.1. The Rad51 protein assembles on the single-strand 3′ extended ends, and the resulting nucleoprotein filament searches the genome for a homologous sequence. The single-strand end then base-pairs with its complement in the homologous donor sequence, displacing the other strand. The 3′ end(s) now serves as a primer for DNA synthesis. The newly synthesized DNA and the donor DNA are disentangled, and the typical outcome that we observe in our system is shown here. The net result is that information from the donor DNA molecule is copied into the recipient DNA molecule, replacing the information that was lost by chromosome breakage. In *S. cerevisiae*, this pathway requires the following genes: *RAD50*, *RAD51*, *RAD52*, *RAD54*, *RAD55*, *RAD54*, *RAD55*, *RAD57*, *RAD59*, *MRE11*, and *XRS2* (the Nibrin protein replaces the XRS2 protein in multicellular eukaryotes) (Paques and Haber, 1999).

P-transposable element. In most instances the P element is inserted in exon 6 of the X-linked *white* locus, causing the w^{hd} allele (Engels *et al.*, 1990). The phenotype of the w^{hd} allele is a bleached white eye color, which is strikingly different from the wild-type deep red eye color. The P-element insertion is in an essential site for *white* gene function; reversion of this allele only results from precise loss of the P element. Excision of the w^{hd} P element occurs when P transposase is supplied by Δ2–3(99B), a stable transposase source located on the third chromosome (Robertson *et al.*, 1988). Repair of the resulting double-strand break by gene conversion in the presence of a homologous donor molecule results in reversion of the eye color phenotype to wild type (Engels *et al.*, 1990). This is because sequence from the donor *white* gene is copied to the recipient *white* gene (Gloor *et al.*, 1991). The donor molecule can be located anywhere in the genome (Engels *et al.*, 1994; Gloor *et al.*, 1991; Nassif *et al.*, 1994); it can even be a *white* gene contained on a nonintegrated plasmid (Keeler *et al.*, 1996). The donor molecule for all these experiments is a modified *white* gene that contains many single-base-pair differences compared to the DNA sequence that flanks the w^{hd} P element (Gloor *et al.*, 1991; Nassif *et al.*, 1994). These differences change restriction

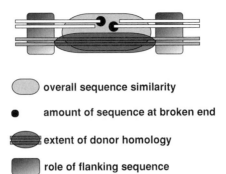

overall sequence similarity

amount of sequence at broken end

extent of donor homology

role of flanking sequence

Figure 4.3. Sequence homology effects on double-strand break repair by gene conversion. Shown here are several distinct ways that sequence homology affects gene conversion.

endonuclease recognition sites in the *white* gene DNA sequence, but do not change the sequence of the white protein. Therefore, the source of the DNA that flanks the w^{hd} P-element excision site in the reverted *white* gene can be determined by restriction mapping of polymerase chain reaction (PCR) amplification products, of appropriate segments of the *white* gene. This provides a convenient method to measure the frequency of gene conversion in *Drosophila*.

The effect of sequence homology and chromosome organization on gene conversion in mitotically dividing, premeiotic germline cells of *Drosophila* is the subject of this review. I will describe a number of experiments that vary mainly in the structure and location of the donor sequence. Figure 4.3 shows several different ways that sequence homology could affect double-strand break repair by gene conversion: sequence identity, extent of homologous sequence on the donor molecule, length of sequence homology on the recipient molecule, sequence location, and chromosome pairing. I will discuss each of these parameters in turn.

II. THE EFFECT OF SEQUENCE IDENTITY ON GENE CONVERSION

One way to investigate the homology requirements for double-strand break repair is to investigate the role of sequence similarity in this process. As shown in Figure 4.3, sequence identity effects extend for substantial distances on both sides of the double-strand break. In all systems, double-strand break repair requires significant sequence similarity between the interacting DNA partner molecules; indeed, double-strand break repair works most efficiently when the two interacting sequences are identical. Thus, there is an inverse relationship between the number of nucleotide mismatches and the frequency of gene conversion.

Nassif and Engels (1993) studied the effect of DNA mismatches on double-strand break repair at the *Drosophila white* locus. In their experiment, the donor and recipient *white* genes were at allelic sites on homologous chromosomes.

There were 15 single-nucleotide differences throughout a 2.8-kbp segment of the *white* gene. Most of the differences were close to the site of the double-strand break; 11 of these nucleotide differences were within a 1-kbp segment, and six were within a 217-bp segment. There was an inverse linear relationship between the number of point mutations and the frequency of gene conversion. Gene conversion frequencies ranged from 18.6% when the donor and recipient were isogenic to 5% when the donor and recipient differed at all 15 sites.

Nassif and Engels (1993) conducted four experiments in which the donor and recipient molecules had the same number of heterozygous sites but the sites were at different positions. Each of these four experiments was conducted in duplicate. In all instances, the gene conversion frequency was similar in the duplicate experiments. Therefore, the location of the sequence difference relative to the double-strand break was not as important as its presence. These results demonstrate the important role that the quality of the sequence match between the donor and recipient plays in double-strand break repair. Studies on mammalian gene targeting show similar effects, where targeting with vectors isogenic with the targeted locus is much more efficient than with nonisogenic vectors (Deng and Capecchi, 1992; te Riele *et al.*, 1992). It is interesting that meiotic recombination in *Drosophila* is not sensitive to the same effects (Hilliker *et al.*, 1994). This suggests that the broken DNA ends that are assumed to initiate meiotic recombination on the recipient molecule (McKim and Hayashi-Hagihara, 1998) do not rely on sequence matching as much as they do in mitotic recombination.

III. THE EFFECT OF THE EXTENT OF DONOR HOMOLOGY ON DOUBLE-STRAND BREAK REPAIR

The second major way that sequence homology requirements have been examined is to investigate the length of sequence homology shared by the donor and the recipient that supports homologous recombination. This has been studied in many different systems and organisms.

A. Determining the *E. coli* MEPS

Shen and Huang (1986) carried out the prototypical experiment in *Escherichia coli*, where they studied recombination between a plasmid and lambda phage. Homologous recombination between these two molecules resulted in co-integration of the plasmid into the phage. The phage containing the plasmid co-integrants could grow on a different host strain, allowing easy detection of the homologous recombination events. These investigators found a linear dependence between the length of homology and the frequency of recombination. Their method did not initiate homologous recombination with a defined DNA break, but rather examined the frequency of homologous recombination between two DNA sequences.

They defined the term MEPS, or minimal effective processing segment, as the length of homology below which the length dependence was no longer linear. The MEPS in E. coli is > 23, < 27 or > 44, < 90 base pairs in the recBC- or recF-dependent recombination pathways. They suggested that the MEPS represented the minimum number of perfectly matched base pairs that were required to initiate recombination. Shen and Huang defined the function $N = L - M + 1$ to give the number of MEPS in a length of homology L between two sequences. In this formula, N is the number of MEPS shared by the sequences and M is the length of the MEPS in the system. Therefore, the frequency of homologous recombination between two DNA sequences is dependent on the number of MEPS shared by the molecules. Sequences with a shared sequence of MEPS + 1 bp would have two MEPS of shared homology, sequences with a shared sequence of MEPS + 2 bp would have three MEPS of shared homology, and so on. They proposed that every base-pair difference destroyed the sequence homology in a region equal in size to the MEPS both upstream and downstream of the sequence difference.

B. Determining the MEPS in *S. cerevisiae*

The MEPS has also been rigorously defined in *Saccharomyces cerevisiae*. Jinks-Robertson et al. (1993) inserted two *ura3* alleles at different sites in the yeast genome; the full-length target allele had a frame-shift mutation and the substrate alleles contained different 3'-end deletions. Homologous recombination between ectopic target and substrate alleles would result in reversion of the yeast strain to URA3. They measured the effect of substrate length on the rate of reversion to the URA+ phenotype, and observed a MEPS of 248 bp. The experiment was repeated with the target and substrate alleles *in cis* and oriented as either inverted or direct repeats, and they observed a similar MEPS. Interestingly, substrates of similar size permitted similar rates of homologous recombination, regardless of the relative positions of the frame-shift mutation and the homology boundary. Indeed, they observed efficient copying of frame-shift mutations within 30 bp of the homology boundary if the substrate molecule was longer than the MEPS. They explained the apparent paradox by suggesting that the extent of sequence shared by the target and substrate alleles was important for the initiation of homologous recombination, but not downstream events. They proposed that heteroduplex DNA could include sequences near the homology boundary very efficiently, after the initiation of homologous recombination.

The MEPS has been estimated in many other organisms, including mammalian cultured cells (200–295 bp) and bacteriophage T4 (50 bp) (Baker et al., 1996; Deng and Capecchi, 1992; Liskay et al., 1987; Metzenberg et al., 1991; Rubnitz and Subramani, 1984; Singer et al., 1982; Watt et al., 1985). It is striking that the eukaryotic MEPS is about four times larger than the prokaryotic MEPS. The reason for this difference is not known, but Thaler and Noordewier (1992) suggest that the apparently much larger MEPS in eukaryotic systems is a safeguard against genome scrambling by recombination between short repeated sequences.

C. Homology requirements for gene conversion in *Drosophila*

The MEPS has not been rigorously determined for double-strand break repair in *Drosophila*. However, two different types of experiments examined the extent of homologous sequence that is required to support *P*-element-induced gene conversion in *Drosophila*.

1. Homology requirements for gene conversion of point mutations

Nassif and Engels (1993) compared the gene conversion frequency of single-nucleotide markers flanked by either a small or a large segment of homology. In the first situation, the nucleotide markers were located on ectopic donor sequences, and the most distal marker had 115 bp of homology between it and the homology boundary with the recipient (Gloor *et al.*, 1991). In the second situation, the same markers were situated on a homolog and were thus allelic to the double-strand break site in the recipeint DNA molecule (Nassif and Engels, 1993). Nassif and Engels found that the conversion frequency of this marker in the limited-homology situation was about 60% of the frequency with infinite homology (compare column 4 in rows 3 and 5 of Table 4.1).

The model for double-strand break repair in this system is one where the gap is enlarged prior to the initiation of gene conversion (Gloor *et al.*, 1991; Nassif and Engels, 1993). In this model, DNA replication accounts for most of the single-base differences copied into the recipient DNA. Furthermore, Nassif

Table 4.1. The Effect of Flanking Sequence on Gene Conversion of a Point Mutation

Distance from dsb (bp)	Extent of flanking homology	Predicted conversion frequency	Observed conversion frequency	Difference (observed − predicted)	Source
884	Infinite	0.3005	0.2949	−1.9%	Nassif and Engels (1993)
827	Infinite	0.3246	0.3462	+6.6%	Nassif and Engels (1993)
884	115	0.3005	0.1831	−39.1%	Gloor *et al.* (1991)
827	172	0.3246	0.1972	−39.3%	Gloor *et al.* (1991)
884	115	0.3005	0.1077	−64.2%	Nassif *et al.* (1994)
827	172	0.3246	0.1385	−57.4%	Nassif *et al.* (1994)
311	Infinite	0.6547	0.5385	−17.8%	Nassif and Engels (1993)
311	688	0.6547	0.6056	−7.5%	Gloor *et al.* (1991)
311	688	0.6547	0.5692	−13.1%	Nassif *et al.* (1994)
600	Infinite	0.44	0.38	−14%	Preston and Engels (1996)
1500	Infinite	0.13	0.14	+7.6%	Preston and Engels (1996)

and Engels assumed that significantly less than 115 bp of homology was required to initiate gene conversion. Under this model, the conversion frequency of any marker in these experiments can be estimated with the formula $f_n = x^n$ (Nassif and Engels, 1993). In this formula, f is the gene conversion frequency, n is the number of base pairs between the marker and the double-strand break. The value of x is a constant that is determined by fitting an exponential decay curve to the observed gene conversion frequency of markers at different distances from the double-strand break. This constant has been determined independently three times, and has values of 0.99855 (Gloor et al., 1991), 0.99873 (Nassif and Engels, 1993), and 0.998635 (Preston and Engels, 1996). This formula estimates the probability of gene conversion initiating at a given position relative to the site of the double-strand break. The conditional probability of the gap being enlarged to a position between the marker and the homology boundary can also be estimated (Nassif and Engels, 1993). With this model, the gene conversion frequency of the marker with 115 bp of homology in the ectopic situation was at least double the conditional probability, although it was only 60% of the frequency observed with infinite flanking homology. Therefore, Nassif and Engels concluded that 115 bp of flanking homology was sufficient to support gene conversion at a frequency similar to that observed with inifinite flanking homology.

A reexamination of their experiment that includes later data from similar experiments does not support this strong conclusion (Gloor et al., 1991; Nassif and Engels, 1993; Nassif et al., 1994; Preston and Engels, 1996). Table 4.1 shows the estimated and observed conversion frequency of these markers in the original experiment and the three other experiments of similar design (Gloor et al., 1991; Nassif and Engels, 1993; Nassif et al., 1994; Preston and Engels, 1996). Note that in the cases where the most distal two markers have either 172 or 115 bp of flanking homology (rows 3–6 in Table 4.1), the observed conversion frequency is at least 39% less than the expected conversion frequency. This is at least double the difference between the expected and observed conversion frequencies for other markers, of which one example is shown in rows 8 and 9 of Table 4.1. The last two rows in Table 4.1 show the estimated and observed conversion frequencies following double-strand break repair between allelic sites at an autosomal locus. In these cases there is effectively inifinite homology between the two interacting DNA molecules. Note the good correspondence between the expected and observed frequencies in this situation. The number of conversions in the ectopic and allelic situations was compared with Fisher's Exact Test to assess the difference between these two data sets. There was a significant difference between the allelic and ectopic conversion frequencies for the markers located 884 bp ($P = 0.008$) and 827 bp ($P = 0.003$), but not the marker located 311 bp ($P = 0.286$) distant from the site of the double-strand break. This shows that the amount of flanking homology required to convert a single base mismatch at the optimal frequency is greater than 172 bp but less than 688 bp.

2. Homology requirements for gene conversion of insertions

Dray and Gloor (1997) examined the amount of homology required to convert a large (8-kbp) insertion of the *yellow* gene from an ectopic donor sequence into a double-strand break at the *white* locus. Nassif *et al.* (1994) had shown previously that the *yellow* gene could be copied from an ectopic donor at a frequency consistent with the conversion frequency of a single base alteration. The *yellow* gene was flanked by 763 bp and 2692 bp of homologous sequence in this experiment. Dray and Gloor (1997) measured the conversion frequency of the *yellow* gene with 0, 25, 51, 375, and 493 bp of flanking homology on one side. In addition, conversion was also examined with unlimited homology. Conversion of the insertion required at least 375 bp of flanking homology, as this amount of homology supported conversion of the *yellow* gene at 50% the maximal observed frequency. Conversion of the *yellow* gene occurred at the same frequency when it was flanked by either 493 bp or unlimited homology, indicating that 500 bp of flanking homology is saturating in this system. It is interesting that the homology requirement for gene conversion of both point mutations and for insertions is similar in this system (see above).

Several other groups have used double-strand break repair to modify the genome by targeted gene insertion; these are reviewed in Gloor and Lankenau (1998) and Lankenau and Gloor (1998). Two experiments are of particular interest. Williams and O'Hare (1996) successfully targeted a cDNA of the *suppressor of forked* gene to the *su(f)* locus with 203 bp and 1549 bp of donor homology on the left and right sides of the recipient double-strand break. Banga and Boyd (1992) injected 50 nt single- or double-stranded donor DNA into embryos in which the w^{hd} P element was excising. They recovered several independent events consistent with those expected for repair of the double-strand break by gene conversion from these donor molecules.

IV. THE EFFECT OF THE EXTENT OF RECIPIENT HOMOLOGY ON DOUBLE-STRAND BREAK REPAIR

Another way to examine the length of sequence required for homologous recombination is to examine the length of sequence required at the broken ends of the DNA molecule, as shown in Figure 4.3. This question is distinct from the one explored above, because the broken end is the site of assembly of the recombination machinery. As outlined below, the large MEPS observed in eukaryotes is at odds with the observation that double-strand break repair occurs in *S. cerevisiae* and in *Drosophila* when the broken end has very little homology to the donor molecule (Banga and Boyd, 1992; Keeler and Gloor, 1997; Lorenz *et al.*, 1995; Manivasakam *et al.*, 1995). The study of P-element transposition and excision

A: Starting components

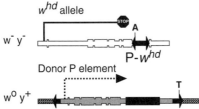

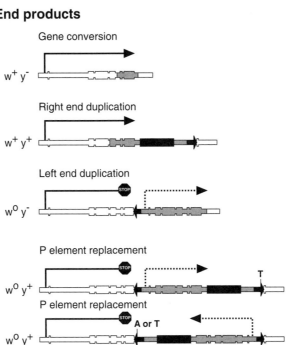

B: End products

Figure 4.4. Gene conversion with P-element ends. The starting components used by Keeler and Gloor are shown in part A. The w^{hd} allele was the recipient in this experiment and an ectopic P{w+ y+} element was the donor. The donor contained a copy of the *yellow* gene flanked by *white* gene sequences. The *white* gene promoter in the donor element expressed the *white* gene poorly, resulting in an orange eye phenotype. In this figure, *white* genes expressed from this promoter are termed w^o, and the dashed line indicates weak transcription from this promoter. The products that were recovered are shown in part B. These diagrams show the structure of the X-linked *white* locus following gene conversion, P-element replacement, or a combination of both. The gene conversion product could be either w+ y−, as shown, or w+ y+, depending on the extent of the gene conversion tract. Right-end duplication products were always w+ y+, because all the sequence between the w^{hd} P-element excision site in exon 6 and the limit of the 3′ untranslated sequence from the donor element were always converted. In a like manner, left-end duplications always had an orange eye color because the *white* gene expressed

products provides an instructive example of the homology required of the broken end to initiate efficient double-strand break repair.

P elements in *Drosophila* are extremely mobile, and several investigators observed the exact replacement of one *P* element with another (Geyer *et al.*, 1988; Gonzy-Treboul *et al.*, 1995; Heslip and Hodgetts, 1994; Heslip *et al.*, 1992; Johnson-Schlitz and Engels, 1993; Keeler and Gloor, 1997; Sepp and Auld, 1999). Geyer *et al.* (1988) first proposed that this replacement occurred by gene conversion. This group started with a *P*-element insertion allele of the *yellow* gene, in which a nonautonomous *P* element was inserted in the promoter for the *yellow* gene. They mobilized the *P* elements in the genome and collected revertants of the *yellow* allele. Interestingly, one of these revertants had a smaller nonautonomous *P* element inserted at the same site and orientation as the original transposon. They mobilized the *P* element in this new allele, and collected flies with a yellow phenotype. Analysis of the *yellow* gene sequence in these new revertants demonstrated the insertion of a nonautonomous *P* element into the same insertion site, but in a different orientation. This nonautonomous *P* element was identical in internal structure to the *P*-element insertion in the original allele.

Keeler and Gloor (1997) examined this phenomenon in more detail. They mobilized the w^{hd} *P* element in the presence of a *P{w+ y+}*; donor element inserted on an autosome, and collected all progeny files with an X-linked white+ or yellow+ phenotype. Figure 4.4 shows the start and end products of their analysis. The DNA of these flies was analyzed to look for gene conversion or *P*-element replacement events at the *white* locus. As expected, they identified a number of independent gene conversions, 10 that were w+ y− and seven that were w+ y+. They also recovered duplications of the *white* locus that included the right end of the donor *P* element inserted into the same site as the w^{hd} *P* element. These so-called right-end duplications result in a wild-type white phenotype and had been recovered previously. A similar number of left end duplications were recovered that had the expected partial white+ phenotype.

in this product was transcribed from the promoter found in the donor element. The coding sequence of the transcript from the endogenous *white* gene promoter is truncated in the middle of exon 6. *P*-element replacements were recovered in both orientations, and in either case had an orange eye color for the same reasons outlined for the left-end duplication products. Recipient and donor sequences corresponding to the *white* gene are shown as white or gray boxes, *yellow* gene sequences are shown as black boxes. Thick boxes in the *white* gene sequence represent exons, and thin boxes represent introns or flanking noncoding sequence. The large black arrows represent *P*-element end sequences. The eye and body color phenotypes that are observed in an otherwise *white* and *yellow* gene null genetic background are shown beside each of the starting components and products. The identity of nuclotide 33 in the donor and recipient *P*-element ends is indicated. Please see the text for a discussion of their significance.

Keeler and Gloor (1997) also searched for P-element replacements. Interestingly, if the same P{w+ y+} donor P element was used, the frequency of P-element replacements was similar to the frequency of gene conversions at the *white* locus. They found 24 P-element replacements in which the donor transposon had exactly replaced the w^{hd} P element. Twenty of these P-element replacements were w+ y+ and three were w+ y−. Keeler and Gloor also recovered one w− y+ replacement in which the *white* gene contained a large deletion. Ten of the replacements were in the same orientation as the original w^{hd} P element, and 14 were in the opposite orientation.

The P-element ends of the w^{hd} P element and the donor P element shared identical sequence except for a polymorphism at nucleotide 33 of the P-element left end. Transposition of a P element, and of other transposons, always results in the maintenance of sequences that are essential to transposition. In the case of P elements, this means that transposition always result in the movement of the first and last 150 bp of the transposon, because these sequences are essential for P-element mobility (Beall and Rio, 1998; Mullins *et al.*, 1989). Therefore, if the polymorphism was derived from the w^{hd} P element in replacements of the same orientation as the w^{hd} P element, then the replacement must have occurred by gene conversion and not by transposition of the donor P element. However, if the polymorphism was derived from the donor P element in those replacements, then the insertion occurred by transposition. The w^{hd} P-element polymorphism occurred in approximately one-third of the same-orientation P-element replacements, demonstrating that these events, at least, were the result of gene conversion. Furthermore, the frequency of P-element replacement was similar to the frequency of gene conversion at the *white* locus, and was several orders of magnitude higher than the frequency of *de novo* P-element insertion into this locus. The elevated frequency of P-element replacement relative to *de novo* insertion at the same locus is a common feature of this process. Therefore, Keeler and Gloor proposed that the simplest explanation for P-element replacement was gene conversion and not transposition.

An experiment by Johnson-Schlitz and Engels (1993) provides additional information on this topic. They observed that a donor chromosome containing a *white* gene isogenic with the sequence flanking the w^{hd} P element supported gene conversion at a frequency of about 19%, following excision of the w^{hd} P element. This group then used allelic donors that included partial P-element ends, and observed that the frequency of gene conversion was greater than 40% when the donor contained at least 17 bp of sequence from one P-element end. Furthermore, homology in the *white* gene sequence flanking the P-element end was not important for this elevated conversion frequency. Johnson-Schlitz and Engels concluded that these allelic donor sequences were competing with the sister chromatid as donors for gene conversion. Furthermore, they concluded that the fragment of P-element end sequence, and not the flanking *white* gene sequence, was the one competing with the sister chromatid. They suggested that the quality of the match between the broken ends and the donor sequence was assessed prior to end processing, and that double-strand break repair proceeded without end processing if the interacting sequences were nearly identical. In this model, the unexcised P element on the sister chromatid would serve as the preferred donor for repair of the double-strand break.

The above results imply that only the inverted terminal repeat sequences at the ends of the P element are required to initiate and complete double-strand break repair by gene conversion. These repeats are only 31 bp long; furthermore, P-element excision results in a 17-nucleotide 3' extended end composed of part of each terminal repeat (Beall and Rio, 1997). Therefore, P-element replacement by gene conversion uses only the terminal 17 nucleotides to conduct an efficient and specific homology search. It is perhaps not surprising that 17 nucleotides of terminal sequence can initiate some specific double-strand break repair, but the high efficiency of this repair is extremely surprising. Small sequences at the termini of DNA can be used to initiate a homology search in other systems. For example, highly specific targeted integration via homologous recombination occurs in yeast when each end of an introduced DNA sequence has as little as 30 bp of sequence homology with the target site (Manivasakam et al., 1995). However, in this instance, the frequency of targeted integration is reduced about 3 orders of magnitude from the optimal frequency. Furthermore, targeted integration of DNA into mammalian cells also requires several kilobases of homologous sequence at each terminus (Deng and Capecchi, 1992). Such high efficiency suggests that the P-element ends may have a special ability to attract the machinery required for homologous recombination.

V. THE EFFECT OF DONOR LOCATION ON DOUBLE-STRAND BREAK REPAIR

There is a large effect of donor location on the frequency of double-strand break repair by gene conversion following excision of the w^{hd} P element. The relative order is sister chromatid > homologue > ectopic (Engels et al., 1994).

The most efficient configuration for repair of the w^{hd} P-element excision by gene conversion uses a sister chromatid donor (Johnson-Schlitz and Engels, 1993). Repair of these double-strand breaks by gene conversion accounts for at least two-thirds of the products. Such repair results in re-insertion of the w^{hd} P element into the *white* locus. This frequency is an underestimate because a significant number of the w^{hd} P elements undergo more than one round of excision (Engels et al., 1990). One reason that the sister chromatid is used so efficiently is that the donor sequence contains P-element ends. As we have seen above, these ends are a powerful attractant for the broken ends following P-element excision. However, there is another reason for the efficient use of a sister chromatid donor. Several lines of evidence now indicate that one of the most important roles for double-strand break repair is the repair of double-strand breaks during S phase (Haber, 1999a, 2000a, 2000b; Holmes and Haber, 1999; Johnson and Jasin, 2000; Kuzminov, 1995; Kuzminov and Stahl, 1999). These breaks occur when a DNA replication fork encounters a single-strand nick in one of the template strands. When this occurs, replication on the unbroken strand proceeds as usual, but replication on the broken strand stops. Exonucleolytic processing of the double-strand end generates a 3′ single-strand end that invades the intact, newly replicated strand. This invasion makes a new DNA replication fork, and allows DNA replication to bypass the single-strand nick. Gene conversion using a sister chromatid, or a just-replicated DNA strand is of such importance to the yeast cell that individual genes specialize in this task (Arbel et al., 1999). It is likely that together these two processes ensure the selection of a sister chromatid donor to repair the double-strand break in the vast majority of cases. As pointed out by Engels, P elements are supremely adapted to use the host DNA repair machinery to increase their copy number by double-strand break repair (Engels, 1992, 1997). They do this by making sure that they transpose when the double-strand break is most likely to be repaired using a sister chromatid donor, and by producing a ready-made 3′ extended end.

An allelic donor is the next most efficient gene conversion template. Up to 20% of the gene conversions recovered after repair of the double-strand break are derived from allelic donors (Nassif and Engels, 1993). This conversion efficiency is obtained only if the donor and recipient alleles are isogenic for sequence flanking the w^{hd} P element. The frequency of gene conversion drops dramatically if the donor allele is on a balancer chromosome (Engels et al., 1990). Therefore, the correct pairing of the donor and recipient chromosomes is an important contributor to efficient donor recognition during double-strand break repair. This will be explored in the next section.

Ectopic donors are also utilized at a significant frequency, and this frequency depends on the physical location of the donor sequence. Engels et al. (1994) investigated the relationship of donor location and gene conversion frequency. They placed the same donor sequence in cis and in trans to the double-strand break made by excision of the w^{hd} P element. They compared the gene conversion rates in these configurations to the rates observed with the same donor sequences inserted at ectopic sites on the autosomes. Engels et al. found that donor sequences located in cis to the double-strand break supported efficient gene conversion. Donor sequences located in cis were used about 2.5-fold more frequently than a donor sequence in trans to the break, and about 6-fold more frequently than a donor sequence located on an autosome. This cis effect on gene conversion frequency was not related to the physical distance between the double-strand break and the donor site. Cis effects on site-specific recombination have been found in both *Drosophila* and in *S. cerevisiae* (Burgess and Kleckner, 1999; Golic and Golic, 1996a). In these cases, the cis effect is strongly dependent on the distance between the interacting sequences. Indeed, the rate of site-specific recombination can be used as an *in vivo* assay for distance between any two DNA molecules (Burgess and Kleckner, 1999). Furthermore, at least in the case of the bacterial recA protein, the prototypical strand-exchange protein, the homology search is conducted by random collisions and not by sliding along the DNA (Adzuma, 1998). It is therefore likely that the cis effect on gene conversion in *Drosophila* is not due to sliding or to an increased local concentration of the donor sequence. This argues that the cis effect on the rate of gene conversion is an intrinsic property of the protein machine that performs double-strand break repair in *Drosophila*.

In addition, not all ectopic insertion sites of donor sequences support equal frequencies of gene conversion (Gloor *et al.*, 1991; Nassif *et al.*, 1994). Ectopic insertions of donor sequences inserted on the autosomes support gene frequencies that range over at least an order of magnitude. Furthermore, there is no apparent relationship between the location of the donor sequence relative to structural chromosomal elements such as centromeres and telomeres. The reason for this wide variation in conversion frequency from identical donor sequences is unknown.

VI. THE EFFECT OF CHROMOSOME PAIRING ON DOUBLE-STRAND BREAK REPAIR

Two separate experiments demonstrate the effect of chromosome pairing on double-strand break repair by gene conversion.

Engels *et al.* examined the effect of global chromosome pairing on double-strand break repair by gene conversion (1990). In this experiment, the donor for double-strand break repair was allelic to the double-strand break, and was located either in the middle of an inversion loop or on a noninverted co-sequential chromosome. In both these instances, the donor sequence was embedded in several megabases of flanking homologous sequence. The frequency of gene conversion was fourfold greater when the donor allele was located on the noninverted chromosome (13.6%) than when the same donor allele was in the middle of the inversion loop (3.0%). Interestingly, this frequency was only slightly greater than the average conversion frequency in experiments that used an ectopic X-linked donor with 2456 bp and 999 bp of flanking homology on the left and right sides (1.5%) (Engels *et al.*, 1994). In addition, the frequency observed with the allelic donor on the inverted chromosome was almost twofold less than with an ectopic donor located *in cis* to the double-strand break (Engels *et al.*, 1994). Therefore, an allelic donor that is located in an inversion loop supports a gene conversion frequency that is on par with an ectopic donor that contains only limited flanking homology. It is noteworthy that crossing over contributes very little to the frequency of gene conversion in these experiments (Engels *et al.*, 1990; Gloor *et al.*, 1991). Therefore the reduction in gene conversion is unlikely to be caused by the failed processing of an intermediate structure. These results suggest that allelic sites are generally paired, or in close association, and that this pairing is an important contributor to double-strand break repair when an allelic donor is present.

A recent experiment by Dray *et al.* (2002) demonstrates that local pairing between allelic sites also contributes to the gene conversion frequency. In this experiment, allelic donor *white* genes where made that had various insertions of ~8000 bp of heterologous DNA inserted ~240 bp distant from the site corresponding to the double-strand break. They observed a gene conversion frequency of ~9% when the donor chromosome contained only single-nucleotide differences from the recipient chromosome. This frequency was similar to that observed by Nassif and Engels for donor chromosomes containing similar numbers of nucleotide differences. The frequency of gene conversion was decreased 6–8-fold when the allelic donor contained an 8000-bp insertion composed of either the *Drosophila yellow* gene or bacteriophage lambda DNA. The frequency of gene conversion was decreased > 4-fold when the large insertion was located *in cis* to the double-strand break, and the donor sequence lacked the insertion. The gene conversion frequency was restored to normal if the same large insertion was located at both the donor and the recipient sites. Interestingly, this effect is observed only with allelic donor sequences. The conversion frequency of ectopic donor sequences with or without the 8-kbp insertion is equivalent. This suggests an active exclusion of the allelic donor sequence from consideration by the homology search machinery when it contains a large insertion or deletion.

Lankenau *et al.* (2000) demonstrated a direct link between proteins important for chromosome architecture, such as the Su(Hw) protein, and double-strand break repair by gene conversion. The Su(Hw) site-specific DNA-binding protein recognizes sets up domains of chromatin by binding to its cognate sites. Such occupied sites act as insulators between genes by preventing transcription signals from one gene acting on another (Roseman *et al.*, 1993). DNA sequences bound by the Su(Hw) protein are thought to associate with each other, and with other insulator sites, in the nucleus (Cai and Shen, 2001; Muravyova *et al.*, 2001). According to this model, the insulator sequences form attachment points for loops of chromatin. Such attachements would unavoidably help constrain the free motion of chromatin in the nucleus.

Lankenau *et al.* compared the frequency of gene conversion in the presence and absence of the Su(Hw) protein with three different donor sequences. They observed that the frequency of double-strand break repair by gene conversion was 1.5–2-fold greater in flies lacking the Su(Hw)-binding protein. This was true whether or not the *white* gene donor sequence was flanked by binding sites for the Su(Hw) protein. In addition, they found that the Su(Hw) site did not affect the frequency of conversion when it was embedded within the donor sequence. The Su(Hw) protein is involved in chromatin organization, but does not have a direct role in double-strand break repair. Lankenau *et al.* (2000) propose that the broken ends have greater freedom in the nucleus, and so can more easily identify ectopic DNA donor sequences. This greater freedom is predicted to cause the increase in gene conversion frequency observed in the absence of this protein.

VII. SUMMARY

The experiments described above show that the homology effects on double-strand break repair by gene conversion are divided into two general phenomena. First, the quality and extent of the sequence match are extremely important. Second, the chromosomal contexts of the donor and recipient sequences affect the ability of the sequences to recognize each other, even when the quality and extent of the sequence match are otherwise sufficient.

A. The role of sequence matching

The quality and extent of the sequence match probably affect at least two distinct stages in gene conversion. First, the RAD51/single-stranded DNA complex conducts a low stringency search for a homologous sequence (Bazemore *et al.*, 1997). Second, the mismatch repair protein MutS assesses the quality of the base pairing between the heteroduplex DNA formed by base pairing between the invading DNA strand and its complement (Evans and Alani, 2000).

The RAD51 protein is a member of the recA/radA family of recombination proteins (Aravind *et al.*, 1999). The recA protein of *E. coli* is the best-characterized member of this family, and its function is representative of the function of other members of this family (Gupta *et al.*, 1998). RecA/Rad51 protein family members bind to single-stranded DNA ends, and form characteristic nucleoprotein filaments in which the single strand is in an extended conformation (Nishinaka *et al.*, 1998). This nucleoprotein filament searches for homologous sequences in duplex DNA by forming a three-stranded parallel triple helix (Hsieh *et al.*, 1990; Rao *et al.*, 1991).

Experiments *in vitro* and *in vivo* demonstrate the relative nonspecificity of the initial homology search by the recA-ssDNA nucleoprotein filament. *In vitro*, Bazemore *et al.* (1997) physically examined the effect of mismatches on the recognition of the target duplex molecule by the nucleoprotein filament. They observed that the nucleoprotein filament recognized the target duplex efficiently when up to 10% of the nucleotides were mismatched, but that a higher proportion of mismatches excluded efficient recognition.

In vivo, the mismatch repair system obscures the role of sequence identity during the homology search by the recA-ssDNA filament. The mismatch repair system checks the quality of the match between the invading DNA single strand and the complementary DNA to which it is bound. This step is responsible for the majority of the discriminatory power observed in homologous recombination (Evans and Alani, 2000). One important protein in this system is the MutS protein (and its eukaryotic homologs, generally known as MSH proteins, or MutS homologs) that bind to mismatched bases. Several different MSH proteins exist in eukaryotes, each with a characteristic ability to recognize different sets of mismatched bases. Binding of the MSH proteins to the mismatch triggers strand-specific repair of the duplex, with the invading DNA strand being the one that is marked as being incorrect. Therefore, repair is directed to correct the invading strand with reference to its complement (Evans and Alani, 2000). The mechanism by which the eukaryotic cell determines the correct DNA strand is currently unknown. In the absence of the MutS protein, recombination between mismatched sequences is much more frequent, demonstrating that the recA-ssDNA nucleoprotein filament also tolerates mismatches during strand displacement *in vivo* (Evans and Alani, 2000; Modrich and Lahue, 1996).

Thus, the initial recognition of the duplex and the subsequent strand invasion steps both contribute to the discrimination of acceptable and unacceptable target molecules. One experiment provides direct proof that the strand pairing activity of the *E. coli* recA protein is sufficient to promote interactions between homologous sequences. Expression of the *E. coli* recA protein increases the frequency of gene targeting at the mouse HPRT locus, without increasing the frequency of nonhomologous recombination (Shcherbakova *et al.*, 2000). Therefore, the recA/single-stranded DNA nucleoprotein filament is able to recognize single copy sequence in a mammalian genome, similarly to that promoted by the RAD51 protein. Therefore, the eukaryotic RAD51 protein does require a greater selectivity than the bacterial recA protein to initiate genetic recombination, although its normal genetic milieu is much more complex.

In vitro, the bacterial recA protein conducts the initial homology search with a small segment of single-stranded DNA, which corresponds to about one turn of the DNA helix (Hsieh *et al.*, 1992). This suggests that the 17-nucleotide 3′ extended end that is generated by P-element excision may truly represent the minimum segment of single-strand sequence that is required for an efficient homology search (Beall and Rio, 1997; Keeler and Gloor, 1997). The P transposase is apparently unique in its ability to generate such a large staggered cut, and to exploit this to increase the copy number of the transposon (Engels, 1996, 1997). The frequency of gene conversion from the sister chromatid, or a homolog with a partial P-element end, is several orders of magnitude greater than is observed in any other system with limiting homology at the broken ends (Johnson-Schlitz and Engels, 1993). It remains to be seen if this highly efficient gene conversion is an intrinsic property of the 17-nt 3′ extended end, or if other host factors are recruited to the break site.

B. The role of chromosomal context

1. Chromosomal context

The chromosomal context has a major effect on the efficiency of gene conversion. The experiments described above demonstrate that different insertion locations of an identical donor sequence supported widely varying frequencies of gene conversion (Gloor et al., 1991; Nassif et al., 1994). The reason for this variation is unknown. Furthermore, as discussed above, the variation is not due simply to the relative distance between the donor and recipient sequences in cis or in trans (Engels et al., 1994). Indeed, unlike for site-specific recombination (Golic and Golic, 1996a), the relative distance between the interacting molecules has no effect.

2. Chromosomal architecture

The previously described observation of Lankenau et al. (2000) that the frequency of gene conversion is generally elevated in the absence of the Su(Hw)-binding protein suggests that a general loosening of nuclear architecture may permit the broken ends to search the genome more efficiently. Thus, some features of nuclear architecture affect gene conversion. This effect could be due to an architectural constraint on either the donor or recipient sequences. The observation that inclusion of a Su(Hw)-binding site inside the donor sequence did not affect the gene conversion frequency of that donor argues that the constraint is not due to anchoring of the donor sequence in the nucleus. Therefore, the Su(Hw) protein probably exerts its effect by anchoring the broken ends at the recipient site. This may explain Lankenau's observation that the frequency of gene conversion was increased for each donor-site insertion in the absence of the Su(Hw) protein, and that the relative gene conversion frequencies supported by each donor was generally preserved. Obviously, further experiments need to be done in this area with mutants in other Drosophila genes encoding proteins involved in nuclear architecture. It may be useful to examine Drosophila homologs of the architectural proteins that are known to be important in mating type switching in S. cerevisiae (Haber, 1998; Paques and Haber, 1999).

3. Chromosome pairing

Proper pairing of homologous chromosomes is the final factor that makes important contributions to the frequency of gene conversion by double-strand break repair. There is strong evidence for the pairing of multiple sites along homologous chromosomes in both *Drosophila* and yeast mitotic cells (Aragon-Alcaide and Strunnikov, 2000; Burgess and Kleckner, 1999; Burgess *et al.*, 1999; Fung *et al.*, 1998; Garcia-Bellido and Wandosell, 1978; Golic and Golic, 1996b; Weiner and Kleckner, 1994). This pairing maintains the homologues in proximity to each other, perhaps promoting DNA repair and other processes. The results discussed above show that homologous chromosomes must be paired both globally and locally for an allelic sequence to be used efficiently as a gene conversion donor (Engels *et al.*, 1990). Strangely, the local pairing requirement is important only for an allelic donor (Dray *et al.*, 2002). An ectopic donor sequence supports an equivalent gene conversion frequency with or without a large insertion (Nassif *et al.*, 1994). The reason for the difference between allelic and ectopic donors in this regard is unknown, but suggests that allelic sequences are actively recruited as gene conversion donors only if pairing is very tight.

Are the same proteins involved in the global pairing of homologs and in the search for homology that occurs at the broken ends? This question has been addressed only in *S. cerevisiae*, where the answer depends on the type of cell. Homolog pairing prior to meiosis is augmented by at least some proteins in the RAD52 epistasis group. Weiner and Kleckner (1994) found significant disruption of premeiotic chromosome pairing in yeast cells mutant for several of these genes. This disruption occurred prior to double-strand break induction by *SPO11*. They proposed that an initial weak interaction between homologs was established prior to meiosis; this weak interaction was converted to a strong one by strand invasion during the double-strand break repair of the *SPO11*-induced double-strand breaks. In contrast, Burgess *et al.* (1999) did not observe any effect on homolog pairing in mitotically dividing yeast cells when the homologous recombination pathway was ablated.

Many experiments using *Drosophila* demonstrate that chromosomes pair in somatic cells throughout the cell cycle, and that local pairing can play an important role in gene regulation (Henikoff and Comai, 1998; Hollick *et al.*, 1997; Marahrens, 1999; Wolffe and Matzke, 1999; Wu and Morris, 1999). The reader is encouraged to investigate the excellent reviews of this topic in this volume. It is exciting to speculate that similar factors are required for chromosome pairing phenomenon such as transvection and for double-strand break repair by gene conversion. This convergence can only lead to a deeper understanding of both phenomena. As always, further experimentation will give us the answers.

Acknowledgments

The author wishes to thank Angela Coveny and Dr. C.-T. Wu for their helpful comments on the manuscript. Financial support for the work in the author's laboratory is from the Medical Research Council of Canada and the Canadian Institutes of Health Research.

References

Adzuma, K. (1998). No sliding during homology search by RecA protein. *J. Biol. Chem.* **273**, 31565–31573.

Aragon-Alcaide, L., and Strunnikov, A. V. (2000). Functional dissection of in vivo interchromosome association in *Saccharomyces cerevisiae*. *Nature Cell Biol.* **2**, 812–818.

Aravind, L., Walker, D. R., and Koonin, E. V. (1999). Conserved domains in DNA repair proteins and evolution of repair systems. *Nucleic Acids Res.* **27**, 1223–1242.

Arbel, A., Zenvirth, D., and Simchen, G. (1999). Sister chromatid-based DNA repair is mediated by RAD54, not by DMC1 or TID1. *EMBO J.* **18**, 2648–2658.

Baker, M. D., Read, L. R., Beatty, B. G., and Ng, P. (1996). Requirements for ectopic homologous recombination in mammalian somatic cells. *Mol. Cell Biol.* **16**, 7122–7132.

Banga, S. S., and Boyd, J. B. (1992). Oligonucleotide-directed site-specific mutagenesis in Drosophila melanogaster. *Proc. Natl. Acad. Sci. USA* **89**, 1735–1739.

Bazemore, L. R., Folta-Stogniew, E., Takahashi, M., and Radding, C. M. (1997). RecA tests homology at both pairing and strand exchange. *Proc. Natl. Acad. Sci. USA* **94**, 11863–11868.

Beall, E. L., and Rio, D. C. (1997). Drosophila *P*-element transposase is a novel site-specific endonuclease. *Genes Dev.* **11**, 2137–2151.

Beall, E. L., and Rio, D. C. (1998). Transposase makes critical contacts with, and is stimulated by, single-stranded DNA at the *P* element termini in vitro. *EMBO J.* **17**, 2122–2136.

Burgess, S. M., and Kleckner, N. (1999). Collisions between yeast chromosomal loci in vivo are governed by three layers of organization. *Genes Dev.* **13**, 1871–1883.

Burgess, S. M., Kleckner, N., and Weiner, B. M. (1999). Somatic pairing of homologs in budding yeast: Existence and modulation. *Genes Dev.* **13**, 1627–1641.

Cai, H. N., and Shen, P. (2001). Effects of cis arrangement of chromatin insulators on enhancer-blocking activity. *Science* **291**, 493–495.

Critchlow, S. E., and Jackson, S. P. (1998). DNA end-joining: From yeast to man. *Trends Biochem. Sci.* **23**, 394–398.

Deng, C., and Capecchi, M. R. (1992). Reexamination of gene targeting frequency as a function of the extent of homology between the targeting vector and the target locus. *Mol. Cell Biol.* **12**, 3365–3371.

Digweed, M., Reis, A., and Sperling, K. (1999). Nijmegen breakage syndrome: Consequences of defective DNA double strand break repair. *Bioessays* **21**, 649–656.

Dray, T., Coveny, A. M., and Gloor, G. B. (2002). The role of local homology on P element induced gene conversion in *Drosophila*. *Genetics*. In Press.

Dray, T., and Gloor, G. B. (1997). Homology requirements for targeting heterologous sequences during P-induced gap repair in *Drosophila melanogaster*. *Genetics* **147**, 689–699.

Engels, W. R. (1992). The origin of P elements in *Drosophila melanogaster*. *Bioessays* **14**, 681–686.

Engels, W. R. (1996). P elements in *Drosophila*. *In* "Transposable Elements" (H. S. a. A. Gierl, ed.), pp. 103–123, Springer-Verlag, Berlin.

Engels, W. R. (1997). Invasions of P elements. *Genetics* **145**, 11–15.

Engels, W. R., Johnson-Schlitz, D. M., Eggleston, W. B., and Sved, J. (1990). High-frequency *P* element loss in *Drosophila* is homolog dependent. *Cell* **62**, 515–525.

Engels, W. R., Preston, C. R., and Johnson-Schlitz, D. M. (1994). Long-range cis preference in DNA homology search over the length of a *Drosophila* chromosome. *Science* **263**, 1623–1625.

Evans, E., and Alani, E. (2000). Roles for mismatch repair factors in regulating genetic recombination [in process citation]. *Mol. Cell Biol.* **20**, 7839–7844.

Fung, J. C., Marshall, W. F., Dernburg, A., Agard, D. A., and Sedat, J. W. (1998). Homologous chromosome pairing in *Drosophila melanogaster* proceeds through multiple independent initiations. *J. Cell Biol.* **141**, 5–20.

Garcia-Bellido, A., and Wandosell, F. (1978). The effect of inversions on mitotic recombination in *Drosophila melanogaster*. *Mol. Gen. Genet.* **161**, 317–321.

Geyer, P. K., Richardson, K. L., Corces, V. G., and Green, M. M. (1988). Genetic instability in *Drosophila melanogaster*: P-element mutagenesis by gene conversion. *Proc. Natl. Acad. Sci. USA* **85**, 6455–6459.

Gloor, G. B., and Lankenau, D. H. (1998). Gene conversion in mitotically dividing cells: A view from *Drosophila*. *Trends Genet.* **14**, 43–46.

Gloor, G. B., Nassif, N. A., Johnson-Schlitz, D. M., Preston, C. R., and Engels, W. R. (1991). Targeted gene replacement in *Drosophila* via P element-induced gap repair. *Science* **253**, 1110–1117.

Golic, K. G., and Golic, M. M. (1996a). Engineering the *Drosophila* genome: Chromosome rearrangements by design. *Genetics* **144**, 1693–1711.

Golic, M. M., and Golic, K. G. (1996b). A quantitative measure of the mitotic pairing of alleles in *Drosophila melanogaster* and the influence of structural heterozygosity. *Genetics* **143**, 385–400.

Gonzy-Treboul, G., Lepesant, J. A., and Deutsch, J. (1995). Enhancer-trap targeting at the Broad-Complex locus of *Drosophila melanogaster*. *Genes Dev.* **9**, 1137–1148.

Gupta, R. C., Golub, E. I., Wold, M. S., and Radding, C. M. (1998). Polarity of DNA strand exchange promoted by recombination proteins of the RecA family. *Proc. Natl. Acad. Sci. USA* **95**, 9843–9848.

Haber, J. E. (1995). *In vivo* biochemistry: physical monitoring of recombination induced by site-specific endonucleases. *Bioessays* **17**, 609–620.

Haber, J. E. (1998). Mating-type gene switching in *Saccharomyces cerevisiae*. *Annu. Rev. Genet.* **32**, 561–599.

Haber, J. E. (1999a). DNA recombination: The replication connection. *Trends Biochem. Sci.* **24**, 271–275.

Haber, J. E. (1999b). DNA repair. Gatekeepers of recombination. *Nature* **398**, 665, 667.

Haber, J. E. (2000a). Partners and pathways repairing a double-strand break. *Trends Genet.* **16**, 259–264.

Haber, J. E. (2000b). Recombination: A frank view of exchanges and vice versa. *Curr. Opin. Cell Biol.* **12**, 286–292.

Haber, J. E., and Leung, W. Y. (1996). Lack of chromosome territoriality in yeast: Promiscuous rejoining of broken chromosome ends. *Proc. Natl. Acad. Sci. USA* **93**, 13949–13954.

Henikoff, S., and Comai, L. (1998). Trans-sensing effects: The ups and downs of being together. *Cell* **93**, 329–332.

Heslip, T. R., and Hodgetts, R. B. (1994). Targeted transposition at the vestigial locus of *Drosophila melanogaster*. *Genetics* **138**, 1127–1135.

Heslip, T. R., Williams, J. A., Bell, J. B., and Hodgetts, R. B. (1992). A P element chimera containing captured genomic sequences was recovered at the vestigial locus in *Drosophila* following targeted transposition. *Genetics* **131**, 917–927.

Hilliker, A. J., Harauz, G., Reaume, A. G., Gray, M., Clark, S. H., and Chovnick, A. (1994). Meiotic gene conversion tract length distribution within the rosy locus of *Drosophila melanogaster*. *Genetics* **137**, 1019–1026.

Hiom, K. (1999). Dna repair: Rad52—The means to an end. *Curr. Biol.* **9**, R446–R448.

Hollick, J. B., Dorweiler, J. E., and Chandler, V. L. (1997). Paramutation and related allelic interactions [see comments]. *Trends Genet.* **13**, 302–308.

Holmes, A. M., and Haber, J. E. (1999). Double-strand break repair in yeast requires both leading and lagging strand DNA polymerases. *Cell* **96**, 415–424.

Hsieh, P., Camerini-Otero, C. S., and Camerini-Otero, R. D. (1990). Pairing of homologous DNA sequences by proteins: Evidence for three-stranded DNA. *Genes Dev.* **4**, 1951–1963.

Hsieh, P., Camerini-Otero, C. S., and Camerini-Otero, R. D. (1992). The synapsis event in the homologous pairing of DNAs: RecA recognizes and pairs less than one helical repeat of DNA. *Proc. Natl. Acad. Sci. USA* **89**, 6492–6496.

Jasin, M. (2000). Chromosome breaks and genomic instability. *Cancer Invest.* **18**, 78–86.

Jinks-Robertson, S., Michelitch, M., and Ramcharan, S. (1993). Substrate length requirements for efficient mitotic recombination in *Saccharomyces cerevisiae*. *Mol. Cell. Biol.* **13**, 3937–3950.

Johnson, R. D., and Jasin, M. (2000). Sister chromatid gene conversion is a prominent double-strand break repair pathway in mammalian cells. *EMBO J.* **19**, 3398–3407.

Johnson-Schlitz, D. M., and Engels, W. R. (1993). P-element-induced interallelic gene conversion of insertions and deletions in *Drosophila melanogaster*. *Mol. Cell Biol.* **13**, 7006–7018.

Karran, P. (2000). DNA double strand break repair in mammalian cells. *Curr. Opin. Genet. Dev.* **10**, 144–150.

Keeler, K. J., Dray, T., Penney, J. E., and Gloor, G. B. (1996). Gene targeting of a plasmid-borne sequence to a double-strand DNA break in *Drosophila melanogaster*. *Mol. Cell Biol.* **16**, 522–528.

Keeler, K. J., and Gloor, G. B. (1997). Efficient gap repair in *Drosophila melanogaster* requires a maximum of 31 nucleotides of homologous sequence at the searching ends. *Mol. Cell Biol.* **17**, 627–634.

Keeney, S., Baudat, F., Angeles, M., Zhou, Z. H., Copeland, N. G., Jenkins, N. A., Manova, K., and Jasin, M. (1999). A mouse homolog of the *Saccharomyces cerevisiae* meiotic recombination DNA transesterase Spo11p. *Genomics* **61**, 170–182.

Keeney, S., Giroux, C. N., and Kleckner, N. (1997). Meiosis-specific DNA double-strand breaks are catalyzed by Spo11, a member of a widely conserved protein family. *Cell* **88**, 375–384.

Kote-Jarai, Z., and Eeles, R. A. (1999). BRCA1, BRCA2 and their possible function in DNA damage response. *Br. J. Cancer* **81**, 1099–1102.

Kuzminov, A. (1995). Collapse and repair of replication forks in *Escherichia coli*. *Mol. Microbiol.* **16**, 373–384.

Kuzminov, A. (1999). Recombinational repair of DNA damage in *Escherichia coli* and bacteriophage lambda. *Microbiol. Mol. Biol. Rev.* **63**, 751–813, table of contents.

Kuzminov, A., and Stahl, F. W. (1999). Double-strand end repair via the RecBC pathway in *Escherichia coli* primes DNA replication. *Genes Dev.* **13**, 345–356.

Lankenau, D. H., and Gloor, G. B. (1998). In vivo gap repair in *Drosophila*: A one-way street with many destinations [in process citation]. *Bioessays* **20**, 317–327.

Lankenau, D. H., Peluso, M. V., and Lankenau, S. (2000). The Su(Hw) chromatin insulator protein alters double-strand break repair frequencies in the *Drosophila* germ line. *Chromosoma* **109**, 148–160.

Liskay, R. M., Letsou, A., and Stachelek, J. L. (1987). Homology requirement for efficient gene conversion between duplicated chromosomal sequences in mammalian cells. *Genetics* **115**, 161–167.

Lorenz, M. C., Muir, R. S., Lim, E., McElver, J., Weber, S. C., and Heitman, J. (1995). Gene disruption with PCR products in *Saccharomyces cerevisiae*. *Gene* **158**, 113–117.

Manivasakam, P., Weber, S. C., McElver, J., and Schiestl, R. H. (1995). Micro-homology mediated PCR targeting in *Saccharomyces cerevisiae*. *Nucleic Acids Res.* **23**, 2799–2800.

Marahrens, Y. (1999). X-inactivation by chromosomal pairing events. *Genes Dev.* **13**, 2624–2632.

McKim, K. S., and Hayashi-Hagihara, A. (1998). mei-W68 in *Drosophila melanogaster* encodes a Spo11 homolog: Evidence that the mechanism for initiating meiotic recombination is conserved. *Genes Dev.* **12**, 2932–2942.

Metzenberg, A. B., Wurzer, G., Huisman, T. H., and Smithies, O. (1991). Homology requirements for unequal crossing over in humans. *Genetics* **128**, 143–161.

Michel, B. (2000). Replication fork arrest and DNA recombination. *Trends Biochem. Sci.* **25**, 173–178.

Modrich, P., and Lahue, R. (1996). Mismatch repair in replication fidelity, genetic recombination, and cancer biology. *Annu. Rev. Biochem.* **65,** 101–133.

Mullins, M. C., Rio, D. C., and Rubin, G. M. (1989). cis-acting DNA sequence requirements for P-element transposition. *Genes Dev.* **3,** 729–738.

Muravyova, E., Golovnin, A., Gracheva, E., Parshikov, A., Belenkaya, T., Pirrotta, V., and Georgiev, P. (2001). Loss of insulator activity by paired Su(Hw) chromatin insulators. *Science* **291,** 495–498.

Nassif, N., and Engels, W. (1993). DNA homology requirements for mitotic gap repair in *Drosophila*. *Proc. Natl. Acad. Sci. USA* **90,** 1262–1266.

Nassif, N., Penney, J., Pal, S., Engels, W. R., and Gloor, G. B. (1994). Efficient copying of nonhomologous sequences from ectopic sites via P-element-induced gap repair. *Trends Biochem. Sci.* **14,** 1613–1625.

Nishinaka, T., Shinohara, A., Ito, Y., Yokoyama, S., and Shibata, T. (1998). Base pair switching by interconversion of sugar puckers in DNA extended by proteins of RecA-family: A model for homology search in homologous genetic recombination. *Proc. Natl. Acad. Sci. USA* **95,** 11071–11076.

Paques, F., and Haber, J. E. (1999). Multiple pathways of recombination induced by double-strand breaks in *Saccharomyces cerevisiae*. *Microbiol. Mol. Biol. Rev.* **63,** 349–404.

Preston, C. R., and Engels, W. R. (1996). P-element-induced male recombination and gene conversion in *Drosophila*. *Genetics* **144,** 1611–1622.

Rao, B. J., Dutreix, M., and Radding, C. M. (1991). Stable three-stranded DNA made by RecA protein. *Proc. Natl. Acad. Sci. USA* **88,** 2984–2988.

Robertson, H. M., Preston, C. R., Phillis, R. W., Johnson-Schlitz, D. M., Benz, W. K., and Engels, W. R. (1988). A stable genomic source of P element transposase in *Drosophila melanogaster*. *Genetics* **118,** 461–470.

Romanienko, P. J., and Camerini-Otero, R. D. (1999). Cloning, characterization, and localization of mouse and human SPO11. *Genomics* **61,** 156–169.

Roseman, R. R., Pirrotta, V., and Geyer, P. K. (1993). The su(Hw) protein insulates expression of the *Drosophila melanogaster white* gene from chromosomal position-effects. *EMBO J.* **12,** 435–442.

Rubnitz, J., and Subramani, S. (1984). The minimum amount of homology required for homologous recombination in mammalian cells. *Mol. Cell Biol.* **4,** 2253–2258.

Sepp, K. J., and Auld, V. J. (1999). Conversion of lacZ enhancer trap lines to GAL4 lines using targeted transposition in *Drosophila melanogaster*. *Genetics* **151,** 1093–1101.

Shannon, M., Richardson, L., Christian, A., Handel, M. A., and Thelen, M. P. (1999). Differential gene expression of mammalian SPO11/TOP6A homologs during meiosis. *FEBS Lett.* **462,** 329–334.

Shcherbakova, O. G., Lanzov, V. A., Ogawa, H., and Filatov, M. V. (2000). Overexpression of bacterial RecA protein stimulates homologous recombination in somatic mammalian cells. *Mutat. Res.* **459,** 65–71.

Shen, P., and Huang, H. V. (1986). Homologous recombination in *Escherichia coli*: Dependence on substrate length and homology. *Genetics* **112,** 441–457.

Singer, B. S., Gold, L., Gauss, P., and Doherty, D. H. (1982). Determination of the amount of homology required for recombination in bacteriophage T4. *Cell* **31,** 25–33.

te Riele, H., Maandag, E. R., and Berns, A. (1992). Highly efficient gene targeting in embryonic stem cells through homologous recombination with isogenic DNA constructs. *Proc. Natl. Acad. Sci. USA* **89,** 5128–5132.

Thaler, D. S., and Noordewier, M. O. (1992). MEPS parameters and graph analysis for the use of recombination to construct ordered sets of overlapping clones. *Genomics* **13,** 1065–1074.

Thompson, L. H., and Schild, D. (1999). The contribution of homologous recombination in preserving genome integrity in mammalian cells. *Biochimie* **81,** 87–105.

Van Dyck, E., Stasiak, A. Z., Stasiak, A., and West, S. C. (1999). Binding of double-strand breaks in DNA by human Rad52 protein. *Nature* **398,** 728–731.

Watt, V. M., Ingles, C. J., Urdea, M. S., and Rutter, W. J. (1985). Homology requirements for recombination in *Escherichia coli. Proc. Natl. Acad. Sci. USA* **82,** 4768–4772.

Weiner, B. M., and Kleckner, N. (1994). Chromosome pairing via multiple interstitial interactions before and during meiosis in yeast. *Cell* **77,** 977–991.

Welcsh, P. L., Owens, K. N., and King, I. (2000). Insights into the functions of BRCA1 and BRCA2. *Trends Genet.* **16,** 69–74.

Williams, C. J., and O'Hare, K. (1996). Elimination of introns at the *Drosophila* suppressor-of-forked locus by P-element-mediated gene conversion shows that an RNA lacking a stop codon is dispensable. *Genetics* **143,** 345–351.

Wolffe, A. P., and Matzke, M. A. (1999). Epigenetics: Regulation through repression. *Science* **286,** 481–486.

Wu, C. T., and Morris, J. R. (1999). Transvection and other homology effects. *Curr. Opin. Genet. Dev.* **9,** 237–246.

5

The Origins of Genomic Imprinting in Mammals

Frank Sleutels
The Netherlands Cancer Institute
1066 CX Amsterdam, The Netherlands

Denise P. Barlow*
ÖAW Institute of Molecular Biology
A5020 Salzburg, Austria

*To whom correspondence should be addressed: E-mail: dbarlow@imb.oeaw.ac.at.

I. INTRODUCTION

Mammals are diploid organisms that inherit a complete chromosome set from each parent. The vast majority of genes are equally expressed from both parental chromosomes. However, in a small subset of genes, a process known as genomic imprinting results in parental-specific gene expression. The imprinted gene is expressed on one parental chromosome, but silent on the other. Parental specific gene expression does not result from genetic changes, but instead results from modifications to DNA or chromatin that are described as "epigenetic." Genomic imprinting is thus an exceptionally good tool to study epigenetic gene regulation because both the active and the silent allele are retained in the same nucleus. Historically, imprinting has been viewed only as a gene-silencing mechanism. However, the actual data obtained from studies of imprinted genes are not compatible with this view and, instead, indicate that imprints were in many cases acquired as a gene-activating mechanism. To accommodate these findings, a model is proposed whereby imprints can activate, or de-repress, genes previously silenced by an epimutation.

II. DEFINITION OF GENOMIC IMPRINTING

Mammals are diploid organisms arising from the fusion of two parental gametes that each donate one set of autosomal chromosomes (22 autosomes in humans, 19 in mice) plus one sex chromosome (X or Y) to an offspring. Diploid cells thus contain two parental copies (or parental alleles) of each autosomal gene, which are predicted to show the same transcription state since for the majority of genes the loss of one copy can be easily tolerated. Genomic imprinting is a rare and unusual process that results in parental specific gene expression of autosomal loci (in males and females) and of loci on the diploid X chromosome (in females) (see reviews by Surani, 1998; Reik and Walter, 2000; Tilghman, 1999; Bartolomei and Tilghman, 1997; Solter, 1998). The transcription of imprinted genes is determined by their parental origin; the two parental alleles can be either maternally expressed and paternally silent or vice versa (maternally silent and paternally expressed). Thus, imprinted genes show haploid expression despite the cell being diploid for the locus. About 50 mammalian imprinted genes have now been identified in humans and mice, with approximately equal numbers of paternal- and maternal-expressed examples (see Beechey and Cattanach, 2000, and Morison *et al.*, 2001, for a full listing and chromosome maps of all known imprinted mouse and human genes and other imprinted phenotypes). The total number of imprinted genes is likely to be relatively small compared to the total number of genes in the genome. Estimates based on genetic experiments in mice and on all the known genetic mutations in mice and humans that, in contrast to imprinted loci, obey Mendel's first law and are functionally equivalent on maternal and paternal inheritance, provide an estimate of 100–200 (Beechey and Cattanach, 2000). However, this number may now be reduced by 60% following the reappraisal of the total number of genes in the mammalian genome (Claverie, 2001).

III. EVOLUTION AND FUNCTION OF IMPRINTING

Genomic imprinting appears to be conserved in all placental mammals (with the caveat that only a limited number of species has yet been tested). It is also present in marsupial mammals where two autosomal genes plus X chromosome inactivation are known to be imprinted (O'Neill *et al.*, 2000; Killian *et al.*, 2000; Graves, 1996). Monotreme or oviparous mammals lack imprinting of two tested autosomal genes, and although they show a partial and tissue-specific X inactivation, it is not yet known if this is random or imprinted (Graves, 1996). Genomic imprinting is thought to be absent from nonmammalian vertebrates but, except for one gene shown not to be imprinted in the chicken (O'Neill *et al.*, 2000), this hypothesis has not been generally tested among all vertebrates. The absence of genomic imprinting in nonmammalian vertebrates is, instead, based on the argument that genomic imprinting imposes a requirement for sexual reproduction in mammals (see Section III.A), and asexual parthenogenetic reproduction has been observed to occur in all nonmammalian vertebrates (Sarvella, 1973; Cuellar, 1977; Cole and Townshend, 1990; Corley-Smith *et al.*, 1996). Genomic imprinting thus likely arose in the vertebrate lineage, after the formation of mammals but prior to the divergence of the marsupial and placental subclasses. However, the presence of parental specific gene expression in the endosperm of angiosperm plants (e.g., *Arabodopsis*) and parental specific chromosome inactivation linked to sex determination in scale insects (*Coccidea*) indicates not only that genomic imprinting evolved several times, but that it serves more than one biological function (Messing and Grossniklaus, 1999; Herrick and Seger, 1999; see also Chapter 6 and reviews by Hurst and McVean, 1998; and Pardo-Manuel de Villena *et al.*, 2000).

Genomic imprinting in mammals today is already known to serve the function of chromosome dosage compensation by inactivating one X chromosome in female mammals, thus equalizing the dosage of X-linked genes between males (XY) and females (XX). X-chromosome inactivation is imprinted so that only the paternal X is inactivated in all tissues of marsupials. In placental mammals, however, imprinted X inactivation is restricted to the placenta and extraembryonic tissues, while somatic tissues show random X inactivation (reviewed by Graves, 1996; Avner and Heard, 2000; see also Chapter 2). Genomic imprinting of autosomal genes in mammals is unlikely to serve the same function of dosage compensation, since males and females have the same number of autosomes. An indication of the other possible function(s) of imprinting in mammals comes from the finding that imprinted genes are present in mammals but absent from all other vertebrate classes. Mammals are very similar to other vertebrates in terms of organism physiology and genetic developmental programs (Ferrier and Holland, 2000). However, in contrast to oviparous vertebrates, mammals depend solely on sexual reproduction, and furthermore, in contrast to oviparous animals that deposit a fixed amount of nutrient per embryo in an egg, mammals do not strictly limit the extent of nutrient transfer between the maternal parent and embryo. All three classes of mammals (monotreme, marsupial, and placental), transfer nutrients and other resources from the maternal parent to the embryo, but only in placental mammals does an open circulatory connection exist between the embryo and the maternal parent *in utero*. All three subclasses of mammals, in addition, have a postnatal growth period in which offspring are fully dependent on maternal milk production. In both pre- and postnatal growth, the mammalian embryo or offspring can reprogram the maternal environment to influence the extent of the nutrient transfer. These and other unique features of mammalian reproduction are listed below, with indications of possible role(s) for imprinted genes.

A. Sexual reproduction

Mammals, alone of all the vertebrates, do not reproduce asexually. The development of imprinted genes that are essential for embryonic survival but show parental-specific silencing will make sexual reproduction essential. Many of the imprinted genes studied by targeted inactivation show an essential developmental function (Table 5.1).

B. Placental development

The placenta is an embryonic tissue necessary for embryonic growth and development. It has been argued that spontaneous activation of eggs leading to parthenogenesis could be dangerous for the development of intrauterine embryonic growth strategies. The use of imprinting to silence maternal genes that regulate placental

Table 5.1. The Function of Imprinted Genes as Determined by Loss of Function Alleles[a]

	Imprinted gene	Loss of function phenotype in Mice	±	References and notes
1	*Igf2*	Growth reduction	+	DeChiara *et al.*, 1990.
2	*Ins2*	Viable—no growth effect		Duvillie *et al.*, 1997. Mice lacking *Ins2* and *Ins1* have nonimprinted growth retardation *Ins1* is not imprinted.
3	*Snrpn/Snurf*	Viable—no growth effect	−	Yang *et al.*, 1997; Tsai *et al.*, 1999.
4	*Snrpr* + *upstream* IC 3'Snrpn-Ube3a region	Growth reduction—lethal Growth reduction—lethal	+	Yang *et al.*, 1997; Tsai *et al.* 1999. These deletions may affect the same gene(s).
5	*Grfl/Ras-Grfl*	Long-term memory Postnatal growth	− +	Brambilla *et al.*, 1997. Itier *et al.*, 1998.
6	*Mest/Peg1*	Growth reduction, nurturing	+	Lefebvre *et al.*, 1998.
7	*Peg3*	Growth reduction, nurturing	+	Li *et al.*, 1999.
8	*Necdin*	Lethal—no growth effect	−	Gerard *et al.*, 1999. Null mice are viable in some genetic backgrounds.
9	*Mas1*	Behavior phenotype	−	Walther *et al.*, 1998. Now known not to be imprinted, Lyle *et al.*, 2000.
10	*GnasX1*	Growth reduction	+	Yu *et al.*, 2000; Peters *et al.*, 1999. Both *GnasX1* and *Gnas* are transcribed in the same direction from different promoters on different parental chromosomes and splice to exon 2.
11	*Gnas*	Growth increase	+	
12	*M6P-Igf2r*	Growth increase—lethal	+	Wang *et al.*, 1994.
13	*H19*	No phenotype	−	Jones *et al.*, 1998. Embryonic growth is increased only if mice also overexpress *Igf2*, Leighton *et al.*, 1995.
14	*Mash2*	Placental defect—lethal	+	Guillemot *et al.*, 1995.
15	*Cdkn1*	Lethal, proliferation defect—no organism growth defect	+	Zhang *et al.*, 1997; Yan *et al.*, 1997
16	*Ube3a*	Motor dysfunction and learning deficit	−	Jiang *et al.*, 1998.
17	*Gtl2*	Growth decrease by transgene insertion upstream to *Gtl2*	+	Schuster-Gossler *et al.*, 1996; Schmidt *et al.*, 2000. Transgene insertion may affect the upstream paternally expressed Dlkl gene.

[a]Paternally expressed imprinted genes are at the top of the table, maternally expressed imprinted genes are shaded boxes at the bottom of the table. The ± column indicates if the phenotype could have a relevance for an imprinting function in any aspect of mammalian development as discussed in the text. See also a similar table in Hurst and McVean (1998).

development and invasion of maternal tissue would prevent the implantation of spontaneously activated nonfertilized eggs. Active placental development and invasion genes, introduced at fertilization by the paternal gamete, would then control pregnancy (trophoblast defense hypothesis; Varmusa and Mann, 1994; Haig, 1994). At this time, none of the paternally expressed imprinted genes studied by targeted inactivation show an essential implantation function, although some, such as *Igf2* and *Peg3*, play a modest role in placental growth (Table 5.1). (N.B.: abbreviations used in the text, and full names of genes, are listed in Table 5.5, below.)

C. Intrauterine embryonic growth

The placental mammalian embryo grows inside the uterus attached via the placenta to the maternal blood supply, where it is nourished and protected. Although the final size of any organism is genetically determined (Conlon and Raff, 1999), there is a wide variation in birthweight that is influenced both by the uterine environment (large uteri will produce large offspring; Snow, 1981), and by the embryo itself (through production of embryonic hormones that manipulate the maternal environment and the extent of maternal nutrient transfer). Birthweight is a crucial parameter for postnatal survival for both polytocous and monotocous (animals with multiple or single offspring in one litter) species. Both underweight and overweight offspring have increased mortality rates (Wilcox and Russell, 1983); however, overweight offspring may present an increased threat to the evolution of intrauterine growth because they endanger the mother and thus also older siblings. Because of the necessity to achieve optimum birthweight, strict dosage control of growth-regulatory genes that may be achieved by silencing one parental allele may have been a necessary requirement for mammalian reproduction (Solter, 1988). In support of this function, many of the imprinted genes studied by targeted inactivation, as well as many of the as yet uncloned but mapped imprinted loci, show growth-regulatory functions (Table 5.1; Beechey and Cattanach, 2000). There is little evidence, however, to indicate that silencing one locus is necessary for tight regulation of gene expression.

The ability of the embryo to influence its own growth at maternal expense is the basis of one of the best-known ideas to explain the evolution of imprinting, the parental conflict hypothesis (Moore and Haig, 1991). This hypothesis proposes that mammalian breeding strategies in which females breed with multiple males will produce a conflict of interest between the two parental genomes. The interests of the maternal genome are best served by equal distribution of her resources among all her offspring; however, the interests of the paternal genome are best served by ensuring that offspring with his genome receive the largest share of maternal nutrients, at the expense of offspring with different paternal genomes. This theory predicts that growth-promoting genes will be paternally expressed, while growth-suppressing genes will be maternally expressed. As can be seen from Table 5.1, there is very strong evidence from the function of imprinted genes to support this hypothesis as at least one of the biological functions of imprinting. However, one prediction of the theory, that the coding sequence of imprinted genes will evolve rapidly (to allow competition between different paternal alleles), is not supported, since the imprinted genes so far examined show a low level of genetic variation (Hurst and McVean, 1998).

D. Avoidance of fetal immune rejection

The direct connection between the maternal and fetal circulation exposes the maternal immune system to fetal antigens. Paternal fetal antigens will be recognized as foreign and stimulate a maternal immune response. Silencing fetal expression of paternal antigens may have been necessary to subvert the maternal immune response during the evolution of intrauterine embryonic growth. However, there is as yet no evidence for this view, and none of the imprinted genes studied by targeted inactivation shows an immune function (Table 5.1). Moreover, it has recently been shown that the fetus actively defends itself against immune attack using nonimprinted genes to suppress maternal immune responses, e.g., by complement inhibition (Xu et al., 2000) and by inhibition of T-cell activation by tryptophan catabolism (Munn et al., 1998).

E. Postnatal nurturing

All newborn mammals are fully dependent on maternal milk for nourishment and maternal behavior for heat and protection in a postnatal period that is of variable length. The neonate is able to stimulate milk production by suckling and other behavior patterns, as well as stimulating appropriate maternal responses by many means, including vocalization (Nowak et al., 2000). Maternal ovulation, and thus the production of new embryos, is also suppressed by suckling in some species. Using the same arguments as in Section III.C, it could be claimed that strict control of genes regulating these behaviors is necessary to limit maternal contribution to the neonate. This would maintain the mother's fitness to care for the present offspring well beyond the weaning period, allowing her to care for previous offspring, as well as ensuring her ability to produce new offspring in the future. Genes affecting nurturing or suckling behavior in the postnatal period would also be selected under the parental conflict theory. Two of the imprinted genes studied by targeted inactivation show a maternal behavioral function (Table 5.1).

The identification of the in vivo function of 17 of the 50 known imprinted genes in mice (Table 5.1) shows that a large majority have growth regulatory functions in late embryonic and postnatal stages. As shown in Table 5.1, growth-promoting effects are shown only by paternally expressed genes (6 cases indicated by +). No paternal-expressed imprinted gene has been shown to have a growth-suppressor effect. In contrast, growth-suppressing effects have been shown only by maternally expressed genes (2 cases—H19 is excluded because of the H19 RNA alone does not generate a growth phenotype; Jones et al., 1998). Because of the similarity between the paired Gtl2 and Dlk1 loci and the paired Igf2 and H19 loci, it is likely, but not yet proven, that the growth decrease associated with a transgene insertion upstream to Gtl2 reflects a change in Dlk1 expression (Table 5.1; Schmidt et al., 2000, discussed in more detail in the Section IX.C). With this caveat in mind, it can also be stated that no maternal-expressed imprinted gene has been shown to have a growth-promoting effect. This finding provides very strong support for the parental conflict theory. However, 8 of the studied genes do not affect growth, which may indicate that the parental conflict theory is false. Conversely, it may also indicate that imprinting serves several different biological function in mammals today, as argued by Hurst and McVean (1998), or even a function not specific to mammals, as argued by Pardo-Manuel de Villena et al. (2000). Any one of the above five listed features of mammalian reproduction may have reinforced selection of genomic imprinting in mammals.

The gene inactivation experiments have provided much information. However, studying a loss of function allele is actually the wrong experiment, since there is a difference between the function of a gene and the significance of imprinting it. A more informative way to examine the biological function of imprinting in mammals would be to change monoallelic imprinted expression to biallelic expression. If no phenotypic consequences result from increasing the gene dosage twofold, then imprinting this gene has no significance for the biological function of imprinting. This is independent of the fact that a loss of function allele shows that the gene has an essential function. Such a gene must be regarded as an "innocent bystander" of the imprinting mechanism (Varmusa and Mann, 1994). "Loss of imprinting" alleles have been generated for two only genes (Igf2 and M6P-Igf2r) by targeted deletion of an imprint control element (ICE). These experiments generated a growth phenotype with a reciprocal effect to that shown by a loss-of-function mutant (Leighton et al., 1995; Caspary et al., 1999; Wutz et al., 2001). However, deletion of an ICE can change expression of more than one gene, and although the phenotype is likely to be due to the two primary imprinted genes regulated by the ICE (respectively, Igf2 and M6P-Igf2r), the effects of the other deregulated genes may influence the phenotype.

IV. ORIGINS OF GENOMIC IMPRINTING

The biological function of genomic imprinting is, however, only one half of the puzzle. The other half is an explanation of the molecular mechanism underlying parental-specific gene silencing. Before we examine the current features of imprinted genes, it may be informative to consider how imprinted genes first arose in the ancestral mammalian population. In considering this, the assumption is made that an ancestral mammalian population lacked the imprinting mechanism but did not lack the genes that are imprinted in the population today. This is likely, since genes imprinted in mammals today have homologs that are not imprinted in nonmammalian vertebrates. Thus, imprinted gene expression is likely to have evolved in a population that initially showed biallelic expression of these genes. In addition, since the majority of genes in the modern day mammalian genome are nonimprinted and show biallelic expression, it is logical to consider the imprinting mechanism initially as one that acts to silence one of the two parental alleles. Models A and B below outline two possible scenarios for acquisition of imprinting by parental-specific silencing of genes that show biallelic expression in an ancestral population.

A. Model A

A genetic "*gain of imprinting*" mutation arose in a single animal that introduced a parental-specific, epigenetic gene-silencing mechanism. An example could be the gain of expression in the gametes of one parent only of a gene silencing protein such as a *de novo* methyltransferase. However, to allow for the spread of the *gain of imprinting* (GOI) mutation in the ancestral population, there must have existed one susceptible gene that could be "imprinted," i.e., a gene whose methylation by the *de novo* methyltransferase reduced expression from one parental allele and so conferred an immediate advantage to the organism. It is important to appreciate that the new GOI allele would not increase in frequency without phenotypic change, and hemizygous RNA expression by itself is unlikely to have been advantageous for the organism (but see Pardo-Manuel de Villena, *et al.*, 2000, for a different point of view). An example of an appropriate susceptible gene could be a dosage-sensitive growth-suppressor or growth-promoter gene. Animals that were effectively hemiyzgous for a growth-suppressor gene because of the GOI mutation would be larger than wild-type littermates and, as a consequence, may be more effective competitors for food and mates, thus allowing the spread of the GOI allele throughout the ancestral population. Animals that were effectively hemiyzgous for a growth-promoter gene would be smaller than littermates, and although it is less easy to suggest an immediate advantage for size reduction, an explanation is needed since many of the imprinted genes listed in Table 5.1 have a growth-promoting effect. There are a few possibilities: first, production of smaller-sized offspring may have been related to the development of a polytocous (multiple offspring in one litter) breeding strategy in mammals. A "small but many" breeding strategy may have had advantages over a "large but few" strategy. Second, small-sized offspring may have provided immediate benefit to the mother by conserving her resources, this may have allowed increased nurturing behavior that would benefit the survival of present and past offspring. A third possibility is that imprinted expression of growth-suppressor genes occurred early in evolution, and imprinted expression of growth-promoting genes occurred in response to reduced levels of suppressor genes. This latter suggestion would require that growth-promoter genes were involved in the same genetic pathways as growth-suppressor genes, as is shown by some of the genes listed in Table 5.1 (e.g., *M6P-Igf2r, Igf2, H19,* and possibly *Grb10;* see Table 5.1 for references). A final suggestion is that the imprint may have altered the expression pattern or level of expression on the active allele of a growth-promoting gene, thereby allowing increased growth despite the complete silence of one allele.

The initial GOI mutation and the nature of the susceptibility of the growth suppressor gene most likely lacked the sophistication of the present-day imprinting mechanism. However, removal of the imprint from the germline of the next generation, and reacquisition according to the sex of the offspring, would be an essential requirement to allow parental specificity. In this model, a different GOI mutation, which would not imprint the same set of genes, would have been acquired later by the reciprocal parental gamete. The acquisition of a GOI mutation suggests the possibility that a large number of existing susceptible genes were simultaneously imprinted and showed hemizygous expression in diploid cells of the ancestral mammal. Although such radical change seems improbable, selection for the GOI mutation by imprinting one key gene that benefited organism fitness would also allow selection for imprinted expression of other genes that have no obvious benefit to the organism. This may explain why some genes imprinted in mammals today appear to have no essential function (Table 5.1); see also later discussion on innocent bystander genes (Section IX.A).

B. Model B

A parental-specific, epigenetic gene-silencing mechanism already existed in ancestral mammals, but served a different purpose. An example could be gamete-specific silencing by *de novo* methyltransferase of infectious exogenous retroviruses and endogenous transposable elements, which would serve as a genome defense function. The maternal gamete would be able to methylate and silence some forms of foreign invasive DNA, while the paternal gamete would be able to methylate and silence complementing forms (Barlow, 1993). With this gamete-specific mechanism in place in the ancestral population, a genetic mutation could arise in a single animal, in a gene or cis-linked regulatory element, that rendered a gene susceptible to this preexisting silencing mechanism. A possible mutation strategy that would serve this purpose would be the integration of a virus or transposon into a dose-sensitive growth-suppressor or growth-promoter gene. The result would be the same as outlined in model A: the mutated allele is silenced by transmission through one gamete only, and heterozygous advantage from haploid expression of the mutated gene in the organism allows spread of the mutated "imprinted allele" throughout the population. An increase in the number of genes subject to imprinting within the framework of model B is likely to have involved acquisition of the susceptible trait by individual genes that were individually tested for function by selective forces.

In both model A and model B, the increased population frequency of the mutated allele resulting from heterozygous advantage cannot lead to homozygous silencing of the gene, since the silencing mechanism is parental-specific. With both models, it is also clear that spread of the first imprinted allele in the ancestral population must have conferred an advantage on an individual organism. However, it is not clear that imprinted genes in modern-day mammals are maintained by the same selective force that drove their initial establishment in ancestral populations.

V. PREDICTIONS FOR AN IMPRINTING MECHANISM

Imprinted genes, in many cases, show parental allele-specific gene silencing that is stable in all cells of the organism. This shows that the imprinting mechanism must be maintained on the same parental allele during cell division. Both the active and silent parental alleles are retained in the nucleus and potentially exposed to the same trans-acting transcription regulators, but only one parental allele is affected; thus the imprinting mechanism must act *in cis* to the chromosome that inherits it. Imprinted genes show the same parental allele-specific expression in succeeding generations (e.g., males containing an imprinted gene that has an active paternal allele and a silent maternal allele will transmit both parental alleles in an active form to their own offspring); this shows that imprints are erased in the germline and then re-formed according to the sex of each new offspring. Imprinted genes can be found in animals with identical maternal and paternal genomes (e.g., inbred mice), although this is not a requirement since outbred populations such as humans also have imprinted genes. This finding shows two further important features of the imprinting mechanism. First, since imprinting can operate with genetically identical parental chromosomes, the mechanism is not genetic and must be epigenetic (i.e., a modification of DNA or chromatin able to regulate gene expression). Second, since the imprinting mechanism can distinguish between genetically identical parental chromosomes, imprints cannot be acquired after the cell becomes diploid and must therefore be acquired when the parental chromosomes are in separate compartments, such as during gametogenesis or just after fertilization prior to pronuclei fusion. Thus, imprints are predicted to be epigenetic modifications that are acquired during gametogenesis by one gamete only, heritable by the same parental allele during mitosis, associated with a cis-acting mechanism able to silence expression, and lastly, reversible in the germline of the next generation.

The degree of silencing of the repressed allele of an imprinted gene can show considerable variation. Some imprinted genes show tissue-specific loss of silencing; e.g., the paternal allele of the mouse *Igf2* gene is not silenced in the choroid plexus, the human *IGF2* gene shows biallelic expression in juvenile liver (DeChiara *et al.*, 1990; Vu and Hoffman, 1994), and the paternal allele of the mouse *M6P-Igf2r* gene is partially activated in adult brain (Schweifer *et al.*, 1997; Hu *et al.*, 1998). Imprinted genes can also show developmental control of imprinted expression. The mouse *M6P-Igf2r* gene shows low-level biallelic expression in preimplantation embryos, while imprinted expression is only seen after embryonic implantation, in parallel with X-chromosome inactivation (Szabo and Mann, 1995; Lerchner and Barlow, 1997; Avner and Heard, 2000). The *Mash2* gene initially also shows biallelic expression before the onset of imprinted expression (Caspary *et al.*, 1998). In contrast, imprinted expression of *Kcnq1* is restricted to the early embryo and placental tissues, and the gene is expressed from both alleles in later tissues (Caspary *et al.*, 1998). Several imprinted genes, e.g., mouse *M6P-Igf2r* and human *CDKN1*, also show a low level of expression of the silent allele in all tissues when samples are analyzed by sensitive means (Wang *et al.*, 1994; Matsuoka *et al.*, 1996). Finally, although imprinted expression is generally highly conserved between mice and humans, several imprinted genes show species variation in imprinted expression. The most dramatic example is *M6P-Igf2r*, which shows full imprinted expression in the mouse embryo and adult, while imprinted expression in humans is limited to postimplantation embryos (20–24 weeks) and, furthermore, is limited to a small proportion of individuals. All tissues at birth and all adult tissues in humans show full biallelic expression of *M6P-Igf2r* (Xu *et al.*, 1993; Kalscheuer *et al.*, 1993; Ogawa *et al.*, 1993). Polymorphic imprinted expression has also been observed for the human *Serotonin-2A receptor* (Bunzel *et al.*, 1998), human *Wilm's tumor* gene (Nishiwaki *et al.*, 1997), and human *IGF2* (Giannoukakis *et al.*, 1996). This variation in the degree of silencing of the repressed allele indicates that the imprinting is not an "all or nothing" mechanism and that the repressed allele may be reactivated under some conditions.

VI. ORGANIZATION AND EPIGENETIC MODIFICATION OF IMPRINTED GENES

According to the models A and B, outlined above, imprinting is considered to arise as a parent-specific gene-silencing mechanism. However, the first results describing the modifications of imprinted genes in the mouse yielded surprising results. The paternally expressed *Igf2* promoter was shown to have potentially active chromatin as well as a somatic methylation imprint upstream of the active paternal allele (Sasaki *et al.*, 1992), while the maternally expressed *M6P-Igf2r* gene was shown to carry a gametic methylation imprint only on the expressed allele (Stöger *et al.*, 1993). If the imprinting mechanism acted by parental allele-specific silencing, the prediction would have been that the silent maternal *Igf2* allele and the silent paternal *M6P-Igf2r* allele should have been modified. The next gene to be studied, the maternally expressed *H19* gene, did, however, support a parental allele-specific silencing mechanism as the basis of imprinting, since it showed specific methylation of the silent paternal promoter (Ferguson-Smith *et al.*, 1993). As further imprinted genes were isolated and their epigenetic modifications studied, several features common to imprinted genes became apparent.

A. Clusters

Imprinted genes are clustered but do not occupy a special chromosome position. Imprinting of the mouse *H19* gene was initially tested because it mapped close to *Igf2* on chromosome 7, the positive result forming the basis of the hypothesis that imprinted genes are clustered and regulated in concert by long-range control elements (Bartolomei *et al.*, 1991). This hypothesis has been spectacularly successful and has allowed the identification of more than half of the known imprinted genes. The extent of clustering in the mouse can be seen in the imprinting maps maintained by Beechey *et al.* (2000), and also from sequence analysis of large regions containing homologous imprinted clusters in human and mice. Mouse chromosome 7 is an extreme example and contains three clusters of imprinted genes mapping to the proximal, central, and distal regions whose homologous regions are separated in the human, respectively, on chromosomes 19, 15 and 11. An approximately 1-Mbp region from the distal mouse chromosome 7 cluster and the same from the homologous human chromosome 11p15 region were sequenced and shown to contain 17 genes, of which 8 showed imprinted expression (Onyango *et al.*, 2000; Engemann *et al.*, 2000). However, it is now known that these 8 imprinted genes are regulated by at least two, and maybe more, distinct imprint control elements (Thorvaldsen *et al.*, 1998; Horike *et al.*, 2000; Mannens and Alders, 1999). Transgenic experiments indicate that clustering may have more relevance for some genes than others. For example, many genes do not maintain their imprinted status as transgenes in ectopic sites (*Igf2*, Lee *et al.*, 1993; *Cdkn1*, John *et al.*, 1999), while others require large tracts of flanking DNA or need to be present in multiple copies (*M6P-Igf2r*, Wutz *et al.*, 1997; *H19*, Pfeifer *et al.*, 1996; *Snrpn*, Blaydes *et al.*, 1999, Shemer *et al.*, 2000; *Xist*, Avner and Heard, 2000).

B. Allelic methylation

DNA methylation fits all the predicted requirements for an imprinting mechanism as outlined above, and the majority of imprinted genes do have parental allele-specific DNA methylation. DNA methylation in mammals modifies cytosine to

5-methyl cytosine in any CpG dinucleotide pair (Bestor, 2000). Other cytosine dinucleotide pairs can be methylated, but this occurs only transiently in very early embryos (Ramsahoye et al., 2000). Methylation patterns are dynamic during mouse development, with two cycles of methylation loss and gain (Oswald et al., 2000). First, methylation is lost during germ cell formation and gained as the gametes mature. Second, after fertilization, methylation is lost by the preimplantation embryo, then regained after embryonic implantation, coincident with gastrulation in the mouse. The allele-specific methylation of imprinted genes has two forms. It can be acquired in the gamete and thus be a candidate for the "imprint," or it can be acquired in the soma and thus be a candidate for maintenance of imprinted expression. For methylation to be a candidate for the "imprint," it must be acquired by one gamete only, survive the preimplantation demethylation wave, and be maintained on the same parental allele in the diploid cell after DNA replication. Gametic methylation imprints with this behavior have been identified on several genes: M6P-Igf2r (Stöger et al., 1993), H19 (Tremblay et al., 1995), RasGrf1 (Shibata et al., 1998), U2af1-rs1 (Shibata et al., 1997), Snrpn (Shemer et al., 1997). Somatic methylation imprints that are acquired after the cell is diploid, and thus cannot be the mark that distinguishes between the parental alleles, have also been shown to occur on many imprinted genes and can also be found on several genes that also carry gametic methylation imprints, e.g., M6P-Igf2r (Stöger et al., 1993). An analysis of mice deficient for the maintenance methyltransferase Dnmt1 gene, and thus deficient for global genomic methylation, has confirmed that DNA methylation plays a major role regulating imprinted expression (Li et al., 1993; see discussion below).

C. CpG islands

A comparative analysis of a 1-Mbp-imprinted region from mouse chromosome 7 and human chromosome 11 prompted the observation that imprinted genes tend to be associated with one or even two CpG islands (Paulsen et al., 2000; Onyango et al., 2000), although a recent review observed that only 88% of imprinted genes have islands, mouse Ins2 being one exception (Reik and Walter, 2000). This is in contrast to the rest of the genes in the mammalian genome, of which approximately 50% have CpG island promoters. CpG islands are defined as stretches of DNA 500–1500 bp long with a CG : GC ratio of more than 0.6, and they are normally found at promoters and contain the 5' end of the transcript (reviewed in Cross and Bird, 1995). Transcription of CpG promoters appears to be regulated by the same panoply of transcription factors as the set of CpG-poor promoters in mammals, but there is little information about their direct regulation. However, it is known that CpG island promoters normally lack TATA boxes, often have multiple transcription starts, tend to be associated with genes with "housekeeping" functions, and are normally unmethylated regardless of expression status. There are also indications that CpG islands act as replication foci in the mammalian genome (reviewed by Antequera and Bird, 1999). The absence of DNA methylation on CpG dinucleotides located inside islands is in contrast to the majority of CpG dinucleotides, which lie outside islands and are invariably methylated. Since methyl-CpG can undergo a spontaneous transition to TpG that is not efficiently repaired, this leaves the mammalian genome with a deficit of CpG dinucleotides compared with GpC dinucleotides, outside islands (Bird, 1980). If methylation is used as a gene-regulatory signal, as is the case for imprinted genes (described in Section VI.B), it makes evolutionary sense to place the important CpGs within islands. For a large number of imprinted genes (but not all), the CpG island promoters are methylated either in the gamete or in the soma (see references Section VI.B). Methylated CpG islands pose two problems. First, it is not clear how CpG dinucleotides are retained in methylated islands, since methyl-CpG/TpG transition will reduce the CpG content. The explanation may lie in the fact that imprinted CpG islands are methylated for only a relatively short time in the germ line (Brandeis et al., 1993), thus they are not exposed for a long time to the hazards of deamination. It is also possible that deamination is reduced following methylation of CpG dinucleotides in CpG islands, or that repair is more efficient. The second problem, which poses one of the major challenges in the imprinting field, is to understand why imprinted CpG islands are methylated when the majority of CpG islands that are linked to nonimprinted genes appear to be actively protected against methylation (Frank et al., 1991; Macleod et al., 1994; and see comments below).

D. Direct repeats

Several but not all imprinted genes contain or are closely linked to a region rich in direct repeats, which shows no homology to the high-copy-interspersed repeat families present in mammalian DNA. The direct repeats are generally found in or close to CpG islands, and their presence but not their sequence can be conserved between mice and humans (Neumann et al., 1995). Imprinted genes containing direct repeats include Xist (Brown et al., 1991), M6P-Igf2r (Stöger et al., 1993; Smrzka et al., 1995), PW71 (Dittrich et al., 1993), Cdkn1 (Hatada and Mukai, 1995), H19 (Neumann et al., 1995; Jinno et al., 1996), Igf2 (Sasaki et al., 1996; Moore et al., 1997), U2af1-rs1 (Shibata et al., 1997), IPW (Wevrick and Francke, 1997), Snrpn (Huq et al., 1997), RasGrf1 (Pearsall et al., 1999; Shibata et al., 1998), Magel2 (Boccaccio et al., 1999), lambdaPenII (Miura et al., 1999), and Peg3 (Li et al., 2000) The presence of direct repeats has been suggested to attract gametic methylation in a sequence-independent manner by inducing DNA to form unusual secondary structure during replication (Neumann et al., 1995). The large number of imprinted genes containing repeats supports this proposal. New results from a comparative sequence analysis of the human and mouse Impact genes also support a role for repeats in attracting methylation. The mouse *Impact* gene contains direct repeats in its CpG island promoter, and is maternally methylated and silenced. In contrast, the human *IMPACT* homologous gene lacks the direct repeats and lacks imprinted methylation and expression (Okamura *et al.*, 2000). However, there are also indications that repeats play no role in attracting the methylation imprint, since it has also been shown that a G-rich repeat element close to the *H19* imprint control element, and a central core from the *U2af1-rs1* CpG island repeat, are not relevant for imprinting (Stadnick *et al.*, 1999; Sunahara *et al.*, 2000). An alternative viewpoint is that differential methylation is achieved by specific sequences that attract or repel methylation and in support of this, a sequence-specific 113-bp element from the *M6P-Igf2r* intron 2 methylated island has been demonstrated to attract methylation in a gamete-specific manner (Birger *et al.*, 1999).

E. Asynchrony

Fluorescent *in situ* hybridization used to label chromosomes during the cell cycle at S phase has identified an apparent asynchrony of replication between the parental alleles of an imprinted gene (Kitsberg *et al.*, 1993). The region showing asynchrony can extend beyond the cluster of imprinted genes to include nonimprinted genes. The asynchrony can go in both directions (maternal early or late) but, in contrast to the behavior of nonimprinted genes, there is no correlation between late replication and silencing (Knoll *et al.*, 1994). In contrast to the fluorescent *in situ* hybridization technique, bromodeoxyuridine incorporation detects replication asynchrony only in highly expressing cells, and shows that the extent of asynchrony between imprinted parental alleles can be less than between loci along a single chromosome. It has been suggested that the asynchrony observed with both these techniques may indicate allelic differences in chromatin structure rather then DNA replication (Kawame *et al.*, 1995; Bickmore and Carothers, 1995). In support of this, increased chromatin packaging on nontranscribed alleles for *Snrpn* and *M6P-Igf2r* has been described (Watanabe *et al.*, 2000). In addition, several imprinted genes have now been shown to have differential chromatin organization over the maternal and paternal promoter of an imprinted gene (Greally *et al.*, 1999; Hark and Tilghman, 1998; reviewed in Feil and Khosla, 1999). However, it is not yet known if chromatin plays a part in acquisition of the imprint, or in the maintenance of imprinted expression, or perhaps has no significance for imprinting.

F. Noncoding RNAs

The identification of the human *XIST* gene, which is essential for imprinted and random X-chromosome inactivation, as a noncoding RNA that lacks a structural function was the first indication that RNAs may possess cis-regulatory properties in mammals (Brown *et al.*, 1991; Marahrens *et al.*, 1997; see also Chapter 3). The first identified imprinted noncoding RNA gene, named *H19*, was noted to have organizational features in common with *Xist*, such as large exons and small introns and association with direct repeats (Bartolomei *et al.*, 1991; Pfeiffer and Tilghman, 1994). Of the 50 known imprinted genes so far identified in mammals, a surprising 27% are noncoding RNAs (see Table 5.2 for references). These 14 imprinted noncoding RNAs can be roughly divided into three categories: (1) "Antisense reciprocal alleles," mostly paternally expressed noncoding RNAs that overlap the silenced allele of a maternally expressed imprinted gene. Examples are **Air**/M6P-Igf2r, **UBE3Aas**/UBE3A, **Kcnq1ot1**/Kcnq1, **Nespas**/Nesp, **Cop2as**/Cop2. The **TsiX**/Xist is an example of a maternally expressed antisense noncoding RNA overlapping the silenced maternal *Xist* allele (which is itself a noncoding RNA), but this parental-specific expression pattern is found only in placental tissues. (2) "Antisense same alleles" in which a noncoding RNA overlaps a protein-coding gene expressed from the same parental allele. The two known examples are also paternally expressed: **Igf2as**/Igf2 and **ZFP127as**/ZFP127. (3) "No overlap detected" alleles include noncoding RNAs that lie close to imprinted protein-coding gene(s) but do not overlap. The examples include two maternally expressed noncoding RNAs linked to the silent allele of a paternally expressed protein-coding gene: **H19**/Igf2 and **Gtl2**/Dlk1. In addition, several paternal-expressed noncoding RNAs have been isolated from the human Prader-Willi/Angelmans syndromic regions: **PWCR1, IPW,** and two incompletely characterized transcripts, **PAR1** and **PAR5**. Of the 14 noncoding RNAs listed in Table 5.2, only two may have a sequence-dependent function. The **Xist** transcript has been shown to be necessary and sufficient for X-chromosome and also autosome inactivation (Wutz *et al.*, 2000). The **PWCR1** RNA is homologous to SNO-guide RNAs that act to modify other RNAs; however, a function has not been genetically demonstrated (de Los Santos *et al.*, 2000).

These noncoding RNAs, particularly the overlapping antisense transcripts, are unusual in the mammalian genome and most share common features. For example, noncoding RNAs mostly show strong expression, many are known to be very long transcripts, several contain interspersed repeats, they also show large exons, and they contain few or no introns. The lack of introns in imprinted genes has been noted previously (Hurst *et al.*, 1996). An extreme example is the *Air* noncoding RNA that overlaps the *M6P-Igf2r* gene; it is 108 kb long, has no identified introns, and appears to be collinear with the genomic sequence (Lyle *et al.*, 2000). Thus, in comparison to mammalian coding genes, noncoding RNAs appear to show reduced ability to splice and reduced ability to terminate. The presence of families of interspersed repeats in noncoding RNAs argues against a sequence specific or structural function. In several cases, a deletion of the promoter of the noncoding RNA has been shown to cause reexpression of the overlapped

Table 5.2. 27% of Known Human and Mouse Imprinted "Genes" Are Noncoding RNAs[a]

Imprinted noncoding RNA	Expressed allele P	M	Overlap	Name, linked coding gene, and genetically determined function (if tested)
1 Nespas (2D)	P		Antisense reciprocal allele	Nesp (NeuroEndocrine Secretory Protein) antisense RNA; Wroe et al., 2000.
2 Copg2as Mitl RNA (6P)	P		Antisense reciprocal allele	Copg2 (COatomer ComPlex subunit Gamma 2) antisense RNA overlaps the 3' end of the upstream Mest1 gene Mitl (Mest-linked Imprinted Transcript 1 in intron 20 of Copg2). Copg2as and Mitl may belong to the same transcript. It is also not excluded that they could derive from Mest1 read-through transcripts. Lee et al., 2000.
3 Pwcr1 (7C)	P		No overlap detected	Prader-Willi Chromosome Region 1/Probable sequence specific function as a SnoRNA; de Los Santos et al., 2000.
4 Zfp127as (7C)	P		Antisense same allele	Zfp127 (Zinc-Finger Protein 127) antisense RNA; Jong et al., 1999.
5 Ipw (7C)	P		No overlap detected	Imprinted in Prader-Willi syndrome; Wevrick and Francke, 1997.
6 PAR1, PAR5 (15q)	P		No overlap detected	Five PAR (Prader-Willi Angelmans Region) RNAs have been isolated only from humans. Pari, 5 are paternally-expressed. However, it is unclear if the PAR RNAs are one or many transcripts. Sutcliffe et al., 1994.
7 UBE3A-AS (15q)	P		Antisense reciprocal allele	UBE3A (Ubiquitin E6-AP protein ligase 3A) antisense RNAs have only been found in humans in the PWS-AS region; Rouguelle et al., 1997.

8	H19 (7D)		M	No overlap detected reciprocal	H19 fetal liver mRNA (closely linked to Igf2); Bartolomei et al., 1991
9	Igf2as (7D)	P		Antisense same allele	Igf2 (insulin-like growth factor 2) antisense RNA; Moore et al., 1997.
10	Kcnqlot (7D)	P		Antisense reciprocal allele	KCNQ1 (Potassium voltage-gated channel, KQT-like subfamily, member 1) overlapping transcript1. Also known as Kvlqt. Smilinich et al., 1999.
11	Gtl2 (12D)		M	No overlap detected reciprocal	Gene Trap Locus 2 (closely linked to Dlk); Schuster-Gossler et al., 1996; Schmidt et al., 2000.
12	Air (17)	P		Antisense reciprocal allele	M6P-Igf2r (insulin-like growth factor 2 receptor) antisense RNA known as Air (Antisense Igf2r RNA); Wutz et al., 1997; Lyle et al., 2000.
13	Xist *placenta (X)	P		Sequence specific function	X chromosome Inactive Specific Transcript, * imprinted only in placenta and extra embryonic tissues. N.B. marsupial Xist is not yet identified. Reviewed by Graves, 1996; Avener and Heard, 2000; Chapter 2.
14	TsiX (X)	P	M	Antisense reciprocal allele	Overlaps and regulates Xist expression (and wins the best-name contest); Lee et al., 1999, 2000.

[a]The 14 known imprinted noncoding RNAs are listed and grouped by linkage groups indicated by the dark and light shading. Chromosome numbers and regions (P, proximal; C, central; D, distal) are listed in column 2. The lane headed "Overlap" indicates three groups. (I) Antisense reciprocal allele: noncoding RNAs that are expressed from one parental allele and overlap a silenced protein coding gene expressed on the other parental allele. (II) Antisense same allele: a noncoding RNA overlaps an active protein coding gene expressed on the same parental allele. (III) No overlap detected: noncoding RNAs that lie close to imprinted coding gene but do not overlap, and are expressed from the opposite parental allele in some cases. See text for further details. Antisense transcripts are indicated in some cases by "as" and in other cases by an independent name as indicated in the last column.

silent gene. This has been shown for the *Air* promoter (Wutz *et al.*, 1997), *TsiX* (Lee *et al.*, 2000), and *KCNQ1OT1* (Horike *et al.*, 2000). However, precise deletion of the *H19* noncoding gene and promoter has no effect on imprinted expression of the linked protein-coding *Igf2* gene in endoderm tissues, although de-repression was seen in mesoderm tissue (Schmidt *et al.*, 1999). Instead, imprinted expression of both *H19* and *Igf2* is regulated by an element lying upstream to *H19* that appears to function as an insulator that binds the CTCF insulator protein in a methylation-dependent manner (Bell and Felsenfeld, 2000; reviewed in Thorvaldsen and Bartolomei, 2000). This finding raises the possibility that noncoding RNAs are not the cause of allele-specific silencing of the linked imprinted gene, but are a consequence of this silencing and may reflect a local alteration in chromatin structure that is permissive for spurious transcription. Thus it is by no means certain that imprinted noncoding RNAs participate directly in allele silencing. The experiments cited above that demonstrate that deletion of a noncoding RNA promoter causes de-repression of a linked protein-coding imprinted gene do not distinguish between a role for the RNA itself and a role for the fragment that was deleted.

G. Homology

The discovery of the widespread existence of RNAi (RNA interference; Bass, 2000; Chapter 11) as a means to silence genes with a similar but altered sequence has raised the question of whether RNAi could be involved in gene silencing in the case of imprinted genes with overlapping noncoding transcripts. RNAi operates through a sense/antisense pairing mechanism and has been shown to exist in a large range of invertebrates and also in two vertebrates, the pregastrulation *Xenopus* embryo (Stancheva and Meehan *et al.*, 2000) and the preimplantation mammalian embryo (Svoboda *et al.*, 2000). There are several good arguments against RNAi action in imprinted allele-specific silencing: the imprinting mechanism is cis-acting, restricted to the nucleus, and can be initiated if a single locus is present in the nucleus. In contrast, RNAi is trans-acting and cytoplasmic. However, it has recently been suggested that RNA silencing can act *in cis*, to repress transgenes in plants (Mette *et al.*, 2000; see also Chapter 8). Thus, in view of the large number of noncoding RNAs linked to imprinted loci, further experiments will be needed to test fully whether RNAi or any form of RNA silencing can be excluded.

The imprinting mechanism does not require the presence of both parental chromosomes, since it occurs in parthenogenetic and androgenetic embryos (i.e., diploid embryos containing only maternal or paternal germline-derived chromosomes; McGrath and Solter, 1984; Surani *et al.*, 1984) and is maintained in uniparental disomic conditions (both copies or part of, one chromosome are derived from one parent). Imprinting also remains parental specific when the copy number of an imprinted locus is altered to one copy, e.g., in deletion syndromes in mice and humans (Beechey *et al.*, 2000) or increased to multiple copies, e.g., in imprinted transgenes (Pfeifer *et al.*, 1996). This indicates that counting or homology interactions between the two parental alleles are unlikely to play a role in imprinting. A chromosomal analysis of the imprinted human *Prader-Willi* syndrome region and a genetic analysis of the imprinted mouse Insulin (*Ins2*) gene did, however, indicate that homologous interactions occurred at these loci (LaSalle and Lalande, 1996; Paldi and Jouvenot, 1999). However, it has also been shown that imprinting in the *Prader-Willi* region and also the *M6P-Igf2r* locus occurs in the absence of obvious homologous interactions (Szabo and Mann, 1996; Nogami *et al.*, 2000). Thus, in view of the known features of the imprinting mechanism, especially its cis-acting mechanism, it is more likely that while homologous interactions are more common than expected in mammalian genes, they are not relevant for imprinting.

Not all of the epigenetic and organizational features associated with imprinted genes are unusual. Many features, such as clustering and CpG island promoters, are found in nonimprinted genes. DNA methylation and replication delay are also typical of nonimprinted genes that show tissue-specific silencing. Noncoding RNAs are more unusual, but are increasingly being found in association with biallelically expressed genes (e.g., *Beta-globin*, Ashe *et al.*, 1997; *Hox 11*, Potter and Branford, 1998; *RPS14*, Tasheva and Roufa, 1995; *bFGF*, Li *et al.*, 1996; *Hsp70*, Murashov and Wolgemuth, 1996; *Nmyc*, Krystal *et al.*, 1990). Direct repeats have not yet been linked to biallelically expressed genes, but it will not be clear if this represents a unique feature of imprinted genes until more of the completed human sequence is available for analysis.

VII. DNA METHYLATION AND GENOMIC IMPRINTS

The study of allele-specific methylation of imprinted genes has in many ways laid the foundation of our current understanding of the imprinting mechanism. In addition to its use to isolate novel imprinted genes (Plass *et al.*, 1996), the identification of regions within imprinted genes that show allele-specific methylation (now referred to as DMRs, *Differentially Methylated Regions*) has led directly to the isolation of critical imprint control elements and of elements that modify imprinted expression. Examples are the *M6P-Igf2r* intron2 DMR (Wutz *et al.*, 1997), *H19* upstream DMR (Thorvaldsen *et al.*, 1998), *Snrpn* promoter DMR (Bielinska *et al.*, 2000), *Kcnq1ot1* intron 10 DMR (Horike *et al.*, 2000), and *Igf2* DMR (Constancia *et al.*, 2000). Furthermore, DMRs have led to the identification of unsuspected imprinted noncoding RNAs transcribed within protein-coding genes (examples are *Air*, Wutz *et al.*, 1997; *Kcnq1ot1*, Smilinich *et al.*, 1999).

A genetic test of the involvement of DNA methylation in imprinting was performed in mice carrying defective alleles of the *Dnmt1* maintenance methyltransferase gene (Li *et al.*, 1993). These mice showed a 70–90% reduction in global genomic DNA methylation and died in the early postimplantation stage (between days 8.5 and 9); however, the expression of several imprinted genes could be analyzed before embryonic death ensued when embryos appeared to display a normal morphology. These experiments have been recently complemented by a conditional gene targeting strategy that removes *Dnmt1 in vitro* in differentiated mouse fibroblasts (MEFs) prepared from E13.5-day postimplantation embryos. GeneChip analysis was used to identify a large number of genes regulated by methylation (Jackson-Grusby *et al.*, 2001). In these reduced methylation conditions (methods 1 and 2 in Table 5.3), three different responses were observed for imprinted protein-coding genes: (1) genes such as *Igf2*, *M6P-Igf2r*, *Kcnq1*, *Dlk1*, *Peg3*, and *Grb10* were repressed, thus both parental alleles are now silent; (2) two genes, *Cdkn1* and *Snrpn*, were active on both parental alleles; and (3) one gene, *Mash2*, showed no change. One type of response was shown by the three noncoding RNAs so far examined: *H19*, *Gtl2*, and *Xist* all showed increased expression, and although this is assumed to result from activation of the silent allele, this has only been shown for *H19* and *Xist*. *Xist* is included in this pattern because the imprinted expression of *Xist* in placenta switches to biallelic expression. A recent report, interestingly, shows that imprinted X inactivation in placental tissues is maintained in the absence of DNA methylation despite the expression of *Xist* from both parental alleles (Sado *et al.*, 2000). Table 5.3 also shows a conflicting result for the *H19* gene between these two assay systems. *H19* has repeatedly been shown to be activated on both alleles in the *Dnmt1* null embryo in different crosses (Li *et al.*, 1993; Caspary *et al.*, 1998), but is repressed in the GeneChip analysis of E13.5 fibroblasts. The reason is not clear but may relate to the fact that these conditional *Dnmt1* fibroblasts are transformed by SV40 or have a *TP53* null phenotype (Jackson-Grusby *et al.*, 2001). A third method (Table 5.3) using demethylating agents has generated results that are broadly similar to those obtained with removal of the Dnmt1 enzyme (El Kharroubi *et al.*, 2000). However, this method also identified a larger number of imprinted genes whose expression was unchanged following treatment with the demethylating agent, including two (*Igf2* and *Snrpn*) that had showed changes in the absence of *Dnmt1*. The reasons for these differences are not yet clear.

Overall, the results in Table 5.3 show very clearly that DNA methylation is involved in the imprinting mechanism at some level. However, the pattern of results is a surprise. Referring back to the two models, A and B, on the origins of imprinting as an allele-specific gene-silencing mechanism, it would be predicted that removal of methylation imprints (or removal of methylation modifications that maintain allele silencing) would lead to expression from both parental

Table 5.3. DNA Methylation and Expression of Imprinted Genes[a]

	Imprinted locus	PC/NC	M+	M− (method 1) (ref. a)	M− (method 2) (ref. a)	M− (method 3) (ref. g)	Ref.
1	Igf2	PC	P	Repressed	Repressed	Unchanged in PG cells	a, e, g
	M6P-Igf2r	PC	M	Repressed	nd	nd	a
	Kcnq1	PC	M	Repressed	nd	nd	b
	Dlk1	PC	P	Repressed	nd	nd	c
	Peg3	PC	P	nd	Repressed	M in PG cells	e, g
	Grb10	PC	M	nd	Repressed	nd	e
2	Cdkn1	PC	M	P+ M	Increased#	P in AG cells	b, e, g
	Snrpn	PC	P	Increased#	nd	Unchanged in AG cells	d, g
3	Mash2	PC	M	M	nd	ne	b
	Grb10	PC	M	nd	nd	Unchanged in AG cells	g
	Rasgrf1	PC	P	nd	nd	Unchanged in AG cells	g
	Sgce	PC	P	nd	nd	Unchanged in AG cells	g
	Zac1	PC	P	nd	nd	Unchanged in PG cells	g
	U2af1	PC	P	nd	nd	Unchanged in AG cells	g
	Peg1/Mest1	PC	P	nd	nd	Unchanged in PG cells	g
4	H19	NC	M	P+ M	Repressed	P in AG cells	a, e, g
	Meg3/Gtl2*	NC	M	Increased#	nd	nd	c
	Xist-imprinted	NC	P	P+ M/P*	nd	nd	a, f, e
	Xist-random		MAE-R	P+ M	Increased		

[a]The effect of DNA methylation has been analyzed using two methods based on removal of the *Dnmt1* enzyme. These two methods are generally in agreement except for the *H19* gene. Method 1 uses allele-specific RNA analysis of *Dnmt1* null embryos at E8.5 days *post coitum*. Method 2 uses GeneCHIP analysis of E13.5 virus transformed or TP53 null MEFs following a conditional deletion of *Dnmt1 in vitro*. Method 3 uses demethylating agents azacytidine or deoxyazacytidine applied to E13.5 androgenetic (AG, diploid for paternal genome) or parthenogenetic (PG, diploid for maternal genome) MEFs. PC, protein coding; NC, noncoding; P, paternal expression; M, maternal expression; M+, expressed allele in the presence of wild-type genomic methylation; M−, expressed allele in the presence of 90% reduced genomic methylation; MAE-R, random monoallelic expression; nd, not determined; ne, not expressed in this cell type. a, Li *et al.*, 1993; b, Caspary *et al.*, 1998; c, Schmidt *et al.*, 2000; d, Shemer *et al.*, 1997; e, Jackson-Grusby *et al.*, 2001; f, Sado *et al.*, 2000. N.B., Tdag51, described as imprinted in reference (e), is not an imprinted gene (Frank *et al.*, 1999). P*, *Xist* expression becomes biallelic in extra embryonic tissue of in E8.5 embryos (e), but imprinted X inactivation is unchanged (f). *The parental alleles were not distinguished. #Increased expression may result from increased transcription of the expressed allele, and may not represent activation of the repressed allele. The results are grouped in four classes as described in the text: (1) protein-coding genes that are activated by DNA methylation, (2) protein-coding genes that are repressed by DNA methylation, (3) protein-coding genes that are unchanged in the reduced DNA methylation conditions, and (4) noncoding RNAs that are repressed by DNA methylation.

alleles. Five of the genes studied (two coding genes, *Cdkn1* and *Snrpn*, and three noncoding RNAs, *H19*, *Gtl2*, and *Xist*) do show this behavior, indicating that methylation suppresses the silent allele of these imprinted genes. However, the behavior of six protein-coding genes, *Igf2*, *M6P-Igf2r*, *Kcnq1*, *Dlk1*, *Peg3*, and *Grb10*, indicates that methylation is needed to maintain expression of the active allele of these imprinted genes. This manner of action of DNA methylation was unexpected, since methylation is normally regarded as a suppressive modification for gene expression (Ng and Bird, 1999; Wolffe and Guschin, 2000).

As described in Section V, it is predicted that genomic imprints are acquired during gametogenesis. Experiments have been performed that transfer early oocyte nuclei and male primordial germ cell nuclei (which are predicted to be "imprint-free") into oocytes, which are then activated and will undergo

Table 5.4. Imprinted Gene Expression from "Imprint-Free" Germ Cell Genomes[a]

Imprinted gene	PC or NC	M/P	a. Expression in diploid parthenogenote embryos derived from late full-grown (fg) oocyte nuclei (fg/fg)	a. Expression in diploid parthenogenote embryos derived from one early nongrowing (ng) oocyte and one full-grown (fg) oocyte (ng/fg)	b. Expression in diploid androgenote embryos derived from diploid male primordial germ cells (ge/gc)
1	Igf2	PC	P	−fg	−fg/−ng	−
2	M6P-Igf2r	PC	M	Not Tested	+fg/−ng	−
3	Peg3	PC	P	−fg	−fg/+ng	+
4	Cdkn1	PC	M	Not tested	+fg/−ng	−
5	Snrpn	PC	P	−fg	−fg/+ng	+
6	Mash2	PC	M	Not tested	Not tested	−*
7	Nnat	PC	P	Not tested	Not tested	+
8	Peg1/Mest	PC	P	−fg	−fg/+ng	+
9	H19	NC	M	Not tested	+fg/+ng	+

[a]This table lists the expression of imprinted genes in postimplantation embryos manipulated by nuclear transplant techniques to contain only the maternal genome (parthenogenote embryos) or only the paternal genome (androgenote embryos). A full-grown oocyte is predicted to represent a stage after imprint acquisition, while the nongrowing oocyte and male primordial germ cell nuclei are predicted to represent stages before imprint acquisition. Three paternally expressed imprinted genes (Peg1/Mest1, Peg3, Snrpn) that are expressed in parthenogenote embryos from the imprint-free "ng" genome are assumed to be repressed by an imprint on the maternal chromosome. Two maternally expressed imprinted genes (M6P-Igf2r, Cdkn1) that are expressed in parthenogenote embryos from the fg, but not the ng, genome are assumed to be activated by an imprint on the maternal chromosome. The androgenote embryos contain only a paternal genome lacking imprints. Expression of maternally expressed imprinted genes such as H19 is assumed to indicate a gene normally repressed by an imprint on the paternal chromosome. Expression of paternally expressed imprinted genes (Peg3, Peg1/Mest1, Nnat, Snrpn) is assumed to indicate genes normally repressed by an imprint on the maternal chromosome. Absence of expression (Igf2, M6P-Igf2r, Cdkn1) is assumed to indicate genes normally activated by an imprint on paternal or maternal chromosome. PC, protein coding; NC, noncoding RNA; M/P, expressed allele in normal differentiated cells. +, expression detected, +fg, expression from the full-grown oocyte genome, +ng, expression from the nongrowing oocyte genome, −fg or −ng, expression not detected. −* the cell type expressing Mash2 may be absent from this embryo type. a. Obata et al., 1998; b, Kato et al., 1999. See text for further explanation of these results.

embryonic development to an early postimplantation stage (Obata *et al.*, 1998; Kato *et al.*, 1999). The resultant embryos are predicted to contain maternal or paternal chromosomes lacking imprints. Analysis of these embryos containing maternal chromosomes (Table 5.4) indicates that some genes are activated by maternal imprints (*M6P-Igf2r* and *Cdkn1*), while some (*Peg3*, *Peg1/Mest*, and *Snrpn*) are silenced by maternal imprints that are acquired in mature full-grown oocytes. The analysis of paternal chromosomes in androgenetic embryos supports the interpretation that (*Igf2*, *M6p-Igf2r*, and *Cdkn1*) are normally activated by an imprint, while other genes (*Peg3*, *Peg1/Mest*, *Nnat*, and *Snrpn*) are predicted to be silenced by an imprint.

In comparing the two models it is expected that genes shown to be activated by imprint acquisition as measured in Table 5.4 should also be shown to be activated by DNA methylation as shown in Table 5.3. *Igf2* and *M6P-Igf2r* behave in both test systems as epigenetically activated genes, but *Cdkn1* does not. Similarly, *Snrpn* behaves as an epigenetically repressed gene in both systems, but *Peg3* does not. The reason for the differences between these two test systems is not fully clear at this moment. However, the important message is that both systems have clearly shown that imprinted genes can be either activated or repressed by epigenetic imprints.

VIII. THE FUNCTION OF METHYLATION IMPRINTS: MODEL C

Imprinted genes have drawn attention to an unexpected role for methylation in regulating gene expression. Methylation imprints as studied in the *Dnmt1*-null model are thus shown to be necessary for expression of the majority of protein-coding genes studied and necessary for repression of all the noncoding RNAs studied so far. In two cases, *Igf2* + *H19*, and *Dlk1* + *Gtl2*, the genes show a reciprocal response to the absence of methylation: the protein-coding gene is silenced, while the linked noncoding RNA is activated. These results were initially interpreted as indicating a form of expression competition between protein-coding genes and noncoding RNAs (Barlow, 1997), and while this may still be a possibility for some coding/noncoding pairs, it has been excluded for *Igf2* and *H19* (Schmidt *et al.*, 1999). However, whatever the mechanism that regulates neighboring or overlapping pairs of imprinted protein-coding and noncoding RNAs, the ability of DNA methylation to *maintain* gene expression requires an explanation that would integrate into a model that explains the origin of imprinting.

Models A and B, described above, outlined a scenario whereby imprinting arose as an allele-specific *silencing* mechanism, acting on genes that were biallelically expressed in the ancestral population. The demonstration that demethylation *silences* many imprinted genes indicates this cannot be the correct interpretation, at least for these genes. Since methylation imprints in modern mammals are

Model C: imprints de-repress epimutations

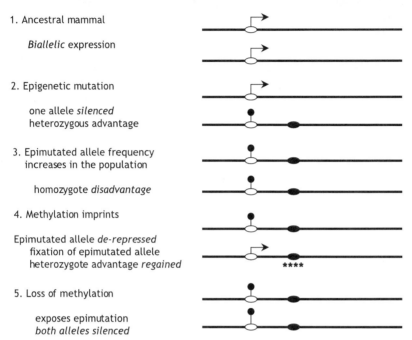

1. Ancestral mammal

   Biallelic expression

2. Epigenetic mutation

   one allele *silenced*
   heterozygous advantage

3. Epimutated allele frequency
   increases in the population

   homozygote *disadvantage*

4. Methylation imprints

   Epimutated allele *de-repressed*
   fixation of epimutated allele
   heterozygote advantage *regained*

5. Loss of methylation

   exposes epimutation
   both alleles silenced

Figure 5.1. Model C: Imprints de-repress epimutations. The steps leading from biallelic expression to parental-specific gene *activation* are displayed. The two parental chromosomes, depicted as black lines, are not distinguished in steps 1, 2, and 3. The white ellipse indicates a gene promoter, the black ellipse indicates the insertion. The direction of transcription is indicated by arrows; a black circle indicates repression. The asterisks indicate the imprint modification (for example, a methylation imprint). In steps 4 and 5, the parental chromosomes are distinguished by the imprint on one chromosome and are depicted with a black or gray line.

needed for expression of the active parental allele, it is logical to assume that this parental allele was silent before acquiring a methylation imprint. Thus imprinting, at least of these genes, must have been acquired in two steps: first the ancestral biallelically expressed gene acquired a silencing mutation, second the silencing mutation acquired a methylation imprint that suppressed its action. Model C (Figure 5.1) suggests a scenario for these two steps that proposes that imprints can act to de-repress epimutations. In model C, it is proposed that an insertion mutation arose in a single gene in a member of the ancestral mammalian population that exerted a cis-acting silencing effect. An example could be an insertion of a retrovirus or retrotransposon or DNA transposon, into a non-exon gene regulatory sequence (step 2, Figure 5.1). The sequence of the target gene mRNA is thus unchanged, but the gene is silenced as a result of the insertion. As with model A and B described in Section IV, hemizygous expression of the mutated gene is proposed to confer an immediate advantage to the organism, such as would occur if the mutated gene was a dose-sensitive growth-suppressor or growth-promoter gene. Animals with hemizygous expression of such a gene could be more competitive in terms of food and mates, because they are either larger and more competitive, or smaller but more numerous, thus allowing the spread of the mutated allele through the ancestral population. Homozygosity for the mutated silent allele is proposed to be disadvantageous but difficult to avoid as the frequency of the mutated allele increased in the population (step 3). The acquisition of an imprint, such as DNA methylation, that acts to suppress the effect of the insertion would restore expression to the mutated allele. Thus the original insertion mutation can be referred to as an "epimutation," since its action would have been to induce epigenetic gene silencing. For imprinted gene expression to occur it is necessary that the imprint is imposed on only one mutated parental allele. Thus model C requires the preexistence of a parental-specific, epigenetic modification mechanism. As described for model B, this could be gamete-specific silencing of retroviruses or transposable elements, which serves a genome defense function. However, in model C (in contrast to model B), gamete-specific modification acts to nullify the effects of the insertion (step 4) and so activate the mutated allele *in cis*. With model C, loss of the imprint can be predicted to result in silencing of both alleles (step 5).

Support for model C, that imprints can suppress the effects of epimutations, comes primarily from the finding that demethylation silences a large number of imprinted protein-coding genes. However, there are also other arguments that can be used to support this model.

1. A prediction of model C is that removal of the insertion mutation that exerts the cis-acting silencing effect (and is proposed to be marked by the DMR) would activate the silenced allele, rendering it equivalent to the active methylated allele. This has been most clearly demonstrated *in vivo* for the M6P-Igf2r gene, where it has been shown that deletion of the intron 2-located DMR restores full expression to the silent paternal allele (Wutz et al., 1997, 2001). Reactivation has also been shown for the silent paternal *KCNQ1* allele after deletion of the intron 10 DMR in a somatic cell hybrid system (Horike et al., 2000). Deletion of the *H19* upstream DMR, however, only restores full expression to *Igf2* when *H19* is also deleted (Leighton et al., 1995), deletion of the DMR leaving *H19* intact results only in partial *Igf2* activation (Thorvaldsen et al., 1998). Finally, mesoderm-specific reactivation of *Igf2* has also been shown following deletion of DMR1 (Constancia et al., 2000).

2. Indirect support for model C comes from the identification of new insertions that can induce epimutations in populations today. A good example is that of spontaneous IAP (Intracisternal A-Particle) retroposon insertions into the mouse *agouti* locus. These insertions have occurred multiple times in the upstream region and in 5′ *introns* in both sense and antisense orientations, and have induced *agouti* overexpression by acting as enhancers or promiscuous promoters. Interestingly, the extent of epigenetic regulation is affected by parental transmission. Paternal transmission correlates with increased methylation and reduced activity of the mutated *agouti* locus (Duhl *et al.*, 1994; Michaud *et al.*, 1994). A second example of an insertional epimutation is the *Mov13* mutation caused by an intronic retrovirus insertion that interferes with tissue-specific regulation of the mouse collagen *Col1a1* gene; however, in this case parental effects have not been noted (Kratochwil *et al.*, 1989).

3. In plants, there has long been abundant evidence that transposon insertions can exert regulatory effects on neighboring genes, as originally described by McClintock (1992). More interesting with regard to supporting the proposal outlined in model C, that imprints can act to de-repress epimutations, is the ability of methylation to mask the effects of transposon insertion in plants (Martienssen, 1998, and references therein). Transposons and retrotransposons can constitute from 10% (e.g., *Arabodopsis*) to 80% (e.g., in maize) of a plant's genome, and there are many descriptions of transposon insertions into essential genes. Some *Robertson's mutator* or *En/Spm* transposon insertions in maize have been described that silence target genes by suppressing transcription or elongation. However, this silencing effect can be suppressed by methylation of the transposon and the target gene is then expressed normally (reviewed in Martienssen, 1998). This action of methylation, to restore expression to a gene following transposon insertion, has clear parallels with model C. In mammals, the genome is also replete with transposons, retrotransposons, retroviruses, and their degenerate copies (for example, 40% of human chromosome 21 and 42% of chromosome 22 is composed of interspersed repeat families; Hattori *et al.*, 2000; Dunham *et al.*, 1999). It has also been shown that absence of methylation leads to a very large increase in transcription of IAP retroposons in the *Dnmt1* null mouse (Walsh *et al.*, 1998). Despite this, similar methylation suppression of transposon mutation effects have not yet been described in mammals.

4. As a final argument in support of model C, there is emerging evidence, particularly with the increasing availability of the genome sequence, that the mammalian genome has evolved a symbiotic relationship with the large numbers of transposons, retrotransposons, and retroviruses that form the repeated DNA comprising a large proportion of the genome (reviewed by Kidwell and Lisch, 1997; Smit, 1999; Luning Prak and Kazazian, 2000). There is a new appreciation that host genomes can make use of transposable elements, e.g., to generate new patterns of tissue-specific gene expression (as described for the *Mov13* mutation

above), to insert new introns, for telomere maintenance (mammalian telomerase is similar to the reverse transcriptase of non-LTR retrotransposons), to mediate V(D)J recombination in B- and T-cell receptors in mammals (the human Recombinase Activating Genes, *Rag 1* and *2*, are structurally and functionally similar to a DNA transposon), to induce syncitium formation in placenta (an *HERV* virus envelope gene is specifically expressed in placenta during syncitium formation) (Mi *et al.*, 2000; and see references in Luning Prak and Kazazian, 2000). The DMRs identified so far in imprinted genes in mammals do not have any sequence similarity to the interspersed repeat families now found in mammals. However, in view of the ability of the genome to co-opt transposable elements for its own benefit, the possibility exists that DMRs represent ancient insertions of mobile elements that have lost their original identity during genome evolution but retained their ability to exert a cis-silencing effect. And, it is this unusual form of gene regulation that has been co-opted for the benefit of the host genome.

IX. INNOCENT BYSTANDERS, FUTURE CHALLENGES

A. Innocent bystanders

Since the discovery of the first mammalian imprinted genes in 1991, a considerable amount of information has been generated concerning the function of imprinted genes and the mechanism imprinting them. This information has been obtained by treating all imprinted genes as the same entity (as in Table 5.1). However, it is possible that while many genes are imprinted, not all are equal. In order to collect relevant information about the function of imprinting and about the imprinting mechanism, we need to discriminate between genes whose imprinted function was selected during mammalian evolution and those "innocent bystander" genes who were carried along for the ride. Innocent bystander genes (Varmuza and Mann, 1994) are genes that show imprinted expression because they lie close to other imprinted genes and are in some manner susceptible. The important but unappreciated test is to examine the effect on the organism of biallelic expression of an imprinted gene. If there is no phenotypic effect, then such a gene would fit the classification of an innocent bystander. The function of innocent bystander genes, their epigenetic modification, structural organization, and features will not be informative in generating a true picture of the function and mechanism of imprinting. The future challenge here is to devise a test that will distinguish the innocent bystanders from the true participants that played a role in the evolution of imprinting in mammals. As mentioned above, this type of test has been performed for only two imprinted genes, *Igf2* (Leighton *et al.*, 1995; Ripoche *et al.*, 1997) and *M6P-Igf2r* (Wutz *et al.*, 2001), and both these genes show a phenotypic effect that is reciprocal to the loss-of-function phenotype.

B. Noncoding RNAs

The results we have so far on the imprinting mechanism, that imprints *activate* the expressed allele for many imprinted protein-coding genes studied (Table 5.3), forces us to pay closer attention to the silent and nonmethylated alleles of these imprinted genes. The silent, nonmethylated alleles of *Igf2*, *M6P-Igf2r*, *Kcnq1*, *Dlk1*, *Peg3*, and *Grb10* are silenced by an epigenetic mechanism that does not use DNA methylation. In four cases, silencing correlates with expression of a noncoding RNA *in cis*; for two of these genes, *M6P-Igf2r* and *Kcnq1*, the noncoding RNA is antisense and arises from an internal promoter, while in the two other genes, *Igf2* and *Dlk1*, the noncoding RNA is sense and approximately 100 kb downstream (Figure 5.2). The future challenge here is to test whether the noncoding RNA is the cause of silencing, or the result. These experiments are technically difficult because we do not know at this moment how to silence a noncoding RNA without changing the sequence of its promoter. And this is the experiment that we would all like to do. Thus, experiments that have deleted the promoter of a noncoding RNA and restored expression of the cis-linked protein-coding gene fail to distinguish between a role for the noncoding RNA and a role for the deleted DNA element (e.g., deletion of the Air promoter, Wutz et al., 1997, and of the *Kcnqort1* promoter, Horike et al., 2000).

A large amount of information has been obtained on the function of the *H19* noncoding RNA in regulating the imprinted expression of the upstream *Igf2* gene (reviewed in Brannan and Bartolomei, 1999; Thorvaldson and Bartolomei, 2000). One key experiment, which deleted the *H19* promoter and gene body but left imprinting of *Igf2* intact in endoderm tissue, provides a strong argument against noncoding RNAs having a role in imprinting. However, this experiment did not fully remove transcription through the deleted *H19* locus and did cause loss of *Igf2* imprinting in mesoderm (Schmidt et al., 1999). In view of these results, it was a surprise to find a second imprinted gene pair with very similar characteristics. The organization of *Dlk1* and *Gtl2* resembles that of *Igf2* and *H19*. Both protein-coding genes (*Igf2* and *Dlk1*) are paternally expressed, while both noncoding RNAs (*H19* and *Gtl2*) are maternally expressed (see references in Tables 5.1 and 5.2). *Igf2* and *Dlk1* are also both separated from their respective noncoding RNAs by approximately 100 kb. And, lastly, in each case the protein-coding and noncoding RNAs are regulated in the same manner by DNA methylation (Table 5.3). This does not represent a locus duplication, since *Igf2* and *H19* have no sequence similarity to *Dlk1* and *Gtl2*, and further work will be needed to determine if the similarities between these two gene pairs indicates any function for the noncoding Gtl2 and H19 RNAs.

The large number of imprinted noncoding RNAs (Table 5.2) have been described as "genes" and mostly treated as entities distinct from their linked protein-coding genes. At this point we do not know if this is a valid distinction, and possibly we should be hesitant to name them as genes. If these noncoding RNAs are cis-regulatory RNAs, there may be some justification in defining them as genes; however, if they are merely the result of silencing a cis-linked protein-coding gene and reflect only promiscuous expression from altered chromatin, they would not fit a definition of a gene.

C. Maternal and paternal imprints

All the models described above (A, B, and C) require the parental gametes to respond differently to an imprinted gene. This indicates that the recognition signal for maternally imprinted genes will be different from that of paternally imprinted genes. The recent identification of the *Dlk1/Gtl2* imprinted pair of protein-coding/noncoding genes and their similarity to the *Igf2/H19* pair contrasts to the five known gene pairs that show the arrangement of an imprinted protein-coding gene overlapped by an antisense noncoding RNA. These two systems may represent paternally and maternally imprinted genes. Figure 5.2A shows the key features of paternally imprinted genes represented by the *Igf2/H19* gene pair. Figure 5.2B shows the key features of maternally imprinted genes represented by the *M6P-Igf2r/Air* transcripts. *Igf2* is a paternally expressed protein-coding gene that lies approximately 100 kb upstream to the *H19* locus (Figure 5.2A). *H19* is a maternally expressed noncoding RNA and has the same transcription orientation. A methylation imprint inherited from the paternal gamete is present on a large CpG-rich element 2 kb upstream to the defined *H19* transcription start, and paternal-specific methylation of this DMR is thought to spread onto the flanking *H19* promoter during preimplantation embryonic development. The methylated paternal *H19* promoter is silent, and the unmethylated *H19* promoter on the maternal chromosome is active (reviewed in Thorvaldson *et al.*, 2000). The *Dlk1/Gtl2* gene pair has the same relative organization and *Gtl2*, similarly, has a paternal-specific methylation mark, but the details have not yet been clarified (Schmidt *et al.*, 2000; Takada *et al.*, 2000). In contrast to these gene pairs, *M6P-Igf2r* is a maternally expressed protein-coding gene that contains the promoter for the *Air* noncoding RNA in intron 2 (Figure 5.2B). The paternally expressed *Air* RNA overlaps the upstream silent *M6P-Igf2r* promoter that is 30 kb distant. A methylation imprint inherited from the maternal gamete is present on a large CpG island in intron 2 that contains the *Air* promoter. The methylated *Air* promoter is silent on the maternal chromosome and active on the paternal chromosome (Lyle *et al.*, 2000; reviewed in Sleutels *et al.*, 2000). The *Kcnq1/Kcnqot1*, *UBE3A/AS-UBE3A* and the *Nesp/as-Nesp* transcripts have a similar expression pattern, and *Kcnq1* also has a similar methylation pattern and regulation (Horike *et al.*, 2000; and references in Table 5.2). Thus, although the data are incomplete, there is a possibility that Figures 5.2A and 5.2B represent

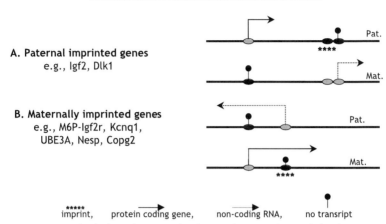

Figure 5.2. A generalized scheme for maternal and paternal imprinted genes. (A) indicates a paternal imprinted gene (the two known examples of a paternally expressed protein-coding gene linked to a maternally expressed non coding RNA are given). (B) indicates a maternal imprinted gene (the five known examples of maternally expressed protein-coding genes overlapped by paternally expressed noncoding RNAs are given). The gray ellipse indicates expressed promoters, the black ellipse indicates repressed promoters. Solid lines indicate transcription of protein-coding genes, dotted lines indicate transcription of noncoding RNAs. The black lollipop indicates repression. Asterisks indicate a methylation (or other epigenetic) imprint. Mat, maternal chromosome (gray line); Pat, paternal chromosome (black line). N.B., three other protein-coding genes, *UBE3A, Nesp,* and *Copg2,* and their respective antisense RNAs, which have been not yet been fully characterized, may also fit into this category.

two distinct parental-specific imprinting systems. Paternal imprints, as depicted in Figure 5.2A, correlate with a silent paternal noncoding RNA and an active paternal protein-coding mRNA. In contrast, maternal imprints, as depicted in Figure 5.2B, correlate with a silent maternal noncoding RNA and an active maternal protein-coding mRNA. As yet, despite the availability of sequence for the DMRs of most of these genes, no sequence homology can be found, even among the separate groups of maternally and paternally methylated DMRs. The future challenge here will be to identify the features common to maternally imprinted genes and those common to paternally imprinted genes. Another major challenge in this area is to identify the molecular pathway leading to *de novo* methylation in the gametes. The methylation enzymes so far identified, *Dnmt1, Dnmt3A, Dnmt3B* (reviewed in Bird, 1999; Bestor, 2000) have not yet been directly tested for their involvement in the acquisition of gamete-specific methylation imprints, although *Dnmt1* is known to be involved in the maintenance of the imprint (Li *et al.*, 1993).

D. Nonmethylation imprinting

The action of DNA methylation on imprinted genes shows that it is used to generate the expression difference between parental alleles, albeit in an unexpected manner. There are, however, two known cases where DNA methylation may not be responsible for the parental-specific difference. Imprinted expression of the autosomal *Mash2* gene in the embryonic placenta is not altered in *Dnmt1*-null mice deficient in global genomic methylation (Caspary *et al.*, 1998; Tanaka *et al.*, 1998). Similarly, Sado *et al.* (2000) have shown that imprinted X inactivation does not change in the placenta of *Dnmt1*-null mice, and these authors have suggested that a distinct methylation-independent imprinting mechanism may exist in extraembryonic tissues. It is worthwhile to note that imprinted X inactivation, which is the only form of chromosome dosage regulation in marsupials, also does not appear to depend on DNA methylation (reviewed in Graves, 1996). Furthermore, it has also recently been shown in marsupials that imprinted expression of the *M6P-Igf2r* gene is not linked to a differentially methylated intronic CpG island. All these results may indicate a distinct imprinting mechanism in marsupials that may be conserved in placental mammals in the placenta for X inactivation and also for some autosomal imprinted genes. The future challenge here is to isolate more marsupial homologs of imprinted genes and to validate the existence of noncoding RNAs in marsupials. The latter is of particular interest, since all attempts to isolate a marsupial *Xist* homolog have so far been unsuccessful (J. Graves, personal communication).

X. CONCLUDING REMARKS

In this review, we have drawn attention to the behavior of the large number of imprinted genes in the absence of DNA methylation, in order to consider how imprinting evolved in mammals. In doing this, we have tried to make two clear distinctions. First, we have distinguished between *how* imprints work to bring about parental-specific expression and *why* imprinted monoallelic expression was selected in the ancestral population. Although model C explains how the imprint works in terms of activating a silenced allele, this should not be confused with the function of imprinting, and we are not arguing that the function of imprinting is to activate silenced genes. The function of imprinting can only be explained by considering why parental allele-specific expression would have been advantageous to the ancestral population, when the alternative would have been biallelic expression. The second distinction is in considering exactly how imprints work to bring about imprinted expression. In this case we have attempted a distinction between the function of the methylation imprint and what the methylation mark actually does. The function of the imprint, is in many cases (as shown in Table 5.3) to activate protein-coding genes, while the methylation mark is proposed to be silencing a linked noncoding RNA. It should be noted that model C does not exclude epimutations targeting positive transcription regulatory sequences. Insertions into promoters or enhancers or boundaries could attract methylation and lead to gene silencing, but in the absence of methylation no effect on expression is seen. In this way, methylation acquired in one gamete would lead to imprinted silencing of a protein-coding gene. It should be noted that for the majority of protein-coding genes, methylation imprints are activating as defined in the *Dnmt1* null model (Table 5.3). However, it should also be noted that the assay in Table 5.4, using "imprint-free" genomes, identified a larger number of protein-coding genes that are silenced by genomic imprints.

At the beginning of this review, models A and B were suggested as ways in which imprinting could be acquired as a gene-silencing mechanism. The actual data obtained from studies of imprinted genes indicate that, for a large number of genes, imprints were acquired as a gene-activating mechanism (as proposed in model C). Our attention is now focused on the silent but un-methylated parental allele of an imprinted gene. This makes imprinting studies much more exciting because, now, hitherto hidden means of epigenetic gene silencing, which do not use DNA methylation, can be revealed. A vast array of epigenetic gene-silencing mechanisms must exist in the mammalian genome to regulate gene expression, and although a lot is known about some mecha-nisms, they likely represent only the tip of the iceberg. Imprinting will have a lot to teach us about how epigenetic gene-regulatory mechanisms function in the genome.

As a parting comment, if model C is the correct interpretation of how im-prints function, it will also throw down a challenge to the proponents/opponents of the popular parental conflict theory (Moore and Haig, 1991). This theory pro-poses that imprinting evolved in mammals because of the conflicting interests of maternal and paternal genes in relation to the transfer of nutrients from the mother to her offspring. The imprinting mechanism is proposed to manipulate embryonic growth by silencing growth-suppressing genes in the paternal germ-line and by silencing growth-promoting genes in the maternal germline. If im-prints are used to derepress epimutations, as outlined in model C, then the parental genomes act in the opposite manner: the maternal genome activates growth-suppressing genes, while the paternal genome activates growth-promoting genes. Although the end result would be the same—maternally expressed growth-suppressing genes and paternally expressed growth-promoting genes—the biolog-ical significance of this may need reappraisal.

Table 5.5. Abbreviations Used in the Text[a]

Air	Antisense Igf2r RNA
as	antisense
bFGF	Basic fibroblast growth factor
Col1a1	Collagen type 1, alpha I chain.
Cop2g	Coatomer protein complex, subunit gamma 2
Cdkn1c	Cyclin-dependent kinase inhibitor 1C (also called p57kip2)
CTCF	CCCTC nucleotide binding factor
Dlk1	Delta-like 1
Dnmt	DNA methyltransferase
DMR	Differentially methylated region
HERV	Human endogenous retro virus
Hsp70	Heat shock protein 70
Igf2	Insulin-like growth factor type 2
Ins2	Insulin II (mice have two insulin genes, humans only one)
GOI	Gain of Imprinting
Grb10	Growth factor receptor-bound protein 10
Gtl2	Gene trap locus 2
H19	a cDNA clone isolated from a fetal Hepatic library
ICE	Imprint Control Element
IPW	Imprinted in Prader-Willi region
KCNQ1	Potassium voltage-gated channel, KQT-like subfamily, member 1
KCNQ1OT1	Kcnq1 overlapping transcript 1
MAGEL2	MAGE-like 2
Mash2	Mus musculus Achaete-Scute homologue 2
Mest	Mesoderm specific transcript
M6P-Ig/2r	Cation-independent Mannose-6-phosphate receptor, also known as Insulin-like growth factor type 2 receptor
Mov13	Molony Virus integration 13
Nesp	Neuroendocrine secretory protein
Nmyc	Neuroblastoma myc (avian myelocytomatosis related)-related oncogene
Par1,5	Prader-Willi/Angelman region 1 and 5
Peg1	Paternally Expressed Gene 1
Peg3	Paternally Expressed Gene 3
PW71	Prader-Willi 71
PWCR1	Prader-Willi critical region 1
RasGrf1	Ras protein-specific guanine nucleotide-releasing factor 1
RNAi	RNA interference
RPS14	Ribosomal protein S14
SNRPN	Small nuclear ribonucleoprotein polypeptide N
TP53	Tumor protein p53
Tsix	X-inactivation-specific transcript-antisense
U2afl-rs1	U2 small nuclear ribonucleoprotein auxiliary factor-related sequence 1
UBE3A	Ubiquitin protein ligase E3A
Xist	X inactive specific transcript
ZFP127	Zinc finger protein 127 also known as ZN127 (now called Makorin 3)

[a]Note that gene loci follow the accepted nomenclature rules and are written in capitals to refer to human gene loci, and in lower case with an initial capital letter, for mouse loci.

Acknowledgments

We thank Andras Paldi, Jenny Graves, Wolf Reik, and Carmen Sapienza for answering questions during the writing of this review; Karl Pfeifer and Hamish Spencer for sending copies of their reviews; and Wolf Reik, Marjori Matzke, Günther Kreil, and Vic Small, for comments on the manuscript. D.P.B. is supported by the Austrian Academy of Sciences, F.S. by the Dutch Cancer Society.

References

Antequera, F., and Bird, A. (1999). CpG islands as genomic footprints of promoters that are associated with replication origins. *Curr. Biol.* **9,** 661–667.

Ashe, H. L., Monks, J., Wijgerde, M., Fraser, P., and Proudfoot, N. J. (1997). Intergenic transcription and transinduction of the human beta-globin locus. *Genes Dev.* **11,** 2494–2509.

Avner, P., and Heard, E. (2000). X-chromosome inactivation: counting, choice and initiation. *Nature Rev. Genet.* **2,** 59–68.

Barlow, D. P. (1993). Methylation and imprinting: from host defense to gene regulation? *Science* **260,** 309–310.

Barlow, D. P. (1997). Competition—A common motif for the imprinting mechanism? *EMBO J.* **16,** 6899–6905.

Bartolomei, M. S., and Tilghman, S. M. (1997). Genomic imprinting in mammals. *Annu. Rev. Genet.* **31,** 493–525.

Bartolomei, M. S., Zemel, S., and Tilghman, S. M. (1991). Parental imprinting of the mouse H19 gene. *Nature* **351,** 153–155.

Bass, B. L. (2000). Double-stranded RNA as a template for gene silencing. *Cell* **101,** 235–238.

Beechey, C. V., Cattanach, B. M., and Selley, R. L. (2000). MRC Mammalian Genetics Unit, Harwell, Oxfordshire. World Wide Web Site: Mouse Imprinting Data and References (URL: http://www.mgu.har.mrc.ac.uk/imprinting/implink.html).

Bell, A. C., and Felsenfeld, G. (2000). Methylation of a CTCF-dependent boundary controls imprinted expression of the Igf2 gene. *Nature* **405,** 482–485.

Bestor, T. H. (2000). The DNA methyltransferases of mammals. *Hum. Mol. Genet.* **9,** 2395–2402.

Bielinska, B., Blaydes, S. M., Buiting, K., Yang, T., Krajewska-Walasek, M., Horsthemke, B., and Brannan, C. I. (2000). De novo deletions of *SNRPN* exon 1 in early human and mouse embryos result in a paternal to maternal imprint switch. *Nature Genet.* **25,** 74–78.

Bickmore, W. A., and Carothers, A. D. (1995). Factors affecting the timing and imprinting of replication on a mammalian chromosome. *J. Cell Sci.* **108,** 2801–2809.

Bird, A. P. (1980). DNA methylation and the frequency of CpG in animal DNA. *Nucleic Acids Res.* **8,** 1499–1504.

Bird, A. P. (1999). DNA methylation de novo. *Science* **286,** 2287–2288.

Birger, Y., Shemer, R., Perk, J., and Razin, A. (1999). The imprinting box of the mouse *Igf2r* gene. *Nature* **397,** 84–88.

Blaydes, S. M., Elmore, M., Yang, T., and Brannan, C. I. (1999). Analysis of murine *Snrpn* and human *SNRPN* gene imprinting in transgenic mice. *Mamm. Genome* **10,** 549–555.

Boccaccio, I., Glatt-Deeley, H., Watrin, F., Roeckel, N., Lalande, M., and Muscatelli, F. (1999). The human *MAGEL2* gene and its mouse homologue are paternally expressed and mapped to the Prader-Willi region. *Hum. Mol. Genet* **8,** 2497–2505.

Brambilla, R., Gnesutta, N., Minichiello, L., White, G., Roylance, A. J., Herron, C. E., Ramsey, M., Wolfer, D. P., Cestari, V., Rossi-Arnaud, C., Grant, S. G., Chapman, P. F., Lipp, H. P., Sturani, E., and Klein, R. (1997). A role for the Ras signalling pathway in synaptic transmission and long-term memory. *Nature* **390,** 281–286.

Brandeis, M., Kafri, T., Ariel, M., Chaillet, J. R., McCarrey, J., Razin, A., and Cedar, H. (1993). The ontogeny of allele-specific methylation associated with imprinted genes in the mouse. *EMBO J.* **12,** 3669–3677.

Brannan, C. I., and Bartolomei, M. S. (1999). Mechanisms of genomic imprinting. *Curr. Opin. Genet. Dev.* **9,** 164–170.

Brown, C. J., Ballabio, A., Rupert, J. L., Lafreniere, R. G., Grompe, M., Tonlorenzi, R., and Willard, H. F. (1991). A gene from the region of the human X inactivation centre is expressed exclusively from the inactive X chromosome. *Nature* **349,** 38–44.

Bunzel, R., Blumcke, I., Cichon, S., Normann, S., Schramm, J., Propping, P., and Nothen, M. M. (1998). Polymorphic imprinting of the *serotonin-2A (5-HT2A) receptor* gene in human adult brain. *Brain Res. Mol. Brain Res.* **59,** 90–92.

Caspary, T., Cleary, M. A., Baker, C. C., Guan, X. J., and Tilghman, S. M. (1998). Multiple mechanisms regulate imprinting of the mouse distal chromosome 7 gene cluster. *Mol. Cell Biol.* **18,** 3466–3474.

Caspary, T., Cleary, M. A., Perlman, E. J., Zhang, P., Elledge, S. J., and Tilghman, S. M. (1999). Oppositely imprinted genes *p57*(Kip2) and *Igf2* interact in a mouse model for Beckwith-Wiedemann syndrome. *Genes Dev.* **13,** 3115–3124.

Claverie, J.-M. (2000). What if there are only 30,000 hman genes? *Science* **16,** 1255–1256.

Cole, C. J., and Townsend, C. R. (1990). Parthenogenetic lizards as vertebrate systems. *J. Exp. Zool. Suppl.* **4,** 174–176.

Conlon, I., and Raff, M. (1999). Size control in animal development. *Cell* **96,** 235–244.

Constancia, M., Dean, W., Lopes, S., Moore, T., Kelsey, G., and Reik, W. (2000). Deletion of a silencer element in *Igf2* results in loss of imprinting independent of *H19*. *Nature Genet.* **26,** 203–206.

Corley-Smith, G. E., Lim, C. J., and Brandhorst, B. P. (1996). Production of androgenetic zebrafish (*Danio rerio*). *Genetics* **142,** 1265–1276.

Cross, S. H., and Bird, A. P. (1995). CpG islands and genes. *Curr. Opin. Genet. Dev.* **5,** 309–314.

Cuellar, O. (1997). Animal parthenogenesis. *Science* **197,** 837–843.

DeChiara, T. M., Efstratiadis, A., and Robertson, E. J. (1990). A growth-deficiency phenotype in heterozygous mice carrying an insulin-like growth factor II gene disrupted by targeting. *Nature* **345,** 78–80.

de Los Santos, T., Schweizer, J., Rees, C. A., and Francke, U. (2000). Small evolutionarily conserved RNA, resembling C/D box small nucleolar RNA, is transcribed from PWCR1, a novel imprinted gene in the Prader-Willi deletion region, which is highly expressed in brain. *Am. J. Hum. Genet.* **67,** 1067–1082.

Dittrich, B., Buiting, K., Gross, S., and Horsthemke, B. (1993). Characterization of a methylation imprint in the Prader-Willi syndrome chromosome region. *Hum. Mol. Genet.* **2,** 1995–1999.

Duhl, D. M., Vrieling, H., Miller, K. A., Wolff, G. L., and Barsh, G. S. (1994). Neomorphic agouti mutations in obese yellow mice. *Nature Genet.* **8,** 59–65.

Dunham, I., Shimizu, N., Roe, B. A., Chissoe, S., Hunt, A. R., and Collins, J. E. *et al.* (1999). The DNA sequence of human chromosome 22. *Nature* **402,** 489–495.

Duvillie, B., Cordonnier, N., Deltour, L., Dandoy-Dron, F., Itier, J. M., Monthioux, E., Jami, J., Joshi, R. L., and Bucchini, D. (1997). Phenotypic alterations in insulin-deficient mutant mice. *Proc. Natl. Acad. Sci. USA* **94,** 5137–5140.

El Kharroubi, A., Piras, G., and Stewart, C. L. (2000). DNA demethylation reactivates a subset of imprinted genes in uniparental mouse embryonic fibroblasts. *J. Biol. Chem.* Dec. 21 [epub ahead of print].

Engemann, S., Strodicke, M., Paulsen, M., Franck, O., Reinhardt, R., Lane, N., Reik, W., and Walter, J. (2000). Sequence and functional comparison in the Beckwith-Wiedemann region: Implications for a novel imprinting centre and extended imprinting. *Hum. Mol. Genet.* **9,** 2691–2706.

Feil, R., and Khosla, S. (1999). Genomic imprinting in mammals: an interplay between chromatin and DNA methylation? *Trends Genet.* **15,** 431–435.

Ferguson-Smith, A. C., Sasaki, H., Cattanach, B. M., and Surani, M. A. (1993). Parental-origin-specific epigenetic modification of the mouse H19 gene. *Nature* **362,** 751–755.

Ferrier, D. E. K., and Holland, P. W. (2000). Ancient origin of the HOX gene cluster. *Nature Rev. Genet.* **2,** 33–38.

Frank, D., Keshet, I., Shani, M., Levine, A., Razin, A., and Cedar, H. (1991). Demethylation of CpG islands in embryonic cells. *Nature* **351,** 239–241.

Frank, D., Mendelsohn, C. L., Ciccone, E., Svensson, K., Ohlsson, R., and Tycko, B. (1999). A novel pleckstrin homology-related gene family defined by Ipl/Tssc3, TDAG51, and Tih1: Tissue-specific expression, chromosomal location, and parental imprinting. *Mamm. Genome* **10,** 1150–1159.

Gerard, M., Hernandez, L., Wevrick, R., and Stewart, C. L. (1999). Disruption of the mouse *Necdin* gene results in early post-natal lethality. *Nature Genet.* **23,** 199–202.

Giannoukakis, N., Deal, C., Paquette, J., Kukuvitis, A., and Polychronakos, C. (1996). Polymorphic functional imprinting of the human *IGF2* gene among individuals, in blood cells, is associated with H19 expression. *Biochem. Biophys. Res. Commun.* **220,** 1014–1019.

Graves, J. A. (1996). Mammals that break the rules: Genetics of marsupials and monotremes. *Annu. Rev. Genet.* **30,** 233–260.

Greally, J. M., Gray, T. A., Gabriel, J. M., Song, L., Zemel, S., and Nicholls, R. D. (1999). Conserved characteristics of heterochromatin-forming DNA at the 15q11-q13 imprinting center. *Proc. Natl. Acad. Sci. USA* **96,** 14430–14435.

Guillemot, F., Caspary, T., Tilghman, S. M., Copeland, N. G., Gilbert, D. J., Jenkins, N. A., Anderson, D. J., Joyner, A. L., Rossant, J., and Nagy, A. (1995). Genomic imprinting of *Mash2,* a mouse gene required for trophoblast development. *Nature Genet.* **9,** 235–242.

Haig, D. (1994). Refusing the ovarian time bomb. *Trends Genet.* **10,** 346–347.

Hark, A. T., and Tilghman, S. M. (1998). Chromatin conformation of the H19 epigenetic mark. *Hum. Mol. Genet.* **7,** 1979–1985.

Hatada, I., and Mukai, T. (1995). Genomic imprinting of CDKN1, a cyclin-dependent kinase inhibitor, in mouse. *Nature Genet.* **11,** 204–206.

Hattori, M., Fujiyama, A., Taylor, T. D., Watanabe, H., Yada, T., Park, H. S., Toyoda, A., Ishii, K., Totoki, Y., Choi, D. K., Soeda, E., Ohki, M., Takagi, T., Sakaki, Y., Taudien, S., Blechschmidt, K., Polley, A., Menzel, U., Delabar, J., Kumpf, K., Lehmann, R., Patterson, D., Reichwald, K., Rump, A., Schillhabel, M., and Schudy, A. (2000). The DNA sequence of human chromosome 21. The chromosome 21 mapping and sequencing consortium. *Nature* **18,** 311–319.

Herrick, G., and Seger, J. (1999). Imprinting and paternal genome elimination in insects. *Results Probl. Cell Differ.* **25,** 41–71.

Horike, S., Mitsuya, K., Meguro, M., Kotobuke, N., Kashwagi, A., Notsu, T., Schulz, T. C., Shirayoshi, Y., and Oshimura, M. (2000). Targeted disruption of the human *Lit1* locus defines a putative imprinting control element playing an essential role in Beckwith-Weidemann syndrome. *Hum. Mol. Genet.* **9,** 2075–2083.

Hu, J. F., Oruganti, H., Vu, T. H., and Hoffman, A. R. (1998). Tissue-specific imprinting of the mouse insulin-like growth factor II receptor gene correlates with differential allele-specific DNA methylation. *Mol. Endocrinol.* **12,** 220–232.

Huq, A. H., Sutcliffe, J. S., Nakao, M., Shen, Y., Gibbs, R. A., and Beaudet, A. L. (1997). Sequencing and functional analysis of the SNRPN promoter: In vitro methylation abolishes promoter activity *Genome Res.* **7,** 642–648.

Hurst, L. D., and McVean, G. T. (1998). Do we understand the evolution of genomic imprinting? *Curr. Opin. Genet. Dev.* **8,** 701–708.

Hurst, L. D., McVean, G., and Moore, T. (1996). Imprinted genes have few and small introns. *Nature Genet.* **12,** 234–237.

Itier, J. M., Tremp, G. L., Leonard, J. F., Multon, M. C., Ret, G., Schweighoffer, F., Tocque, B., Bluet-Pajot, M. T., Cormier, V., and Dautry, F. (1998). Imprinted gene in postnatal growth role. *Nature* **393,** 125–126.

Jackson-Grusby, L., Beard, C., Possemato, R., Tudor, M., Fambrough, D., Csankovszki, G., Dausman, J., Lee, P., Wilson, C., Lander, E., and Jaenisch, R. (2001). Loss of genomic methylation causes p53-dependent apoptosis and epigenetic deregulation. *Nat Genet.* **27**, 31–39.

Jiang, Y. H., Armstrong, D., Albrecht, U., Atkins, C. M., Noebels, J. L., Eichele, G., Sweatt, J. D., and Beaudet, A. L. (1998). Mutation of the Angelman ubiquitin ligase in mice causes increased cytoplasmic p53 and deficits of contextual learning and long-term potentiation. *Neuron.* **21**, 799–811.

Jinno, Y., Sengoku, K., Nakao, M., Tamate, K., Miyamoto, T., Matsuzaka, T., Sutcliffe, J. S., Anan, T., Takuma, N., Nishiwaki, K., Ikeda, Y., Ishimaru, T., Ishikawa, M., and Niikawa, N. (1996). Mouse/human sequence divergence in a region with a paternal-specific methylation imprint at the human *H19* locus. *Hum. Mol. Genet.* **5**, 1155–1161.

John, R. M., Hodges, M., Little, P., Barton, S. C., and Surani, M. A. (1999). A human p57(KIP2) transgene is not activated by passage through the maternal mouse germline. *Hum. Mol. Genet.* **8**, 2211–2219.

Jong, M. T., Gray, T. A., Ji, Y., Glenn, C. C., Saitoh, S., Driscoll, D. J., and Nicholls, R. D. (1999). A novel imprinted gene, encoding a RING zinc-finger protein, and overlapping antisense transcript in the Prader-Willi syndrome critical region. *Hum. Mol. Genet.* **8**, 783–793.

Jones, B. K., Levorse, J. M., and Tilghman, S. M. (1998). *Igf2* imprinting does not require its own DNA methylation or *H19* RNA. *Genes Dev.* **12**, 2200–2207.

Kalscheuer, V. M., Mariman, E. C., Schepens, M. T., Rehder, H., and Ropers, H. H. (1993). The insulin-like growth factor type-2 receptor gene is imprinted in the mouse but not in humans. *Nature Genet.* **5**, 74–78.

Kato, Y., Rideout, W. M., Hilton, K., Barton, S. C., Tsunoda, Y., and Surani, M. A. (1999). Developmental potential of mouse primordial germ cells. *Development* **126**, 1823–1832.

Kawame, H., Gartler, S. M., and Hansen, R. S. (1995). Allele-specific replication timing in imprinted domains: Absence of asynchrony at several loci. *Hum. Mol. Genet.* **4**, 2287–2293.

Kidwell, M. G., and Lisch, D. (1997). Transposable elements as sources of variation in animals and plants. *Proc. Natl. Acad. Sci. USA.* **94**, 7704–11.

Killian, J. K., Byrd, J. C., Jirtle, J. V., Munday, B. L., Stoskopf, M. K., MacDonald, R. G., and Jirtle, R. L. (2000). M6P/IGF2R imprinting evolution in mammals. *Mol. Cell* **5**, 707–716.

Kitsberg, D., Selig, S., Brandeis, M., Simon, I., Keshet, I., Driscoll, D. J., Nicholls, R. D., and Cedar, H. (1993). Allele-specific replication timing of imprinted gene regions. *Nature* **364**, 459–463.

Knoll, J. H., Cheng, S. D., and Lalande, M. (1994). Allele specificity of DNA replication timing in the Angelman/Prader-Willi syndrome imprinted chromosomal region. *Nature Genet.* **6**, 41–46.

Kratochwil, K., von der Mark, K., Kollar, E. J., Jaenisch, R., Mooslehner, K., Schwarz, M., Haase, K., Gmachl, I., and Harbers, K. (1989). Retrovirus-induced insertional mutation in Mov13 mice affects collagen I expression in a tissue-specific manner. *Cell* **57**, 807–816.

Krystal, G. W., Armstrong, B. C., and Battey, J. F. (1990). *N-myc* mRNA forms an RNA-RNA duplex with endogenous antisense transcripts. *Mol. Cell Biol.* **10**, 4180–4191.

LaSalle, J. M., and Lalande, M. (1996). Homologous association of oppositely imprinted chromosomal domains. *Science* **272**, 725–728.

Lee, J. E., Tantravahi, U., Boyle, A. L., and Efstratiadis, A. (1993). Parental imprinting of an *Igf-2* transgene. *Mol. Reprod. Dev.* **35**, 382–390.

Lee, J. T. (2000). Disruption of imprinted X inactivation by parent-of-origin effects at *Tsix*. *Cell* **103**, 17–27.

Lee, J. T., Davidow, L. S., and Warshawsky, D. (1999). *Tsix*, a gene antisense to *Xist* at the X-inactivation centre. *Nature Genet.* **21**, 400–404.

Lee, Y. J., Park, C. W., Hahn, Y., Park, J., Lee, J., Yun, J. H., Hyun, B., and Chung, J. (2000). H. *Mit1/Lb9* and *Copg2*, new members of mouse imprinted genes closely linked to Peg1/Mest(1). *FEBS Lett.* **472**, 230–234.

Lefebvre, L., Viville, S., Barton, S. C., Ishino, F., Keverne, E. B., and Surani, M. A. (1998). Abnormal maternal behavior and growth retardation associated with loss of the imprinted gene *Mest*. *Nature Genet.* **20,** 163–169.

Leighton, P. A., Ingram, R. S., Eggenschwiler, J., Efstratiadis, A., and Tilghman, S. M. (1995). Disruption of imprinting caused by deletion of the *H19* gene region in mice. *Nature* **375,** 34–39.

Lerchner, W., and Barlow, D. P. (1997). Paternal repression of the imprinted mouse M6P-*Igf2r* locus occurs during implantation and is stable in all tissues of the post-implantation mouse embryo. *Mech. Dev.* **61,** 141–149.

Li, A. W., Seyoum, G., Shiu, R. P., and Murphy, P. R. (1996). Expression of the rat *bFGF* antisense RNA transcript is tissue-specific and developmentally regulated. *Mol. Cell Endocrinol.* **118,** 113–123.

Li, E., Beard, C., and Jaenisch, R. (1993). Role for DNA methylation in genomic imprinting. *Nature* **366,** 362–365.

Li, L., Keverne, E. B., Aparicio, S. A., Ishino, F., Barton, S. C., and Surani, M. A. (1999). Regulation of maternal behavior and offspring growth by paternally expressed *Peg3*. *Science* **284,** 330–333.

Li, L. L., Szeto, I. Y., Cattanach, B. M., Ishino, F., and Surani, M. A. (2000). Organization and parent-of-origin-specific methylation of imprinted *Peg3* gene on mouse proximal chromosome 7. *Genomics* **63,** 333–340.

Lyle, R., Watanabe, D., te Vrucht, D., Lerchner, W., Smrzka, O. W., Wutz, A., Schageman, J., Hahner, L., Davies, C., and Barlow, D. P. (2000). The imprinted antisense-M6P-Igf2r-RNA overlaps but does not imprint the flanking *Mas* gene. *Nature Genet.* **25,** 19–21.

Luning Prak, E. T., and Kazazian, H. H. (2000). Mobile elements and the human genome. *Nature Rev. Genet.* **1,** 134–144.

Macleod, D., Charlton, J., Mullins, J., and Bird, A. P. (1994). Sp1 sites in the mouse *aprt* gene promoter are required to prevent methylation of the CpG island. *Genes Dev.* **8,** 2282–2292.

Mannens, M., and Alders, M. (1999). Genomic imprinting: Concept and clinical consequences. *Ann. Med.* **31,** 4–11.

Marahrens, Y., Panning, B., Dausman, J., Strauss, W., and Jaenisch., R. (1997). *Xist*-deficient mice are defective in dosage compensation but not spermatogenesis. *Genes Dev.* **11,** 156–166.

Martienssen, R. (1998). Transposons, DNA methylation and gene control. *Trends Genet.* **14,** 263–264.

Matsuoka, S., Thompson, J. S., Edwards, M. C., Bartletta, J. M., Grundy, P., Kalikin, L. M., Harper, J. W., Elledge, S. J., and Feinberg, A. P. (1996). Imprinting of the gene encoding a human cyclin-dependent kinase inhibitor, CDKN1, on chromosome 11p15. *Proc. Natl. Acad. Sci. USA* **93,** 3026–3030.

McClintock, B. (1992). Nobel Prize Lecture; The significance of responses of the genome to challenge. *In* "The Dynamic Genome, Barbara McClintock's Ideas in the Century of Genetics (N. Fedoroff and D. Botstein, eds.), pp. 361–380. Cold Spring Harbor Press, Cold Spring Harbor, NY.

McGrath, J., and Solter, D. (1984). Completion of mouse embryogenesis requires both the maternal and paternal genomes. *Cell* **37,** 179–183.

Messing, J., and Grossniklaus, U. (1999). Genomic imprinting in plants. *Results Probl. Cell Differ.* **25,** 23–40.

Mette, M. F., Aufsatz, W., van Der Winden, J., Matzke, M. A., and Matzke, A. J. (2000). Transcriptional silencing and promoter methylation triggered by double-stranded RNA. *EMBO J.* **19,** 5194–201.

Mi, S., Lee, X., Li, X. P., Veldman, G. M., Finnerty, H., Racie, L., LaVallie, E., Tang, X. Y., Edourd, P., Howes, S., Keith, J. C., and McCoy, J. M. (2000). Syncitin is a captive retroviral envelope protein involved in placental morphogenesis. *Nature* **403,** 785–789.

Michaud, E. J., van Vugt, M. J., Bultman, S. J., Sweet, H. O., Davisson, M. T., and Woychik, R. P. (1994). Differential expression of a new dominant agouti allele (*Aiapy*) is correlated with methylation state and is influenced by parental lineage. *Genes Dev.* **8,** 1463–1472.

Miura, K., Miyoshi, O., Yun, K., Inazawa, J., Miyamoto, T., Hayashi, H., Masuzaki, H., Yoshimura, S., Niikawa, N., Jinno, Y., and Ishimaru, T. (1999). Repeat-directed isolation of a novel gene preferentially expressed from the maternal allele in human placenta. *J. Hum. Genet.* **44**, 1–9.

Moore, T., Constancia, M., Zubair, M., Bailleul, B., Feil, R., Sasaki, H., and Reik, W. (1997). Multiple imprinted sense and antisense transcripts, differential methylation and tandem repeats in a putative imprinting control region upstream of mouse *Igf2*. *Proc. Natl. Acad. Sci. USA* **94**, 12509–12514.

Moore, T., and Haig, D. (1991). Genomic imprinting in mammalian development: A parental tug-of-war. *Trends Genet.* **7**, 45–49.

Morison, I., Paton, C., and Cleverley, S. D. (2001). The imprinted gene and parent-of-origin effect database. *Nucleic Acids Res.* **29**, 275–276.

Munn, D. H., Zhou, M., Attwood, J. T., Bondarev, I., Conway, S. J., Marshall, B., Brown, C., and Mellor, A. L. (1998). Prevention of allogenic fetal rejection by tryptophan catabolism. *Science* **281**, 1191–1193.

Murashov, A. K., and Wolgemuth, D. J. (1996). Sense and antisense transcripts of the developmentally regulated murine *hsp70.2* gene are expressed in distinct and only partially overlapping areas in the adult brain. *Brain Res. Mol. Brain Res.* **37**, 85–95.

Neumann, B., Kubicka, P., and Barlow, D. P. (1995). Characteristics of imprinted genes. *Nature Genet.* **9**, 12–13.

Ng, H. H., and Bird, A. (1999). DNA methylation and chromatin modification. *Curr. Opin. Genet. Dev.* **9**, 158–163.

Nishiwaki, K., Niikawa, N., and Ishikawa, M. (1997). Polymorphic and tissue-specific imprinting of the human *Wilms tumor* gene, *WT1*. *Jpn. J. Hum. Genet* **42**, 205–211.

Nogami, M., Kohda, A., Taguchi, H., Nakao, M., Ikemura, T., and Okumura, K. (2000). Relative locations of the centromere and imprinted *SNRPN* gene within chromosome 15 territories during the cell cycle in HL60 cells. *J. Cell Sci.* **113**, 2157–2165.

Nowak, R., Porter, R. H., Levy, F., Orgeur, P., and Schaal, B. (2000). Role of mother–young interactions in the survival of offspring in domestic mammals. *Rev. Reprod.* **5**, 153–163.

Obata, Y., Kaneko-Ishino, T., Koide, T., Takai, Y., Ueda, T., Domeki, I., Shiroishi, T., Ishino, F., and Kono, T. (1998). Disruption of primary imprinting during oocyte growth leads to the modified expression of imprinted genes during embryogenesis. *Development* **25**, 1553–1560.

Ogawa, O., McNoe, L. A., Eccles, M. R., Morison, I. M., and Reeve, A. E. (1993). Human insulin-like growth factor type I and type II receptors are not imprinted. *Hum. Mol. Genet.* **2**, 2163–2165.

Okamura, K., Hagiwara-Takeuchi, Y., Li, T., Vu, T. H., Hirai, M., Hattori, M., Sakaki, Y., Hoffman, A. R., and Ito, T. (2000). Comparative genome analysis of the mouse imprinted gene *impact* and its nonimprinted human homolog *IMPACT*: Toward the structural basis for species-specific imprinting. *Genome Res.* **10**, 1878–1889.

O'Neill, M. J., Ingram, R. S., Vrana, P. B., and Tilghman, S. M. (2000). Allelic expression of IGF2 in marsupials and birds. *Dev. Genes Evol.* **210**, 18–20.

Onyango, P., Miller, W., Lehoczky, J., Leung, C. T., Birren, B., Wheelan, S., Dewar, K., and Feinberg, A. P. (2000). Sequence and comparative analysis of the mouse 1-megabase region orthologous to the human 11p15 imprinted domain. *Genome Res.* **10**, 1697–1710.

Oswald, J., Engemann, S., Lane, N., Mayer, W., Olek, A., Fundele, R., Dean, W., Reik, W., and Walter, J. (2000). Active demethylation of the paternal genome in the mouse zygote. *Curr. Biol.* **10**, 475–478.

Paldi, A., and Jouvenot, Y. (1999). Allelic trans-sensing and imprinting. *Results Probl. Cell Differ.* **25**: 271–282.

Pardo-Manuel de Villena, F., de la Casa-Esperon, E., and Sapienza, C. (2000). Natural selection and the function of genome imprinting: Beyond the silenced minority. *Trends Genet.* **16**, 573–579.

Paulsen, M., El-Maarri, O., Engemann, S., Strodicke, M., Franck, O., Davies, K., Reinhardt, R., Reik, W., and Walter, J. (2000). Sequence conservation and variability of imprinting in the Beckwith-Wiedemann syndrome gene cluster in human and mouse. *Hum. Mol. Genet* **22**, 1829–1841.

Pearsall, R. S., Plass, C., Romano, M. A., Garrick, M. D., Shibata, H., Hayashizaki, Y., and Held, W. A. (1999). A direct repeat sequence at the *Rasgrf1* locus and imprinted expression. *Genomics* **55**, 194–201.

Peters, J., Wroe, S. F., Wells, C. A., Miller, H. J., Bodle, D., Beechey, C. V., Williamson, C. M., and Kelsey, G. (1999). A cluster of oppositely imprinted transcripts at the *Gnas* locus in the distal imprinting region of mouse chromosome 2. *Proc. Natl. Acad. Sci. USA* **96**, 3830–3835.

Pfeifer, K., Leighton, P. A., and Tilghman, S. M. (1996). The structural *H19* gene is required for transgene imprinting. *Proc. Natl. Acad. Sci. USA* **93**, 13876–13883.

Pfeifer, K., and Tilghman, S. M. (1994). Allele-specific gene expression in mammals: The curious case of the imprinted RNAs. *Genes Dev.* **8**, 1867–1874.

Plass, C., Shibata, H., Kalcheva, I., Mullins, L., Kotelevtseva, N., Mullins, J., Kato, R., Sasaki, H., Hirotsune, S., Okazaki, Y., Held, W. A., Hayashizaki, Y., and Chapman, V. (1996). Identification of *Grf1* on mouse chromosome 9 as an imprinted gene by RLGS-M. *Nature Genet.* **14**, 106–109.

Potter, S. S., and Branford, W. W. (1998). Evolutionary conservation and tissue-specific processing of Hoxa 11 antisense transcripts. *Mamm. Genome* **9**, 799–806.

Ramsahoye, B. H., Biniszkiewicz, D., Lyko, F., Clark, V., Bird, A. P., and Jaenisch, R. (2000). Non-CpG methylation is prevalent in embryonic stem cells and may be mediated by DNA methyltransferase 3a. *Proc. Natl. Acad. Sci. USA* **97**, 5237–5242.

Reik, W., and Walter, J. (2000). Genomic imprinting: parental influences on the genome. *Nature Rev. Genet.* **2**, 21–32.

Ripoche, M. A., Kress, C., Poirier, F., and Dandolo, L. (1997). Deletion of the H19 transcription unit reveals the existence of a putative imprinting control element. *Genes Dev.* **11**, 1596–1604.

Rougeulle, C., Cardoso, C., Fontes, M., Colleaux, L., and Lalande, M. (1998). An imprinted antisense RNA overlaps UBE3A and a second maternally expressed transcript. *Nature Genet.* **19**, 15–66.

Sado, T., Fenner, M. H., Tan, S. S., Tam, P., Shioda, T., and Li, E. (2000). X inactivation in the mouse embryo deficient for *Dnmt1*: Distinct effect of hypomethylation on imprinted and random X inactivation. *Dev. Biol.* **225**, 294–303.

Sarvella, P. (1973). Adult parthenogenetic chickens. *Nature* **243**, 171.

Sasaki, H., Jones, P. A., Chaillet, J. R., Ferguson-Smith, A. C., Barton, S. C., Reik, W., and Surani, M. A. (1992). Parental imprinting: potentially active chromatin of the repressed maternal allele of the mouse insulin-like growth factor II (*Igf2*) gene. *Genes Dev.* **6**, 1843–1856.

Sasaki, H., Shimozaki, K., Zubair, M., Aoki, N., Ohta, K., Hatano, N., Moore, T., Feil, R., Constancia, M., Reik, W., and Rotwein, P. (1996). Nucleotide sequence of a 28-kb mouse genomic region comprising the imprinted *Igf2* gene. *DNA Res.* **3**, 331–335.

Schmidt, J. V., Levorse, J. M., and Tilghman, S. M. (1999). Enhancer competition between *H19* and *Igf2* does not mediate their imprinting. *Proc. Natl. Acad. Sci. USA* **96**, 9733–9738.

Schmidt, J. V., Matteson, P. G., Jones, B. K., Guan, X. J., and Tilghman, S. M. (2000). The *Dlk1* and *Gtl2* genes are linked and reciprocally imprinted. *Genes Dev.* **14**, 1997–1200.

Schuster-Gossler, K., Simon-Chazottes, D., Guenet, J. L., Zachgo, J., and Gossler, A. (1996). *Gtl2lacZ*, an insertional mutation on mouse chromosome 12 with parental origin-dependent phenotype. *Mamm. Genome* **7**, 20–24.

Schweifer, N., Valk, P. J. M., Delwel, R., Cox, R., Francis, F., Meier-Ewert, S., Lehrach, H., and Barlow, D. P. (1997). Characterization of the C3 YAC contig from proximal mouse chromosome 17 and analysis of allelic expression of genes flanking the imprinted *Igf2r* gene. *Genomics* **43**, 285–297.

Shemer, R., Birger, Y., Riggs, A. D., and Razin, A. (1997). Structure of the imprinted mouse *Snrpn* gene and establishment of its parental-specific methylation pattern. *Proc. Natl. Acad. Sci. USA* **94**, 10267–1072.

Shemer, R., Hershko, A. Y., Perk, J., Mostoslavsky, R., Tsuberi, B. Z., Cedar, H., Buiting, K., and Razin, A. (2000). The imprinting box of the Prader-Willi/Angelman syndrome domain. *Nature Genet.* **26**, 440–443.

Shibata, H., Ueda, T., Kamiya, M., Yoshiki, A., Kusakabe, M., Plass, C., Held, W. A., Sunahara, S., Katsuki, M., Muramatsu, M., and Hayashizaki, Y. (1997). An oocyte-specific methylation imprint center in the mouse *U2afbp-rs/U2af1-rs1* gene marks the establishment of allele-specific methylation during preimplantation development. *Genomics* **44**, 171–178.

Shibata, H., Yoda, Y., Kato, R., Ueda, T., Kamiya, M., Hiraiwa, N., Yoshiki, A., Plass, C., Pearsall, R. S., Held, W. A., Muramatsu, M., Sasaki, H., Kusakabe, M., and Hayashizaki, Y. (1998). A methylation imprint mark in the mouse imprinted gene *Grf1/Cdc25Mm* locus shares a common feature with the *U2afbp-rs* gene: An association with a short tandem repeat and a hypermethylated region. *Genomics* **49**, 30–37.

Sleutels, F., Barlow, D. P., and Lyle, R. (2000). The uniqueness of the imprinting mechanism. *Curr. Opin. Genet. Dev.* **10**, 229–233.

Smilinich, N. J., Day, C. D., Fitzpatrick, G. V., Caldwell, G. M., Lossie, A. C., Cooper, P. R., Smallwood, A. C., Joyce, J. A., Schofield, P. N., Reik, W., Nicholls, R. D., Weksberg, R., Driscoll, D. J., Maher, E. R., Shows, T. B., and Higgins, M. J. (1999). A maternally methylated CpG island in *KvLQT1* is associated with an antisense paternal transcript and loss of imprinting in Beckwith-Wiedemann syndrome. *Proc. Natl. Acad. Sci. USA* **96**, 8064–8069.

Smit, A. F. (1999). Interspersed repeats and other momentos of transposable elements in mammalian genomes. *Curr. Opin. Genet. Dev.* **9**, 657–663.

Smrzka, O. W., Fae, I., Stoger, R., Kurzbauer, R., Fischer, G. F., Henn, T., Weith, A., and Barlow, D. P. (1995). Conservation of a maternal-specific methylation signal at the human *IGF2R* locus. *Hum. Mol. Genet.* **4**, 1945–1952.

Snow, M. H. L. (1981). Growth and its control in early mammalian development. *Br. Med. Bull.* **37**, 221–226.

Solter, D. (1988). Differential imprinting and expression of maternal and paternal genomes. *Annu. Rev. Genet.* **22**, 127–146.

Solter, D. (1998). Imprinting. *Int. J. Dev. Biol.* **42**, 951–954.

Stadnick, M. P., Pieracci, F. M., Cranston, M. J., Taksel, E., Thorvaldsen, J. L., and Bartolomei, M. S. (1999). Role of a 461-bp G-rich repetitive element in *H19* transgene imprinting. *Dev. Genes Evol.* **209**, 239–248.

Stancheva, I., and Meehan, R. R. (2000). Transient depletion of *xDnmt1* leads to premature gene activation in *Xenopus* embryos. *Genes Dev.* **14**, 313–327.

Stöger, R., Kubicka, P., Liu, C. G., Kafri, T., Razin, A., Cedar, H., and Barlow, D. P. (1993). Maternal-specific methylation of the imprinted mouse *M6P-Igf2r* locus identifies the expressed locus as carrying the imprinting signal. *Cell* **73**, 61–71.

Sunahara, S., Nakamura, K., Nakao, K., Gondo, Y., Nagata, Y., and Katsuki, M. (2000). The oocyte-specific methylated region of the *U2afbp-rs/U2af1-rs1* gene is dispensable for its imprinted methylation. *Biochem. Biophys. Res. Commun.* **268**, 590–595.

Surani, M. A. (1998). Imprinting and the initiation of gene silencing in the germ line. *Cell* **93**, 309–311.

Surani, M. A., Barton, S. C., and Norris, M. L. (1984). Development of reconstituted mouse eggs suggests imprinting of the genome during gametogenesis. *Nature* **308**, 548–550.

Sutcliffe, J. S., Nakao, M., Christian, S., Orstavik, K. H., Tommerup, N., Ledbetter, D. H., and Beaudet, A. L. (1994). Deletions of a differentially methylated CpG island at the *SNRPN* gene define a putative imprinting control region. *Nature Genet.* **8**, 52–58.

Svoboda, P., Stein, P., Hayashi, H., and Schultz, R. M. (2000). Selective reduction of dormant maternal mRNAs in mouse oocytes by RNA interference. *Development* **127**, 4147–4156.

Szabo, P. E., and Mann, J. R. (1995). Allele-specific expression and total expression levels of imprinted genes during early mouse development: implications for imprinting mechanisms. *Genes Dev.* **9**, 3097–3108.

Szabo, P. E., and Mann, J. R. (1996). Maternal and paternal genomes function independently in mouse ova in establishing expression of the imprinted genes *Snrpn* and *Igf2r*: No evidence for allelic trans-sensing and counting mechanisms. *EMBO J.* **15**, 6018–6025.

Takada, S., Tevendale, M., Baker, J., Georgiades, P., Campbell, E., Freeman, T., Johnson, M. H., Paulsen, M., and Ferguson-Smith, A. C. (2000). *Delta-like* and *gtl2* are reciprocally expressed, differentially methylated linked imprinted genes on mouse chromosome 12. *Curr. Biol.* **10**, 1135–1138.

Tanaka, M., Puchyr, M., Gertsenstein, M., Harpal, K., Jaenisch, R., Rossant, J., and Nagy, A. (1999). Parental origin-specific expression of *Mash2* is established at the time of implantation with its imprinting mechanism highly resistant to genome-wide demethylation. *Mech. Dev.* **87**, 129–142.

Tasheva, E. S., and Roufa, D. J. (1995). Regulation of human *RPS14* transcription by intronic antisense RNAs and ribosomal protein S14. *Genes Dev.* **9**, 304–316.

Thorvaldsen, J. L., and Bartolomei, M. S. (2000). Mothers setting boundaries. *Science* **288**, 2145–2146.

Thorvaldsen, J. L., Duran, K. L., and Bartolomei, M. S. (1998). Deletion of the H19 differentially methylated domain results in loss of imprinted expression of *H19* and *Igf2*. *Genes Dev.* **12**, 3693–3702.

Tilghman, S. M. (1999). The sins of the fathers and mothers: Genomic imprinting in mammalian development. *Cell* **96**, 185–193.

Tremblay, K. D., Saam, J. R., Ingram, R. S., Tilghman, S. M., and Bartolomei, M. S. (1995). A paternal-specific methylation imprint marks the alleles of the mouse *H19* gene. *Nature Genet.* **9**, 407–413.

Tsai, T. F., Jiang, Y. H., Bressler, J., Armstrong, D., and Beaudet, A. L. (1999). Paternal deletion from *Snrpn* to *Ube3a* in the mouse causes hypotonia, growth retardation and partial lethality and provides evidence for a gene contributing to Prader-Willi syndrome. *Hum. Mol. Genet.* **8**, 1357–1364.

Varmuza, S., and Mann, M. (1994). Genomic imprinting—Defusing the ovarian time bomb. *Trends Genet.* **10**, 118–123.

Vu, T. H., and Hoffman, A. R. (1994). Promoter-specific imprinting of the human insulin-like growth factor-II gene. *Nature* **371**, 714–717.

Walther, T., Balschun, D., Voigt, J. P., Fink, H., Zuschratter, W., Birchmeier, C., Ganten, D., and Bader, M. (1998). Sustained long term potentiation and anxiety in mice lacking the *Mas* protooncogene. *J. Biol. Chem.* **273**, 11867–11873.

Walsh, C. P., Chaillet, J. R., and Bestor, T. H. (1998). transcription of IAP endogenous retroviruses is constrained by cytosine methylation. *Nature Genet.* **20**, 116–117.

Wang, Z. Q., Fung, M. R., Barlow, D. P., and Wagner, E. F. (1994). Regulation of embryonic growth and lysosomal targeting by the imprinted *Igf2/Mpr* gene. *Nature* **372**, 464–467.

Watanabe, T., Yoshimura, A., Mishima, Y., Endo, Y., Shiroishi, T., Koide, T., Sasaki, H., Asakura, H., and Kominami, R. (2000). Differential chromatin packaging of genomic imprinted regions between expressed and non-expressed alleles. *Hum. Mol. Genet.* **9**, 3029–3035.

Wevrick, R., and Francke, U. (1997). An imprinted mouse transcript homologous to the human imprinted in Prader-Willi syndrome (*IPW*) gene. *Hum. Mol. Genet.* **6**, 325–332.

Wilcox, A. J., and Russell, I. T. (1983). Birthweight and perinatal mortality: II. On weight-specific mortality. *Int. J. Epidemiol.* **12**, 319–325.

Wolffe, A. P., and Guschin, D. (2000). Review: Chromatin structural features and targets that regulate transcription. *J. Struct. Biol.* **129**, 102–122.

Wroe, S. F., Kelsey, G., Skinner, J. A., Bodle, D., Ball, S. T., Beechey, C. V., Peters, J., and Williamson, C. M. (2000). An imprinted transcript, antisense to *Nesp*, adds complexity to the cluster of imprinted genes at the mouse *Gnas* locus. *Proc. Natl. Acad. Sci. USA* **97**, 3342–3346.

Wutz, A., and Jaenisch, R. (2000). A shift from reversible to irreversible X inactivation is triggered during ES cell differentiation. *Mol. Cell* **5**, 695–705.

Wutz, A., Smrzka, O. W., Schweifer, N., Schellander, K., Wagner, E. F., and Barlow, D. P. (1997). Imprinted expression of the *Igf2r* gene depends on an intronic CpG island. *Nature* **389**, 745–749.

Wutz, A., Theussl, H. C., Daussman, J., Jaenisch, R., Barlow, D. P., and Wagner, E. F. (2001). Nonimprinted *Igf2r* expression decreases growth and rescues the *Tme* mutation in mice. *Development*, **128**, 1881–1887.

Xu, C., Mao, D., Holers, V. M., Palanca, B., Cheng, A. M., and Molina, H. (2000). A critical role for murine complement regulator Crry in fetomaternal tolerance. *Science* **287**, 498–501.

Xu, Y., Goodyer, C. G., Deal, C., and Polychronakos, C. (1993). Functional polymorphism in the parental imprinting of the human *IGF2R* gene. *Biochem. Biophys. Res. Commun.* **197**, 747–754.

Yan, Y., Frisen, J., Lee, M. H., Massague, J., and Barbacid, M. (1997). Ablation of the CDK inhibitor p57Kip2 results in increased apoptosis and delayed differentiation during mouse development. *Genes Dev.* **11**, 973–983.

Yang, T., Adamson, T. E., Resnick, J. L., Leff, S., Wevrick, R., Francke, U., Jenkins, N. A., Copeland, N. G., and Brannan, C. I. (1998). A mouse model for Prader-Willi syndrome imprinting-centre mutations. *Nature Genet.* **19**, 25–31.

Yu, S., Gavrilova, O., Chen, H., Lee, R., Liu, J., Pacak, K., Parlow, A. F., Quon, M. J., Reitman, M. L., and Weinstein, L. S. (2000). Paternal versus maternal transmission of a stimulatory G-protein alpha subunit knockout produces opposite effects on energy metabolism. *J. Clin. Invest.* **105**, 615–623.

Zhang, P., Liegeois, N. J., Wong, C., Finegold, M., Hou, H., Thompson, J. C., Silverman, A., Harper, J. W., DePinho, R. A., and Elledge, S. J. (1997). Altered cell differentiation and proliferation in mice lacking CDKN1 indicates a role in Beckwith-Wiedemann syndrome. *Nature* **387**, 151–158.

6

Genomic Imprinting during Seed Development

Célia Baroux, Charles Spillane, and Ueli Grossniklaus*
Institute of Plant Biology
University of Zürich
CH-8008 Zürich, Switzerland

*To whom correspondence should be addressed: E-mail: grossnik@botinst.unizh.ch; Fax +41-1-634 8204.

Advances in Genetics, Vol. 46

ABSTRACT

Genomic imprinting allows parent-of-origin specific control over gene expression. Although imprinted genes (or entire chromosomes) are homologous sequences that can be inherited from either parent, they are differentially marked by a heritable epigenetic modification (imprint), which can condition their behavior in term of gene expression. Imprinting-based regulation of entire chromosomes is observed in both insects (paternal genome elimination) and mammals (nonrandom X inactivation). Until recently, it was unknown whether plants possessed a similar epigenetic system discriminating between homologous chromosomes from either paternal or maternal origin. There is now experimental evidence for a genome-wide imprinting phenomenon during seed development in *Arabidopsis*. Genomic imprinting at the gene (or locus) level is observed in both mammals and flowering plants. In maize, only a few allelic variants of several nonessential genes expressed in the endosperm are imprinted. In *Arabidopsis*, gene-specific imprinting has recently been demonstrated for the *MEDEA* (and *FIS2*) gene, which is essential for normal seed development. Unlike the imprinted maize genes, so far all tested *MEA* alleles are subjected to regulation by imprinting. *MEDEA* and *FIS 2* are members of the *FIS* class of genes (*FERTILIZATION INDEPENDENT SEED*) involved in regulation of growth and cell proliferation during seed development. *MEDEA* shares several paradigmatic features with imprinted mammalian genes. The *MEDEA* phenotypes provide empirical support for theories of an intragenomic parental conflict during seed development, whereby imprinting is proposed as a means to differentially balance the selfish interests of each sex's genome during the development of the progeny.   © 2002, Elsevier Science (USA).

I. INTRODUCTION

Selection pressures for developmental plasticity suggest that plants rely extensively on epigenetic regulatory mechanisms (Richards, 1997). Epigenetic mechanisms of regulation can effect a heritable modification in gene expression without any changes in the primary DNA sequence. As such, epigenetic modifications introduce a flexible control mechanism for gene expression, which can occur in addition to the activity of the usual regulatory networks (Bestor, 1994). Epigenetic forms of regulation can occur either during normal development, or as a response to environmental or genetic changes, e.g., following changes in gene dosage, chromosome number, or ploidy levels. Examples of epigenetic regulation in plants include gene silencing (Kaas *et al.*, 1997; Paszkowski and Mittelsten Scheid, 1998; Fagard and Vaucheret, 2000), nucleolar dominance (Pikaard, 2000), paramutation (Chandler *et al.*, 2000), gene dosage compensation (Birchler, 1993;

Matzke and Matzke, 1998), and genomic imprinting (Kermicle and Alleman, 1990, Messing and Grossniklaus, 1999).This review will focus on one particular type of epigenetic regulation that controls gene expression in a parent-of-origin-specific manner, i.e., genomic imprinting and related phenomena.

A. Parent-of-origin effects

Parent-of-origin effects arise from the asymmetry inherent to sexual reproduction, in particular the involvement of two distinct, differentiated gametes (male and female) carrying distinct haploid genetic sets and cytoplasmic contents. Parent-of-origin effects are only apparent during postfertilization development, and can be due to different mechanisms, which affect components essential to zygote or organismal development (Figure 6.1). Transcription of both parental alleles in the zygote does not lead to parent-of-origin differences (Figures 6.1a and 6.1b). By contrast, a temporal asymmetry in the expression of essential maternal and paternal alleles can lead to parent-of-origin effects (Figures 6.1c–6.1g). Because the egg cell provides a large cytoplasmic contribution compared to the sperm cells, it can be expected that disruption of the expression of essential maternal alleles occurring before fertilization will lead to maternal effects (Figures 6.1c–6.1e). This can also occur even if the homologous paternal alleles are expressed in the zygote (Figure 6.1c) or later during development, e.g., the maternal–zygotic transition point T (Figures 6.1d and 6.1e). Asymmetrical expression whereby maternal-origin genes are expressed *de novo* in the zygote, while their paternal counterparts are not, can also lead to parent-of-origin effects (Figures 6.1f and 6.1g). It is necessary to distinguish whether differential accumulation of maternal and paternal transcripts (or proteins) in the embryo is due to the storage/deposition of maternal products (Figure 6.1c), or is due to an active mechanism controlling differential transcription rates of maternally vs paternally derived alleles. While the former mechanism refers to cytoplasmic maternal effects, the latter can typically correspond to genomic imprinting.

In flowering plants, epigenetic regulation via genomic imprinting is potentially more complicated, due to the alternation of generations and double fertilization. Unlike in animals, where the meiotic products differentiate directly into the gametes, the meiotic products of plants undergo several mitoses to form haploid multicellular organisms, the male and female gametophytes. The gametes are formed later in the development of the gametophytes, which is greatly reduced in higher plants. The female gametophyte is a differentiated organism comprising eight haploid nuclei in seven cells: three antipodal cells, two synergid cells, one egg cell, and a homodiploid central cell (Figure 6.2c, see color insert; for review, see Grossniklaus and Schneitz, 1998; Drews et al., 1998). The mature male gametophyte is a tricellular structure comprising two sperm cells contained within

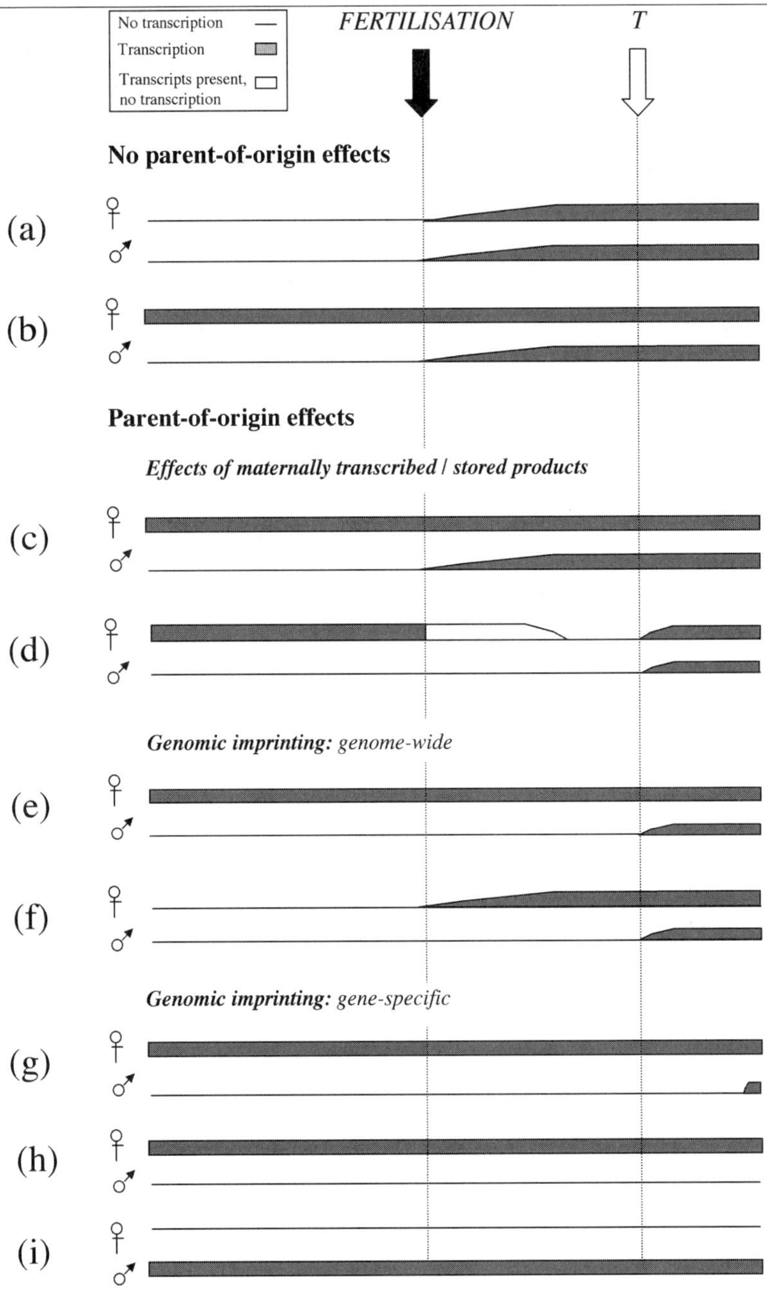
No transcription
Transcription
Transcripts present, no transcription
FERTILISATION
T
No parent-of-origin effects
(a)
(b)
Parent-of-origin effects
Effects of maternally transcribed / stored products
(c)
(d)
Genomic imprinting: genome-wide
(e)
(f)
Genomic imprinting: gene-specific
(g)
(h)
(i)

a vegetative cell (for review, see Mascarenhas, 1989). During double fertilization, one of the sperm cells fuses with the egg cell to give rise to the diploid zygote, whereas the second sperm cell fuses with the homodiploid central cell to give rise to the triploid primary endosperm (Figure 6.2a). Because embryo and endosperm development are closely coordinated in higher plants, parent-of-origin effects can result from the disruption of either embryo or endosperm development. Depending on the species, the relative contribution of embryo and endosperm to seed development varies widely. Thus, parent-of-origin effects may also affect one or the other tissue more severely, depending on the developmental context (Messing and Grossniklaus, 1999).

In addition to the possible parent-of-origin effects described previously (cytoplasmic maternal effect, genomic imprinting), dosage effects play an important role in endosperm development. The triploid endosperm carries two maternal and one paternal copy of the genome. The endosperm is, therefore, particularly prone to be affected by dosage effects. While the amount of a gene product is often directly proportional to the number of functional copies, trans-acting dosage effects, which can be direct or inverse, as well as dosage compensation have been reported (reviewed in Birchler, 1993). Thus, in the endosperm, parent-of-origin effects can also be mediated by dosage effects.

The current view so far in plants is that parent-of-origin effects mostly convey a maternal control during early seed development (Vielle-Calzada *et al.*, 2000; Chaudhury and Berger, 2001; Chaudhury *et al.*, 2001).

Figure 6.1. Parent-of-origin effects on postfertilization gene expression. Schematic representation of mechanisms mediating differential gene expression levels of maternal and the paternal genomes during seed development. Trancriptional levels of both parental genomes are represented, before and after fertilization. In *Arabidopsis*, the transition point (T) corresponds to the mid-globular embryo stage (see Section III.B), but it may vary somewhat depending on the locus. (**a, b**) No parent-of-origin effects occur when both maternal and paternal transcripts are present. (b) The gene product is not required in the zygote; the lag phase during which the paternal transcripts are not equally represented with the maternal transcripts will not trigger parent-of-origin effects. (**c, d**) Parent-of-origin effects mediated by maternally transcribed products deposited before fertilization. In (c), the gene product is required at early stages; the lag phase during which the paternal transcripts are not equally represented to the maternal transcripts can trigger a parent-of-origin effect. In (d), maternal transcripts or products are deposited in the egg; both parental alleles are not expressed in the zygote. (**e–i**) Differential levels and/or rates of transcription of homologous maternal and paternal alleles due to genomic imprinting. (**e, f**) Genomic imprinting can affect gene expression at the entire genome level during the early stages of postfertilization development, until the transition point where the silenced genome becomes derepressed (paternal genome imprinting in *Arabidopsis*, see Section III.B). (**g–i**) Genomic imprinting can specifically control expression of a particular locus (gene-specific imprinting); the gene is expressed exclusively from one parent, the other parental allele being mostly or totally silent (**h, i**); in some instances, the silenced parental allele can be reactivated in a time- or tissue-specific manner (g).

B. Genomic imprinting paradigms

Genomic imprinting results from a mitotically stable epigenetic modification (an imprint) of one of the parental alleles. The imprint leads to differential expression in a parent-of-origin-dependent manner. The epigenetic marking of one of the parental alleles takes place prior to fertilization, presumably during gametogenesis (for review, see Surani, 1998). As a consequence of genomic imprinting, the male and the female genome can be functionally nonequivalent during embryogenesis and later development. Individuals heterozygous for a disruption of an imprinted gene do not show equivalent phenotypes in reciprocal crosses (parent-of-origin effect), because imprinting leads to uniparental functional hemizygosity.

Most data regarding the nature and the establishment of the imprint have come from mammalian studies. In mice, several lines of evidence suggest that, following erasure of the parental imprints, new sex-specific imprints are reestablished during germline differentiation associated with mammalian embryogenesis (Szabo and Mann, 1995; Surani, 1998). The timing of the reestablishment of the imprint differs in male and female germlines: while maternal imprints are established postmeiotically, during oocyte maturation (Obata *et al.*, 1998), paternal imprints are already established in proliferating spermatogonia, prior to meiosis (Ogura *et al.*, 1998). However, *de novo* post-fertilization imprints have been reported (El-Maarri *et al.*, 2001) suggesting that the timing of imprint establishment may differ from locus to locus.

Although the establishment of the imprint correlates with DNA methylation changes, genome-wide methylation changes can also occur without alteration of the imprints. Using embryonic cells derived from primordial germ cells (PGC), Tada *et al.* (1997) showed that demethylation occuring in the PGC is accompanied by the erasure of imprints, and that the reestablishment of sex-specific imprints later, during spermatogonia proliferation and oocyte growth, correlates with remethylation. However, postfertilization demethylation of both maternal and paternal genomes is also observed (Tada *et al.*, 1997; Mayer *et al.*, 2000a; Oswald *et al.*, 2000) and is not correlated with the loss of imprints (Tada *et al.*, 1997). Furthermore, Dnmt3L, a mouse zinc finger protein related to the DNA methyltransferase family 3, is a key regulator of imprint establishment in the gametes, but functions most likely independently of a methyltransferase activity (Aapola *et al.*, 2000; Bourc'his *et al.*, 2001).

C. Genomic imprinting in plants

Genomic imprinting (henceforth referred to as "imprinting") constitutes a largely unknown and intriguing area of plant development, in which interest has burgeoned over the past decade (Haig and Westoby, 1989, 1991; Messing and Grossniklaus, 1999; Kermicle and Alleman, 1990; Alleman and Doctor, 2000;

Grossniklaus *et al.*, 2001; Spillane *et al.*, 2001). The experimental tractability of plant systems will facilitate powerful approaches to elucidate the regulation of genomic imprinting.

Genome- or chromosome-wide imprinting was first described in insects (paternal genome elimination, PGE; Crouse, 1960), and later in marsupials and mammals (nonrandom X inactivation; Cooper *et al.*, 1971; Tagaki and Sasaki, 1975). While it was not widely believed that individual genes could be differentially expressed depending on their parent of origin, Kermicle (1970) was the first to demonstrate such a case, through his study of the *r1*-locus in maize. Genomic imprinting at the chromosomal level in plants was also demonstrated in maize by Lin (1982, 1984). It was subsequently discovered that several loci controlling the development of embryonic and extraembryonic tissues in mice were imprinted (Surani *et al.*, 1984; McGrath and Solter, 1984). These discoveries of chromosome imprinting and locus-specific imprinting constituted a major breakthrough in classical genetics.

Because of imprinting, the parental genomes are nonequivalent, and the contribution of both maternally and paternally derived genomes is necessary for postfertilization development. Many evolutionary theories have been proposed to provide an explanation for genomic imprinting (Hurst and McVean, 1998). Haig and Westoby (1989) proposed that genomic imprinting evolved as a consequence of an intragenomic parental conflict during seed development. The theory was based largely on the results of crosses between lines of different ploidy (interploidy crosses) and chromosome translocation experiments affecting seed size, mainly in maize and potato (Lin, 1982; Johnson *et al.*, 1980). The parental conflict theory was later supported by the mutant phenotypes observed for many (though not all) imprinted genes in mammals (Tilghman, 1999).

Since the discovery of gene-specific imprinting by Kermicle (1970), investigations of imprinting in plants have been renewed. However, imprinting has been conclusively demonstrated at the gene level only for a few cases, mostly for three loci in maize endosperm, and more recently for the *medea* locus in *Arabidopsis*. The current state of research on imprinting in plants raises an ongoing debate regarding whether imprinting takes place in the endosperm only, or in the embryo as well. The study of both the regulation of MEA and genome-wide regulation of paternal gene expression has introduced a different view of imprinting than that based solely on the earlier maize studies, where imprinting was found to be specific to the endosperm.

Here, we review both gene-specific and genome-wide imprinting in plants, and discuss the possible implications in terms of gene regulation during seed development. We will first refer to the historical discoveries of imprinting that highlighted the existence of two types of genomic imprinting, genome-wide and gene-specific, which may be mechanistically quite different. In subsequent sections we will discuss the evidence for both types of imprinting in plants.

II. HISTORICAL OVERVIEW

A. Genome-wide imprinting

1. Chromosomal imprinting and sex determination in insects

a. Paternal genome elimination

The first demonstration of genome-wide imprinting came from the cytogenetic analysis of reproductive stages in the fly *Sciara coprophila*. As reviewed by Herrick and Seger (1999), Metz described as early as 1925 an aberrant meiosis in the male of *S. coprophila* in which one haploid set of autosomal chromosomes (A^P) in the spermatocyte is simply pinched off in a bud, resulting in a sperm cell carrying two paternal X chromosomes ($X^P X^P$) and one autosomal set (A^P). This phenomenon is referred to as paternal genome elimination (PGE). A second PGE happens during cell divisions in the embryo and is due to a failure of chromatid disjunction. As a result, X^P chromosomes are lost in metaphase and are not transmitted to the daughter cells. This loss of X^P chromosomes is the mechanism for sex determination: one X^P is lost in the female embryo, whereas two are lost in the male embryo. A third elimination of a single X^P is observed in the germinal cells of both sexes. The chromosomes eliminated are invariably of paternal origin. This observation led to the conclusion that the paternal chromosomes carry a memory of their parental origin, an imprint, such that the embryo can selectively remove entire sets of paternal chromosome from the genome during development. This chromosomal imprint persists for only one generation and is reset in the next generation.

b. What are the mechanisms mediating the imprint?

Crouse (1960) identified a locus on the *S. coprophila* X chromosome involved in imprinting the paternal set and controlling its paternal-specific elimination. This locus is located near the centromere and acts *in cis* over great distances, thus preventing the entire length of the chromatid to disjoin. As a consequence, the X^P chromatids are not included in the chromosome sets of the daughter cells. Elimination of the autosomal chromosomes relies on different molecular mechanisms, which are poorly understood mechanistically although cytologically well described (reviewed by Goday and Esteban, 2001). Mutational analyses of the Scale insects (coccids such as mealybugs), in which a range of PGE systems are reported, have provided some additional information about the nature of the imprint. In male embryos, dark condensed chromosomes are observed that correspond to heterochromatized paternal chromosomes. This heterochromatic state is acquired either during spermatogenesis or during embryogenesis. Interestingly, it is mitotically heritable (the DNA replicates normally), but the heterochromatized chromosomes are not expressed. This quiescent state is reset in the next generation (reviewed by Herrick and Seger, 1999). It seems that heterochromatin is correlated with an imprinted phenotype. Bongiorni *et al.* (1999) recently showed that in male *Planococcus citri* embryos the heterochromatized region was hypomethylated, compared to euchromatic homologous regions on the maternal chromosomes. However, in female embryos the paternal chromosomes are hypomethylated but not heterochromatized and remain active. Thus, DNA methylation is not responsible for silencing the heterochromatized paternal chromosomes in male embryos, and, therefore, is unlikely to constitute the imprint.

c. Why PGE?

Paternal chromosomal elimination is a process of sex determination in insects (reviewed by Goday and Esteban, 2001). It is intriguing that chromosomal imprinting is used for such a function because this implies that the males are functionally haploid organisms, and as such, male vigor and fitness must be compromized. For the females, PGE is a meiotic drive system, which reinforces maternal chromosome-set transmission at the expense of transmission of the paternal chromosome set (reviewed by Herrick and Seger, 1999). PGE-based imprinting is consistent with theories of parental conflicts proposed for the evolution of imprinting and other parent-of-origin effects in mammals and plants (see below).

2. X inactivation

a. Random and nonrandom X inactivation

X-chromosome inactivation is a process by which one entire X chromosome is silenced in XX mammalian individuals. The mechanism is either random or nonrandom with respect to which X chromosome is silenced. Only the latter case corresponds to an imprint-based inactivation. Random X inactivation was first discovered in the somatic cells of female mammals and has been viewed as a mechanism for achieving dosage compensation (reviewed by Moore *et al.*, 1995). Nonrandom inactivation was later observed in the extraembryonic tissues of mammals. The best-known example is the inactivation of the paternal X chromosome in the extraembryonic membranes of mice (Takagi and Sasaki, 1975). By contrast, in female marsupials, nonrandom inactivation of the (paternal) X chromosome in somatic cells is the rule (reviewed by Graves, 1996).

Nonrandom paternal X inactivation in mice trophectoderm exhibits some flexibility. For instance, even a maternal X chromosome can be inactivated as seen in parthenogenetic embryos derived from two haploid sets of maternal genomes. Nonetheless, in a wild-type situation maternal X chromosomes are typically protected from inactivation (Pereira and Vasques, 2000). It is proposed that the imprint that distinguishes the paternal from the maternal X chromosome confers an enhanced probability of inactivation (rather than a determinate fate of inactivation) to the paternal chromosome in the extraembryonic tissues

(Cattanach and Beechey, 1990). Furthermore, a subset of X-linked genes escapes the silencing mechansim affecting the paternal X chromosome (Tsuchiya and Willard, 2000).

b. What is the imprinting mechanism of paternal X inactivation?

Until recently, it was not known whether the imprint responsible for paternal X inactivation was of paternal or maternal origin. A sex-specific epigenetic mark could be acquired either during male gametogenesis to promote silencing of the paternal X chromosome, or during female gametogenesis to block maternal X-chromosome inactivation. Studies of the mechanisms of random X inactivation in the somatic cells have demonstrated that both the mechanisms for nonrandom and random X inactivation are regulated by the X-inactivation center (*Xic*) located on the X chromosome. *Xic* acts *in cis* to spread the silencing effect (reviewed by Pereira and Vasques, 2000). Silencing during X inactivation is mediated by chromosomal pairing and the establishment of heterochromatic structures (reviewed by Marahrens, 1999). The noncoding RNA *Xist* (belonging to the *Xic* center) is required for the initiation of X inactivation in both imprinted and random X inactivation. Imprinted X inactivation, however, responds to some parentally predetermined factors, in particular, to a maternally imprinted RNA, *Tsix*. In the extraembryonic tissues, the expression of *Tsix*, which is an antagonist of *Xist*, allows maternal X chromosomes to resist inactivation and the paternal X chromosome is therefore inactivated by default (Lee, 2000; Mlynarczyk and Panning, 2000). Although a component of the mechanisms of nonrandom X inactivation has been elucidated, the mode of regulation of *Tsix*, which is itself imprinted, remains poorly understood.

B. Gene-specific imprinting

1. *r1* locus

The two previously described cases provide fascinating illustrations of imprinting mechanisms regulating whole chromosome behavior. Soon after the discoveries of PGE in insects, Kermicle (1970) was the first to demonstrate the regulation by imprinting at a specific locus in maize.

In contrast to animals, sexual reproduction in flowering plants leads to two fertilization products, the embryo and the endosperm. The endosperm is essentially an extraembryonic tissue (Figure 6.2a), supporting growth of the embryo, and an analogy can be drawn with the extraembryonic membranes of mammals (Figure 6.2b). The endosperm is a triploid tissue and would be a genetic twin of the embryo except for its 2:1 maternal:paternal genome ratio (designated 2m:1p, m referring to the maternal and p to the paternal haploid genome), which differs from the 1m:1p ratio in the embryo.

The kernels of some maize varieties display different-colored patterns, ranging from yellow to dark purple, with several intermediate patterns of mottled yellow kernels harboring purple sectors. The dark purple color is due to the production of anthocyanin in a triploid cell layer (aleurone) of the endosperm tissue, beneath the seed coat. Anthocyanin production in maize is controlled by the *r1* (*red color*) locus (Brink, 1956). The *R : r-std* allele of the *r1* locus (designated *R*), confers a solid aleurone pigmentation when *RR* lines are crossed as a female parent with pollen from lines carrying the colorless *r-g* allele (designated *r*) (Figure 6.3a, see color insert). The reciprocal cross, however, results in kernels with mottled pigmentation in the aleurone layer (Figure 6.3b). Thus, the fully colored aleurone phenotype occurs only when the *R* allele is transmitted maternally, indicating a maternal-effect phenotype.

The unusual transmission of the colored aleurone phenotype associated with the *r1* locus prompted Kermicle (1970) to investigate the underlying genetic basis for this phenomenon. He performed a series of crosses with several allelic variants, and manipulated ploidy levels and genome dosage in the endosperm. A summary of the results of the main crosses is provided in Table 6.1. To test

Table 6.1. Maternal Effect on Transmission of Fully Colored Aleurone Phenotype in Maize (Kermicle, 1978)[a]

Female	Male	Endosperm genotype (m:p)	Kernel phenotype
R R	*r r*	*R R:r*	
r r	*R R*	*r r:R*	
r r	*R TR*	*r r:TR TR*	
		r r:R	
R TR	*r r*	TR TR:*r*	
		R R:r	
R R, mdr1 mdr1	*r r, mdr1 mdr1*	*R R, mdr1 mdr1:r, mdr1*	
R R, mdr1 mdr1	*r r, MDR 1 MDR1*	*R R, mdr1 mdr1:r, MDRr1*	

[a] *R* and *r* alleles belong to the *R* class from the *red color* (*r1*) locus that controls aleurone pigmentation. *r* = colorless allele; TR = the *R* allele is carried by the B^{10} chromosome; *mdr1* = mutant allele *maternal derepression of r1*, a trans-acting modifier of *r1* imprinting. Endosperm genotype is indicated with the composition of maternal and paternal genomes, respectively m:p. In the TB-10a line the *R* allele from the long arm of chromosome 10 is located on the B chromosome (B^{10} chromosome) as a result of translocation (Roman, 1947). The B^{10} chromosome fails to disjoin during microspore division, thus resulting in one sperm cell carrying two copies of TR (TR TR) and one with no copy of TR in the TB-10a line. Reciprocal crosses beween *rr* and TB-10a lines produced kernels in which the endosperm contained either two paternal doses of *R* (*rr*:TR TR) or none (*rr*:R).

the possibility that this maternal effect could result from the presence of two maternal copies of the R gene and only one paternal r copy in the endosperm (dosage effect), Kermicle modified the paternal copy number of R using the genetic properties of B chromosomes in maize (details are provided in Table 6.1). Kernels were produced in which the endosperm contained two paternal doses of R in an rr maternal background. All such kernels displayed a mottled aleurone. Thus, the aleurone pigmentation phenotype is correlated with the mode of R transmission and not with dosage effects associated with unequal numbers of chromosomes from the parents in the triploid endosperm. These elegant manipulations allowed Kermicle to demonstrate clearly that the rl locus is imprinted.

Kermicle (1978) subsequently demonstrated that imprinting of the R allele could be suppressed by the introduction of the mdr1 (*maternal derepression of rl*) mutation through the female (Table 6.1), suggesting that the wild-type role of MDR1 is to derepress the expression of the maternal R allele of the rl locus in the central cell. This provided strong evidence that the "imprinted" state of the R allele is the "on" state, with the nonimprinted allele being the "off" state. Kermicle's work constituted a breakthrough in classical genetics, since it was hardly believed at that time that alleles could be differentially expressed depending on their parent of origin. His work also provided possible explanations for previous observations that interploidy crosses and interspecies crosses are often not successful.

2. Imprinting of autosomal genes in the mouse embryo

Experimental embryology and the use of chromosomal translocations, in particular, the experiments of McGrath and Solter (1984) and Surani et al. (1984), initiated scientific research on genomic imprinting in mammals. One-cell-stage embryos or reconstituted eggs were manipulated to create diploid mouse embryos carrying two sets of maternal chromosomes (gynogenote) or two sets of paternal chromosomes (androgenote). However, both gynogenotes and androgenotes died before completion of embryonic development. Any possibility that the lethality could be due to the homozygosity of recessive lethal genes was eliminated by the reconstitution of heterozygous eggs carrying two maternal genomes. Although they contained the normal number of chromosomes, these embryos were unable to complete development. This demonstrated that the paternal and maternal genomes are functionally nonequivalent and indicated that the contributions of both parental genomes are essential for successful completion of embryogenesis in mammals.

Using chromosomal translocations, Cattanach and Kirk (1985) produced a range of mice with maternal or paternal disomy (i.e., with two homologs of a selected chromosome from one parent and none from the other). By studying the development of these disomic mice, a map of chromosomal segments was defined that required both parental contributions for viability. While some regions were lethal when maternally duplicated or paternally deficient, other regions were lethal in the reciprocal configuration. Therefore, several loci essential to postfertilization development are expressed from only one parent. This led to the conclusion that imprinting is established in the germline and controls the expression of some essential genes during embryo development.

At least 40 imprinted genes have been described in mammals, and they are equally divided between the maternally and the paternally repressed classes (reviewed by Reik and Walter, 2001). The regulation of imprinting is poorly understood, but methylation changes are usually correlated with imprinted loci in mammals (Surani, 1998; Feil and Kohsla, 1999). However, changes in chromatin structure also correlate with the parent-of-origin-specific expression of some imprinted mammalian genes (Khosla *et al.*, 1999). Interestingly, most imprinted genes identified to date play a role in growth and cell proliferation. Genomic imprinting of those loci thus raises the question of the selective pressure favoring this epigenetic mode of regulation.

C. Intragenomic parental conflict and the evolution of genomic imprinting

The concept of parent-of-origin-specific gene expression in plants emerged from the research of Kermicle (1970). Lin (1982, 1984) and Johnson *et al.* (1980), and was based on experiments altering either the genetic balance or the parental source of the chromosomes in the endosperm (see Section III.A). Haig and Westoby (1989) interpreted these results to propose a theoretical model aimed at explaining the evolutionary advantage of the triploid endosperm (rather than diploid). In this model of intragenomic parental conflict, the endosperm is the field for the "battle of the sexes." Both the effects of unequal parental genome ratio and parent-of-origin-specific gene expression are considered as a means to defend the opposing selfish interests of the parents for food resources. The embryo proper and the growth of the supportive tissues are expected to be the target of such conflicts. Haig and Westoby (1989) proposed an analogy between the surrounding nutritive tissues of the plant embryo (endosperm) and mammalian embryo (trophectoderm) (Figure 6.2).

In many plants, the endosperm is a highly proliferative extraembryonic tissue, which initially develops as a syncitial structure (at the globular embryo stage, Figure 6.2a) before becoming cellularized (at the heart embryo stage, Figure 6.2a) (for review, see Lopes and Larkins, 1993; Berger, 1999). While the endosperm in plants is the result of a completely independent fertilization event (Figure 6.2a), the trophectoderm in mammals is produced, together with the embryo proper, from a single fertilization event (Figure 6.2b). The egg cleaves to form a blackberry-shaped cluster of cells called the morula (Figure 6.2b). The cells then differentiate into a surrounding outer cell layer, the trophectoderm, and a cluster of cells inside the trophectoderm, called the inner cell mass. The internal intercellular spaces also enlarge to create a central cavity, the blastocoele (Figure 6.2b). The entire embryo proper is derived from the inner cell mass. The trophectoderm is the precursor of part of the placenta and the primitive endoderm and is the earliest component of the system of extraembryonic membranes (Kaufman, 1992).

In plants, the endosperm is thought to be the principal food resource for the embryo, either as a storage tissue for the seedling or as an ephemeral source of nutrients (Lopes and Larkins, 1993). The number of cells in the endosperm determines the nutrient reserves and, thus, endosperm development is correlated directly with the fitness of the seedling (Haig and Westoby, 1991). Theories of parent–offspring conflict thus predict a selection for larger seeds, containing more endosperm tissue. However, such seeds are of greater metabolic cost to the mother and limit the resources available for siblings and future seeds. Therefore, it is expected that some genes expressed in the mother and determining seed size will be selected to restrict seed growth in order to balance the resource demand from the offspring and the cost to the mother. By contrast, it is in the father's interest to produce offspring with higher fitness. Therefore, paternal genes regulating growth are predicted to promote seed size. In order to reflect the specific interests of the mother or the father, parental conflict driven genes involved in regulating seed size must be controlled in a sex-specific manner. In this context, the evolution of genomic imprinting is proposed as a mechanism fulfilling the needs of parent-of-origin-specific gene expression.

Imprinting which supports the intragenomic parental conflict in mammals is illustrated by the differential developmental patterns observed with androgenous and gynogenous mouse embryos (see Section I.D; McGrath and Solter, 1984; Surani *et al.*, 1984). While the former (which express paternal genes) developed a small embryo but produced an excess of extraembryonic tissue (trophectoderm), the latter (which express maternal genes) showed the opposite pattern (normal embryo, trophectoderm deficit). The trophectoderm is a nutritive-supportive tissue, and it can be expected that paternal genes would promote its growth, in order to reallocate nutrients from the mother to the progeny. In contrast, maternal genes would tend to repress its development. Genomic imprinting in mammals may have evolved as a consequence of an intragenomic parental conflict to control the growth of embryonic and extraembryonic tissues (Moore and Haig, 1991; Haig and Graham, 1991; Peterson and Sapienza, 1993) similar to what was proposed for plants (Haig and Westoby, 1989, 1991).

While this theory is consistent with many cases of genomic imprinting in plants and animals, some authors caution regarding the limits of the intragenomic conflict theory (Hurst and McVean, 1998, Spencer *et al.*, 1998). Alternative explanations for the evolution of imprinting have been proposed (Thomas, 1995; Hurst and McVean, 1998).

III. GENOME-WIDE IMPRINTING

Unlike the situation in mammals, some flowering plant species are able to produce viable androgenetic and gynogenetic haploids, either spontaneously, as a
consequence of a mutation (e.g., *indeterminate gametophyte*; Kermicle, 1969),
or following *in vitro* culture (see Kermicle and Alleman, 1990, for review). In
addition, plant embryos can form independently of fertilization in apomictic plants
(Nogler, 1984). For this reason it has been proposed that imprinting does not exist in the embryo or affects only nonessential genes. Alternatively, imprinting
in plants may be flexible enough to allow for the totipotency of maternally or
paternally derived plant cells. If epigenetic programming occurs in the gametes
and introduces a requirement for both parental genomes in postfertilization development, this requirement has to be bypassed in embryos that do not carry
contributions from both parents. For instance, imprinting constraints have to
be relaxed in plants that reproduce asexually via seeds (adventitious embryony,
gametophytic apomixis; Koltunow, 1993; Grossniklaus *et al.*, 2001), or when
regenerating whole plants from somatic cells (somatic embryogenesis) or male
haploid cells (androgenesis) *in vitro*.

Evidence for genome-wide imprinting during seed development has historically been inferred from interploidy crosses in maize (Lin, 1982, 1984) and
Arabidopsis (Redei, 1964; Scott *et al.*, 1998) as well as from crosses between
related species (reviewed by Haig and Westoby, 1991). These crossing experiments
demonstrated the existence of parent-of-origin effects controlling endosperm
development, which can depend on a particular ratio between paternal and maternal genomes (dosage effects) or on the gametic origin of the chromosomes
(genomic imprinting). However, such experiments did not allow any determination of the relative contribution of genomic imprinting, dosage effects or
cytoplasmic maternal effects, to the observed parent-of-origin effects (Figure 6.1
and Section I). Recently, molecular approaches have highlighted a lack of paternal
activity during early seed development in *Arabidopsis*, attributed in part to a
genome-wide imprinting mechanism (Vielle-Calzada *et al.*, 2000).

A. Gene dosage and genomic imprinting in the endosperm

1. Parent-of-origin effects on the development of maize endosperm

Several experiments in maize were performed by Lin (1982, 1984) and by Birchler
and Hart (1987) to alter the ploidy level, and the balance between maternal
and paternal chromosome sets. Lin (1982) took advantage of the *indeterminate
gametophyte* mutation (*ig*; Kermicle, 1971), which can generate a central cell and
egg cell with differing ploidy levels due to supernumerary nuclei. The *ig* mutant was used as a female parent to produce endosperm with an altered parental
genome ratio and ploidy. To enable the determination of ploidy level in cytological

preparations of the endosperm, a male parent was chosen which had a chromosomal translocation (between chromosomes 6 and 10; Roman, 1947), which is easy to score due to the presence of one small and distinct chromosome. An analysis of the size and shape of the kernel progeny from these crosses led to the conclusion that endosperm whose parental genome ratio differed from 2m:1p was not able to complete development. The kernels with unbalanced genome ratios were either abnormal (miniature kernel) or aborted. Normal endosperm development in maize thus requires a strict genetically balanced contribution of both maternal and paternal genomes (2m:1p). Given that an exact dosage of maternal and paternal products has to be maintained, this phenomenon likely involves dosage effects (see Section I).

Does genomic imprinting contribute to parent-of-origin effects during endosperm development? The first set of experiments conducted by Lin (1982, 1984) provided an initial insight but did not provide direct evidence for a role of genomic imprinting in these parent-of-origin effects. Lin employed maize lines of different ploidy levels in a series of crosses to produce grains containing endosperm of higher ploidy level than the normal triploid level. The development of high-ploidy endosperm is influenced by the parental source of the excessive dose of chromosomes. For instance, kernels with an endosperm carrying a tetraploid contribution of the maternal genome (4m) and a diploid set of paternal chromosomes (2p) are normal, but kernels with an endosperm carrying an additional maternal set (5m) and one paternal set only (1p) are abortive. These two types of endosperm are both hexaploid but in the latter (5m:1p) an excess of maternal dose, compared to the former (4m:2p), leads to abortion.

More conclusive evidence for a role of genomic imprinting in endosperm development is provided by chromosomal translocation studies in maize (reviewed by Kermicle and Alleman, 1990), which are similar to those performed in mice (reviewed by Cattanach and Beechey, 1990). A set of maize translocation lines was used in a series of crosses to produce kernels in which the endosperm contains a deletion of a given chromosome arm from the paternal stock. Such kernels have a drastically reduced size. If an additional dose of the corresponding deleted region is transmitted maternally, the defect in kernel size is not compensated. The factors encoded by these chromosomal arms are, thus, not acting in a dosage-dependent manner but are imprinted such that the paternal region is irreplaceable even by its maternal counterpart. Such imprinting effects were shown for eight chromosomes of the maize genome (reviewed by Kermicle and Alleman, 1990).

Kernel development in maize depends on the expression of many genes, which are under different forms of parent-of-origin controls: cytoplasmically stored maternal products, dosage effects, and imprinting. Since both parental genomes are required for proper development, imprinting at the chromosomal level, as demonstrated using translocation studies, must play an important role in conferring parent-of-origin control to endosperm development.

2. Gene dosage and imprinting in *Arabidopsis*

The problem of gene dosage and imprinting in the endosperm has been recently reexamined in *Arabidopsis* (Scott *et al.*, 1998; Adams *et al.*, 2000). Unlike in maize, *Arabidopsis* seed development can tolerate deviations from 2m:1p genome ratio in the endosperm. *Arabidopsis* seeds containing endosperm with ratios of 2m:2p or 4m:1p are viable (Redei, 1964). Self-fertilization of tetraploid or hexaploid plants produces viable seeds of high weight, but crosses altering the maternal-to-paternal genome ratio can induce endosperm and embryo abnormalities. While maternal excess leads to a smaller embryo and a lack of nuclear division in the endosperm, paternal excess promotes the division of the endosperm nuclei and growth of the embryo (Scott *et al.*, 1998). The phenotypes observed depend on the parental dosage ratio, rather than the ploidy per se in the seed. These developmental features indicate that parent-of-origin effects regulate the reciprocal growth of the embryo and the endosperm, but it was not possible with these experiments to distinguish the contribution of genomic imprinting from dosage effects and related phenomena. Unlike in maize, or mammals, approaches based on chromosomal deficiences have not been used to address whether any subchromosomal regions contain imprinted genes. Although γ irradiation can been used to induce chromosomal deficiencies in *Arabidopsis* pollen (Vizir *et al.*, 1994), no tools are yet commonly available for creating deficiencies or duplications of entire chromosome arms such as the translocations to the B chromosome in maize (Roman, 1947).

Although the parent-of-origin interploidy effects are visible in both embryo and endosperm, one of the current hypothesis is that the endosperm is primarily affected by changes in the dosage of imprinted genes Scott *et al.* (1998). However, there is as yet no direct demonstration that the embryo is unaffected by these ploidy changes. This could be tested by, e.g., using genetic mosaic seeds in which the embryo and endosperm genotypes can be manipulated separately (Grossniklaus *et al.*, 2001).

3. Conclusion

From the studies in maize and *Arabidopsis*, it is clear that there are parent-of-origin effects during seed development. However, only the translocation experiments in maize provide direct evidence that entire chromosomal regions contain imprinted genes, and that some epigenetic programming must occur in the gametes, causing the nonequivalence of parental genomes. Imprinting in maize controls several loci involved in regulating seed size, but the absence of cytological data makes it difficult to interpret the underlying causes of the differences in kernel size and to correlate them with differences in cellular proliferation. However, it is possible that the small kernel size resulting from the deficiency in some paternal chromosome arms is due to poor endosperm development. This suggests that some genes exclusively expressed from the father normally promote development of this tissue. This later deduction is consistent with the results of Scott *et al.* (1998) in *Arabidopsis*, in which the parental genome ratio is altered following interploidy crosses: seeds inheriting an excess of paternal genome (e.g., following a cross using a diploid female parent and a tetraploid male parent) display an accelerated growth of the endosperm. Besides, the embryos resulting from this cross grow larger than the embryos inheriting an excess of maternal genome, in a different set of crosses (Scott *et al.*, 1998). This suggests that, as in maize, some paternal genes promote seed growth. Whether genomic imprinting is the major mechanism regulating their expression remains to be determined: interploidy crosses indeed introduce changes in both the parental genome ratio and in the total genome dosage, rendering the interpretation difficult.

A different approach at the gene expression level has been used in *Arabidopsis* to investigate the respective contribution of both parental genomes during seed development, as described in the next section.

B. Genome-wide paternal imprinting in *arabidopsis*

1. Lack of paternal activity during early seed development

Vielle-Calzada *et al.* (2000) discovered a phenomenon of genome-wide differential gene expression depending on the gametic origin of the locus. A collection of enhancer trap insertion lines (transposants) was generated using the maize *Ds* transposable element (Sundaresan *et al.*, 1995) engineered to carry the β-glucuronidase (GUS) reporter gene. A population of insertion lines specifically expressing the GUS reporter gene in the egg cell and/or central cell of the female gametophyte, and/or postfertilization during embryo and endosperm development, were reciprocally crossed to wild-type plants. In the F1 seed progeny, GUS activity was detected postfertilization in the 19 tested lines only if the transposant line was used as a female parent. As a male parent, reporter detection was delayed up to the mid-globular stage. Additional molecular analysis was undertaken to further determine if this was due to: (1) parent-of-origin-specific transgene silencing in the paternal genome, or (2) paternal silencing of endogenous genes. *In situ* transcript detection from the corresponding endogenous genes corroborated the pattern observed in histochemical analyses of GUS reporter expression. Only maternal transcripts were present in the early stages of embryogenesis, while paternal transcripts appeared around the mid-globular stage.

Overall, Vielle-Calzada *et al.* (2000) demonstrated that endogenous genes as well as the reporter genes were under a parent-of-origin control such that maternal and paternal transcripts accumulate differentially in the zygote. There are two possible explanations: the maternal transcripts detected after fertilization consist of maternally deposited products, and both maternal and paternal corresponding genes are not expressed during early embryogenesis (Figure 6.1d); or, a paternal imprint is established during male gametogenesis and prevents transcription of the paternal alleles during early embryogenesis, while the maternal alleles are transcribed (Figure 6.1e or 6.1f). It is likely that both of these processes contribute to the significant absence of paternal gene expression during early seed development.

Several lines of evidence demonstrate that differential gene expression plays a role in this phenomenon. First, the expression patterns of the tested enhancer trap lines show that the GUS activity detected after fertilization in the embryo and/or endosperm is due principally to *de novo* gene expression: (1) several lines with GUS activity before fertilization clearly showed an increase after fertilization; (2) two lines are not expressed prior to fertilization but are strongly induced afterwards, provided the reporter gene is inherited from the mother; (3) one line shows a restricted expression pattern in the two–cell embryo which is not consistent with a cytoplasmically stored product; (4) the GUS protein has a short half-life in these tissues, and essentially no activity can be detected after 12 h if protein synthesis is inhibited (R. Baskar, unpublished). Differential GUS activity depending on the parent of origin of the transgene is therefore, at least in part, a consequence of a mechanism that prevents some paternal loci from being transcribed at similar levels to the counterpart maternal alleles.

The transposon-tagged genes encode a broad range of proteins that play a role in diverse cellular functions. Consequently, this lack of paternal gene expression is not restricted to a cluster of genes or a particular class of genes, and this finding led Vielle-Calzada *et al.* (2000) to make the exciting proposal that a genome-wide imprinting mechanism occurs during early embryogenesis. This does not mean that individual genes could not escape such a mechanism, similarly to the genes escaping X inactivation. It is likely that some genes playing an essential role in early development do not show differential activity, while for the vast majority of genes paternal activity is lacking during early seed development.

It is worth noting from the report of Vielle-Calzada *et al.* (2000) and from independent studies (Sorensen *et al.*, 2001) that the silenced paternal loci in *Arabidopsis* become derepressed later during embryo development. Does such derepression correspond to the transition between maternal control and zygotic control of embryo development, as has been described in mammals? In-depth analysis of gene expression mechanisms in the early stages of mouse embryo development showed that several phases of differential parental gene expression occur. Following fertilization, the egg develops up to the two-cell embryo stage mainly using the maternal stocks of transcripts accumulated in the oocyte (reviewed by Schultz, 1993). A transition occurs at the two-cell stage where activation of zygotic genes is typically observed (zygotic gene activation, ZGA; Schultz, 1993). The paternal genome is transcribed first, while transcription of the maternal genome only begins later, at which stage paternal transcripts are being translated in very high amounts (Nothias *et al.*, 1996). The phenomenon observed by Vielle-Calzada *et al.* (2000) is reminiscent of a mechanism that regulates the timing of zygotic gene activation. The deposition of maternal products, as well as genome-wide paternal silencing, are two mechanisms resulting in a lack of paternal activity, with the consequences that the early stages of seed development are under maternal control.

2. To what extent does genomic imprinting affect paternal gene expression?

Several observations from genetic ablation experiments suggest that some loci in the paternal genome retain transcriptional activity after fertilization, although they are very poorly expressed. The ablation experiments consisted of inducing *de novo* expression of the toxic *BARNASE* ribonuclease gene (Hartley, 1998) in the zygote under the control of a cyclin promoter that is active as soon as fertilization occurs (Ferreira *et al.*, 1994; Baroux *et al.*, 2001). A binary transactivation system (pOp/LhG4 system; Moore *et al.*, 1998) was used for induction of expression. Severe defects in the embryo occurred as a result of *BARNASE* expression soon after the first division of the zygote and led either to embryo arrest in about 40% of the cases (in particular, arrests at the two- to four-cell stage), or to abnormal embryo development with later defects (about 60%; Baroux *et al.*, 2001). Because transactivation of the *BARNASE* gene necessitates both promoter and trans-activating components to be expressed and/or to be accessible, this suggests that transcription of both paternal and maternal transgenes was achieved as early as the second or third division of the zygote. Similar results were obtained in reciprocal crosses. Four transgenic lines were used in these ablation experiments, excluding the possibility that the precocious expression of the paternal genome might be specific to a particular transgene locus or position effect. These results suggest that either the cyclin promoter and the heterologous pOp promoter escapes paternal silencing, or that a very low level of transcription occurs for paternally inherited transgenes.

     In additional experiments in which the GUS reporter gene was substituted for the *BARNASE* gene, GUS expression was not detected in the F1 embryo before a late globular stage, consistent with the observations of Vielle-Calzada *et al.* (2000). By contrast, when both components of the binary system were transmitted through the female genome, *de novo* GUS expression was detected in the zygote. This lack of paternal promoter activity is not specific to a particular endogenous promoter (i.e., from the cyclin gene), because the heterologous pOp promoter also showed the same phenomenon. Altogether, the results suggest that paternally silenced loci retain some basal transcriptional activity but at a minor level compared to the transcriptional activity from the female-derived genome.

3. What is the nature of the imprint mark?

The observations described above clearly indicate that the two parental genomes are not equivalent during the early stages of seed development. What is the epigenetic modification rendering both parental genomes functionally distinct?

a. Chromatin structure

In insects, paternal genome elimination (PGE) has been correlated to dense, heterochromatic, silenced chromosomes. Heterochromatization is an attractive candidate for epigenetic modification rendering certain regions of the paternal genome inaccessible to the general transcription machinery. Heterochromatin is a cytological description of chromatin in one sense (i.e., it stains under certain condition, e.g., with Giemsa) and a molecular genetic description of a chromatin state in another (i.e., late replicating or areas of repressed transcription).

Heterochromatinization is achieved through a modification in chromatin composition, in terms of both proteins and three-dimensional structure. Histones play a structural role in chromatin structure (Wolffe, 1995). Interestingly, some histone variants are specifically expressed during pollen development in *Lilium longiflorum* (Ueda and Tanaka, 1995; Ueda *et al.*, 2000) and tobacco (Prymakowska-Bosak *et al.*, 1999; Xu *et al.*, 1999). Particular isoforms of the core histones H2A, H2B, and H3 progressively accumulate in the elongating and condensing nuclei of generative sperm cells in tobacco (Xu *et al.*, 1999) and *Lilium longiflorum* (Ueda and Tanaka, 1995), suggesting that these histones play a role in chromatin condensation in the male gametes. These proteins are reminiscent of the protamine proteins in mammals that play a similar role in chromatin condensation in the sperm nucleus (Ueda *et al.*, 2000).

Further evidence underlying the importance of chromatin structure in the formation of the male gametes comes from the work of Prymakowska-Bosak *et al.* (1999). Transgenic tobacco plants were constructed that were drastically altered in the stoechiometry of H1 linker histone variants. One of the major consequences was male sterility. This was due to aberrant meiosis in pollen mother cells, mainly because of defects during chromosome assembly and subsequent loss of the chromosomes. One deduction from this study is that a differential requirement for the H1 linker histone exists during male and female gametogenesis. Because H1 histones are particularily enriched in heterochromatin and are thought to stabilize the heterochromatic state (Wolffe, 1995), it is suggested that the heterochromatic pattern is crucial for successful completion of meiosis in the pollen. Heterochromatization is also considered to be a distinct feature of the male genome, in comparison with the female genome. However, whether this heterochromatic state is causally linked to a silencing mechanism of the male genome remains to be demonstrated. In this respect, methods for *in situ* chromosome labeling that allow structural studies of the plant genome promise new insights (De Jong *et al.*, 1999).

b. Methylation and imprinting

In mammals, it is commonly accepted that, following erasure in primordial germ cells, new sex-specific imprints are established in either the male or the female germline (Szabo and Mann, 1995; Surani, 1998). Although it is not clear whether methylation patterns constitute the imprint, dynamic genome-wide methylation changes are observed before and after fertilization. Ovulated mouse oocytes are globally undermethylated compared to the sperm genome. Rapid demethylation of the paternal genome occurs following fertilization, while the maternal genome is gradually demethylated, except at imprinted loci (Mayer *et al.*, 2000a; Oswald *et al.*, 2000). While it is not clear whether such changes in global methylation are linked directly to imprinting in the mouse embryo, it is possible that the asymmetry in demethylation processes may contribute to imprinting (Jaenish, 1997).

Changes in global DNA methylation levels are associated with pollen development in tobacco. In particular, genome-wide demethylation occurs in the sperm cell when pollen grains are mature for pollination (Oakeley *et al.*, 1997; see Section III.B.4). The stage at which such demethylation occurs is concurrent with the limit stage at which regeneration of tobacco plants from pollen tissue culture is possible (androgenesis). In other words, this demethylation event profoundly affects totipotency of these haploid cells, and could also participate in the resetting of epigenetic marks affecting gene expression (Oakeley *et al.*, 1997).

A recent report highlights the role of DNA methylation patterns in parent-of-origin effects observed during seed development (Adams *et al.*, 2000). However, as in mammals, no causal correlation with the observed genomic imprinting can be easily drawn. Using a mutant line reduced in DNA methylation by a *METHYLTRANSFERASE 1* antisense transgene (*asMET1*; Finnegan *et al.*, 1996), Adams *et al.* (2000) carried out reciprocal crosses with *asMET1* diploid and polyploid *Arabidopsis* plants. Seeds from these crosses developed some features reminiscent of seeds normally obtained from crosses with plants of higher-ploidy. For instance, a hypomethylated, haploid paternal genome together with a normal maternal genome creates abnormalities in the endosperm similar to those produced when an excess of maternal dose is introduced using a higher-ploidy parent. The reciprocal cross (a hypomethylated maternal genome together with a normal paternal genome) leads to the opposite effect, i.e., to seeds displaying the features of an excess of paternal dose, as observed in interploidy crosses using a higher-ploidy pollen parent. Thus, hypomethylation is interpreted to be equivalent to the addition of one genome dose of the other sex. One possible explanation for these results is that some imprinted genes, as well as nonimprinted genes, are misregulated in the developing seed following hypomethylation.

There are a number of experimental problems associated with the use of globally hypomethylated inbred *asMET1* lines to ascribe causal effects due to hypomethylation. It is now known that hypomethylation in these lines can be accompanied by redistribution of local methylation profiles such that some loci become hypermethylated, as has been shown for the *superman* and *agamous* loci (Jacobsen *et al.*, 2000). In addition, if *asMET1* has similar effects as *Dnmt1* in mammals, then up to 10% of genes in the plant genome may be misregulated (Jeddeloh *et al.*, 1999; Jackson-Grusby *et al.*, 2001). In such a situation, even in the absence of epistatic effects, separating direct effects of methylation from indirect effects of second-site gene misregulation due to methylation changes is extremely difficult.

c. Compartmentalization and imprinting

An intriguing observation has been made of a nuclear structure allowing the silencing of sex chromosomes during spermatogenesis in mammals. During the early stages of male meiosis, a vesicle at the nuclear periphery of the spermatocyte is formed and captures both X and Y sex chromosomes (McKee and Handel, 1993). This XY body, or sex vesicle, forms an inactive nuclear compartment, while condensation of the autosomal chromosomes goes on. Heterochromatization is correlated with an enrichment of H2A-type histones and subsequent alteration of the nucleosomal structure (Hoyer-Fender *et al.*, 2000). Proteins of the transcription and mRNA maturation machinery are excluded from the XY body (Richler *et al.*, 1994). As a consequence, genes belonging to this XY body are not expressed, while the autosomal chromosomes are. Compartmentalization therefore provides a mechanism for obtaining sex-specific gene expression.

In addition, Mayer *et al.* (2000b) observed a separation of the maternal and paternal chromatin in the nucleus of mouse embryos. This separation was visible up to the four-cell stage. Such structural features might be of functional importance during the early stages of embryogenesis to achieve differential gene expression in a parent-of-origin-dependent manner.

Compartmentalization of whole chromosomes or subcompartmentalization of the parental genomes could be components of the epigenetic mechanisms that govern genome-wide imprinting during postfertilization development. It would be of particular interest to investigate whether the silencing phenomenon observed in the *Arabidopsis* seed is associated with paternal genome compartmentalization. In this context, fine structural studies of the first cytogenetic events following the fertilization process revealed that male and female nucleoli remain physically separated both in the endosperm (at least during the first two or three cell divisions) and in the zygote (J. E. Faure, personal communication). This suggests that, during early seed development, the genetic material from both parents is not yet fully assembled into a nuclear entity containing merged parental chromatin. This compartmentalization may lead to parent-of-origin-specific accessibility to the transcription machinery and contribute to the lack of paternal activity during early development.

DNA methylation, chromatin structure, and genome compartmentalization are all candidates for epigenetic changes which could mediate genomic imprinting. We lack evidence, however, to determine which modification among these, or others, constitutes or maintains the primary imprint. Furthermore, it is now clear from structural and compositional study of the chromatin that DNA methylation, histone composition, higher-order chromatin structure, and possibly chromosome compartmentalization are interrelated (Lewis and Bird, 1991; Wolffe, 1995; Feil and Khosla, 1999).

4. When and where is the imprint established?

Gametogenesis in plants has two major differences to gametogenesis in mammals. First, no germline is predetermined in plants. Cells that will give rise to the male or female gametophyte differentiate from the reproductive meristem, a totipotent group of dividing cells in the adult plant which elaborates all the structures of the flower (both sexual and nonsexual structures). Second, the gametes are formed by meiosis followed by mitotic divisions and differentiation. In plants, the female gametes (the egg cell and the central cell) are associated with sibling, nongametic cells (the synergids and the antipodal cells). During male gametogenesis, the haploid microspore undergoes a mitosis to form a vegetative and a generative cell. A second mitotic division of the generative cell results in the formation of two sperms. The vegetative and sperm cells are enclosed in the pollen coat (reviewed by Mascarenhas, 1989).

If imprints are established during male and female gametogenesis, several possibilities remain open: premeiotic, postmeiotic, premitotic, or postmitotic events could occur. However, if parental imprints are introduced after mitosis in only one of the daughter cells, this would lead to two functionally nonequivalent female gametes or two functionally nonequivalent sperm cells. As mentioned in the previous section, demethylation is observed in the pollen grain after the second meiosis and before the mitosis of the generative cell. Furthermore, additional structural and compositional DNA changes accompany pollen differentiation. These changes are candidate steps for epigenetic erasure and/or reprogramming of the imprints at a genome-wide level, as it occurs in animals. While in plants the male DNA is widely demethylated during sperm cell development, the opposite occurs in mammals, where the DNA is hypermethylated during spermatogonia proliferation. The mouse oocyte, by contrast, is hypomethylated (reviewed by Jaenish, 1997). It would be interesting to determine whether the DNA in the female gametes undergoes such methylation or structural changes, and whether such changes correlate with the establishment of any maternal imprints.

C. Conclusion

In maize and *Arabidopsis*, seed development is largely influenced by parent-of-origin effects resulting from maternally stored products, gene dosage, and genomic imprinting. In this section, we have discussed the current evidence for genomic imprinting in plants, in particular affecting the whole genome. In maize, several studies report the requirement of both parental genomes, mainly for proper endosperm development. In endosperm tissue, the specific requirement for a 2m:1p maternal-to-paternal genome ratio complicates the interpretation of any parent-of-origin effects, because there may be some interplay between genomic imprinting and dosage effects. However, using chromosomal translocations, Lin (1982) clearly showed that the source of the chromosomes is more important than the chromosome dosage *per se*. It is still unclear whether genomic imprinting also influences embryo development in maize, as is the case in *Arabidopsis*, or whether in maize it is a specific mechanism regulating endosperm development only. In the absence of cytological or molecular data, the debate remains open. Genome-wide imprinting may also contribute to the parent-of-origin effects in seed development in *Arabidopsis* observed using interploidy crosses. However, the situation is less clear in *Arabidopsis* than in maize, and additional manipulations are needed to interchange the source of the chromosomes without altering the genome balance in the endosperm and/or the embryo. At the transcriptional level, there is clearly a mechanism favoring gene expression from the maternal genome in the early stages of seed development in *Arabidopsis*, leading to the conclusion that a genome-wide imprinting mechanism must regulate paternal genome activity after fertilization. Although this effect appears to be rather global, that does not mean that every gene is affected. It is likely that some loci escape paternal silencing, especially those playing a crucial role at early stages. Around the mid-globular embryo stage, however, paternal gene expression can be detected, suggesting that a transition occurs derepressing the paternal genome and permitting a transition to zygotic gene expression. The timing of reactivation may differ, however, from locus to locus. This transition also suggests that the imprinted state of the paternal genome is relieved, and this would allow the vast majority of the genes acting during seed development to be turned "on," except for some loci such as *MEDEA*, which are regulated through a gene-specific imprinting process (see Section IV). Together with the reports of DNA modifications accompanying pollen formation, it is tempting to speculate that a genome-wide paternal imprint is introduced during male gametogenesis, which leads to a global lack and/or repression of paternal gene expression. More investigations will be needed to clarify this exciting phenomenon.

IV. GENE-SPECIFIC IMPRINTING AND THE CONTROL OF SEED DEVELOPMENT

At least 40 imprinted genes have been reported in mammals and many of these genes play key roles in growth and differentiation (Tilghman, 1999). While most imprinted chromosomal regions have been described in terms of imprinted genes in mammals, few imprinted loci have been well characterized at the molecular level in plants. Most of the mammalian imprinted loci identified encode essential products for viability and proper development of the organism. By contrast, two distinct categories of imprinted genes are known in plants. The first category comprises allelic variants of four genes expressed in the maize endosperm under parent-of-origin control. These nonessential genes do not affect the viability of the seed. The second category is represented by at least one locus that plays an essential role during seed development. Additionally, the phenotypes of the mutants disrupted in this second class of essential imprinted genes are consistent with the role predicted for imprinted genes by Haig and Westoby's (1989) theory. Imprinted plant genes and their likely function are listed in Table 6.2.

A. Gene-specific imprinting in maize endosperm and during seed development in *arabidopsis*

Maize endosperm was the first object of study of parent-of-origin effects in plants. As pointed out by Haig and Westoby (1991), maize breeders were the first to observe that maternal traits were often dominant in the kernel progeny, but they mostly attributed it to allelic composition rather than to a mechanism that was later termed "imprinting." The conventional view for most maternal effects is that they result from dosage effects (two maternal genomes are present in the endosperm, compared to the single dose of the male genome). Kermicle (1970) designed several experiments based on crosses between distinct maize inbred lines in order to investigate why the parent-of-origin effect on the *red color* phenotype was dependent on its parental origin rather than a result of different dosage in the endosperm (see Section I.C). With the discovery that single genes could be imprinted in plants, increased attention has been brought to genes whose expression displayed a parent-of-origin effect during seed development.

Table 6.2. Imprinted Genes in Plants[a]

Reference	Species	Locus	Allele	Role	Imprinted alleles	Maternal transmission	Methylation associated	Modifier	Type of imprinting
Kermicle, 1970	Maize	r1	R:r:std	Aleurone pigmentation	Maternal alleles: central cell	Fully colored aleurone	√	mdr1	Allele-specific
Lund et al., 1995a	Maize	Zein	19KDa, 22KDa genes W64A inbred	Storage proteins	?	High zein accumulation	√	n.d.	Allele-specific
Lund et al., 1995b	Maize	Tubulin	α-tub2 α-tub4 W64A inbred	Cytoskeleton Cell division	?	High tub RNA accumulation	√	n.d.	Allele-specific
Chaudhuri and Messing, 1994	Maize	dzr1	dzr1 Mo17 inbred	Regulator of 10-kDa zein accumulation	?	Dominant phenotype for low zein accumulation	n.d.	n.d.	Allele-specific
Grossniklaus et al., 1998	Arabidopsis	medea		Cell growth, proliferation during embryo + endosperm development	Paternal allele	Wild-type, normal seed development; mutant, seed abortion	?	ddm1, ddm2 (asMET1)	Locus-specific
Luo et al., 1999 Kiyosue et al., 1999 Castle et al., 1993		(fis1) (f644) (emb173)			?				
Luo et al., 1999	Arabidopsis	fis2		Same as MEA	?	Same as MEA	n.d.	n.d.	n.d.

[a] mdr1 = maternal derepression of r1; fis = fertilisation-independent-seed; fie = fertilisation-independent-endosperm; ddm1 = decrease in DNA methylation 1; asMET1 = METHYLTRANSFERASE antisense. Note: media, fis1, f644 and emb173 an allelic.

1. Allele-specific imprinting in maize endosperm

In addition to the *r1* locus, several genes in maize are differentially expressed in the endosperm depending on their gametic origin. These include one allelic variant of the *dzr1* locus (Chaudhuri and Messing, 1994) and some allelic variants of the zein (Lund *et al.*, 1995a) and the alpha-tubulin genes (Lund *et al.*, 1995b). Although those genes do not affect viability of the seed, they participate in endosperm growth (the alpha-tubulins are a major component of the microtubules; Lund *et al.*, 1995b) and grain filling (zein proteins; Lund *et al.*, 1995a). Zein proteins are storage compounds constituting about 70% of the maize endosperm (Lund *et al.*, 1995a). They play an important role in animal feeding (biological value of the endosperm proteins) as well as in cereal processing (milling and baking qualities of the storage proteins). The level of accumulation of the 10-kDa zein in the endosperm is controlled by the *dzr1* locus (posttranscriptional regulator), and as such, the regulation of this locus has attracted attention (Lund *et al.*, 1995a).

a. *dzr1* gene

Parent-of-origin-specific controls can affect heteroallelic interactions. The *dzr1* gene that encodes a posttranscriptional regulator of the 10-kDa zein accumulation in maize endosperm provides such an example (Chaudhuri and Messing, 1994). The allelic *dzr1* variant from the Mo17 inbred line is associated with low zein accumulation, whereas the BSSS53 inbred accumulates higher levels. When *dzr1*+Mo17 is maternally transmitted to the endosperm following a cross with pollen of the BSSS53 inbred, it exerts a dominant negative effect on the *dzr1*+BSSS53 allele and low levels of zein are found. In the reciprocal cross, hybrid kernels contain high zein accumulation demonstrating that the low zein levels due to the *dzr1*+Mo17 allele are parent-of-origin-specific.

The *dzr1*+Mo17 allele has a unique parent-of-origin behavior since a third allelic variant, *dzr1*+W64A, shows an additive effect when combined with BSSS53, regardless of its parental origin. Additionally, Chaudhuri and Messing (1994) used B-to-A translocations to test whether two copies of paternal *dzr1*+Mo17 allele could mimic the effect of two maternal copies of the same allele (low zein accumulation). Endosperm containing two copies of paternal *dzr1*+Mo17 allele and two maternal copies of *dzr1*+BSSS53 allele accumulate a higher level of zeins than genetically identical endosperms in which the alleles are of opposite parental origin. This shows that the parental origin of *dzr1*+Mo17 determines whether it exerts a dominant negative effect on zein acccumulation.

In conclusion, the *dzr1*+Mo17 allele is regulated by imprinting. Because this imprinting mechanism is specific to only one particular allele, it may not account for a general mechanism of *dzr1* locus regulation.

b. Zein and tubulin

Additional cases of allele-specific imprinting are illustrated by the zein and tubulin genes (Lund *et al.*, 1995a, 1995b). Some alleles of these gene families show differential expression depending on their gametic origin and of their parental inbred line. Expression analysis at the transcript and protein levels demonstrated that the H1, H2 and L1, L2, L3 chains of the 19-kDa and 22-kDa zeins, respectively, as well as α-tubulin-2 and -4 are only expressed in the endosperm when maternally inherited. In all studies, embryos do not express (or poorly express) the zein and tubulin genes, suggesting that the parent-of-origin effects on the expression of these alleles are specific to the endosperm. Furthermore, in both situations, methylation is associated with the silenced state of the paternal copy in the endosperm, leading to the conclusion that the expression of these particular alleles of the zein and tubulin genes are regulated via imprinting. To ensure that detection of the imprinted sequences was not biased by the unequal maternal-to-paternal genome ratio, different DNA preparation mixtures were analyzed from endosperms derived from reciprocal crosses between the imprinted and nonimprinted allelic variants. In these studies, the contribution of gene dosage effects has not yet been fully excluded, but the evidence that the alleles are expressed in the endosperm only if they are demethylated and transmitted via the maternal nuclei strongly favors imprinting as a key regulator of the expression of these genes.

Several common features are evident from the study of the imprinting of the *R-r : std* allele, *dzr-1 Mo17* allele, and different alleles of the zein and tubulin genes. First, these genes are differentially expressed only in the endosperm depending on the sex of origin of the nuclei that transmitted them, namely, the maternal nuclei of the central cell. Second, maternal-specific expression (*R-r : std*, zein, tubulin) or effects (*dzr1*) are observed only for particular maize inbred lines, implying that imprinting is not locus-specific but rather allele-specific. These features constitute the substantive differences to imprinted genes in *Arabidopsis*, as described in the next section.

2. Gene-specific imprinting at the medea locus in *Arabidopsis*

a. The *FIS (FERTILIZATION-INDEPENDENT SEED)* class of genes

Over the past decade, *Arabidopsis* has become a powerful model system for plant developmental geneticists. In particular, the spectacular progresses in large-scale mutagenesis screens provide a tool for genetically dissecting a given developmental pathway. In addition, such mutational approaches are rapidly increasing our knowledge of genes regulating epigenetic phenomena in plants. For instance, the isolation of trans-acting mutations has shed light on the mechanisms of epigenetic silencing affecting some transgene loci (Mittelsten-Scheid *et al.*, 1998). However, the identification of the first candidate imprinted gene(s) in *Arabidopsis* was an indirect consequence of mutagenesis screens for genes affecting female gametophyte development and maternal effects during seed development (Grossniklaus *et al*, 1998c) or fertilization-independent silique elongation (Ohad *et al*, 1996; Chaudhury *et al*, 1997). These *FIS* class mutants enabled a significant advance in the comprehension of genomic imprinting in plants.

Genes of the *FIS* class have been identified in several independent mutational screens and account for three loci thus far described: (*medea*) (allelic to *fis1*, *f644*, and *emb173* loci), *fis2* and *fie* (*fertilization-independent-endosperm*, allelic to *fis3* locus) (Castle *et al.*, 1993; Ohad *et al*, 1996; Chaudhury *et al.*, 1997; Grossniklaus *et al.*, 1998c; Kiyosue *et al.*, 1999; Luo *et al.*, 1999). While the *mea* mutant was isolated in a screen for gametophytic, maternal-effect, embryo-lethal mutations, *emb173* was found in an embryo-mutant screen, and *fis1* and *f644* were isolated by their ability to promote autonomous silique elongation in the absence of pollination. Parent-of-origin maternal effects have been reported for all three loci, but in-depth analysis of genomic imprinting at these loci has been conducted only for the *mea* locus.

b. Parent-of-origin effects controlling MEDEA expression and that of other FIS genes

mea is a mutant disrupted in MEDEA gene and displaying a gametophytic, maternal effect, lethal phenotype in the seed (Figures 6.3c and 6.3d), whereby any seed inheriting a mutant allele from the mother (*mea*[m]) aborts regardless of the paternal contribution (i.e., whether *mea*[p] or MEA[P]). The *mea* mutant phenotype is caused by the disruption of a single gene, MEA, following the insertion of a derivative of the maize *Ds* transposon (Sundaresan *et al.*, 1995). Several experiments allowed Grossniklaus *et al.* (1998c) to rule out the hypothesis that the phenotype could be due to haplo-insufficiency of MEDEA product in the endosperm or to a lethal effect deriving from the absence of the gene product normally deposited in the egg and/or the central cell. While interploidy crosses demonstrated that the defect was not due to dosage effects in the endosperm (Grossniklaus *et al.*, 1998c), *in situ* hybridization analysis demonstrated that only two out of three MEA copies were expressed in the endosperm (Vielle-Calzada *et al*, 1999). Allele-specific RT-PCR assays further confirmed that only the maternally inherited copy of MEA is transcribed both in the embryo and in the endosperm at early stages (Vielle-Calzada *et al.*, 1999).

The two other genes from the *FIS* class (*FIS2*, *FIE*) are also maternally expressed during the early stages of seed development (Luo *et al.*, 2000). The occurrence of reactivation of the paternal copy later during seed development is the critical criterion, which will determine whether these three genes are regulated through a gene-specific imprinting process or through the genome-wide imprinting phenomenon (discussed earlier). Whether MEA is later paternally expressed in both the embryo and endosperm is currently unresolved. RT-PCR techniques detected a paternal transcript signal in a batch of dissected embryos at the torpedo stage, but not in the endosperm (Kinoshita *et al.*, 1999). Nonetheless, expression studies using a reporter gene provided evidence for paternal expression of *MEA* in the chalazal endosperm (a basal region of the endosperm facing the embryo pole; Sorensen *et al.*, 2001) in most cases, and rarely some expression was detected in the embryo (Luo *et al.*, 2000). The fact that the paternal *MEA* allele is reactivated in only a few embryos and at a later stage favors the hypothesis that imprinting regulation at the *medea* locus is independent of the imprinting mechanism governing paternal genome expression (see Section III.B). Reactivation of MEA^P in some embryos further indicates that imprinting may not be stable in all embryos and can break down. The imprinting situation is less clear for *FIS2*, but the studies to date suggest that it is under a gene-specific imprinting control as well. By contrast, *FIE* paternal reactivation corresponds to the transition point observed in previous studies (see Section III.B and Vielle-Calzada *et al.*, 2000), suggesting that its regulation depends on the genome-wide, paternal imprinting process decribed before. In summary, all three genes from the *FIS* class are certainly under parent-of-origin control, such as gene-specific imprinting (*MEA* and probably *FIS2*), or genome-wide imprinting (*FIE*).

c. Imprinting of MEDEA is locus-specific

Since the imprinted maize alleles provided the only examples of gene-specific imprinting in plants, our laboratory was interested in determining if the imprinting observed at the *Arabidopsis medea* locus was ecotype-specific (i.e., a rare allele event). By conducting outcrosses of pollen from 21 *Arabidopsis* ecotype accessions of diverse geographic origin (Figure 6.4) to homozygous *mea* plants (*Landsberg erecta*), we determined that in all cases the MEA^P alleles from the genetically diverse ecotypes were imprinted, typically resulting in 100% seed abortion in the F1 progeny as would be expected if the MEA^P allele was silent. We have since extended this analysis to an additional 50 ecotypes (C. Spillane, H. Hu, and U. Grossniklaus, unpublished results). These studies indicate that the *mea* locus is imprinted in all the ecotypes so far tested across the *Arabidopsis thaliana* gene pool.

d. Imprinting of MEDEA fits the prediction of intragenomic parental conflict theory

Seed abortion in *fis*-class mutants results from a failure in both endosperm and embryo development. It is intriguing that the aborted *mea* embryos exhibit excessive proliferation, leading to embryo enlargement. This is apparently at the expense of endosperm development, which shows a reduced number of peripheral nuclei, but a greater amount of nuclei in the chalazal endosperm (Grossniklaus *et al.*, 1998c; Sorensen *et al.*, 2001). All three *fis*-class mutants also show, to varying

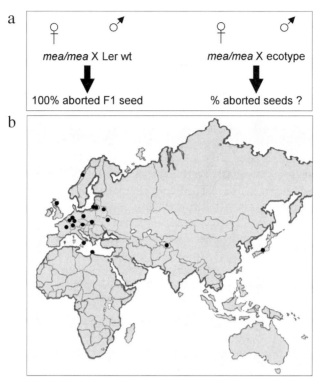

Figure 6.4. The MEA gene is imprinted in a majority of *Arabidopsis* ecotypes. (**a**) To test if all allelic variants of MEA are imprinted, pollen from a range of genetically diverse ecotypes was crossed with *mea/mea* plants. If the MEA^P allele was not imprinted and fully transcribed during early embryogenesis, this test cross would generate 100% normal seeds. (**b**) Geographic origins of 21 genetically diverse *Arabidopsis thaliana* ecotypes tested in which the MEA gene is imprinted.

extents, an ability to induce endosperm development in the absence of fertilization (Grossniklaus and Vielle-Calzada, 1998; Ohad *et al.*, 1996; Chaudhuri *et al.*, 1997; Kiyosue *et al.*, 1999; Luo *et al.*, 1999). It is evident that the *FIS*-class genes are involved in growth and control of cell proliferation during seed development.

It has also been demonstrated that the MEA and FIE proteins interact in a *Polycomb*-like protein complex (Luo *et al.*, 2000; Spillane *et al.*, 2000; Yadegari *et al.*, 2000). *MEDEA* and *FIE* code for proteins that share high similarity with Enhancer of Zeste (E[Z]) and Extra Sex Combs (ESC), respectively, which belong to the *Drosophila Polycomb* group (Grossniklaus *et al.*, 1998c; Ohad *et al.*, 1999). In most organisms studied so far, *PcG* genes are involved in the regulation of cell proliferation and morphogenesis. In *Drosophila*, PcG complexes repress the transcription of homeotic genes by forming complexes that affect higher-order chromatin structures. Similarly, in mammals, PcG homologs maintain the repression of *Hox* genes and also regulate the proliferative response to mitogens during embryo development before gastrulation (Pirotta, 1997). Spillane *et al.* (2000) also found that MEA could bind to ESC in a yeast-two hybrid system. Such conservation of structure and function across the animal and plant kingdoms make it tempting to parallel the role of PcG homologs in mammals with those of MEA and FIE PcG homologs in the control of cell proliferation during embryo and endosperm proliferation.

The genes from the *FIS* class share many features with imprinted genes in mammals, at the structural (PcG homologs), functional (role in cell proliferation), and regulatory (gene-specific imprinting regulation) levels. However, of the three *FIS* class genes, only the *MEA* gene has been conclusively shown to be imprinted at the gene level. More analysis is needed to ascertain the imprinting status of *FIS2*. The imprinted *MEA* gene contrasts strongly with the previously described cases of genomic imprinting in maize. In this respect, the *medea* locus is of particular interest for studying the genetic and molecular basis of imprinting during seed development. Furthermore, the mutant phenotype following gene disruption of *FIS*-class genes, in particular of *MEA*, clearly displays features of an excess of embryo growth, which supports a role of maternally imprinted genes in restricting embryo growth, as predicted by the parental conflict theory (Haig and Westoby, 1991).

B. Imprint mark: establishment and regulation

1. What is the nature of the imprint?

Studies on genomic imprinting require making a clear distinction between the "imprint" itself and the expression state (silenced/expressed) of parental alleles at the imprinted locus (Herrick and Seger, 1999; Sleutels *et al.*, 2000). For instance, in mammals, methylation "imprints" can be found to be associated with the inactive allele (*H19* locus) or the active allele (*IGF2R* locus) (Sleutels *et al.*, 2000).

Differentially methylated regions (DMRs) have been identified for a large number of imprinted mammalian genes (Feil and Khosla, 1999), and assigned a causal role as potential "imprints" (reviewed by Reik and Walter, 2001). However, it may be prudent not to draw too many parallels with imprinting in plants and in mammals (mainly mice and humans), and comparative analogies (e.g., regarding DMRs or imprinting mechanisms) for imprinted loci may not be relevant to explain imprinting mechanisms in all species, which may have evolved different processes.

a. Imprinted alleles in maize endosperm

The nature of the imprint in the maize endosperm remains largely unknown. Analysis of imprinted alleles at the DNA level and using methylation-sensitive restriction enzymes in Southern blotting experiments, has shed some light on possible mechanisms. Strikingly, it was observed that for all imprinted alleles of zein and tubulin genes, the maternal transmission leading to high transcript and protein accumulation was correlated with hypomethylation of the coding region, in comparison with reciprocal hybrids (Lund *et al.*, 1995a, 1995b; Chaudhuri and Messing, 1994). Recently, Alleman and Doctor (2000) report that imprinting of the *R* allele is accompanied by differential methylation of the promoter region of the alleles in the endosperm. DNA methylation is therefore associated with allele-specific imprinting in maize endosperm, and hypomethylation of the promoter and/or coding region seems to accompany the activation of the maternal alleles. Although methylation is an attractive candidate for the imprint mark, more conclusive evidence is required regarding any causal role of methylation in imprinting.

b. DNA methylation and *MEDEA* imprinting

Because methylation is clearly involved in imprinting in mammals, and because the *ddm1* (*DECREASE IN DNA METHYLATION 1*) mutant shows a drastic decrease in global DNA methylation (70% decrease), we determined whether *ddm1* had an effect on the seed abortion phenotype of *mea*.

DDM1 encodes a chromatin remodeling factor of the SWI2/SNF2 family, and the disruption of this gene reduces the overall global DNA methylation level and affects global gene expression patterns (Jeddeloh *et al.*, 1999). This epimutation effect is amplified at each successive generation of inbreeding, as a result of the cumulative action of the *ddm1* mutation. As a consequence, highly inbred lines carry more epigenetic changes than less inbred lines (Jeddeloh *et al.*, 1999). Therefore the capacity of the *ddm1* hypomethylated genome to modify imprinting at the *medea* locus (and thus to have an effect on *mea* mutant seeds) may differ depending on the level of inbreeding the mutant parent has undergone. This is illustrated by two sets of experiments using alternatively low or high-inbred *ddm1* lines.

Double-mutant heterozygous lines were produced that carry a disruption in both the MEA and DDM1 genes. In this work, the *ddm1* mutation was constructed from parental lines with a normally methylated genome. Analysis of such double-mutant lines showed that, in the absence of DDM1 activity, mutant seeds carrying a mutant maternal *mea* allele can survive, provided a wild-type paternal MEA allele is present (Vielle-Calzada *et al.*, 1999). However, a paternal genome from a highly inbred *ddm1* line can rescue *mea*^{m} mutant seeds even in the absence of a wild-type MEA allele (Luo *et al.*, 2000; Yadegari *et al.*, 2000; C. Spillane, D. R. Page, and U. Grossniklaus, unpublished results). This difference with previous results is most likely due to more numerous epigenetic modifications in highly inbred lines. In the latter case, the rescue may be effected by other genes that can substitute for *MEDEA* function, whose activity is modified in the hypomethylated genome of the highly inbred lines. The first experiment, by contrast, suggests that changes induced by *ddm1* can reactivate the expression of MEA[P] and thus break down the imprinting regulation at the *medea* locus.

Similarly, embryos carrying a maternal *fie* mutant allele can be rescued from seed lethality providing a hypomethylated paternal genome from an inbred *ddm1* parent and a wild-type *FIE* allele (Vinkenoog *et al.*, 2000). However, it cannot be concluded that the rescue of the *fie* seeds effect using hypomethylated pollen explicitly requires a *FIE*[P] allele, because (unlike *mea*) it has thus far proved impossible to generate viable *fie/fie* plants. In the case of the *mea* mutant, Vielle-Calzada *et al.* (1999) found that 15 of 15 plants, derived from the enlarged seed progeny of a selfed *mea;ddm1* double heterozygote, were heterozygous for *mea* but homozygous for *ddm1*, indicating that a wild-type paternal *MEA* allele is necessary for the rescue effect.

In addition, the *ddm2* mutant, as well as antisense lines reducing *METHYLTRANSFERASE 1* expression (asMET1), both reduce overall genomic methylation levels (Finnegan *et al.*, 1996, 1998; Ronemus *et al.*, 1996), and similarly to *ddm1*, asMET1 activity in the zygote is able to rescue the *mea*[m] mutant seeds (Luo *et al.*, 2000; C. Spillane, D. R. Page, and U. Grossniklaus, unpublished results).

c. Can we conclude from these experiments that imprinting at the *mea* locus is mediated by DNA methylation?

The rescue of *mea* mutant embryos by hypomethylated genomes is complicated by the lack of direct correlations at the molecular level, such as allele-specific changes in methylation pattern across the *medea* locus, or reactivation of the paternal allele. Initial investigations using methylation-sensitive restriction enzymes to assay for differentially methylated regions (DMRs) across the *medea* locus have so far found no methylation changes (as detected by methylation-sensitive enzymes) in the promoter or coding region of *MEA* (C. Spillane, D. R. Page, and U. Grossniklaus, unpublished).

More experiments are clearly needed to clarify the mechanisms by which the *mea* locus is regulated following fertilization. In particular, it remains to be determined what are the sequences necessary for maternal activation in the seed, paternal silencing, and further reactivation during seed development (Kinoshita *et al.*, 1999). Overall, these results indicate that methylation may play a role in maintenance of the imprint, but they do not provide strong evidence for a causal role for methylation in the establishment or resetting of the imprinting at the *medea* locus.

2. When and where is the imprint established?

The imprinted genes described in maize and *Arabidopsis* differ regarding their pattern of expression and, thus, regarding the tissue specificity of their imprint. While the imprinted alleles described in maize are expressed only in the endosperm, imprinting of *MEA* (and *FIS2*) locus introduce differential expression in the embryo and in the endosperm. Consequently, establishment of the imprint must differ in those two cases, both in the place and time during the development of the male and/or female gametophyte.

a. Female gametophyte differentiation and establishment of the imprint

Gametophytic development is initiated in the adult plant from the reproductive meristem (see Section III.B.4). Among the cells generating the ovule primordia, one diploid cell, the megaspore mother cell, will undergo meiosis to produce four haploid cells (tetrad). While three of these degenerate, the surviving megaspore undergoes three mitotic divisions to produce eight nuclei that will form the seven-celled female gametophyte (Section I.A, and Figure 6.1a; Drews *et al.*, 1998; Grossniklaus and Schneitz, 1998). At some stage during this developmental process, the maternal-specific imprints may be introduced. There is as yet no evidence to determine at which step (premeiotically, postmeiotically, or postmitotically) these epigenetic marks are laid down. The female gametophyte is a highly differentiated structure, which is both polarized and compartmentalized (Figure 6.2a). Cellular differentiation occurs mainly after the last round of mitosis (Grossniklaus and Schneitz, 1998). The polar nuclei migrate from each pole of the embryo sac toward the center of the gametophyte, the antipodal cells as well as the synergid cells form at opposite poles, and the egg cell acquires its shape and position at the micropylar pole. Cell-specific signals and positional cues are involved this differentiation (Grossniklaus and Schneitz, 1998). It is reasonable to expect that the polar nuclei and the nucleus of the egg cell, while genetically identical, could carry some epigenetic differences (Alleman and Kermicle, 1990; Messing and Grossniklaus, 1999).

b. Imprinting in maize endosperm

An interesting feature of the *R* allele is the tissue specificity of its parent-of-origin control. Brink (1956) observed that this class of *R* allele is specifically expressed in the endosperm, and not in the embryo or in the adult tissues. Thus, in this case the imprinting mechanism must activate expression of the *R* allele in the polar nuclei only, and this raises questions about the stage when the epigenetic modification on the *R* allele is introduced or becomes manifest. The same is true for the other maize examples of gene-specific imprinting (Chaudhuri, 1994; Lund *et al.*, 1995a, 1995b). From the discovery of the maternal derepression of *rl* locus, which acts as a regulator of *R*-allele expression in the endosperm (Section II.B.1), it could be deduced that the ground state of the *rl* locus must be "off." By inference from the methylation profile and expression pattern of the imprinted zein and tubulin alleles, a model emerged that the ground state of these alleles was a hypermethylated "off" state, that is, actively turned "on" and demethylated in the central cell or endosperm. The imprinted state of the *R* allele, as well as the alleles of the *dzr1*, zein, and tubulin genes, differ in the embryo and in the endosperm. Therefore, the epigenetic modification has to be introduced after the last mitotic round of division during female gametophyte development, that is, after the formation of the pronuclei of the egg cell and of the central cell. Such a modification could take place either in the central cell after differentiation within the gametophyte (Figure 6.5a), or early during endosperm development. Both possibilities remain open, and there is no evidence to date to distinguish between them.

In addition, these genes are expressed in the endosperm, which is a terminally differentiated tissue and thus does not participate in the next generation. If the imprint resides solely in the endosperm genome, no resetting of the imprinted alleles can occur (Figure 6.5a), and this constitutes an additional and significant difference with imprinting in mammals. In this context, imprinting of the maize alleles may occur in the central cell as a cause or as a consequence of differentiation. In other words, this could mean that endosperm imprinting is a differentiation-based epigenetic modification, which, because it occurs in the gamete, shares the consequences of a gametic-based epigenetic modification. Alternatively, the double dose of chromosomes in the central cell could simply trigger or potentiate imprinting at "imprinting-susceptible" loci, such that only two doses out of three are expressed in the endosperm. This would act as a dosage-compensation mechanism in the endosperm as proposed by Haig and Westoby (1989). In such a situation, genomic imprinting would be a means to balance the amount of a given product in the triploid cell to a level to that present in diploid somatic cells.

c. Regulation of imprinting at the *medea* locus

Imprinting at the *medea* locus has attracted considerable attention and provoked excitement for several reasons. First, the mode of imprinting of this gene is drastically different from the imprinting reported so far in maize (locus-specific vs allele-specific; Vielle-Calzada *et al.*, 1999). Second, imprinting takes place both in the embryo and the endosperm. Third, unlike the imprinted maize alleles but like many imprinted mammalian genes, the imprinting mechanism here regulates an essential gene that governs proliferation during seed development. And fourth, the *FIS*-class genes are able to promote autonomous endosperm development in the absence of fertilization, which is of considerable interest for the engineering of apomictic plants (Grossniklaus *et al.*, 1998a; see Section V.B).

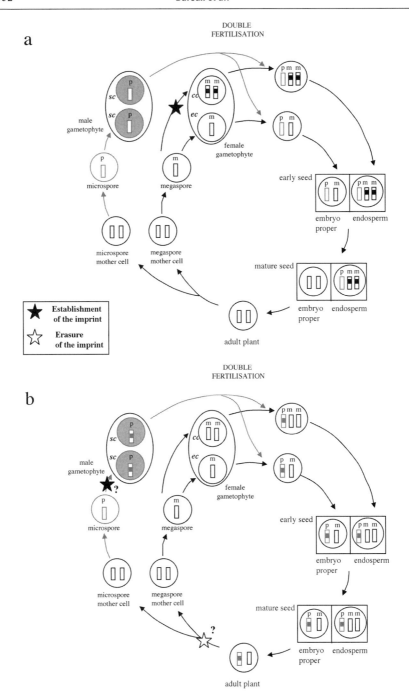

MEDEA is expressed following the mitotic divisions in the eight-nucleated female gametophyte (Vielle-Calzada *et al.*, 1999). After fertilization, *MEDEA* is maternally expressed both in the embryo and in the endosperm. An elegant approach detecting *in situ* nascent transcripts showed that only two of the three *MEDEA* copies were active following fertilization of the central cell. The imprint must therefore be introduced before fertilization, i.e., during gametogenesis. The imprint cue could be established either as a maternal mark leading to the active state of MEA^m, or as a paternal mark repressing MEA^P expression. If it is a maternal imprint, the latter must be introduced postmeiotically and before mitosis. If a paternal imprint was the cause of paternal allele repression, there is no clue indicating the step of establishment. To date, there is no evidence to distinguish between these two possibilities. However, the observation that the paternal MEA^P allele can be reactivated in the chalazal endosperm in seeds at the mid-globular embryo stage, and rarely in the embryo itself (Luo *et al.*, 2000), could indicate that the imprint is a paternal mark, introduced in the sperm cell and repressing MEA^P expression. A working model is proposed Figure 6.5b, in which the cyclical process of imprinting at the *mea* locus during the plant life

---

Figure 6.5. Model for the establishment and resetting of the imprint at the imprinted loci described in Maize and *Arabidopsis*. Schematic representation of the plant life cycle. The main stages of gametophytic development are indicated. The diploid microspore mother cell and megaspore mother cell undergo meiosis to form the haploid microspores and megaspores, respectively. Each of the four microspores and only one of the four megaspores undergo two or three mitotic divisions to produce the mature male gametophyte (pollen) and female gametophyte (embryo sac; see Figure 6.1a), respectively. Only the gametic cells are represented (sc = sperm cell; cc = central cell; ec = egg cell). Mature pollen grains contains two sperm cells, one of which fertilizes the egg cell while the other fertilizes the central cell, to generate the diploid zygote and the triploid endosperm, respectively. Each haploid paternal and maternal genome is represented by a rectangle. The imprint mark at a given locus is indicated schematically by a black box. Hypothetical steps at which the imprint mark may be introduced and reset are indicated. (**a**) Model of allele-specific imprinting in maize endosperm (e.g., *r1 locus, dzr-1*, alleles of the zein and tubulin genes, see Section IV.A.1). A maternal imprint may be established in the central cell and leads to derepression of the maternal copy in the endosperm. Imprinting permits gene expression (from the maternal copy) only in the endosperm. Because the endosperm tissue does not participate in the next generation, no resetting of this imprint occurs during the plant life cycle. This type of allele-specific imprinting observed in maize does not contribute to differential expression in the embryo. (**b**) Model for locus-specific imprinting of *MEDEA* in *Arabidopsis* (see Section IV.A.2). A paternal imprint may be established in the sperm cells and leads to repression of the paternal copy in both the embryo and the endosperm. During seed development, *MEA* is expressed only from the maternal genome, rendering both parental genomes functionally nonequivalent. The imprint has to be reset before the generation of the male and female gametophyte, or it could be reset independently during both developmental processes (possibility not indicated). A new paternal-specific imprint may be reintroduced during male gametogenesis, before or after meiosis.

cycle is represented schematically. In this scheme, one hypothesis is represented in which a paternal imprint is introduced.

d. Conclusion

The imprinted alleles in maize and the imprinted *MEA* gene in *Arabidopsis* are probably regulated by distinct mechanisms. Significantly, the imprints are not introduced in the same time and place (Figures 6.5a and 6.5b). While imprinting of the maize alleles must accompany or participate in the differentiation occurring in the female gametophyte, the imprints at the *medea* locus (and perhaps also for *fis2*) are sex-specific epigenetic marks controlling exclusively expression from the maternal allele in both the embryo and the endosperm.

V. CONCLUSION

A. The significance of imprinting during seed development

Genomic imprinting is a mechanism controlling differential expression of two homologous parental alleles, in a parent-of-origin-dependent manner. This differential expression is achieved through the introduction of an epigenetic modification rendering one parental DNA sequence different from the other, most likely at the structural (e.g., chromatin organization), compositional (e.g., methylation, histone composition), or organisational (e.g., compartmentalization) level, or in a combination of these. Genomic imprinting can occur at the individual gene level, as well as at the entire genome level. One hypothesis is that these two phenomena have different evolutionary significance (see below). Here, we have reviewed the evidence that both types of genomic imprinting occur during seed development. Mainly from recent studies in *Arabidopsis*, a better picture is now emerging of the contribution of genomic imprinting to parent-of-origin effects, which have been observed for many years.

1. Genome-wide imprinting and gene-specific imprinting: two interrelated mechanisms regulating gene expression during seed development

The early stages of embryo and endosperm development are mainly under maternal control: as a consequence of maternally stored products and of genome-wide imprinting that renders the paternal genome contribution mostly underrepresented compared to the maternal genome at the level of gene expression. A transition occurs in seeds at about the mid-globular embryo stage, and this allows derepression of paternal gene expression. From this stage onward, the seed is under the control of zygotic genes. This phenomenon is reminiscent of early development in mouse and *Caenorhabidtis elegans* (Schultz, 1993; Newman-Smith and Rothman, 1998) in which a similar transition from maternal-to-zygotic control occurs. The evolutionary significance of this mechanism remains poorly understood. One hypothesis is that early maternal control constitutes a checkpoint to ensure that no deleterious mutation is transmitted via the mother. Whether the phenomenon observed in *Arabidopsis* refers to a similar process remains to be determined.

A gene-specific imprinting process controls the expression of genes that are essential for seed development in *Arabidopsis*. So far, gene-specific imprinting of one, perhaps two, genes has been demonstrated (*MEA* and *FIS2*). Genes of the *FIS* class repress cell proliferation and growth. They are expressed from the maternal allele during early stage of seed development (*FIE*, *FIS2*) and later (*MEA*). The paternal allele of *MEA* does not contribute to repression of seed growth. If *MEA* is disrupted, seeds inheriting the mutant allele from the mother will show an enlargement of the embryo (see Section IV.A.1). This scenario fits the intragenomic parental conflict theory proposed by Haig and Westoby (1989), in which genes specifically expressed from the mother tend to repress growth of the progeny, ensuring an equal distribution of the resources among the siblings. In contrast, genes specifically expressed from the father tend to promote growth of the seed, to ensure a higher fitness of its progeny. In this theoretical context, genomic imprinting is considered as a means to allow gene expression specifically from the mother and the father, such that each parent can defend its selfish interests in the progeny. *MEA* falls into the class of genes specifically expressed from the mother, through a gene-specific imprinting mechanism, and controlling the rate of cell proliferation in the seed. To date, no specifically paternally-expressed gene has been reported and the field of investigation remains open. There are likely to be more imprinted genes with a function in seed development, as is suggested from interploidy crosses altering the balance of maternal and paternal genomes. Whether imprinting of such genes will be gene-specific or a genome-wide paternal imprinting phenomenon remains to be determined.

What is the significance of allele-specific imprinting of the *r1*, *dzr1*, *mdr1*, tubulin, and zein genes in maize? As proposed before, because these genes are expressed only in the endosperm and do not encode products playing an essential role in seed growth, they are not subject to intragenomic parental conflict. Furthermore, imprinting is allele-specific at these loci and may correspond to an event that appeared in some maize inbred lines. The imprint seems to be introduced in the central cell or in the endosperm, i.e., during or after female gametophyte differentiation. For this reason, we propose that imprinting in this class of genes corresponds to a differentiation-based epigenetic modification during gametophyte development (Messing and Grossniklaus, 1999).

In conclusion, the recent findings highlight two distinct mechanisms of genomic imprinting acting during seed development. It is proposed that one (genome-wide imprinting) may ensure a maternal control of early development, while the other (gene-specific) may have evolved to balance the selfish interests of each parental genome in its progeny (intragenomic parental conflict theory).

2. Allele or locus-specific imprinting?

Spencer (2000) highlights that the evolutionary process most likely affected by imprinting is selection. The imprinted state is unlikely to spread in a population unless it confers some selective advantage (Hurst and McVean, 1998; Pardo-Manuel de Villena *et al*, 2000). The population genetics of imprinted loci has only been empirically explored for a few of the imprinted mammalian loci (Spencer, 2000). The studies to date suggest that polymorphism can exist across populations for imprinted vs nonimprinted alleles, i.e., loci are inactivated in some individuals but not in others. For some loci, the imprinted state is near fixation in the population, whereas in others the nonimprinted state is the most prevalent. For instance, at the imprinted *Igf2* locus, the vast majority of maternally derived *IGF2* alleles are silent across a wide range of mammalian species (mice, rats, pigs, humans) and even in some marsupials (O'Neill *et al.*, 2000). Biallelic expression of *IGF2* is associated with a number of clinical conditions such as Wilms tumor or Beckwith-Wiedeman syndrome. While the majority of *IGF2R* alleles are imprinted in mice (Barlow *et al.*, 1991), only a minority of *IGF2R* alleles are considered to be imprinted in humans (Xu *et al.*, 1993; Ogawa *et al.*, 1993; Kalscheuer *et al.*, 1993; Wutz *et al.*, 1998). The human WT1 gene can be imprinted in some individuals but is biallelically expressed in others (Jinno *et al.*, 1994; Mitsuya *et al.*, 1997). In the case of the imprinted maize loci (*r1*, *dzr1*, zein, and tubulin genes), it is evident that only specific alleles are imprinted. The studies of *MEA* imprinting among various ecotypes (see Section IV.A.2) have indicated that the *medea* locus is imprinted in the majority of ecotypes tested across the *Arabidopsis thaliana* gene pool. We are now conducting a comparative epigenomics approach to determine if the *medea* locus is imprinted in a range of evolutionarily divergent plant species.

3. The evolution of genomic imprinting and gene regulation during postfertilization development

From the study of the imprinted plant genes, it is clear that at least two imprinting mechanisms exist in plants. The mechanisms remain poorly understood, and it is tempting to draw comparative models for imprinting mechanisms between plants (especially for the *medea* locus) and mammals. Since gene-specific imprinting is known only in flowering plants and mammals so far, it is likely that genomic imprinting is a mode of gene regulation that has evolved recently but independently in both organisms. The evolution of imprinting appears to be driven by a parental genome conflict affecting embryo growth in organisms with a placental habit

(Haig and Westoby, 1989, 1991; Haig and Graham, 1991; Haig, 1993). Consistent with this idea, Killian *et al.* (2000) report that, among the mammalian clade, the paradigmatic *Igf2r* locus is not imprinted only in the monotremes which lay eggs. Similarly, the *IGF2* gene is imprinted in eutherians but not in birds, which are oviparous and make no postfertilization contribution to the offspring (O'Neill *et al.*, 2000). Fetal development *in utero* and the development of placental habit are thought to be very important for the evolution of genomic imprinting (Haig, 1993; John and Surani, 1998). In higher plants, the embryo develops inside seeds surrounded by the endosperm, which is thought to be a maternal nutritive tissue (Lopes and Larkins, 1993). A parallel has been proposed between the endosperm and the placenta in mammals for a role in controlling the allocation of the nutrients of the mother to the embryo. Genomic imprinting may have evolved convergently in flowering plants and mammals, and it is likely that studies on imprinting in plants will transcend plant biology and shed some light on imprinting mechanisms in general.

B. Genomic imprinting and apomixis

A major goal in agricultural research is to achieve the production of asexual seeds in a variety of agronomically important crops, in order to fix a desired agronomic trait by generating individuals that are genetically identical to the female parent plant. Apomixis is a form of asexual reproduction through seeds, which could achieve such objectives. Three processes of apomixis allow the formation of an embryo without fertilization and from an unreduced cell lineage. The embryo can develop from an unreduced megaspore mother cell (diplospory) or from an unreduced embryo sac derived directly from a somatic cell (apospory). These two processes occur in the gametophyte, by contrast to adventitious embryony, in which somatic embryos can develop from sporophytic tissues of the ovule (Koltunow, 1993). Although apomictic pathways have been observed in at least 400 plant species, apomixis is not common in agriculturally important crops. For this reason, apomixis research has attracted increased interest in the last decade, and many research efforts are now focused toward the production of apomixis technologies (Grossniklaus *et al.*, 1998b).

In this context, efforts have focused on identifying the genes that could be manipulated in order to induce seed development without fertilization. The interest in the *FIS* class of genes resides on one hand in their ability to induce endosperm development, on the other hand in their regulation by genomic imprinting. Both are key steps to be understood and manipulated in order to genetically engineer asexual seeds (Grossniklaus *et al.*, 1998b). Vinkenoog *et al.* (2000) succeeded in producing endosperm that was almost fully developed in the absence of fertilization in *Arabidopsis*. This was achieved by combining the *fie* mutation and a hypomethylated state of the paternal genome. This shows that these conditions can be sufficient to bypass the requirement of both genomes and initiate some components of asexual seed development.

However, the situation is more complicated in cereals, in which the endosperm is often very sensitive to any modification of both the initial parental genome dosage and the balance between the maternal and paternal genomes (Redei, 1964; Lin, 1982; Haig and Westoby, 1991; Scott *et al.*, 1998; Messing and Grossniklaus, 1999; Grossniklaus *et al.*, 2001). As such, the induction of autonomous endosperm development may not be a successful approach to produce asexual seeds in these species (Grossniklaus *et al.*, 2001). The solution must thus reside elsewhere, as most of the natural apomictic grasses require fertilization of the central cell (pseudogamy; reviewed by Grossniklaus *et al.*, 2001). Pseudogamy may therefore be the preferable way for engineered apomicts to produce asexual but viable seeds. However, the strategic choice of pseudogamy for engineering apomixis in cereals would require tighter control of the fertilization process. Indeed, the transfer of apomixis to cereals would lead to an imbalanced endosperm, which is expected to result in seed abortion (reviewed by Grossniklaus *et al.*, 2001). By contrast, natural apomicts have evolved mechanisms of fertilization that lead to the preservation of the maternal-to-paternal genome ratio in the endosperm (Haig and Westoby, 1991; Messing and Grossniklaus, 1999; Grossniklaus *et al.*, 2001). Transferring pseudogamous systems to cereals will therefore require the engineering of altered fertilization processes or an alleviation of a strict maternal-to-paternal genome ratio.

Acknowledgments

We gratefully acknowledge Baskar Ramamurthy (Institute of Plant Biology, Zurich, Switzerland), Patrick Gallois (University of Manchester, U.K.) and Jean-Emmanuel Faure (ENS Lyon, France) for sharing unpublished results; Urs Jauch for S.E.M. imaging; Jean-Jacques Pittet for helping with the artwork preparation; and Margaret Collinge for critical reading of the manuscript.

References

Aapola, U., Kawasaki, K., Scott, H. S., Ollila, J., Vihinen, M., Heino, M., Shintani, A., Kawasaki, K., Minoshima, S., Krohn, K., Antonarakis, S. E., Shimizu, N., Kudoh, J., and Peterson, P. (2000). Isolation and initial characterization of a novel zinc finger gene, DNMT3L, on 21q22.3, related to the cytosine-5-methyltransferase 3 gene family. *Genomics* **65,** 293–298.

Adams, S., Vinkenoog, R., Spielman, M., Dickinson, H. G., and Scott, R. J. (2000). Parent-of-origin effects on seed development in *Arabidopsis thaliana* require DNA methylation. *Development* **27,** 2493–2502.

Alleman, M., and Doctor, J. (2000). Genomic imprinting in plants: observations and evolutionary implications. *Plant Mol. Biol.* **43,** 147–161.

Barlow, D. P., Stoger, R., Herrmann, B. G., Saito, K., and Schweifer, N. (1991). The mouse insulin-like growth factor type-2 receptor is imprinted and closely linked to the *Tme* locus. *Nature* **349,** 84–87.

Baroux, C., Blanvillain, R., and Gallois, P. (2001). Paternally inherited transgenes are down-regulated but retain low activity during early embryogenesis in *Arabidopsis*. *FEBS Lett.* **509**, 11–16.

Berger, F. (1999). Endosperm development. *Curr. Opin. Plant Biol.* **2**, 28–32.

Bestor, T. H., Chandler, V. L., and Feinberg, A. P. (1994). Epigenetic effects in eukaryotic gene expression. *Dev. Genet.* **15**, 458–462.

Birchler, J. A. (1993). Dosage analysis of maize endosperm development. *Annu. Rev. Genet.* **27**, 181–204.

Birchler, J. A., and Hart, J. R. (1987). Interaction of endosperm size factors in maize. *Genetics* **117**, 309–317.

Bongiorni, S., Cintio, O., and Prantera, G. (1999). The Relationship between DNA methylation and chromosome imprinting in the coccid *Planococcus citri*. *Genetics* **151**, 1471–1478.

Bourc'his, D., Xu, G. L., Lin, C. S., Bollman, B., and Bestor, T. H. (2001). Dnmt3L and the Establishment of Maternal Genomic Imprints. *Science*, in press.

Brink, R. A. (1956). A genetic change associated with the R locus in maize which is directed and potentially reversible. *Genetics* **41**, 872–890.

Castle, L. A., Errampalli, D., Atherton, T. L., Franzmann, L. H., Yoon, E. S., and Meinke, D. W. (1993). Genetic and molecular characterization of embryonic mutants identified following seed transformation in *Arabidopsis*. *Mol. Gen. Genet.* **241**, 504–514.

Cattanach, B. M., and Beechey, C. V. (1990). Autosomal and X-chromosome imprinting. *Dev. Sup.* 63–72.

Cattanach, B. M., and Kirk, M. (1985). Differential activity of maternally and paternally derived chromosome regions in mice. *Nature* **315**, 496–498.

Chandler, V. L., Eggleston, W. B., and Dorweiler, J. E. (2000). Paramutation in maize. *Plant Mol. Biol.* **43**, 121–145.

Chaudhury, A. M., and Berger, F. (2001). Maternal control of seed development. *Semin. Cell Dev. Biol.* **12**, 381–386.

Chaudhury, A. M., Koltunow, A., Payne, T., Luo, M., Tucker, M. R., Dennis, E. S., and Peacock, W. J. (2001). Control of early seed development. *Annu. Rev. Cell Dev. Biol.* **17**, 677–699.

Chaudhuri, S., and Messing, J. (1994). Allele-specific parental imprinting of *dzr1*, a post-transcriptional regulator of zein accumulation. *Proc. Natl. Acad. Sci. USA* **91**, 4867–4871.

Chaudhury, A. M., Ming, L., Miller, C., Craig, S., Dennis, E. S., and Peacock, W. J. (1997). Fertilization-independent seed development in. *Arabidopsis thaliana*. *Proc. Natl. Acad. Sci. USA* **94**, 4223–4228.

Cooper, D. W., Vandeberg, J. L., Sharman, G. B., and Poole, W. E. (1971). Phosphoglycerate kinase polymorphism in kangaroos provide further evidence for paternal X inactivation. *Nature New Biol.* **230**, 155–157.

Crouse, H. V. (1960). The controlling element in sex chromosome behavior in *Sciara*. *Genetics* **45**, 1429–1443.

De Jong, J. H., Franz, P., and Zabel, P. (1999). High resolution FISH in plants—Techniques and applications. *Trends Plant Sci.* **4**, 258–262.

Drews, G. N., Lee, D., and Christensen, C. A. (1998). Genetic analysis of female gametophyte development and function. *Plant Cell* **10**, 5–17.

El-Maarri, O., Buiting, K., Peery, E. G., Kroisel, P. M., Balaban, B., Wagner, K., Urman, B., Heyd, J., Lich, C., Brannan, C. I., Walter, J., and Horsthemke, B. (2001). Maternal methylation imprints on human chromosome 15 are established during or after fertilization. *Nat. Genet.* **27**, 341–344.

Fagard, M., and Vaucheret, H. (2000). (Trans)gene silencing in plants: How many mechanisms? *Annu. Rev. Plant. Physiol. Plant Mol. Biol.* **51**, 167–194.

Feil, R., and Khosla, S. (1999). Genomic imprinting in mammals. An interplay between chromatin and DNA methylation? *Trends Genet.* **15**, 431–435.

Ferreira, P. C. G., Hemerly, A. S., Engler, J. A., Van Montagu, M., Engler, G., and Inze, D. (1994). Developmental expression of the *Arabidopsis* cyclin gene *cyc1At*. *Plant Cell* **6**, 1763–1774.

Finnegan, E. J., Genger, R. K., Peacock, W. J., and Dennis, E. S. (1998). DNA methylation in plants. *Annu. Rev. Plant Physiol. Plant Mol. Biol.* **49**, 223–247.

Finnegan, E. J., Peacock, W. J., and Dennis, E. S. (1996). Reduced DNA methylation in *Arabidopsis thaliana* results in abnormal plant development. *Proc. Natl. Acad. Sci. USA* **93**, 8449–8454.

Goday, C., and Esteban, M. R. (2001). Chromosome elimination in sciarid flies. *Bioessays* **23**, 242–250.

Graves, J. A. M. (1996). Mammals that break the rules: Genetics of marsupials and monotremes. *Annu. Rev. Genet.* **30**, 233–260.

Grossniklaus, U., Koltunow, A., and Van Lookeren Campagne, M. (1998a). A bright future for apomixis. *Trends Plant Sci.* **3**, 415–416.

Grossniklaus, U., Moore, J. M., and Gagliano, W. B. (1998b). Molecular and genetic approaches to understanding and engineering apomixis: *Arabidopsis* as a powerful tool. In "Hybrid Rice" (B. Hardy, ed.), International Rice Research Institute, Manila, The Philippines.

Grossniklaus, U., and Schneitz, K. (1998). The molecular and genetic basis of ovule and megagametophyte development. *Semin. Cell Dev. Biol.* **9**, 227–238.

Grossniklaus, U., Spillane, C., Page, D. R., and Koehler, C. (2001). Genomic imprinting and seed development: Endosperm formation with and without sex. *Curr. Opin. Plant Biol.* **4**, 21–27.

Grossniklaus, U., and Vielle-Calzada, J. P. (1998). Response: Parental conflict and infanticide during embryogenesis. *Trends Plant Sci* **3**, 328.

Grossniklaus, U., Vielle-Calzada, J. P., Hoeppner, M. A., and Gagliano, W. B. (1998c). Maternal control of embryogenesis by MEDEA a *Polycomb* group gene in *Arabidopsis*. *Nature* **280**, 446–450.

Haig, D. (1993). Genetic conflicts in human pregnancy. *Q. Rev. Biol.* **68**, 495–532.

Haig, D., and Graham, C. (1991). Genomic imprinting and the strange case of the insulin-like growth factor II receptor. *Cell* **64**, 1045–1046.

Haig, D., and Westoby, M. (1989). Parent specific gene expression and the triploid endosperm. *Am. Nature* **134**, 147–155.

Haig, D., and Westoby, M. (1991). Genomic imprinting in endosperm: its effect on seed development in crosses between species, and between different ploidies of the same species, and its implications for the evolution of apomixis. *Phil. Trans. R. Soc. Lond.* **333**, 1–13.

Hartley, R. W. (1998). Barnase and Barstar: two small proteins to fold and fit together. *Trends Biotech. Sci.* **14**, 450–454.

Herrick, G., and Seger, J. (1999). Imprinting and paternal elimination in Insects. *Results Probl. Cell Differ.* **25**, 41–71.

Hoyer-Fender, S., Costanzi, C., and Pehrson, J. R. (2000). Histone macroH2A1. is concentrated in the XY-body by the early pachytene stage of spermatogenesis. *Exp. Cell Res* **258**, 254–260.

Hurst, L. D., and McVean, G. T. (1998). Do we understand the evolution of genomic imprinting? *Curr. Opin. Genet. Dev.* **8**, 701–708.

Jackson-Grusby, L., Beard, C., Possemato, R., Tudor, M., Fambrough, D., Csankovszki, G., Dausman, J., Lee, P., Wilson, C., Lander, E., and Jaenisch, R. (2001). Loss of genomic methylation caused p53-dependent apoptosis and epigenetic deregulation. *Nature Biotechnol.* **27**, 31–39.

Jacobsen, S. E., Sakai, H., Finnegan, E. J., Cao, X., and Meyerowitz, E. M. (2000). Ectopic hypermethylation of flower-specific genes in *Arabidopsis*. *Curr. Biol.* **10**, 179–186.

Jaenish, R. (1997). DNA ethylation and imprinting: Why bother? *Trends Genet.* **13**, 323–329.

Jeddeloh, J. A., Stokes, T. L., and Richards, E. J. (1999). Maintenance of genomic methylation requires a SWI2/SNF2–like protein. *Nature Genet.* **22**, 94–97.

Jinno, Y., Yun, K., Nishiwaki, K., Kubota, T., and Ogawa, O. (1994). Mosaic and polymorphic imprinting at the *WT1* gene in humans. *Nature Genet.* **6**, 305–309.

John, R. M., and Surani, M. A. (2000). Genomic imprinting, mammalian evolution and the mystery of egg-laying mammals. *Cell* **101**, 585–588.

Johnson, S. A., den Nijs, T. P. M., Peloquin, S. J., and Hanneman, R. E. (1980). The significance of genic balance to endosperm development in interspecific crosses. *Theor. Appl. Genet.* **57,** 5–9.

Kaas, S. U., Pruss, D., and Wolffe, A.P. (1997). How does DNA methylation repress transcription? *Trends Genet.* **13,** 444–449.

Kalscheuer, V. M., Mariman, E. C., Schepens, M. T., Rehder, H., and Ropers, H. H. (1993). The insulin-like growth factor type-2 receptor gene is imprinted in the mouse but not in humans. *Nature Genet.* **5,** 75–78.

Kaufman, M. H. (1992). "The Atlas of Mouse Development." Springer-Verlag, New York; Academic Press, London.

Kermicle, J. L. (1969). Androgenesis conditioned by a mutation in maize. *Science* **166,** 1422–1424.

Kermicle, J. L. (1970). Dependance of the *R*-mottled aleurone phenotype in maize on mode of sexual transmission. *Genetics* **66,** 69–85.

Kermicle, J. L. (1971). Pleiotropic effects on seed development of the indeterminate gametophyte gene in maize. *Am. J. Bot.* **58,** 1–7.

Kermicle, J. L. (1978). Imprinting of gene action in maize endosperm. *In* "Maize Breeding and Genetics" (D. B. Walden, ed.), pp. 357–371. Wiley, New York.

Kermicle, J. L., and Alleman, M. (1990). Genomic imprinting in maize in relation to the angiosperm life cycle. *Dev. Supp.* 9–14.

Khosla, S., Aitchison, A., Gregory, R., Allen, N. D., and Feil, R. (1999). Parental allele-specific chromatin configuration in a boundary-imprinting-control element upstream of the mouse *H19* gene. *Mol. Cell Biol.* **19,** 2556–2566.

Killian, J. K., Byrd, J. C., Jirtle, J. V., Munday, B. L., Stoskopf, M. K., MacDonald, R. G., and Jirtle, R. L. (2000). M6P/IGF2R imprinting evolution in mammals. *Mol. Cell* **5,** 707–716.

Kinoshita, T., Yadegari, R., Harada, J. H., Goldberg, R. B., and Fisher, R. L. (1999). Imprinting of the *MEDEA* polycomb gene in the *Arabidopsis* endosperm. *Plant Cell* **11,** 1945–1952.

Kiyosue, T., Ohad, N., Yadegri, E., Hannon, M., Dinnery, J., Wells, D., Katz, A., Margossian, L., Harada, J., Goldberg, R. B., and Fisher, R. L. (1999). Control of fertilization-independent endosperm development by the *MEDEA* polycomb gene in *Arabidopsis. Proc. Natl. Acad. Sci. USA* **96,** 4186–4191.

Koltunow, A. M. (1993). Apomixis: embryo-sacs and embryos formed without meiosis or fertlization in ovules. *Plant Cell* **5,** 1425–1437.

Lee, J. T. (2000). Disruption of imprinted X inactivation by parent-of-origin effects at *Tsix. Cell* **103,** 17–27.

Lewis, J., and Bird, A. (1991). DNA methylation and chromatin structure. *FEBS Lett.* **285,** 155–159.

Lin, B-Y. (1982). Association of endosperm reduction with parental imprinting in maize. *Genetics* **100,** 475–486.

Lin, B-Y. (1984). Ploidy barrier to endosperm development in maize. *Genetics* **107,** 103–115.

Lopes, M. A., and Larkins, B. A. (1993). Endosperm origin, development, and function. *Plant Cell* **5,** 1383–1399.

Lund, G., Ciceri, P., and Viotti, A. (1995a). Maternal-specific demethylation and expression of specific alleles of zein genes in the endosperm of *Zea mays* L. *Plant J.* **8,** 571–581.

Lund, G., Messing, J., and Viotti, A. (1995b). Endosperm-specific demethylation and activation of specific alleles of alpha-tubulin genes of *Zea mays* L. *Mol. Gen. Genet.* **246,** 716–722.

Luo, M., Bilodeau, P., Dennis, E. S., Peacock, J., and Chaudhury, A. (2000). Expression and parent-of-origin effects for FIS2, MEA and FIE in the endosperm of developing *Arabidopsis* seeds. *Proc. Natl. Acad. Sci. USA* **94,** 10637–10642.

Luo, M., Bilodeau, P., Koltunow, A., Dennis, E. S., Peacock, W. J., and Chaudhury, A. (1999). Genes controlling fertilization-independent seed development in *Arabidopsis thaliana. Proc. Natl. Acad. Sci. USA* **96,** 296–301.

Marahrens, Y. (1999). X-inactivation by chromosomal pairing events. *Genes Dev.* **13,** 2624–2632.

Mascarenhas, J. P. (1989). The male gametophyte of flowering plants. *Plant Cell* **1**, 657–664.

Matzke, A. J., and Matzke, M. A. (1998). Position effects and epigenetic silencing of plant transgenes. *Curr. Opin. Plant Biol.* **1**, 142–148.

Mayer, W., Niveleau, A., Walter, J., Fundele, R., and Haaf, T. (2000a). Demethylation of the zygotic paternal genome. *Nature* **403**, 501–502.

Mayer, W., Smith, A., Fundele, R., and Haaf, T. (2000b). Spatial separation of parental genomes in preimplantation mouse embryos. *J. Cell Biol.* **148**, 626–634.

McGrath, J., and Solter, D. (1984). Completion of mouse embryogenesis requires both the maternal and paternal genomes. *Cell* **37**, 179–183.

McKee, B. D., and Handel, M. A. (1993). Sex chromosomes, recombination, and chromatin conformation. *Chromosoma* **102**, 71–80.

Messing, J., and Grossniklaus, U. (1999). Genomic imprinting in plants. *Results Probl. Cell Differ.* **25**, 52–70.

Mitsuya, K., Sui, H., Meguro, M., Kugoh, H., Jinno, Y., Niikawa, N., and Oshimura, M. (1997). Paternal expression of *WT1* in human fibroblasts and lymphocytes. *Hum. Mol. Genet.* **6**, 2243–2246.

Mittelsten-Scheid, O., Jakovleva, L., Afsar, K., Maluszynska, J., and Paszkowski, J. (1996). A change of ploidy can modify epigenetic silencing. *Proc. Natl. Acad. Sci. USA* **93**, 7114–7119.

Mlynarczyk, S. K., and Panning, B. (2000). X inactivation: *Tsix* and *Xist* as yin and yang. *Curr. Biol.* **10**, 899–903.

Moore, I., Gälweiler, L., Grosskopf, D., Schell, J., and Palme, K. (1998). A transcription activation system for regulated gene expression in transgenic plants. *Proc. Natl. Acad. Sci. USA* **95**, 376–381.

Moore, T., and Haig, D. (1991). Genomic imprinting in mammalian development: a parental tug-of-war. *Trends Genet.* **7**, 45–49.

Moore, T., Hurst, L. D., and Reik, W. (1995). Genetic conflict and evolution of mammalian X-chromosome inactivation. *Dev. Genet.* **17**, 206–211.

Newman-Smith, E. D., and Rothman, J. H. (1998). The maternal-to-zygotic transition in embryonic patterning of *Caenorhabditis elegans*. *Curr. Opin. Genet. Dev.* **8**, 472–480.

Nogler, G. A. (1984). Gametophytic Apomixis. *In* "Embryology of angiosperms" (B. M. Johri, ed.), pp 475–518. Springer Verlag, Berlin–Heidelberg–New York.

Nothias, J. Y., Miranda, M., and Depamphilis, M. L. (1996). Uncoupling of transcription and translation during zygotic gene activation in the mouse. *EMBO J.* **15**, 5715–5725.

Oakeley, E. J., Podesta, A., and Jost, J. P. (1997). Developmental changes in DNA methylation of the two tobacco pollen nuclei during maturation. *Proc. Natl. Acad. Sci. USA* **94**, 11721–11725.

Obata, Y., Kaneko-Inoshino, T., Koide, T., Takai, Y., Ueda, T., Domeki, I., Shiroishi, T., Ishino, F., and Kono, T. (1998). Disruption of primary imprinting during oocyte growth leads to the modified expression of imprinted genes during embryogenesis. *Development* **125**, 1553–1560.

Ogawa, O., McNoe, L. A., Eccles, M. R., Morison, I. M., and Reeve, A. E. (1993). Human insulin-like growth factor type-I and type-II receptors are not imprinted. *Hum. Mol. Genet.* **2**, 2163.

Ogura, A., Suzuki, O., Tanemura, K., Mochida, K., Kobayashi, Y., and Matsuda, J. (1998). Development of normal mice from metaphase I oocytes fertilized with primary spermatocytes. *Proc. Natl. Acad. Sci. USA* **95**, 5611–5615.

Ohad, N., Margossian, L., Hsu, Y. C., Williams, C., Repetti, P., and Fisher, R. L. (1996). A mutation that allows endosperm development without fertilization. *Proc. Natl. Acad. Sci. USA* **93**, 5319–5324.

Ohad, N., Yadegari, R., Margossian, L., Hannon, M., Michaeli, D., Harada, J. J., Goldberg, R. B., and Fischer, R. L. (1999). Mutations in *FIE*, a WD polycomb group gene, allow endosperm development without fertilization. *Plant Cell* **11**, 407–416.

O' Neill, M. J., Ingram, R. S., Vrana, P. B., and Tilghman, S. M. (2000). Allelic expression of *IGF2* in marsupials and birds. *Dev. Genes Evol.* **210**, 18–20.

Oswald, J., Engemann, S., Lane, N., Mayer, W., Olek, A., Fundele, R., Dean, W., Reik, W., and Walter, J. (2000). Active demethylation of the paternal genome in the mouse zygote. *Curr. Biol.* **10**, 475–478.

Pardo-Manuel de Villena, F., de la Casa-Esperon, E., and Sapienza, C. (2000). Natural selection and the function of genome imprinting. *Trends Genet.* **16**, 573–579.

Paszkowski, J., and Mittelsten Scheid, O. (1998). Plant genes: The genetics of epigenetics. *Curr. Biol.* **8**, 206–208.

Pereira, L. V., and Vasquez, L. R. (2000). X-chromosome inactivation: Lessons from transgenic mice. *Gene* **255**, 363–371.

Peterson, K., and Sapienza, C. (1993). Imprinting the genome: Imprinted genes, imprinting genes, and a hypothesis for their interaction. *Annu. Rev. Genet.* **27**, 7–31.

Pikaard, C. S. (2000). The epigenetics of nucleolar dominance. *Trends Genet.* **16**, 495–500.

Pirrotta, V. (1997). Chromatin-silencing mechanisms in *Drosophila* maintain patterns of gene expression. *Trends Genet.* **13**, 314–318.

Prymakowska-Bosak, M., Przewloka, M. R., Slusarczyk, J., Kuras, M., Lichota, J., Kilianczyk, B., and Jerzmanowski, A. (1999). R linker histones play a role in male meiosis and the development of pollen grains in tobacco. *Plant Cell* **11**, 2317–2330.

Redei, G. P. (1964). Crossing experiences with polyploids. http://www.*Arabidopsis* .org/ais/1964/redei-1964–aagkf.htm.

Reik, W., and Walter, J. (2001). Evolution of imprinting mechanisms: The battle of the sexes begins in the zygote. *Nature Genet.* **27**, 255–256.

Richards, E. J. (1997). DNA methylation and plant development. *Trends Genet.* **13**, 319–323.

Richler, C., Ast, G., Goitein, R., Wahrman, J., Sperling, R., and Sperling, J. (1994). Splicing components are excluded from the transcriptionally inactive XY body in male meiotic nuclei. *Mol. Biol. Cell* **5**, 1341–1352.

Roman, H. (1947). Mitotic non-disjunction in the case of interchanges involving the B-type chromosome in maize. *Genetics* **32**, 391–409.

Schultz, R. M. (1993). Regulation of zygotic gene activation in the mouse. *Bioessays* **15**, 531–538.

Scott R. J., Spielman M., and Bailey J., Dickinson H. G. (1998). Parent-of-origin effects on seed development in *Arabidopsis thaliana*. *Dev.* **125**, 3329–3341.

Sleutels, F., Barlow, D. P., and Lyle, R. (2000). The uniqueness of the imprinting mechanism. *Curr. Opin. Genet. Dev.* **10**, 229–233.

Sorensen, M. B., Chaudhury, A. M., Robert, H., Bancharel, E., and Berger, F. (2001). Polycomb group genes control pattern formation in plant seed. *Curr. Biol.* **11**, 277–281.

Spencer, H. G. (2000). Population genetics and evolution of genomic imprinting. *Annu. Rev. Genet.* **34**, 457–477.

Spencer, H. G., Feldman, M. W., and Clark, A. G. (1998). Genetic conflicts, multiple paternity and the evolution of genomic imprinting. *Genetics* **148**, 893–904.

Spillane, C., MacDougall, C., Stock, C., Koehler, C., Vielle-Calzada, J. P., Nunes, S. M., Grossniklaus, U., and Goodrich, J. (2000). Interaction of the *Arabidopsis* polycomb group proteins FIE and MEA mediates their common phenotypes. *Curr. Biol.* **10**, 1535–1538.

Spillane, C., Vielle-Calzada, J. P., and Grossniklaus, U. (2001). Parent-of-origin effects and seed development: Genetics and er; pigenetics. *In* "Handbook of Transgenic Food" (Y. H. Hui, G. G. Khachatourians, W. K. Nip, and R. Scorza, eds.), Marcel Dekker, New York, in press.

Sundaresan, V., Springer, P., Volpe, T., Haward, S., Jones, J. D. G., Dean, C., and Martienssen, R. (1995). Patterns of gene action in plant development revealed by enhancer trap transposable elements. *Genes Dev.* **9**, 1797–1810.

Surani, M. A. (1998). Imprinting and the initiation of gene silencing in the germ line. *Cell* **93**, 309–312.

Surani, M. A. H., Barton, S. C., and Norris, M. L. (1984). Development of reconstituted mouse eggs suggests imprinting of the genome during gametogenesis. *Nature* **308,** 548–550.

Szabo, P. E., and Mann, J. R. (1995). Biallelic expression of imprinted genes in the mouse germ line: Implications for erasure, establishment, and mechanisms of genomic imprinting. *Genes Dev.* **9,** 1857–1868.

Tada, M., Tada, T., Lefebvre, L., Barton, S. C., and Surani, M. A. (1997). Embryonic germ cells induce epigenetic reprogramming of somatic nucleus in hybrid cells. *EMBO J.* **16,** 6510–6520.

Takagi, N., and Sasaki, M. (1975). Preferential inactivation of the paternally derived X chromosome in the extraembryonic membranes of the mouse. *Nature* **256,** 640–642.

Thomas, J. H. (1995). Genomic imprinting proposed as a surveillance mechanism for chromosome loss. *Proc. Natl. Acad. Sci. USA* **92,** 480–485.

Tilghman, S. M. (1999). The sins of the fathers and mothers: genomic imprinting in mammalian development. *Cell* **96,** 185–193.

Tsuchiya, K. D., and Willard, H. F. (2000). Chromosomal domains and escape from X-inactivation: Comparative X-analysis in mouse and human. *Mamm. Gen.* **11,** 849–854.

Ueda, K., Kinoshita, Y., Xu, Z. J., Ide, N., Ono, M., Akahori, Y., Tanaka, I., and Inoue, M. (2000). Unusual core histones specifically expressed in male gametic cells of *Lilium longiflorum. Chromosoma* **108,** 491–500.

Ueda, K., and Tanaka, I. (1995). The appearance of male gamete-specific histones gH2B and gH3 during pollen development in *Lilium longiflorum. Dev. Biol.* **169,** 210–217.

Vielle-Calzada, J. P., Baskar, R., and Grossniklaus, U. (2000). Delayed activation of the paternal genome during seed development. *Nature* **404,** 91–94.

Vielle-Calzada, J. P., Thomas, J., Spillane, C., Coluccio, A., Hoeppner, M. A., and Grossniklaus, U. (1999). Maintenance of genomic imprinting at the *Arabidopsis medea* locus requires zygotic *DDM1* activity. *Genes Dev.* **13,** 2971–2982.

Vinkenoog, R., Spielman, M., Adams, S., Fisher, R. L., Dickinson, H. G., and Scott, R. J. S. (2000). Hypomethylation promotes autonomous endosperm development and rescues postfertilization lethality in *fie* mutants. *Plant Cell* **12,** 2271–2282.

Vizir, I. Y., Anderson, M. L., Wilson, Z. A., and Mulligan, B. J. (1994). Isolation of deficiencies in the *Arabidopsis* genome by gamma-irradiation of pollen *Genetics* **137,** 1111–1119.

Wolffe, A. (1995). "Chromatin: Structure and function," 2nd ed. Academic Press.

Wutz, A., Smrzka, O. W., and Barlow, D. P. (1998). Making sense of imprinting at the mouse and human *IGF2R* loci. *Novartis Found. Symp.* **214,** 251–259.

Xu, H., Swoboda, I., Bhalla, P. L., and Singh, M. B. (1999). Male gametic cell-specific expression of H2A and H3 histone genes. *Plant Mol. Biol.* **39,** 607–601.

Xu, Y., Goodyer, C. G., Deal, C., and Polychronakos, C. (1993). Functional polymorphism in the parental imprinting of the human *IGF2R* gene. *Biochem. Biophys. Res. Commun.* **197,** 747–754.

Yadegari, R., Kinoshita, T., Lotan, O., Cohen, G., Katz, A., Choi, Y., Katz, A., Nakashima, K., Harada, J. J., Goldberg, R. B., Fischer, R. L., and Ohad, N. (2000). Mutations in the *FIE* and *MEA* genes that encode interacting polycomb proteins cause parent-of-origin effects on seed development by distinct mechanisms. *Plant Cell* **12,** 2367–2381.

7

Long-Distance Cis and Trans Interactions Mediate Paramutation

Vicki L. Chandler,* Maike Stam,† and Lyudmila V. Sidorenko
Department of Plant Sciences
University of Arizona
Tucson, Arizona 85721

I. INTRODUCTION

Paramutation is an interaction between alleles that leads to a mitotically and meiotically heritable change in gene expression. Paramutation was first described in maize (Brink, 1956), and similar phenomena have been observed in other plants (reviewed in Brink, 1973). While paramutation was first described as an interaction between two distinct alleles of the same gene, recent findings indicate

*To whom correspondence should be addressed: E-mail: chandler@ag.arizona.edu; Telephone (520) 626 8725; Fax (520) 621 7186.
†Present address: Department of Developmental Genetics, Vrije Universiteit, De Boelelaan 1087, 1081 HV Amsterdam, The Netherlands.

that paramutation-like interactions can occur between two homologous trans-genes (Meyer *et al.*, 1993), or a transgene and a homologous endogenous gene (Sidorenko and Peterson, 2001). Meiotically heritable changes in gene expression have also been reported in certain animal and fungal systems (reviewed in Klar, 1998). Paramutation has been studied at four genes in maize, and this literature has been reviewed recently (Chandler *et al.*, 2000). In this review we focus on paramutation at two maize loci, *b1* and *p1*, for which the sequences required for paramutation have recently been identified. After reviewing the basic pheno-menology associated with paramutation, we describe recent experiments that have identified cis-acting regions required for paramutation at these two loci, and compare the similarities and differences in results. Finally, we present a model for the mechanism of paramutation that incorporates these recent findings and discuss possible functions for paramutation.

II. BACKGROUND AND DEFINITIONS

Paramutation has been described for four maize genes, all of which encode transcription factors that activate the biosynthesis of colored flavonoid pigments (Figure 7.1). We suspect that paramutation was discovered using these genes because the amount of pigment is exquisitely sensitive to the level of these

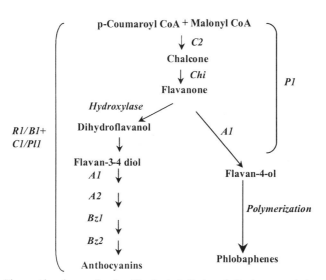

Figure 7.1. Flavonoid pathway of maize. The basic helix-loop-helix factors and the myb factors encoded by the *r1/b1* and *c1/pl1* genes activate all the genes of the pathway required for the synthesis of anthocyanin pigments, while the myb protein P1 activates the subset of these genes required to form phlobaphene pigments. The names of the genes or enzymes for the biosynthetic steps are indicated (Dooner *et al.*, 1991).

regulatory proteins, such that small changes in regulatory protein amounts can produce dramatic changes in the amounts of pigment observed. Two of the genes for which paramutation has been described, *b1* (*booster1*) and *r1* (*red1*), encode functionally duplicate basic helix-loop-helix (bHLH) factors. A third gene showing paramutation, *pl1* (*purple plant1*), and another gene for which paramutation has not been described, *c1* (*colorless1*), encode myb factors. *b1* and *r1* genes are typically expressed in distinct tissues within the seed, seedling, or mature plant (Ludwig and Wessler, 1990). These two genes are homologous, and their map positions are consistent with their being the orthologs in the two species that hybridized to form the ancestral allotetraploid that led to maize (Gaut and Doebley, 1997). The *pl1* or *c1* encoded myb proteins are typically expressed in distinct tissues and are likely to be orthologs (Cone *et al.*, 1993). Pigment is formed where the expression patterns of at least one bHLH and myb protein intersect, as these two classes of proteins interact to activate the pathway (Goff *et al.*, 1992). The fourth gene for which paramutation has been described is *p1* (*pericarp pigmentation1*), which encodes a *myb*-like transcriptional regulator required for the synthesis of the orangy red phlobaphene (flavanoid) pigment in husks, silks, kernel pericarp, cob, and tassel glumes (Grotewold *et al.*, 1991). Extensive studies have indicated that while the C1 and Pl1 proteins require B1 or R1 proteins to activate the anthocyanin pathway, P1 works independently of B1 or R1 to activate the phlobaphene pathway (Grotewold *et al.*, 1994; Sainz *et al.*, 1997; Grotewold *et al.*, 2000). The two pathways share the first few biosynthetic steps (Figure 7.1).

As summarized in detail in a recent review, there are similarities and differences in paramutation at the four loci (Chandler *et al.*, 2000). A common feature of paramutation at all four loci is that paramutation is limited to specific alleles of each gene, and paramutation does not occur between alleles of the different genes. All four examples of paramutation involve an interaction between alleles that leads to a heritable reduction in the expression of one of the alleles. Alleles whose expression becomes altered are termed paramutable, and alleles that induce the change, paramutagenic. Following paramutation, paramutable (sensitive) alleles are termed paramutant (or paramutated) and designated with an apostrophe (i.e., generically, B', R' etc.). Many alleles at each gene do not participate in paramutation; such alleles are referred to in the literature as either neutral or nonparamutagenic. Another similarity is that for two loci (*b1*, *p1*), paramutation is associated with reduced transcription (Patterson *et al.*, 1993; Hollick *et al.*, 2000). Paramutation at *p1* is associated with reduced transcript levels, but nuclear run-on assays have not been done to determine if reduced transcript levels at *p1* are associated with transcriptional or posttranscriptional changes. No studies on *r1* transcript levels or transcription rates have been published. There are also many differences between paramutation at the distinct loci. For example, the reduced expression state is extremely stable at *b1* (Coe, 1966; Patterson *et al.*, 1995), but can change back to higher expression states at *r1* (Styles and Brink, 1966; Brink *et al.*, 1968; Brink, 1973), *pl1* (Hollick and Chandler, 1998; Hollick *et al.*, 1995), and *p1* (Das and Messing, 1994b; Sidorenko and Peterson, 2001). The frequency with which this occurs can vary, depending on the nature of the other allele and genetic background (reviewed in Chandler *et al.*, 2000). Paramutation at *r1* and *p1* correlates with increased DNA methylation in the transcribed or promoter proximal regions (Walker, 1998; Das and Messing, 1994b; Sidorenko and Peterson, 2001), whereas the DNA methylation levels do not change after paramutation within the comparable regions of *b1* and *pl1* (Patterson *et al.*, 1993; Hollick *et al.*, 2000). However, in spite of these differences, paramutation at the different loci is likely to share common mechanistic features, as a mutation in a trans-acting factor that affects *b1* paramutation also affects paramutation at *pl1* and *r1* (Dorweiler *et al.*, 2000).

Experiments have been performed to determine where the sequences required for paramutation are located. The complexity of the *r1* paramutagenic alleles (multiple copies located in tandem arrays) enabled the mapping of the sequences required for paramutagenic activity using unequal crossing over among the repeats (Kermicle *et al.*, 1995; Panavas *et al.*, 1999). These experiments indicated that no particular region was required for paramutagenicity, but that the strength of paramutation correlated with the number of repeats. The sequences required for *pl1* paramutation are tightly linked to the locus (Hollick *et al.*, 1995), but no recombination experiments have been performed to localize the required sequences more precisely. Minimal sequences required for both *p1* and *b1* paramutation have been recently identified and are described in detail below.

III. PARAMUTATION AT *b1* AND *p1*

Figure 7.2 (see color insert) diagrams paramutation at *b1* and *p1*, the focus of this review. At *b1*, the paramutagenic allele is *B′*, which produces a lightly pigmented plant phenotype. *B′* can arise spontaneously from the highly expressed paramutable *B-I* allele (darkly pigmented plants) at a frequency of ∼10% (Coe, 1966). When *B′* is crossed to *B-I* it always changes *B-I* into *B′* (Figure 7.2A), resulting in lightly pigmented plants: no exceptions have been observed in typical inbred and hybrid genetic backgrounds (Coe, 1966). Thus, at *b1*, paramutation is fully penetrant and the resulting *B′* alleles are extremely stable, independent of how the alleles are maintained (homozygous or heterozygous with *B-I* or neutral alleles).

Similar to *b1* paramutation, the paramutable *P1-rr* (red pericarp and red cob) allele can change spontaneously to a lower expression state, *P1-pr* (patterned pericarp and red cob). In contrast to *B-I*, these spontaneous changes are quite rare (∼1 × 10^6; Das and Messing, 1994b). The *P1-pr* epiallele is identical to the *P1-rr* gene in DNA sequence, but is hypermethylated (Das and Messing, 1993) and exhibits a more closed chromatin structure relative to *P1-rr* (Lund *et al.*, 1995). While *B′* exhibits a more closed chromatin structure around the transcription start site (Chandler *et al.*, 2000), there is no detectable hypermethylation relative to *B-I* within the coding region or ~15 kbp upstream (Patterson *et al.*, 1993). The reduced expression state of *P1-pr* is unstable, as germinal reversions of the *P1-pr* state to fully pigmented *P1-rr* occur at a frequency of ~1% (Das and Messing, 1994b). As originally isolated, the *P1-pr* allele was not paramutagenic; it did not induce a heritable change in *P1-rr* in heterozygotes (Das and Messing, 1994a). However, in certain genetic backgrounds, the *P1-pr* allele is paramutagenic, such that when it is crossed with the *P1-rr* stock, the F$_1$ progeny have low pigment and *P1-rr* is heritably changed into *P1-pr* (J. Messing and W. Goettel, personal communication). Thus, at both loci, the alleles can exist in two states. At *b1*, the paramutable allele (*B-I*) is unstable changing into an extremely stable paramutagenic state (*B'*), whereas at *p1*, the paramutagenic allele (*P1-pr*) is unstable changing into a stable paramutable state (*P1-rr*).

A. Identification of sequences required for *p1* paramutation

Paramutation-like effects were observed in the progeny of transgenic maize plants carrying specific regions of the *P1-rr* promoter (Sidorenko and Peterson, 2001). Transgenic lines containing the *GUS* reporter gene fused to the basal promoter, Pb, and one of three different fragments from the 5.2-kbp transcriptional regulatory region of *P1-rr* (Figure 7.3, described further in Section III. C) were crossed with *P1-rr* plants. The P1.2 construct contained a 1.2-kbp enhancer fragment located ~5 kbp upstream and also downstream of the transcription start site, while P1.0 and P2.0 contained the 1.0-kbp enhancer region, or the larger 2.0-kbp region, located ~200 bp upstream of the transcription start site (Figure 7.3C; Sidorenko and Peterson, 2001). Progeny from crosses involving three independent transgenic lines carrying the distal P1.2 enhancer fragment and the *P1-rr* allele showed reduced expression of both the *P1.2b :: GUS* transgene and the endogenous *P1-rr* gene. The reduced expression state of the endogenous gene is referred to as *P1-rr'*. This nomenclature was chosen to distinguish the transgene-induced paramutant state from the spontaneously derived *P1-pr* state. Importantly, *P1-rr'* was observed only in crosses with the P1.2 transgenic lines. No reduced expression of *P1-rr* was observed in progeny from crosses involving transgenic lines that carried the proximal P1.0 enhancer or the larger P2.0 fragment (Sidorenko and Peterson, 2001). All independent transgenic loci are complex, containing multiple copies of the transgene at one or two loci (Sidorenko *et al.*, 2000; Sidorenko and Peterson, 2001).

The *P1-rr'* phenotype was observed in crosses that involved all three independent transgenic lines with the P1.2 construct, but the frequency of *P1-rr'* ears varied between independent crosses with each line (Sidorenko and Peterson, 2001). While not all plants that received the transgene showed the *P1-rr'* phenotype, the *P1-rr'* phenotype was observed only in plants that carried the transgene. The tight correlation between the reduced pigment phenotype and transgene

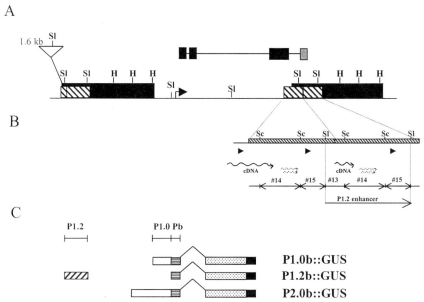

Figure 7.3. Diagram of the *P1-rr* locus and transgenic constructs used to identify sequences required for paramutation. (A) The maize *P1-rr* gene coding sequence is indicated by small black boxes (exons) and a thin line (introns); the alternatively spliced exon is indicated as a gray box. The transcription start site is shown by a bent arrow. The two 5.2-kbp direct repeats flanking the coding sequence are indicated as black bars, and the 1.2-kbp direct repeats are indicated as hatched boxes. The 1.6-kbp transposon present in standard *P1-rr* is indicated as a triangle. Restriction sites are Sl for *Sal*I, H for *Hind*III. (B). The expanded region shows the detailed structure of the 1.2-kbp direct repeats located immediately downstream of the *P1-rr* coding region. This arrangement of sequences is partially repeated ~5 kbp upstream of the coding region. The first large wavy arrow indicates homology to the 3′ portion of the *P1-rr* transcript, while the shorter wavy arrow indicates homology to the *P1-rr* transcript that is present in alternative splicing products (Grotewold *et al.*, 1991). The arrow heads indicate a 64-bp region of homology to the transcribed sequence. The dotted wavy arrows indicate homologies to Stanford EST AI891238 and the Q transcript (Grotewold and Peterson, 1990). Restriction fragments (#13, #14, #15) resulting from *Sal*I + *Sac*I digestion are indicated as double-headed arrows. The location of the 1.2-kbp *Sal*I fragment (the same sequence as that for the P1.2 fragment located upstream of the coding region) is indicated below. Restriction sites are Sc for *Sac*I, S1 for *Sal*I. (C) The P1.0b :: GUS, P2.0b :: GUS and P1.2b :: GUS constructs contain the basal fragment of the *P1-rr* promoter sequence (Pb; horizontally striped box), the *Adh1* gene intron I (bent line), the GUS gene coding region (dotted box), and the *PinII* gene 3′ end (black box). The P1.0b :: GUS and P2.0b :: GUS constructs contain the 1.0-kbp proximal enhancer fragment (open box), while P1.2b :: GUS contains the 1.2-kbp distal enhancer fragment (hatched box). The reversed hatching of the P1.2 fragment indicates its inverted orientation in the P1.2b :: GUS construct.

presence suggested that the transgene might be paramutagenic and its interaction with the paramutable *P1-rr* allele resulted in *P1-rr'*. This hypothesis was confirmed in subsequent crosses as the *P1-rr'* phenotype was highly heritable in the absence of the paramutagenic transgene. These results precisely fit the classical definition of paramutation (Figure 7.2B). Molecular analysis of *P1-rr'* indicated a tight correlation between reduced pigment, increased DNA methylation and decreased transcript levels (Sidorenko and Peterson, 2001).

Importantly, the newly paramutated *P1-rr'* allele was itself paramutagenic. In crosses combining naïve *P1-rr* alleles with *P1-rr'* in plants lacking the transgene, paramutation occurred: *P1-rr* was changed to *P1-rr'* (Sidorenko and Peterson, 2001). The ability to cause paramutation segregated with the *P1-rr'* allele. Taken together, these results demonstrate that a small portion of the transcriptional regulatory region located in ectopic locations is sufficient to initiate paramutation of an endogenous gene. The paramutant state of the endogenous gene is heritable and capable of secondary paramutation. Furthermore, these results suggest a potential link between paramutation and transgene silencing, as transcriptional gene silencing (TGS), which is generally, heritable, can be induced by interaction between a specific transgene with strong silencing capabilities and a second transgene that shares homology within the promoters (Vaucheret, 1993; Vaucheret *et al.*, 1998; Chapter 8).

B. Identification of sequences required for *b1* paramutation

Genetic mapping experiments have demonstrated that the sequences required for *b1* paramutation mapped upstream of the coding region in *B-I* and *B'*. Initially, crosses of *B'* and *B-I* with neutral, nonparamutagenic alleles demonstrated that the ability to participate in paramutation was tightly linked to *B'* and *B-I* (Coe, 1966; Patterson *et al.*, 1995). Further identification of where these sequences lie was achieved by using recombination with the neutral *B-Peru* allele, which is expressed in seeds (Radicella *et al.*, 1992; Selinger *et al.*, 1998) and has some 5' sequences in common with *B'*, but is quite polymorphic further upstream (Patterson *et al.*, 1995). The sequences in common permitted recombination between the alleles, while the polymorphisms enabled mapping of the recombination sites. Recombination within the coding region of the alleles that participate in paramutation (*B'* or *B-I*) and *B-Peru* localized the sequences required for *b1* paramutation (Patterson *et al.*, 1995). All recombinants that retained the upstream regions of *B'* or *B-I* participated in paramutation, whereas recombinants retaining only the coding and 3' regions of *B'* and *B-I* were neutral. The most upstream recombination event occurred within the first exon, demonstrating that sequences within the promoter proximal region or farther upstream were required for paramutation.

A transgenic approach was initiated to try to test whether the available cloned sequences from *B'* were sufficient for paramutation (8.5 kbp upstream

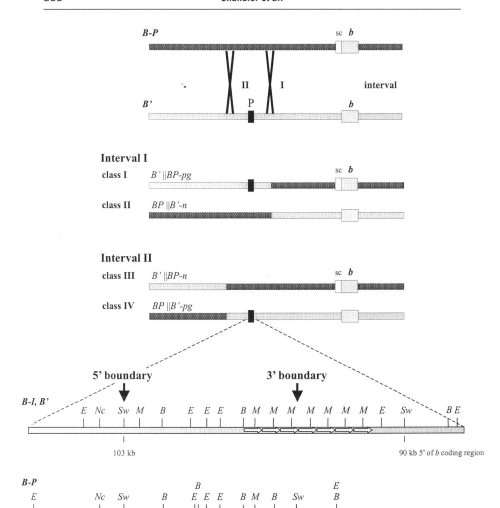

Figure 7.4. Diagram of the strategy to map the sequences required for *b1* paramutation and a restriction map of the defined region. The chromosome carrying the neutral *B-P* allele is darkly dotted and the chromosome carrying the paramutagenic *B′* allele is lightly dotted. The *b1* coding region is indicated by a gray box, the sequence responsible for the purple *B-P* seed color (sc) by a white box, and the sequences required for paramutation (P) by a black box. The *B-P* chromosome contains sequences similar enough to *B′* to allow recombination, but polymorphic enough to allow mapping of the recombination breakpoints. We isolated recombinant alleles between the coding region and P (interval I) and upstream of P (interval II). In interval I, recombinant alleles conferring purple seed pigment (containing the *B-P* coding and promoter proximal region) acquired paramutagenicity (class I), while recombinant alleles conferring colorless seeds (containing the *B′* coding and promoter proximal region) lost their paramutagenicity (class II). In interval II, recombinant alleles conferring purple seed did not gain paramutagenicity (class III), while recombinant

of the coding region, the coding region, and 1.5 kbp downstream of the coding sequences). These sequences from B' were introduced into transgenic plants using biolistics. Eighteen independent lines were generated, but none of these plants showed pigment, even though most of the lines had at least one intact copy of the transgene (V. Chandler, K. Kubo, M. Anderson, unpublished data). Two possibilities could explain the lack of expression: (1) the sequences required for plant expression may not have been within the construct; or (2) there was a very high frequency of transgene silencing. Subsequent crosses were performed to determine if these sequences could participate in paramutation even in the absence of detectable expression. To determine if the transgenes could silence B-I, each line was crossed with B-I plants and the pigment phenotype was examined in the F_1 plants. All plants showed B-I pigment levels, indicating that the transgenes were not paramutagenic. To determine if the transgenes could become paramutagenic by exposure to the endogenous B' allele, each transgenic line (containing a single transgene locus) was first crossed to B', and F_1 plants were subsequently crossed to B-I. If the transgene could become paramutagenic after exposure to B', the expectation would be that plants receiving the transgene would be able to paramutate B-I, observed by reduced amounts of pigment in progeny. That was not observed: only plants that received the endogenous B' allele demonstrated paramutation. No individuals that retained the transgene, but had segregated B' away, were able to paramutate B-I. Thus, the transgenic sequences were unable to participate in paramutation. This negative result could have been obtained for numerous reasons. (1) The sequences required for paramutation may not have been within the introduced construct. (2) All insertions may have occurred within regions nonpermissible for paramutation. (3) The transgenes were not expressed and expression may be required to participate in paramutation. (4) Paramutation at $b1$ requires interacting sequences to be in allelic positions instead of ectopic. In any event, a different approach was needed to further identify the sequences required for $b1$ paramutation.

To further localize the upstream sequences that participate in paramutation, we developed the recombination strategy shown in Figure 7.4. Recombination

alleles conferring colorless seeds did not loose their paramutagenicity (class IV). The most upstream recombination breakpoint in interval I (95 kb upstream of the $b1$ transcription start site) and the most downstream recombination breakpoint in interval II (103 kb upstream of the $b1$ transcription start site) delineate the P region: these are indicated as the 3′ and 5′ boundaries, respectively, in the more detailed structure of this region. An interesting feature of this region is the presence of a sequence that is tandemly repeated seven times in the paramutagenic B' and paramutable B-I alleles (indicated by the white arrows.) In contrast, the neutral B-P allele only contains one copy of this sequence. Key: n for neutral; pg for paramutagenic; || designates a recombinant allele with the parental allele contributing the sequences upstream of the recombination event on the left side of the vertical bars, and the parental allele contributing the sequences downstream of the recombination event, the promoter proximal region and the coding region, on the right side of the bars; E for EcoRI; B for BamHI; M for MluI; Nc for NcoI; Sw for SwaI.

events between B' or B-I and B-$Peru$, upstream of the promoter proximal region, were identified using a combination of phenotypic and molecular markers (M. Stam and V. Chandler, unpublished data). We assumed that B-$Peru$ lacked upstream sequences required for paramutation. These sequences are indicated in Figure 7.4 as a black rectangle and are referred to as the P region. Recombination between the P region and the b coding sequences (interval I) should replace the P sequences in B' with the neutral B-$Peru$ sequences, producing a neutral recombinant allele (Figure 7.4, class II, $BP||B'$-n). The reciprocal event should transfer the P sequences from B' to B-$Peru$, creating a novel paramutagenic allele (Figure 7.4, class I, $B'||BP$-pg). Recombination events within Interval I can be distinguished from parental nonrecombinant alleles based on their pigment phenotypes in the following crossing scheme. When heterozygous B'/B-$Peru$ plants are crossed with B-I/B-I, two parental genotypes segregate: B'/B-I (colorless seeds, light plants) and B-$Peru/B$-I (purple seeds, dark plants). The purple seed phenotype is caused by an insertion in the promoter proximal region of B-$Peru$ relative to B' (sc in Figure 7.4; Radicella *et al.*, 1992; Selinger *et al.*, 1998). The desired recombinant class II ($BP||B'$-n) would have colorless seed (retains B' promoter proximal region and coding sequences), but when heterozygous with B-I, would produce dark purple plants. Dark plants will appear because the alleles would have lost the P sequences and therefore have a region of the B-$Peru$ chromosome that cannot paramutate B-I, resulting in dark F_1 plants. Recombinant class I alleles ($B'||BP$-pg) would have purple seed, but when heterozygous with B-I, would produce only light plants, because they would have gained the P sequences and paramutated B-I. Recombinants within Interval II are not distinguishable from parental nonrecombinant alleles based on pigment phenotypes. Thus, PCR assays with tightly linked molecular markers (Stam *et al.*, 2000) were used to identify recombinant class III and IV alleles (Figure 7.4). All four types of recombinants have been isolated and characterized as discussed in detail below (M. Stam, C. Belele, and V. Chandler, in preparation). These results demonstrate that the paramutagenic B' allele contains sequences required for inducing paramutation and that these sequences can be transferred to a neutral allele, making it paramutagenic. A similar approach was used to isolate recombinants between B-I and B-$Peru$ in which the very weak B-$Peru$ plant expression was substantially increased in certain B-$I||BP$ recombinants. This enabled us to localize an enhancer within B-I that caused increased plant expression (M. Stam, C. Belele, and V. Chandler, in preparation).

The recombination breakpoints were first roughly mapped using DNA blot analyses of gels from standard and pulsed-field gel electrophoresis (Monaco, 1995), which indicated that the recombination breakpoints, and thus the sequences required for paramutation, were quite far upstream (> 50 kbp). We then isolated a BAC (bacterial artificial chromosome) clone that contained ~100 kbp of B' upstream sequences, and used those sequences to compare B', B-I, and B-$Peru$ and to identify more precisely the recombination breakpoints (M. Stam, C. Belele, and V. Chandler, in preparation). These experiments revealed that

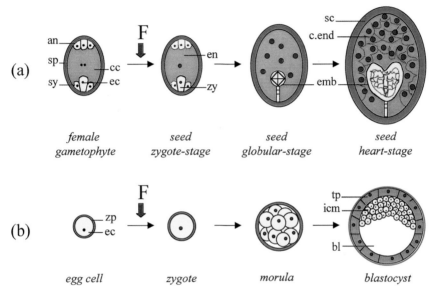

Figure 6.2. Post-fertilisation development of embryonic and extraembryonic tissues in plants and mammals. (a) Seed development in Arabidopsis thaliana: sp, sporophytic tissues (ovule integuments); the embryo sac is constituted by seven cell types: three antipodal cells, two synergids, one egg cell, one central cell. an, antipodal cells sy, synergid cells; ec, egg cell; cc, central cell; emb, embryo proper; c.end, celluarised endosperm. Note that the central cell is homodiploid, and the endosperm is a triploid tissue. The zygote is diploid. (b) Mouse embryo development: zp, zona pellucida; ec, egg cell; tp, trophectoderm; icm, inner cell mass; bl, blastocoel. F, fertilization. Yellow refers to embryonic tissues, orange sectors refer to extra-embryonic tissues. Green refers to maternal surrounding tissue (a) or extracellular matrix (b).

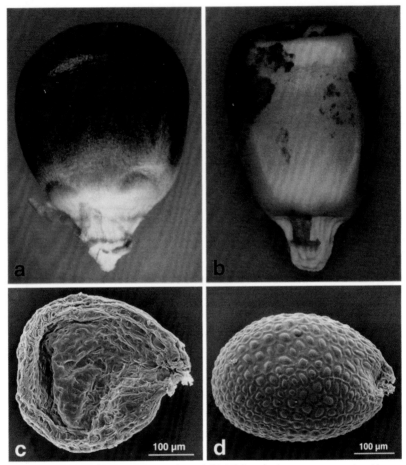

Figure 6.3. Gene-specific imprinting in plants (a,b) Allele-specific imprinting at the rl-locus in maize: The R-r:std allele (designated R allele) encodes a regulator of anthocyanin production in the aleurone, an endosperm cell layer beneath the seed coat. The r-g allele (designated r allele) confers a colorless phenotype to the kernel. Kernels from RR lines are dark purple, and kernels from rr lines are yellow. Kernels derived from reciprocal crosses between RR lines and rr lines display two different phenotypes depending on the parent-of-origin of R: (a) the kernel has a fully colored dark purple phenotype only when R is transmitted through the mother, whereas (b) when R is transmitted through the father, kernels display a mottled phenotype, with purple sectors on the yellow kernel. (c,d) Gene-specific imprinting at MEA locus in Arabidopsis: MEA encodes a Polycomb-group protein and regulates cell proliferation during seed development. Disruption of the MEA gene generates a gametophytic maternal effect embryo-lethal phenotype. Plants heterozygous for the mea mutation produces 50% normal seeds and 50% aborted seeds when self fertilized. Homozygous mea/mea plants can be rescued *in vitro* and produce 100% aborted seeds. Seeds derived from reciprocal crosses between mea/mea and wild-type plants display two different phenotypes depending on the parent-of-origin of mea: (c) seeds abort only when inheriting a mutant mea allele from the mother, while (d) when mea is transmitted through the father, the progeny is wild-type.

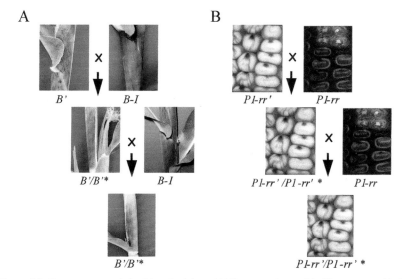

A

B' × B-I

B'/B'* × B-I

B'/B'*

B

Pl-rr' × Pl-rr

Pl-rr'/Pl-rr' * × Pl-rr

Pl-rr'/Pl-rr' *

Figure 7.2. Paramutation of the *b1* and *p1* loci. (A) Paramutation at *b1*. The paramutable *b1* allele *B-I*, which confers dark purple pigment throughout vegetative tissues, is altered with a 100% efficiency into an allele conferring light purple pigment, *B'*, in *B'/B-I* heterozygous plants. The newly paramutated *B'* allele, *B'**, is indistinguishable from *B'*; it paramutates naïve *B-I* alleles in subsequent generations with a 100% frequency. (B) Paramutation at *p1*. The *P1-rr* allele of *p1* confers brick red pigmentation of pericarp, cob, husks, and tassel glumes. *P1-rr'* exhibits reduced pigmentation of all organs with the most dramatic phenotype in the maize kernels as shown. *P1-rr'* can be induced by exposure of the naïve *P1-rr* allele to a transgene carrying the P1.2 enchancer fragment (Sidorenko and Peterson, 2001). The *P1-rr* allele can also spontaneously change to a paramutant state, called *P1-pr* (Das and Messing, 1994b). Both spontaneously derived *P1-pr* and transgene induced *P1-rr'* exhibit phenotypes that are indistinguishable from each other. When *P1-rr'* (shown) or *P1-pr* (not shown) are crossed to *P1-rr*, paramutation occurs.

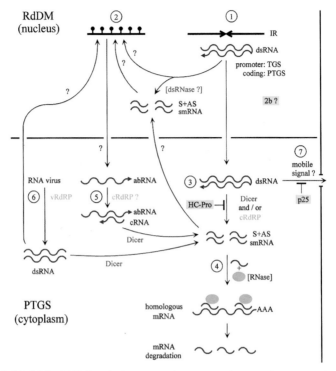

Figure 8.1. Model for RNA-based silencing mechanisms in plants. In the nucleus, dsRNA (red wavy lines) can induce directly, or indirectly via smRNAs, methylation of homologous DNA sequences (RdDM) 2. The most efficient trans-acting dsRNAs are transcribed through IRs ①. If dsRNAs contain promoter sequences, they can trigger TGS via de novo methylation of homologous promoters; dsRNAs that contain coding regions can induce de novo methylation of homologous DNA sequences as well as PTGS of homologous mRNAs in the cytoplasm. DsRNAs trigger degradation of homologous mRNAs in a process involving the production of small sense and antisense RNAs that serve as guides for RNase ③ ④. Cytoplasmic dsRNA can be produced by transcription of IRs ①, or through the action of a cellular ⑤ or viral ⑥ RdRP on aberrant RNAs or viral RNA templates, respectively. An RNA species associated with PTGS, either dsRNA ⑥ or smRNAs ③, can enter the nucleus to trigger RdDM. Enzymes shown in brackets are postulated to exist but have not yet been identified in mutant screens. Question marks indicate activities or alternatives steps that have not yet been completely clarified. Viral suppressors of PTGS (Section I.A.1.a) are shown in yellow boxes. Abbreviations: dsRNA, double stranded RNA; IR, inverted DNA repeat; dsRNase, a ribonuclease that cleaves dsRNA; S, sense, AS, antisense; smRNA, small RNAs; abRNAs, aberrant RNAs; cRdRP, cellular RNA-dependent RNA polymerase; vRdRP viral RNA-dependent RNA polymerase. RdDM, RNA-directed DNA methylation; PTGS, posttranscriptional gene silencing.

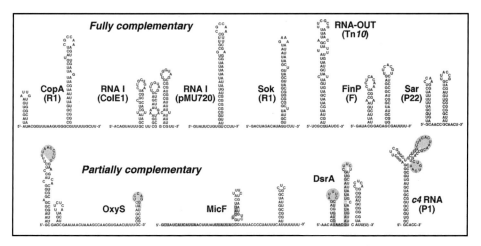

Figure 12.1. Predicted or experimentally determined secondary structures of antisense RNAs. Antisense RNAs in the upper row are completely complementary to target RNAs. In the lower row, regions of target complementarity are indicated by pink color. For DsrA, two different, but overlapping regions of target complementarity are shown (pink and yellow). See text for details.

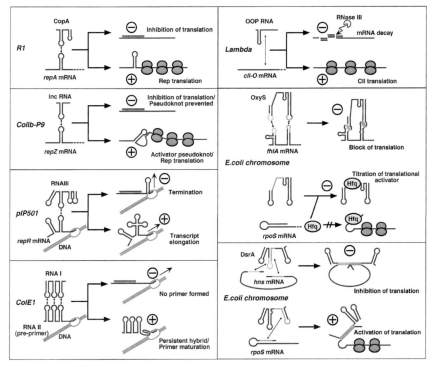

Figure 12.2. Different regulatory mechanisms. Antisense RNAs are shown in red, target RNAs in blue. Interacting sequences (for the RNAs that show only partial complementarity to their targets) are highlighted (OxyS, DsrA). (+) and (-) indicate activation and inhibition, respectively. The left boxes show examples of replication control mechanisms in plasmids. The right boxes show examples of phage and chromosomally trans-encoded regulatory RNAs. See text for details.

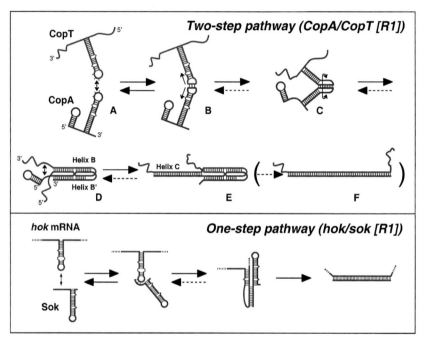

Figure 12.3. One-step and two-step antisense-target RNA binding pathways. Binding pathways of antisense RNAs (red) with their respective target mRNAs (blue) are shown. The upper box shows the proposed binding pathway of CopA to CopT (plasmid R1). Binding initiates by loop-loop contact. The stepwise progression from free RNAs to subsequently more stable intermediates (A–E) is discussed in the text. The arrow between E and F indicates that subsequent duplex formation is slow and biologically not important. The lower box shows the binding pathway of Sok and hok mRNA. Here, binding initiates between the single-stranded 5′-tail of Sok and the hok target stem-loop.

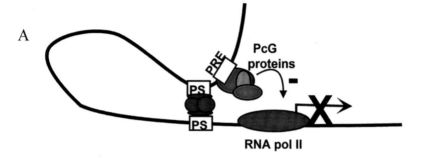

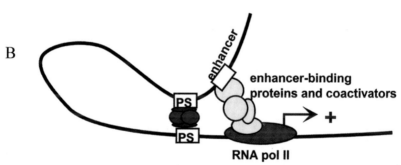

Figure 14.1. Model for pairing-sensitive (PS) sites. Proteins bound to PS sites near the promoter interact with proteins bound to PS sites located near a PRE (A) or an enhancer (B). Note that in this model, the activity of the PRE is dependent on the activity of the flanking PS sites.

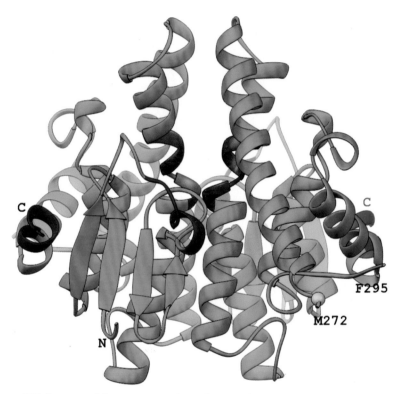

Figure 17.7. Structure of the nitrogen regulation domain of Ure2p. Ure2p is a homodimer with a structure closely similar to glutathione-S-transferases (Umland *et al.*, 2001; Bousset *et al.*, 2001). The figure shows the structure of the C-terminal domain without residues M272 to F295, with one monomer in green and the other in blue (Umland *et al.*, 2001). Deletion of the purple regions increases the frequency of conversion to the prion form while deletion of the red region decreases prion formation. The N-terminus (residue 110) of the molecule is shown as 'N'.

both the paramutagenic B' sequences and an enhancer in B-I mapped within an 8-kbp region located between 95 and 103 kbp upstream of the $b1$ transcription start site (Figure 7.4).

C. Features of the sequences required for paramutation

We have compared the regions required for paramutation at the two loci, $b1$ and $p1$. Interestingly, no significant stretches of sequence similarity are shared between the 1.2-kbp $P1$-rr fragment and the 8 kbp of B' sequences, which are both required for paramutagenicity (L. Sidorenko, M. Stam, and V. Chandler, unpublished data). There are, however, two shared features within each of these regions: both contain enhancer regions and tandem repeats.

The $P1$-rr transcriptional regulatory elements have been identified by a combination of transposon mutagenesis (Athma *et al.*, 1992; Moreno *et al.*, 1992) and functional tests using transient expression assays and transgenic maize plants (Sidorenko *et al.*, 1999, 2000). The results indicate that the 5' regulatory region of $P1$-rr consists of a basal fragment (Pb), including 235 bp of promoter and 326 bp of untranslated leader sequences, and two enhancer elements. One enhancer is contained within a ~1-kbp fragment (P1.0) located immediately upstream of the basal promoter, while another enhancer was found in a 1.2-kbp fragment (P1.2) located ~5 kbp away within repeats that flank the coding region. There is no sequence similarity between P1.0 and P1.2, and only the P1.2-kbp fragment is paramutagenic (Sidorenko and Peterson, 2001). The P1.2 paramutagenic $p1$ sequences are located within small direct repeat sequences within a large 5.2-kbp direct repeat flanking the $p1$ coding region (Figure 7.3A, hatched boxes). The 3' end of the $P1$-rr transcript extends into the downstream 1.2-kbp fragment (Figure 7.3B). This raises the possibility that if this fragment is transcribed from the transgene, posttranscriptional gene silencing could be occurring through interactions of this RNA with the endogenous $p1$ mRNA. Sequence comparison of the P1.2 fragment with GenBank also revealed that it shares sequence similarity to a single EST from the Stanford Genomics Project (accession no. AI891238) and the cDNA clone Q (Grotewold and Peterson, 1990). These two regions partially overlap with each other within the P1.2 fragment (Figure 7.3B). In addition to the above homologies, 64 bp of the 1.2-kb enhancer share homology to part of the transcript as indicated in Figure 7.3B. Fragments #13 and #14 are moderately repetitive in the genome, while fragment #15 is found only at the $p1$ locus (Figure 7.3B). The functional importance of these features for the enhancer or paramutagenic functions of the P1.2 fragment is unclear.

Chromatin studies comparing the $P1$-rr and $P1$-rr' states have not been done, but the reported studies comparing $P1$-rr and $P1$-pr are informative. The $P1$-pr allele showed increased methylation of restriction sites that delimit the P1.2 enhancer fragment (Das and Messing, 1994b). Interestingly, this occurred in both the downstream and upstream sequences that flank the coding region.

Eight DNaseI hypersensitive sites were detected within the *P1-rr* locus (Lund *et al.*, 1995). Six of these sites were hypersensitive in both *P1-rr* and *P1-pr*, while two sites were hypersensitive only in *P1-rr*, not in the silenced *P1-pr*. The two sites that showed differential hypersensitivity between the two alleles are in the upstream and downstream repeats, with each hypersensitivity site located very close to *Sal*I restriction sites that delimit the P1.2 enhancer. The complex nature of the P1.2 transgene precludes examination of DNaseI hypersensitivity in the transgene array, but it will be interesting to determine if the *P1-pr* pattern is also seen in the *P1-rr'* state.

As described above for *b1* paramutation, an 8-kbp region has been identified that contains both the sequences required for paramutation as well as an enhancer required for high expression of *b1* in vegetative plant tissues. Extensive restriction mapping and sequence analyses of the 8-kbp region has revealed no differences between *B-I* and *B'*. However, there is an intriguing difference between *B-I* and *B'* and two neutral alleles. Within both *B'* and *B-I*, the 8-kbp region contains an ~850-kbp region that is repeated seven times in a direct tandem array. The same sequence occurs only once in two neutral alleles, *B-Peru* (Figure 7.4) and *b-K55* (not shown). A recombinant within the 850-bp region between *B-I* and *B-Peru* produced an allele that had lost four of these tandem repeats. This recombinant is highly expressed in vegetative tissues, and its expression is reduced and it becomes paramutagenic when crossed to *B'*, indicating that if these repeats are responsible for the enhancer and paramutation activities, three repeats are sufficient. It is also possible that the sequences immediately upstream of the repeats are responsible for both high expression within *B-I* and paramutation. An indication that the repeats might be involved in paramutation is the observation that the recombinant allele retaining three repeats does not paramutate *B-I* with a 100% efficiency as *B'* does (2 of 57 F_1 plants show no *B-I* paramutation). DNA blot analyses indicate that the 850-bp repeats are unique in the maize genome, and sequence similarity searches do not reveal any matches in Genbank (M. Stam and V. Chandler, unpublished). Given the multiple reports of repeat-induced gene silencing in a variety of organisms and the observation that repeats are often methylated, it will be important to compare DNA methylation levels and chromatin structure within the 850-bp repeats and flanking DNA in *B-I*, *B'*, and *B-Peru*. We would predict that the repeats within *B'* would be in a more condensed chromatin structure, relative to the repeats within *B-I*.

IV. MODELS FOR PARAMUTATION

Models that attempt to explain paramutation must account for the following features. Certain enhancer sequences are associated with sequences required for paramutation in two different genes. The high-expressing alleles can be silenced spontaneously. Trans silencing occurs at a higher frequency than the spontaneous changes to a lower expression state. The distinct expression states are heritably transmitted through mitosis and meiosis. Our model, discussed below, is motivated by the results discussed above, by studies on the β-globin loci in mammals, by models proposed for genomic imprinting at the *Igf2* and *H19* loci in mice, by long-distance enhancers, and models for trans-sensing effects in *Drosophila* (Bell and Felsenfeld, 2000; Bell *et al.*, 2001; Grosveld *et al.*, 1987; Henikoff and Comai, 1998; Reik and Murrell, 2000; Udvardy, 1999).

It is intriguing that at two different genes, *b1*, and *p1*, sequences required for paramutation are localized within the same fragment as enhancer sequences. Clearly, not all enhancer sequences are capable of paramutation, as the P1.0 fragment from *P1-rr* (Figure 7.3C) is a strong enhancer (Sidorenko *et al.*, 2000), yet has no paramutagenic activity (Sidorenko and Peterson, 2001). We propose that at *b1* and *p1*, specific enhancer regions are located very close to or within controlling elements that can mediate epigenetic modification of the enhancer activities. We propose that for both *b1* and *p1* paramutation, the enhancer sequences can exist in two states, functional and nonfunctional. The enhancers may exist in two distinct chromatin states themselves, or they may be adjacent to boundary elements or insulator sequences that can exist in distinct states and influence the activity of the enhancer. In the latter instance it would be the state of these boundary or insulator elements that determines whether the immediately adjacent enhancer region could communicate with promoters located farther away. When the boundary element is functioning the enhancer activity is inhibited, resulting in the reduced expression state of the promoter. When the boundary element is not functioning the enhancer is active, resulting in the higher expression state. Further dissection of the enhancers and tests for boundary elements within the P1.2 *P1-rr* and 8-kbp *B'/B-I* fragments using transgenic approaches should address whether these activities are distinct, overlap, or contained within the same sequences. These approaches, combined with extensive chromatin structure analyses, should determine whether the mechanism is similar to that proposed for the LCR region of the β-globin loci in mammals. The LCR region of the β-globin loci in mammals contains hypersensitive sites located far upstream from the globin genes; some of these sites function as insulator/boundary elements, others function as enhancers (Udvardy, 1999). Another interesting paradigm is seen at the *Igf2* and *H19* loci in mice, which are subject to parent of origin imprinting (reviewed in Chapter 5). An enhancer located ~10 kbp from *H19* and ~100 kbp from *Igf2* functions on one or the other of these genes, depending on the chromatin structure and methylation status of the imprinting control region (ICR) located between the two genes (Bell and Felsenfeld, 2000; Reik and Murrell, 2000). Our current model for the cis effects associated with *b1* paramutation is shown in Figure 7.5A.

One of the most fascinating aspects of paramutation is the allele communication: the ability of an epigenetic state in one allele to induce the same state in the other allele at very high frequencies. Two major models for allele communication are direct interaction between the chromatin complexes formed at each

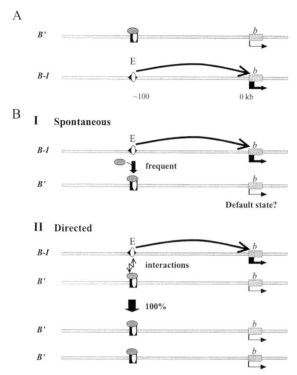

Figure 7.5. Model for *b1* paramutation. The sequences required for paramutation are indicated by
a black box in the paramutagenic B' state and a black diamond in the paramutable *B-I*
state. The sequences required for high expression are indicated by a white oval. The grey
horizontal oval represents chromatin proteins responsible for the condensed B' state.
(A) Model for the cis effects of the enhancer at *b1*. In the condensed B' chromatin state,
the enhancer sequences do not function, whereas in the more open *B-I* chromatin state,
the enhancer is active, indicated with an E for enhancer, and activates *b* transcription.
(B) Model for the trans effects observed with *b1* paramutation. (I) Spontaneous paramutation. The open *B-I* chromatin state is inherently unstable, possibly as the result of the
repeated sequences, attracting chromatin proteins changing the *B-I* into the B' state. The
B' chromatin state is very stable and can be thought of as the default state. The reciprocal
event has not been observed (looking at >50,000 plants). (II) Directed paramutation.
When *B-I* is heterozygous with $B,'$ the paramutation sequences communicate, resulting
in the exchange of chromatin proteins, stably changing *B-I* into a B' state.

allele or interactions between RNA and DNA. In the context of the enhancer
model proposed above, we suggest that the proteins associated with the postulated
condensed chromatin structure of B' may be able to interact with *B-I* protein complexes in trans, inducing a higher-order chromatin structure at *B-I*. It is interesting
that spontaneous changes of *B-I* to B' occur at a much higher frequency when

B-I is homozygous, relative to when it is carried over a neutral allele (Coe, 1966; K. Kubo and V. Chandler, unpublished). Our finding that neutral alleles have only one copy of the 850-bp region reveals a difference between neutral and paramutation-competent alleles that might be responsible for the differences in spontaneous paramutation. If spontaneous paramutation is caused by trans interactions between the repeats, the single copy of the sequence within neutral alleles may be much less capable of interacting relative to the seven copies found in each *B-I* allele in homozygous plants. A diagram of a simple model for trans interactions associated with paramutation at *b1* is shown in Figure 7.5B.

Mitotically heritable trans-sensing phenomena in *Drosophila* provide paradigms for our chromatin model in spite of the observations that the events in *Drosophila* are usually not meiotically heritable. In animals and insects, germ cells are set aside early in development. In contrast, in plants, mitotically heritable changes in expression are often transmitted to progeny when germ cells differentiate from somatic cells late in plant development. Trans-sensing phenomena in *Drosophila* (Henikoff and Comai, 1998) such as transvection (Chapter 13) and pairing-sensitive silencing (Chapter 14) require pairing, as translocations that disrupt somatic pairing, disrupt these phenomena (Dreesen *et al.*, 1991; Smolik-Utlaut and Gelbart, 1987). Translocation of *B-I* or *B'* to other chromosome arms did not disrupt paramutation (Coe, 1966). However, the translocated regions were large such that meiotic pairing was not disrupted, suggesting that somatic pairing might have occurred as well. There is no physical evidence for homolog synapsis in somatic plant cells, except during flower development (Aragon-Alcaide *et al.*, 1996), much later than when paramutation is established (Chandler *et al.*, 1996). However, interactions between homologous chromosomes may be rare or transient, making them difficult to detect microscopically. Only transient interactions are likely to be required between alleles undergoing paramutation, because once the paramutant state is established, it can be maintained in the absence of the paramutagenic allele. The ability of the P1.2 transgene to paramutate the endogenous *P1-rr* allele, together with other studies of homology-dependent gene silencing in plants (Fagard and Vaucheret, 2000), suggest that the genome does have mechanisms for detecting small duplications, which suggests homology-scanning mechanisms. Subnuclear localization could also influence expression. The heterochromatic insertion at *brown*[D] interacts with centromeric heterochromatin sequestering the wild-type gene into a specific heterochromatic compartment affecting its expression (Csink and Henikoff, 1996).

An alternative model to the DNA and protein interactions proposed above is DNA/RNA/protein interactions. A dramatic example of a cis-acting RNA that establishes a repressed state is that of the XIC locus on the inactive X chromosome in mammalian cells (Panning and Jaenisch, 1998; Chapter 2). X-chromosome inactivation involves the stabilization of a nontranslated RNA, which appears to be both necessary and sufficient to confer a chromatin-based mechanism of inactivation on adjacent sequences (Lee and Jaenisch, 1997; Lee *et al.*, 1996; Penny *et al.*, 1996). At the present time it is not known whether the sequences required for paramutation at *p1* or *b1* are transcribed. However, the observation that the P1.2 region does contain sequence similarity to part of the *P1-rr* mRNA and potentially other transcripts makes it a high priority to investigate the role of RNA in the trans interactions.

Another intriguing aspect of paramutation is that the altered expression states are heritable through both mitosis and meiosis. A major difference between paramutation at the *p1* and *b1* loci is the high stability of *P1-rr* (changes to *P1-pr* at one in a million) relative to *B-I* (changes to *B′* at one in 10–100). For both genes, we propose that the most stable state is the state dictated by the DNA sequence and that the more rare, unstable state is caused by altered chromatin structures that can occur stochastically. One possibility is that during DNA replication there is the opportunity to restructure the chromatin. Thus, for the *b1* locus, we propose that the stable *B′* state is the state dictated by the DNA sequence, and that the most common chromatin structure formed by these sequences is one of a higher-order state in which the enhancer is not able to function. We propose that the *B-I* state is a rare state, potentially selected by humans, in which the chromatin is restructured such that the enhancer becomes functional, enabling it to stimulate transcription of the *b1* promoter region located ~95 kbp downstream. Potentially every cell division provides an opportunity for the chromatin to be restructed, which could explain why we often see sectors of *B′* expression throughout the development of *B-I* plants and it is only by selecting for the darkest pigmented plants that we can maintain *B-I* in our stocks. For the *P1-rr* locus we propose that the most common chromatin structure dictated by the DNA sequence is one in which the enhancer fully functions. At the endogenous *P1-rr* allele, it is only rarely that an altered chromatin structure is formed, leading to reduced activity of the enhancer, the *P1-pr* state. Furthermore, we suggest that when the P1.2 enhancer/boundary element region is reintroduced into plants, it more readily forms the higher-order chromatin structure that is not permissive for expression. The dramatic increases in the ability to form the hypothesized repressive chromatin structure in the P1.2 transgene relative to the endogenous gene could be because the sequences are within a complex transgene array, because of where the integration events occur, or because they are separated from other *P1-rr* sequences that function to suppress formation of repressive chromatin.

V. IMPLICATIONS AND POSSIBLE FUNCTIONS

Paramutation has been well documented for only a small number of genes. Paramutation could be much more common, but it has gone undetected because subtle changes in expression are difficult to observe easily at most genes. Alternatively, paramutation could be rare. The ability to examine global gene expression using microarrays will make it possible to determine the number of genes whose expression can change depending on what allele is present. If paramutation does reflect a more general mechanism, what might its function be? Several potential roles have been discussed. One model is that paramutation and other examples of homology-dependent gene silencing may reflect an important cellular mechanism for protection against invasive DNA (Matzke *et al.*, 1996, 2000; Yoder *et al.*, 1997). Another possibility is that the machinery involved in paramutation functions to identify and maintain chromatin boundaries between genes and nearby repetitive sequences. Paramutation may also provide an adaptive mechanism for transmitting favorable expression states to progeny. The latter two possibilities were discussed in a recent review (Chandler *et al.*, 2000).

Independent of its function and the number of genes that paramutate, paramutation provides an excellent system for studying mechanisms involved in establishing and maintaining heritable expression states and allelic communication. Further studies on how the minimal sequences function and on the molecular nature of the trans-acting factors required for paramutation (Dorweiler *et al.*, 2000; Hollick and Chandler, 2001) should reveal how alleles interact in the nucleus to influence the regulation of each other, how heritable expression states are established, and how they are maintained through numerous cell divisions and transmitted to the next generation.

References

Aragon-Alcaide, L., Miller, T., Schwarzacher, T., Reader, S., and Moore, G. (1996). A cereal centromeric sequence. *Chromosoma* **105**, 261–268.

Athma, P., Grotewold, E., and Peterson, T. (1992). Insertional mutagenesis of the maize *P* gene by intragenic transposition of *Ac*. *Genetics* **131**, 199–209.

Bell, A. C., and Felsenfeld, G. (2000). Methylation of a CTCF-dependent boundary controls imprinted expression of the *Igf2* gene. *Nature* **405**, 482–485.

Bell, A. C., West, A. G., and Felsenfeld, G. (2001). Insulators and boundaries: Versatile regulatory elements in the eukaryotic genome. *Science* **291**, 447–449.

Brink, R. A. (1956). A genetic change associated with the *R* locus in maize which is directed and potentially reversible. *Genetics* **41**, 872–889.

Brink, R. A. (1973). Paramutation. *Annu. Rev. Genet.* **7**, 129–152.

Brink, R. A., Styles, E. D., and Axtell, J. D. (1968). Paramutation: directed genetic change. Paramutation occurs in somatic cells and heritably alters the functional state of the *R* locus. *Science* **159**, 161–170.

Chandler, V. L., Eggleston, W. B., and Dorweiler, J. E. (2000). Paramutation in maize. *Plant Mol. Biol.* **43**, 121–145.

Chandler, V., Kubo, K., and Hollick, J. (1996). *b* and *pl* paramutation in maize: heritable transcription states programmed during development. *In* "Epigenetic Mechanisms of Gene Regulation" (V. Russo, R. Martienssen, and A. Riggs, eds.), pp. 289–304. Cold Spring Harbor Laboratory Press, Plainview, NY.

Coe, E. H. J. (1966). The properties, origin and mechanism of conversion-type inheritance at the *b* locus in maize. *Genetics* **53**, 1035–1063.

Cone, K. C., Cocciolone, S. M., Burr, F. A., and Burr, B. (1993). Maize anthocyanin regulatory gene *p1* is a duplicate of *c1* that functions in the plant. *Plant Cell* **5**, 1795–1805.

Csink, A. K., and Henikoff, S. (1996). Genetic modification of heterochromatic association and nuclear organization in *Drosophila*. *Nature* **381**, 529–531.

Das, O. P., and Messing, J. (1993). A heritable mutagenic activity in the maize inbred line A188. *In* "Methods in Molecular Genetics" (K. W. Adolph, ed.), pp. 276–292. Academic Press, Orlando, FL.

Das, O. P., and Messing, J. W. (1994a). A heritable interaction between *P-pr* and *P-rr*. *Maize Genet. Coop. Newslett.* **68**, 80–81.

Das, P., and Messing, J. (1994b). Variegated phenotype and developmental methylation changes of a maize allele originating from epimutation. *Genetics* **136**, 1121–1141.

Dooner, H. K., Robbins, T. P., and Jorgenson, R. A. (1991). Genetic and developmental control of anthocyanin biosynthesis. *Annu. Rev. Genet.* **25**, 173–199.

Dorweiler, J. E., Carey, C. C., Kubo, K. M., Hollick, J. B., Kermicle, J. L., and Chandler, V. L. (2000). *mediator of paramutation l* is required for establishment and maintenance of paramutation at multiple maize loci. *Plant Cell* **12**, 2101–2118.

Dreesen, T. D., Henikoff, S., and Loughney, K. (1991). A pairing-sensitive element that mediates trans-inactivation is associated with the *Drosophila brown* gene. *Genes Dev.* **5**, 331–340.

Fagard, M., and Vaucheret, H. (2000). (Trans)gene silencing in plants: How many mechanisms? *Annu. Rev. Plant Phys. Plant Mol. Biol.* **51**, 167–194.

Gaut, B. S., and Doebley, J. F. (1997). DNA sequence evidence for the segmental allotetraploid origin of maize. *Proc. Natl. Acad. Sci. USA* **94**, 6809–6814.

Goff, S. A., Cone, K. C., and Chandler, V. L. (1992). Functional analysis of the transcriptional activator encoded by the maize *B* gene: Evidence for a direct functional interaction between two classes of regulatory proteins. *Genes Dev.* **6**, 864–875.

Grosveld, F., Van Assendelft, G. B., Greaves, D. R., and Kollias, G. (1987). Position independent, high-level expression of the human β-globin gene in transgenic mice. *Cell* **51**, 975–985.

Grotewold, E., Athma, P., and Peterson, T. (1991). Alternatively spliced products of the maize *P* gene encode proteins with homology to the DNA-binding domain of *myb*-like transcription factors. *Proc. Natl. Acad. Sci. USA* **88**, 4587–4591.

Grotewold, E., Drummond, B. J., Bowen, B., and Peterson, T. (1994). The *myb*-homologous *P* gene controls phlobaphene pigmentation in maize floral organs by directly activating a flavonoid biosynthetic gene subset. *Cell* **76**, 543–653.

Grotewold, E., and Peterson, T. (1990). Isolation of cDNA clones homologous to the *P*-gene flanking regions. *Maize Genet. Coop. Newslett.* **64**, 57.

Grotewold, E., Sainz, M. B., Tagliani, L., Hernandez, J. M., Bowen, B., and Chandler, V. L. (2000). Identification of the residues in the Myb domain of maize C1 that specify the interaction with the bHLH cofactor R. *Proc. Natl. Acad. Sci. USA* **97**, 13579–13584.

Henikoff, S., and Comai, L. (1998). Trans-sensing effects: the ups and the downs of being together. *Cell* **93**, 329–332.

Hollick, J. B., and Chandler, V. L. (1998). Epigenetic allelic states of a maize transcriptional regulatory locus exhibit overdominant gene action. *Genetics* **150**, 891–897.

Hollick, J. B., and Chandler, V. L. (2001). Genetic factors required to maintain repression of a paramutagenic maize *pl1* allele. *Genetics* **157**, 369–378.

Hollick, J. B., Patterson, G. I., Asmundsson, I. M., and Chandler, V. L. (2000). Paramutation alters regulatory control of the maize *pl* locus. *Genetics* **154**, 1827–1838.

Hollick, J. B., Patterson, G. I., Coe, E. H. Jr., Cone, K. C., and Chandler, V. L. (1995). Allelic interactions heritably alter the activity of a metastable maize *pl* allele. *Genetics* **141**, 709–719.

Kermicle, J. L., Eggleston, W. B., and Alleman, M. (1995). Organization of paramutagenicity in *R-stippled* maize. *Genetics* **141**, 361–372.

Klar, A. J. S. (1998). Propagating epigenetic states through meiosis: where Mendel's gene is more then a DNA moiety. *Trends Genet.* **14**, 299–301.

Lee, J. T., and Jaenisch, R. (1997). Long-range cis effects of ectopic X-inactivation centres on a mouse autoscome. *Nature* **386,** 275–279.

Lee, J. T., Strauss, W. M., Dausman, J. A., and Jaenisch, R. (1996). A 450 kb transgene displays properties of the mammalian X-inactivation center. *Cell* **86,** 83–94.

Ludwig, S. R., and Wessler, S. R. (1990). Maize R-gene family-tissue-specific helix-loop-helix proteins. *Cell* **62,** 849–851.

Lund, G., Das, O. P., and Messing, J. (1995). Tissue specific DNaseI-sensitive sites of the maize P gene and their changes upon epimutation. *Plant J.* **7,** 797–807.

Matzke, M. A., Matzke, A. J. M., and Eggleston, W. (1996). Transgene silencing and paramutation: A common response to invasive DNA? *Trends Plant Sci.* **1,** 382–388.

Matzke, M. A., Mette, M. F., and Matzke, A. J. (2000). Transgene silencing by the host genome defense: Implications for the evolution of epigenetic control mechanisms in plants and vertebrates. *Plant. Mol. Biol.* **43,** 401–415.

Meyer, P., Heidmann, I., and Niedenhof, I. (1993). Differences in DNA-methylation are associated with a paramutation phenomenon in transgenic petunia. *Plant J.* **4,** 89–100.

Monaco, A. P. (1995). "Pulsed Field Gel Electrophoresis: A Practical Approach." Oxford University Press, Oxford, U.K.

Moreno, M. A., Chen, J., Greenblatt, I., and Dellaporta, S. L. (1992). Reconstitutional mutagenesis of the maize P gene by short-range Ac transpositions. *Genetics* **131,** 939–956.

Panavas, T., Weir, J., and Walker, E. L. (1999). The structure and paramutagenicity of the R-marbled haplotype of *Zea mays*. *Genetics* **153,** 979–991.

Panning, B., and Jaenisch, R. (1998). RNA and the epigenetic regulation of X chromosome inactivation. *Cell* **93,** 305–308.

Patterson, G. I., Kubo, K. M., Shroyer, T., and Chandler, V. L. (1995). Sequences required for paramutation of the maize b gene map to a region containing the promoter and upstream sequences. *Genetics* **140,** 1389–1406.

Patterson, G. I., Thorpe, C. J., and Chandler, V. L. (1993). Paramutation, an allelic interaction, is associated with a stable and heritable reduction of transcription of the maize b regulatory gene. *Genetics* **135,** 881–894.

Penny, G. D., Kay, G. F., Sheardown, S. A., Rastan, S., and Brockdorff, N. (1996). Requirement for Xist in X chromosome inactivation. *Nature* **379,** 131–137.

Radicella, J. P., Brown, D., Tolar, L. A., and Chandler, V. L. (1992). Allelic diversity of the maize B regulatory gene: Different leader and promoter sequences of two B alleles determine distinct tissue specificities of anthocyanin production. *Genes Dev.* **6,** 2152–2164.

Reik, W., and Murrell, A. (2000). Genomic imprinting. Silence across the border. *Nature* **405,** 408–409.

Sainz, M. B., Grotewold, E., and Chandler, V. L. (1997). Evidence for direct activation of an anthocyanin promoter by the maize C1 protein and comparison of DNA binding by related Myb domain proteins. *Plant Cell* **9,** 611–625.

Selinger, D. A., Lisch, D., and Chandler, V. L. (1998). The maize regulatory gene B-Peru contains a DNA rearrangement that specifies tissue-specific expression through both positive and negative promoter elements. *Genetics* **149,** 1125–1138.

Sidorenko, L. V., Li, X., Cocciolone, S. M., Chopra, S., Tagliani, L., Bowen, B., Daniels, M., and Peterson, T. (2000). Complex structure of a maize Myb gene promoter: functional analysis in transgenic plants. *Plant. J.* **22,** 471–482.

Sidorenko, L. V., and Peterson, T. (2001). Transgene-induced silencing identifies sequences involved in the establishment of paramutation of the maize p1 gene. *Plant Cell* **13,** 319–335.

Sidorenko, L., Li, X., Tagliani, L., Bowen, B., and Peterson, T. (1999). Characterization of the regulatory elements of the maize P-rr gene by transient expression assays. *Plant Mol. Biol.* **39,** 11–19.

Smolik-Utlaut, S. M., and Gelbart, W. M. (1987). The effects of chromosomal rearrangements on the zeste-white interaction in *Drosophila melanogaster*. *Genetics* **116,** 285–298.

Stam, M., Lavin, T., and Chandler, V. L. (2000). *Npi402* and *ncsu1* are identical; *inra1* (*tmp*) maps upstream of the *b* promoter. *Maize Genet. Coop. Newslett.* **74,** 66–67.

Styles, E. D., and Brink, R. A. (1966). The metastable nature of paramutable *R* alleles in maize. I. Heritable enhancement in level of standard R^r action. *Genetics* **54,** 433–439.

Udvardy, A. (1999). Dividing the empire: Boundary chromatin elements delimit the territory of enhancers. *EMBO J.* **18,** 1–8.

Vaucheret, H. (1993). Identification of a general silencer for 19S and 35S promoters in transgenic tobacco plant: 90 bp of homology in the promoter sequences are sufficient for trans-inactivation. *C. R. Acad. Sci. Paris* **316,** 1471–1483.

Vaucheret, H., Beclin, C., Elmayan, T., Feuerbach, F., Godon, C., Morel, J. B., Mourrain, P., Palauqui, J. C., and Vernhettes, S. (1998). Transgene-induced gene silencing in plants. *Plant J.* **16,** 651–659.

Walker, E. L. (1998). Paramutation of the *r1* locus of maize is associated with increased cytosine methylation. *Genetics* **148,** 1973–1981.

Yoder, J. A., Walsh, C. P., and Bestor, T. H. (1997). Cytosine methylation and the ecology of intragenomic parasites. *Trends Genet.* **13,** 335–340.

8

Homology-Dependent Gene Silencing and Host Defense in Plants

Marjori A. Matzke,* Werner Aufsatz, Tatsuo Kanno, M. Florian Mette, and Antonius J. M. Matzke
Institute of Molecular Biology
Austrian Academy of Sciences
A-5020 Salzburg, Austria

*To whom correspondence should be addressed: E-mail: mmatzke@imb.oeaw.ac.at; Fax +43-662-63961-29; Telephone +43-662-63961-26.

Advances in Genetics, Vol. 46
235

ABSTRACT

Analyses of transgene silencing phenomena in plants and other organisms have revealed the existence of epigenetic silencing mechanisms that are based on recognition of nucleic acid sequence homology at either the DNA or RNA level. Common triggers of homology-dependent gene silencing include inverted DNA repeats and double-stranded RNA, a versatile silencing molecule that can induce both degradation of homologous RNA in the cytoplasm and methylation of homologous DNA sequences in the nucleus. Inverted repeats might be frequently associated with silencing because they can potentially interact *in cis* and *in trans* to trigger DNA methylation via homologous DNA pairing, or they can be transcribed to produce double-stranded RNA. Homology-dependent gene silencing mechanisms are ideally suited for countering natural parasitic sequences such as transposable elements and viruses, which are usually present in multiple copies and/or produce double-stranded RNA during replication. These silencing mechanisms can thus be regarded as host defense strategies to foreign or invasive nucleic acids. The high content of transposable elements and, in some cases, endogenous viruses in many plant genomes suggests that host defenses do not always prevail over invasive sequences. During evolution, slightly faulty genome defense responses probably allowed transposable elements and viral sequences to accumulate gradually in host chromosomes and to invade host genes. Possible beneficial consequences of this "foreign" DNA buildup include the establishment of genome defense-derived epigenetic control mechanisms for regulating host gene expression and acquired hereditary immunity to some viruses.

I. INTRODUCTION

Homology-dependent gene silencing (HDGS) is a generic term that has been used to refer to epigenetic silencing effects that rely on recognition of nucleic acid sequence homology at either the DNA or RNA level. Historically, the introduction and use of this term by plant biologists has reflected the diversity of silencing effects observed when multiple copies of transgenes or transgenes with sequence homology to endogenous genes are present in a plant genome (Matzke and Matzke, 1995). The interacting homologous genes can be close together on the same DNA molecule (*cis* inactivation) or on separate chromosomes (*trans* silencing). HDGS can occur at the transcriptional or posttranscriptional level, depending on whether multiple copies of genes share homology in promoters or transcribed sequences, respectively. Transcriptional gene silencing (TGS) results from a block in RNA synthesis, while posttranscriptional gene silencing (PTGS) involves targeted degradation of homologous RNAs in the cytoplasm. Both types of HDGS are often associated with *de novo* DNA methylation, which is usually confined to promoters in cases of TGS and transcribed regions in instances of PTGS (Fagard and Vaucheret, 2000a). The complexity of HDGS effects in plants contrasts to the more restricted phenomenon of RNA interference (RNAi), which was described first in *Caenorhabditis elegans* (Fire *et al.*, 1998), an organism that does not methylate its DNA. RNAi involves targeted destruction of mRNA by homologous double-stranded RNA (dsRNA) and does not appear to lead to any heritable changes in chromatin structure (Chapter 11).

Models for HDGS in plants have traditionally considered TGS and PTGS as mechanistically distinct phenomena that are united primarily through their common requirement for nucleic acid sequence homology. DNA–DNA, RNA–DNA, and RNA–RNA interactions have all been invoked to account for the different silencing effects and sequence-specific genome modifications that have been observed (Kooter *et al.*, 1999; Fagard and Vaucheret, 2000a). These models are being honed in light of recent work demonstrating a central role for dsRNA in many HDGS phenomena. It is now clear that dsRNA is a pivotal player in PTGS and in RNA-directed DNA methylation (RdDM), which can lead to TGS if dsRNAs containing promoter sequences are involved. This convergence on dsRNA as a multipurpose silencing molecule acting in the cytoplasm and at the genome level suggests more extensive mechanistic overlaps among HDGS phenomena than previously assumed. Indeed, most cases of HDGS, including some that were previously believed to result from DNA–DNA pairing, can probably be placed under the new rubric of "RNA silencing" (Baulcombe, 2000; Matzke *et al.*, 2001a).

In this review on HDGS and host defense in plants, we cover RNA silencing phenomena occurring in the cytoplasm and in the nucleus. Despite the potent silencing ability of dsRNA, we argue that a mechanism involving DNA–DNA pairing is still tenable in a subset of HDGS cases that do not seem to be RNA-mediated. We describe the host defense functions of HDGS and show how they can be effective against natural foreign or invasive sequences such as transposable elements (TEs) and viruses. Finally, we argue that a primordial host genome defense system might have been recruited during evolution to establish global epigenetic mechanisms for regulating host gene expression in plants and vertebrates, and consider the possibility that HDGS can generate hereditary resistance to viruses that integrate into host chromosomes.

II. RNA SILENCING

As judged by the number of recent reviews, the field of RNA silencing is thriving. For example, four recent issues of *Current Opinion in Genetics and Development* each contained a review on RNA silencing from the perspective of plant (Matzke *et al.*, 2001a), fungal (Cogoni and Macino, 2000), or animal (Birchler *et al.*, 2000; Plasterk and Ketting, 2000) scientists. Most of the current emphasis is on PTGS and PTGS-like phenomena (RNAi in animals and quelling in *Neurospora crassa*) that involve sequence-specific mRNA degradation induced by homologous dsRNA or other aberrant RNAs (Cogoni and Macino, 2000; Chapter 9). A complete discussion of RNA silencing in plants must also encompass effects at the genome level, as exemplified by the sequence-specific DNA methylation that is the hallmark of RdDM. Whether dsRNA can trigger methylation or chromatin modifications in other organisms is not yet known.

A. Posttranscriptional gene silencing (PTGS)

1. Mechanism of PTGS and related phenomena

Three shared features of PTGS, RNAi, and quelling provide compelling evidence that they involve a common, highly conserved mechanism (Chapter 9 and 11; Plasterk and Ketting, 2000; Matzke *et al.*, 2001a). First, genetic analyses have shown that many of the same protein products are required for all three silencing phenomena. Second, small sense and antisense RNAs (smRNAs) approximately 23 nucleotides (nt) in length that are homologous to the silenced gene have been identified in PTG-silenced plants and an *in vitro* RNAi system in *Drosophila*. Third, mobile silencing signals that can travel through the whole organism in *Caenorhabditis elegans* and in plants and between nuclei in heterokaryotic strains of *Neurospora* have been identified. In plants, advances in understanding the PTGS pathway are coming from studies on PTGS-defective mutants in *Arabidopsis* and on recently discovered viral suppressors of PTGS.

A general outline of RNA silencing is shown in Figure 8.1, see color insert. While many of the details of this pathway remain to be worked out, the essential features of PTGS include a dsRNA from which smRNAs of both polarities are generated. These smRNAs are believed to guide an RNase to the homologous mRNA and tag it for destruction. An RNA species associated with PTGS, presumably either dsRNA or the smRNAs, is able to trigger methylation of homologous DNA sequences. PTGS is thought to occur primarily in the cytoplasm because RNA viruses that can be both inducers and targets of PTGS replicate exclusively in this cellular compartment.

a. Required cellular gene products and virally encoded suppressors of PTGS

Several classes of cellular gene products essential for PTGS, quelling and RNAi have been identified to date in genetic screens (Table 8.1). Only one of these genes, SGS3, which shows no similarity with any known protein, is unique to plants. The identities of the common proteins suggest activities involved in the synthesis and amplification of dsRNA, unwinding of dsRNA, and targeting of mRNAs to the ribosome. Up to now, no genes encoding proteins that could be involved in the RNase complex(es) of PTGS in plants have been identified in mutational analyses. This suggests that genetic approaches might not be sufficient for dissecting the entire PTGS mechanism, particularly those aspects that are essential for normal growth and development (Dalmay et al., 2000).

Virally encoded suppressors of PTGS, which have so far been studied only in plants, and the cellular proteins with which they interact are helping to piece together the PTGS pathway (Table 8.1). Two viral suppressors studied in detail include HC-Pro of potyviruses and the p25 cell-to-cell movement protein of potato virus X (PVX). A cellular calmodulin-related protein, termed rgs-CaM, that interacts with HC-Pro has been identified in a yeast two-hybrid screen. HC-Pro acts upstream of the production of the smRNAs (Mallory et al., 2001) (Figure 8.1③). The PVX p25 protein appears to impede systemic silencing (Voinnet et al., 2000) (Figure 8.1⑦). A third viral suppressor, the 2b protein of cucumber mosaic virus, acts in an unknown way in the nucleus to suppress PTGS (Lucy et al., 2000) (Figure 8.1).

b. Double-stranded RNA and RNA-dependent RNA polymerase

In principle, dsRNA can be produced in a number of ways. In the nucleus, RNA polymerase II (polII) transcription through inverted DNA repeats (IRs) produces a hairpin RNA (Figure 8.1①). PolII transcription and subsequent annealing of separate sense and antisense transcripts from either unlinked loci or from the same locus can generate open dsRNA, which differs from hairpin RNA by being formed through intermolecular (instead of intramolecular) base pairing. Transcribing through IRs appears to be the most efficient means to induce *trans* silencing, presumably because a stable RNA hairpin forms readily (Mette et al., 2000; Smith et al., 2000). Another way to produce dsRNA is through the synthesis of copy antisense RNA from single-stranded sense transcripts (which are probably aberrant in some way; Section II.A.2.c) by an RNA-dependent RNA polymerase (RdRP) (Figure 8.1⑤). The rather vague concept of aberrant RNAs has been invoked in cases of PTGS and quelling, where it is not clear that dsRNA is the direct result of polII transcription (Cogoni and Macino, 2000). Interestingly,

Table 8.1. Proteins Involved in PTGS and PTGS-Like Phenomena

	Cellular proteins	
Protein	Mutant name/organism	Possible roles
RdRP	qde1/*Neurospora*[a] sde1/*Arabidopsis*[b] sgs2/*Arabidopsis*[c] ego1/*C.elegans*[d]	1. Synthesis of cRNA 2. Amplification of dsRNA 3. DNA methylation pathway 4. Synthesis of mobile silencing signal 5. Development 6. Viral defense
eIF2C-like	qde2/*Neurospora*[e] rde1/*C.elegans*[f] ago1/*Arabidopsis*[g]	1. Target PTGS to ribosomes 2. DNA methylation pathway 3. Development
RecQ DNA helicase RNase D-like RNA helicase RNA helicase	qde3/*Neurospora*[h] mut-7/*C.elegans*[i] mut-6/*Chlamydomonas*[j] sde3/*Arabidopsis*[k]	1. Unwinding dsRNA 2. dsRNA degradation 3. Transposon, virus defense 4. Nuclear step?
NMD Proteins	smg2/*C.elegans*[l] smg5/*C.elegans* smg6/*C.elegans*	1. ATPase/helicase/RNA binding (smg2) 2. Dephosphorylation of SMG2 (smg5/smg6)
coiled-coil protein	sgs3/*Arabidopsis*[c]	1. DNA methylation pathway 2. Viral defense
	Viral suppressors	
proteinase	HC-Pro/potyviruses[m,n,o] (TEV, PVY)	1. Interference with small RNA production/accumulation 2. Interference with response to mobile silencing signal 3. Interference with development via RGS-CAM
viral cell-to-cell movement protein	p25/potexviruses (PVX)[p]	1. Interference with production mobile silencing signal
viral long distance movement	2b protein/cucumoviruses[o,q] (CMV, TAV)	1. Interference with nuclear phase of PTGS
	Cellular mediators of viral suppressors	
calmodulin-related protein	rgs-CaM/*Nicotiana* *tabacum*[r]	1. Suppression of PTGS via interaction with HC-Pro 2. Development

Abbreviations: ago1, argonaute1; ego1, enhancer of glp-1; HC-Pro, helper component proteinase; qde, quelling defective; rde, RNA interference deficient; rgr-CaM, regulator of gene silencing-calmodulin-like protein; sde, silencing defective; sgs, suppressor of gene silencing; smg, suppressor with morpho-genetic effect on genitalia

[a]Cogoni and Macino (1999a). [b]Dalmay *et al.* (2000). [c]Mourrain *et al.* (2000). [d]Smardon *et al.* (2000). [e]Catalanotto *et al.* (2000). [f]Tabara *et al.* (1999). [g]Fagard *et al.* (2000). [h]Cogoni and Macini (1999b). [i]Ketting *et al.* (1999). [j]Wu-Scharf *et al.* (2000). [k]Dalmay *et al.* (2001). [l]Domeier *et al.* (2000). [m]Anandalakshmi *et al.* (1998). [n]Kasschau and Carrington (1998). [o]Brigneti *et al.* (1998). [p]Voinnet *et al.* (2000). [q]Lucy *et al.* (2000). [r]Anandalakshmi *et al.* (2000).

screens for mutants defective in PTGS, quelling and RNAi have all recovered genes encoding putative RdRPs (Chapter 9 and 11) (Table 8.1). In plants, there is a twist to the requirement for RdRP, which is encoded by the *SGS2/SDE1* gene. In a study using an *Arabidopsis sde1* mutant, it was found that a cellular RdRP is essential for transgene-induced PTGS but not for RNA virus-induced PTGS, presumably because these viruses encode their own RdRP, which is needed for virus genome replication (Dalmay *et al.*, 2000). The activity of this viral RdRP is a source of dsRNA that is specific to PTGS triggered by viruses (Figure 8.1⑥).

While the requirement for an RdRP in PTGS, RNAi, and quelling is not disputed, the substrates for this enzyme are not entirely clear. As mentioned earlier, an RdRP could be needed to make copy antisense RNA from aberrant sense RNA templates, followed by annealing to form dsRNA. However, an RdRP activity is required even when dsRNA is introduced exogenously to trigger RNAi. It is conceivable that RdRP synthesizes smRNAs using dsRNA as a template (Waterhouse *et al.*, 1998) (Figure 8.1③), which would preserve the initiating dsRNA molecule and provide a means to amplify the silencing signal. It is more likely, however, that the smRNAs are degradation products of dsRNA (Section II.A.1.c). Additional work is required to ascertain the full range of RdRP substrates and activities.

c. Small RNAs

Small sense and antisense RNAs approximately 21–25 nt in length that are homologous to a silenced gene were first found in a PTGS system in plants (Hamilton and Baulcombe, 1999). Similar smRNAs have since been found in many other other HDGS systems (Matzke *et al.*, 2001a). The smRNAs are probably derived primarily from cleavage of dsRNA (Zamore *et al.*, 2000) (Figure 8.1①③) perhaps by an enzyme similar to RNaseIII, which can degrade dsRNA to produce segments of defined size (Bass, 2000; Parrish *et al.*, 2000). Such an RNase has recently been identified by searching for sequence homology to RNaseIII in the *Drosophila* genome. The enyzme has been given the name Dicer, to denote its ability to cleave dsRNA into small fragments of equal size (Bernstein *et al.*, 2001). Dicer homologs exist in mammals, *C. elegans*, *Schizosaccharomyces pombe*, and *Arabidopsis*. No mutants of *dicer* have yet been recovered, but mutations of the *CARPEL FACTORY* gene, which is an *Arabidopsis* homolog of *dicer*, produce defective flowers and display abnormal growth (Jacobsen *et al.*, 1999).

The smRNAs resulting from Dicer cleavage of dsRNA appear to play a key role in the PTGS mechanism by providing the specificity determinant for degradation of single-stranded mRNA by another (as yet unidentified) RNase (Bass, 2000). The proposed guide role of the small RNAs for mRNA degradation has been supported by data showing the association of smRNAs with an RNase in *Drosophila* (Hammond *et al.*, 2000) (Figure 8.1④). Intriguingly, Dicer contains a so-called PAZ domain, which is also present in the ARGONAUTE class of proteins required for PTGS/RNAi. Dicer and ARGONAUTE might interact through the PAZ domain, thus providing a way in which the smRNAs produced by Dicer could be incorporated into a multisubunit RNase complex that degrades single-stranded mRNA (Bernstein *et al.*, 2001; Baulcombe, 2001).

The smRNAs were originally considered attractive candidates for the mobile silencing signal (Hamilton and Baulcombe, 1999) because their diminutive and uniform size would permit relatively unimpeded intercellular movement and systemic transport through the plant vascular system. However, grafting experiments with plants expressing HC-Pro, the viral suppressor of PTGS which abolishes production of the smRNAs but not the mobile silencing signal (Figure 8.1③), suggest the smRNAs are not involved in systemic silencing (Mallory *et al.*, 2001). The mobile signal might consist of dsRNA, which may or may not be complexed with protein (Figure 8.1⑦) (Fagard and Vaucheret, 2000b).

d. PTGS and DNA methylation

PTGS often involves *de novo* methylation of cytosines (Cs) in transcribed DNA sequences, presumably owing to RdDM (Section II.B). Evidence so far suggests that the RNA-triggering RdDM can be produced in the nucleus or in the cytoplasm, in which case it must enter the nucleus to interact with homologous DNA sequences (Jones *et al.*, 1998) (Figure 8.1②③⑥). Further aspects of PTGS and DNA methylation are discussed in Section II.B.3.

2. Natural roles of PTGS

a. Host defense

Long duplex RNAs are normally not abundant in cells, and when present they are usually associated with pathogenic states (Kumar and Carmichael, 1998). Plant RNA viruses, for example, produce dsRNA during genome replication. Some TEs produce RNAs that contain regions of secondary structure (Turker and Bestor, 1997). Multiple copies of TEs could also be integrated into host DNA such that they are transcribed from both DNA strands, which could possibly produce dsRNA (Ketting *et al.*, 1999; Jensen *et al.*, 1999). A system that can recognize and degrade dsRNA is therefore of considerable benefit to the host in the continuous conflict with various types of parasitic sequences. Indeed, it is now well accepted that one natural function of PTGS and related phenomena is host defense to viruses and TEs. This idea was first put forward by Dougherty and co-workers, who found that plant RNA viruses can be both inducers and targets of PTGS (Lindbo *et al.*, 1993). Plant DNA viruses can also be combated by PTGS via targeted degradation of viral transcripts (Covey and Al-Kaff, 2000).

The antiviral function of PTGS has been substantiated by the discovery of viral suppressors of PTGS (Marathe *et al.*, 2000a) (Table 8.1). Counter-defensive strategies are apparently used by many plant viruses (Voinnet *et al.*,

1999). Final confirmation of the antiviral function of PTGS has come from the analysis of *Arabidopsis* PTGS-defective mutants, some of which exhibit a heightened sensitivity to infection by certain viruses (Mourrain *et al.*, 2000). Similarly, TEs are activated in C. *elegans* RNAi mutants, demonstrating the host defense function of a PTGS-like process in this organism (Ketting *et al.*, 1999; Tabara *et al.*, 1999).

b. Development

The first PTGS-defective *Arabidopsis* mutants did not exhibit any obvious developmental abnormalities, suggesting that the PTGS pathway is dispensable for development. This initial conclusion is being rapidly revised, however, as evidence for developmental aberrations in some PTGS mutants begins to accumulate. For example, *Arabidopsis ago-1* mutants are not only deficient in PTGS but also suffer from developmental defects and infertility (Fagard *et al.*, 2000). Mutations in *CARPELFACTORY*, an *Arabidopsis* homolog of Dicer, lead to poor growth and unregulated cell division in floral meristems (Jacobsen *et al.*, 1999). Tobacco plants expressing HC-Pro, the viral suppressor of PTGS, have unusual and distinctive phenotypes, including curled leaves, thick stems, and poor fertility, indicating that suppression of PTGS affects host gene expression. Similar developmental aberrations are caused by overexpression of rgs-CaM, a cellular protein that interacts with HC-Pro (Anandalakshmi *et al.*, 2000), providing the first indication for an endogenous pathway that negatively regulates PTGS. Outside the plant kingdom, C. *elegans ego1* mutants are sterile due to defects in germline development (Smardon *et al.*, 2000).

Even though PTGS as a whole thus appears to be nonessential for development, the phenotypic irregularities observed in a subset of cases where silencing is blocked suggest that development and PTGS/RNAi share common enzymes or pathways. This idea is strengthened by the probable existence of PTGS proteins that have so far escaped detection in mutant screens (Section II.A.1), presumably because they are required for normal growth and development. Determining how PTGS interlaces with plant developmental pathways is one of the most interesting directions for future research on HDGS.

c. RNA surveillance

The phenomenon of PTGS/RNAi/quelling has revealed a new pathway for targeted RNA degradation. Two recent studies, however, place PTGS in the broader realm of RNA surveillance, an evolutionarily conserved system that ensures the fidelity of mRNA synthesis in eukaryotic cells by recognizing and degrading defective transcripts (Hilleren and Parker, 1999). One aspect of RNA surveillance is nonsense mediated decay (NMD), a process in which mRNAs that contain a premature stop codon are selectively degraded. In C. *elegans*, three of six different *smg* genes required for NMD were found to be essential for persistence of RNAi

(Domeier *et al.*, 2000) (Table 8.1). One of these genes, *smg-2*, is homologous to yeast *Upf1*, which encodes an adenosine triphosphatase with RNA-binding and helicase activities.

A DEAH-box RNA helicase (Mut6) that is involved in degrading misspliced and nonpolyadenylated (aberrant) transcripts was shown to be required for transgene and TE silencing in the unicellular green alga *Chlamydomonas reinhardtii* (Wu-Scharf *et al.*, 2000) (Table 8.1). Although it is not yet certain whether this system targets homologous transcripts in *trans* in a PTGS/RNAi-like phenomenon or operates in *cis* only to degrade the aberrant transcripts without affecting homologous RNAs, the detection of sense and antisense RNAs homologous to the silenced transgene and to reactivated TEs supports classical PTGS. In addition, aberrant RNAs, which would include those that are misspliced and nonpolyadenylated, have been suggested as possible substrates for the RdRP that is required for PTGS (Figure 8.1⑤) (Section II.A.1.b). The results from *C. elegans* and *Chlamydomonas* suggest intriguing overlaps between NMD and PTGS/RNAi pathways and provide new handles for unraveling the mechanism of PTGS.

B. RNA-directed DNA methylation (RdDM)

Both PTGS and TGS are associated with *de novo* DNA methylation, which is concentrated in either transcribed regions or promoters, respectively. Because methylation is confined largely to regions of sequence homology between interacting genes, it has long been assumed that sequence-specific signals are involved in directing DNA modification. (A sequence-specific methylation signal refers here to a nucleic acid molecule, DNA or RNA, that guides the DNA methyltransferase [MTase] to homologous nuclear DNA sequences.) While previous models for methylation associated with HDGS frequently invoked DNA–DNA pairing, which triggers methylation of sequence duplications in some fungi (Faugeron, 2000; Chapter 15), RdDM offers an alternative and perhaps more plausible way to induce *trans* silencing and sequence-specific methylation in higher organisms. An RNA molecule is able to diffuse through the nucleoplasm to locate and interact with homologous DNA sequences, regardless of their location in the genome. This model eliminates the need to invoke somatic pairing of unlinked homologous DNA sequences (Section III) in large plant genomes containing substantial repetitive DNA.

1. Discovery and incidence of RdDM

RdDM was first discovered in a viroid system. Viroids are plant pathogens consisting exclusively of noncoding, highly base-paired, rod-shaped RNAs several hundred nucleotides in length. In the original study demonstrating RdDM, viroid cDNAs integrated as transgenes into tobacco nuclear DNA became methylated *de novo* only when the plants were infected with viroids, implicating replicating viroid RNA in DNA modification (Wassenegger *et al.*, 1994). More recently, RdDM of nuclear transgenes has been detected in plants infected with cytoplasmically replicating RNA viruses carrying transgene sequences (Jones *et al.*, 1998, 1999) and in a nonpathogenic transgenic system (Mette *et al.*, 1999). RdDM leads to dense methylation at most symmetrical and nonsymmetrical Cs within the region of homology between the inducing RNA and the target DNA (Pélissier *et al.*, 1999). Target DNA sequences as short as 30 bp can become modified (Pélissier and Wassenegger, 2000).

2. RdDM, double-stranded RNA, and transcriptional gene silencing (TGS)

The ability of viroids and RNA viruses, which replicate via dsRNA intermediates, to provoke RdDM suggested a general requirement for dsRNA in this process. This has been confirmed in a transgenic tobacco TGS system involving promoter RNAs. A dsRNA transcribed from an IR-containing promoter sequences was able to trigger *de novo* methylation and transcriptional silencing of homologous promoters *in trans* (Mette *et al.*, 2000) (Figure 8.1①②). The promoter dsRNA was cleaved to small RNAs approximately 23 nt in length, demonstrating that it entered the same degradation pathway as dsRNAs involved in PTGS. Although it is not yet clear where in the cell the small promoter RNAs are produced, evidence from plants expressing HC-Pro suggests that promoter dsRNA degradation occurs in the nucleus (Mette *et al.*, 2001). It is thus likely that dsRNase activities are present in both the nucleus and the cytoplasm (Figure 8.1).

In another study, an RNA virus vector carrying 35S promoter sequences was able to trigger *de novo* methylation and TGS of nuclear transgenes driven by the 35S promoter in *Nicotiana benthamiana* (Jones *et al.*, 1999). Because RNA viral replication occurs exclusively in the cytoplasm, the RNAs involved in inducing methylation were apparently produced in this cellular compartment and entered the nucleus to modify homologous DNA sequences (Figure 8.1⑥). DsRNAs that can traverse the nuclear envelope bidirectionally thus provide a common molecular link between RdDM, which can result in TGS if promoter sequences are involved, and the RNA degradation step of PTGS (Figure 8.1①③⑥). Although additional work is required to determine the generality of dsRNA-mediated promoter methylation, preliminary studies suggest that endogenous plant promoters can be silenced by promoter dsRNAs (W. Aufsatz and A. Matzke, unpublished results).

Transcribing through an IR is the most efficient means to produce dsRNA that is capable of triggering methylation and silencing *in trans* (Mette *et al.*, 2000) (Figure 8.1①; Figure 8.2B, bottom). It is conceivable, however, that methylation

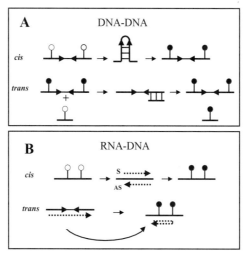

Figure 8.2. DNA and RNA signals triggering *de novo* methylation can potentially act *in cis* and *in trans*. (A) DNA–DNA pairing. (Top) IRs can pair in *cis* to trigger their own methyl-ation. (Bottom) A methylated IR might be able to impose methylation *in trans* on an iden-tical sequence via homologous sequence recognition (Colot *et al.*, 1997). (B) RNA-DNA interactions. (Top) Intermolecular pairing between overlapping sense (S) and antisense (AS) transcripts could produce a dsRNA that triggers DNA methylation of the tran-scribed DNA region *in cis*. (Bottom) RNA hairpins transcribed from an IR can diffuse through the nucleus and methylate homologous DNA sequences *in trans*. RNAs are in-dicated by dotted lines; DNA is depicted by a single straight black line. Methylation is signified by filled circles; unmethylated Cs are indicated by unfilled circles. Homologous DNA sequence recognition or pairing is represented by ladders, which are not intended to specify a mechanism. Pairing steps are shown without lollipops to place emphasis on sequence interactions. Presumably, pairing is not lost during pairing. Abbreviations: IR, inverted DNA repeat; dsRNA, double-stranded RNA.

in cis can be triggered by synthesizing overlapping sense and antisense transcripts from the same DNA molecule (Figure 8.2B, top). Although confirmation of this possibility requires further study, support has come from the observation that methylation can be detected in copies of the nopaline synthase promoter when placed in an antisense (but not sense) orientation relative to the 35S promoter (Mette *et al.*, 2000). *Trans*-acting methylation signals and methylation induced *in cis* by overlapping sense and antisense transcripts might be relevant for genomic imprinting and X-chromosome inactivation in mammals (Section IV).

3. DNA methylation and PTGS

As mentioned previously, *de novo* methylation of protein coding or transcribed regions of transgenes has been detected in many instances of PTGS in plants. How methylation is triggered and whether it is the cause, consequence, or unrelated to PTGS is not known in most cases. The relevance of methylation and whether it is provoked by RdDM or other signals probably varies for different PTGS systems. Methylation is not needed for RNAi in invertebrates such as *C. elegans* that do not methylate their DNA, nor is it required for quelling in *Neurospora*, which has a sparsely methylated genome (Cogoni *et al.*, 1996). In contrast, PTGS in plants can be alleviated when transgene methylation is reduced by drug treatment (Kovarik *et al.*, 2000) or in methylation and chromatin structure mutants of *Arabidopsis* (Morel *et al.*, 2001). In addition, transgene methylation is decreased in PTGS-defective plants mutated in genes encoding RdRP (Elmayan *et al.*, 1998), AGO-1 (Fagard *et al.*, 2000) and SGS-3 (Mourrain *et al.*, 2000) (Table 8.1). These results suggest that the initiation and/or persistence of PTGS requires DNA epigenetic modifications. The nature of this requirement is not yet clear, but it might involve the synthesis of aberrant (truncated or improperly processed) transcripts, which are potential substrates for RdRP (Section II.A.1.b). Aberrant transcripts might be synthesized from methylated DNA templates (Figure 8.1② ③). Dense methylation has been shown to prematurely terminate transcripts in *Ascobolus immersus* (Barry *et al.*, 1993), although it has not yet been shown whether this holds for sparsely methylated DNA templates or for other organisms. Interestingly, even though methylation is dispensable for quelling in *Neurospora*, a nuclear step is suggested by the requirement for a RecQ DNA helicase, QDE3 (Cogoni and Macino, 1999b, 2000; Chapter 9). Continued use of methylation and chromatin structure mutants in *Arabidopsis* (Section V.A.2) will permit an assessment of the general requirements for epigenetic modifications in both PTGS and RNA-based TGS.

4. Mechanism of RdDM

The mechanism of RdDM is unknown but is presumed to require direct RNA-DNA interactions based on sequence homology (Wassenegger, 2000). The recovery of RdDM mutants in *Arabidopsis* will help to determine the mechanism of this process. RNA helicases are potential candidates for the RdDM machinery, since it is probable that RNA–DNA pairing requires single-stranded RNA that is complementary to the target DNA (Pélissier and Wassenegger, 2000).

The minimal DNA target size for RdDM of 30 bp opens the possibility that the smRNAs resulting from dsRNA-cleavage direct DNA methylation. The smRNAs could conceivably guide DNA MTase to unmodified homologous DNA sequences (Figure 8.1③) in a manner similar to what has been proposed for the RNase involved in the mRNA degradation step of PTGS (Figure 8.1④). Conflicting results have been obtained with HC-Pro, which suppresses PTGS by preventing the accumulation of smRNAs. HC-Pro-expressing tobacco plants either did (Mallory *et al.*, 2001) or did not (Llave *et al.*, 2000) maintain methylation of a PTG-silenced transgene. Further studies with HC-Pro using other PTGS systems and plant lines exhibiting RNA-mediated TGS, in which the silenced phenotype is *not* reversed by HC-Pro (Marathe *et al.*, 2000b; Mette *et al.*, 2001), will establish whether smRNAs or dsRNA are required for the initiation and maintenance of RdDM.

An interesting candidate for the MTase involved in RdDM is chromomethylase (CMT). This special form of MTase, which has so far been identified only in plants, contains a chromodomain (Henikoff and Comai, 1998). Chromodomains are present in a variety of chromatin-regulatory proteins, and they are thought to mediate protein–protein interactions. The presence of a chromodomain in CMTs suggests a role for methylating DNA in heterochromatin (Finnegan and Kovac, 2000). Alternatively, the recent finding that the chromodomain of the *Drosophila* histone acetyltransferase MOF is an RNA interaction module (Akhtar *et al.*, 2000) opens up the intriguing possibility that CMTs interact with RNA through the chromodomain to induce RdDM. If CMTs are not involved in RdDM, the interaction of smRNAs by base pairing with DNA could produce RNA–DNA hybrids and single-stranded DNA loops or RNA–DNA triple helices. These unusual structures could attract conventional MTases (Smith *et al.*, 1991).

5. Natural roles of RdDM

Similarly to PTGS, RdDM is a new phenomenon that was discovered first in a transgenic plant system. The extent to which it acts to modify endogenous genes is not yet known. Ascertaining the natural roles and targets of this process awaits the recovery of *Arabidopsis* RdDM-impaired mutants. RdDM could conceivably contribute to host defense by provoking methylation of integrated TEs via their transcripts (Matzke *et al.*, 2000). RdDM might provide the key for understanding viroid pathogenicity, which remains mysterious after more than two decades of study, if plant genes sharing 30 bp or more of sequence homology to viroids are identified (Pélissier and Wassenegger, 2000). Indeed, viroids are believed to be derived from TEs (Kiefer, *et al.*, 1982), which could explain their ability to induce methylation (Matzke *et al.*, 2000).

Clarifying any potential roles for RdDM in normal plant development will also depend on analyses of RdDM-defective mutants. Given that plant development requires proper DNA methylation (Section V.A.1) as well as components of the PTGS pathway (Section II.A.2.b), it is likely that RdDM also participates in the implementation of developmental programs.

C. Advantages of RNA silencing in gene regulation and host defense

HDGS phenomena are expanding the "guide" function of RNA (Caprara and Nielsen, 2000), which must now include targeted RNA degradation (PTGS) and homologous DNA methylation (RdDM). In addition to the ability to target or template sequence-specific modifications via base pairing, an advantageous feature of RNA-based silencing mechanisms is their reversibility: When synthesis of the triggering RNA is discontinued, gene silencing is alleviated. With respect to RdDM, repressing the synthesis of the inducing dsRNA results in loss of methylation from the target DNA in somatic cells (Mette *et al.*, 1999). This suggests that the RNA signal is needed continuously to stimulate *de novo* methylation of target sequences. A constant requirement for an RNA trigger is consistent with the pattern and distribution of C methylation associated with RdDM, which involves dense methylation not only at symmetrical CpG and CpNpG nucleotide groups but also at nonsymmetrical Cs (Section II.B.1). In contrast to methylation at symmetrical Cs, which can persist through the maintenance function of DNA MTase, methylation at nonsymmetrical Cs must be triggered anew following each round of DNA replication, a function which can be fulfilled by RdDM. The reversibility afforded by RdDM generates flexible gene expression patterns, which could underlie the well-known plasticity of plant development and help sessile plants to cope with unpredictable environments.

RNA-based host defenses rely on information in the parasite's own genome. These defensive strategies are therefore able to readily accommodate rapidly evolving parasitic sequences, such as retroelements and RNA viruses, which are replicated by error-prone enzymes (Sijen and Kooter, 2000). Given the selective value of these highly adaptable RNA-based defense strategies, it is not surprising that they have been conserved through evolution.

RNA silencing mechanisms have been studied primarily with transgenes and parasite nucleic acids. Although the extent to which these mechanisms regulate endogenous genes in plants and other eukaryotic organisms is not yet known, it likely that they play a more significant role than is currently appreciated. For example, up to 1200 unknown open reading frames in yeast could correspond to overlapping antisense transcripts (Goffeau, 2000). Noncoding RNAs and natural antisense RNAs are transcribed in plants (Terryn and Rouzé, 2000) and humans (Kumar and Carmichael, 1998; Claverie, 2000). It is presently impossible to predict noncoding genes that might have essential regulatory roles, but this is one important goal of "ribonomics," the RNA analog of proteomics (Doudna, 2000). To identify possible regulatory RNAs, it might also be useful to catalog IRs in genomes (LeBlanc *et al.*, 2000) and determine whether they are transcribed.

III. DNA–DNA PAIRING

In plants, DNA–DNA pairing has long been postulated to occur in various cases of HDGS that involve inactivation and/or methylation of repeated sequences on different chromosomes (Jorgensen, 1992; Matzke and Matzke, 1995; Baulcombe and English, 1996; Bender, 1998). This notion followed from observations of coextensive methylation of unlinked homologous sequences, suggesting that sequence interactions induce DNA modifications. Evidence that DNA–DNA interactions can trigger the modification of linked and unlinked sequence duplications in eukaryotes is provided most convincingly by the repeat-induced point mutation (RIP) and methylation-induced premeiotically (MIP) phenomena observed, respectively, in the filamentous fungi *Neurospora crassa* and *Ascobolus immersus* (Faugeron, 2000; Chapter 15). MIP is a straightforward epigenetic effect in which pairing of duplicated sequences induces dense methylation that is confined largely to the repeated region. These fungal precedents initially fueled arguments for similar DNA-pairing processes as triggers for epigenetic modifications in plants. In addition, a role for DNA–DNA pairing was considered because HDGS effects in plants superficially resembled some pairing-dependent epigenetic phenomena in *Drosophila*, such as dominant-position-effect variegation and transvection (Matzke *et al.*, 1994; Chapter 13). Therefore, prior to the discovery of RdDM, which can also account for sequence-specific methylation (Section II.B), there was much discussion in the plant community about whether homologous DNA pairing acted to trigger gene silencing.

Recently, however, the tide of opinion has changed: DNA–DNA pairing is being discounted in favor of dsRNA–DNA interactions. This shift has come about because some instances of PTGS that were previously believed to result from DNA pairing have been found to involve dsRNA (J. Kooter, personal communication). Moreover, it has become clear that TGS, which was formerly a strong candidate for mediation by DNA pairing, can be triggered by RdDM if dsRNAs containing promoter sequences are synthesized (Section II.B.2). These new results suggest that many examples of both PTGS and TGS can be attributed to dsRNA.

Despite the ascension of dsRNA in the HDGS hierarchy, it remains possible that some types of single-copy or repetitive sequences are able to interact via DNA pairing *in cis* and/or *in trans* to trigger gene silencing and methylation or chromatin modifications. The possible presence of a DNA MTase in *Arabidopsis* that is related to the one required for MIP in *Ascobolus* (Finnegan and Kovac, 2000) helps to keep alive the idea that pairing might also trigger methylation of homologous sequences in plants. Despite considerable speculation about DNA–DNA pairing and gene silencing in plants, there have been few attempts to test the idea. Because several experiments have suggested that IRs can trigger gene silencing via DNA pairing, we focus here on IRs and their roles in different types of HDGS.

A. Inverted repeats (IRs): DNA pairing or dsRNA?

The emphasis on IRs and DNA pairing-mediated silencing in plants follows from structural analyses of silenced transgenes and endogenous genes (Muskens *et al.*, 2000). A comprehensive study revealed that chalcone synthase (*CHS*) transgene loci capable of triggering PTGS in petunia always contained *CHS* transgenes arranged as IRs (Stam *et al.*, 1998). The IRs were densely methylated and re-ported to be transcriptionally silent. Consequently, DNA pairing between a *CHS* transgene IR and endogenous *CHS* genes was invoked as a means to trigger PTGS, possibly by synthesis of aberrant *CHS* RNA from the methylated templates (Stam *et al.*, 1998) (Figure 8.1②③).

The silencing activity of IRs is also illustrated by the endogenous phos-phoribosylanthranilate isomerase (*PAI*) gene family in *Arabidopsis*. In some eco-types, two of members of this family are arranged as an IR, which is densely methylated. This IR is able to trigger methylation and *trans* inactivation of sin-gle copies of the *PAI* gene that reside on a different chromosome (Bender and Fink, 1995). Because methylation is restricted to the regions of sequence identity between the interacting genes, DNA–DNA pairing was proposed to mediate the *trans*-silencing effect (Luff *et al.*, 1999).

These results and others show clearly that IRs are potent silencers. This capability could arise because they are potentially able to pair both *in cis*, to trigger their own methylation, and *in trans* to provoke modification and silencing of unlinked homologous sequences (Figure 8.2A). As discussed previously, IRs can also be transcribed to produce dsRNA, which can trigger RdDM and/or PTGS depending on the sequence composition of the repeated region (Sections II.A and II.B). Unfortunately, it is difficult to distinguish between silencing effects that might be mediated by DNA pairing from those initiated by dsRNA, particularly when the inducing dsRNA is produced in only minute amounts. Conceivably, an IR containing protein-coding regions could trigger PTGS in one of two ways: (1) by being completely transcribed to produce an RNA hairpin that initiates degradation of homologous mRNAs directly (Figure 8.1①); or (2) by DNA pairing to induce methylation of homologous sequences, which can be templates for aberrant transcripts that are potentially substrates for RdRP (Section II.A.1.b) (Figure 8.1②③). To illustrate the difficulties involved, *CHS* dsRNA has indeed been detected recently in the chalcone synthase PTGS system mentioned above (J. Kooter, personal communication), thus overturning the initial conclusion that DNA pairing was involved in this silencing system (Stam *et al.*, 1998).

1. Experimental detection of somatic DNA pairing: transvection-like effects in plants

Detecting DNA-DNA interactions that alter gene activity in somatic plant cells is problematic. Unlike the case in *Drosophila*, where stable associations of homolo-gous chromosomes are observed in somatic cells, cytologically visible pairing is not the rule in most somatic plant nuclei. Progress is being made in localizing complex transgene inserts in plant interphase nuclei by fluorescence *in situ* hybridization (FISH) (Abranches *et al.*, 2000); however, even if close associations between widely separated transgene loci are observed, FISH alone cannot determine whe-ther these interactions between chromosomal domains influence gene expression.

To circumvent these limitations, a transvection system (Chapter 13), which uses *trans* activation (instead of silencing) as a positive measure of physical contact between transgenes on separate chromosomes was established in tobacco (Matzke et al., 2001b). This system assessed the ability of an enhancer on one allele to *trans*-activate an enhancerless promoter on the second allele. Enhancerless alleles were obtained from enhancer-containing alleles *in planta* via Cre-*lox*-mediated recombination. Several single-copy transgene loci did not display any transvection-like effects. One complex transgene locus, however, exhibited behavior that could be interpreted in terms of allelic pairing. Depending on which heteroalleles were present, two distinct transvection-like effects—resulting in either *trans* activation or *trans* silencing—were observed. Because *trans* activation presumably necessitated close physical association of transgene alleles, it could be inferred that *trans* silencing also involved allelic pairing. RNA-mediated silencing could be ruled out because the expected hallmarks of RNA silencing—dsRNAs and small RNAs homologous to the silenced gene—were not detected.

These results provided molecular evidence for allelic pairing that could positively or negatively influence gene expression, perhaps through enhancer interactions. They also suggested that a specific DNA sequence arrangement was required to mediate pairing. Transcriptional regulatory elements have been implicated in homology effects in *Drosophila* (Wu and Morris, 1999; Chapter 13). The complex transgene locus displaying transvection-like effects comprised a single enhancer that occupied a special position in the spacer region of an IR (Figure 8.3A). Conceivably, this sequence configuration could adopt a cruciform structure with the enhancer in the loop (Figure 8.3C), which might be a position favoring *trans* interactions. The lengths of the repeated sequences in the IR and the length of the spacer could determine the frequency of cruciform formation and its stability *in vivo*. While plausible models can be suggested for how this arrangement might trigger either *trans* activation or *trans* silencing, depending on whether only one or both alleles contains an enhancer (Matzke et al., 2001b), the mechanisms involved are not yet known . The information gained from the tobacco system is being used to reproduce the transvection-like effects with easily distinguishable alleles in *Arabidopsis* so that genetic screens can be carried out (A. Matzke and T. Kanno, unpublished results).

2. Multiple epigenetic effects triggered by IRs

Although additional analyses are required, the data obtained so far suggest that IRs can elicit a variety of epigenetic effects, depending on: (1) their length and gene content; (2) the length and content of the spacer region; (3) whether they are transcribed to produce dsRNA; and (4) whether they form DNA secondary structures such as cruciforms *in vivo* (Figure 8.3). What appears certain is that IRs

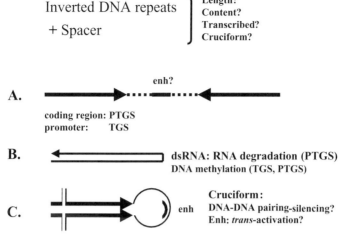

Figure 8.3. Inverted DNA repeats (IRs) can potentially elicit multiple epigenetic effects depending on a number of features. (A) Sequence content. The repeated region can comprise protein coding regions or promoters, which could give rise to PTGS or TGS, respectively. Spacer regions could contain genes or regulatory regions, such as enhancers (enh). (B) Transcriptional activity. Double-stranded RNA (dsRNA) resulting from transcription through an IR containing coding regions can trigger RNA degradation in the cytoplasm (PTGS) or methylation of homologous DNA sequences. If dsRNAs contain promoter sequences, TGS accompanied by *de novo* methylation of homologous promoters can result. (C) Secondary structure. IRs could form cruciforms *in vivo*. Enhancers or other regulatory regions that are present in the loop could be particularly adept at interacting *in trans* with homologous partners. This could lead to either silencing or activation, depending on the sequence content of the spacer.

can be transcribed to produce *trans*-acting dsRNA (Figure 8.2B, bottom) that can trigger DNA methylation and PTGS or TGS depending on the sequence content (Figure 8.3A, B; Sections IIA and IIB). IRs themselves also seem to be able to acquire stable methylation in the absence of dsRNA synthesis (Mette *et al.*, 1999), suggesting that they can pair *in cis* to trigger methylation (Figure 8.2A, top). Less certain is whether an IR is able to pair with homologous DNA sequences *in trans* (Figure 8.2A, bottom) to induce either silencing or activation, depending on the sequence content of the repeated and spacer regions (Figure 8.3C). So far, the tobacco transvection system provides the only molecular evidence for DNA–DNA *trans* interactions involving IRs. These interactions were limited to pairs of transgene complexes inserted in the same ("allelic") chromosomal locations and did not extend to homologous sequences at ectopic locations (Matzke *et al.*, 2001b). These results favor the notion that *trans*-silencing effects involving homologous sequences at unlinked loci are mediated by dsRNA and not by DNA–DNA pairing.

B. Current status of DNA–DNA pairing models

In the previous section, we have gone into detail discussing IRs because there are at least some data supporting their involvement in DNA pairing-mediated epigenetic effects in plants. IRs are not just found with transgene sequences but are common in higher eukaryotic genomes (Gordenin *et al.*, 1993). They must therefore be considered natural sources of epigenetic variability and deserve further study in that context. At present, there are no data to indicate that pairing occurs regularly between alleles lacking repeats or between repeats other than IRs in plants, but it remains possible that certain repeat configurations or highly heterchromatic regions associate in somatic cells. Another possibility is that dispersed repeats present above a certain threshold copy number might engage in frequent, transient contacts, which could maintain a high level of methylation in the repeat population (Section V.B.1).

It will be important to establish tractable experimental systems for analyzing somatic DNA pairing in plants, preferably in *Arabidopsis*, where mutants can be obtained relatively easily. It is also possible that genetic screens to recover mutants of natural silencing phenomena that have been postulated to involve allelic pairing, such as paramutation in maize (Chapter 7), will identify proteins that are obviously involved in homologous DNA interactions.

C. DNA–DNA pairing and host genome defense

In principle, silencing and/or methylation that is triggered by homologous DNA pairing provides an ideal way to protect the host genome from integrated TEs or virus sequences that tend to increase in copy number. RIP and MIP appear to fulfill this host genome defense function in some filamentous fungi (Faugeron, 2000; Chapter 15). IRs can result from TE activity, and it is likely that they become methylated and silenced because they are recognized as products of transposition by the host genome defense. Indeed, several endogenous genes that display *trans*-silencing effects, including the *Arabidopsis PAI* gene family (Section III.A) (Melquist *et al.*, 1999) and semminant *nivea* alleles in snapdragon (Bollman *et al.* 1991), involve IRs that were presumably generated via TE activity. Transient pairing of multicopy dispersed repeats might explain the progressive methylation of TEs observed during plant growth (Brown *et al.*, 1994) or the modification of integrated viral sequences (Section V.B.1).

IV. HOMOLOGY-DEPENDENT GENE SILENCING (HDGS) IN GENOMIC IMPRINTING AND X-CHROMOSOME INACTIVATION IN MAMMALS

We have described how HDGS phenomena in plants result from various nucleic acid sequence interactions, and have specified dsRNA, which can act in the nucleus and the cytoplasm, and pairing of DNA repeats as actual or potential triggers of silencing and methylation. These principles might be applicable to genomic imprinting and X-chromosome inactivation in mammals, which both require DNA methylation. Imprinting and X inactivation have been discussed with respect to foreign DNA (Barlow, 1993), DNA repeats (Reik and Walter, 1998; Lyon, 1998), overlapping sense and antisense RNAs (Sleutels *et al.*, 2000), and chromosome pairing (LaSalle and Lalande, 1996; Duvillie *et al.*, 1998; Marahrens, 1999). These features, which are reminiscent of those associated with silenced transgenes in plants, suggest that similar host defense-derived epigenetic mechanisms and methylation signals might be involved in imprinting and X inactivation.

A. Genomic imprinting

Barlow originally proposed that imprinting mechanisms might have evolved from the host defense function of methylation (Barlow, 1993). The major limitation of this hypothesis is that a host defense mechanism has no inherent requirement to be sexually dimorphic, a characteristic that is the hallmark of imprinting (Spencer *et al.*, 1999). Traditional explanations for the sexual dimorphism of imprinting postulate the activity of DNA MTases (Mertineit *et al.*, 1998) or other factors (Reik and Walter, 1998) that are specific to either the male or female germline. Drawing from HDGS phenomena, we propose an alternative to sex-specific enzymes or factors: a *trans*-acting methylation signal, consisting of either DNA repeats or dsRNA, could originate on one of the sex chromosomes and direct the modification of homologous sequences at an autosomal imprinted locus.

The possibility that sex chromosomes play a role in imprinting through the action of dosage sensitive modifiers has been suggested previously (Sapienza, 1990). In the context of the present discussion, modifiers would be defined as repeats that are present both on a sex chromosome and at an autosomal gene subject to imprinting. These shared repeats would promote interchromosomal communication by means of sequence-specific methylation signals. Sex chromosome repeats could interact *in trans* with homologous repeats on autosomes and trigger methylation either via DNA-DNA pairing (Figure 8.2A) or a diffusible dsRNA transcribed from an IR (Figure 8.2B, bottom). Because the single X chromosome in spermatocytes becomes silent during spermatogenesis (Heard *et al.*, 1997), RNAs originating from X-chromosome repeats would be associated with maternal imprints. Indeed, for most imprinted genes, the methylation imprint is derived from the oocyte (Reik and Walter, 2001). Y-chromosome repeats could in principle establish paternal imprints during gametogenesis; however, these can also be established in somatic cells via the synthesis of paternal antisense RNAs (Section IV.B).

While the imprint imposed *in trans* from a sex chromosome could silence a gene directly, some cases of imprinting are more complex and involve reciprocally imprinted genes or multiple methylation events. An example is imprinting of the

Igf2r gene in mice, which requires two temporally separate but interconnected imprinting steps: one imposed on the maternal allele in oocytes and the second acquired postimplantation on the paternal allele (Chapter 5).

B. *Igf2r* imprinting as a possible example of RNA silencing involving sequential *trans* and *cis* methylation steps

The initial gametic imprint at *Igf2r* consists of methylation in a CpG island/promoter present in an intron of the maternal allele. According to the hypothesis presented here, this methylation could be imposed *in trans* by a dsRNA signal transcribed from X-chromosome repeats that are also present at the intronic promoter of the *Igf2r* gene (Figure 8.4①). Because of the transcriptional inactivation of the single X chromosome during spermatogenesis (Heard *et al.*, 1997), the intronic promoter remains unmethylated on the paternal allele. It is the paternal allele, however, that becomes silenced in somatic cells. This silencing is associated with the synthesis of an antisense transcript, "*Air*," that is initiated in the unmethylated intronic promoter (Lyle *et al.*, 2000) (Figure 8.4②). The promoter of the paternal allele becomes methylated around the time of *Air* synthesis, possibly via a

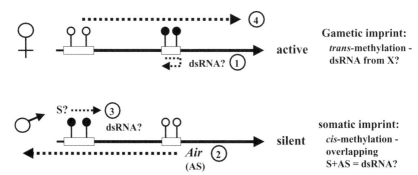

Figure 8.4. An RNA silencing model involving sequential *trans*-and *cis*-methylation steps for imprinting of murine *Igf2r*. An intronic CpG island/promoter (small white box) on the maternal *Igf2r* allele becomes methylated (filled black circles) during oogenesis. This gametic imprint is hypothesized here to be imposed by a *trans*-acting RNA hairpin transcribed from homologous repeats on the X chromosome ①. On the paternal allele, the intronic promoter remains unmethylated (open circles) owing to the silencing of the single X chromosome during spermatogenesis. In postimplantation embryos, a long paternal antisense (AS) RNA, "*Air*," is synthesized from this intronic promoter ②. Hypothetically, if *Air* overlaps with a short *Igf2r* sense (S) RNA for a brief period, an open dsRNA will form that could trigger methylation and silencing *in cis* of the paternal *Igf2r* promoter (large white box) ③. The maternal allele remains active because *Air* synthesis is blocked by the methylation in the intronic promoter ④. RNAs are indicated by dotted lines; double-stranded DNA is depicted by a single heavy black line.

cis-acting dsRNA that forms during a brief period of synthesis of overlapping *Igf2r* sense and *Air* antisense transcripts from the paternal allele (Figure 8.4③). The maternal allele remains active because *Air* transcription is repressed by methylation in the intronic promoter (Figure 8.4④).

Under this hypothesis, *Igf2r* imprinting can be understood as a combination of successive *trans-* and *cis*-acting methylation events that are each triggered by homologous RNA–DNA interactions. Two predictions of this hypothesis are that transcribed repeats homologous to the intronic promoter are present on the X chromosome, and that an dsRNA containing *Igf2r* promoter sequences is generated postimplantation from the paternal allele. Whether other imprinted genes can be accommodated by this scheme remains to be seen, but the facts that the gametic imprint is often in or near promoter regions (Tilghman, 1999) and that antisense RNAs are often synthesized from imprinted loci in somatic cells (Reik and Walter, 1998; Sleutels *et al.*, 2000) suggests that similar principles might be operating in other cases.

C. X-chromosome inactivation and other types of dosage compensation

X-chromosome inactivation involves overlapping sense and antisense RNA (*Xist* and *Tsix*), which could form an RNA duplex (Lee, 2000). Inactivation spreads in *cis* along the X, probably by means of LINE-1 retroelements, which are particularly concentrated on the X chromosome (Bailey *et al.*, 2000). Although the contributions of these noncoding RNAs and L1 repeats to X inactivation and methylation remain to be clarified (Chapter 2), it is intriguing that recurring players have been observed in epigenetic phenomena in plants and in mammals. To further solidify the connections between parasitic sequences, noncoding RNAs, and epigenetic phenomena, it has been suggested that the noncoding RNAs involved in X-chromosome dosage compensation in *Drosophila* are of TE or viral origin (Pannuti and Lucchesi, 2000).

V. HDGS AND THE HOST GENOME DEFENSE: EVOLUTIONARY ASPECTS

We have discussed the actual or potential host defense functions of PTGS (Section II.A.2.a), RdDM (Section II.B.5), and DNA-DNA pairing (Section III.C). It is useful to think in terms of two distinct host defense systems operating against parasitic sequences in plant cells. One is active in the cytoplasm and involves targeted degradation of homologous transcripts (PTGS). The second involves a genome defense characterized by *de novo* DNA methylation that is triggered by sequence-specific methylation signals (i.e., dsRNA-DNA and, possibly, DNA–DNA interactions). The genome defense is most relevant to host evolution because it acts on—and can potentially alter the sequence of—parasitic sequences that have become stable components of host chromosomes.

In this section, we discuss how the genome defense acting on endogenous parasitic sequences might have contributed beneficially to host evolution. In this context, we consider two alternative ways that repeated sequences targeted by the genome defense can evolve: repeats can either remain highly homologous, if this property provides a selective advantage to the host, or they can diverge in sequence, which could produce remnants that might confer a novel function by means of their continued "foreign" character. Both of these possible fates are discussed separately with respect to the evolution of genome defense-derived epigenetic control mechanisms and the acquisition of hereditary resistance to some viruses.

A. Evolution of genome defense-derived epigenetic control mechanisms

1. DNA methylation in genome defense and development

Although DNA methylation is absent in some organisms, it is present in at least some members of all phylogenetic groups, suggesting that it is ancestral (Bestor, 1990; Regev *et al.*, 1998). While the role of DNA methylation in the regulation of eukaryotic gene expression during development remains a contentious issue, the host defense function of methylation is widely acknowledged (Bestor, 2000). DNA methylation is able to silence TEs in plants (Bennetzen, 2000) and mammals (Yoder *et al.*, 1997; Walsh and Bestor, 1999; Cherry *et al.*, 2000). In plants, the host defense function of DNA methylation is substantiated by the observation that some TEs are reactivated in *Arabidopsis* mutants that are defective in DNA methylation (Hirochika *et al.*, 2000; Singer *et al.*, 2001).

Despite the uncertainties regarding the extent to which methylation helps to regulate gene expression in higher eukaryotes, there is no doubt that methylation is essential for proper development in plants and mammals. This requirement is demonstrated by the developmental abnormalities displayed by methylation-deficient mutants of *Arabidopsis* (Richards, 1997) and the embryonic lethality of DNA MTase knockouts in mice (Li *et al.*, 1992; Okano *et al.*, 1999). Moreover, two human pathologies, Rett syndrome and ICF syndrome, are associated with defects in DNA methylation (Robertson and Wolffe, 2000).

2. TGS mutants provide links between methylation and chromatin structure

In eukaryotes that methylate their DNA, changes in methylation and chromatin structure are intimately linked. This was first shown with the *Arabidopsis DDM1*

Table 8.2. Proteins Involved in TGS

Cellular proteins		
Protein	Mutant name/organism	Possible roles
SWI2/SNF2-like protein[a]	ddm1/*Arabidopsis* som1, 4, 5/*Arabidopsis*	1. Maintenance function DNA methylation 2. Chromatin remodeling 3. Development
DNA-methyltransferase[b]	ddm2/*Arabidopsis* met1/*Arabidopsis*	1. Maintenance function in DNA methylation 2. Development
Novel nuclear protein partially similar to the ATPase region of the SWI2/SNF2 family[c]	mom1/*Arabidopsis*	1. Cromatin remodeling 2. Transposon defense? 3. Methylation-independent
Unidentified[d]	hog1/*Arabidopsis*	1. Role in *trans*-TGS 2. Methylation-dependent
Unidentified[d]	sil1/*Arabidopsis* sil2/*Arabidopsis*	1. Role in *cis*-TGS 2. Methylation-independent

Abbreviations: ddm, deficient in DNA methylation; hog, homology-dependent gene silencing; met1, DNA methyltransferase 1; mom, Morpheus' molecule; sil, silencing; som, somniferous
[a] Jeddeloh *et al.* (1999)
[b] Finnegan and Kovac (2000)
[c] Amedeo *et al.* (2000)
[d] Furner *et al.* (1998)

(deficient in DNA methylation) gene, which encodes a SWI2/SNF2 chromatin remodeling factor (Jeddeloh *et al.*, 1999) (Table 8.2). Methylation of rDNA genes and centromeric repeats is greatly reduced in *ddm1* mutants (Vongs *et al.*, 1993). Similarly, mutations in the human gene encoding ATRX, which is involved in chromatin remodeling, cause changes in the methylation pattern of several highly repeated sequences including rDNA arrays, a Y-specific satellite and subtelomeric repeats (Gibbons *et al.*, 2000). Further links between methylation and chromatin structure have been revealed by the ability of histone deacetylases to associate with methyl-DNA-binding proteins (Robertson and Wolffe, 2000) and with DNA MTase (Fuks *et al.*, 2000). Moreover, the possibility that chromodomains might serve generally as RNA interaction modules (Akhtar *et al.*, 2000) suggests that both chromatin-regulatory proteins containing these domains and chromomethylases can target epigenetic modifications to specific regions of the genome via guide RNAs (Section II.B.4). Because of these close molecular connections, it is assumed here that similar signals can elicit changes in both methylation and chromatin structure.

Relationships between chromatin remodeling and methylation are being further defined in TGS-defective mutants in *Arabidopsis* (Table 8.2). These mutants are revealing the existence of both methylation-dependent and methylation-independent pathways of TGS (Vaucheret and Fagard, 2001). In *ddm1* and *hog* mutants, release of TGS is accompanied by loss of methylation from reactivated genes. In contrast, *mom* and *sil* mutants relieve TGS without affecting methylation (Vaucheret and Fagard, 2001). The proteins encoded by *HOG* and *SIL* have not yet been identified. MOM encodes a novel nuclear protein with a region related to SWI2/SNF2 chromatin-remodeling proteins (Amedeo *et al.*, 2000). While *ddm1* mutants display developmental aberrations, no obvious phenotype is associated with the *mom* mutation. This is consistent with the major natural target of *mom*, which consists of sequences derived from the *Athila* retrotransposon family (Steimer *et al.*, 2000). MOM does not act on the natural targets of DDM1 (centromeric and rDNA repeats as well as certain single-copy sequences [Kakutani *et al.*, 1999]), which might explain the more severe phenotype of *ddm1* mutants. An emerging theme from these studies is that different chromatin-remodeling complexes, whose activity may or may not be connected to changes in DNA methylation, can target specific sequences (Jeddeloh *et al.*, 1999; Gibbons *et al.*, 2000). It is anticipated that identification of additional TGS mutants will continue to clarify the relationship between methylation and chromatin structure.

3. Signals for *de novo* methylation

Although DNA methylation is required for genome defense and for development in plants and mammals, much remains to be learned about how methylation is targeted to specific sequences. It is relatively easy to see how preexisting methylation at symmetrical CpG and CpNpG nucleotide groups can be perpetuated during DNA replication by a maintenance MTase activity that prefers hemimethylated DNA as a substrate. Less obvious are the initial signals that provoke *de novo* methylation of an unmodified sequence. We have discussed how transgene HDGS phenomena have revealed that *de novo* DNA methylation can be triggered by sequence-specific methylation signals involving RNA–DNA or DNA–DNA interactions. Although methylation signals based on homologous sequence interactions could in principle play a role in regulating endogenous genes, there is no evidence that such interactions are currently involved in most cases (some exceptions might include certain mammalian imprinted genes; Sections IV.A and IV.B). It is thus likely that other types of signals attract methylation to help regulate promoters of endogenous genes. In this context, it is important to recall the mutagenic potential of DNA methylation, and how this could transform a homology-dependent methylation signal into a different type of signal over evolutionary time.

The precedent for this idea is the RIP phenomenon in *Neurospora* (Chapter 15). RIP silences sequence duplications through both genetic and epigenetic alterations. Initially, RIP generates exclusively C-to-thymine (T) mutations throughout a duplicated region. This mutational bias strongly indicates that mutagenesis occurs via deamination of 5-methylcytosine (m^{5}C) to T. Because the mutations are limited to the duplicated sequences, the methylation that precedes mutagenesis is believed to be induced by DNA–DNA pairing similar to that occurring during MIP in *Ascobolus*. Once a sequence is mutated by RIP, it becomes a hot spot for *de novo* methylation at remaining Cs, even when the sequence is present as a single copy. The signal for *de novo* methylation of single-copy "ripped" sequences appears to be an altered sequence composition (Miao *et al.*, 2000). While RIP greatly accelerates the C-to-T mutation rate, the same process will occur, albeit more slowly, with any methylated sequence, owing to spontaneous deamination of m^{5}C to T (Gonzalgo and Jones, 1997). Therefore, a TE or other repeat can progress from being an identifiable entity that is methylated via homologous sequence interactions with other copies in the genome to an unrecognizable remnant that nevertheless still attracts methylation because of an altered sequence composition. Methylation of such remnants can thus be considered an outcome of the genome defense acting on ancient, degenerate parasitic sequences. Although the existence of such methylation signals in plants and mammalian genes remains to be verified, they might be important contributors to epigenetic regulation of gene expression in these groups (Section V.A.5).

Analyses of epigenetic alleles (epialleles) of endogenous plant genes are confirming the basic foreign nature of signals for *de novo* methylation. The single-copy *SUPERMAN* gene in *Arabidopsis* contains a C[A/T] microsatellite close to the transcription start site. This repeat might be recognized as alien and attract the dense methylation acquired by weak epialleles of this gene (Jacobsen and Meyerowitz, 1997; Jacobsen *et al.*, 2000). A gene involved in flowering, *FWA*, is silenced by methylation at certain stages of development in wild-type plants. *Ddm1*-induced reductions in methylation in *fwa* mutants leads to ectopic expression and delayed flowering. Methylation is believed to be triggered by two direct repeats in the 5′ region of the gene through an HDGS mechanism (Soppe *et al.*, 2000).

Given the high content of repetitive sequences in plant genomes and the proximity of different types of repeats to many plant genes (Bennetzen, 2000), it is likely that further examples of natural epialleles will be identified. Repeats that are close to regulatory regions can fine-tune gene expression levels by provoking varying degrees of epigenetic modifications, thus generating reversible phenotypic variability that could help plants adapt to changeable environments. Some epialleles can also be extraordinarily stable and hence play a significant role in evolution. For example, the *Lcyc* epigenetic mutation that creates natural variation in floral symmetry in toadflax was originally described more than 250 years ago (Cubas *et al.*, 1999).

4. TEs accumulate in plant and vertebrate genomes despite genome defense

Despite the existence of a genome defense that is designed to guard against un-controlled proliferation of invasive sequences, the fact remains that most plant and vertebrate genomes are laden with TEs. Approximately 45% of the euchro-matic portion of the human genome consists of TEs and related sequences (Smit, 1999). At least 50% of the maize genome is derived from TEs (Bennetzen, 2000). Given these enormous TE loads in higher eukaryotic genomes, it is legitimate to question the efficacy of the genome defense.

Possible reasons for the massive accumulation of TEs in plant and ver-tebrate genomes have been presented previously (Matzke *et al.*, 1999). Although the genome defense can in principle suppress TEs by methylation, it cannot be assumed that (1) different types of TEs are equally susceptible to methylation, (2) host defenses are uniformly efficient in all cell types, and (3) all host genomes are similarly susceptible to transposition. During the course of evolution, TEs and their hosts exert a reciprocal influence on each other, with TEs developing ways to evade the genome defense and host genomes varying in their tolerance to transposition (Matzke *et al.*, 1999).

In the context of host genome tolerance, we mention here specifically polyploidy. Polyploidy, which involves whole genome duplications, has been a powerful force in plant and vertebrate evolution. It is estimated that 50–70% of flowering plants have polyploidy in their history (Wendel, 2000). Jawed vertebrates are believed to be ancient octoploids (Spring, 1997). It is likely that the polyploid character of plant and early vertebrate genomes contributed to the accumulation of TEs in these groups. One reason for this is that the duplication of all genetic loci in a polyploid genome buffers against insertional mutagenesis by transposons: if one copy of a gene is disrupted by a TE insertion, there will be a backup copy to compensate. Polyploid genomes thus provide relatively hospitable environments for transposition as well as abundant raw material for genome mod-ification by TEs. In the context of polyploidy, methylation, and TEs, plants and vertebrates share several features. Each group has large populations of TEs in its (polyploid) genomes, which are highly methylated, and methylation has infil-trated into genes. Moreover, methylation is required for proper development in plants and mammals (Section V.A.I).

5. Recruiting the genome defense to regulate host genes in plants and vertebrates

With the above discussion in mind, it becomes possible to envision how the host genome defense might have been harnessed during evolution to regulate host gene expression in plants and vertebrates. The most straightforward way to achieve this is to make host genes or their transcripts appear foreign to the defense machinery. This could occur in the following manner. In the permissive environment afforded by a polyploid genome, active TEs will occasionally integrate into the 5′-flanking regions of individual copies of duplicated genes without disrupting an existing function. A TE insertion could conceivably provide *cis*-acting regulatory sequences that drive a new pattern of expression (Bennetzen, 2000). In addition, because TEs are the primary targets of the genome defense, the TE insertion will attract methylation and/or chromatin modifications to the gene. While this TE insertion might initially become methylated through homologous sequence interactions with other TEs, it could eventually be transformed into a "foreign" remnant that still attracts methylation because of an altered sequence composition (Section V.A.3). In this way, the type of epigenetic silencing associated with the genome defense is imposed on host genes to help repress them in cell types where their protein products are not required. Diversification of gene expression patterns as a consequence of a TE insertion could create developmental or morphological novelties, which would increase the chances that the variant gene becomes fixed in the population (Matzke *et al.*, 2000).

To test the hypothesis that the host genome defense has been recruited to regulate host gene expression in plants and vertebrates, the challenge is to identify a gene family in an intermediate state where it can be shown that (1) A TE remnant in the promoter has modified expression in an evolutionarily meaningful way and (2) the TE remnant attracts methylation that helps silence the gene in cells where it is not required. It is likely that this situation will be found most readily in a relatively recent polyploid, such as tobacco, which has extant diploid ancestors (Figure 8.5).

B. Endogenous viral sequences and hereditary resistance to virus infection

PTGS serves as an antiviral defense that is induced by dsRNA in the cytoplasm (Section II.A.2.a). PTGS acts primarily on RNA viruses, which replicate in the cytoplasm via dsRNA intermediates, and it does not lead to any stable alterations in the host genome. Recent work has provided evidence for a second type of HDGS-based virus resistance that is active against DNA viruses in the nucleus. This form of virus resistance is believed to be conferred by integrated (endogenous) viral sequences and to act at the DNA level. Unlike the inducible and transient nature of PTGS-based resistance, this new kind of virus resistance is acquired gradually over evolutionary time and is associated with permanent changes in the host genome.

1. Homology-dependent virus resistance

Most plant viruses have RNA genomes. Only two groups, caulimoviruses and badnaviruses, possess genomes consisting of double-stranded DNA. Because these

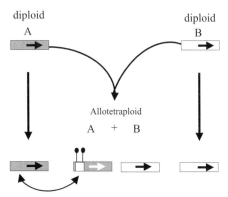

Figure 8.5. Recruiting the genome defense to establish epigenetic control mechanisms in polyploid genomes. Two different but related species, A and B, hybridize to form an allotetraploid. Expression of duplicated genes (short black arrows) can be diversified if a TE jumps into the promoter of one gene copy (white box). This could conceivably change the expression pattern (white arrow) and—because TEs are the primary target of the genome defense—attract methylation to the promoter (filled black circles). Regulatory sequences derived from TEs thus recruit methylation to the promoter. This situation might be found most easily in a relatively recent polyploid with extant diploid progenitors. Changes occurring subsequent to polyploidization can be identified by comparisons to the homologous genes that have been evolving separately in the diploid parents (double-headed arrow).

viruses replicate via an RNA intermediate and require a virus-encoded reverse transcriptase, they are referred to as pararetroviruses, to distinguish them from true retroviruses, which have an RNA genome. Pararetroviruses do not generally encode an integrase function and, unlike retroviruses, integration into the host genome is not required for pararetrovirus replication.

Although integration of viral DNA into host chromosomes occurs regularly in bacteria and animals, there are few reported instances in plants and these involve insertion at only one or a few sites (Hull *et al.*, 2000). Therefore, it came as a surprise to find up to 1000 copies of a previously uncharacterized pararetrovirus inserted into the nuclear DNA of healthy tobacco plants (Jakowitsch *et al.*, 1999). The preferred integration sites of the virus DNA suggested a mechanism involving illegitimate recombination at gaps in the open circular form of the viral genome. The tobacco endogenous pararetroviruses (TEPRVs) were characterized by frame shifts and stop codons, indicating that they were not functional copies. Despite the accumulated mutations, however, many TEPRVs exhibited a high degree of nucleotide sequence similarity. This sequence conservation was remarkable given that TEPRVs appeared to have integrated into host chromosomes prior to the formation of tetraploid tobacco (*N. tabacum*) (Mette *et al.*, 2002), which is believed to have occurred approximately 6 million years ago (Okamuro and Goldberg, 1985).

The maintenance of TEPRV sequence homology over a prolonged period of time suggested selection for a beneficial function. One possibility is that the TEPRVs provide a novel type of homology-dependent virus resistance that acts by triggering silencing of nonintegrated virus genomes (Jakowitsch et al., 1999). This would explain not only the nucleotide sequence conservation of TEPRVs but also the absence of detectable free virus and symptoms of infection in tobacco. This hypothesis is supported by the observation that TEPRV sequences are methylated and silent in tobacco (Mette et al., 2002). Moreover, the TEPRV enhancer is inactive and readily methylated when reintroduced together with a reporter transgene into the tobacco genome. In contrast, the TEPRV enhancer is active and unmethylated when introduced into heterologous species such as *Arabidopsis* and onion, which do not contain TEPRV sequences (Mette et al., 2002). TEPRVs in the tobacco genome thus appear to be able to rapidly trigger silencing of newly integrated sequences derived from TEPRVs.

These data are consistent with a scenario in which the gradual accumulation of TEPRVs in the ancestral tobacco genome led ultimately to homology-dependent silencing and methylation once a certain threshold copy number was reached (Figure 8.6). The estimated copy number of TEPRVs in the tobacco genome could be taken to indicate that approximately 1000 copies were required

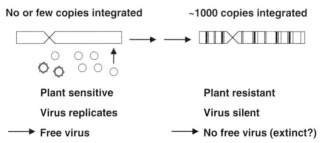

No or few copies integrated **~1000 copies integrated**

Plant sensitive **Plant resistant**

Virus replicates **Virus silent**

⟶ **Free virus** ⟶ **No free virus (extinct?)**

Figure 8.6. Model for acquisition of DNA homology-dependent virus resistance. Initially, when no or few copies of pararetrovirus DNA were integrated into host chromosomes, the virus could replicate and induce symptoms in the plant. At that time, free virus could have been isolated from infected tissues. Gradually, viral sequences invaded the host genome by illegitimate recombination. Once these endogenous viruses reached a threshold copy number, estimated to be ~1000 for tobacco endogenous pararetroviruses, they became collectively methylated and silenced via an HDGS mechanism. The methylated endogenous copies could silence the replication of exogenous virus genomes, thus conferring immunity on the plant. The exogenous virus has since either become extinct or found a new host. Although not yet confirmed, all data are so far consistent with this model.

before collective inactivation was achieved. The methylated TEPRVs could protect against infection by silencing nonintegrated viral genomes via DNA–DNA interactions. Although DNA-mediated resistance remains to be proven, the observations that TEPRVs contain only truncated or defective open reading frames and do not appear to be significantly transcribed argue against the involvement of RNA or interfering viral proteins in virus immunity.

Viral sequences that become stable components of host chromosomes can potentially establish hereditary resistance of the host to future infections with the same virus. Once a plant species acquires immunity, the virus must either find a new host or face extinction. The original pararetrovirus giving rise to TEPRV presumably fell victim to one of these fates, as it remains undetected as a free virus.

2. Comparisons between tobacco endogenous pararetroviruses (TEPRVs) and endogenous retroviruses (ERVs) in animals

Although plant pararetroviruses and animal retroviruses have different types of genomes and modes of replication (Section V.B.1), there are significant parallels between their endogenous forms. Two features of animal endogenous retroviruses (ERVs) are relevant to the present discussion. First, species containing a particular ERV often cannot be superinfected with the corresponding exogenous virus, although the same virus can infect heterologous species. ERVs can thus potentially provide immunity to their exogenous counterparts, as proposed for TEPRVs. Second, many families of human ERVs (HERVs) exist for which no exogenous form is known. The puzzling lack of exogenous virus for many HERVs (which was also observed in the case of TEPRVs) could indicate that successful pathogen control has been achieved by endogenization over the course of evolution (Löwer, 1999). Although ERVs are able to confer resistance through the expression interfering viral proteins (Best *et al.*, 1997; Löwer, 1999), resistance in some cases could be at the DNA level as proposed for TEPRVs. Indeed, the possibility that animal ERVs can provide protection to their exogenous forms through an HDGS or "co-suppression" mechanism has been broached recently (Löwer, 1999), but this idea remains to be tested.

3. EPRVs and ERVs: endogenization and pathogenic potential

The accumulation of EPRVs in plant genomes and ERVs in animal genomes requires that virus infection and integration take place in cells that contribute to the germ line. This is problematic for plants because known DNA viruses are generally excluded from apical meristems, the tissue which gives rise to germ cells (Hull *et al.*, 2000). The pararetrovirus giving rise to TEPRVs must have been an exception, although it is not possible to demonstrate this directly because the free virus has never been observed or isolated. This idea is supported, however, by the finding that the TEPRV enhancer is active preferentially in apical meristems in transgenic *Arabidopsis* (Mette *et al.*, 2002). Recent screens of diverse plant species for TEPRV-like sequences have revealed the existence of other fossil EPRV sequences derived from presumably extinct pararetroviruses that also must have been able to infect meristems (J. S. Heslop-Harrison, personal communication). The lifetime of such viruses might be brief on an evolutionary time scale if their tendency to integrate into host chromosomes in germline cells leads ultimately to host immunity.

Given that germline integrations are probably rare events, the endogenization process leading to virus resistance will occur relatively slowly. If a threshold copy number is necessary to establish DNA homology-dependent resistance (SectionV.B.1), then prior to reaching that threshold, the host remains susceptible to infection by exogenous virus or by unmutated endogenous copies if they are expressed. An example of such an intermediate stage in animals is the mouse mammary tumor virus system, in which endogenization of some subtypes is not yet completed and resistance to the exogenous form has only partially been achieved (Löwer *et al.*, 1996).

In plants, there are two interesting examples of EPRVs causing disease. One concerns banana streak virus (BSV), which—together with TEPRVs—provided one of the first instances of EPRVs (Hull *et al.*, 2000). BSV infections arising mysteriously in presumably virus-free banana tissue cultures were found to be caused by infectious episomal virus created by recombination of integrated BSV copies out of the banana genome. The stress of tissue culture apparently stimulated recombination. Productive infections probably ensued because the endogenous BSV sequences are present at a low copy number in the banana genome and they are not significantly mutated. A second example of pathogenic EPRVs combines features of homology-dependent virus resistance with stress activation to cause infection. When crosses were made between *N. glutinosa*, which carries multiple copies of cryptic TEPRV-like sequences and is symptom-free, and *N. clevelandii*, which is devoid of TEPRVs, the resulting hybrid species, *N. edwardsonii*, exhibited signs of viral infection (Lockhart *et al.*, 2000). Apparently, the silent *N. glutinosa* TEPRV-like sequences were activated only upon formation of the hybrid genome, perhaps owing to epigenetic changes, such as loss of methylation, in the new genomic background. Activation of latent endogenous viruses by demethylation in novel environments might be relevant for xenotransplantation of organs from pigs to humans, which could potentially unleash pig ERVs in the human host (van der Laan *et al.*, 2000).

4. Selective advantages of endogenous virus sequences

Virus resistance conferred by endogenous viral sequences represents a special type of adaptive mutation, and illustrates that so-called junk DNA can provide a host function that has considerable selective value. Although endogenous viruses can furnish protection by means of interfering viral proteins, many cases of resistance might be DNA homology-based and thus an outcome of the host genome defense acting on parasitic sequences. The dynamics of HERV populations and the selective advantage arising from their antiviral function have fascinating implications for human evolution. Indeed, waves of retrovirus infections/integrations and subsequent selection for viral immunity provided by HERVs might have contributed to primate evolution at several stages (Sverdlov, 2000). It is possible that the accumulation of EPRV sequences in plants has contributed to differential survival during virus outbreaks and driven the diversification of related species (Jakowitsch *et al.*, 1999).

We are far from knowing the full impact of endogenous viral sequences on plant and animal evolution and pathology. Information on the abundance and diversity of ERVs and EPRVs should accrue rapidly as more whole genome sequences become available. While comparative analyses between species will be informative, experimental approaches must be devised to assess the evolutionary contributions and pathogenic potential of endogenous viruses.

VI. SUMMARY AND CONCLUSIONS

Plant scientists have made major contributions to research on HDGS. Their discoveries, as well as many of those made by fungal and animal biologists, have all depended on the introduction of "man-made" foreign nucleic acids through transformation or microinjection. Gene silencing phenomena triggered by these "artificial" foreign nucleic acids revealed the existence of host defense mechanisms that limit the proliferation of "natural" foreign or parasitic sequences at the DNA or RNA level. Every type of parasitic sequence found in plants (TEs, viroids, RNA viruses, and DNA viruses) has been found to be a target and/or inducer of some variant of HDGS. The ongoing conflict between parasitic sequences and their hosts has led to major compositional changes in host genomes and to continuous changes in the parasitic sequences themselves, including virus extinctions. Some of the alterations in host genomes have had selective value and hence have contributed to host evolution. Two likely examples of the beneficial consequences of the host genome defense acting on TEs and endogenous viral sequences are the evolution of epigenetic control mechanisms that are essential for plant and vertebrate development, and the acquisition of resistance to viruses that integrate into host chromosomes.

Acknowledgments

Work in our laboratory is supported by the Austrian Fonds zur Förderung der wissenschaftlichen Forschung (grant Z21-MED) and the European Union (contract BIO4-CT96-9253). We thank Vicki

Vance and Hervé Vaucheret for sending papers in press, and Pat Heslop-Harrison and Jan Kooter for communicating unpublished results.

References

Abranches, R., Santos, A. P., Wegel, E., Williams, S., Castilho, A., Christou, P., Shaw, P., and Stöger, E. (2000). Widely separated multiple transgene integration sites in wheat chromosomes are brought together at interphase. *Plant J.* **24**, 1–14.

Akhtar, A., Zink, D., and Becker, P. (2000). Chromodomains are protein-RNA interaction modules. *Nature* **407**, 405–409.

Amedeo, P., Habu, Y., Afsar, K., Mittelsten Scheid, O., and Paszkowski, J. (2000). Disruption of the plant gene MOM releases transcriptional silencing of methylated genes. *Nature* **405**, 203–206.

Anandalakshmi, R., Marathe, R., Ge, X., Herr, J., Mau, C., Mallory, A., Pruss, G., Bowman, L., and Vance, V. B. (2000). A calmodulin-related protein that suppresses posttranscriptional gene silencing in plants. *Science* **290**, 142–144.

Anandalakschmi, R., Pruss, G., Ge, X., Marathe, R., Mallory, A., Smith, T., and Vance, V. B. (1998). A viral suppressor of gene silencing in plants. *Proc. Natl. Acad. Sci. USA* **95**, 13079–13084.

Bailey, J. A., Carrel, L., Chakravarti, A., and Eichler, E. E. (2000). Molecular evidence for a relationship between LINE-1 elements and X chromosome inactivation: the Lyon repeat hypothesis. *Proc. Natl. Acad. Sci. USA* **97**, 6634–6639.

Barlow, D. P. (1993). Methylation and imprinting: From host defense to gene regulation? *Science* **260**, 309–310.

Barry, C., Faugeron, G., and Rossignol, J.-L. (1993). Methylation induced premeiotically in *Ascobolus*: Coextension with DNA repeat lengths and effect on transcript elongation. *Proc. Natl. Acad. Sci. USA* **90**, 4557–4561.

Bass, B. (2000). Double stranded RNA as a template for gene silencing. *Cell* **101**, 235–238.

Baulcombe, D. C. (2000). Unwinding RNA silencing. *Science* **290**, 108–109.

Baulcombe, D. C. (2001). Diced defense. *Nature* **409**, 295–296.

Baulcombe, D. C., and English, J. (1996). Ectopic pairing of homologous DNA and post-transcriptional gene silencing in transgenic plants. *Curr. Opin. Biotechnol.* **7**, 173–180.

Bender, J. (1998). Cytosine methylation of repeated sequences in eukaryotes: the role of DNA pairing. *Trends Biochem. Sci.* **23**, 252–256.

Bender, J., and Fink, G. (1995). Epigenetic control of an endogenous gene family is revealed by a novel blue fluorescent mutant of *Arabidopsis*. *Cell* **83**, 725–734.

Bennetzen, J. L. (2000). Transposable element contributions to plant gene and genome evolution. *Plant Mol. Biol.* **42**, 251–269.

Bernstein, E., Caudy, A., Hammond, S., and Hannon, G. (2001). Role for a bidentate ribonuclease the initiation step of RNA interference. *Nature* **409**, 363–366.

Best, S., Le Tissier, P., and Stoye, J. (1997). Endogenous retroviruses and the evolution of resistance to retroviral infection. *Trends Microbiol.* **5**, 313–339.

Bestor, T. H. (1990). DNA methylation: Evolution of a bacterial immune function into a regulator of gene expression and genome structure in higher eukaryotes. *Phil. Trans. R. Soc. Lond. B* **326**, 179–187.

Bestor, T. H. (2000). The DNA methyltransferases of mammals. *Hum. Mol. Genet.* **9**, 2395–2402.

Birchler, J., Pal Bhadra, M., and Bhadra, U. (2000). Making noise about silence: repression of repeated genes in animals. *Curr. Opin. Genet. Dev.* **10**, 211–216.

Bollman, J., Carpenter, R., and Coen, E. (1991). Allelic interactions at the *nivea* locus of *Antirrhinum*. *Plant Cell* **3**, 1327–1336.

Brigneti, G., Voinnet, O., Li, W., Ji, L.-H., Ding, S.-W., and Baulcombe, D. (1998). Viral pathogenicity determinants are suppressors of gene silencing in *Nicotiana benthamiana*. *EMBO J.* **17**, 6739–6746.

Brown, W. E., Springer, P. S., and Bennetzen, J. L. (1994). Progressive modification of *Mu* transposable elements during development. *Maydica* **39**, 119–126.

Caprara, M. G., and Nielsen, T. W. (2000). RNA: Versatility in form and function. *Nature Struct. Biol.* **7**, 831–833.

Catalanotto, C., Azzalin, G., Macino, G., and Cogoni, C. (2000). Gene silencing in worms and fungi. *Nature* **404**, 245.

Cherry, S, Biniszkiewicz, D., van Parijs, L., Baltimore, D., and Jaenisch, R. (2000). Retroviral expression in embryonic stem cells and hematopoietic stem cells. *Mol. Cell Biol.* **20**, 7419–7426.

Claverie, J.-M. (2000). From bioinfomatics to computational biology. *Genome Res.* **10**, 1277–1279.

Cogoni, C., Irelan, J., Schumacher, M., Schmidhauser, T., Selker, E., and Macino, G. (1996). Transgene silencing of the *al-1* gene in vegetative cells of *Neurospora* is mediated by a cytoplasmic effector and does not depend on DNA-DNA interaction or DNA methylation. *EMBO J.* **15**, 3153–3163.

Cogoni, C., and Macino, G. (1999a). Gene silencing in *Neurospora crassa* requires a protein homologous to RNA-dependent RNA polymerase. *Nature* **399**, 166–169.

Cogoni, C., and Macino, G. (1999b). Posttranscriptional gene silencing in *Neurospora* by a RecQ DNA helicase. *Science* **286**, 2342–2344.

Cogoni, C., and Macino, G. (2000). Post-transcriptional gene silencing across kingdoms. *Curr. Opin. Genet. Dev.* **10**, 638–643.

Colot, V., Maloisel, L., and Rossignol, J.-L. (1997). Interchromosomal transfer of epigenetic states in *Ascobolus*: Transfer of DNA methylation is mechanistically related to homologous recombination. *Cell* **86**, 855–864.

Covey, S. N., and Al-Kaff, N. S. (2000). Plant DNA viruses and gene silencing. *Plant Mol. Biol.* **43**, 307–322.

Cubas, P., Vincent, C., and Coen, E. (1999). An epigenetic mutation responsible for natural variation in floral symmetry. *Nature* **401**, 157–161.

Dalmay, T., Hamilton, A., Rudd, S., Angell, S., and Baulcombe, D. C. (2000). An RNA-dependent RNA polymerase gene in *Arabidopsis* is required for PTGS mediated by a transgene but not by a virus. *Cell* **101**, 543–553.

Dalmay, T., Horsefield, R., Hartig Braunstein, T., and Baulcombe, D. (2001). *SDE3* encodes an RNA helicase required for post-transcriptional gene silencing in *Arabidopsis*. *EMBO J.* **8**, 2069–2077.

Domeier, M. E., Morse, D., Knight, S., Portereiko, M., Bass, B., and Mango, S. (2000). A link between RNA interference and nonsense-mediated decay in *Caenorhabditis elegans*. *Science* **289**, 1928–1930.

Doudna, J. A. (2000). Structural genomics of RNA. *Nature Struct. Biol.* **Suppl.**, Nov. 2000, 954–956.

Duvillie, B., Bucchini, D., Tang, T., Jami, J., and Paldi, A. (1998). Imprinting at the mouse *Ins2* locus: Evidence for *cis-* and *trans-*allelic interactions. *Genomics* **47**, 52–57.

Elmayan, T., Balzergue, S., Béon, F., Bourdon, V., Daubremet, J., Guénet, Y., Mourrainm, P., Palauqui, J., Vernhettes, S., Vialle, T., Wostrikiff, K., and Vaucheret, H. (1998). *Arabidopsis* mutants impaired in cosuppression. *Plant Cell* **10**, 1447–1457.

Fagard, M., Boutet, S., Morel, J., Bellini, C., and Vaucheret, H. (2000). AGO-1, QDE-2 and RDE-1 are related proteins required for PTGS in plants, quelling in fungi and RNAi in animals. *Proc. Natl. Acad. Sci. USA* **97**, 11650–11654.

Fagard, M., and Vaucheret, H. (2000a). (Trans)gene silencing in plants: How many mechanisms? *Annu. Rev. Plant Physiol. Plant Mol. Biol.* **51**, 167–194.

Fagard, M., and Vaucheret, H. (2000b). Systemic silencing signal(s). *Plant Mol. Biol.* **43**, 285–293.

Faugeron, G. (2000). Diversity of homology-dependent gene silencing strategies in fungi. *Curr. Opin. Microbiol.* **3**, 144–148.

Finnegan, E. J., and Kovac, K. A. (2000). Plant DNA methyltransferases. *Plant Mol. Biol.* **43**, 189–201.

Fire, A., Xu, S., Montgomery, M., Kotas, S., Driver, S., and Mello, C. (1998). Potent and specific genetic interference by double stranded RNA in *Caenorhabditis elegans*. *Nature* **391,** 806–811.

Furner, I., Sheikh, A., and Collett, C. (1998). Gene silencing and homology-dependent gene silencing in *Arabidopsis*: Genetic modifiers and DNA methylation. *Genetics* **149,** 651–662.

Fuks, F., Burgers, W., Brehm, A., Hughes-Davies, L., and Kouzarides, T. (2000). DNA methyltransferase Dnmt1 associates with histone deacetylase activity. *Nature Genet.* **24,** 88–91.

Gibbons, R. J., McDowell, T. L., Raman, S., O'Rourke, D. M., Garrick, D., Ayyub, H., and Higgs, D. R. (2000). Mutations in *ATRX*, encoding a SWI/SNF-like protein, cause diverse changes in the pattern of DNA methylation. *Nature Genet.* **24,** 368–371.

Goffeau, A. (2000). Four years of post-genomic life with 6000 yeast genes. *FEBS Lett.* **480,** 37–41.

Gonzalgo, M., and Jones, P. (1997). Mutagenic and epigenetic effects of DNA methylation. *Mutation Res.* **386,** 107–118.

Gordenin, D. A., Lobachev, K. S., Degtyareva, N. P., Malkova, A. L., Perkins, E., and Resnick, M. A. (1993). Inverted DNA repeats: A source of eukaryotic genomic instability. *Mol. Cell Biol.* **13,** 5315–5322.

Hamilton, A., and Baulcombe, D. C. (1999). A species of small antisense RNA in posttranscriptional gene silencing in plants. *Science* **286,** 950–952.

Hammond, S., Bernstein, E., Beach, D., and Hannon, G. (2000). An RNA-directed nuclease mediates PTGS in *Drosophila* cells. *Nature* **404,** 293–296.

Heard, E., Clerc, P., and Avner, P. (1997). X-chromosome inactivation in mammals. *Annu. Rev. Genet.* **31,** 571–610.

Henikoff, S., and Comai, L. (1998). A DNA methyltransferase homolog with a chromodomain exists in multiple polymorphic forms in *Arabidopsis*. *Genetics* **149,** 307–318.

Hilleren, P., and Parker, R. (1999). Mechanisms of mRNA surveillance in eukaryotes. *Annu. Rev. Genet.* **33,** 229–260.

Hirochika, H., Okamoto, H., and Kakutani, T. (2000). Silencing of retrotransposons in *Arabidopsis* and reactivation by the *ddm1* mutation. *Plant Cell* **12,** 357–368.

Hull, R., Harper, G., and Lockhart, B. (2000). Viral sequences integrated into plant genomes. *Trends Plant Sci.* **5,** 362–365.

Jacobsen, S., and Meyerowitz, E. (1997). Hypermethylated *SUPERMAN* epigenetic alleles in *Arabidopsis*. *Science* **277,** 1100–1103.

Jacobsen, S., Running, M., and Meyerowitz, E. (1999). Disruption of an RNA helicase/RNAseIII gene in *Arabidopsis* causes unregulated cell division in floral meristems. *Development* **126,** 5231–5243.

Jacobsen, S., Sakai, H., Finnegan, E. J., Cao, X., and Meyerowitz, E. M. (2000). Ectopic hypermethylation of flower-specific genes in *Arabidopsis*. *Curr. Biol.* **10,** 179–186.

Jakowitsch, J., Mette, M. F., van der Winden, J., Matzke, M. A., and Matzke, A. J. M. (1999). Integrated pararetroviral sequences define a unique class of dispersed repetitive DNA in plants. *Proc. Natl. Acad. Sci. USA* **96,** 13241–13246.

Jeddeloh, J., Stokes, T., and Richards, E. J. (1999). Maintenance of genomic methylation requires a SWI2/SNF2-like protein. *Nature Genet.* **21,** 209–212.

Jensen, S., Gassama, M., and Heidmann, T. (1999). Cosuppression of *I* transposon activity by *I* -containing sense and antisense transgenes. *Genetics* **153,** 1767–1774.

Jones, L., Hamilton, A., Voinnet, O., Thomas, C., Maule, A., and Baulcombe, D. C. (1999). RNA-DNA interactions and DNA methylation in post-transcriptional gene silencing. *Plant Cell* **11,** 2291–2301.

Jones, L., Thomas, C., and Maule, A. (1998). *De novo* methylation and co-suppression induced by a cytoplasmically replicating plant RNA virus. *EMBO J.* **17,** 6385–6393.

Jorgensen, R. (1992). Silencing of plant genes by homologous transgenes. *AgBiotech News Inform.* **4,** 265–273.

Kakutani, T., Munakata, K., Richards, E. J., and Hirochika, H. (1999). Meiotically and mitotically stable inheritance of DNA hypomethylation induced by *ddm1* mutation of *Arabidopsis thaliana*. *Genetics* **151**, 831–838.

Kasschau, C., and Carrington, J. (1998). A counterdefensive strategy of plant viruses: Suppression of posttranscriptional gene silencing. *Cell* **95**, 461–470.

Ketting, R., Haverkamp, T., van Luenen, H., and Plasterk, R. (1999). *Mut-7* of *C elegans*, required for transposon silencing and RNA interference, is a homolog of Werner syndrome helicase and RNaseD. *Cell* **99**, 133–141.

Kiefer, M., Owens, R., and Diener, T. O. (1982). Structural similarities between viroids and transposable genetic elements. *Proc. Natl. Acad. Sci. USA* **80**, 6234–6238.

Kooter, J., Matzke, M. A., and Meyer, P. (1999). Listening to the silent genes: Transgene silencing research identified new mechanisms of gene regulation and pathogen control. *Trends Plant Sci.* **4**, 340–347.

Kovarik, A., Van Houdt, H., Holy, A., and Depicker, A. (2000). Drug-induced hypomethylation of a posttranscriptionally silenced transgene locus of tobacco leads to partial release of silencing. *FEBS Lett.* **467**, 47–51.

Kumar, M., and Carmichael, G. (1998). Antisense RNA: function and fate of duplex RNA in cells of higher eukaryotes. *Microbiol. Mol. Biol. Rev.* **62**, 1415–1443.

LaSalle, J., and Lalande, M. (1996). Homologous association of oppositely imprinted chromosomal domains. *Science* **272**, 725–728.

LeBlanc, M. D., Aspeslagh, G., Buggia, N. P., and Dyer, B. D. (2000). An annotated catalog of inverted repeats of *Caenorhabditis elegans* chromosomes III and X, with observations concerning odd/even biases and conserved motifs. *Genome Res.* **10**, 1381–1392.

Lee, J. T. (2000). Disruption of imprinted X inactivation by parent-of-origin effects at *TSIX*. *Cell* **103**, 17–27.

Li, E., Bestor, T., and Jaenisch, R. (1992). Targeted mutation of the DNA methyltransferase gene results in embryonic lethality. *Cell* **69**, 915–926.

Lindbo, J., Silva-Rosales, L., Proebsting, W., and Dougherty, W. (1993). Induction of a highly specific antiviral state in transgenic plants: Implications for regulation of gene expression and virus resistance. *Plant Cell* **5**, 1749–1459.

Llave, C., Kasschau, K., and Carrington, J. (2000). Virus-encoded suppressor of posttranscriptional gene silencing targets a maintenance step in the silencing pathway. *Proc. Natl. Acad. Sci. USA* **97**, 13401–13406.

Lockhart, B. E., Menke, J., Dahal, G., and Olszewski, N. E. (2000). Characterization and genomic analysis of tobacco vein clearing virus, a plant pararetrovirus that is transmitted vertically and related to sequences integrated in the host genome. *J. Gen. Virol.* **81**, 1579–1585.

Löwer, R. (1999). The pathogenic potential of endogenous retroviruses: Facts and fantasies. *Trends Microbiol.* **7**, 350–356.

Löwer, R., Löwer, J., and Kurth, R. (1996). The viruses in all of us: Characteristics and biological significance of human endogenous retrovirus sequences. *Proc. Natl. Acad. Sci. USA* **93**, 5177–5184.

Lucy, A., Guo, H., Li, W., and Ding, S. (2000). Suppression of posttranscriptional gene silencing by a plant viral protein localized in the nucleus. *EMBO J.* **19**, 1672–1680.

Luff, B., Pawlowski, L., and Bender, J. (1999). An inverted repeat triggers cytosine methylation of identical sequences in *Arabidopsis*. *Mol. Cell* **3**, 505–511.

Lyle, R., Watanabe, D., te Vruchte, D., Lerchner, W., Smrzka, O., Wutz, A., Schageman, J., Hahner, L., Davies, C., and Barlow, D. (2000). The imprinted antisense RNA at the *Igf2r* locus overlaps but does not imprint *Mas1*. *Nature Genet.* **25**, 19–21.

Lyon, M. (1998). X-chromosome inactivation: A repeat hypothesis. *Cytogenet. Cell Genet.* **80**, 133–137.

Mallory, A., Ely, L., Smith, T., Marathe, R., Anandalakshmi, R., Fagard, M., Vaucheret, H., Pruss, G., Bowman, L., and Vance, V. B. (2001). HC-Pro suppression of gene silencing eliminates the small RNAs but not transgene methylation or the mobile signal. *Plant Cell* **13**, 571–583.

Marahrens, Y. (1999). X-inactivation by chromosomal pairing events. *Genes Dev.* **13**, 2624–2632.

Marathe, R., Anandalakshmi, R., Smith, T. H., Pruss, G. J., and Vance, V. B. (2000a). RNA viruses as inducers, suppressors and targets of post-transcriptional gene silencing. *Plant Mol. Biol.* **43**, 295–306.

Marathe, R., Smith, T., Anandalakshmi, R., Bowman, L., Fagard, M., Mourrain, P., Vaucheret, H., and Vance, V. B. (2000b). Plant viral suppressors of post-transcriptional gene silencing do not suppress transcriptional gene silencing. *Plant J.* **22**, 51–59.

Matzke, M. A., and Matzke, A. J. M. (1995). How and why do plants inactivate homologous (trans)genes? *Plant Physiol.* **107**, 679–685.

Matzke, M. A., Matzke, A. J. M., and Mittelsten Scheid, O. (1994). Inactivation of repeated genes—DNA-DNA interaction? *In* "Homologous Recombination and Gene Silencing in Plants" (J. Paszkowski ed.), pp. 271–307. Kluwer, Dordrecht, The Netherlands.

Matzke, M. A., Matzke, A. J. M., Pruss, G. J., and Vance, V. B. (2001a). RNA-based silencing strategies in plants. *Curr. Opin. Genet. Dev.* **11**, 221–227.

Matzke, M. A., Mette, M. F., Aufsatz, W., Jakowitsch, J., and Matzke, A. J. M. (1999). Host defenses to parasitic sequences and the evolution of epigenetic control mechanisms. *Genetica* **107**, 271–287.

Matzke, M. A., Mette, M. F., Jakowitsch, J., Kanno, T., Moscone, E., van der Winden, J., and Matzke, A. J. M. (2001b). Transvection-like effects in plants: Evidence that DNA pairing leads to *trans*-activation or silencing of complex heteroalleles in tobacco. *Genetics* **158**, 451–461.

Matzke, M. A., Mette, M. F., and Matzke, A. J. M. (2000). Transgene silencing by the host genome defense: Implications for the evolution of epigenetic control mechanisms in plants and vertebrates. *Plant Mol. Biol.* **43**, 401–415.

Melquist, S., Luff, B., and Bender, J. (1999). *Arabidopsis* PAI gene rearrangements, cytosine methylation and expression. *Genetics* **153**, 401–413.

Mertineit, C., Yoder, J., Taketo, T., Laird, D., Trasler, J., and Bestor, T. H. (1998). Sex-specific exons control DNA methyltransferase in mammalian germ cells. *Development* **125**, 889–897.

Mette, M. F., Aufsatz, W., van der Winden, J., Matzke, M. A., and Matzke, A. J. M. (2000). Transcriptional gene silencing and promoter methylation triggered by double-stranded RNA. *EMBO J.* **19**, 5194–5201.

Mette, M. F., Kanno, T., Aufsatz, W., Jakowitsch, J., van der Winden, J., Matzke, M. A., and Matzke, A. J. M. (2002). Endogenous viral sequences and their potential contribution to heritable virus resistance in plants. *EMBO J.*, in press.

Mette, M. F., Matzke, A. J. M., and Matzke, M. A. (2001). Resistance of RNA-mediated TGS to HC-Pro, a viral suppressor of PTGS, suggests alternative pathways for dsRNA processing. *Curr. Biol.* **11**, 1119–1123.

Mette, M. F., van der Winden, J., Matzke, M. A., and Matzke, A. J. M. (1999). Production of aberrant promoter transcripts contributes to methylation and silencing of unlinked homologous promoters in *trans*. *EMBO J.* **18**, 241–248.

Miao, V. P., Freitag, M., and Selker, E. U. (2000). Short TpA-rich segments of the ζ-η region induce DNA methylation in *Neurospora crassa*. *J. Mol. Biol.* **300**, 249–273.

Morel, J. B., Mourrain, P., Béclin, C., and Vaucheret, H. (2000). DNA methylation and chromatin structure mutants affect both transcriptional and post-transcriptional transgene silencing in *Arabidopsis*. *Curr. Biol.* **10**, 1591–1594.

Mourrain, P., Béclin, C., Elmayan, T., Feuerbach, F., Godon, C., Morel, J. B., Jouette, D., Lacombe, A. M., Nikic, S., Picault, N., Rémoué, K., Sanial, M., Vo, T. A., and Vaucheret, H. (2000). *Arabidopsis* SGS2 and SGS3 genes are required for posttranscriptional gene silencing and natural virus resistance. *Cell* **101**, 533–542.

Muskens, M. W., Vissers, A. P., Mol, J. N., and Kooter, J. M. (2000). Role of inverted DNA repeats in transcriptional and post-transcriptional gene silencing. *Plant Mol. Biol.* **43**, 243–260.

Okamuro, J. K., and Goldberg, R. B. (1985). Tobacco single-copy DNA is highly homologous to sequences present in the genomes of its diploid ancestors. *Mol. Gen. Genet.* **198**, 290–298.

Okano, M., Bell, D. W., Haber, D., and Li, E. (1999). DNA methyltransferasses Dnmt3a and Dnmt3b are essential for *de novo* methylation and mammalian development. *Cell* **99**, 247–257.

Pannuti, A., and Lucchesi, J. C. (2000). Recycling to remodel: Evolution of dosage-compensation complexes. *Curr. Opin. Genet. Dev.* **10**, 644–650.

Parrish, S., Fleenor, J., Xu, S., Mello, C., and Fire, A. (2000). Functional anatomy of a dsRNA trigger: Differential requirement for the two trigger strands in RNA interference. *Mol. Cell* **6**, 1077–1087.

Pélissier, T., Thalmair, S., Kempe, D., Sänger, H. L., and Wassenegger, M. (1999). Heavy de novo methylation at symmetrical and non-symmetrical sites is a hallmark of RNA-directed DNA methylation. *Nucleic Acids Res.* **27**, 1625–1643.

Pélissier, T., and Wassenegger, M. (2000). A DNA target of 30 bp is sufficient for RNA-directed DNA methylation *RNA* **6**, 55–65.

Plasterk, R. H., and Ketting, R. F. (2000). The silence of the genes. *Curr. Opin. Genet. Dev.* **10**, 562–567.

Regev, A., Lamb, M., and Jablonka, E. (1998). The role of DNA methylation in invertebrates: Developmental regulation or genome defense? *Mol. Biol. Evol.* **15**, 880–891.

Reik, W., and Walter, J. (1998). Imprinting mechanisms in mammals. *Curr. Opin. Genet. Dev.* **8**, 154–164.

Reik, W., and Walter, J. (2001). Evolution of imprinting mechanism: The battle of the sexes begins in the zygote. *Nature Genet.* **27**, 255–256.

Richards, E. J. (1997). DNA methylation and plant development. *Trends Genet.* **13**, 319–322.

Robertson, K., and Wolffe, A. P. (2000). DNA methylation in health and human disease. *Nature Genet. Rev.* **1**, 11–19.

Sapienza, C. (1990). Sex-linked dosage-sensitive modifiers as imprinting genes. *Development* **Suppl.**, 107–113.

Sijen, T., and Kooter, J. (2000). Post-transcriptional gene silencing: RNAs on the attack or on the defense? *BioEssays* **22**, 520–531.

Singer, T., Yordan, C., and Martienssen, R. (2001). Robertson's *Mutator* transposons in *A. thaliana* are regualted by the chromatin-remodeling gene *Decrease in DNA Methylation (DDM1)*. *Genes Dev.* **15**, 591–602.

Sleutels, F., Barlow, D. P., and Lyle, R. (2000). The uniqueness of the imprinting mechanism. *Curr. Opin. Genet. Dev.* **10**, 229–233.

Smardon, A., Spoerke, J., Stacey, S., Klein, M., Mackin, N., and Maine, E. (2000). EGO-1 is related to RNA-directed RNA polymerase and functions in germ-line development and RNA interference in *C. elegans*. *Curr. Biol.* **10**, 169–178.

Smit, A. F. (1999). Interspersed repeats and other mementos of transposable elements in mammalian genomes. *Curr. Opin. Genet. Dev.* **9**, 657–663.

Smith, N. A., Singh, S. P., Wang, M.-B., Stoutjesdijk, P. A., Green, A. G., and Waterhouse, P. M. (2000). Total silencing by intron-spliced hairpin RNAs. *Nature* **407**, 319–320.

Smith, S., Kann, J., Baker, D., Kaplan, B., and Dembek, P. (1991). Recognition of unusual DNA structure by human DNA(cytosine-5)metthytransferase. *J. Mol. Biol.* **217**, 39–51.

Soppe, W., Jacobsen, S., Alonso-Blanco, C., Jackson, J., Kakutani, T., and Koornneef, M. (2000). The late flowering phenotype of *fwa* mutants is caused by gain-of-function epigenetic alleles of a homeodomain gene. *Mol. Cell* **6**, 791–802.

Spencer, H., Clark, A., and Feldman, M. (1999). Reply. *Trends Ecol. Evol.* **14**, 359.

Spring, J. (1997). Vertebrate evolution by interspecific hybridization. Are we polyploid? *FEBS Lett.* **400**, 2–8.

Stam, M., Viterbo, A., Mol, J. N. M., and Kooter, J. M. (1998). Position-independent methylation and transcriptional silencing of transgenes in inverted T-DNA repeats: implications for posttranscriptional silencing of homologous host genes in plants. *Mol. Cell Biol.* **18,** 6165–6177.

Steimer, A., Amedeo, P., Afsar, K., Fransz, P., Mittelsten Scheid, O., and Paszkowski, J. (2000). Endogenous targets of transcriptional gene silencing in *Arabidopsis. Plant Cell* **12,** 1165–1178.

Sverdlov, E. D. (2000). Retroviruses and primate evolution. *BioEssays* **22,** 161–171.

Tabara, H., Sarkissian, M., Kelly, W., Fleenor, J., Grishok, A., Timmons, L., Fire, A., and Mello, C. (1999). The *rde-1* gene, RNA interference, and transposon silencing in C. *elegans. Cell* **99,** 123–132.

Terryn, N., and Rouzé, P. (2000). The sense of naturally transcribed antisense RNAs in plants. *Trends Plant Sci.* **5,** 394–396.

Tilghman, S. (1999). The sins of the fathers and mothers: Genomic imprinting in mammalian development. *Cell* **96,** 185–193.

Turker, M., and Bestor, T. H. (1997). Formation of methylation patterns in the mammalian genome. *Mutation Res.* **386,** 119–130.

van der Laan, L., Lockey, C., Griffeth, B., Frasier, F., Wilson, C., Onions, D., Hering, B., Long, Z., Otto, E., Torbett, B., and Salomon, D. (2000). Infection by porcine endogenous retrovirus after islet xenotransplantation in SCID mice. *Nature* **407,** 90–94.

Vaucheret, H., and Fagard, M. (2001). Transcriptional gene silencing in plants: Targets, inducers and regulators. *Trends Genet.* **17,** 29–35.

Voinnet, O., Lederer, C., and Baulcombe, D. C. (2000). A viral movement protein prevents spread of the gene silencing signal in *Nicotiana benthamiana. Cell* **103,** 157–167.

Voinnet, O., Pinto, Y., and Baulcombe, D. (1999). Suppression of gene silencing: A general strategy used by diverse DNA and RNA viruses of plants. *Proc. Natl. Acad. Sci. USA* **96,** 14147–14152.

Vongs, A., Kakutani, T., Marienssen, R. A., and Richards, E. J. (1993). *Arabidopsis thaliana* DNA methylation mutants. *Science* **260,** 1926–1928.

Walsh, C., and Bestor, T. H. (1999). Cytosine methylation and mammalian development. *Genes Dev.* **13,** 26–34.

Wassenegger, M. (2000). RNA-directed DNA methylation. *Plant Mol. Biol.* **43,** 203–220.

Wassenegger, M., Heimes, S., Riedel, L., and Sänger, H. (1994). RNA-directed *de novo* methylation of genomic sequences in plants. *Cell* **76,** 567–576.

Waterhouse, P. M., Graham, M. W., and Wang, M. B. (1998). Virus resistance and gene silencing in plants can be induced by simultaneous expression of sense and antisense RNA. *Proc. Natl. Acad. Sci. USA* **95,** 13959–13964.

Wendel, J. F. (2000). Genome evolution in polyploids. *Plant Mol. Biol.* **42,** 225–249.

Wu, C.-T., and Morris, J. R. (1999). Transvection and other homology effects. *Curr. Opin. Genet. Dev.* **9,** 237–246.

Wu-Scharf, D., Jeong, B., Zhang, C., and Cerrutti, H. (2000). Transgene and transposon silencing in *Chlamydomonas reinhardtii* by a DEAH-box RNA helicase. *Science* **290,** 1159–1162.

Yoder, J., Walsh, C., and Bestor, T. H. (1997). Cytosine methylation and the ecology of intragenomic parasites. *Trends Genet.* **13,** 335–339.

Zamore, P. D., Tuschl, T., Sharp, P., and Bartel, D. (2000). RNAi: dsRNA directs the ATP-dependent cleavage of mRNA at 21 to 23 nt intervals. *Cell* **101,** 25–33.

9

Quelling in *Neurospora crassa*

Annette S. Pickford, Caterina Catalanotto, Carlo Cogoni, and Giuseppe Macino*
Department of Cellular and Hematologic Biotechnology
Section of Molecular Genetics
Università di Roma "La Sapienza"
00161 Rome, Italy

*To whom correspondence should be addressed: E-mail: macino@bce.med.uniroma1.it.

I. INTRODUCTION

The term homology-dependent gene silencing (HDGS) was introduced by Matzke *et al.* (1994) to include any silencing event that results from the duplication of nucleic acid sequences; however, these events can be divided into those which occur at a transcriptional level (transcriptional gene silencing or TGS) and those which occur at a posttranscriptional level (posttranscriptional gene silencing or PTGS). Quelling in the filamentous fungus *Neurospora crassa* was discovered by Romano and Macino in 1992, as a posttranscriptional mechanism of gene silencing active during vegetative growth. Quelling is distinct from the phenomenon of gene inactivation first reported by Selker *et al.* (1987), originally named rearrangement induced premeiotically (RIP), that is, active during the premeiotic phase of sexual reproduction. Both phenomena were identified as a result of transformation with transgenes which, instead of enhancing gene expression by increasing gene dosage, caused silencing of the duplicated genes. Silencing of gene expression by RIP was subsequently shown to be the result not of rearrangement, but of C-T point mutations in the duplicated homologous sequences, and was consequently renamed repeat-induced point mutation (Cambereri *et al.*, 1989), whereas quelling results in a reduction of mRNA homologous to transgenes.

Gene silencing phenomena similar to quelling have also been discovered in other organisms as a result of the introduction of transgenes or infection by viruses. The posttranscriptional silencing mechanisms which are triggered by the presence of duplicated DNA or RNA sequences, and which result in a reduction of the mRNA homologous to such sequences, have been named co-suppression and RNA interference in plants and animals, respectively.

Silencing in plants was first observed by Napoli *et al.* (1990) and van der Krol *et al.* (1990) in petunia, with genes involved in petal pigmentation. The coordinate silencing of both ectopic transgenes and endogenous homologous genes prompted the name "co-suppression." Co-suppression was demonstrated to be a general phenomenon, as it was also reported in tomato by Smith *et al.* (1990).

The more recent discovery of silencing at a posttranscriptional level in animals was made by Fire *et al.* (1998) in the nematode *Caenorhabditis elegans*. In an attempt to inactivate gene expression by injection of antisense RNA into the animal, it was found that the antisense RNA was no more effective than sense RNA in inducing gene silencing. In fact, it was discovered that small amounts of double-stranded RNA (dsRNA), present as contaminants in *in vitro* RNA preparations, had triggered the observed silencing phenomenon, which was therefore named double-stranded RNA interference (RNAi) (Fire *et al.*, 1998). RNAi was subsequently shown to be active not only in *C. elegans*, but in other animals such as *Drosophila melanogaster* (Kennerdale and Carthew, 1998; Misquitta and Paterson, 1999), *Trypanosoma brucei* (Ngo *et al.*, 1998), and *Planaria* (Sanchez *et al.*, 1999). RNAi has further been shown to be active in *Arabidopsis thaliana* plants (Waterhouse *et al.*, 1998; Chuang and Meyerowitz, 2000) and vertebrates, namely, in mice (Wianny and Zernicka-Goetz, 2000) and in zebrafish (Li *et al.*, 2000; Wargelius *et al.*, 1999).

II. THE DISCOVERY OF QUELLING IN *Neurospora*

The first report of silencing in the vegetative phase of growth in fungi was made by Pandit and Russo (1992) in *Neurospora crassa*. They observed a loss of hygro-mycin resistance as a result of transformation with a plasmid carrying the bacte-rial hygromycin phosphotransferase (*hph*) gene fused to the promoter of the *trpC* gene of *Aspergillus nidulans*. Cytosine methylation of the transgene was correlated with silencing, suggesting possible "position effects" due to the transgene integra-tion site; however, the fact that only stably transformed strains harboring many insertions of the *hph* gene were methylated indicated that methylation may have been due to transgenes being linked as a multicopy sequence rather than their site of insertion. Growth of *Neurospora crassa* in the presence of the cytosine analog 5-azacytidine, which prevents cytosine methylation, resulted in the reactivation of the *hph* gene, further suggesting that DNA methylation is involved in gene silencing. Moreover, gene silencing was found to be reversible, as hygromicin resistance can be recovered during vegetative growth.

While the silencing observed by Pandit and Russo involved transgenic sequences only, Romano and Macino (1992) reported silencing of both trans-forming DNA and homologous endogenous sequences. The new phenomenon observed in *Neurospora crassa* was called "quelling." In the experiments carried out by Romano and Macino (1992), transformation was carried out with constructs of genes involved in the carotenoid biosynthetic pathway, as markers of silenc-ing. Vegetative growth in *Neurospora crassa* is characterized by branching hyphae which form a mycelium that produces asexual spores called conidia. Carotenoids produced both in the mycelium and in conidia confer a bright orange color to *Neurospora*; however, mutation of any one of three genes involved in caroteno-genesis determines an albino phenotype. The *albino* genes (*al-1*, *al-2*, and *al-3*) code for phytoene dehydrogenase (Schmidhauser *et al.*, 1990), phytoene synthase (Schmidhauser *et al.*, 1994), and geranylgeranyl pyrophosphate synthase (Nelson *et al.*, 1989), respectively. These genes are particularly suitable for use as mark-ers of gene silencing, as the albino phenotype is easily visible and facilitates the identification of silenced strains, especially in the screening of large numbers of colonies. Carotenoid content can also be quantified, which is particularly useful to calculate the degree of silencing.

Silencing was observed as a result of transformation of *Neurospora crassa* with plasmids containing constructs of either the *albino-3* (*al-3*) or *albino-1* (*al-1*) gene, demonstrating it to be a general phenomenon, not restricted to a single gene; however, both frequency and severity of silencing varied. Approximately 40% of the progeny of the *al-1* transformants had phenotypes ranging from white to dark yellow, indicating stronger or weaker suppression of gene expression, while only 0.5% of the *al-3* progeny had silenced phenotypes. This difference may be explained by the fact that *al-3* is an essential gene and therefore silenced progeny were not viable. A further demonstration that the frequency of silencing can be variable was reported by Cogoni and Macino (1997a) as a result of the transformation of *Neurospora* with a portion of the *albino-2* (*al-2*) gene. In this case the frequency of silencing was 10%, indicating that the *al-2* gene is less sensistive to silencing. Several factors may determine such variation in frequency. The first is the intrinsic characteristics of the gene determined by its nucleotide sequence. Nucleotide complementarity could lead to the formation of secondary structure in mRNA transcripts, making them more or less resistant to the action of RNAses. Second, the frequency of silencing could also be influenced by the level of gene expression. For example, one might expect that highly expressed endogenous genes could be more resistent to degradation; however, paradoxically, it seems that, at least in plants, highly expressed genes are easier targets for co-suppression, suggesting possible threshold-like effects (Elmayan and Vaucheret, 1996; Jorgensen *et al.*, 1996). Threshold effects are discussed further in Section IV.A.

Further evidence that quelling is a general mechanism, not limited to specific genes in *Neurospora*, was provided by the fact that the *qa-2* gene encoding quinic acid dehydrogenase, used as a genetic marker for transformation in the experiments carried out by Cogoni and Macino (1997a), was also silenced. This was further supported by the reports of quelling of two transcription factors in *Neurospora*, the *white-collar* genes (*wc-1* and *wc-2*) by Ballario *et al.* (1996) and Linden *et al.*, (1997), respectively, and of the *ad-9* gene involved in adenine biosynthesis by Schmidhauser (Cogoni and Macino, 1998, and references therein).

It was shown that inactivation of the *albino* genes was not due to gene disruption, as no rearrangements were found in the endogenous genes, but to a heavy reduction of the steady-state level of mRNA. It was found that over prolonged culture time, 25% of the silenced progeny tended to revert progressively to wild-type or intermediate phenotypes, demonstrating that the silencing of the *albino* genes was reversible. Reversion was unidirectional and correlated with increased levels of steady-state mRNA. Reversion also correlated with the loss of ectopic copies of the *albino* transgenes, probably due to homologous recombination during the vegetative phase (known as mitotic instability), which can result in the deletion of ectopic sequences, especially in the case of tandemly arranged transforming sequences (Selker, 1990), Thus, it seems that quelling per se is not unstable, but that the instability is a consequence of trangenes loss, suggesting that trangenes are not only required for the establishment of quelling, but are also necessary for its maintenance.

The involvement of methylation in quelling was tested. While silenced endogenous *albino* genes were never found to be methylated, it was observed that tandemly arranged transgenic loci whose appearance correlated with the occurrence of gene silencing were frequently methylated, suggesting that methylation of transgenes could somehow be required for silencing. Previous observations in which treatment with 5-azacytidine determined reversion in some silenced transformants seemed to substantiate this hypothesis (Romano and Macino, 1992); however, the subsequent observation of normal levels of quelling in *methylation-deficient* (*dim-2*) *Neurospora crassa* mutants definitively ruled out the involvement of DNA methylation in quelling (Cogoni and Macino, 1997a). The apparent incongruency between the observed reversion of some cases of gene silencing induced by 5-azacytidine and the lack of involvement of DNA methylation in quelling could be explained by the fact that another gene silencing mechanism sensitive to 5-azacytidine is active during the vegetative growth phase of *N. crassa* (C. Cogoni, unpublished results).

Further characterization of the quelling phenomenon exploited the characteristic coenocytic nature of *Neurospora* hyphal cells in which many nuclei share a common cytoplasm. *Neurospora* is defined as a homokaryon when genetically identical nuclei are present in the same cytoplasm, while heterokaryons harbor genetically diverse nuclei. Transformation of multinucleate *Neurospora* spheroplasts often results in the formation of heterokaryons, as not all nuclei incorporate transforming sequences. Since mutations in the *albino* genes are typically recessive, heterokaryons containing both wild-type nuclei and nuclei with mutated *albino* genes present an orange (wild-type) phenotype. In contrast, heterokaryons containing both wild-type nuclei and nuclei with silenced *albino* genes would only be albino in the event that quelling was a dominant trait. An albino phenotype was observed in forced heterokaryons between *al-1* quelled and wild-type nuclei, demonstrating that quelling is indeed a dominant trait, probably mediated by a cytoplasmatically diffusible molecule acting *in trans* (Cogoni *et al.*, 1996).

Also in plants, co-suppression was reported to diffuse from a silenced transgenic stock to a nonsilenced transgenic scion (Palaqui *et al.*, 1997; Voinnet *et al.*, 1997). The phenomenon was called "silencing acquired systemically" (SAS), due to the fact that the silencing signal was able to diffuse through plasmodesmata between cells and via the plant's vascular system to other parts of the plant (Voinnet *et al.*, 1998).

As well as being dominant, Cogoni *et al.* (1996) demonstrated that quelling in *Neurospora* is a posttranscriptional mechanism. The level of unspliced mRNA for the *al-1* gene was the same in both wild-type and silenced strains, while a dramatic reduction of spliced *al-1* mRNA was observed in quelled strains, indicating that quelling acts at a posttranscriptional level and does not influence the rate of transcription. Moreover, the fact that the nuclear levels of *al-1* mRNA were unchanged suggested that the reduction of the mRNA level was due to a degradation process occurring in the cytoplasm. These results were in accordance with those obtained by de Carvalho *et al.* (1992, 1995) in the co-suppression of the β1-3 glucanase gene in tobacco plants and by Smith *et al.* (1990) with the polygalacturanase gene in tomato, as in both cases an increased turnover of RNA of the duplicated genes was observed. Silencing of viral genes in plants was also observed to act at a cytoplasmatic level in transgene-induced resistance to viral infections by Lindbo *et al.* (1993) and van Blokland (1994).

By transformation with various constructs Cogoni *et al.* (1996) also demonstrated that 132 bp of homology to transcribed exonic sequences was still able to trigger quelling, even though at a low frequency. Specific sequences are not required, amino-terminal and carboxy-terminal sequences are equally efficient in triggering quelling, and promoter sequences do not cause quelling. The above characteristics, together with the evidence that the signal for quelling is diffusible, suggested that it may involve an RNA molecule. Grierson *et al.* (1991) proposed that silencing could be a consequence of the unwanted production of transgenic antisense RNA (asRNA) molecules. Such asRNA could be synthesized on a transgenic DNA template integrated in the antisense direction with respect to an endogenous promoter near the integration site. Alternatively, asRNA could originate from the activation of a transgenic cryptic promoter (Baulcombe, 1996). An RNAse protection assay was therefore carried out to identify anti-sense RNA in quelled *Neurospora* strains (Cogoni *et al.*, 1996); however, no asRNA could be detected. To test whether lack of detection may have been due to the highly unstable nature of the asRNA, a highly expressed anti-sense construct was made; however, transformation with this construct did not increase the frequency of quelling, indicating that in *Neurospora* silencing by asRNA does not increment an already functioning process. Instead, the unexpected finding of a chimeric sense RNA transcribed from the promoterless *al-1* transgene suggested that this molecule could have been the silencing trigger molecule. It was thought that transcription of the sense RNA may have occurred by read-through transcription from the promoter of a neighboring gene to the site of integration, or from a cryptic promoter in the plasmid, as vector sequences were identified in the 3' and 5' ends of the unexpected sense RNA. Although a sense RNA would appear to be necessary, it may not be sufficient per se to trigger quelling, as shown by the fact that high levels of a sense transcript driven by the *Aspergillus nidulans* trpC promoter were also found to be present in a nonquelled transformant, indicating that the level of transcription of the exogenous sequence is not a determining factor. It was hypothesized, therefore, that the sense RNA could possess some aberrant characteristic that confers a signal on the sense RNA to trigger the sequence of events that determines a reduction of homologous mRNA in quelled strains.

Flavell *et al.* (1994) proposed that epigenetic modifications of transgenic DNA sequences such as methylation, which can block transcription elongation, could be responsible for the transcription of a preterminated aberrant RNA. Other aberrant characteristics could be represented by covalent modifications and/or formation of a complex with specific proteins. Alternatively, it could be the chimeric nature or abnormal structure of a sense RNA, due respectively to read-through transcription of tandem arrays of transgenes, or to hairpin structures formed by intramolecular bonding in transcripts from inverted repeats of transgenes, which confer on the RNA a quality recognized as a signal for silencing. Evidence of the silencing effect of hairpin RNAs has recently been offered by Smith *et al.* (2000) in tobacco. Almost 100% efficiency of silencing, manifested by immunity to the potato virus Y (PVY), was obtained as a result of transformation of tobacco with a construct of the viral Nia-protease (*Pro*) gene containing a single self-complementary hairpin RNA (hpRNA), while induction of silencing by normal co-suppression and antisense methods resulted in only 7% and 4% of immune transformants. Hairpin dsRNA introduced into *Trypanosoma brucei* has also been found to induce RNAi in up to 100% of transgenic progeny (Tavernarkis *et al.*, 2000).

     The transcript produced from such transgenic sequences could be not only the trigger, but also the diffusible factor responsible for the dominant character of quelling; however, it has still not been demonstrated whether it is the aRNA itself, or some other RNA species produced downstream in the silencing process, that acts as the diffusible molecule.

III. THE QUEST FOR *quelling-defective (qde)* MUTANTS

A mutagenesis approach was used in the quest to isolate cellular factors involved in quelling. As in previous experiments by Romano and Macino (1992), which led to the discovery of quelling, and those of Cogoni *et al.* (1996), which proved that quelling was a posttranscriptional event probably triggered by an aberrant RNA molecule, the carotenogenic gene *al-1* was used as the exogenous transforming sequence in the experiments carried out by Cogoni *et al.* (1997a) in the search for possible *quelling-defective* mutants. It was postulated that mutations affecting a gene required for the quelling mechanism would determine the release of *al-1* silencing, resulting in the recovery of an orange wild-type phenotype. The strategy adopted was to mutagenize a stably transformed *albino* strain, i.e., one that did not present reversion, as is frequently observed in quelled strains (Romano and Macino, 1992), to exclude that the recovery of a wild-type phenotype was due to reversion.

     Nineteen of 100,000 colonies were analyzed for exogenous copy number and possible rearrangements of exogenous DNA that could have accounted for the release from gene silencing. Two groups emerged: four strains had lost exogenous copies, probably the reason for the release from quelling, while the remaining 15 had neither lost copies nor had rearrangements of exogeneous DNA.

A genetic approach was used to classify the *quelling-defective* (*qde*) mutants into complementation groups. As quelling is a dominant trait, it was possible to carry out complementation tests by creating forced heterokaryons between wild-type and quelling-defective strains. Albino phenotypes were predicted in the case of complementation by the wild-type (orange) nuclei producing the *qde* gene products in heterokaryons with *quelling-defective* (orange) nuclei carrying recessive mutations. A heterokaryon between two *qde* strains (orange) was expected to show an albino phenotype if the recessive mutations were in different loci, whereas an orange phenotype would have been shown in the event the two *qde* mutants were alleles.

Three complementation groups emerged from these tests, indicating that three individual genes are involved in the quelling mechanism. The inability to restore quelling in the complementation tests between wild-type strains and members of the group that had lost transgenic copies and produced no transgenic sense RNA demonstrated that a transgenic sense RNA is essential for quelling in the heterokaryon. Confirmation that the *qde* mutants were indeed impaired in quelling was obtained by testing their ability to silence the native *al-2* gene following transformation with a transgenic copy of the *al-2* gene. None of the *qde* mutants presented additional phenotypes to quelling defectiveness, indicating that the *qde* genes are not involved in other essential biological processes.

In an attempt to identify additional loci involved in quelling that may have escaped the initial screening, more than 50 additional independent *qde* mutants were tested for complementation with strains from each of the three previously identified complemenation groups; however, all were found to belong to the previously identified complementation groups (C. Cogoni, unpublished results). Due to the method of screening adopted, the 50 *qde* mutants were identified based on their ability to completely release *al-1* gene silencing (i.e., only colonies presenting a full orange color were selected). However, mutations in genes that encode other, partially or fully redundant factors for quelling which do not determine a completely reverted phenotype, together with silencing factors that also have essential biological functions, are expected to escape from this type of screening.

IV. ISOLATION OF THE GENES INVOLVED IN QUELLING IN *Neurospora*

The insertional mutagenesis approach was used to isolate the *qde* genes. Taking advantage of the fact that in *Neurospora crassa* plasmids are integrated at random into the genome, transformation with a plasmid was used as a simple tool to perform insertional mutagenesis. The same *al-1* stably silenced transformant strain used for UV mutagenesis by Cogoni *et al.* (1997a) to identify the *qde* genes was used as a recipient for insertional mutagenesis. Insertional mutants were isolated by their ability to grow on media selective for transformants only and by their quelling-defective (orange) phenotype.

By complementation analysis with the three previously identified groups of *qde* strains, the mutants were assigned to each of the three complementation groups. Genomic DNA flanking the site of insertion of the transgene was used as a probe to identify cosmids containing genomic DNA capable of complementing each of the mutants. The three *qde* genes are described below in the order in which they were isolated.

A. The *qde-1* gene

Of the three classes of *quelling-defective* (*qde*) genes identified in *Neurospora crassa* (Cogoni *et al.*, 1997a), *qde-1* was the first to be isolated (Cogoni and Macino, 1999a). It encodes a protein of 1402 amino acids with significant homology to an RNA-dependent RNA polymerase (RdRP) isolated in tomato (Schiebel *et al.*, 1998). The region of homology does not involve the entire protein sequence, but is restricted to the carboxy-terminal portion, probably defining a conserved functional domain. Homology between the two proteins strongly supports the involvement of an RdRP in PTGS (Lindbo *et al.*, 1993). The importance of the cloning of the *qde-1* gene resides in the fact that this was the first experimental evidence linking a gene demonstrated to be involved in gene silencing and a biochemical function proposed to play a role in PTGS.

In vitro experiments in tomato have demonstrated that RdRP can use both single-stranded RNA and single-stranded DNA as a substrate for RNA transcription. The insensitivity of RdRP to α-amanitin and actinomycin D excluded the possibility that the enzyme was a DNA-dependent RNA polymerase. Moreover, RNA synthesis may be carried out either with or without primer extension (Schiebel *et al.*, 1993a, 1993b). The apparent versatility of this RdRP enzyme makes it a good candidate to satisfy the various roles attributed to RdRP in different PTGS models. Furthermore, it has been reported that the steady-state level of RdRP mRNA increases as a result of viroid infection in tomato (Schiebel *et al.*, 1998), and similarly in *Neurospora*, the mRNA level of *qde-1* is greater in quelled strains (Cogoni and Macino, 1999), suggesting that the surveillance system that activates PTGS in the presence of invasive nucleic acids determines a transcriptional control of RdRP required for the subsequent sequence-specific mRNA degradation (Baulcombe, 1999).

Several *qde-1* homologous genes have been identified in various species of plants and animals. Seven homologs are present in *Arabidopsis*, of which one, SDE1/SGS2, has been demonstrated to be involved in gene silencing and possibly in virus-induced PTGS (Mourrain *et al.*, 2000; Dalmay *et al.*, 2000). A *qde-1* homolog has also been identified in the yeast *Saccharomyces pombe*, indicating that a PTGS mechanism may exist in this organism. In *Caenorhabditis elegans*, the *ego1* locus encodes a protein homologous to QDE-1 that is involved not only in RNAi (pointing out the correlation of this phenomenon with PTGS and quelling), but also in germline developmental programs (confirming studies on the *qde-2* gene and its homologs that suggest that some steps of gene silencing and developmental pathways are in common). It is interesting to point out that *ego-1* expression is limited to germinal cell lines and that *ego-1* mutants are defective in RNAi of maternal genes only, while genes expressed in the zygote remain sensitive to gene silencing (Smardon *et al.*, 2000).

The exact role of RdRP in quelling, co-suppression, and RNAi has still not been completely clarified because both the substrate and the product of RdRP enzymatic activity remain to be identified. A possible substrate for RdRP could be a transgenic RNA transcript. Recognition by the RdRP enzyme of such a transcript could be due to its aberrant characteristics. Various hypotheses have been advanced as to possible factors that may determine aberrancy. These include pretermination, covalent modifications, and read-through transcription determining a chimeric transcript and secondary structure. The "double-strandedness" of hairpin RNA could be the aberrant characteristic that signals RdRP to use such RNA as a template for polymerisation in plants (Kooter *et al.*, 1999).

The discovery by Hamilton and Baulcombe (1999) of small antisense 21–23-nt RNAs associated with co-suppression in *Arabidopsis* plants led to the hypothesis that RdRP could be involved in the generation of such molecules. This is supprted by the finding that these small RNAs are depleted in an RdRP mutant in *Arabidopsis* (Dalmay *et al.*, 2000). RdRP could generate small asRNAs in two possible ways: first, small asRNA could be synthesized directly on a single-stranded aRNA template by discontinuous polymeration; second, RdRP could convert single-stranded aRNA to a large double-stranded RNA by continous polymerization. The large dsRNA formed thereby could then be processed in small RNA molecules by a dsRNA endonuclease (see Figure 9.1). The latter hypothesis could be supported by the fact that in *in vitro Drosophila* extracts, large dsRNA molecules are processed into 21–23-nt RNAs (Zamore *et al.*, 2000). Also in *Drosophila*, a temporal correlation between the appearance of small RNAs and gene silencing induced by dsRNA has been observed (Yang *et al.*, 2000). Strikingly, target mRNA has been found to be degraded by endonucleolytic cleavage at the same 21–23-nt intervals, and it has therefore been hypothesized that these small RNAs could act as "guide" molecules to mediate the degradation of homologous mRNAs (Zamore *et al.*, 2000). Moreover, the small RNA molecules have been observed to be associated with ribonuclease enzymatic activity (Hammond *et al.*, 2000).

B. The *qde-3* gene

The finding that the 1955-amino acid protein encoded by *qde-3* was homologous to a RecQ-like DNA helicase was an important discovery assigning, as in the case of

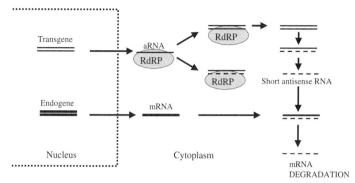

Figure 9.1. Possible mode of action of the RNA-dependent RNA polymerase encoded by the *qde-1* gene. RdRP either synthesizes a single cRNA molecule or many short cRNA molecules on an aberrant RNA template, which is subsequently cleaved into 21–23-nt dsRNA molecules (thin lines). The short RNAs mediate quelling by guiding the degradation of homologous mRNAs.

qde-1, a specific function to a factor involved in the quelling mechanism (Cogoni and Macino, 1999b). Homology from bacteria to humans demonstrates that the RecQ DNA helicase functions are highly conserved. Moreover, some organisms have several homologs: as many as five RecQs have been identified in humans, four in *C. elegans* (Kitao *et al.*, 1998, and references therein), and six RecQ proteins are expressed in *Arabidopsis* (Hartung *et al.*, 2000). All RecQ proteins are characterized by seven highly conserved DNA helicase domains, but homologs vary in size depending on the length of the N- and C-terminal nonhelicase regions. Although homology between the RecQs of different organisms is strictly confined to the helicase domains, some homology may be found in the long terminal regions, with the N-terminal region often characterized by stretches of acidic amino acids. The interaction of the nonhelicase regions with different accessory proteins may confer substrate specifity in organisms where several homologs exist.

The main role of RecQ DNA helicases is maintenance of genome stability, with activities in DNA repair, regulation of illegitimate recombination, and processing of DNA at replication forks. Defects in RecQ genes are responsible for various deleterious effects ranging from hyperrecombination to premature aging and a predisposition to cancer, characteristic of the human genetic disorders known as Werner, Bloom, and Rothmund-Thomson syndromes. The recent discovery of the *qde-3* gene, linked to posttranscriptional gene silencing, has added a new function to the possible roles of DNA helicases.

The demonstration of the interaction of RecQ helicases with topoisomerases and the fact that *qde-3* mutants are sensitive to type I topoisomerase inhibitors suggest a role for *qde-3* in quelling in resolving complex DNA structures between repeated transgenes to allow transcription into an aberrant RNA molecule that triggers the sequence of events that leads to silencing of homologous mRNAs. The N-terminal region of the yeast RecQ homolog *sgs-1* has been demonstrated to interact with topoisomerase III (Gangloff *et al.*, 1994; Bennett *et al.*, 2000) to form an enzymatic complex which suppresses hyperrecombination between repeated sequences. There is also evidence in the fission yeast *Schizosaccharomyces pombe* of a similar functional interaction between the RecQ homolog *rqh-1* and a type III topoisomerase. The carboxy-terminal end of topoisomerase II, essential for faithful chromosome segregation during replication, has also been shown to interact with *sgs-1* (Watt *et al.*, 1995). Brosh *et al.* (2000) have shown a functional interaction between the single-stranded DNA-binding (SSB) protein hRPA and BLM and between topoisomerase III and BLM in humans, and likewise Shen *et al.* (1998) have demonstrated interaction and between SSBs and the WRN protein. QDE-3 is a large helicase, but lacks the N-terminal domain present in the WRN protein with 3'-5' DNA exonuclease activity, hypothesized for the removal of damaged DNA prior to repair. It cannot be excluded, however, that RecQ DNA helicases that do not have the RNAseD-like exonuclease domain in the same gene can interact *in trans* with an exonuclease. This is supported by a recent finding in *Arabidopsis* of the interaction between two of its RecQ homologs, one of which encodes an RNAseD-like exonuclease, demonstrating that the two functions of helicase and exonuclease may act *in trans*, at least in plants (Hartung *et al.*, 2000). The discovery in *C. elegans* of homology between the *mut-7* gene involved in transposon silencing and the WRN exonuclease domain encoding a RNAseD-like exonuclease (Ketting *et al.*, 1999), and in *Neurospora crassa* of homology between QDE-3 involved in quelling and a RecQ DNA helicase (Cogoni and Macino, 1999c), could be futher evidence of *trans* interaction between helicase and exonuclease functions in PTGS. As shown in Figure 9.2, QDE-3, together with a topoisomerase, could therefore be important in remodeling chromatin by resolving complex DNA structures such as Holliday junctions (Holliday and Pugh, 1975) or cruciforms between inverted repeated sequences, to permit transcription. Such DNA–DNA interactions could also lead to methylation and consequently to the transcription of an aberrant RNA to trigger quelling.

C. The *qde-2* gene

While the function of *qde-1* and *qde-3* can be envisaged in a model of the action of the *Neurospora qde* genes in quelling, the collocation of *qde-2* is more difficult due to the fact that little is known regarding both the biochemistry and the function of the protein encoded by this gene. The *qde-2* locus is a member of a large novel gene family, conserved from plants to animals, involved in developmental pathways as well as in silencing processes. Sequence aligments of the *qde-2* gene family reveal a variable N-terminal portion and a conserved C-terminal region that may be involved in some specific function, probably containing a novel functional

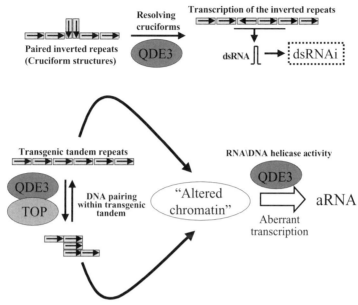

Figure 9.2. Possible mode of action of the DNA-helicase encoded by the *qde-3* gene. The DNA helicase may resolve complex DNA structures formed as a result of pairing of inverted repeated DNA sequences or transgenic tandem repeats, to enable transcription of an aberrant RNA.

domain (Catalanotto *et al.*, 2000). Interestingly, a *qde-2* homolog was identified in *S. pombe*, but not in *S. cerevisiae*. An explanation suggested by Aravind *et al.* (2000) is that functionally linked genes involved in PTGS in *S. cerevisiae* have been co-eliminated. In fact, *S.cerevisiae* lacks both QDE1 and QDE2 homologs.

In *Arabidopsis* there are seven *qde-2* homologs (Fagard *et al.*, 2000). The first and better characterized of these is *AGO1* (*ARGONAUTE-1*), so called for the squidlike aspect of its mutants (Bohmert *et al.*, 1998). The molecular function of AGO1 is not known, but its mutant phenotype involves both leaves and flowers. It has been demonstrated that AGO-1 is involved in PTGS, as it influences the expression level of transgenes (Mourrain *et al.*, 2000). Moreover, in an *AGO-1* mutant background, methylation of transgenic sequences is significantly reduced, indicating a role for AGO-1 in a control step of PTGS that involves DNA methylation (Fagard *et al.*, 2000). Another *qde-2* homolog in *Arabidopsis*, ZWILLE/PINHEAD, is required for the correct development of primary and axillary shoot apical meristems. This locus is responsible not for differentiation, but to maintain the central cells of the embryonic shoot apical meristem in an undifferentiated state (Moussian *et al.*, 1998), and is responsible for the formation of axillary meristems in postembryonic phases (Lynn *et al.*, 1999).

In *Drosophila*, the protein transcribed by the *piwi* gene seems to be a determinant in germline stem cell self-renewing division and maintenance, in both males and females, probably through a somatic signaling pathway; moreover, the presence of *piwi* mRNA in the germline is important during development in embryogenesis (Cox *et al.*, 1998). Again in flies, the STING protein controls the posttranscriptional expression level of the repetitive X-linked *Stellate* locus in the germlines of both sexes. Furthermore, this protein controls male fertility, meiotic disjunction, and meiotic drive (Schmidt *et al.*, 1998). Recently, a protein belonging to this family was cloned in rat, but with different cell-dependent localizations (either in the Golgi apparatus or in the endoplasmic reticulum). The GERp95 protein presents strong homology with AGO1, ZWILLE, and PIWI (Cikalus *et al.*, 1999).

Rabbit eIF2C (previously named Co-eIF-2A) was purified as a member of the translation machinary from a reticulocyte lysate (Chakravarty *et al.*, 1985). The eIF2C protein is implicated in the initiation of protein synthesis, stimulating the formation of a ternary complex between Met-tRNA, GTP, and the eukaryotic initiation factor eIF2 (Zou *et al.*, 1998).

Lastly, in the nematode *C. elegans* there are at least 23 genes that can be grouped in the *qde-2*-like gene family, one of which, *rde-1*, is implicated in double-stranded RNA interference. Although normally viable and fertile, *rde-1* mutants are completely defective in RNAi directed at both maternally and zygotically expressed genes (Tabara *et al.*, 1999). Interestingly, transposon mobility (Tabara *et al.*, 1999) is unaffected by loss of the RDE-1 function, suggesting that this gene (together with *rde-4*) is not required for the initiation of the silencing when the signal is an invasive DNA molecule. It can be assumed that common steps of specific mRNA degradation could be activated by two different pathways: either by an RDE-1-independent pathway triggered by invasive DNA molecules such as transposons or transgenes, or by an alternative pathway induced by dsRNA molecules, such as genomes of viruses or viral RNA intermediates or transgenic dsRNA, in which RDE-1 activity in essential. The fact that *rde-1* seems to be involved in the establishment of dsRNA-directed gene silencing was confirmed by experiments carried out by Grishok *et al.* (2000), showing that *rde-1* is essential for initiation of heritable RNAi, but is not required for the subsequent transmission of the phenomenon.

The large number of QDE2 homologs constitutes a family of genes that embraces functions in both PTGS and developmental pathways. QDE2 and RDE1 appear to be involved in PTGS only, while ZWILLE and PIWI control stem-cell division and maintenance, with AGO1 and STING functioning as a link between the two processes. We suggest that either common steps exist between the two processes, or the same enzymes are involved in the two different biological programs. Alternatively, a reasonable assumption could be that a posttranscriptional

Table 9.1. Genes of the QDE2 Family and Their Involvement in PTGS and/or Development

Gene	Organism	PTGS	Development	Other function
ago-1	*Arabidopsis thaliana*	X	X	
rde-1	*Caenorabditis elegans*	X		
qde-2	*Neurospora crassa*	X		
zll	*Arabidopsis thaliana*		X	
eIF2C	rabbit			Translation initiation factor
sting	*Drosophila melanogaster*	X	X	
piwi	*Drosophila melanogaster*		X	
Gerp95	rat			

regulation mechanism acts during development to control the expression of genes involved in devlopment. The homology of QDE2, RDE1, and AGO1 with eIF2C may lead one to believe that an association between PTGS and translational processes exists, but transgene-induced PTGS in plants (Holtorf *et al.*, 1999) and RNAi in *Drosophila* (Zamore *et al.*, 1999) are not blocked either by initiation or by elongation translation inhibitors. Due to the fact that eIF2C organizes the interaction between RNA (Met-tRNA) and protein (eIF2) molecules in a ternary complex, it would be tempting to speculate that QDE2 in *Neurospora*, RDE1 in *C. elegans*, and AGO1 in *Arabidopsis* could act as mediators in PTGS through RNA–RNA or RNA–protein interactions. A list of homologous *qde-2* genes is shown in Table 9.1.

V. A UNIFIED MODEL FOR QUELLING, CO-SUPPRESSION, AND RNA INTERFERENCE

Comparisons between co-suppression in plants and quelling in *Neurospora* have been important in the formulation of PTGS models, and although some differences do exist, a majority of common features have been revealed, supporting the notion that PTGS is indeed a conserved phenomenon. Common features include the following.

- Gene specifity is observed.
- Both endogenous and transgenes are silenced.
- Duplication of parts only of coding sequences can induce silencing.
- Homology of promoter sequences is not required.
- The introduction of trangenes per se is not sufficient to induce gene silencing.
- Tandem repeats or inverted repeats of transgenes are stronger inducers of silencing than single copies.

- The transcription of transgenes even at a low level seems to be necessary.
- Ongoing protein synthesis is not necessary.
- Varying degrees of the reduction of steady-state mRNA is a consequence of a posttranscriptional event.
- Both mechanisms are reversible in a stepwise manner.
- Both mechanisms are mediated by a *trans*-acting molecole.

See Cogoni and Macino (1997b) and Depicker and Montagu (1997) for reviews and references therein.

Many of the above features are common also to the recently discovered phenomenon of double-stranded RNA interference (RNAi). The main features of RNAi, reviewed in Tabara *et al.* (1999), are the following.

- dsRNA containing exonic sequences is necessary for RNAi.
- Small quantities of dsRNA can induce RNAi in a nonstochiometric fashion.
- dsRNA determines a reduction of homologous endogenous mRNA.
- RNAi acts systemically.

The common features that quelling shares with co-suppression, RNAi and transposon silencing in plants and animals, makes a general model for gene silencing feasible (Cogoni and Macino, 1999c, 2000; Plasterk and Ketting, 2000; Ketting and Plasterk, 2000; Jensen *et al.*, 1999; Sijen and Kooter, 2000). The numerous genes identified in PTGS in both plants and animals, homologous to the *Neurospora crassa qde* genes (see Table 9.2), not only reinforce a unified model, but indicate that PTGS is basically similar in all organisms, probably having evolved from an ancestral gene silencing mechanism aimed at protecting the genome from invading nucleic acid molecules.

A model to explain quelling in *Neurospora crassa* is proposed in which the *qde-3* DNA helicase could be involved in the initial recognition process, maybe unwinding cruciform or other complex DNA structures which could be formed as a result of DNA–DNA pairing of inverted transgenic repeats and/or multiple copies of transgenes, to permit transcription. The aberrant sense RNA molecule transcribed from the exogenous DNA sequence, due to its inherent aberrant characteristics, could be recognised by *qde-1* RdRP and made into a double-stranded RNA molecule either by continuous polymerization or by the synthesis of many 21–23-mer asRNA molecules. These dsRNAs may then be subjected to the endonucleolytic action of a dsRNAse that cleaves the dsRNA into small 21–25-nt RNA species. These small RNAs could mediate the degradation of homologous endogenous mRNAs, acting as "guide" molecules in a complex with an RNAse. Furthermore, the small RNAs could act as primers on endogenous homologous mRNAs to enable RdRP to create double-stranded mRNA molecules which are subsequently processed, activating a positive feedback loop. In this way, additional

Table 9.2. Genes Involved in PTGS in Various Organisms with Putative Functions and Probable Roles

Neurospora crassa	Caenorhabditis elegans	Arabidopsis thaliana	Drosophila melanogaster	Putative function	Probable role	References
qde-1 [1]	ego-1 [2]	sgs-2/sde-1 [3]		RdRP		[1] Cogoni and Macino, 1999a [2] Smardon et al., 2000 [3] Mourrain et al., 2000 Dalmay, et al., 2000
qde-2 [4]	rde-1 [5]	ago-1 [6]		Unknown	Initiation of silencing activity	[4] Catalanotto, et al., 2000 [5] Tabara et al., 1999 [6] Bohmert, et al., 1998
			Dicer [7]	RNAse III	dsRNA processing	[7] Bernstein et al., 2001
	rde-2 [8]			Not cloned	Effector	[8] Tabara et al., 1999
	rde-3 [9]			Not cloned		[9] Tabara et al., 1999
	rde-4 [10]			Unknown	Initiation of silencing activity	[10] Tabara et al., 1999
	mut-2 [11]			Not cloned		[11] Tabara et al., 1999
	mut-6 [12]			Not cloned		[12] Tabara et al., 1999
	mut-7 [13]			RNAse D DNA helicase	Effector	[13] Ketting et al., 1999
qde-3 [14]						[14] Cogoni and Macino, 1999c
		sgs-1 [15]		Unknown		[15] Elmayan et al., 1998
		sgs-3 [16]		Unknown		[16] Mourrain et al., 2000

small RNA molecules could be produced proportionally to the abundance of the target mRNA, leading to the formation of an increased number of RNA-directed RNAse complexes, thus increasing the strength of silencing.

VI. SUPPORTING EVIDENCE FOR THE PROPOSED MODEL

The proposed model is in accordance with several hypotheses that have been advanced to explain various aspects of PTGS. Both quantitative and qualitative features of the RNA trigger molecule may influence silencing, and these aspects, together with the possible roles of RNAse and methylation, are discussed in the following paragraphs.

A. A threshold of mRNA

The threshold model originally proposed to explain co-suppression in plants (Jorgensen, 1991) was based on the notion that a threshold level of transcription of a transgene or virus was the eliciting factor to trigger gene silencing (Smith *et al.*, 1990). Lindbo *et al.* (1993) proposed that RNA concentrations above a threshold level induce RdRP to generate asRNAs involved in co-suppression. Several observations appear to support this model: (1) frequently, transgenes driven by strong promoters induce a higher frequency of silencing; (2) an increased silencing frequency was observed in homozygous with respect to hemizygous plants (Elmayan and Vaucheret, 1996). However, other experimental data are in contrast with these findings. Weakly or negligibly transcribed transgenes can efficiently induce PTGS (Stam *et al.*, 1998), and silencing of endogenous *chs* genes in petunia can be induced by promoter-less *chs* transgenes (van Blokland *et al.*, 1994). The above model proposed for quelling in *Neurospora* could also embrace a threshold hypothesis, i.e., only when abundant target RNA is available can a positive feedback loop be established and an efficient level of silencing be reached.

B. The possible nature of the diffusible signal for silencing

In previous models proposed by Cogoni and Macino (1997b), Stam *et al.*, (1997) and Voinnet *et al.*, (1998), the aberrant characteristics of the RNA transcript have been considered as constituting the signal to trigger the sequence of events which lead to silencing of both transgenic and homologous endogenous sequences. This would explain why not all transgenes are able to induce silencing, why only certain RNAs are the target for RdRPs, and why quelling affects only a percentage of transformants in *Neurospora*. The fact that gene silencing is diffusible from cell to cell in plants and nematodes and through the syncytium of *Neurospora* hyphae has suggested that the aberrant RNA molecule could act not only as the trigger, but also as the diffusible signal of gene silencing able to induce the cascade of events leading to the establishment of the silenced state, even at a distance, far from the site of initiation.

However, the nature of the silencing signal is still completely obscure. It could be equally possible that RNA molecules derived from transgenic aberrant RNA could work as a diffusible signals. For instance, the small double-stranded RNA species could act not only as "guide RNAs" in a protein complex capable of carrying out the degradation of homologous mRNAs in the cell in which they are generated, but could also travel from cell to cell to initiate the posttranscriptional process in neighbouring cells (Zamore *et al.*, 2000, Sharp and Zamore, 2000).

C. The role of RNAse in the degradation step

Hypothetical models proposed by Bass (2000) and Yang *et al.* (2000) to explain RNAi in *Drosophila* are based on evidence found in *Drosophila* by Hammond *et al.* (2000) and Tuschl *et al.* (1999) of the presence of an RNAse enzyme associated to the small guide RNAs. The models suggest that the double-stranded RNA is degraded by an RNAse that remains associated with the double-stranded 25-mer RNAs that are formed as a result of degradation and which determine the specificity for the degradation of the homologous mRNAs. Zamore *et al.* (2000) have demonstrated that ATP is necessary for the various steps in RNAi. In the strand exchange model proposed by Bass (2000), a hypothetical "RNAi nuclease," with a dsRNA-binding domain, ribonuclease domains, and an RNA helicase domain, binds to the complex formed between the guideRNA and homologous mRNA, to permit strand exchange between the sense strand of the guide RNA and the mRNA and its subsequent degradation. ATP would be necessary in this process either to unwind secondary structure of the mRNA or for cleavage. These models also envisage the creation of the small dsRNAs as an amplification step; however, no involvement of an RdRP is foreseen.

The most recent evidence supporting a model envisaging a multicomponent nuclease bound to the guide RNAs as the degradation mediator of mRNA is the identification in *Drosophila* of an enzyme with both a helicase domain and RNAseIII motifs. The ATP-dependent enzyme called Dicer can bind dsRNA to create 22-mer RNAs, and it has been proposed that this could constitute the first part of a two-step process (Bernstein *et al.*, 2001). The second stage would be the base pairing of the 22-mer RNA with homologous mRNA with the RNAse part of the complex responsible for cleavage of mRNAs. Interestingly, the *Dicer* gene sequence presents homology with the PAZ domain found among genes of the Piwi/Argonaute/Zwille-pinhead family of which *qde-2* and *rde-1*, both involved in gene silencing, are members (Cerutti *et al.*, 2000). It has been proposed in *Drosophila* that interaction between an Argonaute-like gene and *Dicer*, by means of their PAZ domains, could explain the formation of the RNA degradation complex (Baulcombe, 2001).

D. Maintenance of PTGS

Although the key events of posttranscriptional gene silencing in plants, animals, and fungi appear to take place in the cytoplasm, several reports suggest that nuclear events involving the homologous sequence could play an important role, especially regarding the activation and maintenance of the silenced state. Jones *et al.* (1999) have found evidence that PTGS induced by viruses is accompanied by an increased level of methylation of the nuclear homologous genes. It has been proposed that sequence-specific methylation can be induced by an RNA-directed DNA methylation (RdDM) mechanism. Indeed, clear evidence of the ability of viroid RNA to cause methylation of homologous sequences has been demonstrated in plants (Wassenegger *et al.*, 1994). Even though the role of DNA methylation associated with PTGS is still not clear, it has been proposed that interactions between the aberrant, viral or small RNAs and endogenous DNA may play a part in the maintenance of the silenced state (Wassenegger and Pélissier, 1998, 2000). In this view, methylation and/or chromatin modifications caused by an RNA-directed mechanism could be important in establishing and/or sustaining an altered epigenetic state essential for the maintenance of silencing, perhaps inducing or enhancing the production of aberrant RNA trigger molecules. Such a maintenance mechanism could be especially significant in SAS phenomena.

VII. THE SEARCH FOR ADDITIONAL PTGS COMPONENTS

Despite the fact that exhaustive screening of *Neurospora quelling-defective* mutants did not reveal any other loci involved in quelling, various experimental approaches may be used in the identification of components that participate in PTGS.

Components of gene silencing mechanisms that are active in protein complexes may be identified by two-hybrid assays, and this could be especially suitable in the case of molecules such as the *qde-2* gene product, the role of which is still unknown.

The comparative genomics approach which envisages the systematic analysis of coordinated gene loss may be used to predict new proteins implicated in silencing mechanisms. For example, homologs to all three *Neurospora qde* genes have been identified in the yeast *S. pombe*, whereas *S. cerevisiae* appears to have lost both the *qde-1* and *qde-2* homologs, retaining a single function encoded by *sgs-1*, homologous to *qde-3*. It has been hypothesized that the complete set of genes necessary for PTGS are present in *S. pombe*, but has been lost in *S. cerevisiae* (Aravind *et al.*, 2000).

Several plant viral proteins have been found to interfere with both the establishment and the maintenance of PTGS, probably interacting with specific cellular components of the silencing machinery. Thus, such viral proteins could be used as a means to isolate new host PTGS components (Brigneti *et al.*, 1998; Kasschau and Carrington, 1998; Voinnet *et al.*, 1999). Indeed, by using such an approach a calmodulin-related gene interacting with the viral protein HC-Pro has recently been isolated (Anandalakshmi *et al.*, 2000).

VIII. CONCLUSIONS

It is now becoming clear that the PTGS phenomena identified in various organisms originate from an ancestral surveillance system aimed at maintaining genome integrity by protection from attack by invasive DNA (such as trasposons and "selfish" repetitive DNA) and RNA (viruses and viroids). The notion that all PTGS phenomena have evolved from an ancestral defense mechanism is also supported by the fact that plants defective in PTGS are more susceptible to viral infections, suggesting that the same silencing mechanism operates on both transgenic and viral RNA. The extreme importance of these silencing mechanisms is shown by the fact that they are responsible not only for genomic surveillance, but also for the regulation of development and cell-differentiation, shown, for example, by the role of the *rde* genes in *C. elegans*. Moreover, an RNA silencing mechanism active in germline cells in animals and in meristems in plants would be a means of blocking transmission to progeny of viral infections.

Although the existence of homology-dependent gene silencing mechanisms has been casually discovered as a result of the experimental introduction of transgenes or dsRNA in various organisms, this discovery may also have important potential repercusions in functional genomics and in both medical and applied biotechnological fields. Despite the fact that genome sequencing projects are producing complete genomic data for several organisms, the biological function of a majority of genes remains unknown. The use of gene silencing as an instrument in reverse genetic analysis may be an effective tool in predicting gene function. For example, by using RNAi to knock out the expression of 90% of the genes located on chromosome I of *C. elegans*, it has been possibile to rapidly identify the function of more than 300 individual genes, with respect to the 70 previously known functions (Fraser *et al.*, 2000). Similarly, genes involved in cell division have been identified on chromosome III of *C. elegans* by Gönczy *et al.* (2000). Recently, the regulatory interconnection between important plant functions has been identified as a result of viral-induced gene silencing in tobacco (Burton *et al.*, 2000). Medical applications of silencing phenomena can be far reaching: from gaining insight into cancer biology by the utilization of RNAi to study genes involved in cell growth and division in *Drosophila*, to the development of vectors that can resist host defense, important for somatic gene therapy (Bestor, 2000). Finally, the understanding of the fine tuning of each step of PTGS is of vital importance in agriculture to improve transformation technology. The ability to control and maybe shut down the PTGS machinery in plants could be useful to allow more reliable expression of introduced transgenes carrying beneficial traits (Bruening, 1998).

The discovery of the genes involved in quelling in *Neurospora crassa* has been a landmark in the understanding of posttranscriptional gene silencing in this organism. The homology of these genes with those identified in other silencing phenomena, such as co-suppression and RNA interference, underlines the fact that these processes may have a common origin, and due to their importance in maintaining genome integrity have been highly conserved during evolution.

Acknowledgments

This work was supported by a grant from the Fondazione Cenci Bolognetti.

References

Anandalakshmi, R., Marathe, R., Ge, X., Herr, J. M. Jr., Mau, C., Mallory, A., Bowman, L., and Vance, V. B. (2000). A calmodulin-related protein that suppresses post-transcriptional gene silencing in plants. *Science* **290,** 142–144.

Anandalakshmi, R., Pruss, G. J., Ge, X., Marathe, R., Mallory, A. C., Smith, T. H., and Vance, V. B. (1998). A viral suppressor of gene silencing in plants. *Proc. Natl. Acad. Sci. USA* **95,** 13079–13084.

Aravind, L., Watanabe, H., Lipman, D. J., and Koonin, E. V. (2000). Lineage-specific loss and divergence of functionally linked genes in eukaryotes. *Proc. Natl. Acad. Sci. USA* **97,** 11319–11324.

Ballario, P., Vittorioso, P., Magrelli, A., Talora, C., Cabibbo, A., and Macino, G. (1996). White colar 1, a central regulator of blue-light responses in *Neurospora crassa. EMBO J.* **15,** 1650–1657.

Bass, B. L. (2000). Double-stranded RNA as a template for gene silencing. *Cell* **101,** 235–238.

Baulcombe, D. C. (1996). RNA as a target and an initiator of post-transcriptional gene silencing in transgenic plants. *Plant Mol. Biol.* **32,** 79–88.

Baulcombe, D. (1999). Viruses and gene silencing in plants. *Arch. Virol. Suppl.* **15,** 189–201.

Baulcombe, D. (2001). Dicer defence. *Nature* **409,** 295–296.

Bennett, R. J., Noirot-Gros, M. F., and Wang, J. C. (2000). Interaction between yeast Sgs1 helicase and DNA topoisomerase III. *J. Biol. Chem.* **275,** 26898–26905.

Bernstein, E., Caudy, A. A., Hammond, S. M., and Hannon, G. J. (2001). Role for a bidentate ribonuclease in the initiation step of RNA interference. *Nature* **409,** 363–366.

Bestor, T. H. (2000). Gene silencing as a threat to the success of gene therapy. *J. Clinic. Invest.* **105,** 409–411.

Brosh, R. M., Li, J. L., Kenny, M. K., Karow, J. K., Cooper, M. P., Kureekattil, R., Hickson, I. D., and Bohr, V. A. (2000). Replication protein Aphisically interacts with the Bloom's syndrome protein and stimulates its helicase activity. *J. Biol. Chem.* **275,** 23500–23508.

Bosher, J. M., and Labouesse, M. (2000). RNA interference: genetic wand and genetic watchdog. *Nature Cell Biol.* **2,** 31–36.

Brigneti, G., Voinnet, O., Li, W. X., Ji, L. H., Ding, S. W., and Baulcombe, D. C. (1998). Viral pathogenicity determinants are suppressors of transgene silencing in *Nicotiana benthamiana. EMBO J.* **17,** 6739–6746.

Bruening, G. (1998). Plant gene silencing regularized. *Proc. Natl. Acad. Sci. USA.* **95,** 13349–13351.

Bohmert, K., Camus, I., Bellini, C., Bouchez, D., Caboche, M., and Benning, C. (1998). AGO1 defines a novel locus of Arabidopsis controlling leaf development. *EMBO J.* **17,** 170–180.

Burton, R. A., Gibeaut, D. M., Bacic, A., Findlay, K., Roberts, K., Hamilton, A., Baulcombe, D. C., and Fincher, G. B. (2000). Virus-induced silencing of a plant cellulose synthase gene. *Plant Cell.* **12**, 691–706.

Cambareri, E. B., Jensen, B. C., Schabtach, E., and Selker, E. U. (1989). Repeat-induced G-C to A-T mutations in Neurospora. *Science* **244**, 1571–1575.

Carrington, J. C. (2000). Moving targets. *Nature* **408**, 150–151.

Catalanotto, C., Azzalin, G., Macino, G., and Cogoni, C. (2000). Gene silencing in worms and fungi. *Nature* **404,** 24.

Cerutti, l., Mian, N., and Bateman, A. (2000). *TIBS* **25,** 481–482.

Cikalus, D. E., Tahbaz, N., Hendricks, L. C., DiMattia, G. E., Hansen, D., Pilgrim, D., and Hobman, T. C. (1999). GERp95, a membrane-associated protein that belongs to a family of proteins involved in stem cell differentiation. *Mol. Biol. Cell.* **10**, 3357–3372.

Chakravarty, I., Bagchi, M. K., Roy, R., Banerjee, A. C., and Gupta, N. K. (1985). Protein synthesis in rabbit reticulocytes. *J. Biol. Chem.* **260**, 6945–6949.

Chuang, C. F., and Meyerowitz, E. M. (2000). Specific and heritable genetic interference by double-stranded RNA in *Arabidopsis thaliana. Proc. Natl. Acad. Sci. USA* **97**, 4985–4990.

Cogoni, C., Irelan, J. T., Schumacher, M., Schmidhauser, T., Selker, E. U., and Macino, G. (1996). Transgene silencing of the *al-1* gene in vegetative cells of *Neurospora* is mediated by a cytoplasmic effector and does not depend on DNA-DNA interactions or DNA methylation. *EMBO J.* **15**, 3153–3163.

Cogoni, C., and Macino, G. (1997a). Isolation of quelling-defective (*qde*) mutants impaired in post-transcriptional transgene-induced gene silencing in *Neurospora crassa. Proc. Natl. Acad. Sci. USA* **94**, 10233–10238.

Cogoni, C., and Macino, G. (1997b). Conservation of transgene-induced post-transcriptional gene silencing in plants and fungi. *Trends in Plant Sci.* **2**, 438–443.

Cogoni, C., and Macino, G. (1998). Quelling: transgene-induced silencing in *Neurospora crassa.* In "Cellular Integration of Signalling Pathways in Plant Development" (NATO ASI Series), Vol. H 104, pp. 103–112. Springer-Verlag, Berlin, Heidelberg.

Cogoni, C., and Macino, G. (1999a). Gene silencing in *Neurospora crassa* requires a protein homologous to RNA-dependent RNA polymerase. *Nature* **399**, 166–169.

Cogoni, C., and Macino, G. (1999b). Homology-dependent gene silencing in plants and fungi: a number of variations on the same theme. *Curr. Opin. Microbiol.* **2**, 657–662.

Cogoni, C., and Macino, G. (1999c). Post-transcriptional gene silencing *in Neurospora* by a RecQ DNA helicase. *Science* **286**, 2342–2343.

Cogoni, C., and Macino, G. (2000). Gene silencing across kingdoms. *Curr. Opin. Genet. Devel.* **10**, 638–643.

Cox, D. N., Chao, A., Baker, J., Chang, L., Qiao, D., and Lin, H. (1998). A novel class of evolutionarily conserved genes defined by piwi are essencial for stem cell self-renewal. *Genes & Dev.* **12**, 3715–3727.

Dalmay, T., Hamilton, A., Rudd, S., Angell, S., and Baulcombe, D. C. (2000). An RNA-dependent RNA polymerase gene in *Arabidopsis* is required for post-transcriptional gene silencing mediated by a transgene but not by a virus. *Cell* **101**, 543–53.

de Carvalho, F., Gheysen, G., Kushnir, S., Van Montagu, M., Inzè, D., and Castresana, C. (1992). Suppression of β-1,3-glucanase transgene expression in homozygous plants. *EMBO J.* **11**, 2595–2602.

de Carvalho, F., Niebel, F., Frendo, P., Van Montagu, M., and Cornelissen, M. (1995). Post-transcriptional cosuppression of β-1, 3-glucanase genes does not affect accumulation of transgene nuclear mRNA. *Plant Cell* **7**, 347–358.

Depicker, A., and Van Montagu, M. (1997). Post-transcriptional gene silencing in plants. *Curr. Opinion Cell. Biol.* **9**, 373–82.

Dernburg, A. F., Zalevsky, J., Colaiácovo, P., and Villeneuve, A. M. (2000). Transgene-mediated cosuppression in the C. elegans germ line. *Genes & Dev.* **14,** 1578–1583.

Domeier, M. E., Morse, D. P., Knight, S. W., Portereiko, M., Bass, B. L., and Mango, S. E. (2000). A link between RNA interference and nonsense-mediated decay in Caenorhabditis elegans. *Science* **289,** 1928–1931.

Elmayan, T., and Vaucheret, H. (1996). Expression of single copies of a strongly expressed 35S transgene can be silenced post-transcriptionally. *Plant J.* **9,** 787–797.

Elmayan, T., Balzergue, S., Beon, F., Bourdon, V., Daubremet, J., Guenet, Y., Mourrain, P., Palauqui, J. C., Vernhettes, S., Vialle, T., Wostrikoff, K., and Vaucheret, H. (1998). *Arabidopsis* mutants impaired in cosuppression. *Plant Cell* **10,** 1747–1758.

Fagard, M., Boutet, S., Morel, J. B., Bellini, C., and Vaucheret, H. (2000). AGO1, QDE-2, and RDE-1 are related proteins required for post-transcriptional gene silencing in plants, quelling in fungi, and RNA interference in animals. *Proc. Natl. Acad. Sci.USA* **97,** 11650–11654.

Fire, A., Xu, S., Montgomery, M. K., Kostas, S. A., Driver, S. E., and Mello, C. C. (1998). Potent and specific genetic interference by double-stranded RNA in *Caenorhabditis elegans*. *Nature* **391,** 806–811.

Flavell, R. B. (1994). Inactivation of gene expression in plants as a consequence of specific sequence duplication. *Proc. Natl. Acad. Sci. USA* **91,** 3490–3496.

Fraser, A. G., Kamath, R. S., Zipperlen, P., Martinez-Campos, M., Sohrmann, M., and Ahringer, J. (2000). Functional genomic analysis of C. elegans chromosome I by systematic RNA interference. *Nature* **408,** 325–330.

Gangloff, S., McDonald, J. P., Bendixen, C., Aurthur, L., and Rothstein, R. (1994). The yeast type I topoisomerase Top3 interacts with Sgs1, a DNA helicase homolog: a potential eukaryotic reverse gyrase. *Mol. Cell. Biol.* **14,** 8391–8398.

Gonczy, P., Echeverri, G., Oegema, K., Coulson, A., Jones, S. J., Copley, R. R., Duperon, J., Oegema, J., Brehm, M., Cassin, E., Hannak, E., Kirkham, M., Pichler, S., Flohrs, K., Goessen, A., Leidel, S., Alleaume, A. M., Martin, C., Ozlu, N., Bork, P., and Hyman, A. A. (2000). Functional genomic analysis of cell division in C. elegans using RNAi of genes on chromosome III. *Nature* **408,** 331–336.

Grierson, D., Fray, R. G., Hamilton, A. J., Smith, C. J. S., and Watson, C. F. (1991). Does co-suppression of sense genes in transgenic plants involve anti-sense RNA? *Trends Biotechnol.* **9,** 122–123.

Grishok, A., Tabara, H., and Mello, C. C. (2000). Genetic requirements for inheritance of RNAi in C. elegans. *Science* **287,** 2494–2497.

Hamilton, A. J., and Baulcombe, D. C. (1999). A novel species of small antisense RNA in posttranscriptional gene silencing in plants. *Science*. **286,** 950–952.

Hammond, S. M., Bernstein, E., Beach, D., and Hannon, G. (2000). An RNA-directed nuclease mediates post-transcriptional gene silencing in *Drosophila* cell extracts. *Nature* **404,** 293–296.

Hartung, F., Plchová, H., and Puchta, H. (2000). Molecular characterisation of RecQ homologues in *Arabidopsis thaliana*. *Nucleic Acids Research* **28,** 4275–4282.

Holliday, R., and Pugh, J. E.. (1975). DNA modification mechanisms and gene activity during development. *Science* **187,** 226–232.

Holtorf, H., Schob, H., Kunz, C., Waldvogel, R., and Meins, F. Jr. (1999). Stochastic and nonstochastic post-transcriptional silencing of chitinase and β1,3-glucanase genes involves increased RNA turnover-possible role for ribosome-indipendent RNA degradation. *Plant Cell* **11,** 471–484.

Jensen, S., Gassama, M. P., and Heidmann, T. (1999). Taming of transposable elements by homology-dependent gene silencing. *Nat. Genet.* **21,** 209–212.

Jones, A. L., Hamilton, A. J., Voinnet, O., Thomas, C. L., Maule, A. J., and Baulcombe, D. C. (1999). RNA-DNA interaction and DNA methylation in post-transcriptional gene silencing. *Plant Cell* **11,** 2291–2301.

Jorgensen, R. (1991). How do genes intoact with homologous plant genes? *Trends Biotechnol.* **9,** 266–267.

Jorgensen, R. A., Cluster, P. D., English, J., Que, Q., and Napoli, C. A. (1996). Chalcone synthase co-suppression phenotypes in petunia flowers: comparison of sense vs. antisense constructs and single copy vs. complex T-DNA sequences. *Plant Mol. Biol.* **31,** 957–973.

Kasschau, K. D., and Carrington, J. C. (1998). A counter defensive strategy of plant viruses: suppression of post-transcriptional silencing. *Cell* **95,** 461–470.

Kennerdell, J. R., and Carthew, R. W. (2000). Heritable gene silencing in *Drosophila* using double-stranded RNA. *Nat. Biotechnol.* **18,** 896–898.

Ketting, R. F., Haverkamp, T. H. A., van Luenen, H. G. A. M., and Plasterk, R. H. A. (1999). *mut-7* of *C. elegans,* required for transposon silencing and RNA interference, is a homolog of Werner syndrome helicase and RNaseD. *Cell.* **99,** 133–141.

Ketting, R. F., and Plasterk, R. H. (2000). A genetic link between co-supression and RNA interference in *C. elegans. Nature* **404,** 296–298.

Kitao, S., Ohsugi, I., Ichikawa, K., Goto, M., Furuichi, Y., and Shimamoto, A. (1998). Cloning of the new human helicase genes of the RecQ family: Biological significance of multiple species in higher eukaryotes. *Genomics* **54,** 443–452.

Kooter, J. M., Matzke, M. A., and Meyer, P. (1999). Listening to the silent genes: transgene silencing, gene regulation and pathogen control. *Trends Plant Sci.* **4,** 340–347.

Li, Y. X., Farrell, M. J., Liu, R., Mohanty, N., and Kirby, M. L. (2000). Double-stranded RNA injection produces null phenotypes in zebrafish. *Dev. Biol.* **217,** 394–405.

Lindbo, J. L., Silva-Rosales, L., Proebsting, W. M., and Dougherty, W. G. (1993). Induction of a highly specific antiviral state in transgenic plants: implications for regulation of gene expression and virus expression. *Plant Cell* **5,** 1749–1759.

Linden, H., and Macino, G. (1997). White collar 2, a partner in blue-light signal transduction, controlling expression of light-regulated genes in *Neurospora crassa. EMBO J.* **16,** 98–109.

Lynn, K., Fernandez, A., Aida, M., Sedbrook, J., Tasaka, M., Masson, P., and Barton, K. (1999). The PINHEAD/ZWILLE gene acts pleiotropically in Arabidopsis development and has overlapping functions with the ARGONAUTE1 gene. *Development* **126,** 469–481.

Matzke, M. A., Matzke, A. J. M., and Mittelsten Scheid, O. (1994). Inactivation of repeated genes – DNA-DNA interaction? *In* "Homologous Recombination and Gene Silencing in Plants" (J. Paszkowski, Ed.), pp. 271–307. Kluwer, Dordrecht.

Misquitta, L., and Paterson, B. M. (1999). Targeted disruption of gene function in *Drosophila* by RNA interference (RNA-i): a role for nautilus in embryonic somatic muscle formation. *Proc. Natl. Acad. Sci. USA* **96,** 1451–1456.

Mourrain, P., Beclin, C., Elmayan, T., Feuerbach, F., Godon, C., Morel, J. B, Jouette, D., Lacombe, A. M., Nikic, S., Picault, N., Remoue, K., Sanial, M., Vo, T. A., and Vaucheret, H. (2000). *Arabidopsis* SGS2 and SGS3 genes are required for post-transcriptional gene silencing and natural virus resistance. *Cell* **101,** 533–542.

Moussian, B., Schoof, H., Haecker, A., Jürgens, G., and Laux, T. (1998). Role of the ZWILLE gene in the regulation of central shoot meristem cell fate during *Arabidopsis* embryogenesis. *EMBO J.* **17,** 1799–1809.

Napoli, C., Lemieux, C., and Jorgensen, R. (1990). Introduction of a chalcone synthase gene into Petunia results in reversible co-suppression of homologous genes in trans. *Plant Cell* **2,** 279–289.

Nelson, M. A., Morelli, G., Carratoli, A., Romano, N., and Macino, G. (1989). Molecular cloning of a *Neurospora crassa* carotenoid biosynthetic gene (*albino-3*) regulated by blue light and the products of white collar genes. *Mol. Cell Biol.* **9,** 1271–1276.

Ngo, H., Tschudi, C., Gull, K., and Ullu, E. (1998). Double-stranded RNA induces mRNA degradation in *Trypanosoma brucei. Proc. Natl. Acad. Sci. USA* **95,** 14687–14692.

Pandit, N. N., and Russo, V. E. (1992). Reversible inactivation of a foreign gene, hph, during the asexual cycle in Neurospora crassa transformants. *Mol. Gen. Genet.* **234,** 412–422.

Palauqui, J. C., Elmayan, T., Pollien, J. M., and Vaucheret, H. (1997). Systemic acquired silencing: transgene-specific post-transcriptional silencing is transmitted by grafting from silenced stocks to non-silenced scions. *EMBO J.* **16**, 4738–4745.

Plasterk, R. H. A., and Ketting, R. F. (2000). The silence of the genes. *Curr. Opin. Gen. Dev.* **10**, 562–567.

Romano, N., and Macino, G. (1992). Quelling: transient inactivation of gene expression in *Neurospora crassa* by transformation with homologous sequences. *Mol. Microbiol.* **6**, 3343–3353.

Ruiz, M. T., Voinnet, O., and Baulcombe, D. C. (1998). Initiation and maintenance of virus-induced gene silencing. *Plant Cell* **6**, 937–946.

Sanchez Alvarado, A., and Newmark, P. A. (1999). Double-stranded RNA specifically disrupts gene expression during planarian regeneration. *Proc. Natl. Acad. Sci. USA* **96**, 5049–5054.

Schmidhauser, T., Lauter, F. R., Russo, V. E. A., and Yanofsky, C. (1990). Cloning, sequence and photoregulation of *al-1*, a carotenoid biosynthetic gene of *Neurospora crassa*. *Mol. Cell.Biol.* **10**, 5064–5070.

Schmidhauser, T. J., Lauter, F., Schumacher, M., Zhou, W., Russo, V. E. A., and Yanofsky, C. (1994). Characterisation of *al-2*, the Pohytoene synthase Gene of *Neurospora crassa*. *J. Biol. Chem.* **269**, 12060–12066.

Schiebel, W., Haas, B., Marinkovic, S., Klanner, A., and Sänger, H. L. (1993a). RNA-directed RNA polymerase from tomato leaves. I: Purification and Physical Properties. *J. Biol. Chem.* **268**, 11851–11857.

Schiebel, W., Haas, B., Marinkovic, S., Klanner, A., and Sänger, H. L. (1993b). RNA-directed RNA polymerase from tomato leaves. II. Catalytic *in vitro* Properties. *J. Biol. Chem.* **268**, 11858–11867.

Schiebel, W., Pelissier, T., Riedel, L., Thalmeir, S., Schiebel, R., Kempe, D., Lottspeich, F., Sänger, H. L., and Wassenegger, M. (1998). Isolation of an RNA-directed RNA polymerase-specific cDNA clone from tomato. *Plant Cell* **10**, 1–16.

Schmidt, A., Palumbo, G., Bozzetti, M. P., Tritto, P., Pimpinelli, S., and Schäfer, U. (1998). Genetic and molecular characterization of sting, a gene involved in crystal formation and meiotic drive in the male germ line of *Drosophila* melanogaster. *Genetics* **151**, 749–760.

Shen, J. C., Gray, M. D., Oshima, J., and Loeb, L. A. (1998). Characterization of Werner syndrome protein DNA helicase activity: directionality, substrate dependence and stimulation by replication protein A. *Nucleic Acids Research* **26**, 2879–2885.

Selker, E. U., Cambareri, E. B., Jensen, B. C., and Haack, K. R.. (1987). Rearrangement of duplicated DNA in specialized cells of *Neurospora*. *Cell.* **51**, 741–752.

Selker, E. U. (1990). Premeiotic instability of repeated sequences in *Neurospora crassa*. *Annu. Rev. Genet.* **24**, 579–613.

Sharp, P. A., and Zamore, P. D. (2000). Molecular biology. RNA interference. *Science* **287**, 2431–2433.

Sijen, T., and Kooter, J. M. (2000). Post-transcriptional gene-silencing: RNAs on the attack or on the defense? *Bioessays* **22**, 520–531.

Smardon, A., Spoerke, J. M., Stacey, S. C., Klein, M. E., Mackin, N., and Maine, E. M. (2000). EGO-1 is related to RNA-directed RNA polymerase and functions in germ-line development and RNA interference in C. elegans. *Curr. Biol.* **10**, 169–178.

Smith, C. J. S., Watson, C. F., Bird, C. R., Ray, J., Schuch, W., and Grierson, D. (1990). Expression of a truncated tomato polygalacturonase gene inhibits expression of the endogenous gene in transgeneic plants. *Mol. Gen. Genet.* **224**, 477–481.

Smith, N. A., Singh, S. P., Wang, M. B., Stoutjesdijk, P. A., Green, A. G., and Waterhouse, P. M. (2000). Total silencing by intron-spliced hairpin RNAs. *Nature* **407**, 319–320.

Stam, M., Mol, J. N. M., and Kooter, J. M. (1997). The silence of genes in transgenic plants. *Annal. Botany* **79**, 3–12.

Stam, M., de Bruin, R., van Blokland, R., van der Hoorn, R. A., Mol, J. N., and Kooter, J. M. (2000). Distinct features of post-transcriptional gene silencing by antisense transgenes in single copy and inverted T-DNA repeate *loci*. *Plant J.* **21**, 27–42.

Tabara, H., Sarkissian, M., Kelly, W. G., Fleenor, J., Grishok, A., Timmons, L., Fire, A., and Mello, C. C. (1999). The rde-1 gene, RNA interference, and transposon silencing in C. elegans. *Cell* **99**, 123–132.

Tavernarakis, N., Wang, S. L., Dorovkov, M., Ryazanov, A., and Driscoll, M. (2000). Heritable and inducible genetic interference by double-stranded RNA encoded by transgenes. *Nature Genetics* **24**, 180–183.

Tuschl, T., Zamore, P. D., Lehmann, R., Bartel, D. P., and Sharp, P. A. (1999). Targeted mRNA degradation by double-stranded RNA *in vitro. Genes Dev.* **13**, 3191–3197.

van Blokland, R., Van der Geest, N., Mol, J. N. M., and Koote, J. M. (1994). Transgene-mediated suppression of chalcone synthase expression in *Petunia hybrida* results from an increase in RNA turnover. *Plant J.* **6**, 861–877.

van der Krol, A. R., Mur, L. A., Beld, M., Mol, J. N., and Stuitje, A. R. (1990). Flavonoid genes in *Petunia:* addition of a limited number of gene copies may lead to a suppression of gene expression. *Plant Cell* **2**, 291–299.

Vaucheret, H., Beclin, C., Elmayan, T., Feurbach, F., Godon, C., Morel, J-B., Mourrainm, P., Palaqui, J-C., and Vernhettes, S. (1998). Transgene-induced gene silencing in plants. *Plant J.* **16**, 651–659.

Voinnet, O., and Baulcombe, D. C. (1997). Systemic signaling in gene silencing. *Nature* **389**, 553.

Voinnet, O., Vain, P., Angell, S., and Baulcombe, D. C. (1998). Systemic spread of sequence-specific transgene RNA degradation in plants is initiated by localized introduction of ectopic promoterless DNA. *Cell* **95**, 177–187.

Voinnet, O., Lederer, C., and Baulcombe, D. C. (2000). A viral movement protein prevents spread of the gene silencing signal in *Nicotiana benthamiana. Cell* **103**, 157–167.

Voinnet, O., Pinto, Y. M., and Baulcombe, D. C. (1999). Suppression of gene silencing: a general strategy used by diverse DNA and RNA viruses of plants. *Proc. Natl. Acad. Sci. USA* **96**, 14147–14152.

Wargelius, A., Ellingsen, S., and Fjose, A. (1999). Double-stranded RNA induces specific developmental defects in zebrafish embryos. *Biochem. Biophys. Res. Commun.* **263**, 156–161.

Wassenegger, M., Heimes, S., Riedel, L., and Ranger, H. L. (1994). RNA-directed *de novo* methylation of genomic sequences in plants. *Cell* **76**, 567–576.

Wassenegger, M., and Pelissier, T. (1998). A model for RNA-mediated gene silencing in higher plants. *Plant Mol. Biol.* **37**, 349–362.

Wassenegger, M. (2000). RNA-directed DNA methylation. *Plant. Mol. Biol.* **43**, 203–220.

Waterhouse, P. M., Graham, M. W., and Wang, M. B. (1998). Virus resistance and gene silencing in plants can be induced by simultaneous expression of sense and antisense RNA. *Proc. Natl. Acad. Sci. USA* **95**, 177–187.

Watt, P. M., Louis, E. J., Borts, R. H., and Hickson, D. (1995). Sgs1: a eukaryotic homolog of *E. coli* RecQ that interacts with topoisomerase II *in vivo* and is required for faithful chromosome segregation. *Cell* **81**, 253–260.

Wianny, F., and Zernicka-Goetz, M. (2000). Specific interference with gene function by double-stranded RNA in early mouse development. *Nature Cell Biol.* **2**, 70–75.

Yang, D., Lu, H., and Erickson, J. W. (2000). Evidence that processed small dsRNAs may mediate sequence-specific mRNA degradation during RNAi in drosophila embryos. *Curr Biol.* **10**, 1191–1200.

Zamore, P. D., Tuschl, T., Sharp, P. A., and Bartel, D. P. (2000). RNAi: double-stranded RNA directs the ATP-dependent cleavage of mRNA at 21 to 23 nucleotide intervals. *Cell* **101**, 25–33.

Zou, C., Zhang, Z., Wu, S., and Osterman, J. C. (1998). Molecular cloning and characterization of rabbit eIF2C protein. *Gene* **211**, 187–194.

10

Non-Mendelian Inheritance and Homology-Dependent Effects in Ciliates

Eric Meyer* and Olivier Garnier
Molecular Genetics Laboratory (CNRS UMR 8541)
Ecole Normale Supérieure
75005 Paris, France

---

*To whom correspondence should be addressed: E-mail: emeyer@wotas.ens.fr.

Advances in Genetics, Vol. 46
Copyright 2002, Elsevier Science (USA).
0065-2660/02 $35.00

ABSTRACT

Ciliates are single-celled eukaryotes that harbor two kinds of nuclei. The germline micronuclei function only to perpetuate the genome during sexual reproduction; the macronuclei are polyploid, somatic nuclei that differentiate from the micronuclear lineage at each sexual generation. Macronuclear development involves extensive and reproducible rearrangements of the genome, including chromosome fragmentation and precise excision of numerous internal sequence elements. In *Paramecium* and *Tetrahymena*, homology-dependent maternal effects have been evidenced by transformation of the vegetative macronucleus with germline sequences containing internal eliminated sequences (short single-copy elements), which can result in a specific inhibition of the excision of the homologous elements during development of a new macronucleus in the sexual progeny of transformed clones. Furthermore, transformation of the *Paramecium* maternal macronucleus with cloned macronuclear sequences can specifically induce new fragmentation patterns or internal deletions in the zygotic macronucleus. These experiments show that the processing of many germline sequences in the developing macronucleus is sensitive to the presence and copy number of homologous sequences in the maternal macronucleus. The generality and sequence specificity of this *trans*-nuclear, epigenetic regulation of rearrangements suggest that it is mediated by pairing interactions between zygotic sequences and sequences originating from the maternal macronucleus, presumably RNA molecules. Alternative macronuclear versions of the genome can be maternally inherited across sexual generations, suggesting a molecular model for some of the long-known cases of non-Mendelian inheritance, and in particular for the developmental determination and maternal inheritance of mating types in *Paramecium tetraurelia*. © 2002, Elsevier Science (USA).

I. INTRODUCTION

A. Non-Mendelian inheritance in ciliates

Long before the advent of molecular genetics, ciliates enjoyed a period of success during which they were strongly associated with non-Mendelian phenomena, and in particular with the question of cytoplasmic inheritance (Nanney, 1985; Preer, 1993). They had become the favorite models of a few biologists, such as T. M. Sonneborn, who believed that Morgan's chromosome theory fell short of explaining all heredity and could not by itself—with no role credited to the cytoplasm—account for complex phenomena such as development and evolution (Harwood, 1985). The early discovery of mating types (Sonneborn, 1937) had made *Paramecium* one of the first single-celled organisms in which genetic analyses could be conducted. Although many characters were found to follow Mendelian inheritance and were ascribed to nuclear genes, it was observed from the very beginning that a number of hereditary characters did not behave in a Mendelian way (Preer, 1993, 2000). One of these non-Mendelian characters, which could hardly have been ignored by the *Paramecium* geneticist, was the mating type itself. Opposite mating types were shown to develop from identical genotypes, and in some species mating types even proved to be cytoplasmically transmitted to sexual progeny. Other well-studied cases of cytoplasmic inheritance included the "killer" trait (production of a toxin that killed other cells) and the serotypes (antigenic variants of the cell surface).

To explain these observations, Sonneborn and others proposed around 1945 that such characters were determined by "plasmagenes," cytoplasmic particles endowed with the genetic properties of reproduction and mutability (Sonneborn, 1948, 1949). Not all plasmagenes, however, were held to be completely independent from nuclear genes; indeed, Mendelian mutations could abolish their maintenance in the cytoplasm, and some plasmagenes even appeared to *originate* from nuclear genes. In its most general form, the plasmagene theory assumed that phenotypic characters can be determined by both nuclear genes and plasmagenes, and that their inheritance patterns in crosses depend on which component differs in the strains analysed (Sonneborn, 1950). Accumulating evidence, however, soon revealed that different mechanisms were at work in different cases, and the search for a unifying theory was abandoned. The killer trait was found to be determined by the presence of an endosymbiotic bacterium multiplying in the cytoplasm, and serotypes were rationalized as the stable inheritance of alternative patterns of expression for a set of nuclear genes encoding variant surface proteins. Unresolved cases of cytoplasmic inheritance, such as that governing *Paramecium* mating types, were confined to the periphery of mainstream research by the strong nuclear hegemony that had built on the remarkable success of Mendelian genetics. As a result of this initial emphasis on cytoplasmic inheritance, ciliates were increasingly perceived as being "anomalous" organisms, and despite Sonneborn's advocacy of the view that the peculiarities of any organism could only help define the universal principles of life (Schloegel, 1999), *Paramecium* genetics largely fell into oblivion (Preer, 1997).

But ciliate research survived, as these organisms continued to prove useful models in other domains, such as cellular morphogenesis or membrane excitability. In the late 1970s, molecular biology began to revive some interest in the genetics of ciliates through a variety of fundamental discoveries, including telomere structure, genome-wide DNA rearrangements, self-splicing introns, deviant genetic codes, telomerase, histone acetyltransferase, and novel histone modifications. Interestingly, some of the new molecular findings again revealed puzzling non-Mendelian phenomena, which have now been characterized as homology-dependent effects. As the Mendelian era is culminating with the first complete sequences of eukaryotic genomes, it is no longer possible to ignore the fact that dividing cells must inherit, in addition to the genome, essential regulatory information that is not to be found in the genome sequence itself, and much excitement is being redirected at some poorly understood epigenetic mechanisms of gene regulation. Homology-dependent effects, in particular, cannot be accounted for by classical paradigms of molecular genetics; yet there are now indications that their widespread occurrence in eukaryotes reflects the evolutionary conservation of a sophisticated machinery. It is the purpose of this review to present available information on homology-dependent effects in ciliates, and their relevance to known cases of non-Mendelian inheritance. The diversity and novel aspects of ciliate effects have the potential of exerting a significant impact in this domain. As in many eukaryotes, ciliate homology effects can lead to the specific silencing of almost any gene during vegetative growth, but in addition they have been shown to participate in the programming of genome rearrangements that occur during development. More generally, the evidence for maternal effects directing an epigenetic programming of the zygotic genome through homology-dependent mechanisms may turn out to be of interest in other systems. For a proper understanding of these maternal effects, some introduction to ciliate biology is necessary.

B. The ciliate life cycle

Ciliates are a monophyletic group of unicellular organisms belonging to the Alveolates, one of the major phyla that emerged at about the same time as plants and the metazoan/fungi clade during the "Big Bang" of eukaryotic evolution (Philippe *et al.*, 2000). One of their most distinctive features is a unique system of separation of germline and somatic functions, which takes the form of two different nuclear lineages coexisting in the cytoplasm of each cell. The germline micronucleus is a diploid nucleus, which divides by mitosis and is transcriptionally silent during vegetative growth. Its only genetic function is to undergo meiosis and produce gametic nuclei during sexual events. In contrast, the somatic macronucleus is a large, highly polyploid nucleus which divides by a nonmitotic process and is responsible for vegetative transcription. It governs the cell phenotype but is lost during sexual events; its genetic material is not transmitted to sexual progeny. Although the numbers of nuclei of each type vary in different species, all nuclei in one cell always originate from a single, diploid zygotic nucleus.

Sexual processes are usually induced by starvation. The sequence of nuclear reorganization is similar in all species with minor variations, and is described in Figure 10.1 for the *P. aurelia* species (see Sonneborn [1974] for full details). An

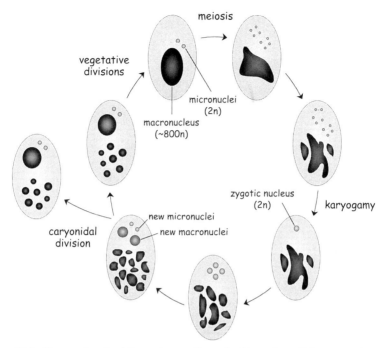

Figure 10.1. The sexual cycle of *Paramecium aurelia* species. Vegetative cells have two micronuclei and one macronucleus. After meiosis of the two micronuclei, seven of the eight haploid products (small circles) degenerate and the surviving one divides once more, yielding two genetically identical gametic nuclei (two small circles with a line between them). In conjugation, one of the gametic nuclei is exchanged with the partner cell before caryogamy; in autogamy, the two gametic nuclei fuse together. In both cases, the diploid zygotic nucleus then undergoes two rounds of mitotic divisions. Of the four products, two become the new micronuclei, while the other two begin to enlarge and develop into new macronuclei. Rearrangements begin after a few replication cycles and are completed by the first cell fission (caryonidal division). During the caryonidal division, the new micronuclei divide by mitosis while the new macronuclei segregate without division to the two daughter cells, called caryonides, and start dividing only at the second cell fission. Each meiotic cell thus gives rise to two caryonidal clones, which derive from distinct events of macronuclear development. Starting shortly after meiosis of the micronuclei, the parental macronucleus is progressively fragmented into about 30 pieces, which segregate randomly to daughter cells during the first vegetative fissions and then disintegrate.

important point is that meiosis of the micronuclei eventually results in two genetically identical gametic nuclei in each cell, as only one haploid product survives and this undergoes an additional division. If the cell is engaged in conjugation with a partner of compatible mating type, karyogamy occurs after the reciprocal exchange of one of the two gametic nuclei, resulting in identical zygotic nuclei in the two conjugants. In the absence of a partner, some species can undergo autogamy, a self-fertilization process by which the two identical gametic nuclei fuse together. In both cases new micronuclei and new macronuclei develop from mitotic products of the diploid zygotic nucleus. The parental macronucleus is not immediately degraded: although DNA replication is rapidly inhibited, transcription continues actively throughout the development of the new macronuclei. The parental genome thus contributes ~80% of the RNA synthesized during the first two cell cycles in *Paramecium* (Berger, 1973).

The genetic consequences of conjugation and autogamy are illustrated in Figure 10.2A by a cross between two homozygous *Paramecium* cells differing by a pair of alleles. Because the exchange of gametic nuclei always makes the two conjugating cells become identical in genotype (see the heterozygous F1

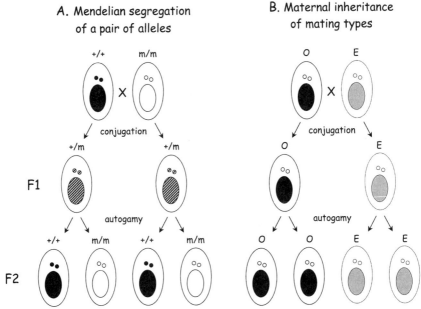

Figure 10.2. Inheritance patterns in *Paramecium aurelia*. (A) Mendelian segregation of a pair of alleles in a cross between homozygous mutant (*m/m*) and wild-type (+/+) cells. Conjugation results in identical zygotic nuclei in the two F1 exconjugants (genotype *m*/+). Autogamy of the F1 clones yields entirely homozygous F2 progeny that have a 50% chance of receiving each of the parental alleles. (B) Maternal inheritance of mating types in *P. tetraurelia*. Mating types O and E follow a cytoplasmic pattern of inheritance, but are determined by the differentiated macronuclei. Maternal inheritance results from the recurrent programming of the zygotic macronucleus by the maternal macronucleus during development. Adapted with permission from Figure 2 in E. Meyer and S. Duharcourt (1996), *J. Eukaryot. Microbiol.*, **43**(6):453–461.

exconjugants in Figure 10.2A), any phenotypic difference between F1 clones is indicative of non-Mendelian effects. In *Paramecium*, there is almost no exchange of cytoplasm between conjugating cells under normal conditions. This makes the system ideally suited for the detection of cytoplasmically inherited characters, such as those encoded in mitochondria. However, a cytoplasmic pattern of inheritance can also reflect maternal effects on zygotic development, exerted through the cytoplasm by the maternal macronucleus (see Figure 10.2B and Section I.D). Note that ciliates are hermaphrodites; "maternal" refers here to the cytoplasmic parent of each progeny cell. Finally, genetic analyses are greatly facilitated in *Paramecium* by the occurrence of autogamy, as this process generates an entirely homozygous genotype in just one sexual generation. F2 individuals obtained by autogamy of a heterozygous F1 clone have a 50% chance of receiving each parental allele, which translates into a characteristic 1:1 segregation of characters in the F2 population (Figure 10.2A).

C. Developmental genome rearrangements

Developmentally regulated rearrangements occur in a variety of other eukaryotes, but rarely on a scale as massive as that seen in ciliates (for reviews, see Coyne et al., 1996; Klobutcher and Herrick, 1997; Prescott, 1994; Yao et al., 2002). During the development of a new macronucleus, the genome is amplified to a high ploidy level ($\sim$50n in *Tetrahymena*, $\sim$1000n in *P. tetraurelia*), and rearranged at a large number of sites ranging from 6000 in *Tetrahymena* to perhaps as many as 100,000 in hypotrichous ciliates. The process results in the elimination of a sizable fraction of the genome, which varies from 10–15% to 95% in different species. Virtually all transposable elements and repeated sequences belong to the eliminated fraction. A recent pilot survey of the *P. tetraurelia* macronuclear genome ($\sim$10^8 bp) revealed a very high gene density (Dessen et al., 2001), leading to an estimate of at least 20,000 genes and possibly many more. As a rule, rearrangements are highly reproducible from one sexual generation to the next. They occur at specific time periods, after some initial amplification of the diploid genome but before the final ploidy level is reached. Although little is known about the molecular mechanisms involved, a link between heterochromatin formation and DNA elimination is suggested by the identification of two chromodomain proteins that are required for rearrangements in *Tetrahymena*. There appears to be significant differences in mechanistic details among ciliates; here we shall limit ourselves to a brief overview of the two main types of rearrangements that are observed in all species.

The first type is the precise excision of internal eliminated sequences (IESs). Both coding and noncoding regions of the micronuclear genome contain various sequence elements that reside between two short direct repeats and are deleted during macronuclear development. Flanking sequences are rejoined with high precision, leaving one of the repeats in the macronuclear sequence. The excision of each IES usually results in a unique junction sequence in the mature macronucleus, although a reproducible set of alternative junctions have been described in *Tetrahymena* and for a few *Paramecium* IESs. IESs can be either relatively large transposable elements, mostly Tc1/mariner family members with copy numbers ranging from dozens to tens of thousands per haploid genome, or short, single-copy, noncoding elements that are also very numerous (about 50,000 different elements in the *Paramecium* haploid genome). One class of single-copy IESs, called TA IESs because they are invariably bounded by two 5′-TA-3′ repeats, is present in widely divergent ciliates. Although they contain no other invariant sequence motif, a short, loosely conserved consensus is present at their ends, adjacent to the TA boundaries. Point mutations at some consensus positions, or in the TA repeats, were shown to abolish excision. The resemblance of this consensus to the extremities of inverted repeats of Tc1/mariner transposons has led to the hypothesis that single-copy IESs represent ancient, degenerate copies of such transposons that are flanked by the duplication of a TA integration site. Developmental excision of both transposon-like and TA IESs was shown to generate covalently closed circular molecules with a species-specific junction structure.

The second type of rearrangement observed in all species is the fragmentation of germline chromosomes into smaller molecules, which are healed by *de novo* telomere addition. The average size of these "macronuclear chromosomes" varies considerably among species, from ~600 kb in *Tetrahymena thermophila* to about 2 kb in hypotrichous ciliates such as *Euplotes crassus*, in which most macronuclear chromosomes contain a single gene. Mechanisms for fragmentation also appear to be quite diverse. In both *Tetrahymena* and *Euplotes*, the process is directed by short *cis*-acting sequences, but only in the latter are telomeric repeats added at a fixed number of nucleotides from the *cis*-acting sequence element. In contrast, no putative *cis*-acting consensus sequence has been identified in *Paramecium aurelia* species. In these species, the process appears to be associated with the elimination of significant lengths of germline-limited sequences around fragmentation sites. Macronuclear telomeres are formed by addition of telomeric repeats at multiple random positions within regions that are 0.8–2 kb in length. Although these telomere addition regions do not contain any conserved sequences, they are reproducibly used each time a new macronucleus develops. Additional heterogeneity is seen for some macronuclear chromosome ends, which have a choice of several alternative telomere addition regions, spaced 2–20 kb apart. All of this heterogeneity is observed within each developing macronucleus. Furthermore, some fragmentation events are facultative: in these cases the deletion of germline sequences can lead either to chromosome fragmentation or to the joining of the sequences flanking the eliminated segment. These internal deletions differ from IES excision by their larger size and the variability of breakpoints (A. Le Mouël *et al.*, unpublished). In two cases where they have been examined, the germline sequences that were eliminated at fragmentation sites were found to contain different Tc1/mariner transposons and, for one of them, a 69-bp minisatellite (A. Le Mouël, unpublished; O. Garnier, unpublished). As both transposable elements and satellites are commonly associated with heterochromatin in eukaryotic genomes, this finding suggests that chromosome fragmentation in *P. aurelia* may result from the elimination of heterochromatin through a developmentally regulated, non-sequence-specific mechanism.

D. Mating type determination in *P. tetraurelia:* maternal inheritance of developmental alternatives

More than 60 years after the discovery of mating types in *P. aurelia*, the molecular basis for the expression of alternative types is still unknown. Nevertheless, a considerable sum of experimental observations has led to a detailed description of the developmental process of mating type determination. The summary given below shows that the maternal macronucleus can direct differentiation of the zygotic macronucleus for stable alternative characters, providing an interesting precedent for the maternal homology-dependent effects presented in Section II (for reviews, see Sonneborn, 1974, 1977).

Each species of the *P. aurelia* group has two complementary mating types, O and E. Both types can be produced in entirely homozygous wild-type strains, indicating that they are not determined by genotypic differences. Rather, they correspond to alternative differentiated states of the macronucleus. Mating type is determined in an irreversible manner during a critical period of macronuclear development and remains unchanged throughout vegetative growth, until the macronucleus is lost at the following conjugation or autogamy and replaced by a new one.

Of special interest here is the cytoplasmic pattern of inheritance of mating types at sexual reproduction, a feature that is observed in roughly half of the *P. aurelia* species, including *P. tetraurelia* (see Figure 10.2B). After conjugation of an O cell with an E cell, the two new macronuclei developing in the cytoplasm of the O parent are almost always determined for O, while those developing in the E parent cell are almost always determined for E (Nanney, 1957). Likewise, autogamy produces a zygotic macronucleus that is determined for the same mating type as the maternal macronucleus. Numerous experiments, including the transfer of cytoplasm between cells at the appropriate stage of development, have shown that the E maternal macronucleus produces a cytoplasmic factor that causes the zygotic macronucleus to be determined for E, hence the capacity to produce, in turn, the same factor at the following sexual event. This E-determining factor was shown to act only at a specific stage of macronuclear development and to have no effect on mature macronuclei. As no evidence for an O-determining cytoplasmic factor was obtained, mating type O appears to be the default developmental alternative. Although recurrent determination of the zygotic macronucleus by the maternal macronucleus gives a cytoplasmic pattern of inheritance, this is more properly called maternal inheritance, to distinguish it from true cytoplasmic inheritance.

The timing and irreversibility of mating type determination suggest that it could be achieved during macronuclear development by an alternative rearrangement of the genome. Some support for this possibility came from studies of the only Mendelian mutation known to affect mating type determination in *P. tetraurelia*, the *mtF^E* mutation. All other mating type mutations restrict homozygotes to the *expression* of type O during sexual reactivity, but do not affect *determination* (Brygoo, 1977; Byrne, 1973). In contrast, *mtF^E* is a pleiotropic mutation which affects the process of macronuclear development, making determination for E constitutive in homozygotes, regardless of the maternal mating type (Brygoo and Keller, 1981a), and also causing stable differentiation for several unrelated mutant characters (Brygoo and Keller, 1981b). Among *mtF^E*-induced developmental abnormalities, one was shown to be the failure to excise an IES located in the G gene, which encodes a nonessential surface antigen, resulting in a nonfunctional form of the gene in the macronucleus (Meyer and Keller, 1996). This suggests that the *mtF* gene product is involved in the excision of a subset of germline IESs, and that the *mtF^E* pleiotropy reflects rearrangement defects at a number of unrelated loci. Constitutive determination for mating type E may thus result from the inability to rearrange a putative gene controlling mating type.

If type O is indeed a default state determined by the excision of an IES from a mating type gene, could the maternal E-determining cytoplasmic factor specifically block excision of this IES in wild-type cells? Again, indirect evidence was obtained through the study of the *mtF^E* pleiotropy. Indeed, like mating type E, some (possibly all) of the *mtF^E*-induced characters can be maternally transmitted to sexual progeny, even when the wild-type *mtF* allele is reintroduced in the cytoplasmic lineage by conjugation and made homozygous by autogamy (Brygoo and Keller, 1981b). These mutant characters are further inherited in subsequent generations but can revert with relatively high frequencies at each autogamy and revert independently from each other. Thus, they appear to be determined by epigenetic alterations arising during macronuclear development, each of which is self-reproduced from maternal to zygotic macronucleus. A striking example, which can be described in molecular terms, is the retention of the G gene IES in the macronuclear genome. Once induced by the *mtF^E* mutation, the character is stably maintained through an indefinite number of autogamies in genetically wild-type cells (Meyer and Keller, 1996). Like mating types, it is also maternally inherited at conjugation (Duharcourt *et al.*, 1995). The next section shows that this results from a homology-dependent control of IES excision that is exerted by the maternal macronucleus.

II. REGULATION OF IES EXCISION BY HOMOLOGY-DEPENDENT MATERNAL EFFECTS

A. An IES inhibiting its own excision in *P. tetraurelia*

The genetic analysis presented above showed that wild-type *P. tetraurelia* cell lines can be induced to retain a particular IES located in the G gene during macronuclear development. The IES$^+$ (retention) and IES$^-$ (excision) states of the macronucleus are alternative characters that show maternal inheritance in breeding analyses, confirming that they are not attributable to any genotypic difference. If developmental excision of the IES in the zygotic macronucleus is controlled by the maternal macronucleus, it could either be induced by maternal copies of the correctly rearranged G gene, or be inhibited by maternal copies of the IES-retaining G gene. To answer this question, the macronucleus of IES$^+$ or IES$^-$ cells was transformed by direct microinjection of plasmids containing a fragment of the G coding sequence in either its micronuclear (IES$^+$) or macronuclear (IES$^-$) versions (Duharcourt *et al.*, 1995). After being introduced into the *Paramecium* macronucleus, any DNA molecule above ~4 kb can be stably maintained at a wide range of copy numbers during vegetative growth, replicating autonomously as extrachromosomal monomers and multimers after *de novo* addition of telomeres. Transformed clones were grown and allowed to undergo autogamy, which leads to the loss of the transformed maternal macronucleus, and excision of the IES was examined in the new macronucleus of sexual progeny. While the IES$^-$ plasmid proved unable to induce excision in the progeny of transformed IES$^+$ cells, the IES$^+$ plasmid resulted in the retention of the IES in the progeny of transformed IES$^-$ cells (Figure 10.3).

    The effect of the IES$^+$ plasmid was clearly dependent on its copy number in the maternal macronucleus: low copy numbers resulted in a partial effect, i.e., the retention of the IES on only a fraction of the ~1000 copies of the genome in the postautogamous macronucleus. The IES sequence itself is responsible for the phenomenon, since the IES$^-$ plasmid, which contained the same length of flanking sequences from the G gene, had no detectable effect, even at high copy numbers. Furthermore, a plasmid containing only the IES, but no flanking sequences, also caused IES retention in sexual progeny from transformed cells, although with a reduced efficiency (see Section II.B). In contrast, an internal deletion removing two-thirds of the 222-bp IES sequence completely abolished the effect.

    The IES was still retained after additional autogamies of first-generation IES$^+$ cells, showing that injection of the IES is sufficient to turn the wild-type cell line into a permanent IES$^+$ cell line. Inspection of a few other IESs in the same gene or in other genes showed them to be correctly excised. Furthermore, IES$^+$ cells have no obvious phenotype, indicating that the vast majority of germline

A. IES⁻ cell line

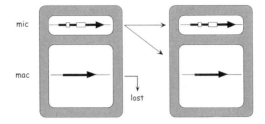

B. IES⁺ cell line

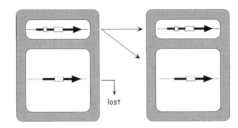

C. IES⁻ cell line transformed with IES⁺ plasmid

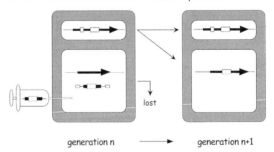

generation n ⟶ generation n+1

Figure 10.3. Maternal inhibition of IES excision. (A) In the IES⁻ (wild-type) cell line, all IESs (open boxes) are removed from the G gene (arrow) during macronuclear differentiation. (B) In the IES⁺ cell line, developmental excision of the 222-bp IES is specifically inhibited, although the germline genome is entirely wild type. (C) Transformation of the IES⁻ macronucleus with a plasmid containing the 222-bp IES specifically inhibits excision of this IES after autogamy. Adapted from Figure 2 in *Cell* **87,** Meyer, E., and Duharcourt, S., Epigenetic programming of developmental genome rearrangements in ciliates, pp. 9–12, Copyright © 1996, with permission from Elsevier Science.

IESs are not affected. Thus, when present in the maternal macronucleus, this IES specifically causes the retention of the homologous germline IES during development of a new macronucleus.

B. A homology-dependent maternal effect

Do other IESs show similar maternal effects? Another study addressed this question by transforming the macronucleus of wild-type cells with large segments of micronuclear DNA containing the G gene or another nonessential surface antigen gene, A, belonging to the same multigene family (Duharcourt *et al.*, 1998). The injected segments contained 6 and 9 IESs, respectively. Thirteen of these 15 IESs were examined in the postautogamous progeny of transformed cells, and five were found to be retained on part or all of the copies of the new macronuclear genomes. Specificity between the two genes was complete: each micronuclear gene induced the retention of some of the IESs located in its own homologous copy (two for G, three for A), but did not affect any of the IESs in the other gene. For three of these five IESs, the test of plasmids containing only one IES showed that each of them acts only on itself and does not affect other IESs in the same gene. A control DNA fragment containing most of the macronuclear G gene had no effect on any of the IESs tested. The sequence of IESs retained in the macronuclear genome was found to be identical to the germline sequence, although two of the IESs contain smaller, internal IESs that were fully excised from the new macronuclear copies. These small internal IESs are also excised when excision of the larger IESs is blocked by point mutations at their boundaries, rather than by the maternal epigenetic effect (Mayer and Forney, 1999; Mayer *et al.*, 1998).

Quantitative analyses showed that the fraction of copies of the new macronuclear genome that retain an IES increases with the copy number of the IES-containing DNA fragment in the maternal macronucleus. However, different IES sequences were clearly retained with different efficiencies. In the most efficient case, 50% retention was obtained with a maternal copy number of ~0.4 copies per haploid genome (cphg), i.e., ~400 copies per macronucleus, and 100% retention with ~1 cphg; in the least efficient case, 50% retention was not reached with ~6 cphg. For a given IES, the efficiency was further shown to depend on the length of flanking sequences linked to the IES sequence carried by the DNA fragment in the maternal macronucleus. While 50% retention, in the most efficient case, required ~40 cphg of a plasmid containing only the IES sequence, this level was obtained with ~3 cphg of a plasmid that also contained some 600 bp of flanking sequences, and with only ~0.4 cphg of the 12-kb DNA fragment containing the entire micronuclear gene. The latter value is close to the maximum efficiency (~0.3 cphg in this case), which was observed for IES-retaining endogenous chromosomes in the maternal macronucleus.

These quantitative analyses were performed after mass cultivation of pools of 40 different caryonidal clones (clones deriving from single events of macronuclear development) picked from the postautogamous progeny of each transformed clone. Although the *average* fraction of new macronuclear copies retaining the IES is a fairly regular function of the maternal IES copy number, a great variability was observed among individual caryonidal clones deriving from the same transformed clone. Caryonidal clones showing 0% or 100% retention were common in mass populations giving an average figure of 50%, although individual clones showing partial retention were also found. Thus IES retention in each developing macronucleus appears to be a stochastic event, the probability of which is determined by the copy number of the IES in the maternal macronucleus.

C. Epigenetic inhibition of IES excision in *Tetrahymena thermophila*

A very similar phenomenon has been described in *Tetrahymena thermophila* (Chalker and Yao, 1996). Transformation of the vegetative macronucleus with either one of two IESs that are closely linked in the germline genome, the M and R elements, resulted in the retention of the homologous IES in the macronuclear genome of sexual progeny after conjugation of transformed clones. Here again the effect was quite specific, although parental copies of the M element were also found to interfere weakly with the excision of R. In addition, the presence of IES-bearing constructs in the parental macronucleus appeared to increase the frequency of aberrant excision events (i.e., deletions using novel endpoints), which are rarely seen in the wild type. As was observed in *Paramecium*, IES sequences alone were sufficient for the effect, and flanking macronuclear sequences had no effect at all. Conjugation of IES-retaining F1 progeny yielded F2 clones that still retained the IES in their macronuclear genome, showing that IES retention had become a non-Mendelian, heritable trait.

In this study, the fraction of IES-retaining copies in the new macronucleus did not appear to correlate with the copy number of IES-bearing constructs in the parental macronucleus. As the *Tetrahymena* macronucleus does not have the capacity to maintain introduced DNA fragments as autonomously replicating molecules, macronuclear transformation relies on an rDNA-based vector which allows the replacement of the endogenous rDNA minichromosome by a modified version carrying the sequence of interest. Introduced IESs, therefore, had the same high copy number as the rDNA, which is overamplified during macronuclear development and then maintained at ~180 cphg, or ~9000 copies per macronucleus. Such a high copy number did not result in complete IES retention in the developing macronucleus. As in the *Paramecium* effect, the level of IES retention was very variable, even among different F1 progeny of the same transformed clones, but most of these progeny contained a substantial fraction of correctly rearranged copies. Although IES-retaining F1 clones had at most 50 copies of the IES in their macronucleus (the normal macronuclear ploidy), the same variability and incomplete retention was also observed in the F2 generation. No correlation was apparent between IES copy numbers in the various F1 and derived F2 macronuclei. However, an average correlation could have been masked by the large stochastic component of the effect, as only a few F2 progeny were analysed for each F1. Since the efficiencies of the various *Paramecium* IESs to cause sequence retention were shown to depend on their sequence context, the efficiency of rDNA-linked IES copies in *Tetrahymena* may also be inherently smaller than the efficiency of chromosomal copies of the same elements.

One interesting difference between the *Tetrahymena* and *Paramecium* effects is that conjugation of IES-transformed cells with normal cells results in IES retention in the progeny of *both* mates in *Tetrahymena*, whereas the IES^+ and IES^- characters show maternal inheritance in *Paramecium*. This may be related to the fact that conjugating cells form a cytoplasmic bridge large enough for complete cytoplasmic mixing to occur before the two conjugants separate in *Tetrahymena*, but not in *Paramecium*. The cytoplasmic "dominance" observed in *Tetrahymena* therefore confirms that the effect of the IES-containing parental macronucleus on the developing zygotic macronucleus is mediated through the cytoplasm.

D. Models

Although they both belong to the class Oligohymenophorea, *Paramecium* and *Tetrahymena* are fairly distant evolutionarily (Tourancheau *et al.*, 1998). Their IESs show a number of differences: *Tetrahymena* IESs are bounded by short direct repeats that are variable in sequence, as compared to the invariant *Paramecium* TA, and their excision appears frequently to use a discrete set of alternative boundaries; they have never been found within coding sequences; they are fewer and longer than *Paramecium* IESs. Nevertheless, the homology-dependent maternal effects leading to the retention of specific zygotic IESs in these two species are remarkably similar, suggesting the conservation of an ancient mechanism, which may also exist in other ciliates.

Among the 13 *Paramecium* IESs tested, there is no obvious difference in size, base composition, or position within the genes between the five that show the maternal effect and the eight that do not. Four of the five IESs showing the effect begin with the sequence 5′-TATT . . . -3′ at both ends, where the T at the 4th position is a rare variant in the general consensus 5′-TAYAGYNR . . . -3′ (Duharcourt *et al.*, 1998). However the significance of this deviation is unclear, as the 5th one has very standard ends. The phenomenon is not limited to IESs located in surface antigen genes, as one of those showing the effect is inserted upstream of the gene's promoter. Furthermore, two of two *Tetrahymena* IESs tested showed the effect, suggesting it may be observed for a large number of other IESs in both genomes.

The most puzzling aspect of this *trans*-nuclear effect is its sequence specificity. The limits of this specificity have not been fully explored, but minor sequence differences, such as a few mutations present in allelic sequences, or the deletion of 28-bp and 29-bp internal IESs in two 370-bp IESs, do not hinder recognition of homologous target sequences (Duharcourt *et al.*, 1998). How can a sequence introduced in one nucleus specifically affect the excision of the homologous sequence in another nucleus? Two different types of models have been proposed. In the first type, IES copies in the maternal macronucleus act by sequestering sequence-specific protein factors that are required for excision in the developing macronucleus. Because it is highly unlikely that each IES requires a different factor, it was proposed, on the basis of the *Tetrahymena* study, that different IESs use different subsets of a smaller number of factors, which could explain the weak effect of the M element on excision of R. However, this model still requires an unreasonably large number of different factors if, as suggested by the *Paramecium* study, a significant fraction of the ~50,000 IESs in the genome can inhibit their own excision with high specificity. Furthermore, such factors would have to bind the inner portion of IES sequences, since constructs carrying only the consensus-bearing extremities, or flanking macronuclear sequences, have no effect by themselves. This is difficult to reconcile with the higher efficiency of molecules containing longer flanking sequences, and with the rapid, apparently unconstrained, evolution of the inner part of IES sequences (Scott *et al.*, 1994a).

In the second type of model, sequence specificity is achieved by pairing interactions between homologous nucleic acids. However, the parental macronucleus, or its fragments, have never been observed to fuse with the developing macronucleus, and there is ample evidence that its DNA never gets incorporated into the new macronucleus. Indeed, the well-documented Mendelian segregation of alleles during autogamy of heterozygotes shows that only one of the two maternal alleles is present in the zygotic macronucleus; similarly, injected exogenous sequences, such as plasmid vectors, are maintained in the maternal macronucleus but cannot be recovered from sexual progeny. Thus, one has to assume that IES copies, presumably RNA molecules, are exported from the maternal macronucleus to the developing macronucleus, where they could pair with homologous sequences of the zygotic genome. The stronger effect of longer IES flanking sequences in the constructs may then be explained by differences in pairing efficiency. In support of this model, copies of the M and R elements present in the maternal macronuclear chromosomes were recently shown to be transcribed during conjugation in *Tetrahymena* (Chalker and Yao, 2001). This nongenic transcription appears to be determined by internal promoter sequences, since maternal macronuclear flanking sequences are not detectably transcribed when IESs are excised from the chromosomes. IES-driven transcription of chromosomal flanking sequences, however, might somehow be responsible for the weak effect of M on the excision of R, as these elements are separated only by 2.7 kb. In *Paramecium*, transcription of maternal IESs has not been tested, but the maternal macronucleus is known to be transcriptionally active throughout the development of the new macronucleus. Maternal transcription level and RNA stability might certainly be some of the factors that determine the characteristic inhibition efficiency of each IES.

One version of the pairing model postulated that IESs are in fact constitutively excised from zygotic chromosomes during macronuclear development, leaving a gap that is repaired by polymerization with a homologous template originating from the maternal macronucleus. Since IES excision involves DNA cleavage at the ends of the element in *Tetrahymena* (Saveliev and Cox, 1995) as well as in *Paramecium* (M. Bétermier and A. Gratias, personal communication), IESs could be resynthesized in the new macronucleus by copying maternal IES-containing templates. It should be noted that such a mechanism is not likely to account for the formation of correct macronuclear junctions after IES excision, as these junctions were observed to form spontaneously after autogamy of cells that were completely devoid of junction templates: Indeed, IES+ *Paramecium* cell lines can be made to return to normal IES excision when the copy number of the IES-containing chromosomal region is first reduced in the maternal macronucleus, through the experimental induction of macronuclear deletions (see Section III.C), to the point where it is no longer sufficient to cause IES retention on all copies of the zygotic macronucleus (Duharcourt *et al.*, 1995). Template-independent formation of macronuclear junctions is also consistent with the reported lack of maternal transcription of these junctions in *Tetrahymena* (Chalker and Yao, 2001). The remaining possibility, that templated repair is involved in cases of IES *retention*, was disproved by injecting IES sequences that were marked by the introduction of a restriction site or by allelic mutations: IESs retained in the new macronuclear genome were copied from the germline genome and not from maternal copies (Duharcourt *et al.*, 1995, 1998).

Thus, the hypothesized pairing interactions apparently cause IES retention by inhibiting endonucleolytic cleavage or recognition of IES sequences. Direct protection of zygotic IESs from the excision machinery by the pairing of maternal transcripts is unlikely, as one would then expect small internal IESs to be retained concomitantly with the larger ones in which they are located. Maternal transcripts may instead induce some epigenetic modification of germline IESs, which in turn would block excision. The modification of specific nucleotides or dinucleotides could affect the excision of each IES to a different extent, depending on the precise location of modified sites relative to IES boundaries. Such a highly localized modification could allow small internal IESs to be excised even when excision of the larger ones in which they reside is inhibited. N6-methyladenine is known to occur in *Tetrahymena* and *Paramecium* (Cummings *et al.*, 1974; Gorovsky *et al.*, 1973; Karrer and VanNuland, 1998), but no role for this modification in the developmental regulation of genome rearrangements has yet been demonstrated.

Other possible mechanisms are suggested by the recent finding that the zygotic M and R IESs, as well as other *Tetrahymena* germline-limited elements, are transcribed at an early stage of macronuclear development, well before they are excised (Chalker and Yao, 2001). Transcription of the M element was shown to occur from both strands, giving rise to nonpolyadenylated transcripts that are heterogeneous in size, at least a fraction of which originate or terminate in flanking sequences. Brief treatment of cells with actinomycin D during early macronuclear development resulted in impaired IES excision, leading the authors to propose that transcription of zygotic IESs is necessary for their excision, either because it establishes a chromatin structure that allows access to the excision machinery, or because the transcripts themselves are used to guide the excision machinery to the IESs. In both cases, the presence of IES copies in the parental macronucleus could inhibit excision by mechanisms akin to homology-dependent gene silencing. Parental IES transcripts (which appear to be qualitatively different from zygotic transcripts, although they have not been fully characterized) could be responsible for the transcriptional silencing of zygotic IESs, which would reduce their accessibility; alternatively, parental transcripts could cause the degradation of zygotic transcripts by posttranscriptional mechanisms, which would affect the recognition of IESs to be excised.

Whatever the mechanism may be, the homology-dependent effect through which a maternal IES can inhibit its own excision in the zygotic macronucleus provides cells with a simple way to transmit a variety of alternative macronuclear versions of the genome to sexual progeny. For one of the *Paramecium* IESs studied (51G4404), the efficiency of the maternal effect is such that the presence of the IES on a given fraction of copies of the macronuclear genome will result in a *greater* fraction of IES-retaining copies in the macronucleus of the following sexual generation (Duharcourt *et al.*, 1995). The dynamics of the system thus imply that, over the course of many sexual generations in a cytoplasmic lineage, a stable equilibrium can be reached only for two alternative states, IES$^-$ (100% excision) and IES$^+$ (0% excision). This is reminiscent of the stable maternal inheritance of mating types O and E in *P. tetraurelia*. The same fundamental asymetry is seen in both systems: IES excision and determination for O both appear to be default developmental pathways, while IES retention and determination for E both require a specific cytoplasmic signal produced by the maternal macronucleus. The hypothesis that mating type is determined by the regulated excision of an IES is further supported by the pleiotropic effects of the Mendelian mutation mtF^E, which impairs excision of IES 51G4404 regardless of the IES$^-$ or IES$^+$ state of the maternal macronucleus, and similarly makes determination for E constitutive.

III. REGULATION OF ZYGOTIC GENOME AMPLIFICATION AND CHROMOSOME FRAGMENTATION BY HOMOLOGY-DEPENDENT MATERNAL EFFECTS

A. Maternal inheritance of alternative chromosome fragmentation patterns

The first evidence for an epigenetic regulation of the other type of developmental genome rearrangement, the fragmentation of germline chromosomes, came from the study of d48, a variant cell line of *P. tetraurelia*. Originally isolated in a screen for mutants unable to express the A surface antigen, the d48 cell line indeed showed this defect, but genetic analyses surprisingly indicated that it did not carry any germline mutation. Although the A expression defect was found to be due to the complete absence of the A gene in the macronucleus, the defect followed a cytoplasmic, rather than Mendelian, pattern of inheritance in crosses with the wild type (Epstein and Forney, 1984). Nuclear transplantation experiments directly confirmed that the d48 micronuclear genome is entirely wild type, and that vegetative micronuclei are not determined in any way for that character: the replacement of the d48 vegetative micronucleus with a wild-type micronucleus did not prevent maternal transmission of the A gene deletion to sexual progeny, and the d48 micronucleus, when transplanted into a wild-type cell, gave rise after autogamy to a new macronucleus which contained the A gene (Harumoto, 1986; Kobayashi and Koizumi, 1990).

In contrast, injection of wild-type vegetative macronucleoplasm into the d48 vegetative macronucleus was shown to rescue the A gene defect in the injected clone, as well as in its postautogamous progeny, thus causing a permanent reversion of d48 to wild type (Harumoto, 1986). The cytoplasm from wild-type cells, when transferred to d48 cells at an early stage of macronuclear development, also proved able to prevent deletion of the A gene, but only when it was taken from donor cells that were themselves undergoing autogamy (Koizumi and Kobayashi, 1989). In the absence of any reported evidence for a d48-determining cytoplasmic factor, these experiments collectively suggest that the A gene is deleted during development of the d48 macronucleus as a consequence of the lack of a cytoplasmic factor transiently produced by the wild-type, A-containing maternal macronucleus, which determines the maintenance of the gene in the new macronucleus.

In the wild type, the A gene is located near the end of a macronuclear chromosome; developmental chromosome fragmentation results in the addition of telomeres in one of three alternative regions located 8, 13, and 26 kb downstream of the gene, all three forms being represented in each macronucleus. Molecular characterization of the d48 macronuclear deletion revealed that the gene is lost as part of a larger terminal deletion that also removes all downstream sequences, the d48 telomere forming in a single region located at the 5′ end of the gene

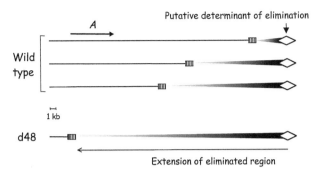

Figure 10.4. The macronuclear A-gene locus in the wild-type and d48 cell lines. The A-gene coding sequence is shown as an arrow. Hatched boxes represent the telomere addition regions. The open lozenge on the right symbolizes a putative germline sequence element determining elimination of this region during macronuclear development.

(Figure 10.4) (Forney and Blackburn, 1988). The d48 cell line thus provides evidence that alternative chromosome fragmentation patterns can be maternally inherited in genetically wild-type cells. However, it should be noted that the alternative pattern is not simply the result of chromosome breakage occurring upstream of the gene instead of at one of the wild-type downstream positions, as one would then expect the A gene to be present on a different macronuclear chromosome. Even if breakage occurred at both upstream and downstream positions, the gene would be present on 13- to 34-kb molecules, which should be long enough to be maintained at normal copy numbers in the macronucleus.

An alternative view is that chromosome fragmentation results from the imprecise elimination of a variable length of downstream sequences, which can be extended to include the A gene in d48. If elimination is determined by a heterochromatin-like structure of germline-limited sequences, the wild-type telomere addition regions could represent alternative boundaries of heterochromatin spreading from downstream sequences toward the A gene. In d48, heterochromatin would spontaneously spread further to include the gene, perhaps because of the attractive presence of four nearly identical 210-bp tandem repeats in the middle of its coding sequence. In support of the gene's intrinsic propensity to be deleted, the use of upstream probes has recently revealed that the wild-type macronucleus does in fact contain, in addition to the forms already described, a limited number of copies of a shorter chromosome form which lacks the gene, ending with telomeres in the d48 region (unpublished results). In this view, the maternal cytoplasmic factor that is necessary for amplification of the gene to wild-type copy numbers would act by *preventing* the spontaneous formation of heterochromatin over the gene.

The d48 alternative fragmentation pattern thus appears to result from a maternal effect that primarily controls the amplification or deletion of the A-gene genomic region. Can such maternal effects be evidenced in other genomic regions? Scott *et al.* (1994b) have addressed this question by taking advantage of a mutant strain carrying a Mendelian deletion of the B gene, another surface antigen gene that is located close to the end of a macronuclear chromosome in the wild type. In this strain the B gene is completely absent from the micronuclei, and therefore is also missing in the macronucleus. The mutant micronuclei were replaced with a wild-type micronucleus, resulting in a cell line analogous to d48, which contained the B gene in its micronucleus but not in its macronucleus. When transplanted clones were allowed to undergo autogamy, the wild-type micronucleus gave rise to a new macronucleus in which the B gene was deleted. Like the d48 deletion, this macronuclear deletion was maternally transmitted to subsequent sexual generations. Because there is no reason to suspect that the macronucleus of the Mendelian mutant lacked anything else than the B gene, this experiment suggests that the B gene is amplified to a wild-type copy number in the developing macronucleus only if it is already present in the maternal macronucleus.

B. Homology-dependent rescue of maternally inherited macronuclear deletions

If amplification of the A gene in the developing macronucleus is determined by its presence in the maternal macronucleus, it can be assumed that the wild-type cytoplasmic factor that allows zygotic amplification is produced by the maternal A-gene copies themselves. Indeed, transformation of the d48 vegetative macronucleus with the cloned A gene, followed by induction of autogamy, was shown to restore A-gene amplification in the developing macronucleus of sexual progeny, thus resulting in a permanent rescue of the d48 defect (Figure 10.5) (Jessop-Murray *et al.*, 1991; Koizumi and Kobayashi, 1989; You *et al.*, 1991). However, the cytoplasmic factor cannot be the protein product of the gene, nor its full-length mRNA, as even truncated copies of the gene were able to rescue the d48 defect when present in the maternal macronucleus. Furthermore, clones transformed with the entire gene often express the A surface antigen throughout vegetative growth, whereas production of the rescuing cytoplasmic factor was shown to be restricted to the period of nuclear reorganization.

Two studies have attempted to identify specific regions of the A gene that are responsible for the maternal rescue effect by transforming the d48 macronucleus with different subfragments, and then testing the capacity of postautogamous progeny to express the A gene, which requires its presence in the new macronucleus (Kim *et al.*, 1994; You *et al.*, 1994). Rescuing activity appears to be spread over most of the ~8-kb coding sequence, as several nonoverlapping fragments showed the effect. In contrast, the G gene from *P. primaurelia*, which is 78% identical to A over the coding sequence, had no rescue activity. Quantitative measures of the rescue efficiency, based on the fraction of progeny cells that

A. Wild-type macronucleus

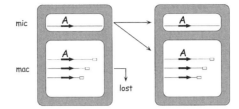

B. d48 macronucleus

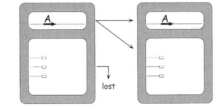

C. Transformed d48 macronucleus

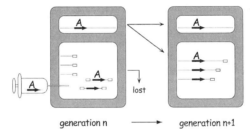

generation n   ⟶   generation n+1

Figure 10.5. Developmental processing of the A gene in *P. tetraurelia* during autogamy. (A) Wild-type cell line. Open boxes represent telomere addition regions. (B) d48 cell line. Although the germline genome is wildtype, the macronuclear genome shows a terminal deletion of the A gene. (C) Transformation of the d48 macronucleus with an A-gene plasmid rescues the macronuclear deletion after autogamy. Adapted from Figure 1 in *Cell* **87**, Meyer, E., and Duharcourt, S., Epigenetic programming of developmental genome rearrangements in ciliates, pp. 9–12, Copyright © 1996, with permission from Elsevier Science.

could be induced to express the A gene, indicated that different fragments have different efficiencies, but conflicting results were obtained for some fragments. Interestingly, longer fragments were generally more efficient, and co-transformation with two different fragments resulted in a better rescue than each fragment alone. Constructs that contained the gene's promoter appeared to have a poor rescue efficiency. However, these measures should be taken with caution, as the ability of sexual progeny to express antigen A is not a direct measure of A-gene amplification levels. Furthermore, the copy numbers of the different fragments in the transformed clones were not measured, and this was later shown to be a critical parameter in such maternal effects (see Section III.D).

The sequence specificity of the maternal rescue effect was examined using a genetically wild-type cell line that carried maternally inherited macronuclear deletions of both the A and B genes (Scott *et al.*, 1994b). These genes are members of the same multigene family and their coding sequences are 74% identical overall, although the central portion is much more divergent than distal portions. Transformation of the maternal macronucleus with the A gene restored amplification of the A gene in the developing macronucleus, but not that of the B gene; similarly, the B gene restored only its own amplification. On the other hand, the macronuclear deletion of the A gene could be rescued by transformation with a different allele of the A gene, showing 97% identity (Forney *et al.*, 1996). Thus, the maternal rescue of macronuclear deletions is a homology-dependent process which does not require any specific sequence within the genes, but requires a minimum level of sequence identity.

C. Homology-dependent induction of macronuclear deletions

Transformation of the maternal macronucleus can also have opposite consequences on the copy number of zygotic sequences in the new macronucleus. This opposite effect was first observed with the G surface antigen gene, which lies ~5 kb from the end of a macronuclear chromosome, in the sibling species *P. primaurelia*. When the wild-type macronucleus was transformed with a high copy number of the cloned G gene, autogamy of the transformed clones resulted in the complete deletion of the gene in the new macronucleus (Meyer, 1992). Like the d48 deletion, the induced deletion also removed downstream sequences up to the wild-type telomere, the new telomere addition region being located upstream of the gene and overlapping the 5′ end of the injected sequence (Figure 10.6). The induced deletion also showed maternal inheritance in conjugation with the wild type or after a second autogamy, although its heritability was weaker than that of the d48 deletion: many second-generation clones regained variable lengths of G-gene sequences on a fraction of macronuclear copies. Further work showed that similar terminal deletions can be obtained in sexual progeny by transforming the wild-type macronucleus with nonoverlapping subfragments of the G-gene region, indicating that the deletion effect does not depend on a particular sequence in that region (Meyer *et al.*, 1997). However, some differences were noted in the sizes of deletions induced by different fragments (see Figure 10.6).

The maternal deletion effect is not restricted to subtelomeric surface antigen genes, as it could be reproduced at two other macronuclear subtelomeric regions that did not contain such genes (unpublished results). Furthermore, transformation with an arbitrarily chosen fragment internal to a macronuclear

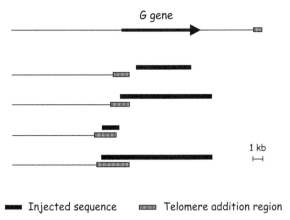

■ Injected sequence   ▥ Telomere addition region

Figure 10.6. Induced terminal deletions in the G-gene region. The top line shows the single telomere addition region (hatched box) downstream of the G gene in the wild-type macronucleus. Black boxes below represent the fragments injected into the maternal macronucleus, and the hatched boxes the telomere addition regions observed in the sexual progeny of transformed clones. Adapted with permission from Figure 6 in E. Meyer and S. Duharcourt (1996), *J. Eukaryot. Microbiol.* **43**(6):453–461.

chromosome, ~80 kb away from the nearest telomere, was shown to induce *internal* deletions of the homologous sequence in sexual progeny, albeit only on a fraction of new macronuclear copies (Meyer *et al.*, 1997). A number of macronuclear molecules carrying these internal deletions were cloned and sequenced to study the deletion endpoints. Alignment with the full-length macronuclear sequence, which is never deleted in the wild type, revealed that deletions occurred between short (3–12 bp) direct repeats, one of which was left in the macronuclear sequence. Strikingly, these repeats were made almost exclusively of alternating Ts and As. Twenty-four sequenced deletion endpoints were clustered in a few hotspots, located on either side of the sequence that had been injected into the maternal macronucleus. These hotspots were the same in several caryonidal clones derived independently from transformed clones, although they were mostly outside the injected sequence and were found to coincide precisely with short segments showing the highest density of 5′-TA-3′ dinucleotides. These TA-rich short segments themselves coincided with intergenic regions, which are very short in the *Paramecium* genome, so that individual genes appear to be the smallest segments that can be deleted. These results show that the maternal deletion effect does not necessarily lead to chromosome fragmentation. However, maternally induced deletions could be facilitated when the injected sequence is close to a wild-type macronuclear telomere, i.e., close to "spontaneously" eliminated germline sequences, in which case the two regions programmed for elimination appear to be bridged, resulting in a single deletion event removing the whole region. It is not known whether induced internal deletions would show the same maternal inheritance as terminal macronuclear deletions in subsequent sexual generations.

The surface antigen multigene family of *P. tetraurelia* was used to study different parameters that could affect the efficiency of the maternal deletion effect, such as the structure and copy number of the injected sequence and its level of sequence identity with the targeted zygotic sequence (O. Garnier, unpublished results). To this end, a ~10-kb fragment containing the A gene was injected at various copy numbers either as a circular or prelinearized plasmid, or as a linear fragment with *Paramecium* macronuclear telomeres at each end. By quantifying both the copy numbers of the injected A gene in the maternal macronucleus and the resulting A gene copy numbers in the macronucleus of postautogamous progeny, it was concluded that deletion efficiency is not strongly affected by the structure of injected sequences, but is highly dependent on their copy numbers. Under 20 cphg (~20,000 copies per macronucleus) in the maternal macronucleus, the injected sequence induced little or no reduction of the A-gene copy number in the zygotic macronucleus; above 40 cphg, all macronuclear copies of the A gene were deleted. The newly induced macronuclear deletion of the A gene showed the same stable maternal inheritance in further sexual generations as the original d48 deletion.

In the same experiments, zygotic amplification of paralogous genes G, B, and C was also examined in the sexual progeny of A-transformed clones. These genes respectively show 78%, 74%, and 55% of overall sequence identity with A, although identity levels can be locally higher in any pair. Like A, the G and B genes could be completely deleted with more than 40 cphg of the A gene in the maternal macronucleus, but in intermediate cases where the A gene was only partially deleted, they were less affected than A. Copy numbers of the more divergent C gene were not affected, even at much higher maternal A-gene copy numbers. Thus, the maternal deletion effect is a homology-dependent effect that is induced only at high copy numbers. The effect of the length of the injected sequence was not fully explored, but comparison with a previous analysis of the effects of short (213–851 bp) G-gene fragments (Duharcourt *et al.*, 1995) suggests that deletions are more efficiently induced by large fragments.

An unexpected link with the expression of the injected gene was also uncovered (O. Garnier, unpublished results). The quantitative analysis presented above used an allele of the A gene (A^{29}), which is rarely expressed in the transformed clones. Rare cases of A^{29}-transformed clones that did express the A antigen did not show the strong deletion effect that was expected from their A^{29} copy numbers. Similarly, the A^{51} allele, which is ~97% identical to A^{29} but is readily expressed after injection, consistently showed a much lower deletion efficiency. Constructs in which normal expression of the A^{51} allele was made impossible by the introduction of a frameshift in the coding sequence, or by the truncation of its 3' end, had an increased deletion efficiency, similar to that of the unexpressed

A^{29} allele. Thus, the stable expression of the injected gene appears to be incompatible with the maternal deletion effect. However, inactivating the A^{51} allele by removing its promoter did not significantly increase its deletion efficiency. Intriguingly, another promoterless construct containing an internal fragment of the A^{51} coding sequence also had a reduced deletion efficiency.

D. Resolving the paradox: rescue vs induction of macronuclear deletions

While both the rescue effect and the deletion effect clearly show that maternal macronuclear sequences can influence the copy number of homologous sequences in the developing macronucleus, it is unclear why transformation promotes amplification in some experiments and induces deletion in others. How can these conflicting observations be reconciled? It should first be pointed out that possible outcomes of these experiments are limited by their very design: in the d48 cell line, the only effect of transformation that *can* be observed is the restoration of amplification, but in the wild type it is deletion. Thus, if the same A-gene fragment were capable of inducing both effects, the conclusion drawn would simply depend on which recipient cell line was used. Furthermore, in both sets of experiments some of the tested fragments were found to be more efficient than others, and the most efficient fragments may not be the same for both effects. Although the parameters that determine the efficiencies of both effects are not fully understood, it is interesting to note that promoter-containing fragments of the A^{51} allele generally performed poorly in the d48 rescue, whereas they were the most efficient in inducing deletions in the wild type. Methodological differences also probably play a role. For instance, when the whole A gene was used in rescue experiments, transformants were usually screened on the basis of A gene expression, which involves a convenient immunological test. However, this would screen out any high-copy-number transformant likely to show a strong deletion effect in sexual progeny, as the deletion effect was shown to be incompatible with the expression of the injected gene.

The idea that copy number can influence the type of effect to occur was tested by transforming the macronucleus of a d48-like strain with a wide range of copy numbers of the rarely expressed A^{29} allele (O. Garnier, unpublished results). Very low copy numbers (0.1 cphg) were sufficient to induce a significant rescue of zygotic A gene amplification, and wild-type amplification levels were obtained for maternal copy numbers between 0.6 and 10 cphg. However, further increasing the maternal copy number resulted in a sharp decrease of zygotic copy numbers, and the A gene was again completely deleted in the new macronucleus above 40 cphg in the maternal macronucleus. This experiment showed conclusively that both effects can be induced by injecting the same fragment in the same cell line, and that they are determined primarily by the copy number of the injected sequence.

E. Models

Like the maternal effect leading to the inhibition of the excision of zygotic IESs, both of the homology-dependent maternal effects that modulate the copy number of zygotic sequences are *trans*-nuclear effects. For this reason, attempts to explain the sequence specificity of copy number modulation encounter precisely those problems also encountered by attempts to explain the sequence specificity of the IES effect. Considering the large number of different DNA fragments and subfragments that were shown to induce homology-dependent effects, models involving the production or titration of sequence-specific protein factors by sequences injected in the maternal macronucleus, resulting in a specific effect in the developing macronucleus, are very unlikely. Here again, we are left with the only alternative that sequence information is conveyed by maternal nucleic acids, which, as argued in the case of the IES effect, are presumably RNA molecules. This would imply that almost any DNA fragment can be transcribed, but this is not unreasonable, as a study of the vegetative homology-dependent gene silencing phenomenon (Ruiz *et al.*, 1998) has shown that promoterless coding sequences can indeed be transcribed after injection into the macronucleus.

In both effects, RNA copies of the sequences injected into the maternal macronucleus could exert their effects by pairing with homologous zygotic sequences in the developing macronucleus. It remains to be explained how these pairing interactions could either promote amplification or induce the deletion of the targeted sequence. In the rescue effect, very low copy numbers of maternal A gene sequences are sufficient to allow A-gene amplification. It is tempting to believe that RNA copies of the injected sequences are functionally equivalent to the cytoplasmic factor required for zygotic amplification of the A gene in wild-type cells, which appears to be produced by maternal A-gene copies. The timing of production of the wild-type cytoplasmic factor, which is limited to sexual events, indicates that it is distinct from the normal A-gene mRNA, which is known to be continuously produced during vegetative growth in A-expressing cells. A specialized transcription system would also be implied by the fact that normal A-gene amplification occurs in wild-type cells expressing a different surface antigen, as the expression of different surface antigen genes is mutually exclusive, and regulated primarily at the transcriptional level. The rescuing activity of nonoverlapping subfragments of the A gene indicates that pairing of maternal transcripts over the whole gene is not necessary, although their additive effects suggest that this would further enhance A-gene amplification.

We have proposed that, in the absence of maternal copies of the A gene, a heterochromatin-like structure determined by downstream germline-limited sequences is able to spread upstream to include the A gene and ultimately result in A-gene deletion. Preliminary results indicate that in the d48 cell line, the A gene is first amplified for at least a few replication cycles during early macronuclear development before being eliminated by an active mechanism (O. Garnier, unpublished results). The timing of this elimination coincides roughly with that of chromosome fragmentation at another genomic site (A. Le Mouël, personal communication). The pairing of maternal transcripts could block heterochromatin spreading, either directly or because it first induces some epigenetic modification such as DNA methylation. In contrast to the rescue effect, the deletion effect requires a large excess of maternal A-gene sequences. Could an excess of the maternal transcripts that would otherwise rescue a deletion result in opposite consequences? Although this cannot be excluded, another possibility is that an excess of maternal A-gene sequences leads to the production of qualitatively different RNA molecules, with opposite effects on zygotic sequences. For instance, double-stranded RNA molecules could be produced from head-to-head multimers of the injected sequences, which represent only a fraction of the total copy number. This would be consistent with the higher deletion efficiency of promoter-containing fragments.

The idea that qualitatively different deletion-inducing transcripts could be double-stranded molecules further suggests an alternative mechanism for the deletion effect. Instead of pairing directly with zygotic sequences, double-stranded RNA could result in deletions indirectly, by preventing rescuing transcripts (such as the wild-type cytoplasmic factor) from promoting amplification. In many other eukaryotes, double-stranded RNA has been shown to induce the specific degragation of single-stranded homologous transcripts, providing a simple explanation for the inactivation of rescuing transcripts. This model would explain why an excess of maternal A-gene sequences is functionally equivalent to their complete absence. Some support for this hypothesis comes from the study of homology-dependent gene silencing in *Paramecium* (Ruiz et al., 1998). High-copy number transformation of the wild-type macronucleus with promoterless and terminatorless coding sequences indeed results in the specific disappearance of endogenous homologous transcripts during vegetative growth of the transformed clones. Vegetative silencing was further shown to be associated with the production of aberrantly sized transcripts from both strands of the injected sequence. Although the exact role of these aberrant transcripts remains to be determined, in at least one case the silencing mechanism was shown to be posttranscriptional (Galvani and Sperling, 2001), raising the possibility of an RNAi-like mechanism. A connection between the maternal deletion effect and the vegetative silencing phenomenon is suggested by the observation that expression of the injected A gene is incompatible with the deletion effect. Most of the deletion-inducing fragments of the A gene were indeed shown to silence the expression of the endogenous A gene during vegetative growth of the transformed clones, before autogamy is induced (O. Garnier, unpublished results). These observations support the idea that similar transcripts may be responsible for both effects.

IV. CONCLUSIONS

Studies of the developmentally regulated rearrangements of the genome in ciliates have generally been based on the assumption that rearrangement patterns are determined entirely by the sequence of the germline genome. Classical paradigms of molecular biology have inspired the view that the reproducibility of these patterns can be explained by *cis*-acting sequence elements directing the action of *trans*-acting protein factors. While a few such *cis*-acting elements have indeed been identified, the comparison of germline micronuclear sequences with their macronuclear rearranged counterparts has not been very successful in explaining the reproducibility of IES excision in many species, or of chromosome fragmentation in *Paramecium*. We have reviewed the available evidence for an epigenetic programming of rearrangements, which provides an alternative type of explanation for their reproducibility: genetic analyses, as well as macronuclear transformation experiments, reveal that homology-dependent maternal effects play an important role in specifying rearrangement patterns. The mechanistic models discussed for each of these effects are necessarily very speculative, as nothing is known of their biochemistry. The most robust conclusion is probably that the sequence specificity of these *trans*-nuclear effects cannot be explained by the production or titration of sequence-specific protein factors, and therefore implies a direct comparison of homologous nucleic acids through pairing interactions. As in other cases of *trans*-nuclear or *trans*-cellular homology-dependent effects (Grishok *et al.*, 2000; Strauss, 1999; van West *et al.*, 1999; other chapters in this volume), the cross-talk between different genomes is best explained by RNA molecules moving in and out of nuclei to convey sequence information.

For convenience, maternal homology-dependent effects were here classified according to the type of rearrangement they affect, IES excision or chromosome fragmentation. However, in *Paramecium*, chromosome fragmentation appears to result from the elimination of specific germline sequences, in a process that differs from IES excision only by the larger size of the deleted segments, the variability of deletion endpoints, and the frequent lack of rejoining of flanking sequences. In this respect, it is interesting to note that the maternal effect leading to inhibition of IES excision is formally analogous to the maternal effect resulting in the amplification of genomic regions located close to macronuclear telomeres, such as the A-gene region. In both effects, the presence of normal copy numbers of these sequences in the maternal macronucleus is sufficient to determine their maintenance in the macronuclear genome of sexual progeny. These effects may therefore draw on similar mechanisms. There is clear evidence that a cytoplasmic factor produced by maternal sequences actively promotes A-gene maintenance in the developing macronucleus; such a factor is also likely to mediate IES retention. In both cases, it was hypothesized that maternal transcripts somehow protect homologous sequences of the zygotic genome from deletion, even though the mechanisms through which these sequences would otherwise be deleted may be different.

Another interesting connection was made between the maternal deletion effect and the phenomenon of homology-dependent gene silencing (Ruiz *et al.*, 1998). Both effects can be induced by transformation of the macronucleus with high copy numbers of a nonfunctional gene, which results in the silencing of homologous endogenous genes during vegetative growth of the transformed clones, and also in the deletion of these genes after autogamy, during the development of a new macronucleus. Both effects can be observed with many different genes, and the requirements of copy number and sequence similarity with the targeted genes appear to be similar. Both effects are likely to be RNA-mediated and may therefore rely on similar mediators. Interestingly, the study of homology-dependent gene silencing in many other systems has led to the idea that these effects evolved as a mechanism of defense against transposable elements (Matzke *et al.*, 2000; Plasterk and Ketting, 2000). Such mechanisms theoretically provide cells with a simple way to silence any transposable element on the sole basis of its copy number in the genome, without prior knowledge of its sequence. Homology-dependent mechanisms indeed are responsible for the silencing of retrotransposons in *Drosophila* (Jensen *et al.*, 1999; Malinsky *et al.*, 2001), and perhaps also in mammals (Whitelaw and Martin, 2001). The transposon connection is certainly relevant in ciliates; it has been proposed that the developmental system of genome-wide rearrangements is the result of a co-evolution process which allows the host genome to be purged of all transposable elements before it is expressed (Klobutcher and Herrick, 1997). If any sequence can be deleted in the developing macronucleus in response to an excess copy number in the maternal macronucleus, the developmental elimination of transposable elements may similarly be determined by homology-dependent effects, on the basis of their high copy numbers in the germline genome itself.

Whatever its mechanisms and evolutionary origins, the developmental remodeling of the genome appears to involve a comparison of the zygotic genome to be rearranged with the maternal macronuclear genome, which allows alternative rearrangement patterns to be transmitted to sexual progeny through the cytoplasm, i.e., independently from the germline genome. This system could account for many of the long-known cases of non-Mendelian inheritance. In particular, it suggests an attractive model of mating-type determination and inheritance in *P. tetraurelia*, which illustrates its potential usefulness in maintaining phenotypic polymorphism within populations of genetically identical cells. Because changes in rearrangement patterns occur much more frequently than Mendelian mutations, the maternal inheritance of alternative macronuclear editions may also help cell populations to adapt to a rapidly changing environment without having to alter their germline genome, which can thus be kept optimized for the long term.

Acknowledgments

We wish to thank M. Bétermier, A. Galvani, A. Gratias, A. Le Mouël, and L. Sperling for communicating results prior to publication. We are indebted to J. Beisson, M. Morange, and all members of the lab for helpful comments on the manuscript. O. Garnier is a recipient of a fellowship from the Fondation pour la Recherche Médicale. Work in our lab was supported by the Association pour la Recherche sur le Cancer (grant # 5733), the Centre National de la Recherche Scientifique (Programme Génome), the Ministère de l'Education Nationale, de la Recherche et de la Technologie (Programme de Recherche fondamentale en Microbiologie et Maladies infectieuses et parasitaires), and the Comité de Paris de la Ligue Nationale centre le Cancer (grant #75/01-RS/73).

References

Berger, J. D. (1973). Nuclear differentiation and nucleic acid synthesis in well-fed exconjugants of *Paramecium aurelia. Chromosoma* **42**, 247–268.

Brygoo, Y. (1977). Genetic analysis of mating-type differentiation in *Paramecium tetraurelia. Genetics* **87**, 633–653.

Brygoo, Y., and Keller, A. M. (1981a). Genetic analysis of mating type differentiation in *Paramecium tetraurelia*. III. A mutation restricted to mating type E and affecting the determination of mating type. *Dev. Genet.* **2**, 13–22.

Brygoo, Y., and Keller, A. M. (1981b). A mutation with pleiotropic effects on macronuclearly differentiated functions in *Paramecium tetraurelia. Dev. Genet.* **2**, 23–34.

Byrne, B. C. (1973). Mutational analysis of mating type inheritance in syngen 4 of *Paramecium aurelia. Genetics* **74**, 63–80.

Chalker, D. L., and Yao, M. C. (1996). Non-Mendelian, heritable blocks to DNA rearrangement are induced by loading the somatic nucleus of *Tetrahymena thermophila* with germ line-limited DNA. *Mol. Cell. Biol.* **16**, 3658–3667.

Chalker, D. L., and Yao, M.-C. (2001). Non-genic, bi-directional transcription precedes and may promote developmental DNA deletion in *Tetrahymena thermophila. Genes Dev.* **15**, 1287–1298.

Coyne, R. S., Chalker, D. L., and Yao, M. C. (1996). Genome downsizing during ciliate development: Nuclear division of labor through chromosome restructuring. *Annu. Rev. Genet.* **30**, 557–578.

Cummings, D. J., Tait, A., and Goddard, J. M. (1974). Methylated bases in DNA from *Paramecium aurelia. Biochim. Biophys. Acta* **374**, 1–11.

Dessen, P., Zagulski, M., Gromadka, R., Plattner, H., Kissmehl, R., Meyer, E., Bétermier, M., Shultz, J. E., Linder, J. U., Pearlman, R. E., Kung, C., Forney, J., Satir, B. H., Van Houten, J. L., Keller, A.-M., Froissard, M., Sperling, L., and Cohen, J. (2001). *Paramecium* genome survey: A pilot project. *Trends Genet.* **17**, 306–308.

Duharcourt, S., Butler, A., and Meyer, E. (1995). Epigenetic self-regulation of developmental excision of an internal eliminated sequence on *Paramecium tetraurelia. Genes Dev.* **9**, 2065–2077.

Duharcourt, S., Keller, A. M., and Meyer, E. (1998). Homology-dependent maternal inhibition of developmental excision of internal eliminated sequences in *Paramecium tetraurelia. Mol. Cell. Biol.* **18**, 7075–7085.

Epstein, L. M., and Forney, J. D. (1984). Mendelian and non-Mendelian mutations affecting surface antigen expression in *Paramecium tetraurelia. Mol. Cell. Biol.* **4**, 1583–1590.

Forney, J. D., and Blackburn, E. H. (1988). Developmentally controlled telomere addition in wild-type and mutant Paramecia. *Mol. Cell. Biol.* **8**, 251–258.

Forney, J. D., Yantiri, F., and Mikami, K. (1996). Developmental controlled rearrangement of surface protein genes in *Paramecium tetraurelia. J. Eukaryot. Microbiol.* **43**, 462–467.

Galvani, A., and Sperling, L. (2001). Transgene-mediated post-transcriptional gene silencing is inhibited by 3′ non-coding sequences in *Paramecium. RNA* **29**, 4387–4394.

Gorovsky, M. A., Hattman, S., and Pleger, G. L. (1973). (6 N)methyl adenine in the nuclear DNA of a eucaryote, *Tetrahymena pyriformis. J. Cell Biol.* **56,** 697–701.

Grishok, A., Tabara, H., and Mello, C. C. (2000). Genetic requirements for inheritance of RNAi in *C. elegans. Science* **287,** 2494–2497.

Harumoto, T. (1986). Induced change in a non-Mendelian determinant by transplantation of macronucleoplasm in *Paramecium tetraurelia. Mol. Cell. Biol.* **6,** 3498–3501.

Harwood, J. (1985). The erratic career of cytoplasmic inheritance. *Trends Genet.* **1,** 298–300.

Jensen, S., Gassama, M. P., and Heidmann, T. (1999). Taming of transposable elements by homology-dependent gene silencing. *Nature Genet.* **21,** 209–212.

Jessop-Murray, H., Martin, L. D., Gilley, D., Preer, J. R., Jr., and Polisky, B. (1991). Permanent rescue of a non-Mendelian mutation of *Paramecium* by microinjection of specific DNA sequences. *Genetics* **129,** 727–734.

Karrer, K. M., and VanNuland, T. A. (1998). Position effect takes precedence over target sequence in determination of adenine methylation patterns in the nuclear genome of a eukaryote, *Tetrahymena thermophila. Nucleic Acids Res.* **26,** 4566–4573.

Kim, C. S., Preer, J. R., Jr., and Polisky, B. (1994). Identification of DNA segments capable of rescuing a non-Mendelian mutant in *Paramecium. Genetics* **136,** 1325–1328.

Klobutcher, L. A., and Herrick, G. (1997). Developmental genome reorganization in ciliated protozoa: The transposon link. *Prog. Nucleic Acid Res. Mol. Biol.* **56,** 1–62.

Kobayashi, S., and Koizumi, S. (1990). Characterization of Mendelian and non-Mendelian mutant strains by micronuclear transplantation in *Paramecium tetraurelia. J. Protozool.* **37,** 489–492.

Koizumi, S., and Kobayashi, S. (1989). Microinjection of plasmid DNA encoding the A surface antigen of *Paramecium tetraurelia* restores the ability to regenerate a wild-type macronucleus. *Mol. Cell. Biol.* **9,** 4398–4401.

Malinsky, S., Bucheton, A., and Busseav, I. (2000). New insights on homology-dependent silencing of I factor containing ORF1 in. *Drosophila melanogaster. Genetics* **156,** 1147–1155.

Matzke, M. A., Mette, M. F., and Matzke, A. J. (2000). Transgene silencing by host genome defense: Implications for the evolution of epigenetic control mechanisms in plants and vertebrates. *Plant Mol. Biol.* **43,** 401–415.

Mayer, K. M., and Forney, J. D. (1999). A mutation in the flanking 5′-TA-3′ dinucleotide prevents excision of an internal eliminated sequence from the *Paramecium tetraurelia* genome. *Genetics* **151,** 597–604.

Mayer, K. M., Mikami, K., and Forney, J. D. (1998). A mutation in *Paramecium tetraurelia* reveals functional and structural features of developmentally excised DNA elements. *Genetics* **148,** 139–149.

Meyer, E. (1992). Induction of specific macronuclear developmental mutations by microinjection of a cloned telomeric gene in *Paramecium primaurelia. Genes Dev.* **6,** 211–222.

Meyer, E., Butler, A., Dubrana, K., Duharcourt, S., and Caron, F. (1997). Sequence-specific epigenetic effects of the maternal somatic genome on developmental rearrangements of the zygotic genome in *Paramecium primaurelia. Mol. Cell. Biol.* **17,** 3589–3599.

Meyer, E., and Keller, A. M. (1996). A Mendelian mutation affecting mating-type determination also affects developmental genomic rearrangements in *Paramecium tetraurelia. Genetics* **143,** 191–202.

Nanney, D. L. (1957). Mating type inheritance at conjugation in Variety 4 of *Paramecium aurelia. J. Protozool.* **4,** 89–95.

Nanney, D. L. (1985). Heredity without genes: Ciliate explorations of clonal heredity. *Trends Genet.* **1,** 295–298.

Philippe, H., Guermot, A., and Moreira, D. (2000). The new phylogeny of eukaryotes. *Curr. Opin. Genet. Dev.* **10,** 596–601.

Plasterk, R. H., and Ketting, R. F. (2000). The silence of the genes. *Curr. Opin. Genet. Dev.* **10,** 562–567.

Preer, J. R. (1993). Nonconventional genetic systems. *Perspect. Biol. Med.* **36,** 395–419.

Preer, J. R. (1997). Whatever happened to *Paramecium* genetics? *Genetics* **145,** 217–225.

Preer, J. R. (2000). Epigenetic mechanisms affecting macronuclear development in *Paramecium* and *Tetrahymena. J. Eukaryot. Microbiol.* **47,** 515–524.

Prescott, D. M. (1994). The DNA of ciliated protozoa. *Microbiol. Rev.* **58,** 233–267.

Ruiz, F., Vayssie, L., Klotz, C., Sperling, L., and Madeddu, L. (1998). Homology-dependent gene silencing in *Paramecium. Mol. Biol. Cell* **9,** 931–943.

Saveliev, S. V., and Cox, M. M. (1995). Transient DNA breaks associated with programmed genomic deletion events in conjugating cells of *Tetrahymena thermophila. Genes Dev.* **9,** 248–255.

Schloegel, J. J. (1999). From anomaly to unification: Tracy Sonneborn and the species problem in Protozoa, 1954–1957. *J. Hist. Biol.* **32,** 93–132.

Scott, J., Leeck, C., and Forney, J. (1994a). Analysis of the micronuclear B type surface protein gene in *Paramecium tetraurelia. Nucleic Acids Res.* **22,** 5079–5084.

Scott, J. M., Mikami, K., Leeck, C. L., and Forney, J. D. (1994b). Non-Mendelian inheritance of macronuclear mutations is gene specific in *Paramecium tetraurelia. Mol. Cell. Biol.* **14,** 2479–2484.

Sonneborn, T. M. (1937). Sex, sex inheritance and sex determination in *Paramecium aurelia. Proc. Natl. Acad. Sci. USA* **23,** 378–385.

Sonneborn, T. M. (1948). The determination of hereditary antigenic differences in genically identical *Paramecium* cells. *Proc. Natl. Acad. Sci. USA* **34,** 413–418.

Sonneborn, T. M. (1949). Beyond the gene. *Am. Scientist* **37,** 33–59.

Sonneborn, T. M. (1950). Partner of the genes. *Sci. Am.* **November,** 2–11.

Sonneborn, T. M. (1975). *Paramecium aurelia. In* "Handbook of Genetics" (R. King, ed.), pp. 469–594. *Plenum,* New York.

Sonneborn, T. M. (1977). Genetics of cellular differentiation: Stable nuclear differentiation in eucaryotic unicells. *Annu. Rev. Genet.* **11,** 349–367.

Strauss, E. (1999). RNA molecules may carry long-distance signals in plants. *Science* **283,** 12–13.

Tourancheau, A. B., Villalobo, E., Tsao, N., Torres, A., and Pearlman, R. E. (1998). Protein coding gene trees in ciliates: Comparison with rRNA-based phylogenies. *Mol. Phylogenet. Evol.* **10,** 299–309.

van West, P., Kamoun, S., van 't Klooster, J. W., and Govers, F. (1999). Internuclear gene silencing in *Phytophthora infestans. Mol. Cell* **3,** 339–348.

Whitelaw, E., and Martin, D. I. (2001). Retrotransposons as epigenetic mediators of phenotypic variation in mammals. *Nature Genet.* **27,** 361–365.

Yao, M.-C., Duharcourt, S., and Chalker, D. L. (2002). Genome-wide rearrangements of DNA in ciliates. *In* "Mobile DNA II" (N. Craig, R. Craigie, M. Gellert, and A. Lanbomitz, eds.), pp. 730–758. American Society for Microbiology, Washington, DC.

You, Y., Aufderheide, K., Morand, J., Rodkey, K., and Forney, J. (1991). Macronuclear transformation with specific DNA fragments controls the content of the new macronuclear genome in *Paramecium tetraurelia. Mol. Cell. Biol.* **11,** 1133–1137.

You, Y., Scott, J., and Forney, J. (1994). The role of macronuclear DNA sequences in the permanent rescue of a non-Mendelian mutation in *Paramecium tetraurelia. Genetics* **136,** 1319–1324.

11

RNAi (Nematodes: *Caenorhabditis elegans*)

Alla Grishok
Program in Molecular Medicine
University of Massachusetts Medical School
Worcester, Massachusetts 01605

Craig C. Mello
Howard Hughes Medical Institute
Program in Molecular Medicine
University of Massachusetts Medical School
Worcester, Massachusetts 01605

Advances in Genetics, Vol. 46
Copyright 2002, Elsevier Science (USA).
All rights reserved.
0065-2660/02 $35.00

VI. RNAi and Development
   References

ABSTRACT

RNA interference in *Caenorhabditis elegans* is a type of homology dependent post-transcriptional gene silencing induced by dsRNA. In this chapter we describe the history of the discovery of RNAi, its systemic nature, inheritance, and connection to other homology-dependent silencing phenomena like co-suppression and transcriptional gene silencing. We discuss RNAi-deficient mutants in *C. elegans* as well as characterized components of the RNAi, pathway, the molecular mechanism of RNAi, and its possible role in development and immunity.   © 2002, Elsevier Science (USA).

I. INTRODUCTION/HISTORY

The natural regulation of gene function by antisense RNAs is documented in diverse organisms, such as bacteria (Stolt and Zillig, 1993; van Biesen *et al.*, 1993; Delihas, 1995), *Caenorhabditis elegans* (Lee *et al.*, 1993; Reinhart *et al.*, 2000), and mammals (Hastings *et al.*, 2000; Li and Murphy, 2000), including regulation of noncoding *Xist* RNA by antisense *Tsix* (Lee and Lu, 1999). Antisense transcripts have also been detected and implicated in the regulation of gene expression in *Drosophila* (Akhmanova *et al.*, 1997) and plants (Terryn and Rouze, 2000).

Antisense RNA was first used experimentally by Izant and Weintraub (1984, 1985) to induce a sequence-specific block of mRNA expression in tissue culture cells. Soon thereafter the technique found applications in a variety of other systems, including *Xenopus* oocytes (Harland and Weintraub, 1985), *Drosophila* embryos (Rosenberg *et al.*, 1985), and mouse oocytes (Strickland *et al.*, 1988). In *C. elegans*, antisense RNAs expressed from transgenes were shown to be effective in blocking the expression of two muscle genes *unc-22* and *unc-54* (Fire *et al.*, 1991). However, inconsistency in the effectiveness of antisense approach from gene to gene or from one application to another in many systems led to the perception that this methodology was somewhat unreliable for inhibiting gene function. This perception was dramatically altered for *C. elegans* in 1995, when Guo and Kemphues found that the microinjection of antisense RNA corresponding to the *par-1* gene induced a strikingly accurate *par-1* loss of function phenotype. Surprisingly, they found that control preparations of *par-1* sense RNA also induced a *par-1* loss of function phenotype (Guo and Kemphues, 1995).

Despite the apparent lack of strand specificity, the use of antisense RNA to inhibit gene function rapidly gained acceptance as the number of C. *elegans* genes that could be silenced with this technique continued to grow. Two additional observations suggested that something more than a simple concentration-dependent pairing between the antisense RNA and the mRNA must be involved in the interference process in C. *elegans*. The first of these surprising observations came with the discovery that the interference effect could be inherited for at least two generations after the injection of RNA (Mello, unpublished observations). The second came with the discovery that interference could spread from the site of injection into the other tissues in the organism (Fire *et al.*, 1998). Together these findings led C. *elegans* researchers to coin a new name for the methodology: RNA interference, or simply RNAi.

The mystery of RNAi took on a helical twist with the discovery that dsRNA was at least 10 times more effective than preparations of sense or antisense RNA (Fire *et al.*, 1998). This discovery was of great importance because it led rapidly to applications of dsRNA to silence genes in other organisms, including plants (Waterhouse *et al.*, 1998), trypanosomes (Ngo *et al.*, 1998), flies (Kennerdell and Carthew, 1998), planaria (Sanchez-Alvarado and Newmark, 1999), hydra (Lohmann *et al.*, 1999), and mouse embryos (Wianny and Zernicka-Goetz, 2000).

The improved efficiency of dsRNA made possible new methods for inducing RNAi. These included simply soaking worms in dsRNA solutions (Tabara *et al.*, 1998), feeding worms *Escherichia coli* expressing a dsRNA segment of a target gene (Timmons and Fire, 1998), or driving dsRNA expression from a transgene (Tabara *et al.*, 1999; Tavernarakis *et al.*, 2000). These new methods in turn opened new doors for the application of RNAi. The applications included genetic screens for mutants resistant to RNAi (Tabara *et al.*, 1999) and, more recently, genome-wide applications of RNAi to inhibit C. *elegans* genes systematically (Fraser *et al.*, 2000; Gonczy *et al.*, 2000; Piano *et al.*, 2000).

Studies on the mechanism of RNAi in C. *elegans* have identified similarities to posttranscriptional gene silencing (PTGS) mechanisms previously described in plants and *Neurospora*. The similarities and differences between RNAi and other homology-dependent silencing phenomena, such as co-suppression and transgene silencing, have been recognized both within a given organism (Tabara *et al.*, 1999; Dernburg *et al.*, 2000; Ketting and Plasterk, 2000) and between different organisms (see reviews by Montgomery and Fire, 1998; Fire, 1999; Grant, 1999; Sharp, 1999, 2001; Wolffe and Matzke, 1999; Hunter, 1999, 2000; Bass, 2000; Bosher and Labouesse, 2000; Cogoni and Macino, 2000; Gura, 2000; Maine, 2000; Marx, 2000; Plasterk and Ketting, 2000; Carthew, 2001). In this review we will focus on RNAi in C. *elegans* and discuss connections to other silencing phenomena.

II. RNAi AND OTHER SILENCING MECHANISMS IN *C. elegans*

A. RNAi-deficient mutants

The remarkable features of RNAi suggest that at least several distinct mechanisms may exist in the animal to facilitate this process. It is likely that specific mechanisms underlie (1) the spreading or transport of dsRNA or a secondary agent within and between tissues, (2) the formation and inheritance of an active interfering agent, and, of course, (3) the interference process itself. To identify genes required for these mechanisms in C. *elegans*, Tabara and colleagues (1999) cultured populations of mutagenized worms on E. *coli* expressing a dsRNA corresponding to a segment of an essential C. *elegans* gene. Wild-type animals feeding on this E. *coli* strain produced inviable embryos due to RNA interference with the essential gene (targeted via the ingested dsRNA). In contrast, mutants resistant to RNAi escaped interference and survived. Mutants identified in this way were named RNAi-deficient or *rde* mutants.

This screen proved very powerful as a means for selecting mutations in RNAi pathway genes, and three classes of mutants were found. The first and largest class included mutants resistant to RNAi by feeding but sensitive to injected dsRNA, suggesting that these mutants may either be weak mutants or may be defective in the uptake or transport of RNA from the intestine. This class of mutants has yet to be characterized further. The second class consisted of six mutants defining two genes, *rde-1* and *rde-4*, which are absolutely required for RNAi. Although *rde-1* and *rde-4* mutants are completely resistant to RNAi, they exhibit no other obvious phenotypes. In contrast, the third and final class, comprised of 14 mutants and defining six complementation groups, was deficient in RNAi targeting germline genes but remained sensitive to RNAi targeting several somatic genes. Members of this third class also exhibited several additional phenotypes, including temperature-dependent sterility, a high incidence of males, and mobilization of transposons in the germline (Tabara et al., 1999), which are the features of mutator mutants (Collins et al., 1987; Ketting et al., 1999).

B. RNAi and transposon silencing

The observation that transposons are mobilized in several of the RNAi-deficient strains was exciting, as it suggested a possible *in vivo* function for RNAi—a defense against transposons. Indeed, Ketting and colleagues (1999) showed that many of the mutator strains they had identified in a screen for mutants with increased transposition were also resistant to RNAi. They proposed a model for transposon silencing via RNAi initiated by transposon-derived dsRNA. This simple model was complicated, however, by the fact that the strongest RNAi deficient mutants, *rde-1* and *rde-4*, do not exhibit transposon mobilization (Tabara et al., 1999), indicating that there is some mechanistic distinction between the two types of silencing. The simplest interpretation of these findings is that the mutator class of RNAi mutants disrupts a step common to both RNAi and transposon silencing (see Sections II.C, III.B, and IV.A).

A peculiar aspect of the mutator class of *rde* mutants is that they exhibit only partial loss of RNAi and primarily in the germline. One explanation may be that redundant genes exist that carry out the functions of these genes in other tissues. Alternatively, some of these mutations might disrupt RNAi indirectly, by causing an excess of an unrelated dsRNA to accumulate in the germline. Indeed, an unrelated dsRNA can, under some conditions, render wild-type worms resistant to a second dsRNA, suggesting that there is a saturable step involved in the process (Parrish *et al.*, 2000). In a recent study of posttranscriptional gene silencing in *Chlamydomonas*, it was shown that both transposon and transgene silencing are affected by a mutation in the *mut-6* gene, which encodes a DEAH-box RNA helicase (Wu-Scharf *et al.*, 2000). The authors demonstrate that aberrant transcripts accumulate in the *mut-6* mutant. Thus, disruption of an unrelated RNA degradation pathway or the upregulation of genes expressing natural dsRNA might indirectly cause mutant strains to become partially resistant to RNAi. Cloning and future analysis of more of the mutator class of *rde* mutants should shed light on whether these genes are direct or indirect effectors of RNAi.

C. RNAi and co-suppression

Co-suppression was first discovered in plants (Napoli *et al.*, 1990; van der Krol *et al.*, 1990) as a silencing phenomenon in which a transgene bearing an extra copy of a cellular gene initiates a posttranscriptional silencing of both the transgene and the endogenous copy of the corresponding gene. Co-suppression was subsequently shown to occur in fungi and called quelling (Romano and Macino, 1992). Recent work has clearly demonstrated that this phenomenon is related to RNAi. For example, the *qde-2* gene, which is essential for co-suppression in *Neurospora*, is a homolog of *C. elegans rde-1*, and another *rde-1* homolog, *ago1*, is important for posttranscriptional gene silencing in *Arabidopsis* (Fagard *et al.*, 2000).

But how similar are RNAi and co-suppression? For example, do both mechanisms involve a dsRNA trigger? Although in some cases of PTGS in plants the initiation of silencing was correlated with dsRNA expression from the transgenes (Waterhouse *et al.*, 1998; Smith *et al.*, 2000), the presence of dsRNA has not been documented in most cases. The initiation of PTGS by bombardment of plants with gold particles containing promoterless DNA (Voinnet *et al.*, 1998) also indicates differences in mechanism, as injection of DNA into *C. elegans* does not induce systemic silencing. However, both in plants and in *Neurospora*, mutations in genes that encode proteins with homology to RNA-dependent RNA polymerases (RdRP) completely abolish PTGS induced by transgenes (Cogoni and Macino, 1999a; Dalmay *et al.*, 2000; Mourrain *et al.*, 2000), suggesting that dsRNA plays a role in co-suppression as well. Nevertheless, it is not clear yet at which step the target RNA gets copied into dsRNA. The prevailing model assumes that full-length mRNA molecules are copied into dsRNA, which act as an initiator of mRNA degradation (Dalmay *et al.*, 2000; Voinett *et al.*, 2000). However, it is still possible that transgene-encoded aberrant or excess RNA gets into the degradation machinery first and partial products of degradation get copied into dsRNA molecules which mark more transgene-specific RNAs for degradation and ensure amplification and maintenance of the process.

In *C. elegans* it is now clear that co-suppression, which appears to occur only in the germline, is genetically distinct from RNAi. In elegant genetic studies, Dernburg and colleagues (2000) and Ketting and Plasterk (2000) have recently shown that co-suppression is independent of *rde-1* activity but is dependent on the activity of other members of the mutator class of *rde* mutants. These findings suggest that whatever function *rde-1* provides in RNAi must be provided by some other gene in co-suppression (Figure 11.1). Considering that *rde-1* homologs are involved in co-suppression in at least two other species, it seems likely that a *C. elegans* homolog of *rde-1* will be involved in co-suppression (see Sections III.B and V.A). It also remains possible that *rde-1* is involved in co-suppression in *C. elegans* but that some redundant gene exists that can mediate co-suppression when *rde-1* is not functional. Whatever the specific explanation might be, it is clear that the factors mediating co-suppression are not sufficient for carrying out RNAi in the absence of *rde-1*.

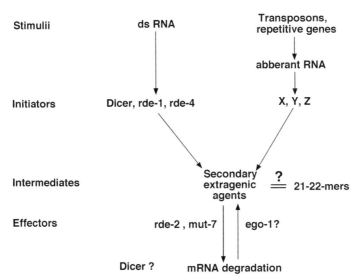

Figure 11.1. Model for RNAi and other PTGS-like silencing pathways in *C. elegans*.

D. RNAi and transgene silencing

In C. *elegans*, transgenes are readily expressed in somatic tissues and can be transmitted for generations with undiminished expression. In contrast, transgenes that drive expression in the germline are rapidly silenced. Transgene silencing is an extremely interesting phenomenon that is, from a practical perspective, terribly frustrating for researchers who would like to stably express a transgene. Germline silencing of transgenes appears to occur at a transcriptional level, as it has been shown to depend on the function of C. *elegans* genes homologous to members of the *Drosophila* Polycomb group, thought to regulate chromatin structure (Holdeman *et al.*, 1998; Kelly and Fire, 1998; Korf *et al.*, 1998).

In plants, co-suppression has been correlated with the methylation of the silenced gene (Wassenegger *et al.*, 1994; Jones *et al.*, 1998). In addition, transcriptional silencing can be triggered in plants via dsRNA targeting a promoter region (Mette *et al.*, 2000). Although a connection between RNAi and transcriptional silencing has not been firmly established in C. *elegans*, a significant degree of transgene de-silencing does occur when silent transgenes are crossed into strains homozygous for mutator class *rde* mutants (Tabara *et al.*, 1999). Interestingly, silencing was not strongly suppressed unless the animals were cultured at an elevated temperature, culture condition that renders the mutator *rde* mutants nearly completely sterile. Thus, the mutator-class *rde* genes seem to function in a temperature-dependent process that is essential for both fertility and gene silencing. It is tempting to speculate that transgene silencing in C. *elegans* involves a posttrancriptional step necessary for intitiation and/or maintenance of a silent chromatin state.

III. SPREADING AND PERSISTENCE OF RNAi

A. Systemic nature of RNAi

One of the most intriguing features of RNAi is its ability to spread to other tissues after exposure to dsRNA via injection, soaking, or feeding. Appreciation of this systemic nature of RNAi came with the observation that it is not necessary to inject dsRNA directly into the target tissue. For example, injection of dsRNA into the intestine or body cavity was found to cause interference in the germline and even among the progeny of the injected animal (Fire *et al.*, 1998). In fact, the intestine appears to be the best injection site for dsRNA even when the target mRNA is expressed in the germline (Mello, unpublished observations). If RNAi represents an immunity mechanism against a dsRNA pathogen (see Section III.C), the gut would likely be an ideal place to prepare the initial defense. Viruses encountered while feeding might be ingested, their RNA fragmented and then transported into the body cavity for circulation to potentially infected tissues, where they might block viral mRNA expression.

Little is known about the uptake and transport of dsRNA in C. elegans. So far, none of the rde mutants that have been characterized prohibits the transport of dsRNA or its derivatives from the intestine or body cavity into the germline. Conceivably, mutations in genes involved in transport could be found within the first and largest class of rde mutants (Tabara et al., 1999), those resistant to RNAi by feeding but sensitive to injection.

Systemic transport of RNAi in C. elegans is often related to a similar phenomenon in plants, wherein PTGS can spread via a sequence-specific agent (Palauqui et al., 1997; Palauqui and Vaucheret, 1998). For plants it has been shown that DNA sequences corresponding to the 5′ end of the target gene caused systemic silencing that was able to target the 3′ portion of the gene not present in the trigger DNA (Voinnet et al., 1998). This suggests that the systemic signal is produced by copying the target mRNA, possibly via a RdRP. In C. elegans, the systemic transport of the injected material (dsRNA or its derivative) and its persistence to the F1 generation occurs in the absence of the target gene (Tabara et al., 1999; our unpublished data). Although we cannot exclude that long-term persistence of RNAi requires target-dependent amplification of the signal, this has not yet been observed in C. elegans. It will be very exciting in the future to learn more about the mechanisms involved in the systemic transport of nucleic acids into and out of cells in C. elegans and other organisms.

B. Inheritance of RNAi

As we mentioned above, exposure to dsRNA causes interference both in the exposed animal and, via transport to the germline, can also cause interference in the exposed animal's progeny. This is a truly remarkable process, as in some cases the targeted gene is completely silent in the progeny of an injected animal more than a week after the initial injection of dsRNA. Interestingly, although males respond to RNAi and can inherit RNAi from their mothers, male animals exposed to dsRNA do not transmit interference to their progeny. This suggests that some process that occurs in the hermaphrodite but not in the male is required for uptake of the interfering agent by the germ cells. One could imagine that because the sperm has a relatively small cytoplasmic compartment, there simply is not enough cytoplasm to store the interfering agent in the sperm. However, as we will describe below, males can pass the interfering agent to their progeny, provided that they inherited the interfering agent from their mother (Grishok et al., 2000). Thus, males can transmit but cannot create the inherited interfering agent.

While inheritance of RNAi for one generation is remarkable enough, for at least several genes, germline transmission of interference is possible for two or more generations (Grishok et al., 2000; our unpublished data). The ability to passage the interfering agent for two generations permits a genetic analysis of the inheritance phenomenon. Such a study has demonstrated that inheritance can occur even when the target gene is completely absent in the gamete. Thus, the inherited agent is not associated with, or dependent in any way, on the target locus or associated chromatin (Grishok *et al.*, 2000). Interestingly, interference in response to the inherited factor exhibits a distinct set of genetic requirements from interference induced directly by the trigger RNA. Specifically, the *rde-1* and *rde-4* genes were dispensable for RNAi in response to the inherited agent. The simplest interpretation of this result is that the RDE-1 and RDE-4 products are involved in an upstream step of RNAi that either processes the trigger RNA or ensures that its derivatives become engaged with downstream factors required for targeting the mRNA. This model (Figure 11.1) is consistent with the apparent dedication of RDE-1 and RDE-4 to interference triggered by dsRNA. In this model, distinct triggers that share a common set of downstream effectors are required for co-suppression and transposon silencing.

Recent biochemical studies of RNAi in *Drosophila* tissue culture have shown that the *Drosophila* Argonaute-2 protein, an RDE-1 homolog, co-purifies with a complex that can degrade the target mRNA (Bernstein *et al.*, 2001). Thus, in *Drosophila* an RDE-1 homolog may play a more downstream role than the one predicted from the above genetic analysis of the *rde-1* gene. Perhaps a distinct RDE-1 homolog functions upstream of Argonaute 2 in *Drosophila* and is required for processing the trigger RNA. Alternatively, although the genetic studies indicate that continued expression of *rde-1* is not necessary for RNAi induced by the inherited agent, it remains possible that RDE-1 protein becomes a stable component of the inherited agent and is also involved in targeting. In such a model, the continued expression of *rde-1* gene would not be required to mediate interference as RDE-1, and the processed trigger RNA could exist in a stable inherited complex. Future biochemical and genetic studies should shed more light on the composition of the inherited interfering agent and on the sequence of events that process the trigger and lead to targeting.

C. RNAi and immunity

Both *rde-1* and *rde-4* null mutants are completely deficient for RNAi and apparently have no other phenotypes (Tabara *et al.*, 1999). If animals lacking these genes are wild type in appearance and development, what then might be the physiological relevance of these genes? The most attractive model is that these genes (and RNAi in general) may be part of an immunity mechanism. In the wild, *C. elegans* may encounter dsRNA viruses that are absent under laboratory culture conditions; *rde-1* and *rde-4* could confer immunity against such pathogens.

The idea that RNAi might represent a form of acquired sequence-specific immunity is very attractive. *C. elegans* adults live only a few days and, thus, immunity developed in an individual animal may not be of much benefit. However, if an infected animal could pass immunity to its offspring, as can occur with RNAi (see Section III.B), this would likely confer a very considerable selective benefit. It seems likely that co-suppression, transposon silencing, and transgene silencing may all represent related forms of sequence-directed immunity that share a mechanism but have evolved to respond to independent triggers. In such a model, *rde-1* and *rde-4* would function to trigger this immunity mechanism in response to dsRNA. If this model is correct, then *rde-1* and *rde-4* must be very good at what they do: as yet, there are no known viruses that infect C. *elegans*. It will be interesting in the future to identify and challenge these mutant strains with potential pathogens to see if they are compromised in any way.

IV. MOLECULAR MECHANISM OF RNAi

A. Initiation of PTGS: dsRNA and RNAi

There is a confusion in the literature concerning the distinctions between different types of posttranscriptional silencing mechanisms, such as co-suppression and RNAi. And yet the genetic analysis of these phenomena in C. *elegans* suggests that they must have distinct mechanisms and employ distinct but overlapping sets of genes (Dernberg *et al.*, 2000; Grishok *et al.*, 2000; Ketting and Plasterk, 2000). The confusion, no doubt, reflects the fact that all of these mechanisms are likely to get their sequence specificity through base pairing between a guide RNA and the target mRNA. Indeed, once targeting occurs, the mechanisms may thereafter be indistinguishable, perhaps involving self- renewing reactions that produce more targeting complexes with each round of mRNA degradation (Figure 11.1).

RNAi in C. *elegans* is by definition initiated by dsRNA (injected, fed, or expressed from transgenes) (Fire *et al.*, 1998). The most upstream components responding to dsRNA are *rde-1* and *rde-4* (Grishok *et al.*, 2000), and the activities of these two genes are totally dispensable for the initiation of other silencing phenomena in C. *elegans*, such as transposon silencing (Tabara *et al.*, 1999), co-suppression (Dernburg *et al.*, 2000; Ketting and Plasterk, 2000), and transcriptional silencing of germline transgenes (Tabara *et al.*, 1999). The work in C. *elegans* clearly suggests that co-suppression and transposon silencing must have initial triggers other than dsRNA. However, it is likely that dsRNA also functions downstream as a common factor involved in the renewal, amplification, and targeting of all PTGS phenomena. Thus, the defining feature of each form of PTGS is likely to be the initiation signal (or trigger). These triggers could be aberrant RNA, RNA accumulated above a threshold, the repetitive nature of the DNA, or the modification of DNA or chromatin structure. Involvement of RecQ DNA helicase family member, *qde-3*, in posttranscriptional gene silencing in *Neurospora* may indicate that initiation of this process starts with DNA–DNA interactions facilitated by the activity of *qde-3* (Cogoni and Macino, 1999b). Clearly, an area of considerable importance for future study will be defining the sequence of events that initiates RNAi and other forms of PTGS.

B. Role of small 21–22-nt RNA species in RNAi

In 1999, Hamilton and Baulcombe discovered what seems likely to be the common currency of PTGS mechanisms, a species of small 21–25-nt RNAs associated with PTGS in plants (Hamilton and Baulcombe, 1999). These small RNAs have now been found associated with RNAi in *Drosophila* (Hammond *et al.*, 2000; Zamore *et al.*, 2000) and RNAi in *C. elegans* (Parrish *et al.*, 2000). Moreover, small RNAs derived from a transgene expressing promoter-specific dsRNA have been correlated with transcriptional gene silencing in plants (Mette *et al.*, 2000). Conceivably, these small RNAs serve as sequence-specific guide RNAs that direct distinct protein complexes to the target RNA or DNA. Presently, however, presence of small RNAs in plants has been shown only in plants actively involved in the process of target RNA degradation, and hence, the initiation role of small RNAs in PTGS has not been confirmed yet (Hamilton and Baulcombe, 1999; Dalmay *et al.*, 2000; Voinnet *et al.*, 2000).

　　　The best evidence for a direct role for small RNAs in the initiation of RNAi comes from biochemical studies. For example, it has been shown *in vitro* that dsRNA is processed into small 21–22mer RNAs prior to targeted mRNA degradation (Zamore *et al.*, 2000). In cultured Drosophila S2 cells, small RNAs were shown to co-purify with a nuclease activity that degrades the target mRNA (Hammond *et al.*, 2000). Furthermore, in *Drosophila* embryos the appearance of small RNAs derived from a dsRNA trigger were correlated with disappearance of the target RNA (Yang *et al.*, 2000). Direct evidence for the sufficiency of 21–22-nt RNA fragments as mediators of RNAi has been obtained using *Drosophila* lysates wherein chemically synthesized 21–22 dsRNA molecules with overhanging 3′ ends were shown to be efficient triggers for target mRNA cleavage (Elbashir *et al.*, 2001b). The observation that 3′ overhangs stimulate RNAi is consistent with the idea that an RNase III-like processing reaction is involved in processing the trigger RNA. Indeed, as discussed below, a recent study has identified an RNAse III-type enzyme from *Drosophila* that may process dsRNA into the small RNAs involved in RNAi (Bernstein *et al.*, 2001). Duplexes of small 21-nt RNAs have been recently shown to suppress gene expression in cultured mammalian cells (Elbashir *et al.*, 2001a).

　　　Biochemical studies analyzing the degradation of target RNA have shown that the target RNA is cleaved with a periodicity roughly corresponding to 20–23 nt (Zamore *et al.*, 2000), consistent with the idea that small RNAs may guide degradation. In experiments with synthetic small dsRNA triggers, the target RNA cleavage site was located near the center of the region covered by the 22-nt guide RNA, and there was no nucleotide preference for this reaction (Elbashir *et al.*, 2001b). Although both strands of small 21–22-nt guide RNAs have been identified in targeting complexes, each complex appears to have an asymmetric character and either contains only one of the strands or allows only one of the strands to guide target RNA cleavage (Elbashir *et al.*, 2001b).

C. Dissection of the trigger RNA

Several studies have examined structural and sequence requirements for the trigger RNA. In *C. elegans*, dsRNA of several hundred base pairs in length is normally used to achieve optimal knock-out phenotypes, and similar length requirements have been observed in other systems (Ngo *et al.*, 1998; Tuschl *et al.*, 1999; Hammond *et al.*, 2000). In a recent study examining length requirements for the RNAi trigger in *C. elegans*, dsRNA molecules as short as 26 bp were found to induce interference but showed much higher concentration requirements (Parrish *et al.*, 2000). However, this study did not examine small RNAs with 3′ overhangs similar to those shown to be effective in *Drosophila* extracts. Perhaps this structure will improve the efficiency of targeting via small synthetic RNAs *in vivo*. Indeed, it would be interesting to see if such small RNAs might bypass certain of the *rde* mutants in *C. elegans* to cause interference or cause interference in the organisms that apparently lack RNAi.

The degree of similarity required between the dsRNA and target RNA in *C. elegans* was also examined by Parrish and colleagues (2000). Efficient degradation of a target RNA required 96% sequence identity with the trigger dsRNA. Perhaps not surprisingly, targets lacking segments of at least 23 nucleotides of perfect identity were not efficiently degraded. To determine the relative importance of sense and antisense strands in the trigger, dsRNA with mismatches or chemical modifications have been tested in *Drosophila* (Yang *et al.*, 2000) and *C. elegans* (Parrish *et al.*, 2000). These studies concluded that modifications in the antisense strand of dsRNA are less well tolerated than are modifications of the sense strand. These findings strongly support the notion that the antisense strand determines the target specificity. These findings also suggest that direct amplification of the introduced dsRNA is not necessary for induction of interference.

D. Targets of RNAi

In the initial studies of RNAi in *C. elegans*, it was shown that intron and promoter sequences are not effective in causing RNAi (Fire *et al.*, 1998), which argued against transcription or RNA processing as the targets for RNAi. In most cases RNAi seems to target only sequences present in the mature mRNA. For example, individual cistrons within a multicistronic gene can be targeted separately (Montgomery *et al.*, 1998). However, in one study, two genes in an operon were inactivated by injection of dsRNA corresponding to one of the genes and also by dsRNA corresponding to an intron within the operon (Bosher *et al.*, 1999). This finding suggests that the pre-mRNA can be targeted by RNAi. Indeed, *in situ* hybridization studies in C. *elegans* have shown that both nuclear and cytoplasmic levels of target RNA can be reduced by RNAi (Montgomery *et al.*, 1998). These studies suggest that RNAi may occur in the nucleus or in both the nucleus and cytoplasm.

Several studies suggest that RNAi in other systems can target mature mRNAs. For example, in the early mouse embryo the mature *c-mos* mRNA was susceptible to RNAi (Wianny and Zernicka-Goetz, 2000). Also, mature mRNAs corresponding to *bicoid* and *hunchback* homologs were successfully knocked out in the fly, *Megaselia abdita* (Stauber *et al.*, 2000). Finally, in *Drosophila* extracts, mature capped and polyadenylated mRNAs are efficiently degraded (Tuschl *et al.*, 1999). These studies imply that the mature cytoplasmic mRNA can serve as a target for RNAi. Interestingly however, in C. *elegans* mature maternal mRNAs can apparently co-exist and remain functional even when sufficient interfering RNA is present in the oocyte to direct interference in the next generation (Grishok *et al.*, 2000). This observation indicates that either mature maternal RNA in C. *elegans* are protected from RNAi, or some component necessary for RNAi initiation is not present in the oocyte.

Another intriguing aspect of RNAi in C. *elegans* is that certain genes are relatively resistant to interference (Rappleye *et al.*, 1999; Shin *et al.*, 1999; Mello, unpublished observations). Although this seems to be especially true for genes expressed in the nervous system (Fire, 1999), it is not clear that this is a problem of tissue specificity, as in some cases genes expressed in the same cell exhibit different sensitivities to RNAi. It will be interesting in the future to examine what factors render certain genes resistant to RNAi.

V. COMPONENTS OF THE RNAi PATHWAY

A. The RDE-1 gene family

The *rde-1* gene is a member of a large gene family with members in plants, fungi, D. *melanogaster*, and mammals (Tabara *et al.*, 1999). Members of this family have the greatest degree of similarity within the carboxy-terminal 300 amino acids of the protein, which has been referred to as the PIWI box (Cox *et al.*, 1998) or PIWI domain (Cerutti *et al.*, 2000), after a *Drosophila* member of the gene family. Cerutti and colleagues (2000) identified a second, apparently transportable motif in the proteins of the RDE-1 family—Piwi, Argonaute, and Zwille—and called the motif PAZ. They also found this motif in *Arabidopsis* gene CAF (Carpel Factory); (Jacobsen *et al.*, 1999). The CAF protein is related to proteins in *Drosophila* and C. *elegans*, including the *Drosophila* Dicer protein discussed below. The PAZ domain consists of two small regions of similarity that extend from amino acid 390 to 421 in RDE-1. Although the functions of the PAZ and PIWI domains are not known, it is likely that these domains are important for RDE-1 function. The existing alleles of *rde-1* include three stop-codon alleles that truncate the protein before or within the PIWI domain. The allele (*ne219*) contains a single amino acid substitution that changes a highly conserved glutamate residue to a lysine within the PAZ domain. The N-terminal domain of *rde-1* is the least conserved region in the protein, but some members of the family exhibit sequence similarity that extends to the very N-terminus.

To date, the biochemical function of *rde-1* gene family members has not been elucidated. One homolog, eIF2C, was purified as a possible translation factor (Zou *et al.*, 1998) from rabbit reticulocyte lysates. As mentioned above, members of this family have been implicated in development and silencing in diverse organisms. In *D. melanogaster*, a homolog of *rde-1*, *aubergine/sting*, is implicated in regulation of embryonic development (Schupbach and Wieschaus, 1991; Schmidt *et al.*, 1999). It plays a role in the posttranscriptional silencing of the *Stallete* locus (Schmidt *et al.*, 1999), and also in activating the translation of the *oskar* mRNA (Wilson *et al.*, 1996). Another *Drosophila rde-1* homolog, *piwi*, is important for stem cell maintenance (Cox *et al.*, 1998). In *Arabidopsis thaliana*, mutations in the gene of the same family, *argonaute1* (*ago1*), lead to small-size plants and defects in plant architecture (Bohmert *et al.*, 1998). *Ago1* has recently been shown to play a role in PTGS (Fagard *et al.*, 2000). Another member of the *rde-1* family in *Arabidopsis*—*pinhead/zwille*—has been shown to have role in the formation of central shoot meristem (Moussian *et al.*, 1998). This gene appears to have a partially overlapping function with *ago1* in controlling expression of the *shoot meristemless* (STM) homeobox-containing gene required for shoot apical meristem formation and maintenance (Lynn *et al.*, 1999). It has been proposed that Pinhead and Argonaute might promote translation of specific mRNAs such as STM (Lynn *et al.*, 1999). In the rat, a close homolog of Argonaute1, pGERp95, is localized to the cytoplasmic side of the ER and Golgi (Cikaluk *et al.*, 1999). Finally, the *Neurospora* gene *qde-2* is required for a co-suppression (Catalanotto *et al.*, 2000). QDE-2, like RDE-1, does not have any obvious essential functions for viability or development.

In the genome sequence of *C. elegans* there are at least 23 sequences related to *rde-1*, with both the PAZ and PIWI domains. Inactivation by RNAi in *C. elegans* of the two *rde-1* homologs most similar to *Drosophila piwi* (*prg-1* and *prg-2*) leads to defects in germline stem cell production and reduced brood size, similarly to defects reported for *Drosophila piwi* mutants (Cox *et al.*, 1998). Perhaps members of the *rde-1* family perform related functions in distinct silencing and/or developmental mechanisms. For example, since co-suppression and transposon silencing do not require *rde-1*, it seems plausible that one or more of the worm *rde-1* homologs will prove to be required for these silencing mechanisms. It is also possible that *rde-1* has functions in co-suppression as well as development but that these activities are redundantly specified by other family members.

B. Role of MUT-7, EGO-1, and other components of RNAi

Another cloned gene important for RNAi as well as transposon and transgene silencing is *mut-7*. The MUT-7 protein has homology to a 3′-5′ exonuclease motif found in RNAseD and the Werner syndrome helicase (Ketting *et al.*, 1999). Its homology to ribonuclease D has led to models wherein MUT-7 functions in RNAi to degrade the target RNA (Ketting *et al.*, 1999; Lin and Avery, 1999; Bosher and Labouesse, 2000). If MUT-7 is the enzyme that degrades the target RNA, then there must be a related function encoded by another gene, as MUT-7 does not seem to be necessary for RNAi in somatic cells.

The *ego-1* gene encodes a protein essential for fertility in *C. elegans* and is related to RNA-dependent RNA polymerase. The *ego-1* mutant is defective in RNAi for only a small number of genes expressed in the germline (Smardon *et al.*, 2000). In both *Neurospora* and *Arabidopsis*, genes encoding proteins with homology to RdRP were among the first discovered to play role in the PTGS phenomena (Cogoni and Macino, 1999a; Dalmay *et al.*, 2000; Mourrain *et al.*, 2000). The apparently limited role of *ego-1* in RNAi may be explained by the possible existence of redundant genes, as there are three more genes homologous to RdRP in the *C. elegans* genome. The involvement of an RNA polymerase could explain the remarkably robust and long-lasting interference response involved in RNAi.

The *dicer* gene of *Drosophila* is a recent newcomer to the list of cloned genes involved in RNAi (Bernstein *et al.*, 2001). Dicer appears to be a multi-functional protein with a helicase domain, two dsRNA-binding domains, and two RNAse III domains. Interestingly, this protein also contains the PAZ motif found in members of the *rde-1* family and in the *Arabidopsis* gene CAF (Jacobsen *et al.*, 1999; Cerutti *et al.*, 2000). Bernstein and colleagues (2001) have shown that Dicer can process dsRNA into the small 22-nucleotide RNAs which are thought to be the sequence-specific factors involved in both co-suppression in plants (Hamilton and Baulcombe, 1999) and RNAi in *Drosophila* (Zamore *et al.*, 2000). Previous work in *Drosophila* cultured cells demonstrated that these small RNAs co-purify with an enzyme complex that can degrade a target RNA, and, moreover, the RNA moiety was necessary for the sequence specificity of the complex (Hammond *et al.*, 2000). At present, it is not known if Dicer is involved in both trigger dsRNA processing and degrading the template RNA. Although homologs of Dicer exist in other animals and plants, it is not yet known if these genes have any function in RNAi or other gene silencing mechanisms.

One study has linked components of nonsense-mediated decay in *C. elegans* to the persistence of the interference effect (Domeier *et al.*, 2000).

Nonsense-mediated decay (NMD) is a mechanism conserved from yeast to vertebrates wherein messages with premature stop codons are degraded. Domeier and colleagues (2000) have reported that several, but not all, of the C. *elegans* NMD mutant strains recover more rapidly from RNAi. Since all of the strains are initially sensitive to RNAi, this finding suggests that some components of the NMD pathway play a role in the maintenance of RNAi.

It seems that new factors involved in RNAi and other silencing mechanisms are now being reported at a steadily increasing pace. In C. *elegans* there is much more work to do. The screens performed to date have clearly not identified all the components involved in RNAi in C. *elegans*. Indeed, gene products important for fertility or viability were excluded by the design of the original screens for *rde* mutants. New genetic screens as well as protein interaction screens and biochemical studies will be necessary to identify all the factors that mediate RNAi.

VI. RNAi AND DEVELOPMENT

Considering the sophisticated mechanism and apparently ancient origin of RNAi, one would expect to find related mechanisms employed in natural developmental or cellular functions. In C. *elegans* at least two genes, *lin-4* and *let-7*, are known to encode small 21–22-nucleotide antisense RNAs (Lee *et al.*, 1993; Reinhart *et al.*, 2000). Both of these genes regulate stage-specific developmental events and belong to the heterochronic pathway (Ambros and Horvitz, 1984). The most obvious defects in *lin-4* and *let-7* include reiteration of larval molts in adult worms. The *lin-4* and *let-7* RNAs are complementary to the sequences present in the 3′ UTR of transcripts derived from genetically downstream genes, and appear to downregulate the translation of their respective target mRNAs (Olsen and Ambros, 1999). Both genes appear to be expressed as ~70-nucleotide-long precursors predicted to form dsRNA stem loop structures. The *let-7* gene appears to be conserved in metazoans, as is one of its target mRNAs, encoded by the gene *lin-41* (Pasquinelli *et al.*, 2000).

Could the *let-7* and *lin-4* small RNAs share similarities in their formation and activity with small RNAs that form in response to RNAi? A recent study (Grishok *et al.*, 2001) provides evidence that supports such a connection. First of all, RNA interference targeting a worm homolog of the *Drosophila* gene *Dicer* induces a heterochronic phenotype that exhibits striking similarities to aspects of both the *lin-4* and *let-7* loss of function phenotypes. Similarly, two C. *elegans* homologs of *rde-1*, more closely related to *Arabidopsis ago1* than to *rde-1*, also exhibit interference phenotypes strikingly similar to *let-7* and *lin-4* mutants. These phenotypes include a clear reiteration of the larval molting patterns. Also, interference with these functions leads to derepression of reporter genes carrying the *lin-4* and *let-7* target sites in their 3′ UTRs. Finally, interference with CeDicer and *C. elegans* homologs of Argonaute 1 leads to the accumulation of unprocessed forms of the *lin-4* and *let-7* precursor RNAs.

One explanation for the above findings is that the processing of *let-7* and *lin-4* precursor RNAs relies on the same mechanism that processes the trigger dsRNA in the initiation of RNAi. Consistent with this idea, RNAi targeting CeDicer appears to diminish RNAi (Grishok *et al.*, 2001). This finding suggests that CeDicer is able to trigger interference targeting itself in *C. elegans*. Thus, CeDicer appears to have a role in both the developmental regulation of *lin-4* and *let-7* and in RNA interference. Recent study by Zamore and co-workers implicated *Drosophila* Dicer and its mammalian homolog in the processing of *let-7* precursor (Hutvágner *et al.*, 2001).

Despite the similarities in the upstream portions of the processing mechanisms, it seems likely that there are distinct downstream effectors. First of all, RNAi triggers rapid and nearly complete destruction of the target mRNA (Montgomery *et al.*, 1998; Tuschl *et al.*, 1999), whereas *lin-4* and *let-7* block mRNA expression without reducing the target mRNA levels (Wightman *et al.*, 1993; Moss *et al.*, 1997; Olsen and Ambros, 1999; Slack *et al.*, 2000). Furthermore, RNAi depends on near-perfect complementarity between the target sequence and the trigger RNA, while in contrast, both *lin-4* and *let-7* have only partial identity with their target RNAs. Both *lin-4* and *let-7* are predicted to form two regions of Watson-Crick base pairing flanking a central bulge of 2–3 nucleotides. Little is known about the sequence requirements for *lin-4* and *let-7* function. Perhaps the bulged region provides access for unknown sequence-specific RNA-binding proteins that are involved in mediating translational repression on the mRNAs.

Additional RNAi-induced phenotypes of the two *rde-1* homologs (Grishok *et al.*, 2001), including embryonic defects, suggest that more genes in *C. elegans* might be regulated by small RNAs. Also, similar developmental abnormalities in *ago1* and *caf* mutants in *Arabidopsis* suggest that a pathway including small regulatory RNAs might also exist in plants. In conclusion, it seems that much of the RNA world is yet to be discovered.

References

Akhmanova, A., Kremer, H., Miedema, K., and Hennig, W. (1997). Naturally occurring testis-specific histone H3 antisense transcripts in Drosophila. *Mol. Reprod. Dev.* **48**, 413–420.

Ambros, V., and Horvitz, H. R. (1984). Heterochronic mutants of the nematode *Caenorhabditis elegans*. *Science* **226**, 409–416.

Bass, B. L. (2000). Double-stranded RNA as a template for gene silencing. *Cell* **101**, 235–238.

Bernstein, E., Caudy, A. A., Hammond, S. M., and Hannon, G. J. (2001). Role for a bidentate ribonuclease in the initiation step of RNA interference. *Nature* **409**, 363–366.

Bohmert, K., Camus, I., Bellini, C., Bouchez, D., Caboche, M., and Benning, C. (1998). AGO1 defines a novel locus of Arabidopsis controlling leaf development. *EMBO J.* **17**, 170–180.

Bosher, J. M., Dufourcq, P., Sookhareea, S., and Labouesse, M. (1999). RNA interference can target pre-mRNA: Consequences for gene expression in a *Caenorhabditis elegans* operon. *Genetics* **153,** 1245–1256.

Bosher, J. M., and Labouesse, M. (2000). RNA interference: Genetic wand and genetic watchdog. *Nature Cell Biol.* **2,** E31–E36.

Carthew, R. W. (2001). Gene silencing by double-stranded RNA. *Curr. Opin. Cell Biol.* **13,** 244–248.

Catalanotto, C., Azzalin, G., Macino, G., and Cogoni, C. (2000). Gene silencing in worms and fungi. *Nature* **404,** 245.

Cerutti, L., Mian, N., and Bateman, A. (2000). Domains in gene silencing and cell differentiation proteins: The novel PAZ domain and redefinition of the Piwi domain. *Trends. Biochem. Sci.* **25,** 481–482.

Cikaluk, D. E., Tahbaz, N., Hendricks, L. C., DiMattia, G. E., Hansen, D., Pilgrim, D., and Hobman, T. C. (1999). GERp95, a membrane-associated protein that belongs to a family of proteins involved in stem cell differentiation. *Mol. Biol. Cell* **10,** 3357–3372.

Cogoni, C., and Macino, G. (1999a). Gene silencing in *Neurospora crassa* requires a protein homologous to RNA-dependent RNA polymerase. *Nature* **399,** 166–169.

Cogoni, C., and Macino, G. (1999b). Posttranscriptional gene silencing in *Neurospora* by a RecQ DNA helicase. *Science* **286,** 2342–2344.

Cogoni, C., and Macino, G. (2000). Post-transcriptional gene silencing across kingdoms. *Curr. Opin. Genet. Dev.* **10,** 638–643.

Collins, J., Saari, B., and Anderson, P. (1987). Activation of a transposable element in the germ line but not the soma of *Caenorhabditis elegans*. *Nature* **328,** 726–728.

Cox, D. N., Chao, A., Baker, J., Chang, L., Qiao, D., and Lin, H. (1998). A novel class of evolutionarily conserved genes defined by piwi are essential for stem cell self-renewal. *Genes Dev.* **12,** 3715–3727.

Dalmay, T., Hamilton, A., Rudd, S., Angell, S., and Baulcombe, D. C. (2000). An RNA-dependent RNA polymerase gene in *Arabidopsis* is required for posttranscriptional gene silencing mediated by a transgene but not by a virus. *Cell* **101,** 543–553.

Delihas, N. (1995). Regulation of gene expression by trans-encoded antisense RNAs. *Mol. Microbiol.* **15,** 411–414.

Dernburg, A. F., Zalevsky, J., Colaiacovo, M. P., and Villeneuve, A. M. (2000). Transgene-mediated cosuppression in the C. *elegans* germ line. *Genes Dev.* **14,** 1578–1583.

Domeier, M. E., Morse, D. P., Knight, S. W., Portereiko, M., Bass, B. L., and Mango, S. E. (2000). A link between RNA interference and nonsense-mediated decay in *Caenorhabditis elegans*. *Science* **289,** 1928–1931.

Elbashir, S. M., Harborth, J., Lendeckel, W., Yalcin, A., Weber, K., and Tuschl, T. (2001a). Duplexes of 21-nucleotide RNAs mediate RNA interference in cultured mammalian cells. *Nature* **411,** 494–498.

Elbashir, S. M., Lendeckel, W., and Tuschl, T. (2001b). RNA interference is mediated by 21- and 22-nucleotide RNAs. *Genes Dev.* **15,** 188–200.

Fagard, M., Boutet, S., Morel, J. B., Bellini, C., and Vaucheret, H. (2000). AGO1, QDE-2, and RDE-1 are related proteins required for post-transcriptional gene silencing in plants, quelling in fungi, and RNA interference in animals. *Proc. Natl. Acad. Sci. USA* **97,** 11650–11654.

Fire, A. (1999). RNA-triggered gene silencing. *Trends Genet.* **15,** 358–363.

Fire, A., Albertson, D., Harrison, S. W., and Moerman, D. G. (1991). Production of antisense RNA leads to effective and specific inhibition of gene expression in C. *elegans* muscle. *Development* **113,** 503–514.

Fire, A., Xu, S., Montgomery, M. K., Kostas, S. A., Driver, S. E., and Mello, C. C. (1998). Potent and specific genetic interference by double-stranded RNA in *Caenorhabditis elegans*. *Nature* **391,** 806–811.

Fraser, A. G., Kamath, R. S., Zipperlen, P., Martinez-Campos, M., Sohrmann, M., and Ahringer, J. (2000). Functional genomic analysis of *C. elegans* chromosome I by systematic RNA interference. *Nature* **408**, 325–330.

Gonczy, P., Echeverri, G., Oegema, K., Coulson, A., Jones, S. J., Copley, R. R., Duperon, J., Oegema, J., Brehm, M., Cassin, E., Hannak, E., Kirkham, M., Pichler., S., Flohrs, K., Goessen, A., Leidel, S., Alleaume, A. M., Martin, C., Ozlu, N., Bork, P., and Hyman, A. A. (2000). Functional genomic analysis of cell divisions in *C. elegans* using RNAi of genes on chromosome III. *Nature* **408**, 331–336.

Grant, S. R. (1999). Dissecting the mechanisms of posttranscriptional gene silencing: Divide and conquer. *Cell* **96**, 303–306.

Grishok, A., Pasquinelli, A. E., Conte, D., Li, N., Parrish, S., Ha, I., Baillie, D. L., Fire, A., Ruvkun, G., and Mello, C. C. (2001). Genes and mechanisms related to RNA interference regulate expression of the small temporal RNAs that control *C. elegans* developmental timing. *Cell* **106**, 23–34.

Grishok, A., Tabara, H., and Mello, C. C. (2000). Genetic requirements for inheritance of RNAi in *C. elegans*. *Science* **287**, 2494–2497.

Guo, S., and Kemphues, K. J. (1995). par-1, a gene required for establishing polarity in *C. elegans* embryos, encodes a putative Ser/Thr kinase that is asymmetrically distributed. *Cell* **81**, 611–620.

Gura, T. (2000). A silence that speaks volumes. *Nature* **404**, 804–808.

Hamilton, A. J., and Baulcombe, D. C. (1999). A species of small antisense RNA in posttranscriptional gene silencing in plants. *Science* **286**, 950–952.

Hammond, S. M., Bernstein, E., Beach, D., and Hannon, G. J. (2000). An RNA-directed nuclease mediates post-transcriptional gene silencing in *Drosophila* cells. *Nature* **404**, 293–296.

Harland, R., and Weintraub, H. (1985). Translation of mRNA injected into *Xenopus* oocytes is specifically inhibited by antisense RNA. *J. Cell Biol.* **101**, 1094–1099.

Hastings, M. L., Ingle, H. A., Lazar, M. A., and Munroe, S. H. (2000). Post-transcriptional regulation of thyroid hormone receptor expression by cis-acting sequences and a naturally occurring antisense RNA. *J. Biol. Chem.* **275**, 11507–11513.

Holdeman, R., Nehrt, S., and Strome, S. (1998). MES-2, a maternal protein essential for viability of the germline in *Caenorhabditis elegans*, is homologous to a *Drosophila* Polycomb group protein. *Development* **125**, 2457–2467.

Hunter, C. P. (1999). Genetics: A touch of elegance with RNAi. *Curr. Biol.* **9**, R440–R442.

Hunter, C. P. (2000). Gene silencing: Shrinking the black box of RNAi. *Curr. Biol.* **10**, R137–R140.

Hutvágner, G., McLachlan, J., Pasquinelli, A. E., Balint, E., Tuschl, T., and Zamore, P. D. (2001). A cellular function for the RNA-interference enzyme Dicer in the maturation of *let-7* small temporal RNA. *Science* **293**, 834–838.

Izant, J. G., and Weintraub, H. (1984). Inhibition of thymidine kinase gene expression by anti-sense RNA: A molecular approach to genetic analysis. *Cell* **36**, 1007–1015.

Izant, J. G., and Weintraub, H. (1985). Constitutive and conditional suppression of exogenous and endogenous genes by anti-sense RNA. *Science* **229**, 345–352.

Jacobsen, S. E., Running, M. P., and Meyerowitz, E. M. (1999). Disruption of an RNA helicase/RNAse III gene in *Arabidopsis* causes unregulated cell division in floral meristems. *Development* **126**, 5231–5243.

Jones, A. L., Thomas, C. L., and Maule, A. J. (1998). De novo methylation and co-suppression induced by a cytoplasmically replicating plant RNA virus. *EMBO J.* **17**, 6385–6393.

Kelly, W. G., and Fire, A. (1998). Chromatin silencing and the maintenance of a functional germline in *Caenorhabditis elegans*. *Development* **125**, 2451–2456.

Kennerdell, J. R., and Carthew, R. W. (1998). Use of dsRNA-mediated genetic interference to demonstrate that frizzled and frizzled 2 act in the wingless pathway. *Cell* **95**, 1017–1026.

Ketting, R. F., Haverkamp, T. H., van Luenen, H. G., and Plasterk, R. H. (1999). Mut-7 of *C. elegans*, required for transposon silencing and RNA interference, is a homolog of Werner syndrome helicase and RNaseD. *Cell* **99,** 133–141.

Ketting, R. F., and Plasterk, R. H. (2000). genetic link between co-suppression and RNA interference in *C. elegans*. *Nature* **404,** 296–298.

Korf, I., Fan, Y., and Strome, S. (1998). The Polycomb group in *Caenorhabditis elegans* and maternal control of germline development. *Development* **125,** 2469–2478.

Lee, J. T., and Lu, N. (1999). Targeted mutagenesis of *Tsix* leads to nonrandom X inactivation. *Cell* **99,** 47–57.

Lee, R. C., Feinbaum, R. L., and Ambros, V. (1993). The *C. elegans* heterochronic gene lin-4 encodes small RNAs with antisense complementarity to lin-14. *Cell* **75,** 843–854.

Li, A. W., and Murphy, P. R. (2000). Expression of alternatively spliced FGF-2 antisense RNA transcripts in the central nervous system: Regulation of FGF-2 mRNA translation. *Mol. Cell Endocrinol.* **162,** 69–78.

Lin, R., and Avery, L. (1999). RNA interference. Policing rogue genes. *Nature* **402,** 128–129.

Lohmann, J. U., Endl, I., and Bosch, T. C. (1999). Silencing of developmental genes in *Hydra*. *Dev. Biol.* **214,** 211–214.

Lynn, K., Fernandez, A., Aida, M., Sedbrook, J., Tasaka, M., Masson, P., and Barton, M. K. (1999). The *PINHEAD/ZWILLE* gene acts pleiotropically in *Arabidopsis* development and has overlapping functions with the *ARGONAUTE1* gene. *Development* **126,** 469–481.

Maine, E. M. (2000). A conserved mechanism for post-transcriptional gene silencing? *Genome Biol.* **1,** 10181–10184.

Marx, J. (2000). Interfering with gene expression. *Science* **288,** 1370–1372.

Mette, M. F., Aufsatz, W., van der Winden, J., Matzke, M. A., and Matzke, A. J. (2000). Transcriptional silencing and promoter methylation triggered by double-stranded RNA. *EMBO J.* **19,** 5194–5201.

Montgomery, M. K., and Fire, A. (1998). Double-stranded RNA as a mediator in sequence-specific genetic silencing and co-suppression. *Trends Genet.* **14,** 255–258.

Montgomery, M. K., Xu, S., and Fire, A. (1998). RNA as a target of double-stranded RNA-mediated genetic interference in *Caenorhabditis elegans*. *Proc. Natl. Acad. Sci. USA* **95,** 15502–15507.

Moss, E. G., Lee, R. C., and Ambros, V. (1997). The cold shock domain protein LIN-28 controls developmental timing in *C. elegans* and is regulated by the lin-4 RNA. *Cell* **88,** 637–646.

Mourrain, P., Beclin, C., Elmayan, T., Feuerbach, F., Godon, C., Morel, J. B., Jouette, D., Lacombe, A. M., Nikic, S., Picault, N., Remoue, K., Sanial, M., Vo, T. A., and Vaucheret, H. (2000). *Arabidopsis* SGS2 and SGS3 genes are required for posttranscriptional gene silencing and natural virus resistance. *Cell* **101,** 533–542.

Moussian, B., Schoof, H., Haecker, A., Jurgens, G., and Laux, T. (1998). Role of the *ZWILLE* gene in the regulation of central shoot meristem cell fate during *Arabidopsis* embryogenesis. *EMBO J.* **17,** 1799–1809.

Napoli, C., Lemieux, C., and Jorgensen, R. (1990). Introduction of a chalcone synthase gene into *Petunia* results in reversible co-suppression of homologous genes in trans. *Plant Cell* **2,** 279–289.

Ngo, H., Tschudi, C., Gull, K., and Ullu, E. (1998). Double-stranded RNA induces mRNA degradation in *Trypanosoma brucei*. *Proc. Natl. Acad. Sci. USA* **95,** 14687–14692.

Olsen, P. H., and Ambros, V. (1999). The lin-4 regulatory RNA controls developmental timing in *Caenorhabditis elegans* by blocking LIN-14 protein synthesis after the initiation of translation. *Dev. Biol.* **216,** 671–680.

Palauqui, J. C., Elmayan, T., Pollien, J. M., and Vaucheret, H. (1997). Systemic acquired silencing: Transgene-specific post-transcriptional silencing is transmitted by grafting from silenced stocks to non-silenced scions. *EMBO J.* **16,** 4738–4745.

Palauqui, J. C., and Vaucheret, H. (1998). Transgenes are dispensable for the RNA degradation step of cosuppression. *Proc. Natl. Acad. Sci. USA* **95,** 9675–9680.

Parrish, S., Fleenor, J., Xu, S., Mello, C., and Fire, A. (2000). Functional anatomy of a dsRNA trigger. Differential requirement for the two trigger strands in RNA interference. _Mol. Cell_ **6,** 1077–1087.

Pasquinelli, A. E., Reinhart, B. J., Slack, F., Martindale, M. Q., Kuroda, M. I., Maller, B., Hayward, D. C., Ball, E. E., Degnan, B., Muller, P., Spring, J., Srinivasan, A., Fishman, M., Finnerty, J., Corbo, J., Levine, M., Leahy, P., Davidson, E., and Ruvkun, G. (2000). Conservation of the sequence and temporal expression of let-7 heterochronic regulatory RNA. _Nature_ **408,** 86–89.

Piano, F., Schetterdagger, A. J., Mangone, M., Stein, L., and Kemphues, K. J. (2000). RNAi analysis of genes expressed in the ovary of _Caenorhabditis elegans_. _Curr. Biol._ **10,** 1619–1622.

Plasterk, R. H., and Ketting, R. F. (2000). The silence of the genes. _Curr. Opin. Genet. Dev._ **10,** 562–567.

Rappleye, C. A., Paredez, A. R., Smith, C. W., McDonald, K. L., and Aroian, R. V. (1999). The coronin-like protein POD-1 is required for anterior-posterior axis formation and cellular architecture in the nematode _Caenorhabditis elegans_. _Genes Dev._ **13,** 2838–2851.

Reinhart, B. J., Slack, F. J., Basson, M., Pasquinelli, A. E., Bettinger, J. C., Rougvie, A. E., Horvitz, H. R., and Ruvkun, G. (2000). The 21–nucleotide let-7 RNA regulates developmental timing in _Caenorhabditis elegans_. _Nature_ **403,** 901–906.

Romano, N., and Macino, G. (1992). Quelling: Transient inactivation of gene expression in _Neurospora crassa_ by transformation with homologous sequences. _Mol. Microbiol._ **6,** 3343–3353.

Rosenberg, U. B., Preiss, A., Seifert, E., Jackle, H., and Knipple, D. C. (1985). Production of pheno-copies by Kruppel antisense RNA injection into _Drosophila_ embryos. _Nature_ **313,** 703–706.

Sanchez Alvarado, A., and Newmark, P. A. (1999). Double-stranded RNA specifically disrupts gene expression during planarian regeneration. _Proc. Natl. Acad. Sci. USA_ **96,** 5049–5054.

Schmidt, A., Palumbo, G., Bozzetti, M. P., Tritto, P., Pimpinelli, S., and Schafer, U. (1999). Genetic and molecular characterization of sting, a gene involved in crystal formation and meiotic drive in the male germ line of _Drosophila melanogaster_. _Genetics_ **151,** 749–760.

Schupbach, T., and Wieschaus, E. (1991). Female sterile mutations on the second chromosome of _Drosophila melanogaster_. II. Mutations blocking oogenesis or altering egg morphology. _Genetics_ **129,** 1119–1136.

Sharp, P. A. (1999). RNAi and double-strand RNA. _Genes Dev._ **13,** 139–141.

Sharp, P. A. (2001). RNA interference—2001. _Genes Dev._ **15,** 485–490.

Shin, T. H., Yasuda, J., Rocheleau, C. E., Lin, R., Soto, M., Bei, Y., Davis, R. J., and Mello, C. C. (1999). MOM-4, a MAP kinase kinase kinase-related protein, activates WRM-1/LIT-1 kinase to transduce anterior/posterior polarity signals in _C. elegans_. _Mol. Cell_ **4,** 275–280.

Slack, F. J., Basson, M., Liu, Z., Ambros, V., Horvitz, H. R., and Ruvkun, G. (2000). The lin-41 RBCC gene acts in the _C. elegans_ heterochronic pathway between the let-7 regulatory RNA and the LIN-29 transcription factor. _Mol. Cell_ **5,** 659–669.

Smardon, A., Spoerke, J. M., Stacey, S. C., Klein, M. E., Mackin, N., and Maine, E. M. (2000). EGO-1 is related to RNA-directed RNA polymerase and functions in germ-line development and RNA interference in _C. elegans_. _Curr. Biol._ **10,** 169–178.

Smith, N. A., Singh, S. P., Wang, M. B., Stoutjesdijk, P. A., Green, A. G., and Waterhouse, P. M. (2000). Total silencing by intron-spliced hairpin RNAs. _Nature_ **407,** 319–320.

Stauber, M., Taubert, H., and Schmidt-Ott, U. (2000). Function of bicoid and hunchback homologs in the basal cyclorrhaphan fly _Megaselia_ (Phoridae). _Proc. Natl. Acad. Sci. USA_ **97,** 10844–10849.

Stolt, P., and Zillig, W. (1993). Antisense RNA mediates transcriptional processing in an archaebac-terium, indicating a novel kind of RNase activity. _Mol. Microbiol._ **7,** 875–882.

Strickland, S., Huarte, J., Belin, D., Vassalli, A., Rickles, R. J., and Vassalli, J. D. (1988). Antisense RNA directed against the 3′ noncoding region prevents dormant mRNA activation in mouse oocytes. _Science_ **241,** 680–684.

Tabara, H., Grishok, A., and Mello, C. C. (1998). RNAi in C. elegans: Soaking in the genome sequence. _Science_ **282,** 430–431.

Tabara, H., Sarkissian, M., Kelly, W. G., Fleenor, J., Grishok, A., Timmons, L., Fire, A., and Mello, C. C. (1999). The rde-1 gene, RNA interference, and transposon silencing in C. elegans. Cell **99**, 123–132.

Tavernarakis, N., Wang, S. L., Dorovkov, M., Ryazanov, A., and Driscoll, M. (2000). Heritable and inducible genetic interference by double-stranded RNA encoded by transgenes. Nature Genet. **24**, 180–183.

Terryn, N., and Rouze, P. (2000). The sense of naturally transcribed antisense RNAs in plants. Trends Plant Sci. **5**, 394–396.

Timmons, L., and Fire, A. (2000). Specific interference by ingested dsRNA. Nature **395**, 854.

Tuschl, T., Zamore, P. D., Lehmann, R., Bartel, D. P., and Sharp, P. A. (1999). Targeted mRNA degradation by double-stranded RNA in vitro. Genes Dev. **13**, 3191–3197.

van Biesen, T., Soderbom, F., Wagner, E. G., and Frost, L. S. (1993). Structural and functional analyses of the FinP antisense RNA regulatory system of the F conjugative plasmid. Mol. Microbiol. **10**, 35–43.

van der Krol, A. R., Mur, L. A., Beld, M., Mol, J. N., and Stuitje, A. R. (1990). Flavonoid genes in Petunia: Addition of a limited number of gene copies may lead to a suppression of gene expression. Plant Cell **2**, 291–299.

Voinnet, O., Lederer, C., and Baulcombe, D. C. (2000). A viral movement protein prevents spread of the gene silencing signal in Nicotiana benthamiana. Cell **103**, 157–167.

Voinnet, O., Vain, P., Angell, S., and Baulcombe, D. C. (1998). Systemic spread of sequence-specific transgene RNA degradation in plants is initiated by localized introduction of ectopic promoterless DNA. Cell **95**, 177–187.

Wassenegger, M., Heimes, S., Riedel, L., and Sanger, H. L. (1994). RNA-directed de novo methylation of genomic sequences in plants. Cell **76**, 567–576.

Waterhouse, P. M., Graham, M. W., and Wang, M. B. (1998). Virus resistance and gene silencing in plants can be induced by simultaneous expression of sense and antisense RNA. Proc. Natl. Acad. Sci. USA **95**, 13959–13964.

Wianny, F., and Zernicka-Goetz, M. (2000). Specific interference with gene function by double-stranded RNA in early mouse development. Nature Cell Biol. **2**, 70–75.

Wightman, B., Ha, I., and Ruvkun, G. (1993). Posttranscriptional regulation of the heterochronic gene lin-14 by lin-4 mediates temporal pattern formation in C. elegans. Cell **75**, 855–862.

Wilson, J. E., Connell, J. E., and Macdonald, P. M. (1996). aubergine enhances oskar translation in the Drosophila ovary. Development **122**, 1631–1639.

Wolffe, A. P., and Matzke, M. A. (1999). Epigenetics: Regulation through repression. Science **286**, 481–486.

Wu-Scharf, D., Jeong, B., Zhang, C., and Cerutti, H. (2000). Transgene and transposon silencing in Chlamydomonas reinhardtii by a DEAH-box RNA helicase. Science **290**, 1159–1162.

Yang, D., Lu, H., and Erickson, J. W. (2000). Evidence that processed small dsRNAs may mediate sequence-specific mRNA degradation during RNAi in Drosophila embryos. Curr. Biol. **10**, 1191–1200.

Zamore, P. D., Tuschl, T., Sharp, P. A., and Bartel, D. P. (2000). RNAi: Double-stranded RNA directs the ATP-dependent cleavage of mRNA at 21 to 23 nucleotide intervals. Cell **101**, 25–33.

Zou, C., Zhang, Z., Wu, S., and Osterman, J. C. (1998). Molecular cloning and characterization of a rabbit eIF2C protein. Gene **211**, 187–194.

12

Antisense RNAs in Bacteria and Their Genetic Elements

E. Gerhart H. Wagner*
Institute of Cell and Molecular Biology
Biomedical Center
Uppsala University
751 24 Uppsala, Sweden

Shoshy Altuvia
Department of Molecular Genetics and Biotechnology
The Hebrew University—Hadassah Medical School
91120 Jerusalem, Israel

Pascale Romby
UPR 9002, CNRS-IBMC
Institut de Biologie Moléculaire et Cellulaire
67084 Strasbourg Cedex, France

*To whom correspondence should be addressed: E-mail: gerhart.wagner@icm.uu.se; Telephone: +46 18 4714866; Fax: +46 18 530396.

Copyright 2002, Elsevier Science (USA).
0065-2660/02 $35.00

ABSTRACT

Antisense RNA-mediated regulation is widespread in bacteria. Most antisense RNA control systems have been found in plasmids, phages, and transposons. Fewer examples were identified in bacterial chromosomes. This chapter summarizes our current knowledge about antisense RNAs with respect to their occurrence, their biological roles, and their diverse mechanisms of action. Examples of *cis*- or *trans*-encoded antisense RNAs are discussed, and their properties compared. Most antisense RNAs are posttranscriptionally acting inhibitors of target genes, but a few examples of activator antisense RNAs are known. The implications of RNA structure on topologically and kinetically favored binding pathways are addressed, and solutions that have evolved to permit productive interactions between intricately folded RNAs are discussed. Finally, we describe how particular properties of individual antisense/target RNA systems match their respective biological roles. © 2002, Elsevier Science (USA).

I. INTRODUCTION—ANTISENSE PRINCIPLE AND GENE REGULATION

Homology effects—in a broad sense—underlie a variety of effects that alter gene expression in all organisms. The common denominator in all these diverse biological phenomena is some form of interaction between nucleic acids, based either on homology/sequence similarity or on complementarity. Antisense RNAs are a class of regulators of gene expression that act on target RNAs via sequence complementarity. In the vast majority of the known cases this entails inhibition of target RNA function. In a few cases, antisense RNAs can activate gene expression. In other cases, in particular in eukaryotes, a variety of processes make use of RNA sequence complementarity to *mediate* or *facilitate* biological processes (RNA editing, splicing, etc.). However, in the context of this article, a bona-fide antisense RNA is an independent regulator of one or more target genes, whose effect is exerted posttranscriptionally. Furthermore, we will be concerned only with systems in which antisense RNAs play biological roles in nature, and disregard all cases in which antisense RNAs have artificially been introduced to interfere with gene expression.

Following this definition, very few cases in eukaryotes (Hildebrandt and Nellen, 1992; Reinhart et al., 2000; Wightman et al., 1993) and archea (Stolt and Zillig, 1993a) qualify, while the overwhelming majority of natural antisense RNA-regulated systems are found in bacteria, in particular in their accessory genetic elements—plasmids, phages, and transposons (Wagner and Simons, 1994). From the examples of bacterial antisense RNAs known, it can be concluded that they are generally small (most often 55–150 nt long), diffusible, and untranslated. They are in addition characterized by distinct secondary and tertiary structures, and particular motifs appear to correlate with regulatory performance (Figure 12.1, see color insert). Contrary to earlier concepts, our current understanding is that the high specificity and efficiency displayed by antisense RNA control systems cannot be explained by simple sequence complementarity-dependent hybridization, but is rather a consequence of interactions between intricately designed three-dimensional structures of the interactant RNAs.

This review will focus on the biology of a number of well-characterized bacterial systems, give an overview of the mechanisms of action used by antisense RNAs, discuss the particular properties of antisense RNAs with respect to structure, binding kinetics, and topology, and set some characteristic features in context with their biological functions. Reviews on some aspects of bacterial antisense RNA have been published previously (Delihas, 1995; Eguchi et al., 1991; Nordström and Wagner, 1994; Wagner and Brantl, 1998; Wagner and Simons, 1994; Zeiler and Simons, 1998).

II. NATURALLY OCCURRING ANTISENSE RNA CONTROL SYSTEMS

The first natural antisense RNAs were discovered in 1981. Tomizawa and co-workers showed that a small, plasmid-encoded regulator controlled the copy number of a plasmid, ColE1 (Tomizawa and Itoh, 1981; Tomizawa et al., 1981). The Nordström group identified antisense RNA-mediated control in a low-copy-number plasmid called R1 (Stougaard et al., 1981). In subsequent years, it became apparent that the negative control circuits that operate to "count" plasmids in a cell, and that correct deviations from the default copy number either use so-called iteron control (not covered here, but see Chattoraj, 2000) or antisense RNAs as the principal regulatory elements. Since much of our current understanding of antisense RNA functions and mechanisms is derived from several well-characterized plasmids, a substantial part of this review will be devoted to these systems.

Table 12.1. Prokaryotic Antisense RNAs, Occurrence, Mechanisms, and Biological Roles

Antisense RNA[a]	Biological function	Target(s)	Mechanism	Comments	References[b]
Plasmids					
IncFII relatives—CopA RNA	Replication control	*repA* mRNA	Translation inhibition		Light and Molin, 1982; Womble et al., 1985
IncB and Inclα relatives—RNAI	Replication control	*repZ* mRNA	Translation inhibition, inhibition of activator pseudoknot		Hama et al., 1990; Praszkier et al., 1991
ColE1 relatives—RNAI	Replication control	RNAII (preprimer)	Primer maturation	Rom protein involvement	Tomizawa et al., 1981
pT181 and pIP501 relatives—RNAI	Replication control	*repC* mRNA	Transcription attenuation	Unusually stable antisense RNA	Brantl et al., 1993; Novick et al., 1989
ColE2 relatives—RNAI	Replication control	*rep* mRNA	Translation inhibition		Takechi et al., 1994
pLS1	Replication control	*cop-rep* mRNA	Translation inhibition		del Solar and Espinosa, 1992
R1162—ct RNA	Replication control	*repI* mRNA	Translation inhibition		Kim and Meyer, 1986
pVT736-1—Cop RNA	Replication control	*rep* mRNA	Translation inhibition (?)		Galli and Leblanc, 1995
R1 and relatives—Sok RNA	Postsegregational killing	*mok/hok* mRNA	Translation inhibition	Activity in absence of gene locus	Gerdes et al., 1997
pAD1—RNAII	Postsegregational killing	RNAI	Translation inhibition	First case in a Gram-positive host	Greenfield et al., 2000
IncFI/FII relatives—FinP RNA	Control of conjugation	*traJ* mRNA	Translation inhibition	Antisense RNA stabilized by FinO protein	Frost et al., 1989; Koraimann et al., 1991
Transposons					
IS10—RNA-OUT	Transposition	RNA-IN (*tnp* mRNA)	Translation inhibition		Ma and Simons, 1990
IS30—RNA-C	Transposition	*tnp* mRNA	Inhibition of translation elongation (?)		Arini et al., 1997

	Function	Target	Mechanism	Notes	Reference[a],[b]
Bacteriophages					
λ—OOP RNA	Lysis/lysogeny switch	*cII* mRNA	mRNA stability		Krinke and Wulff, 1990
P1/P4—*c4/cI* RNA	Lysis/lysogeny switch	*icd-ant* mRNA	Transcription termination		Citron and Schuster, 1990; Sabbattini et al., 1995
P22—Sar RNA	Lysis/lysogeny switch	*arc-ant* mRNA	Translation inhibition		Liao et al., 1987
P22—Sas RNA	Superinfection override	*sieB-esc* mRNA	Switch of translation start sites		Ranade and Poteete, 1993
ΦH-T$_{ant}$	Lysis/lysogeny switch	T1 RNA	RNA processing		Stolt and Zillig, 1993
Bacterial					
E. coli MicF	Stress responses, osmoregulation	*ompF* mRNA	Translation inhibition, mRNA stability (?)		Aiba et al., 1987
E. coli OxyS	Oxidative stress	*fhlA* mRNA *rpoS* mRNA	Translation inhibition Translation inhibition/sequestration of Hfq	Regulator of ~40 genes, acts as an antimutator	Altuvia et al., 1998; Zhang et al., 1998
E. coli DrsA	Stress responses	*rpoS* mRNA *hns* mRNA	Translation activation Translation inhibition/mRNA stability	Two (or more) targets, one activated	Sledjeski et al., 1996
E. coli DicF	Cell division	*ftsZ* mRNA	Translation inhibition		Tetart and Bouche, 1992
E. coli Sof	Cell killing	*gef* mRNA	Translation inhibition	Encoded by prophage Chromosomal *hok/sok* homolog	Poulsen et al., 1991
S. aureus RNAIII	Virulence	*hla* mRNA	Activation of translation	Activator antisense RNA	Morfeldt et al., 1995; Novick et al., 1993
V. anguillarum RNAα	Iron transport regulation	*fatA/fatB* mRNAs	mRNA stability		Salinas et al., 1993
B. subtilis tRNAs	aminoacyl-tRNA synthetase regulation	aa-tRNA synthetase mRNAs	Transcription antitermination		Grundy and Henkin, 1994
C. acetobutylicum p3 RNA	Nitrogen regulation (?)	*glnA* mRNA	Translation inhibition (?)		Jansen et al., 1990

[a]Prototypical examples are given; see text for details.
[b]Additional references can be found in the text.

Moving ahead in time from the early days of antisense RNA discovery, we can conclude that these regulators of gene expression do not represent a rare phenomenon but rather are widespread. To illustrate the plethora of different regulatory systems, a necessarily incomplete list of examples is given in Table 12.1. Clearly, inhibition mechanisms used by antisense RNAs vary considerably. Binding pathways differ, and many systems accomplish inhibition without full RNA/RNA pairing. In particular, the finding that a class of so-called riboregulators, i.e., small RNAs with regulatory function, mostly act as antisense RNAs, but do so with only partial target complementarity, broadens our view on the versatility of RNA-mediated regulation. Thus, one can distinguish antisense RNA systems that are either *cis* or *trans* encoded. In the former case, the same DNA sequence encodes both the antisense and the target RNA. These RNAs are transcribed in an overlapping fashion from both strands and, hence, are fully complementary. In the latter case, the gene encoding the regulator is located at a locus different from that encoding the target(s), and thus the RNAs exhibit only partial complementarity.

III. ANTISENSE RNA MECHANISMS AND BIOLOGICAL CONTEXT

In the simplest case, an antisense RNA is fully complementary to its target. Since the target usually is an mRNA, the most likely mechanism of inhibition would involve the formation of an antisense/target RNA duplex; if such a structure were present at the ribosomal loading site, translation would be blocked. Such mechanisms do indeed operate. However, antisense RNAs are involved in even more sophisticated activities. This will be exemplified by biological control systems in which antisense RNAs, e.g., inhibit *or* activate translation, induce premature target RNA termination, facilitate mRNA decay, or inhibit formation of a mature replication primer (Table 12.1).

A. Plasmid systems

Bacterial plasmids share a problem with all other selfish elements. Their presence is (normally) a burden for host cells, and thus the host would benefit from elimination of the intruder. Large naturally occurring low-copy-number plasmids are—within the limits of feasible experiments—never lost, even in the absence of selective pressure (Nordström and Aagaard-Hansen, 1984). This is surprising if one considers the statistically expected loss rates. Thus, it is clear that ingenious maintenance functions have evolved: plasmids carry several genetic "packages" that contribute to stable inheritance (Bingle and Thomas, 2001).

The problem solved by all copy number control systems can be stated as follows. Fluctuations in the number of plasmids among individual cells are unavoidable. Great decreases or increases are deleterious; lower copy numbers increase the risk of plasmid loss (Nordström *et al.*, 1984; Wagner and Brantl, 1998), and uncontrolled high replication rates may lead to "runaway" (Uhlin and Nordström, 1978), which often kills the host. In the groups of plasmids discussed here, antisense RNAs are the key elements that control replication. These antisense RNAs are constitutively synthesized and metabolically unstable (for exceptions, see below), and thus any change in plasmid concentration will be reflected in corresponding concentration changes of the regulator. These concentration changes are "sensed," leading to altered replication frequencies; i.e., antisense RNA-mediated replication control works through a negative control circuit—increased copy number entails increased antisense RNA concentration, which in turn inhibits a function that is rate-limiting for replication. Vice versa, a drop in copy number decreases inhibition and thus results in increased replication. This scenario holds true for all plasmids described below, although the underlying mechanisms differ.

Plasmids often encode additional systems that aid stable maintenance. Partition loci encode a centromere-like system that ensures segregation of at least one plasmid copy to each of the daughter cells (Gerdes *et al.*, 2000; Hiraga, 1992). Host killing systems (also called postsegregational killing systems) encode a "latent potential" to kill cells. As long as plasmids are present, cells survive, whereas plasmid-less daughter cells are killed. Killer systems can be proteic (killer and antidote are proteins; Jensen and Gerdes, 1995) or antisense RNA-regulated (Gerdes *et al.*, 1997; see below). Finally, a problem which indirectly can cause plasmid loss is dimerization. When two plasmid copies undergo homologous recombination and thus form a dimer, the doubled probability of replication of this molecule (due to the presence of two origins of replication) will eventually result in dimer-only cells ("dimer catastrophe"; Summers *et al.*, 1993). Having fewer molecules per cell statistically increases plasmid loss. Thus, plasmid-encoded site-specific recombination systems resolve this potentially dangerous situation by converting dimers back into monomers. In at least one case, the *cer* recombination system of ColE1, a small RNA, Rcd, appears to be involved in delaying cell division, thereby increasing the time allotted for the site-specific recombinase to act. Whether or not Rcd is an antisense RNA, and if so, the identity of its targets, is at present unknown (Patient and Summers, 1993; Sharpe *et al.*, 1999).

Many bacterial plasmids, in addition to the ones discussed below, use antisense RNAs for regulation (e.g., del Solar *et al.*, 1995; del Solar and Espinosa, 1992, 2000; Galli and Leblanc, 1995; Sugiyama and Itoh, 1993; Takechi *et al.*, 1994). However, due to space limitations, we will focus on the best-characterized ones, with an emphasis on differences in mechanisms.

1. Replication control in R1-related plasmids

A great number of low-copy-number plasmids, belonging to different incompatibility groups (FII, IFc, FIII, and others), exhibit a similar genetic organization. Possibly the best-studied case with respect to antisense control is R1, discussed here as a representative of this group. R1 resides in *Escherichia coli* and closely related bacteria. Its basic replicon, defined as the region containing all genes and sites required for wild-type copy number control (Rosen *et al.*, 1981), carries *oriR1* (replication origin) and the *repA*, *tap*, *copA*, and *copB* genes. The RepA protein is the plasmid-encoded initiator, rate-limiting for replication. Thus, by controlling its synthesis, the replication frequency can be regulated. Regulation occurs on two levels. One control element is a repressor protein, CopB, and is not discussed here (Dong *et al.*, 1985; Riise *et al.*, 1982). The second, main control element is an antisense RNA, CopA. CopA is unstable (Söderbom and Wagner, 1998), untranslated, and highly structured (Wagner and Nordström, 1986; Figure 12.1). Its target, denoted CopT, is located in the leader region of the *repA* mRNA (Light and Molin, 1982). Binding of CopA to CopT sterically blocks initiation of translation of the *tap* leader peptide reading frame (Malmgren *et al.*, 1996), and also results in RNaseIII-dependent cleavage, which, however, has only minor effects on control (Blomberg *et al.*, 1990). Translation of *tap* is required for translation of *repA* (translational coupling), because a stable RNA structure sequesters the *repA* translation initiation site (Blomberg *et al.*, 1992, 1994; Wu *et al.*, 1992). Hence, CopA inhibits *repA* translation via inhibition of *tap* translation. Genetic evidence has shown that copy-number-up mutations map to a region of the *copA* gene that encodes the loop of the major CopA stem-loop structure (Brady *et al.*, 1983; Givskov and Molin, 1984). Many of these mutations also change incompatibility properties; i.e., a single base change leads to the inability of mutant CopA to recognize and inhibit a wild-type CopT target encoded by a co-resident plasmid. These drastic effects on target binding rate and specificity, modifying copy number and incompatibility groups, respectively, pinpoint the central importance of the CopA and CopT loop regions for the rate-limiting step in binding and the efficiency of control (Givskov and Molin, 1984; Persson *et al.*, 1988, 1990a, 1990b).

As will become apparent, the theme of interaction between highly structured antisense and target RNAs, initiating by defined loop–loop contacts, is a recurrent one, valid for most of the cases discussed here. With respect to the consequences of antisense RNA binding, R1 is one of many systems in which the ultimate effect is that of translational inhibition (Table 12.1, Figure 12.2, see color insert).

2. Replication control in IncB/IncIα-related plasmids

A family of low-copy-number plasmids, with representatives classified as IncB, IncIα, IncZ, IncK, and IncL/M, is similar to the R1-family but displays interesting differences in its mechanism of regulation . This is illustrated by the two best-studied cases, ColIb-P9 (IncIα plasmid) and pMU720 (IncB). A gene corresponding to *copB* of R1 is absent. In these plasmids, inhibition of *rep* expression involves a long-distance RNA pseudoknot (Asano *et al.*, 1991a, 1991b; Wilson *et al.*, 1993). As in R1, a leader peptide reading frame, *repY* (in ColIb-P9), must be translated to permit RepZ protein synthesis (Hama *et al.*, 1990a; Praszkier *et al.*, 1992). However, the main function of *repY* translation is to disrupt a stem-loop at the *rep* ribosome loading site. This in turn permits the formation of a short helix between the target loop and the disrupted stem, located ∼100 nt apart. This long-distance pseudoknot activates *repZ* translation. The indispensable copy-number regulator is a ∼70-nt antisense RNA, Inc RNA/RNAI, encoded in a position equivalent to that of *copA* (Praszkier *et al.*, 1989; Shiba and Mizobuchi, 1990). Its function is twofold: interaction with its target blocks translation of *repY*, and since the site of interaction involves the set of nucleotides required for pseudoknot formation, activation is blocked as well (Asano and Mizobuchi, 1998a; Wilson *et al.*, 1997). That is, this family of plasmids uses antisense RNA for translational inhibition, but with an unusual twist: inhibition works through prevention of an activation pseudoknot structure (Figure 12.2).

3. Replication control in ColE1-related plasmids

ColE1 is the prototype of many closely related high-copy-number plasmids. Its replication control circuit has been investigated in great detail (for a review, see Eguchi *et al.*, 1991). ColE1 does not require any plasmid-encoded proteins for regulation of its copy number. Host RNA polymerase synthesizes a preprimer, RNAII, of ∼550 nt in length. During synthesis, this transcript undergoes specific conformational changes which are required for activity (Masukata and Tomizawa, 1984, 1986; Polisky *et al.*, 1990; Wong and Polisky, 1985). The active conformation of this RNA forms a persistent hybrid which involves two regions of contact between RNAII and the DNA in the *ori* region (Masukata and Tomizawa, 1990). The RNA in the hybrid is cleaved by RNase H and converted to a mature primer for replication (Itoh and Tomizawa, 1980, 1982; Masukata and Tomizawa, 1986, 1990). DNA polymerase I uses the newly generated 3'-OH and extends it, starting leading strand synthesis.

The negative control element is RNAI, a 108-nt-long antisense RNA complementary to the 5'-most region of RNAII (Figure 12.1; Lacatena and Cesareni, 1981; Tomizawa and Itoh, 1981). This RNA consists of three stem-loops and an unstructured 5'-tail (Tamm and Polisky, 1983). The sites of mutations affecting copy number and incompatibility properties identify here as well the loop regions as the most important determinants for specificity and rate of binding (Lacatena and Cesareni, 1983; Muesing *et al.*, 1981). Binding of RNAI to RNAII prevents the refolding of the nascent preprimer; the structure formed upon binding is incompatible with persistent RNAII-DNA hybrid formation, and hence primer maturation is prevented. A peculiarity of this system is an activity window: RNAI can bind to nascent RNAII of all lengths (provided that the latter contains the target region), however, it can only block primer formation when it interacts with target RNA of 100–150 nt in length (Tomizawa, 1986). Binding at a later stage does not result in inhibition.

Taken together, the ColE1-plasmid family illustrates an antisense RNA mechanism that does *not* affect expression of a gene, but rather the activity of a target RNA by induction of a nonfunctional conformation (Figure 12.2).

4. Replication control in pT181- and pIP501-related plasmids of Gram-positive bacteria

Replication control of some plasmids residing in Gram-positive bacteria involves transcriptional attenuation, a regulatory mechanism so far not found in plasmids of Gram-negative hosts. Novick and co-workers were the first to describe this control mode in the rolling-circle-replicating plasmid pT181 (Novick *et al.*, 1989). Subsequently, the same mechanism was demonstrated for the theta-replicating plasmids pIP501 (Brantl *et al.*, 1993) and pAMβ1 (Le Chatelier *et al.*, 1996). Regulation works as follows. The nascent *rep* mRNA can adopt two mutually exclusive conformations. One conformation permits transcription throughout the *rep* gene, allowing for the synthesis of the replication rate-limiting Rep protein. The alternative conformation is induced when the antisense RNA binds to the target region—as in the ColE1 case, this must occur within a short time window in order to be effective. This alternative folding results in premature termination of transcription immediately upstream of the *rep* reading frame, and thus no Rep protein is synthesized. *In vitro* work has confirmed these conclusions and added structural and kinetic details (Brantl and Wagner, 1994, 2000). Thus, the antisense RNA mechanism used by these plasmids affects gene expression by aborting elongation of a target transcript (Figure 12.2).

5. Postsegregational killing systems

The prototype of antisense RNA-controlled host killing loci is *hok/sok* of R1. This system has been elucidated in minute detail by the Gerdes group (Gerdes *et al.*, 1997). Many *hok/sok*-like loci have been identified, and all conform to the same general theme (Nielsen *et al.*, 1991). The *hok/sok* locus of plasmid R1 encodes a small protein, Hok (host killing), which causes damage to the bacterial membrane leading to cell death (Gerdes *et al.*, 1986). Translation of an overlapping reading frame, *mok* (modulator of killing) is required for *hok* expression (Thisted and Gerdes, 1992). The antidote is the Sok (suppressor of killer) antisense RNA of 67 nt (Figure 12.1). Initially, the way by which such a system could work seemed paradoxical: killing occurs when plasmids are no longer present, i.e., in the *absence* of the locus that is responsible for this phenotype. It is now clear that peculiar features of the *hok* mRNA explain how this is possible. This message accumulates as an extremely compact RNA, with many stable stem-loop structures and a flush 5′-3′-end pairing. This confers both metabolic stability (long half-life) and an inert state; the mRNA is neither translated nor a target for Sok (Franch and Gerdes, 1996). At a low rate, processing near the 3′-end generates a truncated mRNA which is *both* translationally active *and* able to bind the inhibitor (Franch *et al.*, 1997). These properties resolve the paradox and lead to the following model. In plasmid-carrying cells, an active mRNA generated by truncation will be attacked by Sok, followed by inhibition of translation and decay of the message. By contrast, plasmid-free segregants will rapidly experience a loss of Sok due to its metabolic instability, whereas the reservoir of inactive but activatible mRNA persists. Thus, when 3′ truncation occurs, the absence of the inhibitor permits *hok* translation, and the host cell is killed. Mechanistically, inhibition in systems of the *hok/sok* family probably occurs by a combination of effects exerted at the level of translation and enhanced message degradation. Interestingly, a number of *hok/sok* homologs have been found in bacterial chromosomes (Pedersen and Gerdes, 1999). Their biological functions are at present still obscure, but may be related to stress responses (Gerdes, 2000).

Recently, the first plasmid-borne killer system was identified in a Gram-positive, *Enterococcus faecalis*. The *par* stability locus of plasmid pAD1 encodes a toxic protein, Fst, and an antisense RNA, RNAII. The overall properties of this system mirror that of *hok/sok*, although it differs in detail. The genes encoding toxin and antidote are transcribed convergently and overlap in a bi-directional terminator. However, the two RNAs contain two additional short regions of complementarity due to two repeats encoded far apart. These two motifs, in the mRNA, are located close to the *fst* ribosome-binding site (RBS). Interaction between antisense and target RNA within these two complementary regions, and in the complementary terminators, suffices to block ribosome binding and thus *fst* translation (Greenfield *et al.*, 2000; Greenfield and Weaver, 2000). As in *hok/sok*, the toxin mRNA is stable and initially translationally inert, whereas the antisense RNA is unstable, consistent with the requirements of a system that must be able to become active in the absence of its locus. Mechanistically, *par* appears to blend properties of *cis*- and *trans*-encoded systems. Two segments of the interactant RNAs are fully complementary (the terminators), and two other complementary segments are encoded by spatially separated DNA regions.

6. Conjugation

Conjugal transfer of plasmids such as F and R1 involves a large number of genes, encoded by the Tra-operon (Firth *et al.*, 1996). Initiation of this complex process requires synthesis of an activator, TraJ. An antisense RNA, FinP (Figure 12.1), and a protein, FinO, cooperate to repress *traJ* expression (Koraimann *et al.*, 1996; van Biesen and Frost, 1994). The *finP* gene overlaps with the leader region of *traJ*. FinP appears to block translation of *traJ* by binding to its RBS. FinO confers intracellular stability to the otherwise unstable FinP RNA, probably by shielding it from the action of degrading enzymes (Ghetu *et al.*, 1999; Jerome *et al.*, 1999). Possibly, FinO also enhances the binding rate of FinP to its target (van Biesen and Frost, 1994). As in most of the previously discussed examples, mutations affecting the activity of FinP reside mostly in the two loop regions of this RNA (Koraimann *et al.*, 1991, 1996; van Biesen *et al.*, 1993).

B. Transposons

Transposons and IS (insertion sequence) elements are mobile genetic entities. Transposition frequencies have to be tightly regulated, since frequent jumping would lead to accumulation of these elements, ultimately compromising the genomic integrity of the host. The enzyme required for mobility is called transposase, and control of its synthesis can occur by different mechanisms, one of which involves antisense RNA. IS*10*, the active element of transposon Tn*10*, encodes an antisense RNA denoted RNA-OUT. RNA-OUT partially overlaps the *tnp* mRNA (RNA-IN). The metabolically stable ~70-nt antisense RNA (Figure 12.1) consists of one single stem-loop with a destabilized upper stem, and appears to be generated by RNA processing (Figure 12.1; Case *et al.*, 1989). The 5′-most segment of RNA-IN is bound within the hexa-loop sequence of RNA-OUT (Kittle *et al.*, 1989). After initiation of binding, helix progression unwinds the RNA-OUT stem-loop, entailing full pairing of the two RNAs, thereby sequestering the *tnp* RBS (Ma and Simons, 1990). Additionally, the antisense/target RNA duplex is degraded by RNase III, although this does not contribute significantly to control (Case *et al.*, 1990). A special feature of *tnp* mRNA that further reduces its translatability is the potential to form a fold-back inhibition (*fbi*) structure by pairing between the nascent 3′-end and the 5′-region. When only one copy of the transposon is present, antisense control is barely operating. When multiple copies are present, RNA-OUT efficiently downregulates *tnp* expression (Simons and Kleckner, 1983). The antisense mode of action in IS*10* is that of translational inhibition.

Another transposon, Tn*30*, also uses antisense RNA to control *tnp* expression. Here, binding appears to occur within the *tnp* coding region of the mRNA. It was proposed that duplex formation could block elongating ribosomes (Arini *et al.*, 1997).

C. Bacteriophages

Many bacteriophages have the option of two different lifestyles, lytic cycle or lysogeny. The former will yield progeny phage, the latter implies maintenance as a plasmid or integration into the host chromosome. Regulation of lysis/lysogeny switches is extraordinarily complex. Generally, repressor-operator control dominates control decisions but, in many cases, an additional level of control operates, involving antisense RNAs. Most likely, these RNAs serve to fine-tune protein-mediated control. One might also consider them safety devices that prevent inadvertent escape from the lysogenic state.

1. Lambda OOP RNA induces facilitated mRNA decay

In the coliphage lambda, CII is a key protein in the establishment of the lysogenic state. A 77-nt-long antisense RNA, OOP, is encoded between the *cII* and *O* gene sequences, overlapping the 3-end of the *cII* mRNA. When OOP binds to its target site, formation of an RNA duplex that extends into the 3′-part of the *cII* coding region creates a substrate for RNase III (Figure 12.2). Cleavage by this enzyme initiates the rapid decay of the *cII* mRNA segment, carried out by other RNases (Krinke *et al.*, 1991; Krinke and Wulff, 1990). In support of this, overexpression of OOP *in trans* prohibits lysogeny of an incoming lambda phage (Krinke and Wulff, 1987).

2. P1, P4, and related phages: co-transcription of antisense and sense RNAs

An unusual theme is illustrated by the process of maintenance of the lysogenic state in a group of phages related to P1, P7, and P4. In P1 and P7, lysogeny is primarily achieved by repressor control. An antirepressor (Ant) can mediate the switch to the lytic cycle. The *ant* mRNA contains an upstream reading frame, *icd*, which is translationally coupled to *ant*. Furthermore, its translation on the *nascent* mRNA is required for passage of RNA polymerase through a downstream region of the DNA which specifies a transcription terminator (Riedel *et al.*, 1993). Ant synthesis is inhibited by the 77-nt *c4* antisense RNA which, oddly, is co-transcribed with *ant*, upstream of *icd* (Citron and Schuster, 1990). The *c4* RNA is excised from the 5′-segment of the mRNA in a process involving RNase P, and accumulates as a free, *trans*-acting regulator (Citron and Schuster, 1992). Two single-stranded loops in *c4* RNA are complementary to two sequence motifs overlapping the *icd* RBS (Figure 12.1). Interaction blocks *icd* translation and induces Rho-dependent termination immediately downstream (Biere *et al.*, 1992). Consequently, the antisense RNA aids in maintaining the lysogenic state. Additionally, *c4* thereby functions as a superinfection immunity element, preventing productive lytic infection by a phage with complementary target motifs.

    A similar mechanism is supported for P4 and its relatives (Sabbattini *et al.*, 1995, 1996). P4 can be present as an integrated prophage, but also as a replicating plasmid. Control of the replication genes is exerted by *cI* RNA, which conforms, with respect to location of its gene and secondary structure, to *c4* RNA of P1. This RNA is 69 nt long, and its excision from the mRNA depends on RNase P and PNPase (Forti *et al.*, 1995; Piazza *et al.*, 1996). Here, sequence elements in *cI* are complementary to motifs located upstream *and* downstream within the mRNA. Interaction with the upstream motif induces termination of transcription upstream of *cI*, that is, *cI* is autoregulated. Binding to the downstream site induces premature termination, shutting off replication (Briani *et al.*, 2000).

3. Bacteriophage P22: Sar and Sas

Similar to the above cases, the *Salmonella* phage P22 requires silencing of *ant* expression to maintain the lysogenic state. The main repressor is the Mnt protein, but additional downregulation of Ant synthesis depends on Sar, an antisense RNA of 69 nt that folds into a structure consisting of two stem-loops (Jacques and Susskind, 1991; Schaefer and McClure, 1997a). The *arc-ant* mRNA (mRNA encoding both the Arc and Ant regulators) is the target of Sar; complementarity extends throughout the intergenic region between these two reading frames, and binding sequesters the *ant* Shine-Dalgarno region (Liao *et al.*, 1987; Wu *et al.*, 1987), repressing translation.

Superinfecting lambdoid phages can overcome immunity of a resident prophage by a second antisense RNA-mediated mechanism. The SieB (superinfection exclusion) protein of P22 mediates abortion of lytic superinfections (Ranade and Poteete, 1993a). Esc (escape factor) is a truncated SieB protein translated from an alternative start site. Normally, SieB is in excess and Esc in limited supply. Thus, Esc cannot inactivate SieB by subunit mixing. Upon entry, a superinfecting P22 produces a small RNA, Sas. Sas RNA binds to its target which overlaps the *sieB* RBS, and induces a translational switch that shifts the balance of synthesis in favor of Esc (Ranade and Poteete, 1993b), resulting in an override of exclusion.

4. Antisense RNA in the Third Kingdom: phage ϕH

The first and, to our knowledge, only case of antisense RNA control in archea was found in a phage, ϕH, of the extreme halophile *Halobacterium salinarum* (Stolt and Zillig, 1993a). When phage ϕH is in a lysogenic state, it accumulates an antisense RNA, T_{ant}, which is complementary to T1, an early lytic mRNA. In response to T_{ant}, T1 RNA levels are decreased. Degradation of T1 has been shown to be dependent on cleavage by a host-encoded structure-specific ds/ss-RNase. The biological consequences of this control circuit are still elusive (Stolt and Zillig, 1993b).

D. Chromosomally encoded antisense systems

Compared to the vast number of antisense RNAs discovered in accessory elements, the harvest in bacterial chromosomes has been poor. In addition, antisense RNAs in plasmids, phages and transposons are *cis*-encoded, whereas most chromosomally encoded antisense RNAs are *trans*-encoded (Altuvia and Wagner, 2000; Delihas, 1995; Wassarman *et al.*, 1999). This is intriguing, since at a first glance this feature suggests a drawback: mutations in either antisense or target sequences may result in loss of regulation. On the other hand, for riboregulators that act on multiple targets, being *trans*-encoded would be beneficial. Interestingly, all *trans*-encoded antisense RNAs appear to be involved in stress responses.

1. MicF

MicF, the first chromosomally encoded antisense RNA identified, is a regulator whose synthesis is induced by a variety of environmental stresses, such as changes in temperature, low osmolarity, redox stress, and others (Andersen *et al.*, 1989; Gidrol and Farr, 1993; Ramani *et al.*, 1994). The target of this 93-nt RNA is the *ompF* mRNA, whose product is one of the two major outer membrane porins in *Escherichia coli*; OmpF and OmpC are inversely regulated with respect to each other, resulting in changed membrane properties as an adaptive response to stress. When MicF binds to its target, *ompF* mRNA, a partial duplex is formed that blocks initiation of translation and promotes degradation of the mRNA (Delihas, 1997; Schmidt *et al.*, 1995). Secondary structures of MicF (Figure 12.1), *ompF* target, and the complex appear to be conserved in related bacteria (Schmidt *et al.*, 1995). Indicative of further complexity, several proteins that affect transcription levels are integrated in MicF-dependent regulation, e.g., Lrp, H-NS, and SoxRS (Chou *et al.*, 1993; Ferrario *et al.*, 1995; Suzuki *et al.*, 1996).

2. OxyS

OxyS and DsrA (see below) are both riboregulators that act on multiple targets, and do so by different modes of action. OxyS is a pleiotropic regulator of more than 40 genes in *E. coli*. The expression of *oxyS* is strongly induced by oxidative stress (Altuvia *et al.*, 1997). The mechanism of regulation of two targets, *rpoS* (stationary-phase Sigma factor) and *fhlA* (transcriptional activator of formate metabolism) has been examined. OxyS is a 109-nt RNA consisting of two major and one minor stem-loop structures, and a long unstructured region (Figure 12.1). Interestingly, different segments of the RNA act on the two known targets. In the *rpoS* case, OxyS inhibits translation by sequestration of a protein, Hfq (host factor Q), which is required to unfold and thereby activate the 5′-segment of *rpoS* mRNA; the unstructured region in OxyS appears to bind Hfq (Zhang *et al.*, 1997, 1998). In the case of *fhlA*, the two major loop regions base-pair to two noncontiguous regions within the 5′-part of the *fhlA* mRNA. Both kissing complexes contribute to inhibition; one contact occludes the *fhlA* RBS, and the other one probably stabilizes the inhibitory complex (Argaman and Altuvia, 2000). Thus,

OxyS is an example of a riboregulator that inhibits more than one target mRNA, by an antisense mechanism and by protein titration, respectively (Figure 12.2).

3. DsrA

DsrA also acts on several target genes (Lease and Belfort, 2000b). This 87-nt RNA (Figure 12.1) was discovered through its effect on polysaccharide capsule synthesis in *E. coli* (Sledjeski and Gottesman, 1995). Also, DsrA action is more complex than expected from a conventional antisense RNA. Current knowledge indicates that DsrA acts on at least two targets, *rpoS* and *hns* (encoding a transcriptional modulator). More fascinating, DsrA works through antisense mechanisms in both cases, but in one (*rpoS*) translation becomes *activated*, while in the other (*hns*) it is *inhibited* (Lease *et al.*, 1998; Majdalani *et al.*, 1998; Sledjeski *et al.*, 1996). Recent work indicates that different parts of the RNA act on each target. In *rpoS* mRNA, a stem-loop structure blocks the RBS. Interaction between a 5′-segment of DsrA and one half of the stem-loop induces opening of the *rpoS* RBS region and permits translation. Two adjacent sequence segments within the middle stem-loop of DsrA appear to interact with 5′- and 3′-regions in the *hns* mRNA (Lease and Belfort, 2000a). This results in inhibition of translation and destabilization of the message. By homology searches, similar 3′/5′ interactions have been proposed for two additional mRNAs, making them putative targets although a confirmation of this proposal awaits experimental support (Lease and Belfort, 2000a). Hence, DsrA is a *trans*-encoded antisense RNA capable of *both* activating and repressing different target genes, respectively (Figure 12.2).

4. Activation by antisense RNA: RNAIII in regulation of virulence in *Staphylococcus aureus*

Virulence in *Staphylococcus aureus* is under global control by the *agr* (accessory gene regulator) locus. RNAIII (514 nt), the key effector encoded within *agr*, has multiple functions: it acts as a message by encoding *hld* (δ-hemolysin) and as a regulator—by *activating* several virulence-associated genes (secreted toxins and enzymes) or by *repressing* others (cell-surface proteins) (Novick *et al.*, 1993). Its regulatory function is independent of the *hld* reading frame or its translation. On most target genes, RNAIII appears to affect transcription by an as yet unknown mechanism (Tegmark *et al.*, 1998). Possibly, RNAIII acts by sequestration of *trans*-acting proteins, since the RNA forms hairpin structures that may create protein-binding sites (Benito *et al.*, 2000; Novick *et al.*, 1993; Tegmark *et al.*, 1998). Alternatively, RNAIII may indirectly affect transcription through inhibition of mRNAs which encode transcription/regulatory factors.

Interestingly, RNAIII also acts as an activator antisense RNA; binding of the 5′-region of RNAIII to a complementary region encompassing the RBS of the *hla* (α-hemolysin) mRNA competes with an inhibitory structure within the target, so that translation is activated (Morfeldt *et al.*, 1995).

5. ... and various others

Over the years, a number of additional chromosomally encoded antisense RNAs have been reported to control gene expression (Table 12.1). Most of these are less well characterized. A (probably incomplete) summary is given here. (1) The *ftsZ* gene, a major player in initiation of cell division, is subject to complex control. One gene, encoded by a defective prophage of *E. coli*, is *dicF*. Its product, the 53-nt DicF RNA, is partially complementary to the *ftsZ* mRNA, inhibiting its translation and thereby blocking cell division (Tetart and Bouche, 1992). (2) In *Clostridium acetobutylicum*, the *glnA* (glutamine synthetase) gene is nitrogen-regulated. It appears that part of this control is exerted by a nitrogen-induced *cis*-encoded antisense RNA whose target is the *glnA* RBS region (Fierro-Monti *et al.*, 1992; Janssen *et al.*, 1990). (3) A chromosomally encoded killer function with features similar to *hok/sok* is encoded by the *gef* locus in *E. coli*. The antisense RNA, Sof, is complementary to the RBS of the *mok*-equivalent and thereby prevents killing by a translational block (Poulsen *et al.*, 1991). The biological significance of this is still elusive. (4) The *Acetobacter methanolicus* phage Acm1 encodes a *trans*-acting regulatory antisense RNA of 97 nt, designated *lbi* (LPS biosynthesis-interfering) RNA. When expressed in *E. coli*, the RNA appears to form intracellular RNA duplexes with mRNA, affecting conversion to rough-type LPS (Mamat *et al.*, 1995). A recent report indicates that *lbi* RNA also may control D-galactan II biosynthesis in *Klebsiella pneumoniae*. So far, it is not clear whether this occurs by an antisense RNA mechanism (Warnecke *et al.*, 2000). (5) Filamentous cyanobacteria, e.g., *Calotrix* 7601, differentiate hormogonia in response to environmental change. During differentiation, the *gvpABC* operon is expressed and synthesizes products which are needed for gas vesicle formation. An unusually long ~400-nt RNA is encoded within the operon, but in antisense orientation. Differential accumulation of this RNA in differentiated/undifferentiated cells suggests that it may be involved in regulation of the *gvpABC* operon, although experimental support is lacking (Csiszàr *et al.*, 1987). (6) Finally, a plasmid-encoded antisense RNA system is considered here since its biological function is *not* related to plasmid lifestyle, but rather to the virulence properties of the host. Regulation of iron transport in the fish pathogen *Vibrio anguillarum* requires several genes located on a virulence plasmid. Two genes encoding transport proteins, FatA and FatB, are downregulated by a *trans*-encoded antisense RNA, RNAα, of ~400 nt. Synthesis of this RNA is induced under iron-rich conditions. The formation of partial duplexes between RNAα and *fatA/fatB* mRNA is believed to result in mRNA destabilization (Chen and Crosa, 1996; Salinas *et al.*, 1993; Waldbeser *et al.*, 1995).

A peculiar regulatory mechanism that relies on very short sequence complementarity operates in the control of most aminoacyl-tRNA synthetase genes in *Bacillus subtilis* (Condon *et al.*, 1996; Henkin, 1994; Putzer *et al.*, 1995). Starvation for a cognate amino acid induces transcription antitermination within the leader of the corresponding synthetase mRNA. The leader can adopt two alternative conformations, one entailing termination, the other readthrough. Uncharged cognate tRNA binds to two sequence/structure elements in the leader via its anticodon and NCCA3'-sequences, stabilizes the antiterminator structure, and promotes readthrough.

IV. ANTISENSE STRUCTURE AND KINETICS OF BINDING

The efficiency of an antisense RNA acting on its complementary target is determined by two parameters, the binding rate constant and the intracellular concentration. From a number of studies it is clear that the rate of formation of a stable complex between the interacting RNAs rarely follows saturation kinetics (e.g., Persson *et al.*, 1990a; Tomizawa, 1985). This specific difference to, e.g., a repressor protein binding to its DNA target, implies that great changes in antisense RNA concentration range are reflected in almost linearly correlated changes in inhibition rates, whereas increases in concentration of a repressor eventually saturate DNA binding and, hence, inhibition. That is, inhibition by a repressor protein (usually) is determined by its binding strength (K_D); inhibition by antisense RNAs depends on binding rate. An abundance of data supports the critical importance of RNA structure features as determinants of binding rate as well as of specificity.

A. Antisense and target RNA structure

Probably more that 100 antisense RNA secondary structures have been predicted from their sequences, using folding algorithms. Experimentally supported structures have been determined by enzymatic and chemical probing for few of these. Due to space limitations, we discuss here a few examples that may serve to illustrate general patterns. Almost all antisense RNAs consist of one, two, or three stem-loop structures (Figure 12.1). Most often, one of these serves as the recognition stem-loop used for initiation of target binding. In addition, many antisense RNAs use their Rho-independent terminator stem-loops as recognition elements, which restricts the structural degrees of freedom these RNA segments can adopt. Loops can vary in size, though preferred antisense RNA loop sizes range from 5 to 7 nt (Hjalt and Wagner, 1992). Recent publications have suggested that particular U-turn structure motifs are present in almost all antisense loops or their complements (Asano *et al.*, 1998; Franch and Gerdes, 2000; Franch *et al.*, 1999). U-turns are defined loop structures in which the bases following the conserved U residue are presented in a preformed A-helical conformation, with bases directed outwards. The ubiquitous presence of this motif (which was previously found in all anticodon loops of tRNAs, where decoding takes place) suggests its superior properties in promoting rapid and specific interaction. The upper stems regions of the major recognition stem-loops tend to be destabilized by bulges and internal loops, a feature whose significance will be addressed below. In addition, single-stranded regions are present as 5′-tails or as unstructured middle regions. The importance of these regions lies in the stabilization of antisense/target complexes formed and in allowing the conversion of kissing complexes to stable, inhibitory complexes.

Structures of target RNAs have also been determined by biochemical experiments (Asano and Mizobuchi, 1998b; Brantl and Wagner, 2000; Masukata and Tomizawa, 1986; Öhman and Wagner, 1989; Thisted et al., 1995; Wilson et al., 1997). In some cases, notably RNAII of ColE1, mutant phenotypes have aided in the elucidation of complex structures and folding pathways (Masukata and Tomizawa, 1984; Polisky et al., 1990). Target RNA structures can be intrinsically difficult to assess. In bacteria, the time frame of regulation is very short. Antisense RNA will bind to nascent target RNA. This "moving target" concept raises the question of what length of target RNA to choose when conducting a secondary structure probing in vitro. Nevertheless, secondary structure models have been proposed for relevant segments of target RNAs. In most cases, the stem-loops involved in initial recognition by the antisense RNAs are mirror images of the antisense stem-loops.

B. Mutations affect binding rate and specificity

Mutations in antisense/target RNA genes change nucleotides in both RNAs simultaneously so that complementarity is maintained. Therefore, regulation is not abolished but may be changed. This is because binding rates can be altered, and thus the specific inhibitory activity of the antisense RNA is affected. Strikingly, antisense RNA systems in nature operate at the upper limit of their inhibitory potential. Mutations can at best maintain binding rate—but change target specificity.

The central role of loop sequences is indicated by many copy number up-mutations in plasmids (Asano and Mizobuchi, 1998a; Brady et al., 1983; Givskov and Molin, 1984; Muesing et al., 1981; Polisky et al., 1985; Wu et al., 1985). In plasmids, single changes in subsets of loop nucleotides create new incompatibility groups, which can be viewed as a "speciation" event. The mutant antisense RNA regulates its own target RNA, but fails to cross-react with the parental wild-type target RNA, and vice versa. Therefore, single-base-change mutant plasmids can most often coexist with the wild-type plasmid (plasmid compatibility). This phenotype supports that antisense/target RNA complexes are initiated at the loops; single-base mismatches result in failure to proceed through the normal binding pathways.

Many different, distantly related plasmids encode antisense RNAs of identical secondary structure and even sequence identity in their recognition loops, but display sequence differences within their upper stem segments (Kolb *et al.*, 2001; Praszkier *et al.*, 1991). Consequently, heterologous antisense/target RNA pairs can form initial, reversible loop–loop complexes, but productive binding is prevented, since subsequent steps are abolished by the mismatches encountered upon helix progression. Thus, upper stem sequence differences often result in plasmid compatibility.

C. RNA–RNA Binding Kinetics

Complementary nucleic acids hybridize in solution; short unstructured DNA or RNA oligomers form dimers with association rate constants of $\sim10^6$ M^{-1} s^{-1} (Craig *et al.*, 1971; Pörschke and Eigen, 1971). Duplexes are initiated by the reversible formation of a helix nucleus of 2–3 bp, followed by addition of base pairs (zipping up). The critical step is the addition of the first base pair after nucleation, which is aided by stacking upon the structurally prearranged initiating helix. Thereafter, zipping up is rapid and proceeds to completion (all-or-none model). Complex formation between complementary anticodons in pairs of tRNAs gave similar values for the association rate constant (Grosjean *et al.*, 1976).

Antisense and target RNAs interact as folded RNAs, and thus are subject to topological and kinetic barriers that are absent in the simpler model systems. *In vitro* studies in a variety of antisense/target RNA systems have shown that the overall binding rate constants are remarkably similar, usually $\sim10^6$ M^{-1} s^{-1}. That is, stable complex formation occurs at rates comparable to that in the model studies, indicating similar or identical rate-limiting steps. Nevertheless, zipping up does not occur in most systems and, instead, only topologically permitted structures are formed. The structures of critical binding intermediates and their association and dissociation rates have been analyzed by gel-shift analyses, binding competition, RNase protection, and surface plasmon resonance. K_D-values of the extended kissing complexes of CopA/CopT (R1) and Inc/RepZ (ColIb-P9) have been determined to be ~8 nM (Asano and Mizobuchi, 1998a; Hjalt and Wagner, 1995; Kolb *et al.*, 2000a; Nordgren *et al.*, 2001). The association rate constants of these complexes were $\sim10^6$ M^{-1} s^{-1}, i.e., close to the value obtained for stable complex formation. Thus, when helix formation has initiated, subsequent steps are unimpeded, and rapid progression commits the interactants to stable complex formation. Thus, it is the rate of formation, rather than the affinity of a loop–loop complex, which determines inhibition. This is corroborated by loop size mutations that increase the affinity of the extended kissing complex, yet result in slower binding rate and poor inhibition *in vivo* (Hjalt and Wagner, 1992; Nordgren *et al.*, 2001). Conversely, when mutations stabilize upper stems, this affects the rate of formation of base pairs subsequent to the initiating contact, and overall binding rate constants decrease (Siemering *et al.*, 1993; Hjalt and Wagner, 1995; Jain, 1995).

Tomizawa has analyzed the stepwise conversion of initial RNAI/RNAII binding intermediates to progressively more stable structures (Eguchi *et al.*, 1991; Eguchi and Tomizawa, 1991; Tomizawa, 1985, 1990a, 1990b). A reversible complex, C_χ, whose structure is not known but which is likely to involve a single pair of stem-loops, initiates binding. By kinetic inhibition and RNase protection studies, the rate constant of formation of the more stable complex C^* was determined at 6×10^6 M^{-1} s^{-1} and, subsequently, a kissing complex possible involving all three RNAI loops is formed with $\sim 3 \times 10^6$ M^{-1} s^{-1}. Finally, stable complex formation (complex Cs) occurs at a rate constant of $\sim 10^6$ M^{-1} s^{-1}. Thus, the high rates characteristic of early steps are almost maintained throughout the binding pathways.

Very little is known about the binding kinetics of *trans*-acting antisense RNAs. In the OxyS/*fhlA* target interaction, the total complementarity of 7 and 9 nt, respectively, is restricted to two loop regions. The complex formed *in vitro* is moderately stable (K_D of 25 nM), and its association rate constant is 5×10^5 M^{-1} s^{-1} (Argaman and Altuvia, 2000). This suggests that inhibition by OxyS occurs almost as rapidly as in the above systems in spite of the inability to form irreversible complexes. Given that the calculated half-life of the OxyS/*fhlA* RNA complex is ~ 1 min, and the intracellular concentration of induced OxyS exceeds 100 nM, inhibition is efficient.

Several other systems have been analyzed for antisense/target RNA-binding kinetics, but cannot be covered here (e.g., Brantl and Wagner, 1994; Jain, 1995; Kittle *et al.*, 1989; Schaefer and McClure, 1997a, 1997b; Sugiyama and Itoh, 1993; Thisted *et al.*, 1994; van Biesen *et al.*, 1993; Zeiler and Simons, 1998).

D. Binding Pathways and Topology Problems

When folded RNAs interact at loop sequences, a problem arises. Simple helix progression in both directions is topologically impossible, since torsional stress accumulates as both RNAs twist around each other to form double-stranded RNA. Therefore, loop–loop initiating systems require a subsequent interaction at a distal site in order to circumvent this limitation (two-step pathway). In systems in which an unstructured RNA end initiates binding within the loop of its counterpart, a duplex can form by strand exchange (one-step pathway). In the case of partially complementary (*trans*-encoded) RNAs, binding probably initiates in loops, but the paths to the inhibitory complexes are not quite understood. In the OxyS case, two kissing complexes are formed, and no further steps are required for inhibition. Examples of two-step and one-step pathways are given below.

1. Two-step pathways

Binding pathways have been proposed for RNAI/RNAII of ColE1, CopA/CopT of R1, Inc RNA and its target of ColIb-P9/pMU720, and some others. Tomizawa and co-workers have shown that binding initiates between one or two loop pairs (out of three). A reversible, unstable kissing complex was inferred, but its structure was not observed experimentally. Subsequent loop–loop contacts result in progressively more stable complexes (Eguchi and Tomizawa, 1990, 1991; Tomizawa, 1990a, 1990b). RNAI lacking its 5′-tail is arrested at this stage, but inhibits primer formation *in vitro* and *in vivo*, which implies that a full RNA duplex is not required for control (Lin-Chao and Cohen, 1991; Tomizawa, 1984, 1990b). Nontruncated RNAI proceeds by interaction between the 5′-tail and its complement in RNAII. *In vitro* RNase susceptibility data indicate that, starting from this 5′-contact, the duplex is formed very slowly, concomitant with stepwise loss of loop–loop contacts and unfolding of the stem regions (Tomizawa, 1984). Due to structural complexity, intermediates have been difficult to analyze. By enzymatic probing, complexes between single complementary stem-loops have been studied. It was concluded that all seven loop bases are base-paired to each other, creating a coaxial stack of the two stems bent at the loop–loop helix (Eguchi and Tomizawa, 1990, 1991). NMR studies of such a pair of stem-loops suggested the same properties (Lee and Crothers, 1998; Marino *et al.*, 1995). However, this complex carried a loop-sequence inversion which, in its biological context, inactivated the antisense RNA.

In plasmid R1, binding initiates within a subset of nucleotides (CGCC in CopA) in the hexanucleotide loop of the major CopA stem-loop, and its complement in CopT (Figure 12.3B, see color insert for Figure 12.3A–F). This short helix is rapidly extended into the upper stem regions, facilitated by the destabilizing presence of bulges. In the absence of bulges, stable complexes are not formed, and CopA is inactive *in vivo* (Hjalt and Wagner, 1995). The moderately stable complex formed is denoted as an extended kissing complex (Figure 12.3D). It consists of two intermolecular helices (B, B′) and two intramolecular helices in a four-helix junction structure, connected by single-stranded loop regions— the nucleotides that initiated binding are again rendered single-stranded (Kolb *et al.*, 2000b). Helix extension from the loops is directional: helix B (9 bp) must form first (Figure 12.3C), then helix B′ (6 bp) can be formed from the freed halves of the upper stems (Figure 12.3D). Congruent with this, mutant CopA regulates wild-type CopT (and vice versa) *in vitro* or *in vivo* when mismatches are present in helix B′, whereas activity is lost when helix B cannot form (Kolb *et al.*, 2000a). Unidirectional helix progression causes a side-by-side alignment of the CopA and CopT stem regions. This enables the single-stranded middle regions of both RNAs to hybridize and form a stabilizer helix of 30 bp. Thus, the stable CopA/CopT complex consists of two *intra*molecular and three *inter*molecular helices (Figure 12.3E). The structures of the extended kissing complex and the stable complex are supported by extensive enzymatic and chemical probing data, mutational analysis, and molecular modeling (Kolb *et al.*, 2000b). Notably, full duplex formation does not occur appreciably *in vitro* (Malmgren *et al.*, 1997), and both the extended kissing complex and the stable complex suffice for inhibition *in vitro* and *in vivo* (Malmgren *et al.*, 1996; Wagner *et al.*, 1992).

The IncIα and IncB group plasmids follow very similar binding pathways (e.g., Asano and Mizobuchi, 2000; Siemering *et al.*, 1994). The antisense RNAs carry only one major stem-loop with a hexanucleotide loop, a destabilized upper stem, and a 5'-tail. The 6-base loop sequence is identical in all IncB-related plasmids, and is also identical to that of R1. Yet, productive heterologous interaction does not occur (i.e., no incompatibility) because of sequence differences in the upper stem regions. As in CopA/CopT, Inc RNA first binds its target within the loop region. Subsequently, helix progression unfolds the upper stems, resulting in a four-helix junction structure. The proposed secondary structure resembles that of the CopA/CopT complex, but differs in the position of the junction (Asano and Mizobuchi, 2000). Recent data suggest that antisense/target RNA complexes of CollIb-P9, R1, and, by analogy, of many other distantly related plasmids, may share the same overall topology including the position of the junction (Kolb *et al.*, 2001). Thus, initiation by loop–loop contact is the preferred solution for efficient inhibitors, but subsequent conversion to stable, inhibitory complexes must follow topologically possible routes. In the CopA/CopT pathway, but also others, a progression through a hierarchy of distinct intermediates requires that topological stress can be kept at a minimum.

2. One-step pathways

The topological problems described above are essentially absent when a single-stranded tail of one RNA initiates binding to a loop region in the other. This is the case in Tn*10* and *hok/sok*. The unstructured Sok 5'-tail binds to a structured target loop in *hok* mRNA (Figure 12.3). The target loop carries a U-turn structure motif that has been demonstrated to be important for maintaining high binding rates (Franch *et al.*, 1999). Helix progression from the Sok 5'-end unwinds the helical stem of the target, forming a short-lived intermediate and committing the RNAs to irreversible and rapid formation of a duplex. Binding of RNA-OUT and RNA-IN of Tn*10* occurs in a similar fashion, although here it is the unstructured 5'-tail of the target RNA that binds to the loop of the antisense RNA. Mutational analyses have shown that upper-stem destabilizing elements are required for antisense RNA activity, suggesting that helix propagation is driven by a thermodynamically favored strand exchange from the imperfectly base-paired upper stem of RNA-OUT to the perfectly base-paired intermolecular RNA-IN/RNA-OUT duplex (Jain, 1995; Kittle *et al.*, 1989).

V. MATCHING ANTISENSE RNA PROPERTIES WITH BIOLOGICAL ROLES

Antisense RNAs have evolved to fulfill biological functions, and thus their properties reflect the particular requirements of the regulatory system in question. Thus, it may be instructive to view the differences displayed in light of the functions these regulators have to carry out in the cell.

A. The role of structure

Antisense RNA structures present short single-stranded regions for initial interaction with target RNAs, most often in loop regions characterized by particular structure motifs. This appears favorable, since recognition between the complementary RNAs becomes restricted to a small subset of nucleotides within both RNAs. This entails that "wrong starts" are unlikely, and most encounters will result in rapid progression to stable, inhibitory complexes. Hence, inhibition rates are maximized. If antisense and target RNAs were mostly unstructured, formation of helix nuclei of 2–3 base pairs would, statistically, occur even between regions that are not flanked by continuous complementarity, and thus dissociation would be required before binding can initiate again. Thus, hairpins in which a loop structure is presented for interaction are a preferred feature, but the topological problems of propagating intermolecular helices from a loop–loop contact present an obstacle. However, antisense and target RNAs have evolved to cope with this: loop–loop helices can easily be extended into destabilized upper stems, and hence initial high rates of binding can be maintained throughout the following binding steps.

Less is known about target structure. During transcription, RNAs can potentially adopt many conformations (moving target) and interconvert between alternative structures, due to the continuous addition of sequences that can participate in folding. However, many target RNAs form structure modules that persist almost irrespective of the length of the target RNA (Masukata and Tomizawa, 1986; Öhman and Wagner, 1989). Most often these are the ones required for binding of the antisense RNA. In some cases, target sequences have evolved a capacity to form mutually exclusive alternative conformations, one of which presents the antisense target, the other one formed after binding of the antisense RNA (Brantl and Wagner, 1994, 2000; Masukata and Tomizawa, 1986; Novick *et al.*, 1989)].

B. Short or long half-life

Differences in the metabolic stability of antisense RNAs suggest optimization for function and economy. Most chromosomally encoded *trans*-acting antisense regulators, and a few of the antisense RNAs of phages and transposons, are stable

(Pepe *et al.*, 1994; Wassarman *et al.*, 1999; Wrobel *et al.*, 1998). Their half-lives, often 10–60 min *in vivo*, by far exceed the average half-life of, e.g., bacterial mRNAs. This is in line with their functions: maintaining an appropriate concentration in a cell at the lowest possible energetic cost appears sensible. RNAs such as OxyS, DsrA, and MicF, in addition, are under transcriptional control, and accumulate only when appropriate environmental signals are present. High levels of these RNAs have to be maintained until the stress response ceases.

On the other hand, antisense RNAs involved in plasmid copy number regulation, and in postsegregational killing systems, are very unstable (half-lives ~1 min; Dam Mikkelsen and Gerdes, 1997; Söderbom *et al.*, 1997; Söderbom and Wagner, 1998; Xu *et al.*, 1993). This is explained by their particular properties. In order to measure plasmid copy numbers, high RNA turnover is crucial, so that antisense RNA concentrations can be tightly correlated with fluctuating copy numbers (e.g., according to the inhibitor dilution model; Pritchard *et al.*, 1969). Thus, the short half-life of these antisense RNAs is a requirement of the regulatory circuit. The observed decay rates must represent a trade-off between metabolic cost associated with continuous, high rate of synthesis of the inhibitor and the need for high turnover. For the *hok/sok* killer system, the same applies: killing relies on rapid decay of Sok. An exception to the rule is represented by pIP501, which encodes an unusually stable copy number regulator antisense RNA (~30-min half-life; Brantl and Wagner, 1996). As discussed in Brantl and Wagner (1998), stable regulator antisense RNAs are predicted to be unable to correct downward fluctuations of plasmid copy number. However, recent work has shown that the dimeric CopR repressor protein of pIP501 can compensate for this potentially detrimental effect (Brantl and Wagner, 1997).

The determinants of antisense RNA stability are still elusive. At a first glance, there are no structural features that enable us to predict the half-life of these regulators, although it is known that RNase E, RNase II, and PNPase are involved in decay of these RNAs. Most often, RNase E cleavage initiates decay, and studies of RNAI (ColE1), CopA (R1) and Sok (*hok/sok* system) have shown that poly(A)-polymerase (formerly: PcnB) is required for their rapid decay (Dam Mikkelsen and Gerdes, 1997; Söderbom *et al.*, 1997; Söderbom and Wagner, 1998; Xu *et al.*, 1993). For some stable antisense RNAs, the absence of RNase E cleavage sites and/or oligo(A) tails may increase stability.

C. Single or multiple targets

Cis-acting antisense RNAs act on single targets. *Trans*-acting antisense RNAs are only partially complementary to targets. Therefore, it is conceivable that one antisense RNA may carry several antisense determinants. From an admittedly limited number of bacterial regulatory RNAs, two have been reported to act on more than one target. These are OxyS and DsrA (see Sections III.D.2 and

III.D.3). At present, a number of new putative riboregulators in *E. coli* are under investigation (Argaman, Hershberg, Vogel, Bejerano, Wagner, Margalit and Altuvia, unpublished). A challenging question to be tackled concerns whether many more *trans*-acting antisense RNAs play significant roles in biological physiology, and whether they are primarily regulators of single or multiple target genes.

D. Protein involvement

Binding of antisense to target RNAs in the absence of any protein occurs *in vitro*. Binding rates *in vivo* are unknown, but values obtained *in vitro* can quantitatively account for control *in vivo*, as shown for the ColE1 system (Brenner and Tomizawa, 1991). Even inhibition has been reproduced in purified assay systems, e.g., transcriptional attenuation of pIP501 and pT181, block of translation in R1 and Tn*10*, and inhibition of primer formation in ColE1 (Brantl and Wagner, 1994, 2000; Ma and Simons, 1990; Malmgren *et al.*, 1996; Tomizawa, 1990a, 1990b). Nevertheless, it cannot be excluded that proteins might modulate binding of antisense RNA to target RNA, affect an inhibitory event, or promote formation of a full RNA duplex, when present in living cells. At present, most proteins known to affect inhibition carry out functions that only indirectly affect antisense RNA control. For example, mutations in genes encoding degradation enzymes result in changes in antisense RNA half-life and thus intracellular concentration.

Experimentally supported direct interactions between specific proteins and antisense RNAs are rare. Plasmid ColE1 encodes a protein, Rom, whose activity affects copy number, although it is dispensable (Twigg and Sherratt, 1980). Rom dimers bind to RNAI/RNAII kissing complexes and decrease their dissociation rate (Eguchi and Tomizawa, 1990; Predki *et al.*, 1995). Nevertheless, though the effect of Rom on complex stability is more than 100-fold, its effect on the overall rate of binding is small (Eguchi and Tomizawa, 1990, 1991; Tomizawa, 1990a), consistent with its minor effect on copy number. In control of conjugation, the FinO protein protects the FinP RNA from degradation by blocking the access of RNase E (Jerome *et al.*, 1999). For MicF, interaction with a cellular protein of ~70 kDa has been reported, but the significance of this interaction is yet unclear (Andersen and Delihas, 1990). Thus, current knowledge suggests that antisense RNA mechanisms essentially rely on RNA–RNA interactions for their biological effects.

E. Beyond intuition and experiments

For an understanding of complex regulatory pathways, experimental determinations of the key parameters is required. Most important, one needs to know intracellular concentrations of the reactants, and rate constants for binding between antisense and target RNA. Most often this is not sufficient to model the properties of a regulatory circuit. Several attempts have been made to use experimental data as a basis for modeling of ColE1 plasmid replication (Brendel and Perelson, 1993; Brenner and Tomizawa, 1991), and subsequent work by the Ehrenberg group has extended such analysis for both ColE1 and R1 (Ehrenberg, 1996; Ehrenberg and Sverredal, 1995; Paulsson *et al.*, 1998, 2000; Paulsson and Ehrenberg, 1998, 2000). As it turns out, mathematical modeling of control circuits can indicate properties that are otherwise not obvious. Optimization strategies that must operate for any selfish element, such as a plasmid, have to balance metabolic burden and segregational stability. Regulation operates in individual cells and not in an idealized large volume. Thus, the numbers of antisense RNAs, target RNAs, and plasmids fluctuate greatly in each cell, and therefore regulation has to be modeled mesoscopically rather than macroscopically. New models have already suggested interesting solutions that may be employed by plasmids, and, moreover, are testable. For instance, it has been proposed that plasmid loss is dependent on whether regulation follows hyperbolic or exponential curves. In special cases, even hyperbolic regulation can result in very precise control of replication, and a clocklike behavior is expected (Paulsson and Ehrenberg, 2000). The challenging task for future years is to enhance the communication between experimentally and mathematically inclined scientists in order to increase our understanding of regulation in a biologically relevant setting.

VI. CONCLUDING REMARKS

In this review we have attempted to give a comprehensive overview over the biological roles of antisense RNAs in bacteria, and of the various modes of regulation. A striking observation is that so many antisense RNA systems operate in accessory genetic elements, whereas the number of known chromosomally encoded (mostly *trans*-acting) antisense RNAs is still small. We predict that a systematic search, based on combinations of computational, biochemical, and genetic approaches will, in the near future, result in the identification of a considerable number of novel riboregulators. A first step in this direction is the very recent discovery, independently reported by two groups, of more than 20 novel small RNA-encoding genes in *E. coli* whose biological roles remain to be characterized (Argaman *et al.*, 2001; Wassarman *et al.*, 2001).

Binding pathways, kinetics, and structural constraints imposed on RNA/RNA interactions have been studied in a few cases. For example, it has become clear that full RNA duplex formation most often is unimportant for regulation, yet our understanding of the intricate details of most systems is still in its infancy. We anticipate that high-resolution structures of antisense and target RNAs, and of their complexes, will increase our understanding of structure/function relationships. Finally, though complexes between antisense and target RNAs have been studied in the test tube, nothing is as yet known about their conformations *in vivo*. It will be a challenging task to map secondary structures, and possibly infer higher-order structures, of functional complexes formed in living cells.

Acknowledgments

The authors acknowledge helpful discussions with the lab members of their research groups. G.W. and S.A. acknowledge grants from the Human Frontier Science Program. G.W. is grateful for financial support by the Swedish Natural Science Research Council. P.R is supported by grants from the Centre National de la Recherche Scientifique (CNRS).

References

Aiba, H., Matsuyama, S., Mizuno, T., and Mizushima, S. (1987). Function of *micF* as an antisense RNA in osmoregulatory expression of the *ompF* gene in *Escherichia coli*. *J. Bacteriol.* **169**, 3007–3012.

Altuvia, S., and Wagner, E. G. H. (2000). Switching on and off with RNA. *Proc. Natl. Acad. Sci. USA* **97**, 9824–9826.

Altuvia, S., Weinstein-Fischer, D., Zhang, A., Postow, L., and Storz, G. (1997). A small, stable RNA induced by oxidative stress: Role as a pleiotropic regulator and antimutator. *Cell* **90**, 43–53.

Altuvia, S., Zhang, A., Argaman, L., Tiwari, A., and Storz, G. (1998). The Escherichia coli OxyS regulatory RNA represses *fhlA* translation by blocking ribosome binding. *EMBO J.* **17**, 6069–6075.

Andersen, J., and Delihas, N. (1990). *micF* RNA binds to the 5′ end of *ompF* mRNA and to a protein from . *Escherichia coli Biochemistry* **29**, 9249–9256.

Andersen, J., Forst, S. A., Zhao, K., Inouye, M., and Delihas, N. (1989). The function of *micF* RNA. *micF* RNA is a major factor in the thermal regulation of OmpF protein in *Escherichia coli*. *J. Biol. Chem.* **264**, 17961–17970.

Argaman, L., and Altuvia, S. (2000). *fhlA* repression by OxyS RNA: Kissing complex formation at two sites results in a stable antisense-target RNA complex. *J. Mol. Biol.* **300**, 1101–1112.

Argaman, L., Hershberg, R., Vogel, J., Bejerano, G., Wagner, E. G. H., Margalit, H., and Altuvia, S. (2001). Novel small RNA-encoding genes in the intergenic regions of *Escherichia coli*. *Curr. Biol.* **11**, 941–950.

Arini, A., Keller, M. P., and Arber, W. (1997). An antisense RNA in IS30 regulates the translational expression of the transposase. *Biol. Chem.* **378**, 1421–1431.

Asano, K., Kato, A., Moriwaki, H., Hama, C., Shiba, K., and Mizobuchi, K. (1991a). Positive and negative regulations of plasmid ColIb-P9 *repZ* gene expression at the translational level. *J. Biol. Chem.* **266**, 3774–3781.

Asano, K., and Mizobuchi, K. (1998a). Copy number control of Inclα plasmid ColIb-P9 by competition between pseudoknot formation and antisense RNA binding at a specific RNA site. *EMBO J.* **17**, 5201–5213.

Asano, K., and Mizobuchi, K. (1998b). An RNA pseudoknot as the molecular switch for translation of the *repZ* gene encoding the replication initiator of Inclα plasmid ColIb-P9. *J. Biol. Chem.* **273**, 11815–11825.

Asano, K., and Mizobuchi, K. (2000). Structural analysis of late intermediate complex formed between plasmid ColIb-P9 Inc RNA and its target RNA. How does a single antisense RNA repress translation of two genes at different rates? *J. Biol. Chem.* **275**, 1269–1274.

Asano, K., Moriwaki, H., and Mizobuchi, K. (1991b). An induced mRNA secondary structure enhances *repZ* translation in plasmid ColIb-P9. *J. Biol. Chem.* **266**, 24549–24556.

Asano, K., Niimi, T., Yokoyama, S., and Mizobuchi, K. (1998). Structural basis for binding of the plasmid ColIb-P9 antisense Inc RNA to its target RNA with the 5′-rUUGGCG-3′ motif in the loop sequence. *J. Biol. Chem.* **273**, 11826–11838.

Benito, Y., Kolb, F. A., Romby, P., Lina, G., Etienne, J., and Vandenesch, F. (2000). Probing the structure of RNAIII, the *Staphylococcus aureus agr* regulatory RNA, and identification of the RNA domain involved in repression of protein A expression. *RNA* **6**, 668–679.

Biere, A. L., Citron, M., and Schuster, H. (1992). Transcriptional control via translational repression by c4 antisense RNA of bacteriophages P1 and P7. *Genes Dev.* **6**, 2409–2416.

Bingle, L. E., and Thomas, C. M. (2001). Regulatory circuits for plasmid survival. *Curr. Opin. Microbiol.* **4**, 194–200.

Blomberg, P., Engdahl, H. M., Malmgren, C., Romby, P., and Wagner, E. G. H. (1994). Replication control of plasmid R1: Disruption of an inhibitory RNA structure that sequesters the *repA* ribosome-binding site permits *tap*-independent RepA synthesis. *Mol. Microbiol.* **12**, 49–60.

Blomberg, P., Nordström, K., and Wagner, E. G. H. (1992). Replication control of plasmid R1: RepA synthesis is regulated by CopA RNA through inhibition of leader peptide translation. *EMBO J.* **11**, 2675–2683.

Blomberg, P., Wagner, E. G. H., and Nordström, K. (1990). Control of replication of plasmid R1: The duplex between the antisense RNA, CopA, and its target, CopT, is processed specifically *in vivo* and *in vitro* by RNase III. *EMBO J.* **9**, 2331–2340.

Brady, G., Frey, J., Danbara, H., and Timmis, K. N. (1983). Replication control mutations of plasmid R6–5 and their effects on interactions of the RNA-I control element with its target. *J. Bacteriol.* **154**, 429–436.

Brantl, S., Birch-Hirschfeld, E., and Behnke, D. (1993). RepR protein expression on plasmid pIP501 is controlled by an antisense RNA-mediated transcription attenuation mechanism. *J. Bacteriol.* **175**, 4052–4061.

Brantl, S., and Wagner, E. G. H. (1994). Antisense RNA-mediated transcriptional attenuation occurs faster than stable antisense/target RNA pairing: An *in vitro* study of plasmid pIP501. *EMBO J.* **13**, 3599–3607.

Brantl, S., and Wagner, E. G. H. (1996). An unusually long-lived antisense RNA in plasmid copy number control: *In vivo* RNAs encoded by the streptococcal plasmid pIP501. *J. Mol. Biol.* **255**, 275–288.

Brantl, S., and Wagner, E. G. H. (1997). Dual function of the *copR* gene product of plasmid pIP501. *J. Bacteriol.* **179**, 7016–7024.

Brantl, S., and Wagner, E. G. H. (2000). Antisense RNA-mediated transcriptional attenuation: An *in vitro* study of plasmid pT181. *Mol. Microbiol.* **35**, 1469–1482.

Brendel, V., and Perelson, A. S. (1993). Quantitative model of ColE1 plasmid copy number control. *J. Mol. Biol.* **229**, 860–872.

Brenner, M., and Tomizawa, J. (1991). Quantitation of ColE1-encoded replication elements. *Proc. Natl. Acad. Sci. USA* **88**, 405–409.

Briani, F., Ghisotti, D., and Dehò, G. (2000). Antisense RNA-dependent transcription termination sites that modulate lysogenic development of satellite phage P4. *Mol. Microbiol.* **36**, 1124–1134.

Case, C. C., Roels, S. M., Jensen, P. D., Lee, J., Kleckner, N., and Simons, R. W. (1989). The unusual stability of the IS*10* anti-sense RNA is critical for its function and is determined by the structure of its stem-domain. *EMBO J.* **8**, 4297–4305.

Case, C. C., Simons, E. L., and Simons, R. W. (1990). The IS*10* transposase mRNA is destabilized during antisense RNA control. *EMBO J.* **9**, 1259–1266.

Chattoraj, D. K. (2000). Control of plasmid DNA replication by iterons: No longer paradoxical. *Mol. Microbiol.* **37**, 467–476.

Chen, Q., and Crosa, J. H. (1996). Antisense RNA, *fur*, iron, and the regulation of iron transport genes in *Vibrio anguillarum*. *J. Biol. Chem.* **271**, 18885–18891.

Chou, J. H., Greenberg, J. T., and Demple, B. (1993). Posttranscriptional repression of *Escherichia coli* OmpF protein in response to redox stress: Positive control of the *micF* antisense RNA by the *soxRS* locus. *J. Bacteriol.* **175**, 1026–1031.

Citron, M., and Schuster, H. (1990). The c4 repressors of bacteriophages P1 and P7 are antisense RNAs. *Cell* **62**, 591–598.

Citron, M., and Schuster, H. (1992). The c4 repressor of bacteriophage P1 is a processed 77 base antisense RNA. *Nucleic Acids Res.* **20**, 3085–3090.

Condon, C., Putzer, H., and Grunberg-Manago, M. (1996). Processing of the leader mRNA plays a major role in the induction of *thrS* expression following threonine starvation in Bacillus subtilis. *Proc. Natl. Acad. Sci. USA* **93**, 6992–6997.

Craig, M. E., Crothers, D. M., and Doty, P. (1971). Relaxation kinetics of dimer formation by self complementary oligonucleotides. *J. Mol. Biol.* **62**, 383–401.

Csiszàr, K., Houmard, J., Damerval, T., and Tandeau de Marsac, N. (1987). Transcriptional analysis of the cyanobacterial *gvpABC* operon in differentiated cells: Occurrence of an antisense RNA complementary to three overlapping transcripts. *Gene* **60**, 29–37.

Dam Mikkelsen, N., and Gerdes, K. (1997). Sok antisense RNA from plasmid R1 is functionally inactivated by RNase E and polyadenylated by poly(A) polymerase I. *Mol. Microbiol.* **26**, 311–320.

del Solar, G., Acebo, P., and Espinosa, M. (1995). Replication control of plasmid pLS1: Efficient regulation of plasmid copy number is exerted by the combined action of two plasmid components, CopG and RNA II. *Mol. Microbiol.* **18**, 913–924.

del Solar, G., and Espinosa, M. (1992). The copy number of plasmid pLS1 is regulated by two *trans*-acting plasmid products: The antisense RNA II and the repressor protein, RepA. *Mol. Microbiol.* **6**, 83–94.

del Solar, G., and Espinosa, M. (2000). Plasamid copy number control: An ever-growing story. *Mol. Microbiol.* **37**, 492–500.

Delihas, N. (1995). Regulation of gene expression by *trans*-encoded antisense RNAs. *Mol. Microbiol.* **15**, 411–414.

Delihas, N. (1997). Antisense *micF* RNA and 5'-UTR of the target *ompF* RNA: Phylogenetic conservation of primary and secondary structures. *Nucleic Acids Symp. Ser.* 33–35.

Dong, X., Womble, D. D., Luckow, V. A., and Rownd, R. H. (1985). Regulation of transcription of the *repA1* gene in the replication control region of IncFII plasmid NR1 by gene dosage of the *repA2* transcription repressor protein. *J. Bacteriol.* **161**, 544–551.

Eguchi, Y., Itoh, T., and Tomizawa, J. (1991). Antisense RNA. *Annu. Rev. Biochem.* **60**, 631–652.

Eguchi, Y., and Tomizawa, J. (1990). Complex formed by complementary RNA stem-loops and its stabilization by a protein: Function of ColE1 Rom protein. *Cell* **60**, 199–209.

Eguchi, Y., and Tomizawa, J. (1991). Complexes formed by complementary RNA stem-loops. Their formations, structures and interaction with ColE1 Rom protein. *J. Mol. Biol.* **220**, 831–842.

Ehrenberg, M. (1996). Hypothesis: Hypersensitive plasmid copy number control for ColE1. *Biophys. J.* **70**, 135–145.

Ehrenberg, M., and Sverredal, A. (1995). A model for copy number control of the plasmid R1. *J. Mol. Biol.* **246**, 472–485.

Ferrario, M., Ernsting, B. R., Borst, D. W., Wiese, D. E. n., Blumenthal, R. M., and Matthews, R. G. (1995). The leucine-responsive regulatory protein of Escherichia coli negatively regulates transcription of *ompC* and *micF* and positively regulates translation of *ompF*. *J. Bacteriol.* **177**, 103–113.

Fierro-Monti, I. P., Reid, S. J., and Woods, D. R. (1992). Differential expression of a *Clostridium acetobutylicum* antisense RNA: Implications for regulation of glutamine synthetase. *J. Bacteriol.* **174**, 7642–7647.

Firth, N., Ippen-Ihler, K., and Skurray, R. A. (1996). Structure and function of the F factor and mechanism of conjugation. *In* "*Escherichia coli* and *Salmonella*" (F. C. Neidhard, R. Curtiss III, J. L. Ingraham, E. C. C. Lin, K. B. Low , B. Magasanik, W. S. Reznikoff, M. Schaechter, and H. E. Umbarger, eds.), vol. **2**, pp. 2377–2401. ASM Press, Washington, DC.

Forti, F., Sabbattini, P., Sironi, G., Zangrossi, S., Dehò, G., and Ghisotti, D. (1995). Immunity determinant of phage-plasmid P4 is a short processed RNA. *J. Mol. Biol.* **249**, 869–878.

Franch, T., and Gerdes, K. (1996). Programmed cell death in bacteria: Translational repression by mRNA end-pairing. *Mol. Microbiol.* **21,** 1049–1060.

Franch, T., and Gerdes, K. (2000). U-turns and regulatory RNAs. *Curr. Opin. Microbiol.* **3,** 159–164.

Franch, T., Gultyaev, A. P., and Gerdes, K. (1997). Programmed cell death by hok/sok of plasmid R1: Processing at the *hok* mRNA 3′-end triggers structural rearrangements that allow translation and antisense RNA binding. *J.Mol. Biol.* **273,** 38–51.

Franch, T., Petersen, M., Wagner, E. G. H., Jacobsen, J. P., and Gerdes, K. (1999). Antisense RNA regulation in prokaryotes: Rapid RNA/RNA interaction facilitated by a general U-turn loop structure. *J. Mol. Biol.* **294,** 1115–1125.

Frost, L., Lee, S., Yanchar, N., and Paranchych, W. (1989). *finP* and *fisO* mutations in FinP anti-sense RNA suggest a model for FinOP action in the repression of bacterial conjugation by the Flac plasmid JCFL0. *Mol. Gen. Genet.* **218,** 152–160.

Galli, D. M., and Leblanc, D. J. (1995). Transcriptional analysis of rolling circle replicating plasmid pVT736-1: Evidence for replication control by antisense RNA. *J. Bacteriol.* **177,** 4474–4480.

Gerdes, K. (2000). Toxin-antitoxin modules may regulate synthesis of macromolecules during nutritional stress. *J. Bacteriol.* **182,** 561–572.

Gerdes, K., Gultyaev, A. P., Franch, T., Pedersen, K., and Mikkelsen, N. D. (1997). Antisense RNA-regulated programmed cell death. *Annu. Rev. Genet.* **31,** 1–31.

Gerdes, K., Moller-Jensen, J., and Bugge Jensen, R. (2000). Plasmid and chromosome partitioning: Surprises from phylogeny. *Mol. Microbiol.* **37,** 455–466.

Gerdes, K., Rasmussen, P. B., and Molin, S. (1986). Unique type of plasmid maintenance function: Postsegregational killing of plasmid-free cells. *Proc. Natl. Acad. Sci. USA* **83,** 3116–3120.

Ghetu, A. F., Gubbins, M. J., Oikawa, K., Kay, C. M., Frost, L. S., and Glover, J. N. (1999). The FinO repressor of bacterial conjugation contains two RNA binding regions. *Biochemistry* **38,** 14036–14044.

Gidrol, X., and Farr, S. (1993). Interaction of a redox-sensitive DNA-binding factor with the 5′-flanking region of the *micF* gene in *Escherichia coli. Mol. Microbiol.* **10,** 877–884.

Givskov, M., and Molin, S. (1984). Copy mutants of plasmid R1: Effects of base pair substitutions in the copA gene on the replication control system. *Mol. Gen. Genet.* **194,** 286–292.

Greenfield, T. J., Ehli, E., Kirshenmann, T., Franch, T., Gerdes, K., and Weaver, K. E. (2000). The antisense RNA of the *par* locus of pAD1 regulates the expression of a 33-amino-acid toxic peptide by an unusual mechanism. *Mol. Microbiol.* **37,** 652–660.

Greenfield, T. J., and Weaver, K. E. (2000). Antisense RNA regulation of the pAD1 par postsegregational killing system requires interaction at the 5′ and 3′ ends of the RNAs. *Mol. Microbiol.* **37,** 661–670.

Grosjean, H., Soll, D. G., and Crothers, D. M. (1976). Studies of the complex between transfer RNAs with complementary anticodons. I. Origins of enhanced affinity between complementary triplets. *J. Mol. Biol.* **103,** 499–519.

Grundy, F. J., and Henkin, T. M. (1994). Conservation of a transcription antitermination mechanism in aminoacyl-tRNA synthetase and amino acid biosynthesis genes in gram-positive bacteria. *J. Mol. Biol.* **235,** 798–804.

Hama, C., Takizawa, T., Moriwaki, H., and Mizobuchi, K. (1990a). Role of leader peptide synthesis in *repZ* gene expression of the ColIb-P9 plasmid. *J. Biol. Chem.* **265,** 10666–10673.

Hama, C., Takizawa, T., Moriwaki, H., Urasaki, Y., and Mizobuchi, K. (1990b). Organization of the replication control region of plasmid ColIb-P9. *J. Bacteriol.* **172,** 1983–1991.

Henkin, T. M. (1994). tRNA-directed transcription antitermination. *Mol. Microbiol.* **13,** 381–387.

Hildebrandt, M., and Nellen, W. (1992). Differential antisense transcription from the Dictyostelium EB4 gene locus: Implications on antisense-mediated regulation of mRNA stability. *Cell* **69,** 197–204.

Hiraga, S. (1992). Chromosome and plasmid partition in *Escherichia coli. Annu. Rev. Biochem.* **61,** 283–306.

Hjalt, T., and Wagner, E. G. H. (1992). The effect of loop size in antisense and target RNAs on the efficiency of antisense RNA control. *Nucleic Acids Res.* **20**, 6723–6732.

Hjalt, T. A., and Wagner, E. G. H. (1995). Bulged-out nucleotides in an antisense RNA are required for rapid target RNA binding *in vitro* and inhibition *in vivo*. *Nucleic Acids Res.* **23**, 580–587.

Itoh, T., and Tomizawa, J. (1980). Formation of an RNA primer for initiation of replication of ColE1 DNA by ribonuclease H. *Proc. Natl. Acad. Sci. USA* **77**, 2450–2454.

Itoh, T., and Tomizawa, J. (1982). Purification of ribonuclease H as a factor required for initiation of *in vitro* ColE1 DNA replication. *Nucleic Acids Res.* **10**, 5949–5965.

Jacques, J. P., and Susskind, M. M. (1991). Use of electrophoretic mobility to determine the secondary structure of a small antisense RNA. *Nucleic Acids Res.* **19**, 2971–2977.

Jain, C. (1995). IS*10* antisense control *in vivo* is affected by mutations throughout the region of complementarity between the interacting RNAs. *J. Mol. Biol.* **246**, 585–594.

Janssen, P. J., Jones, D. T., and Woods, D. R. (1990). Studies on *Clostridium acetobutylicum glnA* promoters and antisense RNA. *Mol. Microbiol.* **4**, 1575–1583.

Jensen, R. B., and Gerdes, K. (1995). Programmed cell death in bacteria: Proteic plasmid stabilization systems. *Mol. Microbiol.* **17**, 205–210.

Jerome, L. J., van Biesen, T., and Frost, L. S. (1999). Degradation of FinP antisense RNA from F-like plasmids: The RNA-binding protein, FinO, protects FinP from ribonuclease E. *J. Mol. Biol.* **285**, 1457–1473.

Kim, K., and Meyer, R. J. (1986). Copy-number of broad host-range plasmid R1162 is regulated by a small RNA. *Nucleic Acids Res.* **14**, 8027–8046.

Kittle, J. D., Simons, R. W., Lee, J., and Kleckner, N. (1989). Insertion sequence IS*10* anti-sense pairing initiates by an interaction between the 5′ end of the target RNA and a loop in the anti-sense RNA. *J. Mol. Biol.* **210**, 561–572.

Kolb, F. A., Engdahl, H. M., Slagter-Jäger, J. G., Ehresmann, B., Ehresmann, C., Westhof, E., Wagner, E. G. H., and Romby, P. (2000a). Progression of a loop-loop complex to a four-way junction is crucial for the activity of a regulatory antisense RNA. *EMBO J.* **19**, 5905–5915.

Kolb, F. A., Malmgren, C., Westhof, E., Ehresmann, C., Ehresmann, B., Wagner, E. G. H., and Romby, P. (2000b). An unusual structure formed by antisense-target RNA binding involves an extended kissing complex with a four-way junction and a side-by-side helical alignment. *RNA* **6**, 311–324.

Kolb, F. A., Westhof, E., Ehresmann, B., Ehresmann, C., Wagner, E. G. H., and Romby, P. (2001). Four-way junctions in antisense RNA-mRNA complexes involved in plasmid replication control: A common theme? *J. Mol. Biol.* **309**, 605–614.

Koraimann, G., Koraimann, C., Koronakis, V., Schlager, S., and Högenauer, G. (1991). Repression and derepression of conjugation of plasmid R1 by wild-type and mutated *finP* antisense RNA. *Mol. Microbiol.* **5**, 77–87.

Koraimann, G., Teferle, K., Markolin, G., Woger, W., and Högenauer, G. (1996). The FinOP repressor system of plasmid R1: Analysis of the antisense RNA control of *traJ* expression and conjugative DNA transfer. *Mol. Microbiol.* **21**, 811–821.

Krinke, L., Mahoney, M., and Wulff, D. L. (1991). The role of the OOP antisense RNA in coliphage lambda development. *Mol. Microbiol.* **5**, 1265–1272.

Krinke, L., and Wulff, D. L. (1987). OOP RNA, produced from multicopy plasmids, inhibits lambda *cII* gene expression through an RNase III-dependent mechanism. *Genes Dev.* **1**, 1005–1013.

Krinke, L., and Wulff, D. L. (1990). RNase III-dependent hydrolysis of lambda *cII-O* gene mRNA mediated by lambda OOP antisense RNA. *Genes Dev.* **4**, 2223–2233.

Lacatena, R. M., and Cesareni, G. (1981). Base pairing of RNA I with its complementary sequence in the primer precursor inhibits ColE1 replication. *Nature* **294**, 623–626.

Lacatena, R. M., and Cesareni, G. (1983). Interaction between RNA1 and the primer precursor in the regulation of ColE1 replication. *J. Mol. Biol.* **170**, 635–650.

Le Chatelier, E., Ehrlich, S. D., and Janniere, L. (1996). Countertranscript-driven attenuation system of the pAMβ1 *repE* gene. *Mol. Microbiol.* **20**, 1099–1112.

Lease, R. A., and Belfort, M. (2000a). A trans-acting RNA as a control switch in *Escherichia coli*: DsrA modulates function by forming alternative structures. *Proc. Natl. Acad. Sci. USA* **97**, 9919–9924.

Lease, R. A., and Belfort, M. (2000b). Riboregulation by DsrA RNA: *trans*-actions for global economy. *Mol. Microbiol.* **38**, 667–672.

Lease, R. A., Cusick, M. E., and Belfort, M. (1998). Riboregulation in *Escherichia coli*: DsrA RNA acts by RNA:RNA interactions at multiple loci. *Proc. Natl. Acad. Sci. USA* **95**, 12456–12461.

Lee, A. J., and Crothers, D. M. (1998). The solution structure of an RNA loop-loop complex: The ColE1 inverted loop sequence. *Structure* **6**, 993–1005.

Liao, S. M., Wu, T. H., Chiang, C. H., Susskind, M. M., and McClure, W. R. (1987). Control of gene expression in bacteriophage P22 by a small antisense RNA. I. Characterization *in vitro* of the Psar promoter and the sar RNA transcript. *Genes Dev.* **1**, 197–203.

Light, J., and Molin, S. (1982). The sites of action of the two copy number control functions of plasmid R1. *Mol. Gen. Genet.* **187**, 486–493.

Lin-Chao, S., and Cohen, S. N. (1991). The rate of processing and degradation of antisense RNAI regulates the replication of ColE1-type plasmids *in vivo*. *Cell* **65**, 1233–1242.

Ma, C., and Simons, R. W. (1990). The IS*10* antisense RNA blocks ribosome binding at the transposase translation initiation site. *EMBO J.* **9**, 1267–1274.

Majdalani, N., Cunning, C., Sledjeski, D., Elliott, T., and Gottesman, S. (1998). DsrA RNA regulates translation of RpoS message by an anti-antisense mechanism, independent of its action as an antisilencer of transcription. *Proc. Natl. Acad. Sci. USA* **95**, 12462–12467.

Malmgren, C., Engdahl, H. M., Romby, P., and Wagner, E. G. H. (1996). An antisense/target RNA duplex or a strong intramolecular RNA structure 5′ of a translation initiation signal blocks ribosome binding: The case of plasmid R1. *RNA* **2**, 1022–1032.

Malmgren, C., Wagner, E. G. H., Ehresmann, C., Ehresmann, B., and Romby, P. (1997). Antisense RNA control of plasmid R1 replication. The dominant product of the antisense RNA-mRNA binding is not a full RNA duplex. *J. Biol. Chem.* **272**, 12508–12512.

Mamat, U., Rietschel, E. T., and Schmidt, G. (1995). Repression of lipopolysaccharide biosynthesis in *Escherichia coli* by an antisense RNA of *Acetobacter methanolicus* phage Acm1. *Mol. Microbiol.* **15**, 1115–1125.

Marino, J. P., Gregorian, R. S. J., Csankovszki, G., and Crothers, D. M. (1995). Bent helix formation between RNA hairpins with complementary loops. *Science* **268**, 1448–1454.

Masukata, H., and Tomizawa, J. (1984). Effects of point mutations on formation and structure of the RNA primer for ColE1 DNA replication. *Cell* **36**, 513–522.

Masukata, H., and Tomizawa, J. (1986). Control of primer formation for ColE1 plasmid replication: Conformational change of the primer transcript. *Cell* **44**, 125–136.

Masukata, H., and Tomizawa, J. (1990). A mechanism of formation of a persistent hybrid between elongating RNA and template DNA. *Cell* **62**, 331–338.

Morfeldt, E., Taylor, D., von Gabain, A., and Arvidson, S. (1995). Activation of alpha-toxin translation in *Staphylococcus aureus* by the *trans*-encoded antisense RNA, RNAIII. *EMBO J.* **14**, 4569–4577.

Muesing, M., Tamm, J., Shepard, H. M., and Polisky, B. (1981). A single base-pair alteration is responsible for the DNA overproduction phenotype of a plasmid copy-number mutant. *Cell* **24**, 235–242.

Nielsen, A. K., Thorsted, P., Thisted, T., Wagner, E. G. H., and Gerdes, K. (1991). The rifampicin-inducible genes *srnB* from F and *pnd* from R483 are regulated by antisense RNAs and mediate plasmid maintenance by killing of plasmid-free segregants. *Mol. Microbiol.* **5**, 1961–1973.

Nordgren, S., Slagter-Jäger, J. G., and Wagner, E. G. H. (2001). Real time kinetic studies of interaction between folded antisense and target RNAs using surface plasmon resonance. *J. Mol. Biol.* **310,** 1125–1134.

Nordström, K., and Aagaard-Hansen, H. (1984). Maintenance of bacterial plasmids: Comparison of theoretical calculations and experiments with plasmid R1 in *Escherichia coli. Mol. Gen. Genet.* **197,** 1–7.

Nordström, K., Molin, S., and Light, J. (1984). Control of replication of bacterial plasmids: Genetics, molecular biology, and physiology of the plasmid R1 system. *Plasmid* **12,** 71–90.

Nordström, K., and Wagner, E. G. H. (1994). Kinetic aspects of control of plasmid replication by antisense RNA. *Trends Biochem. Sci.* **19,** 294–300.

Novick, R. P., Iordanescu, S., Projan, S. J., Kornblum, J., and Edelman, I. (1989). pT181 plasmid replication is regulated by a countertranscript-driven transcriptional attenuator. *Cell* **59,** 395–404.

Novick, R. P., Ross, H. F., Projan, S. J., Kornblum, J., Kreiswirth, B., and Moghazeh, S. (1993). Synthesis of staphylococcal virulence factors is controlled by a regulatory RNA molecule. *EMBO J.* **12,** 3967–3975.

Öhman, M., and Wagner, E. G. H. (1989). Secondary structure analysis of the RepA mRNA leader transcript involved in control of replication of plasmid R1. *Nucleic Acids Res.* **17,** 2557–2579.

Patient, M. E., and Summers, D. K. (1993). ColE1 multimer formation triggers inhibition of *Escherichia coli* cell division. *Mol. Microbiol.* **9,** 1089–1095.

Paulsson, J., Berg, O. G., and Ehrenberg, M. (2000). Stochastic focusing: Fluctuation-enhanced sensitivity of intracellular regulation. *Proc. Natl. Acad. Sci. USA* **97,** 7148–7153.

Paulsson, J., and Ehrenberg, M. (1998). Trade-off between segregational stability and metabolic burden: A mathematical model of plasmid ColE1 replication control. *J. Mol. Biol.* **279,** 73–88.

Paulsson, J., and Ehrenberg, M. (2000). Molecular clocks reduce plasmid loss rates: The R1 case. *J. Mol. Biol.* **297,** 179–192.

Paulsson, J., Nordström, K., and Ehrenberg, M. (1998). Requirements for rapid plasmid ColE1 copy number adjustments: A mathematical model of inhibition modes and RNA turnover rates. *Plasmid* **39,** 215–234.

Pedersen, K., and Gerdes, K. (1999). Multiple *hok* genes on the chromosome of *Escherichia coli. Mol. Microbiol.* **32,** 1090–1102.

Pepe, C. M., Maslesa-Galic, S., and Simons, R. W. (1994). Decay of the IS*10* antisense RNA by 3′ exoribonucleases: evidence that RNase II stabilizes RNA-OUT against PNPase attack. *Mol. Microbiol.* **13,** 1133–1142.

Persson, C., Wagner, E. G. H., and Nordström, K. (1988). Control of replication of plasmid R1: Kinetics of *in vitro* interaction between the antisense RNA, CopA, and its target, CopT. *EMBO J.* **7,** 3279–3288.

Persson, C., Wagner, E. G. H., and Nordström, K. (1990a). Control of replication of plasmid R1: Formation of an initial transient complex is rate-limiting for antisense RNA-target RNA pairing. *EMBO J.* **9,** 3777–3785.

Persson, C., Wagner, E. G. H., and Nordström, K. (1990b). Control of replication of plasmid R1: Structures and sequences of the antisense RNA, CopA, required for its binding to the target RNA, CopT. *EMBO J.* **9,** 3767–3775.

Piazza, F., Zappone, M., Sana, M., Briani, F., and Deho, G. (1996). Polynucleotide phosphorylase of *Escherichia coli* is required for the establishment of bacteriophage P4 immunity. *J. Bacteriol.* **178,** 5513–5521.

Polisky, B., Tamm, J., and Fitzwater, T. (1985). Construction of CoIE1 RNA1 mutants and analysis of their function *in vivo. Basic Life Sci.* **30,** 321–333.

Polisky, B., Zhang, X. Y., and Fitzwater, T. (1990). Mutations affecting primer RNA interaction with the replication repressor RNA I in plasmid ColE1: Potential RNA folding pathway mutants. *EMBO J.* **9,** 295–304.

Pörschke, D., and Eigen, M. (1971). Co-operative non-enzymic base recognition. 3. Kinetics of the helix-coil transition of the oligoribouridylic–oligoriboadenylic acid system and of oligoriboadenylic acid alone at acidic pH. *J. Mol. Biol.* **62,** 361–381.

Poulsen, L. K., Refn, A., Molin, S., and Andersson, P. (1991). The gef gene from *Escherichia coli* is regulated at the level of translation. *Mol. Microbiol.* **5,** 1639–1648.

Praszkier, J., Bird, P., Nikoletti, S., and Pittard, J. (1989). Role of countertranscript RNA in the copy number control system of an IncB miniplasmid. *J. Bacteriol.* **171,** 5056–5064.

Praszkier, J., Wei, T., Siemering, K., and Pittard, J. (1991). Comparative analysis of the replication regions of IncB, IncK, and IncZ plasmids. *J. Bacteriol.* **173,** 2393–2397.

Praszkier, J., Wilson, I. W., and Pittard, A. J. (1992). Mutations affecting translational coupling between the rep genes of an IncB miniplasmid. *J. Bacteriol.* **174,** 2376–2383.

Predki, P. F., Nayak, L. M., Gottlieb, M. B., and Regan, L. (1995). Dissecting RNA-protein interactions: RNA-RNA recognition by Rop. *Cell* **80,** 41–50.

Pritchard, R. H., Barth, P. T., and Collins, J. (1969). Control of DNA synthesis in bacteria. *Symp. Soc. Gen. Microbiol.* **19,** 263–297.

Putzer, H., Grunberg-Manago, M., and Springer, M. (1995). Bacterial aminoacyl-tRNA synthetases: Genes and regulation of expression. *In* "tRNA: Structure, Biosynthesis and Function" (D. Söll and U. RajBhandari, eds.), pp. 293–333. ASM Press, Washington, DC.

Ramani, N., Hedeshian, M., and Freundlich, M. (1994). *micF* antisense RNA has a major role in osmoregulation of OmpF in *Escherichia coli.* *J. Bacteriol.* **176,** 5005–5010.

Ranade, K., and Poteete, A. R. (1993a). Superinfection exclusion (*sieB*) genes of bacteriophages P22 and lambda. *J. Bacteriol.* **175,** 4712–4718.

Ranade, K., and Poteete, A. R. (1993b). A switch in translation mediated by an antisense RNA. *Genes Dev.* **7,** 1498–1507.

Reinhart, B. J., Slack, F. J., Basson, M., Pasquinelli, A. E., Bettinger, J. C., Rougvie, A. E., Horvitz, H. R., and Ruvkun, G. (2000). The 21-nucleotide let-7 RNA regulates developmental timing in *Caenorhabditis elegans.* *Nature* **403,** 901–906.

Riedel, H. D., Heinrich, J., and Schuster, H. (1993). Cloning, expression, and characterization of the *icd* gene in the immI operon of bacteriophage P1. *J. Bacteriol.* **175,** 2833–2838.

Riise, E., Stougaard, P., Bindslev, B., Nordström, K., and Molin, S. (1982). Molecular cloning and functional characterization of a copy number control gene (*copB*) of plasmid R1. *J. Bacteriol.* **151,** 1136–1145.

Rosen, J., Ryder, T., Ohtsubo, H., and Ohtsubo, E. (1981). Role of RNA transcripts in replication incompatibility and copy number control in antibiotic resistance plasmid derivatives. *Nature* **290,** 794–797.

Sabbattini, P., Forti, F., Ghisotti, D., and Dehò, G. (1995). Control of transcription termination by an RNA factor in bacteriophage P4 immunity: Identification of the target sites. *J. Bacteriol.* **177,** 1425–1434.

Sabbattini, P., Six, E., Zangrossi, S., Briani, F., Ghisotti, D., and Dehò, G. (1996). Immunity specificity determinants in the P4-like retronphage phi R73. *Virology* **216,** 389–396.

Salinas, P. C., Waldbeser, L. S., and Crosa, J. H. (1993). Regulation of the expression of bacterial iron transport genes: Possible role of an antisense RNA as a repressor. *Gene* **123,** 33–38.

Schaefer, K. L., and McClure, W. R. (1997a). Antisense RNA control of gene expression in bacteriophage P22. I. Structures of *sar* RNA and its target, *ant* mRNA. *RNA* **3,** 141–156.

Schaefer, K. L., and McClure, W. R. (1997b). Antisense RNA control of gene expression in bacteriophage P22. II. Kinetic mechanism and cation specificity of the pairing reaction. *RNA* **3,** 157–174.

Schmidt, M., Zheng, P., and Delihas, N. (1995). Secondary structures of *Escherichia coli* antisense *micF* RNA, the 5'-end of the target *ompF* mRNA, and the RNA/RNA duplex. *Biochemistry* **34,** 3621–3631.

Sharpe, M. E., Chatwin, H. M., Macpherson, C., Withers, H. L., and Summers, D. K. (1999). Analysis of the ColE1 stability determinant Rcd. *Microbiology* **145,** 2135–2144.

Shiba, K., and Mizobuchi, K. (1990). Posttranscriptional control of plasmid ColIb-P9 *repZ* gene expression by a small RNA. *J. Bacteriol.* **172,** 1992–1997.

Siemering, K. R., Praszkier, J., and Pittard, A. J. (1993). Interaction between the antisense and target RNAs involved in the regulation of IncB plasmid replication. *J. Bacteriol.* **175,** 2895–2906.

Siemering, K. R., Praszkier, J., and Pittard, A. J. (1994). Mechanism of binding of the antisense and target RNAs involved in the regulation of IncB plasmid replication. *J. Bacteriol.* **176,** 2677–2688.

Simons, R. W., and Kleckner, N. (1983). Translational control of IS*10* transposition. *Cell* **34,** 683–691.

Sledjeski, D., and Gottesman, S. (1995). A small RNA acts as an antisilencer of the H-NS-silenced *rcsA* gene of *Escherichia coli. Proc. Natl. Acad. Sci. USA* **92,** 2003–2007.

Sledjeski, D. D., Gupta, A., and Gottesman, S. (1996). The small RNA, DsrA, is essential for the low temperature expression of RpoS during exponential growth in *Escherichia coli. EMBO J.* **15,** 3993–4000.

Söderbom, F., Binnie, U., Masters, M., and Wagner, E. G. H. (1997). Regulation of plasmid R1 replication: PcnB and RNase E expedite the decay of the antisense RNA, CopA. *Mol. Microbiol.* **26,** 493–504.

Söderbom, F., and Wagner, E. G. H. (1998). Degradation pathway of CopA, the antisense RNA that controls replication of plasmid R1. *Microbiology* **144,** 1907–1917.

Stolt, P., and Zillig, W. (1993a). Antisense RNA mediates transcriptional processing in an archaebacterium, indicating a novel kind of RNase activity. *Mol. Microbiol.* **7,** 875–882.

Stolt, P., and Zillig, W. (1993b). Structure specific ds/ss-RNase activity in the extreme halophile *Halobacterium salinarium. Nucleic Acids Res.* **21,** 5595–5599.

Stougaard, P., Molin, S., and Nordström, K. (1981). RNAs involved in copy-number control and incompatibility of plasmid R1. *Proc. Natl. Acad. Sci. USA* **78,** 6008–6012.

Sugiyama, T., and Itoh, T. (1993). Control of ColE2 DNA replication: *In vitro* binding of the antisense RNA to the Rep mRNA. *Nucleic Acids Res.* **21,** 5972–5977.

Summers, D. K., Beton, C. W., and Withers, H. L. (1993). Multicopy plasmid instability: The dimer catastrophe hypothesis. *Mol. Microbiol.* **8,** 1031–1038.

Suzuki, T., Ueguchi, C., and Mizuno, T. (1996). H-NS regulates OmpF expression through micF antisense RNA in *Escherichia coli. J. Bacteriol.* **178,** 3650–3653.

Takechi, S., Yasueda, H., and Itoh, T. (1994). Control of ColE2 plasmid replication: Regulation of Rep expression by a plasmid-coded antisense RNA. *Mol. Gen. Genet.* **244,** 49–56.

Tamm, J., and Polisky, B. (1983). Structural analysis of RNA molecules involved in plasmid copy number control. *Nucleic Acids Res.* **11,** 6381–6397.

Tegmark, K., Morfeldt, E., and Arvidson, S. (1998). Regulation of *agr*-dependent virulence genes in *Staphylococcus aureus* by RNAIII from coagulase-negative staphylococci. *J. Bacteriol.* **180,** 3181–3186.

Tetart, F., and Bouche, J. P. (1992). Regulation of the expression of the cell-cycle gene *ftsZ* by DicF antisense RNA. Division does not require a fixed number of FtsZ molecules. *Mol. Microbiol.* **6,** 615–620.

Thisted, T., and Gerdes, K. (1992). Mechanism of post-segregational killing by the *hok*/*sok* system of plasmid R1. Sok antisense RNA regulates *hok* gene expression indirectly through the overlapping *mok* gene *J. Mol. Biol.* **223,** 41–54.

Thisted, T., Sorensen, N. S., and Gerdes, K. (1995). Mechanism of post-segregational killing: Secondary structure analysis of the entire Hok mRNA from plasmid R1 suggests a fold-back structure that prevents translation and antisense RNA binding. *J. Mol. Biol.* **247,** 859–873.

Thisted, T., Sørensen, N. S., Wagner, E. G. H., and Gerdes, K. (1994). Mechanism of post-segregational killing: Sok antisense RNA interacts with Hok mRNA via its 5′-end single-stranded leader and competes with the 3′-end of Hok mRNA for binding to the *mok* translational initiation region. *EMBO J.* **13,** 1960–1968.

Tomizawa, J. (1984). Control of ColE1 plasmid replication: The process of binding of RNA I to the primer transcript. *Cell* **38**, 861–870.

Tomizawa, J. (1985). Control of ColE1 plasmid replication: Initial interaction of RNA I and the primer transcript is reversible. *Cell* **40**, 527–535.

Tomizawa, J. (1986). Control of ColE1 plasmid replication: Binding of RNA I to RNA II and inhibition of primer formation. *Cell* **47**, 89–97.

Tomizawa, J. (1990a). Control of ColE1 plasmid replication. Interaction of Rom protein with an unstable complex formed by RNA I and RNA II. *J. Mol. Biol.* **212**, 695–708.

Tomizawa, J. (1990b). Control of ColE1 plasmid replication. Intermediates in the binding of RNA I and RNA II. *J. Mol. Biol.* **212**, 683–694.

Tomizawa, J., and Itoh, T. (1981). Plasmid ColE1 incompatibility determined by interaction of RNA I with primer transcript. *Proc. Natl. Acad. Sci. USA* **78**, 6096–6100.

Tomizawa, J., Itoh, T., Selzer, G., and Som, T. (1981). Inhibition of ColE1 RNA primer formation by a plasmid-specified small RNA. *Proc. Natl. Acad. Sci. USA* **78**, 1421–1425.

Twigg, A. J., and Sherratt, D. (1980). Trans-complementable copy-number mutants of plasmid ColE1. *Nature* **283**, 216–218.

Uhlin, B. E., and Nordström, K. (1978). A runaway-replication mutant of plasmid R1drd-19: Temperature-dependent loss of copy number control. *Mol. Gen. Genet.* **165**, 167–179.

van Biesen, T., and Frost, L. S. (1994). The FinO protein of IncF plasmids binds FinP antisense RNA and its target, *traJ* mRNA, and promotes duplex formation. *Mol. Microbiol.* **14**, 427–436.

van Biesen, T., Söderbom, F., Wagner, E. G. H., and Frost, L. S. (1993). Structural and functional analyses of the FinP antisense RNA regulatory system of the F conjugative plasmid. *Mol. Microbiol.* **10**, 35–43.

Wagner, E. G. H., Blomberg, P., and Nordström, K. (1992). Replication control in plasmid R1: Duplex formation between the antisense RNA, CopA, and its target, CopT, is not required for inhibition of RepA synthesis. *EMBO J.* **11**, 1195–1203.

Wagner, E. G. H., and Brantl, S. (1998). Kissing and RNA stability in antisense control of plasmid replication. *Trends Biochem. Sci.* **23**, 451–454.

Wagner, E. G. H., and Nordström, K. (1986). Structural analysis of an RNA molecule involved in replication control of plasmid R1. *Nucleic Acids Res.* **14**, 2523–2538.

Wagner, E. G. H., and Simons, R. W. (1994). Antisense RNA control in bacteria, phages, and plasmids. *Annu. Rev. Microbiol.* **48**, 713–742.

Waldbeser, L. S., Chen, Q., and Crosa, J. H. (1995). Antisense RNA regulation of the *fatB* iron transport protein gene in *Vibrio anguillarum*. *Mol. Microbiol.* **17**, 747–756.

Warnecke, J. M., Nitschke, M., Moolenaar, C. E., Rietschel, E. T., Hartmann, R. K., and Mamat, U. (2000). The 5′-proximal hairpin loop of *lbi* RNA is a key structural element in repression of D-galactan II biosynthesis in *Klebsiella pneumoniae* serotype O1. *Mol. Microbiol.* **36**, 697–709.

Wassarman, K. M., Repoila, F., Rosenow, C., Storz, G., and Gottesman, S. (2001). Identification of novel small RNAs using comparative genomics and microarrays. *Genes Dev.* **15**, 1637–1651.

Wassarman, K. M., Zhang, A., and Storz, G. (1999). Small RNAs in *Escherichia coli*. *Trends Microbiol.* **7**, 37–45.

Wightman, B., Ha, I., and Ruvkun, G. (1993). Posttranscriptional regulation of the heterochronic gene *lin-14* by *lin-4* mediates temporal pattern formation in *C. elegans*. *Cell* **75**, 855–862.

Wilson, I. W., Praszkier, J., and Pittard, A. J. (1993). Mutations affecting pseudoknot control of the replication of B group plasmids. *J. Bacteriol.* **175**, 6476–6483.

Wilson, I. W., Siemering, K. R., Praszkier, J., and Pittard, A. J. (1997). Importance of structural differences between complementary RNA molecules to control of replication of an IncB plasmid. *J. Bacteriol.* **179**, 742–753.

Womble, D. D., Dong, X., Luckow, V. A., Wu, R. P., and Rownd, R. H. (1985). Analysis of the individual regulatory components of the IncFII plasmid replication control system. *J. Bacteriol.* **161**, 534–543.

Wong, E. M., and Polisky, B. (1985). Alternative conformations of the ColE1 replication primer modulate its interaction with RNA I. *Cell* **42,** 959–966.

Wrobel, B., Herman-Antosiewicz, A., Szalewska-Palasz, S., and Wegrzyn, G. (1998). Polyadenylation of *oop* RNA in the regulation of bacteriophage lambda development. *Gene* **212,** 57–65.

Wu, R., Wang, X., Womble, D. D., and Rownd, R. H. (1992). Expression of the *repA1* gene of IncFII plasmid NR1 is translationally coupled to expression of an overlapping leader peptide. *J. Bacteriol.* **174,** 7620–7628.

Wu, R. P., Womble, D. D., and Rownd, R. H. (1985). Incompatibility mutants of IncFII plasmid NR1 and their effect on replication control. *J. Bacteriol.* **163,** 973–982.

Wu, T. H., Liao, S. M., McClure, W. R., and Susskind, M. M. (1987). Control of gene expression in bacteriophage P22 by a small antisense RNA. II. Characterization of mutants defective in repression. *Genes Dev.* **1,** 204–212.

Xu, F., Lin-Chao, S., and Cohen, S. N. (1993). The *Escherichia coli pcnB* gene promotes adenylylation of antisense RNAI of ColE1–type plasmids *in vivo* and degradation of RNAI decay intermediates. *Proc. Natl. Acad. Sci. USA* **90,** 6756–6760.

Zeiler, B., and Simons, R. W. (1998). Antisense RNA structure and function. *In* "RNA Structure and Function" (R. W. Simons and M. Grunberg-Manago, eds.), pp. 437–464. Cold Spring Harbor Laboratory Press, Cold Spring Harbor, NY.

Zhang, A., Altuvia, S., and Storz, G. (1997). The novel *oxyS* RNA regulates expression of the sigma S subunit of *Escherichia coli* RNA polymerase. *Nucleic Acids Symp. Ser.* 27–28.

Zhang, A., Altuvia, S., Tiwari, A., Argaman, L., Hengge-Aronis, R., and Storz, G. (1998). The OxyS regulatory RNA represses *rpoS* translation and binds the Hfq (HF-I) protein. *EMBO J.* **17,** 6061–6068.

Transvection in *Drosophila*

James A. Kennison* and Jeffrey W. Southworth
Section on Drosophila Gene Regulation
Laboratory of Molecular Genetics
National Institute of Child Health and Human Development
National Institutes of Health
Bethesda, Maryland 20892

ABSTRACT

Pairing-dependent interallelic complementation was first described for the *Ultrabithorax* gene of the bithorax-complex in *Drosophila* by Lewis and cited as an example of a new phenomenon that Lewis called the "trans-vection effect." Several different kinds of pairing-dependent gene expression have been observed in *Drosophila*, and it is now clear that a variety of different molecular mechanisms

*To whom correspondence should be addressed: E-mail: Jim-Kennison@nih.gov.

probably underlie the changes in gene expression that are observed after disrupting chromosome pairing. Transvection in the bithorax-complex appears to result from the ability of *cis*-regulatory elements to regulate transcription of the promoter on the homologous chromosome. The same phenomenon appears to be responsible for pairing-dependent interallelic complementation at numerous other genes in *Drosophila*. Some transvection effects are dependent on the presence of wild-type or specific mutant forms of the protein encoded by the *zeste* trans-regulatory gene, but other transvection effects are *zeste*-independent. The ease with which chromosome aberrations can disrupt transvection also varies widely among different genes.   © 2002, Elsevier Science (USA).

I. INTRODUCTION

> One is tempted to suggest that if homologous maternal and paternal chromosomes in the same cell ever exert any influence on each other, such that it is manifest in the heredity of the offspring, there is more opportunity for such influence in these flies than in cases where pairing of homologous chromosomes occurs but once in a generation. Possibly experiments in cross-breeding of flies may bring out some interesting facts in heredity (Stevens, 1908).

Although predicted by Nettie Stevens in 1908, the ability of chromosome pairing to alter gene expression has been intensely studied only in recent years. Many of our current ideas can be traced to a single remarkable paper by E. B. Lewis (1954), which is cited as precedent by almost every publication on pairing-dependent phenomena within the last 50 years. In his 1954 paper, Lewis demonstrated the ability of chromosome pairing to alter the phenotype of specific mutant heterozygotes and referred to the phenomenon as the "trans-vection effect." In this review, we would first like to examine the ideas that led Lewis to design experiments to examine the effects of chromosome pairing. Much of this was outlined earlier by Lewis himself (1992). We will then describe several other examples of pairing-dependent gene expression in *Drosophila* and try to summarize what is known about the molecular mechanisms responsible. We have not included any of the work on transgenes in *Drosophila*, as this is covered extensively in Chapter 14.

II. HISTORICAL BACKGROUND TO BITHORAX-COMPLEX TRANSVECTION

The discovery of transvection begins with two fundamental and related problems in genetics. What is a gene, and how do new genes arise? The best-characterized gene with multiple alleles in the early days of genetics was the *white* gene of

Drosophila. Between 1910 and 1917, seven phenotypically distinguishable alleles of *white* were isolated (Morgan, 1910, 1912; Safir, 1913, 1916; Hyde, 1916; Lancefield, 1918). Heterozygous females with 19 of the 21 possible pairwise combinations of different alleles were synthesized between 1913 and 1920. Their progeny were then examined for evidence of recombination between these alleles. No recombinants were found among 51,661 progeny (Morgan and Bridges, 1913; Safir, 1913, 1916; Hyde, 1916, 1920). Based largely on these data, it was concluded that genes were not divisible by recombination and that recombination occurred only between genes. This dogma dominated genetics for the next 30 years. For example, in their textbook of genetics, Sturtevant and Beadle (1939) describe a cross with females heterozygous for two different alleles of *white*, w^1 and w^a. They state that "since these two genes occupy the same locus, no crossing over is possible between them."

      The first example of recombination between two alleles of the same gene was described by Oliver (1940) for two alleles of the *lozenge* gene of *Drosophila*. Oliver suggested that the results were due either to unequal exchange, or that the two *lozenge* alleles were actually mutant for two closely linked genes with similar functions. The latter suggestion derived from speculation about the origin of new genes by gene duplication. In a brief communication describing duplications, Bridges (1919) had suggested that the most significant bearing of duplications would be "upon the idea of evolution of chromosome groups." This idea was expanded independently by Bridges (1935, 1936a, 1936b) and Muller (1935a, 1935b, 1936). Both suggested that once a duplication had arisen, the original and the duplicated gene could now diverge in function by the accumulation of different mutations. This would eventually lead to two (or more) genes with related, but slightly different functions. Since many of the duplications were thought to arise by unequal crossing over, producing small tandem duplications, the duplicated genes would probably remain closely linked in the genome. Grüneberg (1937) listed several pairs of closely linked mutations in *Drosophila* that had similar phenotypes, including the *bithorax* and *bithoraxoid* mutations in the bithorax complex, which were to play important roles in the discovery of transvection. Closely linked and functionally related mutations between which meiotic recombination was observed were later called "pseudoalleles." Morgan and his colleagues had originally used the word "pseudo-allelism" in a different context to describe the failure of a deficiency to complement a "dissimilar, non-allelomorphic but neighboring mutant" (Morgan *et al.*, 1938). Several dominant *vestigial* mutations were lethal when homozygous and failed to complement the *scabrous* and *l(2)C* mutations, even though the *scabrous* and *l(2)C* mutations complemented each other and complemented recessive *vestigial* mutations. The *scabrous*, *l(2)C*, and recessive *vestigial* mutations also had different mutant phenotypes. The dominant *vestigial* mutations were shown to be associated with large chromosomal deletions that probably removed the wild-type alleles of *vestigial*, *scabrous*, and *l(2)C*. To indicate that the failure of complementation was due to deletion and not mutation of the *scabrous* and *l(2)C* genes, Morgan called it pseudo-allelism. McClintock (1944) later described a case in corn in which two complementing mutants with similar phenotypes, *pyd* (pale-yellow seedlings) and *yg2* (yellow-green plants), were closely linked and uncovered by some of the same chromosomal deletions. In this case, she used the word "pseudo-allelic" to describe the relationship between the two similar and closely linked genes.

Lewis was an undergraduate in Oliver's laboratory at the time that Oliver was recovering recombinants between *lozenge* alleles. After moving to Pasadena to begin his graduate work with Sturtevant, Lewis decided to investigate evolution of new genes by gene duplication using closely linked mutations with similar phenotypes. He chose a number of genes for his initial studies, including *Star*, *Stubble*, and *bithorax*. These were chosen because each gene had at least one dominant allele and one recessive allele with related but somewhat different phenotypes. For each, Lewis was able to recover recombinants between the dominant and recessive alleles (Lewis, 1945, 1948). Since recombination was believed to occur only between genes, Lewis renamed $Star^{recessive}$ as *asteroid*, $Stubble^{recessive2}$ as $stubbloid^2$, and $bithorax^{Dominant}$ as *Ultrabithorax* to indicate that the dominant and recessive mutations were in different but closely linked genes. Following McClintock, Lewis called these closely linked mutations "pseudo-alleles" (1948). In contrast to McClintock's pseudoalleles, which complemented each other, Lewis's pseudoalleles failed to complement. For example, flies heterozygous for the pseudoalleles $bithorax^{34e}$ and *Ultrabithorax* did not look like either single heterozygote, but more closely resembled the $bithorax^{34e}$ homozygotes. Lewis suggested that the failure of the *Drosophila* pseudoalleles to complement was due to some type of "position effect," and he later used the term "position pseudoalleles" to indicate the failure to complement as expected (Lewis, 1951). In 1967, Lewis began referring to the five known position pseudoalleles of *bithorax* collectively as the "bithorax complex" (Lewis, 1967).

That the location of a gene in the genome could affect its function (position effect) was first realized by Sturtevant (1925) in his studies on the *Bar* mutation. Muller also became increasingly interested in position effects from his studies on X-ray-induced chromosome rearrangements (Muller, 1930, 1935c, 1941). Lewis (1950) reviewed the extensive evidence for position effects on gene function. Both Lewis (1951, 1955) and Pontecorvo (1950) proposed similar models of position effects to explain the failure of some pseudoalleles to complement. Figure 13.1 summarizes these models for a group of three related pseudogenes, *G1*, *G2*, and *G3*. These three genes are proposed to derive by duplication from a single gene, and then diverge by separate mutations to catalyze three successive steps in a chain of biochemical steps to convert an initial substance S0 to a final substance S3. The product of each gene is produced and acts only locally on the chromosome, with little diffusion between the alleles on homologous chromosomes. The wild-type phenotype would require the synthesis of the final

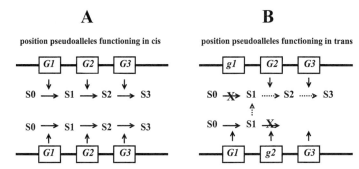

Figure 13.1. Model for the functioning of position pseudoalleles *in cis* and *in trans*. (A) illustrates functioning of the G1, G2, and G3 position pseudoalleles *in cis*. Each position pseudoallele produces an enzyme that functions in a sequential pathway that converts substance S0 to substance S3. The product of G1 converts S0 to S1, the product of G2 converts S1 to S2, and the product of G3 converts S2 to S3. The sequential steps occur primarily *in cis* because diffusion of S1 and S2 between homologous chromosomes is limited. (B) illustrates functioning of the same position pseudoalleles *in trans*. The g1 mutation on the upper chromosome blocks the conversion of S0 to S1 *in cis*. The g2 mutation on the lower chromosome blocks the conversion of S1 to S2 *in cis*. Limited diffusion of S1 substance from the lower chromosome to the upper allows some conversion of S1 to S2, and S2 to S3 by the products of the G2 and G3 position pseudoalleles of the upper chromosome (indicated by the dashed arrows). This provides some complementation between the g1 and g2 mutations.

substance (S3). A mutation in G1 on one chromosome (*g1*) would prevent the synthesis of S1 (and the synthesis of S2 and S3 by subsequent steps) by that chromosome, while a mutation in G2 (*g2*) would prevent synthesis of S2 (and the synthesis of S3 by the subsequent step) by that chromosome. Thus, the *g1/g2* heterozygote would be unable to produce the final substance S3 and have a mutant phenotype.

For the *bithorax*[34e] and *Ultrabithorax* position pseudoalleles, the actual phenotypes were even more complicated than predicted by the model. The *bithorax*[34e]/*Ultrabithorax* heterozygotes did not fail to completely complement as predicted, but showed some complementation. Lewis suggested that this partial complementation was due to some diffusion of the product substances (S1, S2, and S3 in the model) from one chromosome to the homologous chromosome. This diffusion would be facilitated by the pairing of homologous chromosomes that was known to occur in mitotic cells of *Drosophila* (Stevens, 1908; Metz, 1916). Lewis thus expected that disruption of somatic pairing in *bithorax*[34e]/*Ultrabithorax* heterozygotes would block the diffusion of product substances between homologous chromosomes and prevent the partial complementation observed in those heterozygotes. To disrupt chromosome pairing, Lewis used either existing chromosome aberrations or induced new chromosome aberrations that carried either the *bithorax*^{34e} or *Ultrabithorax* mutation. He used the symbol *R(a)* to indicate that the chromosome carrying the mutant *a* also carried a chromosome aberration. *R(bithorax*^{34e}*)/Ultrabithorax* or *bithorax*^{34e}*/R(Ultrabithorax)* heterozygous flies were expected to have more extreme mutant phenotypes than *bithorax*^{34e}*/Ultrabithorax* heterozygous flies. Further, *R(bithorax*^{34e}*)/R(Ultrabithorax)* flies (in which normal pairing is restored) should resemble the *bithorax*^{34e}*/Ultrabithorax* heterozygous flies. Since many of the chromosome aberrations were not viable when homozygous, Lewis also used different chromosome aberrations with similar breakpoints (*R1* and *R2*) to restore normal pairing in *R1(bithorax*^{34e}*)/R2(Ultrabithorax)* flies. The results were as predicted. A heterozygous chromosome aberration that was expected to reduce pairing of the *bithorax*^{34e} and *Ultrabithorax* mutations also reduced the ability of the mutations to partially complement. Restoration of normal pairing in the *R1/R2* trans-heterozygotes also restored the ability to complement (Lewis, 1954). While Lewis's model for transvection was consistent with knowledge at the time, rapid discoveries in genetics and molecular biology soon presented problems.

Lewis (1955) called the model in Figure 13.1 the "genetic interpretation" of position pseudoalleles and an alternative model advocated by Pontecorvo (1952) the "functional interpretation." In the functional interpretation, position pseudoalleles were viewed as mutations at different sites within the same gene. Recombination between position pseudoalleles then meant that recombination could occur within a single gene as well as between different genes. Support for the functional interpretation came largely from studies in various fungi, in which large numbers of meiotic products could be examined on selective media to identify rare recombinants. In 1952, both Lewis and Pontecorvo reexamined recombination between different *white* alleles in *Drosophila* (Lewis, 1952; Mackendrick and Pontecorvo, 1952). By examining larger numbers of progeny from females heterozygous for two different alleles, recombinants between several different *white* alleles were finally observed. Both Lewis and Pontecorvo used these same results as evidence supporting their respective interpretations. Later, Benzer's extensive fine structural analyses of the rII region of the bacteriophage T4 (Benzer, 1955, 1957) clearly demonstrated recombination between noncomplementing mutations within a single gene and strongly supported Pontecorvo's functional interpretation of position pseudoalleles.

The model in Figure 13.1 for *cis* functioning of position pseudoalleles became less likely with the discovery that the protein products of most genes were synthesized in the cytoplasm. Lewis (1963) suggested that position pseudoalleles might be a special case in which the products were synthesized at the chromosome and functioned only briefly. This led to the suggestion that position pseudoalleles might function as regulatory RNA molecules that diffused primarily *in cis* along the chromosome from the site of production (Jack and Judd, 1979; Lewis, 1985; Micol and Garcia-Bellido, 1988; illustrated in Figure 13.2A).

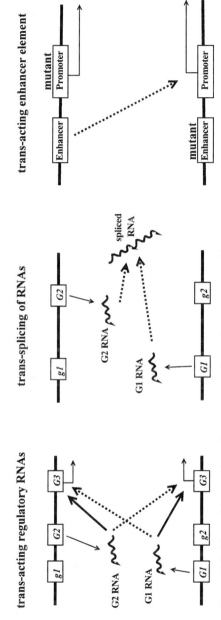

Figure 13.2. Three of the possible mechanisms proposed to explain transvection. (A, trans-acting regulatory RNAs) illustrates the production of regulatory RNAs by the G1 and G2 position pseudoalleles. The G1 and G2 RNAs are both required to regulate expression of G3, function primarily *in cis* (solid arrows), but can diffuse slowly to function in trans (dashed arrows). The *g1* and *g2* mutations prevent the production of the G1 and G2 RNAs, respectively. (B, *trans* splicing of RNAs) illustrates two RNAs that are produced by the G1 and G2 genes and then spliced into the functional RNA at the right. Again, the *g1* and *g2* mutations prevent the production of the G1 and G2 RNAs, respectively. (C, *trans*-acting enhancer element) indicates by the dashed arrow the ability of an enhancer element on one chromosome to activate transcription of the promoter on the homologous promoter. This is facilitated by the mutant promoter *in cis*.

Trans splicing of RNAs (Judd, 1979; illustrated in Figure 13.2B) and template switching of RNA polymerase between homologous genes (Babu and Bhat, 1980) were also suggested as molecular bases for transvection. However, the observation that *Ultrabithorax* mutations mapped at opposite ends of the transcription unit failed to complement made *trans* splicing or template switching unlikely (Akam *et al.*, 1984; Lewis, 1985). Ashburner (1977) suggested that transvection might also be due to changes in chromatin structure that were somehow transferred to the homologous gene. With the characterization of "enhancer" elements (modular *cis*-regulatory elements that control transcription) in the early 1980s, it was also suggested that transvection might result from enhancer elements acting *in trans* on the homologous promoter (Pirrotta *et al.*, 1985; Zachar *et al.*, 1985; illustrated in Figure 13.2C). Enhancer elements have been shown to act *in trans* in plasmid-based systems in *Drosophila* (Rothberg *et al.*, 1991) and in other organisms (Dunaway and Dröge, 1989; Müller and Schaffner, 1990). Molecular analyses support the idea that the *Ultrabithorax* transvection described by Lewis results from enhancer elements acting in trans (the model illustrated in Figure 13.2C). The *bithorax, anterobithorax, postbithorax,* and *bithoraxoid,* position pseudoalleles are *cis*-regulatory mutations in the *Ultrabithorax* gene that delete or block specific enhancer elements (reviewed in Duncan, 1987). *Ultrabithorax cis*-regulatory elements do appear to activate or repress transcription of the promoter on the homologous chromosome in some genotypes (Micol *et al.*, 1990; Martínez-Laborda *et al.*, 1992; Goldsborough and Kornberg, 1996; Casares *et al.*, 1997).

III. TRANSVECTION, *TRANS*-SENSING, AND HOMOLOGY EFFECTS

Lewis introduced the term transvection to describe "the position effect that is revealed by modifying the trans-heterozygote by means of chromosomal rearrangements" (Lewis, 1954). Since the original observation by Lewis, several phenomena that might be pairing-dependent in *Drosophila* have been called transvection. These pairing-dependent phenomena appear to act by several different molecular mechanisms. To avoid confusion, Tartof and Henikoff (1991) suggested that the term "trans-sensing" be used to include all pairing-dependent phenomena, and recommended that the term transvection be retained for pairing-dependent effects that closely resembled the original *Ultrabithorax* example described by Lewis. We find this a useful distinction, although Wu and Morris (1999) suggested that when the terms *trans*-sensing and transvection are rigorously defined, they are synonymous. In any case, we will first describe several examples of *trans*-sensing phenomena in *Drosophila* that appear to involve different molecular mechanisms from the *Ultrabithorax* example. Then we will describe examples of transvection that closely resemble *Ultrabithorax* transvection.

A. Dominant *trans*-silencing

Muller (1930) described several radiation-induced dominant eye color mutations. Most are dominant alleles of the *brown* gene. The dominant alleles of *brown* are all associated with chromosome aberrations that bring heterochromatin to the vicinity of the *brown* gene (Slatis, 1955a, 1955b; Talbert *et al.*, 1994). This heterochromatin is able to inactivate both the *brown* gene *in cis* on the chromosome aberration, and any wild-type *brown* genes that are able to pair with that *brown* gene (Henikoff and Dreesen, 1989; Dreesen *et al.*, 1991; Talbert *et al.*, 1994; Henikoff *et al.*, 1995). The molecular mechanism responsible appears to be the relocation of *brown* to a different region of the nucleus mediated by heterochromatin–heterochromatin interactions (Csink and Henikoff, 1996; Dernburg *et al.*, 1996). Several other dominant mutations may act through a similar mechanism. At least four different dominant *Punch* eye color mutations are associated with chromosome aberrations that place the *Punch* gene close to centric heterochromatin (Lindsley and Zimm, 1992). Oliver (1943) showed that one of these dominant *Punch* mutations was suppressed when heterozygous to a translocation, suggesting that the dominant inactivation of *Punch* by the original chromosome aberration is pairing dependent. In a third example, a chromosome aberration that placed the *karmoisin* eye color gene close to centric heterochromatin behaved as a dominant *karmoisin* allele (Henikoff, 1979); all other known *karmoisin* alleles are recessive (Lindsley and Zimm, 1992). It was not determined whether the dominant *karmoisin* mutation was pairing-dependent. Finally, dominant eye defects associated with *In(2LR)40d*, a chromosome aberration that has one breakpoint in centric heterochromatin, were suppressed by further rearrangements (Hinton, 1950), suggesting pairing dependence. The dominant eye defects associated with *In(2LR)40d* included a roughening of the facets of the eye, mottling of the pigment of the eye, and the occurrence of black spots on the surface of the eye. The gene responsible for these defects was not identified.

B. Dominant *trans*-derepression

A different molecular mechanism is probably responsible for pairing-dependent derepression of the *Sex combs reduced* gene in imaginal tissues. This pairing-dependent derepression is associated with all chromosome aberrations (except deficiencies) that have a breakpoint within a 70-kb portion of the *Sex combs reduced* gene (Hazelrigg and Kaufman, 1983; Pattatucci and Kaufman, 1991; Southworth and Kennison, unpublished). The chromosome aberrations cause derepression of the wild-type homolog. This derepression can be blocked by chromosome aberrations that are expected to disrupt pairing. We believe that the derepression occurs by interference with interaction of *cis*-regulatory elements in the *Sex combs reduced* gene whose normal functions are to interact *in cis* to facilitate the maintenance of silencing (Southworth and Kennison, unpublished).

C. The *cubitus interruptus* gene and the "Dubinin" effect

Dubinin and Sidorov (1934a, 1934b) described a genetic phenomenon involving the fourth chromosome mutation *cubitus interruptus* (*ci*). This is sometimes called the "Dubinin" effect (Lewis, 1950). Dubinin and Sidorov crossed 19 translocations involving the fourth chromosome to the recessive *ci* mutation. Ten of these translocations failed to completely complement the *ci* mutation, that is, the *ci* mutation behaved as a dominant when heterozygous to the translocations. Dubinin (1935) and Sturtevant and Dobzhansky (Dobzhansky, 1936) tested another 29 translocations, of which 12 also failed to complement the *ci* mutation. Flies homozygous for some of these translocations could be recovered; the homozygotes did not have a *ci* mutant phenotype, suggesting that the *ci* gene was not mutated on the translocation chromosomes (Dubinin and Sidorov, 1934b). Several of the translocations were also examined in haplo-IV flies that had only the translocated fourth chromosome (the fourth chromosome is small enough that these aneuploids survive). Again, if the *ci* gene in the translocated chromosome were altered, these flies were expected to have the *ci* mutant phenotype; they did not (Dubinin and Sidorov, 1934b). Sidorov (1941) and Stern and Heidenthal (1944) irradiated chromosomes carrying the *ci* mutation and recovered translocations carrying the *ci* mutation *in cis*. Many of these translocations behaved as though they now carried a dominant *ci* mutation, i.e., flies heterozygous for a translocation often had a *ci* mutant phenotype, even though they carried a wild-type *ci* allele on the normal fourth chromosome. The recessive *ci* mutation became dominant when the fourth chromosome carrying it was involved in a translocation. Although Ephrussi and Sutton (1944) pointed out that these results suggested a role for chromosome pairing in the phenomenon, the effects of translocations on *ci* expression continued to be primarily cited as an example of a position effect. Although studied extensively by Stern and his colleagues (Stern and Heidenthal, 1944; Stern *et al.*, 1946a, 1946b; Stern and Kodani, 1955), a simple model to explain the results remained elusive.

A simple explanation for the Dubinin effect was finally proposed by Locke and Tartof (1994). They proposed that the recessive *ci* mutation is not a loss-of-function mutation, but causes misexpression of the *ci* gene due to loss or inactivation of a negative *cis*-regulatory element. When able to pair with a wild-type *ci* gene, the negative *cis*-regulatory element in the wild-type gene can repress the mutant *ci* gene in trans. Disruption of pairing in translocation heterozygotes (which should be independent of whether the *ci* mutation is carried by the wild-type or translocated chromosome) prevents the negative *cis*-regulatory element of the wild-type gene from repressing *in trans*. The *ci* mutant gene is derepressed, causing the *ci* mutant phenotype. Molecular and developmental analyses support this model. The *ci* phenotype of recessive *ci* mutations does appear to result from derepression of the *ci* gene (Schwartz *et al.*, 1995; Slusarski *et al.*, 1995). Recessive *ci* mutations map to a region upstream of the *ci* transcription unit that includes a *cis*-regulatory element that can repress a reporter gene in a transgenic assay (Locke and Tartof, 1994; Schwartz *et al.*, 1995; Slusarski *et al.*, 1995).

D. Transvection similar to Lewis's original bithorax-complex example

The remaining examples that we would like to describe all resemble the original example of transvection presented by Lewis. These all have the property that a^1/a^2 *trans*-heterozygotes have a more normal phenotype than either a^1/a^1 or a^2/a^2 homozygotes. This is also a property of interallelic complementation (Crick and Orgel, 1964). Transvection is distinguished from interallelic complementation by its pairing dependence. In transvection, $R(a^1)/a^2$ *trans*-heterozygotes (where R indicates a chromosome aberration with a breakpoint that is not within gene a, but is close enough to disrupt pairing of the homologs) have a different phenotype than a^1/a^2 *trans*-heterozygotes. $R(a^1)/R(a^2)$ *trans*-heterozygotes (in which normal pairing is restored) should resemble the a^1/a^2 *trans*-heterozygotes in phenotype. In interallelic complementation, chromosome rearrangements that disrupt pairing have no effect on the phenotype of *trans*-heterozygotes. Pairing dependence has been examined either by introducing known chromosome aberrations into the a^1/a^2 *trans*-heterozygotes by meiotic recombination, or by recovering irradiated a^1 mutant chromosomes heterozygous to a^2 and looking for new chromosome aberrations that alter the a^1/a^2 phenotype. From these $R(a^1)/a^2$ *trans*-heterozygotes, the new R chromosome aberration can be recovered and studied.

Several examples of pairing-dependent polytene chromosome puffs in the larval salivary gland are known. Ashburner (1967, 1969, 1970a, 1970b) described a chromosome puff at 64C that was only formed in larvae heterozygous or homozygous for a third chromosome from the vg6 strain. In the vg6 heterozygotes, the vg6 chromosome almost always formed a puff at 64C, but the non-vg6 homolog only formed a puff when the chromosomes were paired. When the chromosomes failed to pair in the 64C region, the non-vg6 chromosome did not form a puff at 64C. Similar results were described for a chromosome puff at 22B8-9 that formed only in homozygotes or heterozygotes for a second chromosome from the *ast ho ed dp cl* mutant strain (Ashburner, 1970a, 1970b). Korge (1977, 1981) studied a third polytene chromosome puff at 3C. This polytene chromosome puff is the site of transcription for the *Sgs-4* gene, which is expressed at high levels in the salivary gland at the time that the polytene chromosome puff is present. *Sgs-4* alleles were characterized from several different "wild-type" laboratory strains. Most *Sgs-4* alleles (such as the alleles from the Berlin or FM1 strains) formed polytene chromosome puffs at 3C and produced *Sgs-4* proteins when homozygous or hemizygous. The *Sgs-4* allele from the Hikone-R strain did not form a polytene chromosome puff at 3C or produce *Sgs-4* proteins when homozygous or hemizygous, but it did form the 3C puff and produce *Sgs-4* proteins when heterozygous to the *Sgs-4* alleles from the Berlin and FM1 strains. The formation of the 3C puff in the Hikone-R chromosome, however, was dependent on chromosome pairing in the heterozygotes. Molecular analysis showed that the *Sgs-4* allele in the Hikone-R chromosome has a 51-bp deletion in the 5′-*cis*-regulatory region necessary for formation of the 3C puff (Muskavitch and Hogness, 1980; McGinnis *et al.*, 1983; Korge *et al.*, 1990).

Transvection at the *decapentaplegic* (*dpp*) locus was first demonstrated by Gelbart (1982) for *dpp^{ho}* (currently called *dpp^{d-ho}*), which is a deletion of *cis*-regulatory sequences 3′ to the *dpp* transcription unit (Irish and Gelbart, 1987; Masucci *et al.*, 1990), and *dpp^4* (currently called *dpp^{d4}*), which is a missense mutation in the open reading frame (Wharton *et al.*, 1996). Flies homozygous for *dpp^{ho}* survived, but their wings were held out from the body in an abnormal position. Flies homozygous for *dpp^4* did not survive. The *dpp^{ho}*/*dpp^4* *trans*-heterozygous flies survived and held their wings in a normal position. When also heterozygous for chromosome aberrations that should disrupt chromosome pairing of the *dpp* genes, the *dpp^{ho}*/*dpp^4* *trans*-heterozygous flies had the same phenotype as the *dpp^{ho}* homozygotes, i.e., they held their wings in an abnormal position. Thus, complementation between *dpp^{ho}* and *dpp^4* is pairing-dependent.

Pairing-dependent complementation between alleles of the *eyes absent* (*eya*) gene has also been demonstrated (Leiserson *et al.*, 1994). The *eya^2* mutation is a 322-bp deletion in a *cis*-regulatory region 5′ to the *eya* transcription unit; *eya^4* has a transposable element inserted in the 5′-UTR (Zimmerman *et al.*, 2001). Flies homozygous for the *eya^2* mutation had no eyes and flies homozygous for the *eya^4* mutation had eyes that were severely reduced in size. These two alleles largely complemented each other so that *eya^2*/*eya^4* *trans*-heterozygous flies had eyes that were almost normal in size (about three-fourths of normal size). This complementation was pairing-dependent; chromosome aberrations that should disrupt pairing of the *eya* alleles blocked complementation. Further, using two different chromosome aberrations with similar breakpoints (*R1* and *R2*), pairing should no longer be disrupted. As expected, the *R1*(*eya^2*)/*R2*(*eya^4*) flies had eyes of almost normal size, similar in size to those of *eya^2*/*eya^4* flies.

Transvection has also been observed for three other homeotic genes in addition to *Ultrabithorax*. These are the *abdominal-A* (Lewis, 1985; Jijakli and Ghysen, 1992), *Abdominal-B* (Lewis, 1985; Hendrickson and Sakonju, 1995; Hopmann *et al.*, 1995; Sipos *et al.*, 1998), and *Sex combs reduced* (Southworth and Kennison, unpublished observations) genes. For example, the *iab-2^{D24}* and *iab-4,5^{DB}* *cis*-regulatory mutations of *abdominal-A* showed pairing-dependent complementation (Lewis, 1985). The *iab-5^{C7}* and *iab-7^{D14}* *cis*-regulatory mutations of *Abdominal-B* also showed pairing-dependent complementation (Lewis, 1985). For the *Sex combs reduced* gene, we have observed pairing-dependent activation of the promoter by an imaginal leg enhancer on the homologous chromosome (Southworth and Kennison, unpublished). We observed *trans* activation by the imaginal leg enhancer only when the *Sex combs reduced* promoter *in cis* was deleted.

Several different pairing-dependent phenomena have been observed at the *white* locus of *Drosophila,* including interactions of *zeste* mutations with wild-type alleles of *white* (reviewed in Jack and Judd, 1979; Judd, 1988; Wu and Goldberg, 1989; Pirrotta, 1991), interactions between the *white*^{DZL} allele and wild-type *white* alleles (Bingham, 1980; Bingham and Zachar, 1985), and interallelic complementation involving the *spotted* (*sp*) *cis*-regulatory alleles of *white* (Lewis, 1956; Babu and Bhat, 1980). While clearly related, these pairing-dependent interactions between *white* alleles may occur by different molecular mechanisms. Complementation between the *white*^{sp} alleles and other *white* alleles, such as *white*^{1} and *white*^{ch}, most closely resembles the example of transvection described by Lewis for *Ultrabithorax*. The *white*^{sp} alleles map to a 5′-*cis*-regulatory region that includes an eye-specific enhancer (Davidson *et al.*, 1985; Pirrotta *et al.*, 1985; Levis *et al.*, 1985).

The last example of transvection that we would like to describe in detail is the *yellow* gene. This includes some of the most elegant molecular analyses of transvection. Interallelic complementation involving the *yellow*^{2} mutation was first described in 1935 (Stone, 1935), but it was not recognized that this resulted from transvection until 1990 (Geyer *et al.*, 1990). The *yellow*^{2} mutation is caused by the insertion of a *gypsy* transposable element between the *yellow* promoter and *yellow cis*-regulatory enhancer elements 5′ to the transcription unit. The *gypsy* insertion acts as an insulator to block interactions between the enhancer elements and the promoter *in cis*. Deletion of the *yellow* promoter on the homologous chromosome allows the *yellow* enhancer elements on the deleted chromosome to activate the promoter on the *yellow*^{2} mutant chromosome in trans. This *trans* activation was shown to be pairing-dependent (Geyer *et al.*, 1990). Transvection at the *yellow* locus does not require the gypsy insertion, but was also observed between the *yellow*^{82f29} allele (which deletes the upstream enhancer elements) and the *yellow* alleles that delete the *yellow* promoter (Morris *et al.*, 1998).

Morris and colleagues (Morris *et al.*, 1999) used targeted gene replacement to make specific mutations of the *yellow* promoter to determine which promoter elements need to be inactivated *in cis* in order to allow the enhancer elements to interact with the promoter on the homolog. Mutations of two core promoter elements (the TATA box and the initiator) were constructed and tested for their ability to facilitate transvection of the *yellow*^{2} and *yellow*^{82f29} alleles. Mutation of either core promoter element facilitated transvection, but not to the same extent as the large promoter deletions. The double mutant of both the TATA box and initiator elements facilitated transcription better than either single mutant, but still not to the same extent as the deletions. This suggests that other elements (in addition to the TATA box and initiator) are involved in transvection at the *yellow* gene.

One recent surprising observation is that activation of the promoter on the *yellow*[2] mutant chromosome can occur by a second transvection mechanism in addition to activation by enhancer elements *in trans*. The *yellow*[3c3] mutation, which deletes the *yellow* promoter and one of the 5′-enhancer elements, interacts with the *yellow*[2] mutant chromosome to allow the enhancer elements on the *yellow*[2] mutant chromosome to bypass the gypsy insulator block and activate the *yellow* promoter *in cis* (Morris *et al.*, 1998).

In addition to the examples that we have already discussed, interallelic complementation that may be caused by transvection has been reported for the *Notch* (Siren and Portin, 1988), *pointed* (Scholz *et al.*, 1993), *Gpdh* (Gibson *et al.*, 1999), *wingless* (Neumann and Cohen, 1996; Buratovich *et al.*, 1997), *hedgehog* (Lee *et al.*, 1992), *vestigial* (Alexandrov and Alexandrova, 1987; Williams *et al.*, 1990a, 1990b, 1991), and *engrailed* (Condie and Brower, 1989) loci. For *vestigial* and *engrailed*, efforts to demonstrate pairing dependence failed (Williams *et al.*, 1990a; Gustavson *et al.*, 1996), but the other loci remain to be further examined.

IV. *TRANS*-ACTING *zeste* PROTEINS AND TRANSVECTION

The *zeste* protein is a *trans*-acting factor required for some transvection (reviewed in Wu and Goldberg, 1989: Pirrotta, 1991). It was once believed to be the Rosetta stone of transvection, but if true, it still needs to be translated. The *zeste* gene was first identified by a gain-of-function mutation, *zeste*[1], with unusual properties (Gans, 1953). Females homozygous for the X-linked *zeste*[1] mutation have yellow eyes, while hemizygous males have eyes of a normal red color. The yellow eye color of *zeste*[1] females is due to a reduction in expression of the *white* genes, which occurs only if the *white* genes can pair (Jack and Judd, 1979; Gelbart and Wu, 1982). The repression of paired *white* genes requires the gain-of-function *zeste*[1] mutation; loss-of-function *zeste* mutations (such as the *zeste*[a] allele) have no pairing-related effect on wild-type *white* alleles. Lewis isolated two X-linked mutations that blocked *Ultrabithorax* transvection (Lewis, 1959). These two mutations were later shown to be *zeste* loss-of-function alleles (Kaufman *et al.*, 1973). Zeste is not essential for viability; flies that completely lack the *zeste* gene can survive and appear almost completely normal, except that the eyes are a dull reddish-brown color instead of the normal bright red (Goldberg *et al.*, 1989). Zeste proteins are DNA-binding transcriptional activators (Biggin *et al.*, 1988) and aggregate *in vitro* to form large multimeric complexes (Bickel and Pirrotta, 1990). These properties suggested that *zeste* proteins might bind to specific sequences in enhancer elements and promoters. Aggregation of *zeste* proteins would bring the enhancers and promoters closer to each other, facilitating transvection if the enhancers and promoters were on homologous chromosomes (Pirrotta, 1991; Qian *et al.*, 1992).

For many examples of transvection, the role of *zeste* has not been examined. Where the role of *zeste* has been examined, the conclusions are not always clear. Transvection of some combinations of alleles at the *Ultrabithorax* locus (such as *bithorax³⁴ᵉ/Ultrabithorax*) was blocked by most loss-of-function *zeste* mutations (including *zesteᵃ*, *zesteᵃᵉ⁽ᵇˣ⁾*, *zesteᵃ⁶⁹⁻²*, *zesteᵃ⁶⁹⁻³*, and *zeste¹¹ᴳ³*), but not by the *zesteᵛ⁷⁷ʰ* loss-of-function mutation or the *zeste¹* gain- of-function mutation (Kaufman *et al.*, 1973; Gelbart and Wu, 1982; Green, 1984). Transvection of other combinations of alleles (such as *bithorax³⁴ᵉ/Ultrabithoraxᴹˣ⁶* and *bithorax³⁴ᵉ/Ultrabithorax¹⁹⁵ʳˣ¹*) was not blocked by the *zesteᵃ⁶⁹⁻²* or *zesteᵃ⁶⁹⁻³* mutation (other *zeste* alleles were not tested) (Martínez-Laborda *et al.*, 1992). The *zesteᵃ* mutation blocked transvection of the *whiteˢᵖᵒᵗᵗᵉᵈ* allele (Babu and Bhat, 1980), but had no effect on paired wild-type *white* alleles (Gans, 1953; Kaufman *et al.*, 1973; Jack and Judd, 1979). In contrast, the *zeste¹* mutation repressed paired wild-type *white* alleles (Jack and Judd, 1979; Gelbart and Wu, 1982), but did not block transvection of *whiteˢᵖᵒᵗᵗᵉᵈ* (Babu and Bhat, 1980). Transvection of *decapentaplegic* was blocked by all *zeste* mutations tested (*zesteᵃ*, *zesteᵃ⁶⁹⁻²*, *zeste¹¹ᴳ³*, and *zeste¹*), but only if the flies were also heterozygous for a chromosome aberration that weakly disrupted transvection in a *zeste⁺* background (Gelbart and Wu, 1982). Effects of the *zeste* mutations were observed in other *decapentaplegic* mutant genotypes that were structurally normal, but the phenotypes of these mutant genotypes were not shown to be pairing-dependent (Gelbart and Wu, 1982). Transvection of *eyes absent* was blocked by the *zesteᵃᵉ⁽ᵇˣ⁾* mutation, but no other *zeste* mutations were tested (Leiserson *et al.*, 1994). Transvection of *Abdominal-B* was not blocked by any of the *zeste* mutations tested (including *zesteᵃ*, *zesteᵃᵉ⁽ᵇˣ⁾*, *zesteᵃ⁶⁹⁻²*, *zesteᵃ⁶⁹⁻³*, *zeste¹¹ᴳ³*, *zeste¹*, and *zesteᵛ⁷⁷ʰ*) (Hendrickson and Sakonju, 1995; Hopmann *et al.*, 1995; Sipos *et al.*, 1998). Transvection of *Sex combs reduced* was also not blocked by any of the *zeste* mutations tested (*zesteᵃ*, *zeste¹¹ᴳ³*, or *zeste¹*) (Pattatucci and Kaufman, 1991; Southworth and Kennison, unpublished observation).

V. "CRITICAL REGIONS" FOR TRANSVECTION

In his original experiments characterizing transvection, Lewis found that most chromosome aberrations that prevented transvection had at least one breakpoint in a "critical region" for transvection (Lewis, 1954). For the *bithorax* pseudoalleles, the critical region was fairly large, including almost the entire euchromatic region of the chromosome between *Ultrabithorax* and the centromere. Lewis suggested that somatic pairing is initiated in the proximal part of the chromosome arm and then proceeds distally. A chromosome aberration would inhibit somatic pairing most effectively distal to its breakpoint in that chromosome arm. The critical regions for the *dpp* and *eya* genes are fairly large (Gelbart, 1982; Leiserson *et al.*, 1994), but the critical region for the *Abdominal-B* gene appears to be very small (Hendrickson and Sakonju, 1995; Hopmann *et al.*, 1995; Sipos *et al.*, 1998). The critical region for *zeste*[1] repression of paired *white* genes is also very small (Gans, 1953; Green, 1967; Smolik-Utlaut and Gelbart, 1987). Surprisingly, when the paired *white* genes were located proximal to *Ultrabithorax* but distal to the breakpoints of chromosome aberrations that disrupted *Ultrabithorax* transvection, *zeste*[1] repression was not affected (Smolik-Utlaut and Gelbart, 1987). Similar results were observed for paired *white* genes proximal to *dpp* (Smolik-Utlaut and Gelbart, 1987). How can chromosome aberrations affect transvection of *Ultrabithorax* (or *dpp*) without affecting the pairing of *white* genes in between? A possible solution to this paradox has been proposed (Golic and Golic, 1996; Gubb *et al.*, 1997). They presented evidence that the length of the cell cycle may affect the extent of somatic chromosome pairing. Tissues in which the cells divide more rapidly may be more sensitive to disruptions of pairing by chromosome aberrations. Since *zeste*[1] repression of paired *white* genes and transvection of *Ultrabithorax* occur in different tissues and at different times during development, the difference in their sensitivity to the same chromosome aberrations may be due to differences in the lengths of the cell cycle.

While tremendous progress (as evidenced by the collection of reviews in this volume) has been made in the past few years in understanding the mechanisms underlying homology effects in many different organisms, much work remains for the future.

Acknowledgments

We would like to thank Judy Kassis for many helpful discussions and for comments on the manuscript. Jim Kennison also thanks Dan Lindsley and Claude Hinton for sparking his interest in the history of *Drosophila* genetics.

References

Akam, M. E., Moore, H., and Cox, A. (1984). *Ultrabithorax* mutations map to distant sites within the bithorax complex of *Drosophila*. *Nature* **309,** 635–637.

Alexandrov, I. D., and Alexandrova, M. V. (1987). A new *nw* allele and interallelic complementation at the *vg* locus of *Drosophila melanogaster*. *Drosophila Information Service* **66,** 11–12.

Ashburner, M. (1967). Gene activity dependent on chromosome synapsis in the polytene chromosomes of *Drosophila melanogaster*. *Nature* **214,** 1159–1160.

Ashburner, M. (1969). Patterns of puffing activity in the salivary gland chromosomes of *Drosophila*. IV. Variability of puffing patterns. *Chromosoma (Ber.)* **27,** 156–177.

Ashburner, M. (1970a). The genetic analysis of puffing in polytene chromosomes of *Drosophila*. *Proc. R. Soc. Lond. B* **176,** 319–327.

Ashburner, M. (1970b). A prodromus to the genetic analysis of puffing in *Drosophila*. *Cold Spring Harbor Symp. Quant. Biol.* **35,** 533–538.

Ashburner, M. (1977). Happy birthday—puffs! *Chromosomes Today* **6**, 213–222.

Babu, P., and Bhat, S. (1980). Effect of *zeste* on *white* complementation. *In* "Development and Neurobiology of Drosophila" (O. Siddiqi, P. Babu, L. M. Hall, J. C. Hall, eds.), pp. 35–40. Plenum, New York.

Benzer, S. (1955). Fine structure of a genetic region in bacteriophage. *Proc. Natl. Acad. Sci. USA* **41**, 344–354.

Benzer, S. (1957). The elementary units of heredity. *In* "The Chemical Basis of Heredity" (W. D. McElroy and B. Glass, eds.), pp. 70–93. Johns Hopkins University Press, Baltimore.

Bickel, S., and Pirrotta, V. (1990). Self-association of the *Drosophila zeste* protein is responsible for transvection effects. *EMBO J.* **9**, 2959–2967.

Biggin, M. D., Bickel, S. E., Benson, M., Pirrotta, V., and Tjian, R. (1988). Zeste encodes a sequence-specific transcription factor that activates the *Ultrabithorax* promoter in vitro. *Cell* **53**, 713–722.

Bingham, P. (1980). The regulation of white locus expression: a dominant mutant allele at the white locus of *Drosophila melanogaster*. *Genetics* **95**, 341–353.

Bingham, P. M., and Zachar, Z. (1985). Evidence that two mutations, w^{DZL} and z^1, affecting synapsis-dependent genetic behavior of *white* are transcriptional regulatory mutations. *Cell* **40**, 819–825.

Bridges, C. B. (1919). Duplications. *Anat. Rec.* **15**, 357–358.

Bridges, C. B. (1935). Salivary chromosome maps with a key to the banding of the chromosomes of *Drosophila melanogaster*. *J. Hered.* **26**, 60–64.

Bridges, C. B. (1936a). The bar "gene" a duplication. *Science* **83**, 210–211.

Bridges, C. B. (1936b). Genes and chromosomes. *The Teaching Biologist* **November 1936**, 17–23.

Buratovich, M. A., Phillips, R. G., and Whittle, J. R. S. (1997). Genetic relationships between the mutations *spade* and *Sternopleural* and the *wingless* gene in *Drosophila* development. *Dev. Biol.* **185**, 244–260.

Casares, F., Bender, W., Merriam, J., and Sanchez-Herrero, E. (1997). Interactions of *Drosophila Ultrabithorax* regulatory regions with native and foreign promoters. *Genetics* **145**, 123–137.

Condie, J. M., and Brower, D. L. (1989). Allelic interactions at the *engrailed* locus of *Drosophila*: engrailed protein expression in imaginal discs. *Dev. Biol.* **135**, 31–42.

Crick, F. H. C., and Orgel, L. E. (1964). The theory of inter-allelic complementation. *J. Mol. Biol.* **8**, 161–165.

Csink, A. K., and Henikoff, S. (1996). Genetic modification of heterochromatin association and nuclear organization in *Drosophila*. *Nature* **381**, 529–531.

Davison, D., Chapman, C. H., Wedeen, C., and Bingham, P. M. (1985). Genetic and physical studies of a portion of the *white* locus participating in transcriptional regulation and in synapsis-dependent interactions in *Drosophila* adult tissues. *Genetics* **110**, 479–494.

Dernburg, A. F., Broman, K. W., Fung, J. C., Marshall, W. F., Philips, J., Agard, D. A., and Sedat, J. W. (1996). Perturbation of nuclear architecture by long-distance chromosome interactions. *Cell* **85**, 745–759.

Dobzhansky, Th. (1936). Position effects on genes. *Biol. Rev.* **11**, 364–384.

Dreesen, T. D., Henikoff, S., and Loughney, K. (1991). A pairing-sensitive element that mediates *trans*-inactivation is associated with the *Drosophila brown* gene. *Genes Dev.* **5**, 331–340.

Dubinin, N. P. (1935). Discontinuity and continuity in the structure of the hereditary materials. *Trudy Din. Razv.* **10**, 345–360.

Dubinin, N. P., and Sidorov, B. N. (1934a). Relation between the effect of a gene and its position in the system. *Biol. Zh.* **3**, 307–331.

Dubinin, N. P., and Sidorov, B. N. (1934b). Relation between the effect of a gene and its position in the system. *Am. Naturalist* **68**, 377–381.

Dunaway, M., and Dröge, P. (1989). Transactivation of the *Xenopus* rRNA gene promoter by its enhancer. *Nature* **341**, 657–659.

Duncan, I. M. (1987). The bithorax complex. *Annu. Rev. Genet.* **21**, 285–319.

Ephrussi, B., and Sutton, E. (1944). A reconsideration of the mechanism of position effect. *Proc. Natl. Acad. Sci. USA* **30**, 183–197.

Gans, M. (1953). Etude genetique et physiologique du mutant *z* de *Drosophila melanogaster*. *Bull. Biol. Fr. Belg.* **38**, 1–90.

Gelbart, W. M. (1982). Synapsis-dependent allelic complementation at the decapentaplegic gene complex in *Drosophila melanogaster*. *Proc. Natl. Acad. Sci. USA* **79**, 2636–2640.

Gelbart, W. M., and Wu, C.-T. (1982). Interactions of *zeste* mutations with loci exhibiting transvection effects in *Drosophila melanogaster*. *Genetics* **102**, 179–189.

Geyer, P. K., Green, M. M., and Corces, V. G. (1990). Tissue-specific transcriptional enhancers may act in *trans* on the gene located in the homologous chromosome: The molecular basis of transvection in Drosophila. *EMBO J.* **9**, 2247–2256.

Gibson, J. B., Reed, D. S., Bartoszewski, S., and Wilks, A. V. (1999). Structural changes in the promoter region mediate transvection at the *sn*-glycerol-3–phosphate dehydrogenase gene of *Drosophila melanogaster*. *Biochem. Genet.* **37**, 301–315.

Goldberg, M. L., Colvin, R. A., and Mellin, A. F. (1989). The *Drosophila zeste* locus is nonessential. *Genetics* **123**, 145–155.

Goldsborough, A. S., and Kornberg, T. B. (1996). Reduction of transcription by homologue asynapsis in *Drosophila* imaginal discs. *Nature* **381**, 807–810.

Golic, M. M., and Golic, K. G. (1996). A quantitative measure of the mitotic pairing of alleles in *Drosophila melanogaster* and the influence of structural heterozygosity. *Genetics* **143**, 385–400.

Green, M. M. (1967). Variegation of the eye color mutant zeste as a function of rearrangements at the white locus in *Drosophila melanogaster*. *Biol. Z.* **86**, 221–220.

Green, M. M. (1984). Genetic instability in *Drosophila melanogaster*: Transpositions of the white gene and their role in the phenotypic expression of the *zeste* gene. *Molec. Gen. Genet.* **194**, 275–278.

Grüneberg, H. (1937). Gene doublets as evidence for adjacent small duplications in *Drosophila*. *Nature* **140**, 932–932.

Gubb, D., Roote, J., Trenear, J., Coulson, D., and Ashburner, M. (1997). Topological constraints on transvection between *white* genes within the transposing element *TE35B* in *Drosophila melanogaster*. *Genetics* **146**, 919–937.

Gustavson, E., Goldsborough, A. S., Ali, Z., and Kornberg, T. B. (1996). The *Drosophila engrailed* and *invected* genes: Partners in regulation, expression and function. *Genetics* **142**, 893–906.

Hazelrigg, T., and Kaufman, T. C. (1983). Revertants of dominant mutations associated with the *Antennapedia* gene complex of *Drosophila melanogaster*: cytology and genetics. *Genetics* **105**, 581–600.

Hendrickson, J. E., and Sakonju, S. (1995). Cis and trans interactions between the *iab* regulatory regions and *abdominal-A* and *Abdominal-B* in *Drosophila melanogaster*. *Genetics* **139**, 835–848.

Henikoff, S. (1979). Position effects and variegation enhancers in an autosomal region of *Drosophila melanogaster*. *Genetics* **93**, 105–115.

Henikoff, S., and Dreesen, T. D. (1989). Trans-inactivation of the *Drosophila brown* gene: Evidence for transcriptional repression and somatic pairing dependence. *Proc. Natl. Acad. Sci. USA* **86**, 6704–6708.

Henikoff, S., Jackson, J. M., and Talbert, P. B. (1995). Distance and pairing effects on the *brown*[Dominant] heterochromatic element in *Drosophila*. *Genetics* **140**, 1007–1017.

Hinton, T. (1950). A correlation of phenotypic changes and chromosomal rearrangements at the two ends of an inversion. *Genetics* **35**, 188–205.

Hopmann, R., Duncan, D., and Duncan, I. (1995). Transvection in the *iab-5,6,7* region of the bithorax complex of *Drosophila*: Homology independent interactions in *trans*. *Genetics* **139**, 815–833.

Hyde, R. R. (1916). Two new members of a sex-linked multiple (sextuple) allelomorph system. *Genetics* **1**, 535–580.

Hyde, R. R. (1920). Segregation and recombination of the genes for tinged, blood, buff, and coral in *Drosophila melanogaster*. *Indiana Acad. Sci. Rep.* **1921**, 291–300.

Irish, V. F., and Gelbart, W. M. (1987). The decapentaplegic gene is required for dorsal-ventral patterning of the *Drosophila* embryo. *Genes Dev.* **1**, 868–879.

Jack, J. W., and Judd, B. H. (1979). Allelic pairing and gene regulation: A model for the *zeste-white* interaction in *Drosophila melanogaster*. *Proc. Natl. Acad. Sci. USA* **76**, 1368–1372.

Jijakli, H., and Ghysen, A. (1992). Segmental determination in *Drosophila* central nervous system: Analysis of the *abdominal-A* region of the bithorax complex. *Int. J. Dev. Biol.* **36**, 93–99.

Judd, B. H. (1979). Allelic complementation and transvection in *Drosophila melanogaster*. *ICN-UCLA Symp. Mol. Cell Biol.* **15**, 107–115.

Judd, B. H. (1988). Transvection: Allelic cross talk. *Cell* **53**, 841–843.

Kaufman, T. C., Tasaka, S. E., and Suzuki, D. T. (1973). The interaction of two complex loci, *zeste* and *bithorax* in *Drosophila melanogaster*. *Genetics* **75**, 299–321.

Korge, G. (1977). Direct correlation between a chromosome puff and the synthesis of a larval saliva protein in *Drosophila melanogaster*. *Chromosoma (Berl.)* **62**, 155–174.

Korge, G. (1981). Genetic analysis of the larval secretion gene *Sgs-4* and its regulatory chromosome sites in *Drosophila melanogaster*. *Chromosoma (Berl.)* **84**, 373–390.

Korge, G., Heide, I., Sehnert, M., and Hofmann, A. (1990). Promoter is an important determinant of developmentally regulated puffing at the *Sgs-4* locus of *Drosophila melanogaster*. *Dev. Biol.* **138**, 324–337.

Lancefield, D. E. (1918). Three mutations in previously known loci. *Am. Naturalist* **52**, 264–269.

Lee, J. J., Von Kessler, D. P., Parks, S., and Beachy, P. A. (1992). Secretion and localized transcription suggest a role in positional signaling for products of the segmentation gene *hedgehog*. *Cell* **71**, 33–50.

Leiserson, W. M., Bonini, N. M., and Benzer, S. (1994). Transvection at the *eyes absent* gene of *Drosophila*. *Genetics* **138**, 1171–1179.

Levis, R., Hazelrigg, T., and Rubin, G. M. (1985). Separable *cis*-acting control elements for expression of the *white* gene of *Drosophila*. *EMBO J.* **4**, 3489–3499.

Lewis, E. B. (1945). The relation of repeats to position effect in *Drosophila melanogaster*. *Genetics* **30**, 137–166.

Lewis, E. B. (1948). Pseudoallelism in *Drosophila melanogaster*. *Genetics* **33**, 113–113.

Lewis, E. B. (1950). The phenomenon of position effect. *Adv. Genet.* **3**, 73–115.

Lewis, E. B. (1951). Pseudoallelism and gene evolution. *Cold Spring Harbor Symp. Quant. Biol.* **16**, 159–174.

Lewis, E. B. (1952). The pseudoallelism of *white* and *apricot* in *Drosophila melanogaster*. *Proc. Natl. Acad. Sci. USA* **38**, 953–961.

Lewis, E. B. (1954). The theory and application of a new method of detecting chromosomal rearrangements in *Drosophila melanogaster*. *Am. Naturalist* **88**, 225–239.

Lewis, E. B. (1955). Some aspects of position pseudoallelism. *Am. Naturalist* **90**, 73–89.

Lewis, E. B. (1956). An unstable gene in *Drosophila melanogaster*. *Genetics* **41**, 651–651.

Lewis, E. B. (1959). New mutants—Report of E. B. Lewis. *Drosophila Information Service* **33**, 96–96.

Lewis, E. B. (1963). Genes and developmental pathways. *Am. Zool.* **3**, 33–56.

Lewis, E. B. (1967). Genes and gene complexes. *In* "Heritage from Mendel" (R. A. Brink, ed.), pp. 17–47. University of Wisconsin Press, Madison.

Lewis, E. B. (1985). Regulation of the genes of the bithorax complex in *Drosophila*. *Cold Spring Harbor Symp. Quant. Biol.* **50**, 155–164.

Lewis, E. B. (1992). Clusters of master control genes regulate the development of higher organisms. *JAMA* **267**, 1524–1531.

Lindsley, D. L., and Zimm, G. G. (1992). "The Genome of *Drosophila melanogaster*." Academic, San Diego, CA.

Locke, J., and Tartof, K. D. (1994). Molecular analysis of *cubitus interruptus* (*ci*) mutations suggests an explanation for the unusual *ci* position effects. *Mol. Gen. Genet.* **243**, 234–243.

MacKendrick, M. E., and Pontecorvo, G. (1952). Crossing-over between alleles at the *w* locus in *Drosophila melanogaster. Experientia* **8,** 390–391.

Martínez-Laborda, A., González-Reyes, A., and Morata, G. (1992). *Trans* regulation in the *Ultrabithorax* gene of *Drosophila:* Alterations in the promoter enhance transvection. *EMBO J.* **11,** 3645–3652.

Masucci, J. D., Miltenberger, R. J., and Hoffmann, F. M. (1990). Pattern-specific expression of the *Drosophila decapentaplegic* gene in imaginal disks is regulated by 3′ *cis*-regulatory elements. *Genes Dev.* **4,** 2011–2023.

McClintock, B. (1944). The relation of homozygous deficiencies to mutations and allelic series in maize. *Genetics* **29,** 478–502.

McGinnis, W., Shermoen, A. W., Heemskerk, J., and Beckendorf, S. K. (1983). DNA sequence changes in an upstream DNase I-hypersensitive region are correlated with reduced gene expression. *Proc. Natl. Acad. Sci. USA* **80,** 1063–1067.

Metz, C. W. (1916). Chromosome studies in the Diptera. II. The paired association of chromosomes in the Diptera and its significance. *J. Exp. Zool.* **21,** 213–279.

Micol, J. L., Castelli-Gair, J. E., and Garcia-Bellido, A. (1990). Genetic analysis of transvection effects involving *cis*-regulatory elements of the *Drosophila Ultrabithorax* gene. *Genetics* **126,** 365–373.

Micol, J. L., and Garcia-Bellido, A. (1988). Genetic analysis of "transvection" effects involving *Contrabithorax* mutations in *Drosophila melanogaster. Proc. Natl. Acad. Sci. USA* **85,** 1146–1150.

Morgan, T. H. (1910). Sex-limited inheritance in *Drosophila. Science* **32,** 120–122.

Morgan, T. H. (1912). Further experiments with mutations in eye-color of *Drosophila:* The loss of the orange factor. *J. Acad. Nat. Sci. Phila.* **15,** 321–346.

Morgan, T. H., and Bridges, C. B. (1913). Dilution effects and bicolorism in certain eye colors of *Drosophila. J. Exp. Zool.* **15,** 429–466.

Morgan, T. H., Bridges, C. B., and Schultz, J. (1938). Constitution of the germinal material in relation to heredity. *Carnegie Inst. Washington Year Book* **37,** 304–309.

Morris, J. R., Chen, J. L., Geyer, P. K., and Wu, C. T. (1998). Two modes of transvection: Enhancer action in trans and bypass of a chromatin insulator in cis. *Proc. Natl. Acad. Sci. USA* **95,** 10745.

Morris, J. R., Geyer, P. K., and Wu, C. T. (1999). Core promoter elements can regulate transcription on a separate chromosome in trans. *Genes Dev.* **13,** 253–258.

Muller, H. J. (1930). Types of visible variations induced by X-rays in *Drosophila. J. Genet.* **22,** 299–334.

Muller, H. J. (1935a). The origination of chromatin deficiencies as minute deletions subject to insertion elsewhere. *Genetica* **17,** 237–252.

Muller, H. J. (1935b). A viable two-gene deficiency phaenotypically resembling the corresponding hypomorphic mutations. *J. Hered.* **26,** 469–478.

Muller, H. J. (1935c). The position effect as evidence of the localization of the immediate products of gene activity. *Int. Congr. Physiol.* 15 *Trudy* **1935,** 286–289.

Muller, H. J. (1936). Bar duplication. *Science* **83,** 528–530.

Muller, H. J. (1941). Induced mutations in *Drosophila. Cold Spring Harbor Symp. Quant. Biol.* **9,** 151–165.

Müller, H.-P., and Schaffner, W. (1990). Transcriptional enhancers can act in trans. *Trends Genet.* **6,** 300–304.

Muskavitch, M. A. T., and Hogness, D. S. (1980). Molecular analysis of a gene in a developmentally regulated puff of *Drosophila melanogaster. Proc. Natl. Acad. Sci. USA* **77,** 7362–7366.

Neumann, C. J., and Cohen, S. M. (1996). *Sternopleural* is a regulatory mutation of *wingless* with both dominant and recessive effects on larval development of *Drosophila melanogaster. Genetics* **142,** 1147–1155.

Oliver, C. P. (1940). A reversion to wild-type associated with crossing-over in *Drosophila melanogaster. Proc. Natl. Acad. Sci. USA* **26,** 452–454.

Oliver, C. P. (1943). A case of change in dominance of a dominant mutant in *Drosophila melanogaster. Anat. Rec.* **87,** 461–461.

Pattatucci, A. M., and Kaufman, T. C. (1991). The homeotic gene *Sex combs reduced* of *Drosophila melanogaster* is differentially regulated in the embryonic and imaginal stages of development. *Genetics* **129**, 443–461.

Pirrotta, V. (1991). The genetics and molecular biology of *zeste* in *Drosophila melanogaster*. *Adv. Genet.* **29**, 301–348.

Pirrotta, V., Steller, H., and Bozzetti, M. P. (1985). Multiple upstream regulatory elements control the expression of the *Drosophila white gene*. *EMBO J.* **4**, 3501–3508.

Pontecorvo, G. (1950). New fields in the biochemical genetics of micro-organisms. *Biochem. Soc. Symp.* **4**, 40–50.

Pontecorvo, G. (1952). Genetical analysis of cell organization. *Symp. Soc. Exp. Biol.* **6**, 218–229.

Qian, S., Varjavand, B., and Pirrotta, V. (1992). Molecular analysis of the *zeste-white* interaction reveals a promoter-proximal element essential for distant enhancer-promoter communication. *Genetics* **131**, 79–90.

Rothberg, I., Hotaling, E., and Sofer, W. (1991). A *Drosophila Adh* gene can be activated in *trans* by an enhancer. *Nucleic Acids Res.* **19**, 5713–5717.

Safir, S. R. (1913). A new eye-color mutation in *Drosophila*. *Biol. Bull.* **25**, 45–45.

Safir, S. R. (1916). Buff, a new allelomorph of white eye color in *Drosophila*. *Genetics* **1**, 584–590.

Scholz, H., Deatrick, J., Klaes, A., and Klämbt, C. (1993). Genetic dissection of *pointed*, a *Drosophila* gene encoding two ETS-related proteins. *Genetics* **135**, 455–468.

Schwartz, C., Locke, J., Nishida, C., and Kornberg, T. B. (1995). Analysis of *cubitus interruptus* regulation in *Drosophila* embryos and imaginal disks. *Development* **121**, 1625–1635.

Sidorov, B. N. (1941). Study on the nature of changes caused by structural chromosome mutations. *C. R. Acad. Sci. USSR* **31**, 390–391.

Sipos, L., Mihály, J., Karch, F., Schedl, P., Gausz, J., and Gyurkovics, H. (1998). Transvection in the *Drosophila Abd-B* domain: Extensive upstream sequences are involved in anchoring distant cis-regulatory regions to the promoter. *Genetics* **149**, 1031–1050.

Siren, M., and Portin, P. (1988). Effect of transvection on the expression of the *Notch* locus in *Drosophila melanogaster*. *Heredity* **61**, 107–110.

Slatis, H. M. (1955a). Position effects at the *brown* locus in *Drosophila melanogaster*. *Genetics* **40**, 5–23.

Slatis, H. M. (1955b). A reconsideration of the *brown*-dominant position effect. *Genetics* **40**, 246–251.

Slusarski, D. C., Motzny, C. K., and Holmgren, R. (1995). Mutations that alter the timing and pattern of *cubitus interruptus* gene expression in *Drosophila melanogaster*. *Genetics* **139**, 229–240.

Smolik-Utlaut, S. M., and Gelbart, W. M. (1987). The effects of chromosomal rearrangements on the *zeste-white* interaction in *Drosophila melanogaster*. *Genetics* **116**, 285–298.

Stern, C., and Heidenthal, G. (1944). Materials for the study of the position effect of normal and mutant genes. *Genetics* **30**, 197–205.

Stern, C., MacKnight, R. H., and Kodani, M. (1946a). The phenotypes of homozygotes and hemizygotes of position alleles and of heterozygotes between alleles in normal and translocated positions. *Genetics* **31**, 598–619.

Stern, C., Schaeffer, E. W., and Heidenthal, G. (1946b). A comparison between the position effects of normal and mutant alleles. *Proc. Natl. Acad. Sci. USA* **32**, 26–33.

Stern, C., and Kodani, M. (1955). Studies on the position effect at the *cubitus interruptus* locus of *Drosophila melanogaster*. *Genetics* **40**, 343–373.

Stevens, N. M. (1908). A study of the germ cells of certain Diptera. *J. Exp. Zool.* **5**, 359–379.

Stone, W. (1935). Allelomorphic phenomena. *Drosophila Information Service* **4**, 62–63.

Sturtevant, A. H. (1925). The effects of unequal crossing over at the *bar* locus in *Drosophila*. *Genetics* **10**, 117–147.

Sturtevant, A. H., and Beadle, G. W. (1939). "An Introduction to Genetics." Saunders, Philadelphia.

Talbert, P. B., LeCiel, C. D. S., and Henikoff, S. (1994). Modification of the *Drosophila* heterochromatic mutation *brown^{Dominant}* by linkage alterations. *Genetics* **136**, 559–571.

Tartof, K. D., and Henikoff, S. (1991). Trans-sensing effects from *Drosophila* to humans. *Cell* **65,** 201–203.

Wharton, K., Ray, R. P., Findley, S. D., Duncan, H. E., and Gelbart, W. M. (1996). Molecular lesions associated with alleles of *decapentaplegic* identify residues necessary for TGF-β/BMP cell signaling in *Drosophila melanogaster. Genetics* **142,** 493–505.

Williams, J. A., Atkin, A. L., and Bell, J. B. (1990a). The functional organization of the vestigial locus in *Drosophila melanogaster. Mol. Gen. Genet.* **221,** 8–16.

Williams, J. A., Scott, I. M., Atkin, A. L., Brook, W. J., Russell, M. A., and Bell, J. B. (1990b). Genetic and molecular analysis of vg^U and vg^W: Two dominant vg alleles associated with gene fusions in *Drosophila. Genetics* **125,** 833–844.

Williams, J. A., Bell, J. B., and Carroll, S. B. (1991). Control of *Drosophila* wing and haltere development by the nuclear *vestigial* gene product. *Genes Dev.* **5,** 2481–2495.

Wu, C.-T., and Goldberg, M. L. (1989). The *Drosophila zeste* gene and transvection. *Trends Genet.* **5,** 189–194.

Wu, C. T., and Morris, J. R. (1999). Transvection and other homology effects. *Curr. Opin. Genet. Dev.* **9,** 237–246.

Zachar, Z., Chapman, C. H., and Bingham, P. (1985). On the molecular basis of transvection effects and the regulation of transcription. *Cold Spring Harbor Symp. Quant. Biol.* **50,** 337–346.

Zimmerman, J. E., Bui, Q. T., Lu, H., and Bonini, N. M. (2001). Molecular genetic analysis of *Drosophila eyes* absent mutants reveals an eye enhancer element. *Genetics* **154,** 237–246.

14 Pairing-Sensitive Silencing, Polycomb Group Response Elements, and Transposon Homing in *Drosophila*

Judith A. Kassis
National Institute of Child Health and Human Development
National Institutes of Health
Bethesda, Maryland 20892

ABSTRACT

Regulatory DNA from a diverse group of *Drosophila* genes causes silencing of the linked reporter gene mini-*white* in the P-element vector *CaSpeR*. This silencing can occur in flies heterozygous for the P-element construct but is often enhanced in flies homozygous for the construct. In *Drosophila*, somatic chromosomes are

Copyright 2002, Elsevier Science (USA).
0065-2660/02 $35.00

paired and this pairing is important for the enhancement of silencing in most cases. Thus, this type of silencing has been called pairing-sensitive silencing. Many of the DNA fragments that cause pairing-sensitive silencing are regulatory elements required for the activity of the Polycomb group of transcriptional repressors (Polycomb group response elements, PREs). However, some PREs do not appear to cause pairing-sensitive silencing, and some fragments of DNA that cause pairing-sensitive silencing do not appear to act as PREs. I suggest that many PREs are composite elements of sites important for silencing and sites important for "pairing" or bringing together distant DNA elements. Both activities may be required for PRE function. In a related phenomenon, fragments of DNA included within P-element vectors can cause those transposons to insert in the genome near the parent gene of the included DNA (transposon homing). I suggest that DNA fragments that cause transposon homing or pairing-sensitive silencing are bound by protein complexes that can interact to bring together distant DNA fragments. © 2002, Elsevier Science (USA).

I. INTRODUCTION

A commonly used vector for making transgenic *Drosophila* is the P-element-based vector *CaSpeR* (Pirrotta, 1988). *CaSpeR* includes a minigene for the selectable marker *white*; expression of the *white* minigene causes mutant white-eyed flies to have colored eyes. The eye color is sensitive to the dose of *white*, i.e., the higher the levels of *white* mRNA, the darker the eye color. At saturation, a red, wild-type color is observed. In addition, the *white* gene is cell autonomous. Thus, different cells in the eye can have different colors if *white* expression levels differ. Mini-*white* lacks most of the regulatory DNA present at the endogenous *white* locus. It contains only a minimal *white* promoter that is expressed at a low level in its ground state and produces yellow-eyed flies (Kellum and Schedl, 1991). However, the eye color of *CaSpeR* transgenic flies varies tremendously with the site of insertion, since flanking genomic enhancers and silencers influence its expression. These properties of the mini-*white* gene make it a sensitive marker of gene expression and led to the discovery of pairing-sensitive silencing and related phenomena.

In 1991 we reported that a particular fragment of *cis*-regulatory DNA from the segmentation gene *engrailed* had an unusual effect on the expression of the mini-*white* gene in *CaSpeR* (Kassis *et al.*, 1991). Normally, flies homozygous for a given *CaSpeR* insertion have a darker eye color than heterozygotes. However, when a particular *engrailed* DNA fragment was included in that transposon, homozygotes often had a lighter eye color than heterozygotes. Thus, the *engrailed* DNA caused the mini-*white* gene to be repressed in homozygotes. Chromosomes are somatically paired in *Drosophila* (Metz, 1916; Lifschytz and Harevan, 1982), thus, in the homozygous state the two *CaSpeR* insertions would be near each other in the genome. Duplicating the transposon on the same chromosome also led to mini-*white* silencing. Silencing did not occur when two insertions were located on different chromosomes. I called this type of repression "pairing-sensitive suppression" (Kassis, 1994), since the repression of *white* was dependent on two copies of the transposon in close proximity in the genome—either *in cis* (on the same chromosome) or *in trans* (on homologous chromosomes). We originally proposed that "pairing-sensitive" (PS) sites might be involved in mediating interactions between regulatory DNAs located throughout the 70-kb *engrailed* locus; i.e., PS sites located near the *engrailed* promoter might allow it to interact with distant silencers and enhancers (Kassis *et al.*, 1991; Kassis, 1994). Since our initial report, DNA fragments from many genes have been found to cause mini-*white* silencing in homozygotes and/or heterozygotes (Table 14.1). Many of these DNA fragments are "Polycomb group response elements" (PREs). PREs are regulatory DNA necessary for the action of the Polycomb group (PcG) of transcriptional repressors (Simon *et al.*, 1993; Chan *et al.*, 1994; reviewed in Pirrotta, 1997a, 1997b). Here I attempt a comprehensive review of the DNA fragments that cause silencing of mini-*white* in *CaSpeR*. I will describe the different types of silencing observed and explore the relationships between PREs and PS sites. The data suggest that not all PREs are PS sites, and vice versa. However, the data suggest that PS sites may potentiate the action of PREs and may be integral parts of these elements. I will also discuss another related phenomenon, transposon homing. Both pairing-sensitive silencing and transposon homing may be due to the formation of protein complexes that can bring together distant DNA sites.

II. THREE TYPES OF mini-*white* SILENCING

As pointed out in Table 14.1, three types of mini-*white* silencing are observed: variegation, patterning, and pairing-sensitive silencing. In variegation, mini-*white* silencing occurs in heterozygotes and is not homogenous; pigmentation occurs in patches that are variable from one eye to the next. This type of silencing was first observed with a fragment of DNA from the PcG gene, *polyhomeotic* (Fauvarque and Dura, 1993). Patterning gives partially pigmented eyes in which one portion of the eye is regularly and reproducibly more highly pigmented than others, as if mini-*white* is repressed only in a section of the eye. The pattern observed is dependent on the location of the transposon in the genome. Patterned eyes can be seen in homozygotes, in heterozygotes, or in both. In pairing-sensitive silencing, the eye colors of homozygous flies are lighter than those of heterozygous flies. Patterned and variegated eyes can become lighter or completely silenced in homozygous flies. For all these phenomena, the degree of silencing observed is highly dependent on

Table 14.1. Regulatory DNA Reported to Cause Pairing-Sensitive Silencing

Regulatory DNA	Variegated[a]	Patterned[b]	PSS[c]	Eye color sensitive to PcG mutations[d]	PRE in embryos[e]	PHO binding sites[f]	pho sensitive[g]
engrailed							
2.6-kb	no	yes	yes	no	yes	yes	ND
Fragment 8[h]	no	yes	yes	ND	yes	yes	yes
Fragment 5[h]	no	yes	yes	ND	ND	yes	ND
polyhomeotic	yes	yes	yes	yes	ND	yes	ND
escargot	no	no	yes	ND	ND	yes	ND
even-skipped	yes	yes	yes	ND	ND	yes	ND
Sex combs reduced							
SSRN + 8.2-kb *Xba*I	yes	no	yes	yes	ND[e]	yes	no[g]
SSRN+ 5.5-kb *Hind* III	no	no	yes	no	ND	yes	ND
10-kb *Xba*I	yes	no	no	yes	yes	yes	no[g]
Probosipedia							
0.58-kb	yes	yes	yes	no	ND	yes	ND
pbZR	yes	yes	no	no	ND	yes	ND
2.1-kb enhancer + pbZR	yes	yes	yes	no	ND	yes	ND
iab-2 (1.7) enhancer	NR	NR	yes	no	no	yes	yes
bxd PRE							
2212H6.5	yes	NR	NR	yes	yes	yes	ND
1.5-kb *Eco*RI-*Sty*I	yes	NR	yes[k]	ND	yes	yes	ND
HS × 3[i]	yes	no	yes[k]	ND	ND	no	ND
HH2 × 4[i]	yes	no	yes[k]	ND	ND	no	ND
HA × 4[i]	yes	no	yes[k]	ND	ND	no	ND
AB × 6[i]	no	yes	yes	yes	no	yes	ND
BP × 6[i]	yes	no	yes	yes	yes	yes	ND
PF × 4[i]	yes	no	yes	yes	no	yes	ND

Mcp							
2.9-kb	yes	yes	yes	yes	yes	yes	yes
810-bp core[j]	no	no	no	ND	yes	yes	ND
core + ftz DNA[j]	no	yes	yes	yes	ND	yes	yes
iab-7 PRE	yes	yes	yes	yes	yes	yes	yes
iab-8 PRE	yes	NR	yes	ND	yes	yes	ND

NR, not reported; ND, not done. [a] Eye color variegated in heterozygotes. Pigmentation occurs in patches that are variable from one eye to the next. [b] Patterned eyes are those in which one portion of the eye is regularly and reproducibly more highly pigmented than others. This can occur in either homo- or heterozygotes. For *engrailed* DNA, this occurs in about 10% of insertion sites. [c] PSS-pairing-sensitive silencing. The eye color of homozygotes is lighter than heterozygotes. A fragment is said to cause pairing-sensitive silencing if mini-*white* is silenced in greater than 20% of chromosomal insertion sites. [d] Heterozygous mutations in PcG genes cause an increase in eye color of insertions at a high percentage of chromosomal insertion sites. Effect of *pho* mutations is listed separately. [e] PcG-dependent silencing of a reporter construct in embryos. Although the activity of the 8.2-kb *XbaI* fragment from *Scr* was tested in embryos, it was not tested in a vector designed to test for PRE function. [f] At least one consensus binding site for PHO present in the DNA. [g] Eye color is darker in a *pho* mutant. For the *engrailed* fragment, effects were not seen in $pho^1/+$ heterozygotes, but were observed in pho^1/pho^{cv} flies. The *Scr* fragments were tested only in *pho* heterozygotes. [h] Fragment 8 and Fragment 5 contain nonoverlapping subsets of the 2.6-kb *engrailed* fragment. [i] These fragments contain subsets of the 1.5-kb EcoRI-StyI fragment. These fragments were tested as multimers (i.e., HS × 3 means that three copies of the HS fragment were present in CaSpeR). [j] This is a subset of the 2.9-kb fragment. The *Mcp* core is required for the activity of the 2.9-kb fragment. Although the *Mcp* core has no silencing activity on its own, when combined with *ftz* (or *yellow*) regulatory sequences, silencing is observed. [k] Homozygotes have variegated eyes. Complete silencing is not observed.

References: *engrailed*: Kassis et al., 1991; Kassis, 1994; Brown et al., 1998; J. Americo, M. Fujioka, M. Whiteley, J. B. Jaynes, and J. A. Kassis, in preparation. *polyhomeotic*: Fauvarque and Dura, 1993. *escargot*: Kassis, 1994. *even-skipped*: Fujioka et al., 1999; M. Fujioka and J. Jaynes, personal communication. *Sex combs reduced*: Gindhart and Kaufman, 1995. *Proboscipedia*: Kapoun and Kaufman, 1995. *bxd* PRE: Chan et al., 1994; Horard et al., 2000; Sigrist and Pirrotta, 1997; V. Pirrotta, personal communication. *Mcp*: Muller et al., 1999; M. Muller, personal communication; Bustaria et al., 1997. *iab-7* PRE: Hagstrom et al., 1997; Cavalli and Paro 1999; Mishra et al., 2001. *iab-8* PRE: Barges et al., 2000; Zhou et al., 1999.

the insertion site in the genome. At some insertion sites, no silencing is observed, whereas at other sites, complete silencing is observed. Thus, silencing of mini-*white* is highly subject to the influence of flanking genomic DNA.

III. mini-*white* SILENCING BY PREs

PcG genes have been implicated in all types of mini-*white* silencing. PcG genes are necessary for the heritable maintenance of the silenced state of many different genes in *Drosophila* (reviewed in Paro, 1993; Kennison, 1995; Simon, 1995; Bienz and Muller, 1995; Pirrotta, 1997a, 1997b; Hagstrom and Schedl, 1997). Eleven PcG genes have been characterized at the molecular level (reviewed in Pirrotta, 1997a, 1997b; Yamamoto *et al.*, 1997; Brown *et al.*, 1998; Sinclair *et al.*, 1998), and seven are known to encode chromatin-associated proteins. Antibody staining of polytene chromosomes showed that many chromatin-associated PcG proteins bind to 80–100 loci in the genome (Zink and Paro, 1989; DeCamillis *et al.*, 1992; Rastelli *et al.*, 1993; Martin and Adler, 1993; Lonie *et al.*, 1994; Carrington and Jones, 1996; Peterson *et al.*, 1997; Sinclair *et al.*, 1998). Although many of the binding sites for different PcG proteins overlap, some do not, implying the existence of different complexes comprised of subsets of the PcG proteins (Rastelli *et al.*, 1993; Strutt and Paro, 1997). Consistent with this, biochemical evidence now exists for at least three different types of PcG protein complexes (reviewed in Satijn and Ottie, 1999; Shao *et al.*, 1999; Ng *et al.*, 2000). In addition, different PcG mutants have different loss of function phenotypes (Breen and Duncan, 1986; Soto *et al.*, 1995), suggesting that each protein may be involved in regulating a different subset of genes. Several models exist to explain the mode of action of the PcG proteins. These include compacting the chromatin (Paro, 1990), interfering with promoter-enhancer interactions (Pirrotta and Rastelli, 1994), formation of an inactive promoter–silencer complex (Bienz and Müller, 1995), sequestering target genes into nuclear compartments (Paro, 1993), repositioning of nucleosomes (Pirrotta, 1997a, 1997b), and creating a repressed chromatin state (Shao *et al.*, 1999).

Within homeotic genes, fragments of DNA (PREs) have been identified that are necessary for maintenance of a repressed transcriptional state (Simon *et al.*, 1993; Chan *etal.*, 1994; reviewed in Pirrotta 1997a, 1997b). PREs have been identified in reporter constructs by their ability to render the expression of reporter genes responsive to PcG genes (Busturia and Bienz, 1993; Simon *et al.*, 1993; Chan *et al.*, 1994; Gindhardt and Kaufman, 1995; Poux *et al.*, 1996; Busturia *et al.*, 1997; Hagstrom *et al.*, 1997). The activity of one PRE has been demonstrated *in situ*. Deletion of the *iab-7* PRE from the bithorax complex caused derepression of homeotic gene expression (Mihaly *et al.*, 1997). PREs are thought to recruit PcG proteins and act as nucleation sites for PcG protein function (Chan *et al.*, 1994; Müller, 1995).

PcG proteins are physically associated with PREs. This has been shown by two methods. First, when present in reporter constructs in trangenic *Drosophila*, PREs create an additional chromosomal binding site for PcG proteins at the site of insertion of the reporter construct (Zink *et al.*, 1991; Chan *et al.*, 1994; Lonie *et al.*, 1994; Chiang *et al.*, 1995; Peterson *et al.*, 1997; Fritsch *et al.*, 1999). Second, chromatin immunoprecipitation experiments show that PcG proteins are concentrated at PREs (Strutt *et al.*, 1997). Finally, it should be noted that PREs are required throughout development for silencing to be maintained (Busturia *et al.*, 1997). This suggests that a physical association between PREs and PcG proteins is required for silencing.

Chan, Rastelli, and Pirrotta (1994) first reported that a PRE from the homeotic gene *Ultrabithorax* caused silencing of mini-*white* in *CaSpeR* vectors. Mini-*white* variegation was observed in heterozygous flies in about 50% of chromosomal insertion sites. In some lines, pigmentation was present in only a few ommatidia, suggesting almost complete silencing. In fact, by using another selectable marker, *Hsp70-neo*, these investigators recovered white-eyed flies showing that mini-*white* can be completely silenced at some chromosomal insertion sites in the heterozygous state. We have not observed this kind of extreme silencing in heterozygotes with *engrailed* PS sites, even with the 2.6-kb fragment that contains two strong PS sites (Kassis *et al.*, 1991; Kassis, 1994; and unpublished observations). This highlights a difference in the silencing activity of the *engrailed* fragment and the *bxd* PRE.

Like mini-*white* silencing, the ability of a PRE to maintain repression of a reporter gene is dependent on the position of insertion in the genome (Simon *et al.*, 1993; Chan *et al.*, 1994; Poux *et al.*, 1996). Interestingly, Chan *et al.* (1994) noted a correlation between mini-*white* variegation in adult eyes and PRE activity in embryos and suggested that both activities might be caused by PcG proteins. They therefore tested the effects of PcG mutations on mini-*white* variegation and found a dramatically increased eye color in flies homozygous for a temperature-sensitive $E(z)$ allele (a PcG gene). This suggested that PcG protein complexes may be responsible for mini-*white* variegation. Since then, studies on the *iab-7*, *iab-8*, and *Mcp* PREs have confirmed that PREs cause mini-*white* variegation in heterozygotes (Cavalli and Paro, 1999; Muller *et al.*, 1999; Zhou *et al.*, 1999). For *iab-7* and *Mcp* PREs, this variegation has been shown to be suppressed by mutations in the PcG genes (Cavalli and Paro, 1999; Muller *et al.*, 1999). The *bxd*, *iab-7*, *iab-8*, and *Mcp* PREs have also been shown to cause pairing-sensitive silencing of the mini-*white* promoter in *CaSpeR* (Sigrist and Pirrotta, 1997; Tillib *et al.*, 1999; Horard *et al.*, 2000; Hagstrom *et al.*, 1997; Barges *et al.*, 2000; Muller *et al.*, 1999).

Dissection of a core 1.5-kb *bxd* PRE led to the identification of six nonoverlapping fragments of DNA that, when multimerized, could suppress mini-*white* in *CaSpeR* in either heterozygotes or homozygotes (Horard *et al.*, 2000; see Table 14.1). The presence of 6 PS sites within a 1.5-kb fragment might explain why the *bxd* PRE so efficiently silences mini-*white* in heterozygotes. One of these fragments, BP, was tested as a single copy and found to suppress mini-*white* in a much lower percentage of insertion sites (35% of sites with one copy vs 95% with 6 copies).

Three of the *bxd* PRE subfragments were tested for their ability to cause PcG-dependent silencing of reporter constructs in embryos. Only the BP element had this activity. The BP fragment also created a new Polycomb-binding site in polytene chromosomes from larval salivary glands suggesting that BP could recruit PcG complexes to DNA in larval tissues as well. Interestingly, the PF element also created a new Polycomb-binding site in polytene chromosomes but did not act as a PRE in embryos (Horard *et al.*, 2000). These data suggest that the PF fragment may act as a PRE in larva but not in embryos. Several lines of evidence suggest that PcG silencing may not be equivalent in larva and embryos (Struhl and Brower, 1982; Ingham, 1984; Poux *et al.*, 2000). Thus, different DNA fragments may act as PREs at different developmental stages.

IV. EFFECT OF MUTATIONS IN POLYCOMB GROUP GENES ON mini-*white* SILENCING

If mini-*white* silencing is due to repression by PcG proteins, then removal of those proteins should lead to an increase in eye color. Because mutations in most Polycomb group genes are homozygous lethal, the effects of removal of one copy on either mini-*white* variegation in heterozygotes or on pairing-sensitive silencing in homozygotes has been examined. Mini-*white* silencing caused by some DNA fragments was suppressed by heterozygous mutations in PcG genes (Table 14.1). In these instances, mini-*white* silencing was suppressed at many, but not all, chromosomal insertion sites (Hagstrom *et al.*, 1997; Muller *et al.*, 1999; Tillib *et al.*, 1999; Horard *et al.*, 2000). For other DNA fragments, heterozygous mutations in PcG genes had no effect or only affected a small number of chromosomal insertion sites (Kassis, 1994; Kapoun and Kaufman, 1995a; Gindhardt and Kaufman, 1995). For the *polyhomeotic* DNA-induced variegation, the results are puzzling. While *polyhomeotic* mutations suppressed both variegation and PS silencing as expected; *Polycomb* mutations enhanced silencing in heterozygotes but had no effect on pairing-sensitive silencing (Fauvarque and Dura, 1993). These results are puzzling because Polycomb and polyhomeotic are both thought to act as transcriptional repressors and may act in the same protein complex (Shao *et al.*, 1999).

Since there are different types of PcG protein complexes, there might be different PREs for the recruitment of each complex. Thus, we might be able to identify DNA fragments that render mini-*white* silencing sensitive to only a subset of PcG genes. In general, this has not been observed (note one exception in Tillib *et al.*, 1999). Instead, the chromosomal insertion site seems to dictate which PcG mutations effect silencing (Kassis, 1994; Hagstrom *et al.*, 1997; Muller *et al.*, 1999;

Horard *et al.*, 2000). This suggests that the silencing of mini-*white* at a particular chromosomal location is the result of an interaction between proteins bound to flanking genomic PREs and the PREs present in the transposon (as proposed in Pirrotta, 1997a, 1997b). Since mini-*white* silencing at particular chromosomal insertion sites can be sensitive to mutations in only a subset of PcG genes, it seems that not all PcG proteins contribute to silencing of all PREs. However, the extreme position dependence of PRE silencing in transgenes, along with the clustering of PS sites within PREs suggests that PREs do act independently.

V. mini-*white* SILENCING BY OTHER ELEMENTS

DNA fragments that cause pairing-sensitive silencing of mini-*white* were iden- tified near the promoters from the segmentation gene *engrailed*, the PcG gene *polyhomeotic*, the developmentally regulated gene *escargot*, and the homeotic gene *probosipedia* (Kassis *et al.*, 1991; Fauvarque and Dura, 1993; Kassis, 1994; Kapoun and Kaufman, 1995a). PS sites were also found in a *probosipedia* intron, the *iab-2(1.7)* enhancer, and downstream of the *eve* transcription unit (Kapoun and Kaufman, 1995a; Shimell *et al.*, 2000; Fujioka *et al.*, 1999). Within the *Sex combs reduced* gene, three large fragments of regulatory DNA have been shown to silence mini-*white* in different ways (see Table 14.1; Gindhardt and Kaufman, 1995). While two of these fragments may be PREs, the other may not be. Sim- ilarly, the *iab-2 (1.7)* fragment is an embryonic enhancer and did not act as a PRE in a reporter construct in embryos (Shimell *et al.*, 2000). The *prososcipedia* intron fragment also acted as an enhancer, and both the enhancer and PS silenc- ing activity of this intron depended on the presence of the *probosipedia* promoter (Kapoun and Kaufman 1995a, 1995b). These data suggest that different types of regulatory elements have the ability to act as pairing-sensitive silencers. These different regulatory elements may share core components.

VI. IS THE *engrailed* PS FRAGMENT A PRE?

Like the *bxd* PRE, the 2.6-kb *engrailed* fragment contains multiple PS sites (Kassis, 1994). However, unlike the *bxd* PRE, the *engrailed* fragment does not cause mini-*white* variegation in a significant percentage of insertions in either heterozygotes or homozygotes. Rather, about 90% of heterozygous lines have a homogenous eye color and about 10% have a patterned eye color. Pairing-sensitive silencing occurs at about 50% of insertion sites, and is independent of patterning (some lines with patterned eyes are silenced in homozygotes while others are not; Kassis *et al.*, 1991; Kassis, 1994; unpublished observations). Heterozygous PcG mutations have effects on silencing from only a few *engrailed-CaSpeR* insertions. This might suggest that the *engrailed* fragment is not a PRE. However, other data suggest that it is (see below). *engrailed* expression in embryos is relatively insensitive to the loss of one PcG protein (Moazed and O'Farrell 1990). Thus, it may not be too surprising that the eye colors of *engrailed-CaSpeR* flies are insensitive to loss of one dose of a PcG gene.

In chromatin-immunoprecipitation experiments Strutt and Paro (1997) found that PcG proteins were associated with the 2.6-kb fragment of *engrailed* DNA, suggesting that it may be a PRE. Therefore, we tested whether the 2.6-kb *engrailed* fragment could act as a Polycomb-dependent silencer in *Drosophila* embryos and it did confer Polycomb-dependent silencing to a *bxd-Ubx-lacZ* reporter (J. Americo, M. Fujioka, M. Whiteley, J. B. Jaynes, and J. A. Kassis, in preparation). This suggests that, at least in combination with the regulatory sequences present in this reportor construct, the *engrailed* fragment can function as a PRE. Since the *bxd* enhancer appears to contain weak PRE activity (Müller and Bienz, 1991), it might be that PS sites present in the *engrailed* fragment potentiate the activity of the weak PRE present in the reporter construct.

VII. *Pleiohomeotic,* PREs, AND PS SITES

Many PREs and PS sites contain binding sites for the PcG gene *pleiohomeotic* (*pho*; Mihaly *et al.*, 1998; Brown *et al.*, 1998). *pho* encodes a zinc-finger DNA-binding protein homologous to the ubiquitous transcription factor YY1 (Brown *et al.*, 1998). All identified PREs and PS sites with the exception of three subfragments of the *bxd* PRE (see Table 14.1), contain matches to the Pho binding consensus (G/t)CCATN(T/a)(T/g/c) (Table 14.1; Hyde-DeRuyscher *et al.*, 1995)). In the three cases examined, Pho-binding sites were found to be required for pairing-sensitive silencing (Table 14.1; Brown *et al.*, 1998; Shimell *et al.*, 2000; Mishra *et al.*, 2001). In addition, *pho* mutations suppressed mini-*white* silencing (both variegation and pairing-sensitive silencing) induced by the *Mcp* PRE, *engrailed* fragment 8, the *iab-2 (1.7)* enhancer, and the *iab-7* PRE (Muller *et al.*, 1999; Brown *et al.*, 1998; Shimell *et al.*, 2000; Cavalli and Paro, 1999; Mishra *et al.*, 2001). Finally, Pho-binding sites were required for PcG-dependent silencing of a reporter gene in imaginal disks (Fritsch *et al.*, 1999). Recent experiments suggest that these Pho-binding sites are not required for PRE activity in embryos (Poux *et al.*, 2000). However, additional Pho sites are present in this reporter construct, so it cannot be concluded that Pho sites are not required for PRE function in embryos until all of the Pho sites have been mutated. Still, the data presented in Poux *et al.* (2000) raise the possibility that *pho* may only be required for PRE function postembryonically.

The available data suggest that *pho* activity is required at the majority of identified PREs and PS sites (Table 14.1). However, Pho-binding sites are not sufficient for PRE or PS site function (Brown *et al.*, 1998). In fact, our preliminary data suggest that *engrailed* fragment 8 contains binding sites for at least five proteins

(L. E. Brown and J. A. Kassis, unpublished data). Mutation of any one of these five sites leads to a loss in the pairing-sensitive silencing activity of fragment 8 (J. Americo, M. Fujioka, M. Whiteley, J. B. Jaynes, and J. A. Kassis, in preparation). One of these five sites is the sequence GAGAG, which binds the protein GAGA factor (Soeller *et al.*, 1993). GAGA factor is thought to be important for the function of the *bxd* PRE (Horard *et al.*, 2000). GAGA-binding sites are required for the pairing-sensitive silencing activity of the *iab-7* PRE (Mishra *et al.*, 2001). In addition, mutations in the gene that encodes GAGA factor (*Trithorax-like*) have been shown to suppress mini-*white* silencing by the *iab-7* PRE (Hagstrom *et al.*, 1997; Mishra *et al.*, 2001).

VIII. NOT ALL PS SITES ARE PREs AND VICE VERSA

Not all DNA fragments that cause pairing-sensitive silencing of mini-*white* act as PREs in embryos. Notably, the *iab-2(1.7)* PS site from *abdominal-A* and the 2.1-kb PS site from *proboscipedia* both contain, or are tightly linked to enhancer elements (Shimell *et al.*, 2000; Kapoun and Kaufman, 1995a, 1995b). In fact, the *iab-2(1.7)* PS site could not act as a PRE in a reporter construct (Shimell *et al.*, 2000). Similarly, not all PREs act as pairing-sensitive silencers of mini-*white*. The first intron of *engrailed* was shown to be associated with PcG proteins in chromatin-immunoprecipitation assays (Strutt and Paro, 1997) suggesting it may be a PRE. However, it does not cause pairing-sensitive silencing of mini-*white* (Kassis *et al.*, 1991). Also, although the *Mcp* core fragment was sufficient to act as a PRE in reporter constructs embryos, it required additional sequences to act as a pairing-sensitive silencer of mini-*white* in adult eyes (Busturia *et al.*, 1997; Muller *et al.*, 1999). These additional sequences could come from three different sources: flanking sequences, *ftz* or *yellow* regulatory DNA (Muller *et al.*, 1999). I would like to point out that the embryonic reporter constructs used to test for PRE activity also contain additional regulatory sequences that may contribute to the activity of the tested fragment. Thus, it is clear that PREs and PS sites are not simple elements but are made up of binding sites for multiple proteins. As proposed by Pirrotta (1997a, 1997b), core elements may need to interact with flanking, weaker elements to yield a functional PRE.

IX. PAIRING-SENSITIVE SILENCING HAS NOT BEEN DEMONSTRATED IN THE EMBRYO

Pairing-sensitive silencing is not limited to mini-*white* expression in the adult eye. Pairing-sensitive silencing of the *yellow* and mini-*white* promoters have been found to occur in the adult epidermis and the testis (Mallin *et al.*, 1998; Muller *et al.*, 1999; Hagstrom *et al.*, 1997). Pairing-sensitive silencing may also occur in larval tissues (Kassis, 1994). In contrast, pairing-sensitive silencing has not been shown to occur in embryos (Kassis *et al.*, 1991; Muller *et al.*, 1999). In these cases, early embryonic enhancers for *engrailed* and *ftz* DNA were linked to *engrailed* PS sites and the *Mcp* PRE, respectively. The expression driven by these enhancers was not silenced in homozygous embryos, even in lines where mini-*white* was completely silenced in homozygous eyes. This might mean that regulatory proteins necessary for pairing-sensitive silencing are not present in the embryo. Alternatively, positive regulatory sequences present in these constructs may have acted before silencing could be established. It has been proposed that genes activated early in development cannot be inactivated by PcG genes (Poux *et al.*, 1996).

X. TRANSPOSON HOMING

P-element-based transposons insert in a nonselective manner throughout the *Drosophila* genome. Although there are hotspots for P-element insertions, up to 25% of vital *Drosophila* genes are mutable by P-transposons (Spradling *et al.*, 1999). Certain fragments of regulatory DNA dramatically alter the insertion site specificity of P-transposons. DNA from *engrailed*, *linotte*, and the *bithorax* complex cause P-transposons to insert preferentially within or near their parent gene (Hama *et al.*, 1990; Kassis *et al.*, 1992; Taillebourg and Dura, 1999; Bender and Hudson, 2000). This type of selective insertion has been called transposon homing (Hama *et al.*, 1990). A 1.6-kb fragment of *linotte*, when included in a P-transposon, caused insertion into the *linotte* gene in 20% of transposition events (Taillebourg and Dura, 1999). Transposons containing at least 3.4-kb of *engrailed* regulatory DNA inserted in the *engrailed/invected* complex in 7 of 20 events. The 2.6-kb fragment of *engrailed* discussed in this chapter caused transposons to home to *engrailed* in approximately 5% of insertion events (Kassis *et al.*, 1992). A 7-kb fragment from the middle of the bithorax complex caused transposons to insert in the bithorax complex at a frequency of 18% following injection (Bender and Hudson, 2000). Insertions into *engrailed* and the bithorax complex were regional, i.e., they occurred throughout the 100-kb *engrailed/invected* complex and the 300-kb bithorax complex, although there were hotspots for insertion within these broad regions (Hama *et al.*, 1990; Whiteley and Kassis, 1996; Bender and Hudson, 2000). For *linotte*, insertions occurred in a very narrow region within the same 1.6-kb fragment that caused homing (Taillebourg and Dura, 1999). For all three of these cases, proteins bound to the transposons are thought to recognize proteins bound to the endogenous locus concentrating the transposon in the region of the locus. Transposition is then thought to occur by the usual mechanism (Hama *et al.*, 1990; Kassis *et al.*, 1992; Taillebourg and Dura, 1999; Bender and Hudson, 2000). The broad distribution of insertion sites within *engrailed/invected* and the bithorax complex might indicate that there is more than one target fragment located within these complexes.

What are the homing fragments? The *engrailed* fragment contains at least two PS sites and perhaps these are responsible for the homing. However, we generated over 100 lines with fragment 8-containing transposons and saw no evidence of homing (J. Americo, M. Fujioka, and J. A. Kassis, unpublished data) suggesting that a single PS site is not sufficient for homing. The 7-kb bithorax fragment contains a boundary/insulator element (Bender and Hudson, 2000), and this might be responsible for homing. This is intriguing in light of the recent data that suggests that insulator elements bring DNA to particular regions of the nucleus (Gerasimova *et al.*, 2000). It is unknown whether other bithorax complex boundary elements cause transposon homing, since these elements have been assayed in the *CaSpeR* vector using *white* as the selectable marker (Hagstrom *et al.*, 1997; Barges *et al.*, 2000). The mini-*white* gene is silenced when inserted into the bithorax complex thus most homing transformants would not have been recovered (Bender and Hudson, 2000; M. Muller, personal communication). The function of the *linotte* regulatory sequences that mediate homing is unknown, but these sequences do not appear to include a PRE (Taillebourg and Dura, 1999).

XI. PREFERENTIAL INSERTION OF PRE-CONTAINING TRANSPOSONS NEAR ENDOGENOUS PREs

Favauque and Dura (1993) found that the same *polyhomeotic* regulatory DNA that suppresses mini-*white* expression causes preferential transposon insertion near endogenous PREs; 11 of 19 variegating lines were localized in the same polytene chromosomal sections as an endogenous Polycomb binding site. *P[engrailed]* transposons, containing the 2.6-kb fragment discussed here, were reported to insert near genes expressed in stripes at a high frequency (Kassis *et al.*, 1992). An analysis of the chromosomal insertions sites of 27, independent, nonselected inserts showed that 55% of *P[engrailed]* insertion sites were near endogenous Polycomb protein binding sites in polytene chromosomes (data from Kassis *et al.*, 1991, constructs A through F). Further, Chiang *et al.* (1995) found that DNA fragments from the bithorax complex that contained Polycomb-binding sites also inserted preferentially near endogenous Polycomb-binding sites. Although these data suggest that PS sites and PREs cause transposons to insert near endogenous PREs, the numbers of lines analyzed in each case were small and sometimes represented a preselected subset of the total lines obtained.

XII. CONCLUSIONS

In this chapter I have discussed DNA fragments that silence mini-*white* expression in adult eyes in the P-based transposon *CaSpeR*. Three types of silencing are observed: variegation, patterning, and pairing-sensitive silencing. In pairing-sensitive silencing, the eye color of homozygotes is lighter than that of heterozygotes. Fragments of DNA that mediate pairing-sensitive silencing are called "PS sites." The characteristics of mini-*white* silencing differ for some DNA fragments. For example, the *bxd* PRE causes variegated eye colors in heterozygotes while the *engrailed* fragment does not. However, both of these DNAs cause pairing-sensitive silencing of mini-*white* and therefore both can be called PS sites. The *bxd* PRE may be a composite element made up of PS sites and fragments of DNA that cause mini-*white* silencing, a PRE. The *engrailed* fragment, on the other hand, may not be able to act as a silencer on its own, but may potentiate the action of PREs if cloned near one either in a reporter construct or if one is nearby in the genome. An alternative explanation for their differing effects on mini-*white* silencing could be that that the *engrailed* fragment is a PRE for a different class of PcG proteins than the *bxd* PRE. However, I suggest that PS sites may be "transcriptionally neutral," i.e., they may have the ability to bring together or stabilize interactions between proteins bound to distantly located DNA fragments, but not to repress or activate mini-*white* transcription on their own (Figure 14.1A, see color insert). In this model, PREs and PS sites are separable elements. Protein complexes bound to PS sites interact with complexes bound to PS sites near the promoter. This interaction brings the PRE near the promoter and allows it to silence transcription. In this model, PS sites are required for the activity of a PRE. Thus, PS sites could restore the activity of a weak or inactive PRE in a reporter construct. Therefore, PS sites might appear indistinguishable from a PRE. PS sites could also be used to bring an enhancer close to a promoter to allow transcriptional activation (Figure 14.1B, see color insert).

If PS sites are "neutral elements," then their ability to silence mini-*white* should be dependent on flanking genomic sequences. This is true. For *engrailed* PS sites, silencing occurs at about half of transposon insertion sites and the degree of silencing is dependent on the insertion site. Complete silencing is achieved at only a small percentage of insertion sites (Kassis *et al.*, 1991; Kassis, 1994; Kassis, unpublished). I suggest that in the presence of flanking, positive, genomic regulatory sequences, protein complexes bound to PS sites may interact with activating proteins and promote transcriptional activation. However, when repressor proteins are bound to flanking genomic sequences they may interact with proteins bound to PS sites and silence the mini-*white* promoter. The stability of the silencing complex is enhanced by the formation of two repressive complexes on homologous chromosomes causing pairing-sensitive silencing.

What is the evidence that PS sites can bring together distant DNA fragments? First, although most pairing-sensitive silencing occurs between insertions located near each other in the genome, interactions between insertions located hundreds of kilobases away have been observed (for one example, see Kassis *et al.*, 1991). Second, the *Mcp* PRE promoted interactions between transposons located on different chromosomes (Muller *et al.*, 1999). Finally, the data suggest that PS sites cause preferential insertion of transposons near PREs in the genome. It will be interesting to learn the biochemical nature of the protein complexes that mediate these remarkable phenomena.

Acknowledgments

I am indebted to Jim Kennison for many interesting discussions and stimulating ideas. I thank all the authors in this field for the interesting papers and review articles and Vince Pirrotta and Martin Muller for discussions. I apologize for any omissions or misstatements of data. Finally, I thank Jim Kennison and Lesley Brown for help in preparing this manuscript.

References

Barges, S., Mihaly, J., Galloni, M., Hagstrom, K., Müller, M., Shanower, G., Schedl, P., Gyurkovics, H., and Karch, F. (2000). The *Fab-8* boundary defines the distal limit of the bithorax complex *iab-7* domain and insulates *iab-7* from initiation elements and a PRE in the adjacent *iab-8* domain. *Development* **127,** 779–790.

Bender, W., and Hudson, A. (2000). P element homing to the *Drosophila* bithorax complex. *Development* **127,** 3981–3992.

Bienz, M., and Müller, J. (1995). Transcriptional silencing of homeotic genes in *Drosophila*. *Bioessays* **17,** 775–784.

Breen, T. R., and Duncan, I. M. (1986). Maternal expression of genes that regulate the Bithorax complex of *Drosophila melanogaster*. *Dev. Biol.* **118,** 442–456.

Brown, L. J., Mucci, D., Whiteley, M., Dirksen, M.-L., and Kassis, J. A. (1998). The *Drosophila* Polycomb group gene *pleiohomeotic* encodes a DNA binding protein with homology to the transcription factor YY1. *Mol. Cell* **1,** 1057–1064.

Busturia, A., and Bienz, M. (1993). Silencers in *abdominal-B*, a homeotic *Drosophila* gene. *EMBO J.* **12,** 1415–1425.

Busturia, A., Wightman, C. D., and Sakonju, S. (1997). A silencer is required for maintenance of transcriptional repression throughout *Drosophila* development. *Development* **124,** 4343–4350.

Carrington, E. A., and Jones, R. S. (1996). The *Drosophila Enhancer of zeste* gene encodes a chromosomal protein: Examination of wild-type and mutant protein distribution. *Development* **122,** 4073–4083.

Cavalli, G., and Paro, R. (1999). Epigenetic inheritance of active chromatin after removal of the main transactivator. *Science* **286,** 955–958.

Chan, C.-S., Rastelli, L., and Pirrotta, V. (1994). A *Polycomb* response element in the *Ubx* gene that determines an epigenetically inherited state of repression. *EMBO J.* **13,** 2553–2564.

Chiang, A., O'Conner, M. B., Paro, R., Simon, J., and Bender, W. (1995). Discrete Polycomb-binding sites in each parasegmental domain of the bithorax complex. *Development* **121,** 1681–1689.

DeCamillis, M., Cheng, N., Pierre, D., and Brock, H. (1992). The *polyhomeotic* gene of *Drosophila* encodes a chromatin protein that shares polytene chromosome-binding sites with Polycomb. *Genes Dev.* **6,** 223–232.

Fauvarque, M.-O., and Dura, J.-M. (1993). *Polyhomeotic* regulatory sequences induce developmental regulator-dependent variegation and targeted *P*-element insertions in *Drosophila*. *Genes Dev.* **7,** 1508–1520.

Fritsch, C., Brown, J. L., Kassis, J. A., and Mueller, J. (1999). The DNA-binding Polycomb group protein Pleiohomeotic mediates silencing of a homeotic gene. *Development* **126,** 3905–3913.

Fujioka, M., Emi-Sarker, Y., Yusibova, G. L., Goto, T., and Jaynes, J. B. (1999). Analysis of an *even-skipped* rescue transgene reveals both composite and discrete neuronal and early blastoderm enhancers, and multi-stripe positioning by gap gene repressor gradients. *Development* **126,** 2527–2538.

Gerasimova, T. I., Byrd, K., and Corces, V. G. (2000). A chromatin insulator determines the nuclear localization of DNA. *Mol. Cell* **6,** 1025–1035.

Gindhardt, J. G., and Kaufman, T. C. (1995). Identification of *Polycomb* and *trithorax* group responsive elements in the regulatory region of the *Drosophila* homeotic gene *Sex combs reduced. Genetics* **139,** 797–814.

Hagstrom, K., Muller, M., and Schedl, P. (1997). A Polycomb and GAG dependent silencer adjoins the Fab-7 boundary in the *Drosophila* Bithorax complex. *Genetics* **146,** 1365–1380.

Hagstrom, K. and Schedl, P. (1997). Remembrance of things past: Maintaining gene expression patterns with altered chromatin *Curr. Opin. Genet. Dev.* **7,** 814–821.

Hama, C., Ali, Z., and Kornberg, T. B. (1990). Region-specific recombination and expression are directed by portions of the *Drosophila engrailed* promoter. *Genes Dev.* **4,** 1079–1093.

Horard, B., Tatout, C., Poux, S., and Pirrotta, V. (2000). Structure of a Polycomb response element and in vitro binding of Polycomb group complexes containing GAGA factor. *Mol. Cell Biol.* **20,** 3187–31397.

Hyde-DeRuyscher, R. P., Jennings, E., and Shenk, T. (1995). DNA binding sites for the transcriptional activator/repressor YY1. *Nucleic Acids Res.* **23,** 4457–4465.

Ingham, P. W. (1984). A gene that regulates the Bithorax complex differentially in larval and adult cells of *Drosophila. Cell* **37,** 815–823.

Kapoun, A. M., and Kaufman, T. C. (1995a). Regulatory regions of the homeotic gene *proboscipedia* are sensitive to chromosomal pairing. *Genetics* **140,** 643–658.

Kapoun, A. M., and Kaufman, T. C. (1995b). A functional analysis of 5′, intronic and promoter regions of the homeotic gene *proboscipedia* in *Drosophila melanogaster. Development* **121,** 2127–2141.

Kassis, J. A. (1994). Unusual properties of regulatory DNA from the *Drosophila engrailed* gene: *Three* "pairing-sensitive" sites within a 1.6-kb region. *Genetics* **136,** 1025–1038.

Kassis, J. A., Noll, E., VanSickle, E. P., Odenwald, W. F., and Perrimon, N. (1992). Altering the insertional specificity of a *Drosophila* transposable element. *Proc. Natl. Acad. Sci. USA* **89,** 1919–1923.

Kassis, J. A., VanSickle, E. P., and Sensabaugh, S. M. (1991). A fragment of *engrailed* regulatory DNA can mediate transvection of the *white* gene in *Drosophila. Genetics* **128,** 751–761.

Kellum, R., and Schedl, P. (1991). A position-effect assay for boundaries of higher order chromosomal domains. *Cell* **64,** 941–950.

Kennison, J. A. (1995). The Polycomb and trithorax group proteins of *Drosophila*: Trans-regulators of homeotic gene function. *Annu. Rev. Genet.* **29,** 289–303.

Lifschytz, E., and Hareven, D. (1982). Heterochromatin markers: Arrangement of obligatory heterochromatin, histone genes and multisite gene families in the interphase nucleus of *D. melanogaster. Chromosoma* **86,** 443–455.

Lonie, A., D'Andrea, R., Paro, R., and Saint, R. (1994). Molecular characterization of the *Polycomblike* gene of *Drosophila melanogaster*, a *trans*-acting negative regulator of homeotic gene expression. *Development* **120,** 2629–2636.

Mallin, D. R., Myung, J. S., Patton, J. S., and Geyer, P. K. (1998). Polycomb group repression is blocked by the *Drosophila* suppressor of Hairy-wing[su(Hw)] insulator. *Genetics* **148,** 331–339.

Martin, E. C., and Adler, P. N. (1993). The Polycomb group gene *Posterior Sex Combs* encodes a chromosomal protein. *Development* **117,** 641–655.

Metz, C. W. (1916). Chromosome studies in the Diptera. II. The paired association of chromosomes in the Diptera, and its significance. *J. Exp. Zool.* **21,** 213–279.

Mihaly, J., Hogga, I., Gausz, J., Gyurkovics, H., and Karch, F. (1997). *In situ* dissection of the Fab-7 region of the bithorax complex into a chromatin domain boundary and a Polycomb-response element. *Development* **124,** 1809–1820.

Mihaly, J., Mishra, R. K., and Karch, F. (1998). A conserved sequence motif in Polycomb-response elements. *Mol. Cell* **1,** 1065–1066.

Mishra, R. K., Mihaly, J., Barges, S., Spierer, A., Karch, R., Hagstrom, K., Schweinsberg, S. E., and Schedl, P. (2001). The *iab-7* Polycomb response element maps to a nucleosome-free region of chromatin and requires both GAGA and pleiohomeotic for silencing activity. *Mol. Cell Biol.* **21,** 1311–1318.

Moazed, D., and O'Farrell, P. H. (1992). Maintenance of the *engrailed* expression pattern by Polycomb group genes in *Drosophila*. *Development* **116,** 805–810.

Müller, J. (1995). Transcriptional silencing by the Polycomb protein in Drosophila embryos. *EMBO J.* **14,** 1209–1220.

Müller, J., and Bienz, M. (1991). Long range repression conferring boundardies of *Ultrabithorax* expression in the *Drosophila* embryo. *EMBO J.* **10,** 3147–3155.

Muller, M., Hagstrom, K., Gyurkovics, H., Pirrotta, V., and Schedl, P. (1999). The Mcp element from the *Drosophila melanogaster* Bithorax complex mediates long-distance regulatory interactions. *Genetics* **153,** 1333–1356.

Ng, J., Hart, C. M., Morgan, K., and Simon, J. A. (2000). A *Drosophila* ESC-E(Z) protein complex is distinct from other Polycomb group complexes and contains covalently modified ESC. *Mol. Cell Biol.* **20,** 3069–3078.

Paro, R. (1990). Imprinting a determined state into the chromatin of *Drosophila*. *Trends Genet.* **6,** 416–421.

Paro, R. (1993). Mechanism of heritable gene expression during development of Drosophila. *Curr. Opin. Cell. Biol.* **5,** 999–1005.

Peterson, A. J., Kyba, M., Bornemann, D., Morgan, K., Brock, H., and Simon, J. (1997). A domain shared by the Polycomb group proteins Scm and ph mediates heterotypic and homotypic interactions. *Mol. Cell Biol.* **17,** 6683–6692.

Pirrotta, V. (1988). Vectors for P-mediated transformations in *Drosophila*. *In* "Vectors, A Survey Molcular Cloning Vectors and Their Uses" (R. L. Rodriguez and D. T. Denhardt, eds.), pp. 437–456. Buttersworths, Boston.

Pirrotta, V. (1997a). Chromatin-silencing mechanisms in *Drosophila* maintain patterns of gene expression. *Trends Genet.* **13,** 314–318.

Pirrotta, V. (1997b). PcG complexes and chromatin silencing. *Curr. Opin. Gen. Dev.* **7,** 249–258.

Pirrotta, V., and Rastelli, L. (1994). *white* gene expression, repressive chromatin domains and homeotic gene regulation in *Drosophila*. *Bioessays* **16,** 549–556.

Poux, S., Kostic, C., and Pirrotta, V. (1996). Hunchback-independent silencing of late *Ubx* enhancers by a Polycomb group response element. *EMBO J.* **15,** 4713–4722.

Poux, S., McCabe, D., and Pirrotta, V. (2000). Recruitment of components of Polycomb group chromatin complexes in *Drosophila*. *Development* **128,** 75–85.

Rastelli, L., Chan, C. S., and Pirrotta, V. (1993). Related chromosome binding sites for zeste, *suppressors of zeste* and Polycomb group proteins in *Drosophila* and their dependence on *Enhancer of zeste* function. *EMBO J.* **12,** 1513–1522.

Satijn, D. P. E., and Otte, A. P. (1999). Polycomb group protein complexes: Do different complexes regulate distinct target genes? *Biochem. Biophys. Acta* **1447,** 1–16.

Shao, Z., Raible, R., Mollaaghababa, R., Guyon, J. R., Wu, C.-t, Bender, W., and Kingston, R. E. (1999). Stabilization of chromatin structure by PRC1, a Polycomb complex. *Cell* **98,** 37–46.

Shimell, M. J., Peterson, A. J., Burr, J., Simon, J. A., and O'Conner, M. B. (2000). Functional analysis of repressor binding sites in the *iab-2* regulatory region of the *abdominal-A* homeotic gene. *Dev. Biol.* **218,** 38–52.

Sigrist, C. J. A., and Pirrotta, V. (1997). Chromatin Insulator Elements block the silencing of a target gene by the *Drosophila* Polycomb Response Element (PRE) but allow trans interactions between PREs on different chromosomes. *Genetics* **147**, 209–221.

Simon, J. (1995). Locking in stable states of gene expression: Transcriptional control during *Drosophila* development. *Curr. Opin. Cell Biol.* **7**, 376–385.

Simon, J., Chiang, A., Bender, W., Shimell, M. J., and O'Conner, M. (1993). Elements of the *Drosophila* bithorax complex that mediate repression by Polycomb goup products. *Dev. Biol.* **158**, 131–144.

Sinclair, D. A. R., Milne, T. A., Hodgson, J. W., Shellard, J., Salinas, C. A., Kyba, M., Randazzo, F., and Brock, H. W. (1998). The *Additional sex combs* gene of *Drosophila* encodes a chromatin protein that binds to shared and unique Polycomb group sites on polytene chromosomes. *Development* **125**, 1207–1216.

Soeller, W. C., Oh, C. E., and Kornberg, T. B. (1993). Isolation of cDNAs encoding the *Drosophila* GAGA transcription factor. *Mol. Cell Biol.* **13**, 7961–7970.

Soto, M. C., Chou, T., and Bender, W. (1995). Comparison of germline mosaics of genes in the Polycomb group of *Drosophila melanogaster*. *Genetics* **140**, 231–243.

Spradling, A. C., Stern, D., Beaton, A., Rhem, E. J., Laverty, T., Mozden, N., Misra, S., and Rubin, G. M. (1999). The Berkeley Drosophila Genome Project gene disruption project: Single P-element insertions mutating 25% of vital *Drosophila* genes. *Genetics* **153**, 135–177.

Struhl, G., and Brower, D. (1982). Early role of the *esc+* gene product in the determination of segments in *Drosophila*. *Cell* **31**, 285–292.

Strutt, H., Cavalli, G., and Paro, R. (1997). Co-localization of Polycomb protein and GAGA factor on regulatory elements responsible for the maintenance of homeotic gene expression. *EMBO J.* **16**, 3621–3632.

Strutt, H., and Paro, R. (1997). The Polycomb group protein complex of *Drosophila* melanogaster has different compositions at different target genes. *Mol. Cell Biol.* **17**, 6773–6783.

Taillebourg, E., and Dura, J.-M. (1999). A novel mechanism for *P* element homing in *Drosophila*. *Proc. Natl. Acad. Sci. USA* **96**, 6856–6861.

Tillib, S., Petruk, S., Sedkov, Y., Kuzin, A., Fujioka, M., Gogo, T., and Mazo, A. (1999). Trithorax- and Polycomb-group repsonse elements within an *Ultrabithorax* transcription maintenance unit consist of closely situated but separable sequences. *Mol. Cell Biol.* **19**, 5189–5202.

Whiteley, M., and Kassis, J. A. (1997). Rescue of *Drosophila engrailed* mutants with a highly divergent mosquito *engrailed* cDNA using a homing, enhancer-trapping transposon. *Development* **124**, 1531–1541.

Yamamoto, Y., Girard, F., Bello, B., Affolter, M., and Gehring, W. J. (1997). The *cramped* gene of *Drosophila* is a member of the Polycomb-group, and interacts with *mus209*, the gene encoding Proliferating cell nuclear antigen. *Development* **124**, 3385–3394.

Zhou, J., Ashe, H., Burks, C., and Levine, M. (1999). Characterization of the transvection mediating region of the *Abdominal-B* locus in *Drosophila*. *Development* **126**, 3057–3065.

Zhou, J., Barolo, S., Szymanski, P., and Levine, M. (1996). The Fab-7 element of the bithorax complex attenuates enhancer-promoter interactions in the *Drosophila* embryo. *Genes Dev.* **10**, 3195–3201.

Zink, B., Engstrom, Y., Gehring, W. J., and Paro, R. (1991). Direct interaction of the *Polycomb* protein with *Antennapedia* regulatory sequences in polytene chromosomes of *Drosophila melanogaster*. *EMBO J.* **10**, 153–162.

Zink, B., and Paro, R. (1989). *In vivo* binding pattern of a *trans*-regulator of homoeotic genes in *Drosophila melanogaster*. *Nature* **337**, 468–471.

15

Repeat-Induced Gene Silencing in Fungi

Eric U. Selker*

Institute of Molecular Biology
University of Oregon
Eugene, Oregon 97403

I. INTRODUCTION

The notion that the structure and behavior of an organism are determined primarily by the structure and behavior of the organism's genome is well accepted. Lamarckism aside, it is time to contemplate the complementary possibility, namely, that the structure and behavior of a genome reflect, in part, the structure and behavior of the organism. Consider prokaryotes. The relatively compact genomes of bacteria and their viruses presumably reflect the economy required for success in extremely competitive environments. If one considers a thermophilic organism, it is not hard to imagine that the base composition of its genome reflects an adaptation to the extreme environment. The same reasoning should apply to

*To whom correspondence should be addressed: E-mail: selker@molbio.uoregon.edu; Telephone: (541) 346-5193; Fax: (541) 346-5891.

eukaryotes, but presumably some organisms need to be more fastidious "genome keepers" than others. For example, fungi such as *Neurospora crassa* that rely on their ability to rapidly exploit nutrient opportunities (Davis, 2000) may place a premium on maintaining an uncluttered genome. In contrast, organisms with large genomes composed substantially of transposable elements, such as higher plants and animals, must have lifestyles that can tolerate untidy genomes. An appreciation of the forces acting on genomes should give us a better understanding of the variation in eukaryotic genomes, such as the dramatic size differences between the genomes of Arabidopsis (~120 Mb; Arabidopsis Genome Initiative, 2000) and other higher plants (typically >1000 Mb), and between those of the pufferfish (~400 Mb; Elgar *et al.*, 1996) and mammals (~3000 Mb).

Of course genomes are not simply products of their physical environments. Evidence for the importance of "selfish" DNA (e.g., transposons) in evolution is undeniable. Moreover, the existence of genome defense systems, such as repeat-induced point mutation (RIP) in *Neurospora*, methylation-induced premeiotically (MIP) in *Ascobolus immersus*, and posttranscriptional gene silencing (e.g., RNAi) in plants, animals, and fungi, suggests that organisms developed measures to counter perturbations of their genomes. If we wish to study the control of genome structure, it makes sense to look first to organisms that appear to be fastidious genome keepers. Toward this end, I will briefly review RIP and MIP here. Early, more extensive reviews of these phenomena are still pertinent (Rossignol and Faugeron, 1994; Selker, 1990). It is noteworthy that additional processes that should influence genome structure have been more recently discovered in *Neurospora*, including a form of meiotic silencing (Aramayo and Metzenberg, 1996) similar to transvection in *Drosophila* and a form of posttranscriptional gene silencing ("quelling"). The latter is the topic of Chapter 9.

II. DISCOVERY AND BASIC FEATURES OF RIP AND MIP

RIP was discovered in 1986 in the course of research on the control of methylation using a chromosomal region in *Neurospora* called zeta-eta ($\zeta-\eta$; Selker and Stevens, 1985), which incidentally is now known to be a natural relic of RIP (Grayburn and Selker, 1989). Because of clues that the methylation of $\zeta-\eta$ was related to a tandem duplication, we were careful to select single-copy transformants to analyze the potential of various sequences to induce DNA methylation (Selker *et al.*, 1987b; Selker and Stevens 1985, 1987). Both unique and repeated copies of $\zeta-\eta$ sequences induced methylation in vegetative cells. Surprisingly, however, we found that duplicated sequences, independent of their origin or methylation state, displayed extreme instability in the sexual phase of the life cycle (Selker *et al.*, 1987a). Whereas single-copy sequences were stable throughout the life cycle, tandemly repeated sequences frequently suffered deletions, and those that were not deleted almost invariably showed evidence of *de novo* methylation and sequence alterations. Unlinked duplications were also found to be modified, although at lower frequencies; typically ~50% of sexual spores showed alterations of unlinked duplications. Genetic and molecular analyses revealed that the sequence alterations were all G:C to A:T polarized transition mutations and that they occurred in the stage of the sexual phase in which the dikaryotic cells, which form by fusion of cells of opposite mating types, proliferate prior to nuclear fusion and meiosis (Cambareri *et al.*, 1989; Selker *et al.*, 1987a). We proposed that a mechanism (RIP) specifically targets repeated sequences for sequence alterations. Proof that duplications, per se, trigger RIP, and that inactivation occurs in a pairwise manner, came from carefully controlled genetic experiments in which the stability of a sequence was compared in strains differing only in the presence or absence of a second copy of the sequence elsewhere in the genome (Selker and Garrett, 1988). Single copies of a sequence at unlinked sites in separate strains were stable, but they became sensitive to RIP when they were united in a nucleus by crossing. It is important to note that every sizable (e.g., >1 kb) duplication that we know is subject to RIP, unlike the case for some other genome defense systems such as quelling (Cogoni *et al.*, 1996). Nevertheless, every duplication escapes RIP at some frequency (typically less than 1% for a tandem duplication or approximately 50% for an unlinked duplication). Thus duplicated sequences are never "immune" but are not necessarily eliminated immediately. Even duplications of chromosomal segments containing numerous genes are sensitive to RIP (Bhat and Kasbekar, 2001; Perkins *et al.*, 1997). Because RIP reliably destroys both copies of duplicated genes, it has been used by numerous *Neurospora* researchers to specifically inactivate endogenous genes (e.g., see Arganoza *et al.*, 1994; Barbato *et al.*, 1996; Chakraborty *et al.*, 1995; Chang and Staben, 1994; Connerton, 1990; Connerton *et al.*, 1991; da Silva *et al.*, 1996; Ferea and Bowman, 1996; Fincham, 1990; Fincham *et al.*, 1989; Foss *et al.*, 1991; Glass and Lee, 1992; Grad *et al.*, 1999; Kuldau *et al.*, 1998; Marathe *et al.*, 1990; Nelson and Metzenberg, 1992; Perkins *et al.*, 1997; Plesofsky-Vig and Brambl, 1995; Selker *et al.*, 1989; Zhou *et al.* 1998).

After the discovery of RIP, the occurrence of similar, or identical, gene silencing mechanisms was sought in a number of fungi including members of at least the following genera: *Saccharomyces*, *Schizosaccharomyces*, *Aspergillus*, *Schizophillum*, *Coprinus*, *Ustilago*, *Gibberellia*, *Sordaria*, *Podospora*, *Magnaporthe*, *Cochliobolus*, and *Ascobolus*. MIP, a process similar to RIP, but with important differences, was promptly found and characterized by Faugeron, Rossignol, and colleagues in the filamentous ascomycete *Ascobolus immersus* (Barry *et al.*, 1993; Faugeron *et al.*, 1990; Goyon *et al.*, 1988; Rhounim *et al.*, 1992). Evidence of a version of MIP was also found in *Coprinus cinereus* (Freedman and Pukkila, 1993). RIP has not yet been demonstrated outside of *Neurospora crassa* but sequences showing hallmarks of RIP (e.g., numerous C to T mutations especially at CpA dinucleotides) or a similar process have been found in several fungi including

Magnaporthe grisea (Nakayashiki *et al.*, 1999), *Aspergillus fumigatus* (Neuveglise *et al.*, 1996), *Podospora anserina* (Graia *et al.*, 2001; Hamann *et al.*, 2000), *Fusarium* species (Hua-Van *et al.*, 1998; M.-J. Daboussi, personal communication) as well as several species of *Neurospora* (Kinsey *et al.*, 1994; Zhou *et al.*, 2001). Like RIP, MIP detects linked and unlinked sequence duplications during the period between fertilization and karyogamy and inactivates them in a pairwise manner (Faugeron *et al.*, 1990; Fincham *et al.*, 1989; Selker and Garrett, 1988). Unlike RIP, MIP relies exclusively on DNA methylation for inactivation; no evidence of mutations has been found in sequences inactivated by MIP.

III. *DE NOVO* AND MAINTENANCE METHYLATION ASSOCIATED WITH MIP AND RIP

Although DNA methylation is, by definition, integral to MIP, it is not yet known whether the methylation itself actually occurs premeiotically. The alternative possibility is that an unidentified imprint is established premeiotically, which then directs *de novo* methylation later, i.e., during or after meiosis. Support for the direct possibility has come from the observation that an *Ascobolus* gene that shows strong sequence similarities to known methyltransferases, *masc1*, plays a role in MIP (Malagnac *et al.*, 1997). Crosses heterozygous for *masc1* show reduced frequencies of MIP, especially when the duplication is in the nucleus with the defective *masc1* allele. Crosses homozygous for defective *masc1* are sterile. Negative results from efforts to detect methyltransferase activity of the product of *masc1* may be a reflection of its dependence on one or more accessory factors, such as a factor required for recognition of repeated sequences. Interestingly, deletion of *masc1* does not result in any obvious reduction in DNA methylation in vegetative cells. Similarly, mutations in a gene of *Neurospora* that appears to be a homolog of *masc1* does not result in reduced DNA methylation (M. Freitag and E. U. Selker, unpublished).

Evidence for a distinction between methylation induced in the sexual phase and that induced as the vegetative phase of the life cycle of fungi comes from studies in *Neurospora*. Methylation is frequently, but not invariably, found in sequences mutated by RIP. This is consistent with the possibility that methylation is involved in the mechanism of RIP. For example, the C-to-T mutations of RIP might be the result of deamination of methylated cytosines (Selker, 1990). On the other hand, methylation of the modified sequences could simply be a consequence of RIP. Indeed, early studies with the $\zeta-\eta$ region demonstrated that this relic of RIP can induce methylation *de novo* in vegetative cells (Miao *et al.*, 2000; Selker *et al.*, 1987b, 1993). Thus there are two nonexclusive possible explanations for the methylation found associated with sequences subjected to RIP: (1) it could reflect methylation established during RIP (and perhaps integral to its mechanism) and then maintained by some sort of copying mechanism; (2) it could reflect the creation, by RIP, of signals triggering methylation *de novo*. To test the first possibility, a set of methylated *am* alleles generated by RIP were demethylated, either by treatment with the methylation inhibitor 5-azacytidine, or by cloning followed by targeted gene replacement, and then assayed for their potential to trigger *de novo* methylation (Singer *et al.*, 1995b). The alleles showing the highest level of mutagenesis by RIP became immediately remethylated, demonstrating that they had become capable of triggering methylation, as found previously with some relics of RIP such as $\zeta-\eta$ (Cambareri *et al.*, 1991; Selker and Stevens, 1987). Several alleles with lower levels of mutation did not become remethylated, however. This, and a similar observation made with the bacterial antibiotic resistance gene *hph* located between copies of *am* that had been subjected to RIP (Irelan and Selker, 1997), told us that: (1) methylation is induced in the sexual phase of *Neurospora*, consistent with the possibility that methylation is involved in the mechanism of RIP; (2) the trigger for this methylation is different from that operating in vegetative cells; (3) methylation of some sequences is dependent on preexisting methylation, i.e., *Neurospora* is capable of some form of maintenance methylation. Interestingly, this continuously propagated methylation in *Neurospora*, and that resulting from MIP in *Ascobolus*, is not limited to symmetrical sites. Thus maintenance methylation can occur by a mechanism other than that originally proposed by Riggs (1975) and Holliday and Pugh (1975) based on the symmetry of methylated sites (e.g., $5'CpG3'/3'GpC5'$) in higher eukaryotes. This raises some fascinating questions, such as: (1) To what extent does methylation at nonsymmetrical sites depend on methylation at symmetrical sites? (2) Does the methylation found at symmetrical sites in higher and lower eukaryotes depend simply on the methylation status at those sites, as in the original maintenance methylation model? With respect to the latter possibility, it should be noted that the occurrence of "spreading" of methylation in animals (Arnaud *et al.*, 2000; Doerfler *et al.*, 1990; Turker, 1999) and fungi (Miao *et al.*, 2000) suggests that methylation of a site can promote methylation in its vicinity. Results of preliminary experiments suggest that *Neurospora* and *Ascobolus* may have somewhat different mechanisms for propagating methylation. Methylation in *Ascobolus* may simply depend on a framework of methylation at symmetrical sites (G. Faugeron, V. Rocco, A. Hanguehard, A. Grégoire, B. Margolin, J.-L. Rossignol, and E. U. Selker, unpublished), but this does not seem to be the case in *Neurospora* (B. Margolin and E. U. Selker, unpublished). One possibility is that maintenance methylation in *Neurospora*, and perhaps in other organisms, depends on modifications of chromatin such as acetylation or methylation of histones.

Control of *de novo* methylation is not yet well understood in any eukaryote. Although the extent to which principles of methylation discovered in one system will be applicable to others is not yet clear, it is obvious that *Neurospora* offers an excellent system for investigating *de novo* methylation. Sequences that reproducibly trigger *de novo* methylation can be generated simply from any sequence by RIP and such sequences can be dissected, modified, and tested to identify the underlying principles. The most extensive example of this approach used the 1.6-kb $\zeta-\eta$ region (Grayburn and Selker, 1989; Miao *et al.*, 1994, 2000; Selker *et al.*, 1993). Tests were carried out to assess the methylation potential of a variety of fragments of this region, as well as chimeras between $\zeta-\eta$ sequences and the homologous unmutated allele, theta (θ). Synthetic variants were also tested. Contructs were integrated precisely in single copy at the *am* locus on linkage group VR and/or the *his-3* locus on linkage group IR to control for possible confounding effects of random integration including differences in copy number, chromosomal location, and arrangement of the transforming DNA. Some conclusions from this work include: (1) the $\zeta-\eta$ region contains at least two nonoverlapping methylation signals; (2) different fragments of the region can induce different levels of methylation; (3) methylation induced by $\zeta-\eta$ sequences can spread far into flanking sequences; (4) fragments as small as 171 bp can trigger methylation; (5) methylation signals behave similarly, but not identically, at different chromosomal sites; (6) mutation density, per se, does not determine whether sequences become methylated; (7) both A:T-richness and high densities of TpA dinucleotides, typical attributes of methylated sequences in *Neurospora*, appear to promote *de novo* methylation but neither is an essential feature of methylation signals. An important general conclusion is that *de novo* methylation of $\zeta-\eta$ sequences does not simply reflect the absence of signals that prevent methylation; rather, the region contains multiple, positive signals that trigger methylation (Miao *et al.*, 2000). Recent tests of synthetic sequences have identified a variety of simple sequences that can promote *de novo* methylation in *Neurospora* (H. Tamaru and E. U. Selker, in preparation).

Classical genetic approaches to identify elements of the methylation machinery are starting to pay off. Mutant hunts in *Neurospora* have already implicated five genes involved in DNA methylation (Foss *et al.*, 1993, 1995, 1998; Tamaru and Selker, 2001; M. Freitag, A. Hagemann, and E. U. Selker, in preparation). One, *dim-2*, the only eukaryotic gene currently known in which mutations appear to eliminate DNA methylation, encodes a DNA methyltransferase (MTase; Kouzminova and Selker, 2001). The *dim-2* MTase is apparently responsible for methylation at both symmetrical and asymmetrical sites and is required for both *de novo* and maintenance methylation. Curiously, *dim-2* does not play a role in RIP; duplicated sequences are mutated in *dim-2* strains, as usual, but the mutated sequences are not methylated (Kouzminova and Selker, 2001). The *Neurospora* genome sequence is nearly complete, and only one other potential DNA MTase has been found. This MTase homolog does not appear to be involved in methylation in vegetative cells (M. Freitag and E. U. Selker, unpublished). The *Ascobolus masc1* gene is the closest known homolog of this potential MTase gene. Thus it will be interesting to learn whether *mth* plays a role in RIP.

IV. CONSEQUENCES OF RIP AND MIP

The most common outcome of RIP is gene inactivation due to nonsense and/or missense mutations. In addition, RIP can generate functional, or partially functional alleles (e.g., see Barbato et al., 1996; Fincham, 1990; Glass and Lee, 1992). RIP frequently makes just C-to-T or G-to-A changes on a given strand and prefers certain sites (e.g., CpA dinucleotides). Because of peculiarities of codon usage in Neurospora, C-to-T changes on the coding strand are not as serious as C-to-T changes on the noncoding strand (Singer et al., 1995a, 1995b; Watters et al., 1999). Some potentially functional alleles are not expressed, however, due to methylation resulting from RIP. The methylation resulting from RIP and MIP causes a transcriptional block by a mechanism that remains largely unexplored (Barry et al., 1993; Rountree and Selker, 1997). There are several clues, however. Results of nuclear run-on assays using extracts from Neurospora strains bearing various methylated sequences demonstrated that the block is at the level of transcript elongation; initiation of transcription appeared unperturbed (Rountree and Selker, 1997). Considering that methylation does not inhibit transcription in vitro, such as in nuclear extracts, methylation must be having its effect by an indirect mechanism. Two nonexclusive possibilities are being examined: (1) the inhibitory effect results from effects of hypothetical methyl-DNA binding proteins, perhaps similar to those described in higher eukaryotes (e.g., see Hendrich and Bird, 1998); (2) the inhibitory effect results from modifications of chromatin, such as acetylation or methylation of histones. Methyl-DNA binding proteins have been detected in Neurospora (G. Kothe, M. Rountree, and E. U. Selker, unpublished), and there is evidence for chromatin modifications associated with methylation. For example, in Neurospora, the histone deacetylase inhibitor Trichostatin A was found to cause derepression of a gene silenced by methylation, repression of a gene that is activated by methylation of an adjacent transposon, and selective loss of DNA methylation (Selker, 1998), indicating that DNA methylation and protein acetylation are connected in one or more ways (see Dobosy and Selker, 2001).

One can only speculate about the long-term consequences of RIP and MIP. On the one hand, there is good evidence that these processes successfully control transposable elements. The Neurospora genome contains a variety of sequences related to known transposons, but virtually all of them show evidence of RIP, and active transposons are absent from most Neurospora strains (Cambareri et al., 1998; Kinsey et al., 1994; Margolin et al., 1998; Kinsey, 1989; E. U. Selker, unpublished). On the other hand, it seems likely that RIP and MIP limit the evolution of new functions through gene duplication and divergence. Aside from the rDNA, the structurally similar functional genes that have been noted in Neurospora appear to lack segments that are long enough and sufficiently similar to be subject to RIP. Conceivably, sequences altered by RIP might provide useful raw material for the evolution of new genes, but I am unaware of evidence that this has occurred. A survey of genomic sequences that show the hallmarks of RIP has revealed that most are methylated (T. Wolf, B. Margolin, and E. U. Selker, unpublished). It is not known whether such sequences drift, eventually lose their methylation and become functional, or are lost by deletion.

V. CONCLUDING REMARKS

The discovery of RIP provided dramatic evidence that related sequences can interact efficiently regardless of their relative locations in a genome and provided the first example of a genome defense system. To date, RIP and MIP are the only examples of genome defense systems that almost certainly rely on DNA:DNA interactions. Other genome defense systems based in some way on sequence homology have come to light more recently, including transcriptional and post-transcriptional gene silencing in plants, quelling in *Neurospora* and RNAi in animals. These processes do not appear as potent as RIP and MIP in that they do not inactivate all duplications and they do not typically result in permanent silencing. This may be responsible for their broader distribution in nature. The realization that organisms have genome defense systems that can silence and modify selfish DNA should cause us to reconsider the oft-repeated dogma that experience cannot be inherited.

Acknowledgments

I thank Michael Freitag and Jeanne Selker for comments on the manuscript. Work in my laboratory is supported by a grant from the National Institutes of Health (GM35690).

References

Arabidopsis Genome Initiative (2000). Analysis of the genome sequence of the flowering plant *Arabidopsis thaliana*. *Nature* **408,** 796–815.

Aramayo, R., and Metzenberg, R. L. (1996). Meiotic transvection in fungi. *Cell* **86,** 103–113.

Arganoza, M. T., Ohrnberger, J., Min, J., and Akins, R. A. (1994). Suppressor mutants of *Neurospora crassa* that tolerate allelic differences at single or at multiple heterokaryon incompatibility loci. *Genetics* **137,** 731–742.

Arnaud, P., Goubely, C., Pelissier, T., and Deragon, J. M. (2000). SINE retroposons can be used in vivo as nucleation centers for de novo methylation. *Mol. Cell. Biol.* **20,** 3434–3441.

Barbato, C., Calissano, M., Pickford, A., Romano, N., Sandmann, G., and Macino, G. (1996). Mild RIP-an alternative method for in vivo mutagenesis of the albino-3 gene in *Neurospora crassa*. *Mol. Gen. Genet.* **252,** 353–361.

Barry, C., Faugeron, G., and Rossignol, J.-L. (1993). Methylation induced premeiotically in *Ascobolus*: Coextension with DNA repeat lengths and effect on transcript elongation. *Proc. Natl. Acad. Sci. USA* **90,** 4557–4561.

Bhat, A., and Kasbekar, D. P. (2001). Escape from repeat-induced point mutation of a gene-sized duplication in *Neurospora crassa* crosses that are heterozygous for a larger chromosome segment duplication. *Genetics* **157**, 1581–1590.

Cambareri, E. B., Aisner, R., and Carbon, J. (1998). Structure of the chromosome VII centromere region in *Neurospora crassa*: Degenerate transposons and simple repeats. *Mol. Cell. Biol.* **18**, 5465–5477.

Cambareri, E. B., Jensen, B. C., Schabtach, E., and Selker, E. U. (1989). Repeat-induced G-C to A-T mutations in *Neurospora*. *Science* **244**, 1571–1575.

Cambareri, E. B., Singer, M. J., and Selker, E. U. (1991). Recurrence of repeat-induced point mutation (RIP) in *Neurospora crassa*. *Genetics* **127**, 699–710.

Chakraborty, B. N., Ouimet, P. M., Sreenivasan, G. M., Curle, C. A., and Kapoor, M. (1995). Sequence repeat-induced disruption of the major heat-inducible HSP70 gene of *Neurospora crassa*. *Curr. Genet.* **29**, 18–26.

Chang, S., and Staben, C. (1994). Directed replacement of *mt A* by *mt a-1* effects a mating type switch in *Neurospora crassa*. *Genetics* **138**, 75–81.

Cogoni, C., Irelan, J. T., Schumacher, M., Schmidhauser, T. J., Selker, E. U., and Macino, G. (1996). Transgene silencing of the al-1 gene in vegetative cells of *Neurospora* is mediated by a cytoplasmic effector and does not depend on DNA-DNA interactions or DNA methylation. *EMBO J.* **15**, 3153–3163.

Connerton, I. F. (1990). Premeiotic disruption of the *Neurospora crassa* malate synthase gene by native and divergent DNAs. *Mol. Gen. Genet.* **223**, 319–323.

Connerton, I. F., Deane, S. M., Butters, J. A., Loeffler, R. S., and Hollomon, D. W. (1991). RIP (repeat induced point mutation) as a tool in the analysis of P-450 and sterol biosynthesis in *Neurospora crassa*. *Biochem. Soc. Trans.* **19**, 799–802.

da Silva, M. V., Alves, P. C., Duarte, M., Mota, N., Lobo da Cunha, A., Harkness, T. A., Nargang, F. E., and Videira, A. (1996). Disruption of the nuclear gene encoding the 20.8–kDa subunit of NADH: Ubiquinone reductase of *Neurospora* mitochondria. *Mol. Gen. Genet.* **252**, 177–183.

Davis, R. H. (2000). "Neurospora: Contributions of a Model Organism." Oxford University Press.

Dobosy, J. R., and Selker, E. U. (2001). Emerging connections between DNA methylation and histone acetylation. *Cell. Mol. Life Sci.* **58**, 721–727.

Doerfler, W., Toth, M., Kochanek, S., Achten, S., Freisem, R. U., Behn, K. A., and Orend, G. (1990). Eukaryotic DNA methylation: Facts and problems. *FEBS Lett.* **268**, 329–333.

Elgar, G., Sandford, R., Aparicio, S., Macrae, A., Venkatesh, B., and Brenner, S. (1996). Small is beautiful: Comparative genomics with the pufferfish (*Fugu rubripes*). *Trends Genet.* **12**, 145–150.

Faugeron, G., Rhounim, L., and Rossignol, J.-L. (1990). How does the cell count the number of ectopic copies of a gene in the premeiotic inactivation process acting in *Ascobolus immersus*? *Genetics* **124**, 585–591.

Ferea, T. L., and Bowman, B. J. (1996). The vacuolar ATPase of *Neurospora crassa* is indispensable: Inactivation of the *vma-1* gene by repeat-induced point mutation. *Genetics* **143**, 147–154.

Fincham, J. R. (1990). Generation of new functional mutant alleles by premeiotic disruption of the *Neurospora crassa am* gene. *Curr. Genet.* **18**, 441–445.

Fincham, J. R. S., Connerton, I. F., Notarianni, E., and Harrington, K. (1989). Premeiotic disruption of duplicated and triplicated copies of the *Neurospora crassa am* (glutamate dehydrogenase) gene. *Curr. Genet.* **15**, 327–334.

Foss, E. J., Garrett, P. W., Kinsey, J. A., and Selker, E. U. (1991). Specificity of repeat induced point mutation (RIP) in *Neurospora*: Sensitivity of non-*Neurospora* sequences, a natural diverged tandem duplication, and unique DNA adjacent to a duplicated region. *Genetics* **127**, 711–717.

Foss, H. M., Roberts, C. J., Claeys, K. M., and Selker, E. U. (1993). Abnormal chromosome behavior in *Neurospora* mutants defective in DNA methylation. *Science* **262**, 1737–1741; corrections: **267**, 316.

Foss, H. M., Roberts, C. J., and Selker, E. U. (1998). Reduced levels and altered patterns of DNA methylation caused by mutations in *Neurospora crassa*. *Mol. Gen. Genet.* **259,** 60–71.

Freedman, T., and Pukkila, P. J. (1993). *De novo* methylation of repeated sequences in *Coprinus cinereus*. *Genetics* **135,** 357–366.

Glass, N. L., and Lee, L. (1992). Isolation of *Neurospora crassa* A mating type mutants by repeat induced point (RIP) mutation. *Genetics* **132,** 125–133.

Goyon, C., Faugeron, G., and Rossignol, J.-L. (1988). Molecular cloning and characterization of the *met2* gene from *Ascobolus immersus*. *Gene* **63,** 297–308.

Grad, L. I., Descheneau, A. T., Neupert, W., Lill, R., and Nargang, F. E. (1999). Inactivation of the *Neurospora crassa* mitochondrial outer membrane protein TOM70 by repeat-induced point mutation (RIP) causes defects in mitochondrial protein import and morphology. *Curr. Genet.* **36,** 137–146.

Graia, F., Lespinet, O., Rimbault, B., Dequard-Chablat, M., Coppin, E., and Pieard, M. (2001). Genome quality control: RIP (repeat-induced point mutation) comes to *Podospora*. *Mol. Microbiol.* **40,** 586–595.

Grayburn, W. S., and Selker, E. U. (1989). A natural case of RIP: Degeneration of DNA sequence in an ancestral tandem duplication. *Mol. Cell. Biol.* **9,** 4416–4421.

Hamann, A., Feller, F., and Osiewacz, H. D. (2000). The degenerate DNA transposon Pat and repeat-induced point mutation (RIP) in *Podospora anserina*. *Mol. Gen. Genet.* **263,** 1061–1069.

Hendrich B., and Bird, A. (1998). Identification and characterization of a family of mammalian methyl-CpG binding proteins. *Mol. Cell. Biol.* **18,** 6538–6547.

Holliday, R., and Pugh, J. E. (1975). DNA modification mechanisms and gene activity during development. *Science* **187,** 226–232.

Hua-Van, A., Hericourt, F., Capy, P., Daboussi, M. J., and Langin, T. (1998). Three highly divergent subfamilies of the impala transposable element coexist in the genome of the fungus Fusarium oxysporum. *Mol. Gen. Genet.* **259,** 354–362.

Irelan, J. T., and Selker, E. U. (1997). Cytosine methylation associated with repeat-induced point mutation causes epigenetic gene silencing in *Neurospora crassa*. *Genetics* **146,** 509–523.

Kinsey, J. A. (1989). Restricted distribution of the Tad transposon in strains of Neurospora. *Curr. Genet.* **15,** 271–275.

Kinsey, J. A., Garrett-Engele, P. W., Cambareri, E. B., and Selker, E. U. (1994). The *Neurospora* transposon Tad is sensitive to repeat-induced point Mutation (RIP). *Genetics* **138,** 657–664.

Kouzminova, E. A., and Selker, E. U. (2001). *Dim-2* encodes a DNA-methyltransferase responsible for all known cytosine methylation in *Neurospora*. *EMBO J.* **20,** 4309–4323.

Kuldau, G. A., Raju, N. B., and Glass, N. L. (1998). Repeat-induced point mutations in pad-1, a putative RNA splicing factor from *Neurospora crassa*, confer dominant lethal effects on ascus development. *Fungal Genet. Biol.* **23,** 169–180.

Malagnac, F., Wendel, B., Goyon, C., Faugeron, G., Zickler, D., Rossignol, J.-L., Noyer-Weidner, M., Vollmayr, P., Trautner, T. A., and Walter, J. (1997). A gene essential for de novo methylation and development in *Ascobolus* reveals a novel type of eukaryotic DNA methyltransferase structure. *Cell* **91,** 281–290.

Marathe, S., Connerton, I. F., and Fincham, J. R. S. (1990). Duplication-induced mutation of a new *Neurospora* gene required for acetate utilization: Properties of the mutant and predicted amino acid sequence of the protein product. *Mol. Cell. Biol.* **10,** 2638–2644.

Margolin, B. S., Garrett-Engele, P. W., Stevens, J. N., Yen-Fritz, D., Garrett-Engele, C., Metzenberg, R. A., and Selker, E. U. (1998). A methylated *Neurospora* 5S rRNA pseudogene contains a transposable element inactivated by RIP. *Genetics* **149,** 1787–1797.

Miao, V. P., Freitag, M., and Selker, E. U. (2000). Short TpA-rich segments of the zeta-eta region induce DNA methylation in *Neurospora crassa*. *J. Mol. Biol.* **300,** 249–273.

Miao, V. P. W., Singer, M. J., Rountree, M. R., and Selker, E. U. (1994). A targeted replacement system for identification of signals for de novo methylation in *Neurospora crassa*. *Mol. Cell. Biol.* **14**, 7059–7067.

Nakayashiki, H., Nishimoto, N., Ikeda, K., Tosa, Y., and Mayama, S. (1999). Degenerate MAGGY elements in a subgroup of *Pyricularia grisea*: A possible example of successful capture of a genetic invader by a fungal genome. *Mol. Gen. Genet.* **261**, 958–966.

Nelson, M. A., and Metzenberg, R. L. (1992). Sexual development genes of *Neurospora crassa*. *Genetics* **132**, 149–162.

Neuveglise, C., Sarfati, J., Latge, J. P., and Paris, S. (1996). Afut1, a retrotransposon-like element from *Aspergillus fumigatus*. Nucleic Acids Res. **24**, 1428–1434.

Perkins, D. D., Margolin, B. S., Selker, E. U., and Haedo, S. D. (1997). Occurrence of repeat induced point mutation in long segmental duplications of *Neurospora*. *Genetics* **147**, 125–136.

Plesofsky-Vig, N., and Brambl, R. (1995). Disruption of the gene for hsp30, an alpha-crystallin-related heat shock protein of *Neurospora crassa*, causes defects in thermotolerance. *Proc. Natl. Acad. Sci. USA* **92**, 5032–5036.

Rhounim, L., Rossignol, J.-L., and Faugeron, G. (1992). Epimutation of repeated genes in *Ascobolus immersus*. *EMBO J.* **11**, 4451–4457.

Riggs, A. D. (1975). X inactivation, differentiation, and DNA methylation. *Cytogenet. Cell Genet.* **14**, 9–25.

Rossignol, J.-L., and Faugeron, G. (1994). Gene inactivation triggered by recognition between DNA repeats. *Experientia* **50**, 307–317.

Rountree, M. R., and Selker, E. U. (1997). DNA methylation inhibits elongation but not initiation of transcription in *Neurospora crassa*. *Genes Dev.* **11**, 2383–2395.

Selker, E. U. (1990). Premeiotic instability of repeated sequences in *Neurospora crassa*. *Annu. Rev. Genet.* **24**, 579–613.

Selker, E. U. (1998). Trichostatin A causes selective loss of DNA methylation in *Neurospora*. *Proc. Natl. Acad. Sci. USA* **95**, 9430–9435.

Selker, E. U., Cambareri, E. B., Garrett, P. W., Jensen, B. C., Haack, K. R., *et al.* (1989). Use of RIP to inactivate genes in *Neurospora crassa*. *Fungal Genet. Newsl.* **36**, 76–77.

Selker, E. U., Cambareri, E. B., Jensen, B. C., and Haack, K. R. (1987a). Rearrangement of duplicated DNA in specialized cells of *Neurospora*. *Cell* **51**, 741–752.

Selker, E. U., and Garrett, P. W. (1988). DNA sequence duplications trigger gene inactivation in *Neurospora crassa*. *Proc. Natl. Acad. Sci. USA* **85**, 6870–6874.

Selker, E. U., Jensen, B. C., and Richardson, G. A. (1987b). A portable signal causing faithful DNA methylation *de novo* in *Neurospora crassa*. *Science* **238**, 48–53.

Selker, E. U., Richardson, G. A., Garrett-Engele, P. W., Singer, M. J., and Miao, V. (1993). Dissection of the signal for DNA methylation in the $\zeta-\eta$ region of *Neurospora*. Cold Spring Harbor Symp. Quant. Biol. **58**, 323–329.

Selker, E. U., and Stevens, J. N. (1985). DNA methylation at asymmetric sites is associated with numerous transition mutations. *Proc. Natl. Acad. Sci. USA* **82**, 8114–8118.

Selker, E. U., and Stevens, J. N. (1987). Signal for DNA methylation associated with tandem duplication in *Neurospora crassa*. *Mol. Cell. Biol.* **7**, 1032–1038.

Singer, M. J., Kuzminova, E. A., Tharp, A., Margolin, B. S., and Selker, E. U. (1995a). Different frequencies of RIP among early vs. late ascospores of *Neurospora crassa*. *Fungal Genet. Newsl.* **42**, 74–75.

Singer, M. J., Marcotte, B. A., and Selker, E. U. (1995b). DNA methylation associated with repeat-induced point mutation in *Neurospora crassa*. *Mol. Cell. Biol.* **15**, 5586–5597.

Tamaru, H., and Selker, E. U. (2001). A histone H3 methyltransferase controls DNA methylation in Neurospora crassa. *Nature* **414**, 277–283.

Turker, M. S. (1999). The establishment and maintenance of DNA methylation patterns in mouse somatic cells. *Semin. Cancer Biol.* **9,** 329–337.

Watters, M. K., Randall, T. A., Margolin, B. S., Selker, E. U., and Stadler, D. R. (1999). Action of repeat-induced point mutation on both strands of a duplex and on tandem duplications of various sizes in *Neurospora. Genetics* **153,** 705–714.

Zhou, L. W., Haas, H., and Marzluf, G. A. (1998). Isolation and characterization of a new gene, sre, which encodes a GATA- type regulatory protein that controls iron transport in *Neurospora crassa. Mol. Gen. Genet.* **259,** 532–540.

Zhou, Y., Cambareri, E. B., and Kinsey, J. A. (2001). DNA methylation inhibits expression and transposition of the *Neurospora* Tad retrotransposon. *Mol. Gen. Genet.* **265,** 748–754.

The Evolution of Gene Duplicates

Sarah P. Otto*
Department of Zoology
University of British Columbia
Vancouver, British Columbia V6T 1Z4, Canada

Paul Yong
Department of Medicine
University of British Columbia
Vancouver, British Columbia V6T 1Z3, Canada

*To whom correspondence should be addressed: E-mail: otto@zoology.ubc.ca; Telephone: (604) 822-2778; Fax: (604)822-2416.

Advances in Genetics, Vol. 46

ABSTRACT

Gene and genome duplications have given rise to enormous variability among species in the number of genes within their genomes. Gene copies have in turn played important roles in adaptation, having been implicated in the evolution of the immune response, insecticide resistance, efficient protein synthesis, and vertebrate body plans. In this chapter, we discuss the life history of gene duplications, from their first appearance within a population, through the period during which they rise in frequency or disappear, to their long-term fate. At each phase, we discuss the evolutionary processes that have influenced the dynamics of gene duplications and shaped their ultimate roles within a population. We argue that there is no evidence that organisms have evolved strategies to promote gene duplication in order to permit adaptive evolution. In contrast, many mechanisms exist to silence or eliminate duplicated genes, suggesting that selection has acted largely to reduce the rate of gene duplication. We also argue that natural selection has functioned as an effective sieve, increasing the representation of beneficial gene duplicates among those that establish within a population and that play a long-term role in evolution. To refine our understanding of how selection acts on new gene duplications, we provide a model incorporating a single-copy gene, its gene duplicate, and selection either favoring heterozygotes or eliminating deleterious mutations. Although both forms of selection can increase the initial rate of spread of a gene duplicate, the efficacy with which they do so differs dramatically. Heterozygote advantage always increases the rate of spread and can have a large impact. In contrast, masking deleterious mutations never has a large effect on the rate of spread of the duplicate, and this minor effect can be negative as well as positive. In both cases, the degree of linkage between the two gene copies affects the rate of spread of the duplication. Finally, we discuss evolutionary processes that occur over longer periods after a gene duplication has become established within a population. These long-term processes include maintenance, inactivation, and diversification in function. Consideration of each of the short-term and long-term processes affecting duplicated genes illustrates the subtle ways in which selection has acted to shape genomic structure.   © 2002, Elsevier Science (USA).

I. INTRODUCTION

The number of genes within the genome will evolve over time whenever (1) there is heritable variation in copy number (caused by gene duplication, gene loss, and polyploidization events) and (2) individuals that carry variant genomes differ in the number of offspring that they bear, either by chance (drift) or as a result of differences in survival ability and/or reproductive potential (selection).

Recent genomic analyses have clarified the extent to which gene duplications (e.g., Lynch and Conery, 2000; Arabidopsis Genome Initiative, 2000) and genome duplications (e.g., Lundin, 1993; Wolfe and Shields, 1997; Postlethwait *et al.*, 1998; Vision *et al.*, 2000) occur. These studies have found that a remarkably high fraction of genes are closely related to other genes within the genome. For example, the fraction of genes that represent recent duplication events (recent enough to generate recognizable paralogs) is 11.2% in *Haemophilus influenzae*, 28.6% in *Saccharomyces cerevisiae*, 65.0% in *Arabidopsis thaliana*, 27.5% in *Drosophila melanogaster*, and 44.8% in *Caenorabditis elegans* (Arabidopsis Genome Initiative, 2000). These numbers increase substantially if less stringent criteria are used to identify paralogs (compare the above results to those of Rubin *et al.*, 2000). Indeed, it seems likely that a large fraction, if not all, genes within a genome are ultimately related by descent to a small number of genes that arose early in our evolutionary history (Maynard Smith, 1998), although there is some evidence for the *de novo* evolution of short regulatory and signal transduction domains (Chervitz *et al.*, 1998). Understanding how these ancestral genes have given rise, through duplication followed by diversification, to the vast number and array of genes present in extant organisms has fascinated evolutionary biologists for decades (Haldane, 1933; Fisher, 1935; Grant, 1963; Spofford, 1969; Ohno, 1970; Stebbins, 1980). In this chapter, we discuss the evolutionary forces that have acted on gene copy number. We briefly review the extent of variation in copy number and the array of mechanisms by which duplications may arise. We then examine processes that affect the initial spread of duplicates within polymorphic populations. Finally, we consider the ultimate fate of these duplicates over longer periods of evolutionary time.

A. Variation in gene copy number

There is great variation in the total number of genes that organisms carry (Table 16.1). Most bacteria have hundreds to thousands of genes, with a minimum near the 480 genes found in *Mycoplasma*. The range is shifted upward by an order of magnitude among the eukaryotes that have been studied. Among metazoans, vertebrates tend to have more genes than invertebrates, perhaps as a result of two genomic doubling events before ($\sim$500 MYA) and after ($\sim$430 MYA) the divergence of jawless fish (Ohno, 1970; Amores *et al.*, 1998; Postlethwait *et al.*, 1998; Gibson and Spring, 2000; but see Skrabanek and Wolfe, 1998). Beyond such coarse divisions, there is little relationship between the number of genes carried by an organism and its size or complexity (Valentine, 2000). Indeed, the largest number of genes is likely to be found in organisms that have undergone several rounds of polyploidization. Extensive polyploidization has led to organisms with vast genome sizes, such as the fern *Ophioglossum pycnostichum*, with a record number of 1260 chromosomes in the sporophytic (diploid) phase (Löve *et al.*, 1977).

Table 16.1. Number of Protein-Coding Genes Inferred from Completed Genome
Sequencing Projects

Species	Estimated number of genes[a]	Reference
Bacteria		
Mycoplasma genitalum	480	Hutchison *et al.*, 1999
Buchnera sp.	583	Shigenobu *et al.*, 2000
Mycoplasma pneumoniae	677	Tekaia and Dujon, 1999
Rickettsia prowazekii	837	Tekaia and Dujon, 1999
Borrelia burgdorferi	850	Tekaia and Dujon, 1999
Chlamydia trachomatis	894	Tekaia and Dujon, 1999
Treponema pallidum	1,031	Tekaia and Dujon, 1999
Aquifex aeolicus	1,522	Tekaia and Dujon, 1999
Helicobacter pylori	1,577	Tekaia and Dujon, 1999
Haemophilus influenzae	1,709	Fleischmann *et al.*, 1995
Campylobacter jejuni	1,731	Tekaia and Dujon, 1999
Synechocystis sp. PCC6803	3,168	Tekaia and Dujon, 1999
Mycobacterium tuberculosis	3,924	Tekaia and Dujon, 1999
Bacillus subtilis	4,100	Tekaia and Dujon, 1999
Escherichia coli	4,288	Blattner *et al.*, 1997
Archaebacteria		
Methanococcus jannaschii	1,735	Tekaia and Dujon, 1999
Methanobacterium thermoautrophicum	1,871	Tekaia andDujon, 1999
Pyrococcus horikoshii	2,061	Tekaia and Dujon, 1999
Archaeoglobus fulgidus	2,437	Tekaia and Dujon, 1999
Eukaryotes		
Saccharomyces cerevisiae	6,241	Rubin *et al.*, 2000
Arabidopsis thaliana	25,498	Arabidopsis Genome Initiative, 2000
Drosophila melanogaster	13,601	Adams *et al.*, 2000
Caenorhabditis elegans	19,320	C. *elegans* Sequencing Consortium, 1998[b]
Homo sapiens	35,000–120,000[c]	Crollius *et al.*, 2000; Ewing and Green, 2000; Liang *et al.*, 2000

[a]Protein-coding genes.
[b]http://www.sanger.ac.uk/Projects/C_elegans/wormpep/
[c]The number in humans remains so uncertain that geneticists have wagered on its exact value
(http://www.ensembl.org/Genesweep/).

It is almost certainly the case that most variation among organisms in
gene number simply reflects the predilection for gene or genomic doubling in their
ancestors, rather than selection for increased gene number in lineages that are
evolving greater complexity. Recently, through genomic analyses of the function
and similarity of genes, it has become possible to estimate the number of genes in
the "core" proteome. The proteome is defined as the number of protein-coding

genes that are so divergent from one another that they no longer share sequence similarity, which uses the fairly arbitrary benchmark of whether or not we can detect sequence similarity to assess uniqueness. One might expect that counting genes in this way would lead to a strong relationship between gene number and complexity, but this also appears to be false. For example, *Drosophila* has a core proteome of 10,736 (8065) genes, while *C. elegans* has a core proteome of 14,177 (9453) genes (Arabidopsis Genome Initiative, 2000; second estimates in parentheses from Rubin *et al.*, 2000), despite the fact that flies have roughly six times more cell morphotypes than nematodes (Valentine, 2000). Again, the size of the core proteome likely reflects the past tendency toward gene or genome duplication among ancestors distant enough that the duplicated genes have diverged substantially in sequence.

B. Mechanisms of increase in gene copy number

Gene copy number has increased through both duplication of relatively small regions of DNA and through genome-wide duplication events. The mechanisms involved in these processes are quite different, and we briefly review them in turn.

1. Gene duplication

Stretches of DNA may be duplicated in a variety of ways, although the main mechanisms involve mobile elements and unequal crossing over. Replicative transposons, retrotransposons, and retroviruses are elements that encode proteins enabling their own replication and insertion into the host genome. Occasionally, these elements replicate neighboring host genes as well as their own (Lewin, 1994; Baldo and McClure, 1999). Although such "accidents" may be rare per replication event, transposition can occur frequently enough to ensure a substantial rate of production of gene duplicates. Indeed, Charlesworth and Langley (1991) note that, in *Drosophila*, replicative transposition occurs at a rate of approximately 10^{-4} per transposable element per generation, which is orders of magnitude higher than the point mutation rate of a transposable element.

Duplications can also arise through unequal random breakage and reunion between nonhomologous sequences (Maeda and Smithies, 1986). An example is the human haptoglobin genes, where Hp2 is a result of what appears to be an unequal exchange between the fourth intron of Hp1F and the second intron of Hp1S. However, the crossover points show no homology, providing evidence that an unequal nonhomologous breakage and reunion has occurred. More frequently, unequal homologous recombination may occur between repeated elements, including transposable elements (Charlesworth and Langley, 1991), causing either the duplication or loss of intervening sequences. A signature of such a recombination event is a copy of the repeated element between the gene and its duplicate. An example is the human growth hormone and human chorionic somatomammotropin (HGH–HCS) gene cluster, where a SINE sequence resides in between the two genes. The rate of unequal crossing-over can rise dramatically in areas of tandemly repeated genes. For example, Guillemaud *et al.* (1999) examined the rate of unequal recombinants in a tandemly repeated array of *esterase* B genes in *Culex pipiens*, finding new gene amplifications in 4 of 60 mosquitoes (7%). This represents an extremely high rate of gene duplication compared to studies of regions lacking tandem arrays. In *D. melanogaster*, for example, only nine tandem duplications were detected in the *rosy* region (Gelbart and Chovnick, 1979), among 2.25 million offspring examined. We conclude that the mechanisms by which duplicate genes are generated vary greatly, depending on the activity of mobile elements and on the nature of the surrounding sequence.

2. Genome duplication

In addition to gene duplication, whole-scale genome duplication (polyploidization) has been a major source of gene duplicates. The main mechanisms by which polyploidization is achieved have recently been reviewed by Ramsey and Schemske (1998). Typically, polyploidization is the result of an error that occurs at one of three points: (1) mitosis, (2) meiosis, or (3) fertilization (Stebbins, 1980; Ramsey and Schemske, 1998; Otto and Whitton, 2000). Failure of a mitotic cell division can result in polyploidization and is known to occur in both animals and plants. If the failure occurs within the germline or, more generally, within cells that give rise to reproductive tissue, then the polyploid genome may be inherited. However, failure of the first or second reductive division during meiosis is a more usual route to polyploidy. Ramsey and Schemske (1998), reviewing several plant studies, estimated that unreduced gametes comprise 0.56% of pollen from nonhybrid individuals and as much as 27.52% of pollen in hybrids. Errors in fertilization are also a common source of polyploids. Fertilization of an egg by multiple sperm (polyspermy) is the most widespread error in fertilization; in humans, for example, polyspermy is the main mechanism by which triploids (~1–3% of conceptions; McFadden *et al.*, 1993) are produced (Uchida and Freeman, 1985). Another error in fertilization that can occur among asexual species is the occasional fertilization of an unreduced egg by sperm from a related sexual species. This is a likely explanation for the unusually high incidence of polyploidy among animals that are gynogenetic (in which fertilization requires stimulation by sperm that are not incorporated genetically) or hybridogenetic (in which sperm contribute genetically to the zygote, but the paternal genome is lost sometime during development; Vrijenhoek *et al.*, 1989; Otto and Whitton, 2000).

II. EVOLUTIONARY FORCES AFFECTING GENE COPY NUMBER

Given that gene and genome duplications do arise, we now turn to an examination of how evolutionary forces, including selection, mutation, and genetic drift, shape the evolution of gene number. We first discuss the rate at which duplications occur; in particular, we ask how this rate may have been molded by evolution. Second, we examine the processes that affect the frequency of a gene duplicate after it arises. Selection during this period has a very strong influence on whether the duplicate is lost or rises in frequency within the population. Finally, we turn to long-term evolutionary processes that act on a gene duplicate after it is established in a population. These long-term processes determine the ultimate fate of the gene duplicate.

A. The rate of duplication

The completion of several genome sequencing projects has enabled researchers to quantify, in an accurate and unbiased manner, the extent to which the genome is comprised of duplicated sequences. The enormity of the contribution of duplications has recently been assessed in *Arabidopsis thaliana*: 17% of its genome falls within one of 1528 tandemly duplicated arrays, while an even greater percentage (~58%) falls within one of the 24 large duplicated segments that most likely arose via polyploidization (Arabidopsis Genome Initiative, 2000). These numbers are limited to detectable and thus relatively recent events (within the past 200 MY or so; Vision *et al.*, 2000) and hence underestimate the total fraction of genes that have arisen through duplication.

1. Rate of gene duplication

What is the rate at which new gene duplications appear? Lynch and Conery (2000) recently addressed this question by scouring complete genomic sequences for very young gene duplicates (excluding multigene families and transposable elements). They found 10 pairs of genes in *D. melanogaster*, 32 pairs in *S. cerevisiae*, and 164 pairs in *C. elegans* whose level of silent-site divergence was less than 0.01. Using estimated rates of silent-site substitution, they inferred that 31 new duplicates arose per genome per million years in flies, 52 in yeast, and 383 in nematodes. In the absence of selection, one can use the neutral theory (Kimura, 1983) to make two important predictions about gene duplications from these estimates. (1) The rate at which gene duplicates spread throughout a population (= a substitution) should equal the genome-wide rate of gene duplication, ν, regardless of the population size. (2) The probability that any two haploid genomes randomly sampled from a diploid species will differ in terms of the duplicate genes that they carry is $4N_e v/(4N_e v + 1)$. (Aside: N_e is a measure of the size of a population in terms of how much allele frequencies change by chance from generation to generation, i.e., as a result of random genetic drift. A population with a very large census size, N, may nevertheless have a low N_e if only a few individuals ever reproduce. See Crow and Kimura, 1970, for further details.) Given their estimates of v, Lynch and Conery (2000) estimate that roughly half of the genes within the genome are expected to produce duplicates that spread through the population over a time period of 35–350 million years. For scale, mammals and birds first appeared in the fossil record around 200–250 MYA, which is also the time period during which monocots and dicots diverged. From the second prediction of the neutral theory, we expect that a large fraction of individuals in sizeable populations ($N_e > 10^4$) should be heterozygous for a duplication somewhere within their genome. It is, however, unreasonable to assume that new gene duplicates are always neutral (see Section II.B). If duplicated genes are often immediately deleterious, lineages that have survived to be sequenced would have fewer accumulated duplications than expected from the neutral theory. As an extreme example, if all gene duplicates were lethal in the heterozygous condition, then surviving individuals would never carry a gene duplicate, regardless of the rate at which gene duplicates arise. Consequently, the method of Lynch and Conery would underestimate the rate of gene duplication to the extent that selection has eliminated deleterious duplicates.

2. Rate of genome duplication

In addition to high rates of gene duplication, genome duplication has occurred repeatedly, especially in plants. A large fraction of angiosperms are thought to have undergone genomic doubling at some time in their evolutionary history, with estimates ranging from 20–40% (Stebbins, 1938), to 57% (Grant, 1963), to 70% (Goldblatt, 1980; Masterson, 1994), depending on the method used to infer polyploidization events (reviewed in Otto and Whitton, 2000). These estimates are insufficient to gauge the role of polyploidization in genome evolution, however, because they tell us only whether polyploidy has occurred at some point in the past. Groups that have polyploidized once and those that have done so repeatedly are treated equally. Recently, we have developed a new method to estimate the rate of polyploidization based on the excess of species bearing an even number of chromosomes at the gametic or haploid stage. Using this method, we inferred a rate of polyploidization per speciation event of ~2–4% in angiosperms and ~7% in ferns (Otto and Whitton, 2000). Even though polyploidy is much rarer among animals, we identified over 170 independent polyploidization events among insects and vertebrates, with many more known from other invertebrate species (see citations in Otto and Whitton, 2000). Several of these events represent ancient polyploidization events and affect the number of gene copies in a large fraction of animals. Large taxonomic groups of polyploid fish include the Actynopterygii (the highly speciose ray-finned fishes), Catostomidae (the sucker family), Salmonidae, the group *Corydoras–Aspidoras–Brochis* (Callichthyidae, the catfish family), the subfamily Schizothoracinae (Cyprinidae, the carp family), the subgenus *Barbus* (Cyprinidae), and the subgenus *Labeobarbus* (Cyprinidae). Among amphibia, polyploid groups include the Sirenidae and *Xenopus* (Pipidae). Finally, the extensive collinearity of genetic content on different chromosomes within vertebrates (Postlethwait *et al.*, 1998; Gibson and Spring, 2000) supports Ohno's (1970) contention that vertebrates underwent two genomic doubling events early in the paleozoic era.

3. Evolution of the rate of duplication

Even though duplication has played a large role in evolution, it is a logical error to assume that organisms have necessarily evolved strategies to promote duplication per se. Another possibility is that the rate of duplications has evolved to its current level as a side consequence of selection on other aspects of the replicative machinery of the cell, including selection to reduce the rate of point mutations, selection to ensure proper chromosome segregation, and selection to reduce the energy and time costs of DNA replication. Simply put, duplication events may be the consequence of the imperfection of cellular processes involved in DNA replication, and the rate of these errors might reflect a balance of selective forces other than the fitness effects of gene duplications. To demonstrate that organisms have evolved strategies to promote or hinder gene duplications, one must be able to reject the null hypotheses that they have not. One way to reject this null hypothesis is to show that mechanisms have evolved specifically to increase or decrease the rate of gene duplication.

In fact, complex mechanisms have evolved by which genes can promote their duplication, including replicative transposition, reverse transcription, and insertion. These events are made possible by a suite of enzymes, including transposase, resolvase, reverse transcriptase, and integrase (Lewin, 1994), which are encoded by the very elements that become duplicated but otherwise play no essential role in the cell. Nevertheless, these mechanisms almost certainly evolved to promote the duplication of certain genetic elements (transposons, retroposons, and retroviruses) and not as part of a generalized strategy to promote gene duplication on the part of an organism. This claim is supported by the fact that these mechanisms do not duplicate all possible gene sequences. For example, integrase requires a conserved CA near the end of an inverted repeat to allow insertion into the host genome of DNA transcribed from the RNA of retroposons and retroviruses (Lewin, 1994). A generalized mechanism to promote gene duplication would not be so picky.

Conversely, complex mechanisms have evolved to decrease the rate at which gene duplications become established within populations (reviewed by Selker, 1997, 1999, 2002; Wu and Morris, 1999). The most remarkable of these mechanisms is repeat induced point mutation (RIP), whereby duplicated genes are recognized by a poorly understood process and are subsequently subject to a high rate of mutation, which generally inactivates both gene copies. RIPing is only known to occur in *Neurospora,* but several other mechanisms that either recognize and silence duplicated sequences or degrade messenger RNA transcripts of duplicated genes have been uncovered, including methylation induced premeiotically, quelling, transvection, and paramutation (see Wu and Morris, 1999, for a description of these and other homology-dependent silencing processes). In many cases, these mechanisms were discovered because they interfered with expected patterns of expression in transgenic organisms. For example, the *hygromycin phosphotransferase* gene (*hpt*) was introduced into *Arabidopsis.* Initially, this transgene conferred resistance to hygromycin, but this resistance was lost in multiple homozygous lines as a result of duplicate gene silencing without loss of the transgene (Mittelsten Scheid *et al.,* 1991). Many of these mechanisms appear to be taxonomically widespread and general in action, although duplicated genes do vary in their susceptibility to silencing, depending on their sequence and site of insertion (e.g., inverted repeats appear to be particularly prone to silencing; Selker, 1999). It is thought that these homology-associated mechanisms have evolved as defense mechanisms against the proliferation of transposable elements and retroviruses (e.g., Ratcliff *et al.,* 1997; Al-Kaff *et al.,* 1998). The facts that many eukaryotic genomes are riddled with transposable elements (~35% of the human genome may be relics of mobile elements; Wolffe and Matzke, 1999) and that transposition may induce severe mutations (insertions and chromosomal rearrangements) suggest that there may well have been a substantial selective advantage to inactivating such elements.

Similarly, there is no evidence that organisms have evolved mechanisms to promote polyploidization in order to increase the number of genes available for evolution to act upon. As discussed earlier, polyploidization is typically the result of an error in mitosis, meiosis, or fertilization. There are some very unusual reproductive systems in which polyploidization occurs frequently and regularly, suggesting that some mechanism has evolved to promote polyploidization. For example, in some soft-scale insect species (Coccidae: Homoptera: Insecta), males develop from unfertilized haploid eggs that then polyploidize by the fusion of the first two cleavage nuclei (Nur, 1980). Even in these cases, however, the mechanisms promoting polyploidization have not been selected to increase the number of genes, for in fact this number stays constant from one generation to the next. Instead, polyploidization has evolved simply to restore the diploid state without fertilization. On the other hand, it is commonly observed that genome size declines after polyploidization events and that gene expression patterns of polyploid lineages decline toward diploid levels, a process known as rediploidization (Ferris and Whitt, 1977; Grant, 1981; Leipoldt, 1983; Werth and Windham, 1991; Soltis and Soltis, 1993). These observations suggest that mechanisms may have evolved to eliminate genes added by polyploidization. Although rapid and extensive genomic rearrangements and gene silencing or loss can occur following polyploidization (reviewed by Soltis and Soltis, 1993, 1999), this process is poorly understood. Mittelsten Scheid *et al.* (1996) observed gene silencing of a single-copy transgene after a change in ploidy, which they argue is not a homology-dependent process but one that depends on chromosome number or genome size. More work on the mechanisms by which gene silencing and loss occurs following polyploidization promises to clarify several puzzling observations, such as rediploidization.

Thus, the preponderance of evidence suggests that organisms have evolved processes to hinder rather than to promote the establishment of gene duplicates within their genomes. Consequently, we can reject our null hypothesis, but not in favor of the hypothesis that organisms have evolved strategies to promote the generation of gene copies. Rather, we suggest that organisms have evolved strategies to silence or inactivate gene duplicates before they rise to fixation within a population. Why then are duplications so widespread? We argue that the costs of ensuring error-free meiosis, mitosis, fertilization, and recombination would prevent the rate of gene or genome duplication from ever evolving to zero. Similarly, mechanisms that recognize and silence gene duplications must entail several costs, including the time involved in searching out homologous sequences, the energetic costs of silencing (including the costs of producing enzymes that promote silencing), and the risk of silencing critical genes and gene families. If the benefits of eliminating or silencing duplicated genes are not always greater than these costs, then homology-dependent silencing mechanisms may evolve to a point where the benefits and costs balance. Further reductions in the rate of establishment of gene duplicates would then be disadvantageous. One prediction from this line of reasoning is that we would expect stronger homology-dependent gene silencing mechanisms in organisms with a history of infectious spread of mobile elements, because the past benefits to silencing would have been greater, shifting the balance point to a higher overall level of homology-dependent silencing. A similar trade-off is thought to play an important role in the evolution of mutation rates (Sniegowski *et al.*, 2000).

In summary, gene and genome duplication events are commonplace. There is currently no good evidence that organisms have evolved mechanisms to increase the chance that their offspring carry gene or genome duplicates, although mobile elements within organisms clearly have evolved mechanisms to promote their own replication. In contrast, evidence from several species suggests that a variety of mechanisms have evolved to eliminate or silence gene duplicates soon after they arise. Thus, there is every reason to believe that the appearance of duplicated segments within genomes represents nothing more than a series of historical accidents. What happens to these duplicated genes after their appearance is the subject of the next section.

B. The spread of new duplicates

After a new gene duplicate has appeared within a population, its evolution may be divided into two distinct phases, a polymorphic period and a fixed period (Ohta, 1988c). Although evolutionary forces such as selection and mutation may affect both phases, their consequences are different. We consider first the polymorphic period and will return to the fixed period in the following section. During the polymorphic period, the gene duplicate, which is originally present in a single copy, changes in frequency, ultimately becoming lost or fixed within the species. In humans, such polymorphic gene duplicates have been found in both the opsin (Wolf *et al.*, 1999) and ribosomal (Veiko *et al.*, 1996) gene families. If the gene duplicate were selectively neutral, it would have a $1/(2N)$ chance of becoming fixed within a diploid population of census size N, and it would take on average $4N_e$ generations to do so. These estimates assume that gene silencing, mutations, and unequal crossing-over events that silence, inactivate, or knockout the duplicate are rare ($\ll 1/2N$ in frequency); if such events are frequent ($\geq 1/4N$), it is essentially impossible for the entire population ever to contain two active gene copies (proven using equation 8.8.38 of Crow and Kimura, 1970). This assumes that the gene duplication arises only once. In a series of papers, Ohta (1987, 1988b, 1988c) developed models in which duplicates are continuously being produced and destroyed by unequal crossing over. She also found that gene families are unlikely to evolve when duplicate genes are inactivated at an appreciable rate, unless there is positive natural selection favoring duplicates carrying different alleles (Ohta, 1987).

As noted by Walsh (1995), it is often assumed that gene duplications rise to fixation by random genetic drift, such that the gene duplications observed in a genome are a random subset of all the duplications that have occurred. However, selection acting on individuals carrying a duplicated gene has a profound effect on the probability of fixation of the duplicate. If the duplicate alters the fitness of its heterozygous carriers by a factor $1 + s$, through changes in survival, mating ability, or reproductive potential, its fixation probability becomes:

$$\frac{1 - e^{-2sN_e/N}}{1 - e^{-4sN_e}} \tag{1}$$

(Crow and Kimura, 1970). To simplify this discussion, we assume that the population is diploid and that the duplicate is additive in action, so that a homozygote carrying the duplication has an expected fitness of $1 + 2s$ (see Crow and Kimura, 1970, equation 8.8.3.21, to incorporate dominance). As equation (1) confirms, the probability of fixation is always lower than $1/(2N)$ if the gene duplicate reduces fitness (i.e., $s < 0$). Conversely, if the duplicate has a beneficial effect on fitness, its probability of fixation is greater than $1/(2N)$, being approximately $2s$ for small s. Notice that even directly beneficial gene duplicates are not guaranteed to fix within a population; 98% of gene duplicates that confer a 1% fitness advantage will ultimately be lost by chance!

It seems likely that when a gene duplicate does affect fitness, it will most often have a negative effect ($s < 0$). Gene duplicates may insert into other genes or gene regulatory regions, disrupting their function. Gene duplicates may also induce deleterious chromosomal rearrangements as a result of ectopic exchange. Gene duplicates may have been incompletely or inaccurately copied in such a way that the gene product decreases fitness. For instance, gene duplicates that have been reverse transcribed from a processed RNA intermediate may lack introns that regulate gene expression or coordinate alternative splicing, leading to an incorrect expression pattern of the gene products. Furthermore, the timing or level of gene expression may be disturbed by the presence of different regulatory factors near the new position of the duplicated gene. Even when both gene copies are properly expressed, there may be a fitness cost to having increased levels of the gene product. The opposite problem, underexpression, may also arise if the gene duplicate induces inactivation or silencing of both the original and the duplicate copy, as is observed with RIPing and quelling (Selker, 1997). Of course, each such change may increase rather than decrease fitness, but the latter will be more common as long as the duplication causes a relatively large change in gene expression and the original unduplicated gene was functioning well in its current environment. Under these circumstances, the chance that an alteration in gene expression will improve the ability of an individual to survive and reproduce is small. Such an argument applies to mutations of any form and has received both theoretical validation (Fisher, 1930; Kimura, 1983) and empirical support (Mukai *et al.*, 1972; Simmons and Crow, 1977; Deng and Lynch, 1997; Elena, 1997; Vassilieva *et al.*, 2000, but see Shaw *et al.*, 2000).

Even if a small fraction of gene duplicates that arise are beneficial, the subset of duplicates that fix within a population will contain a much higher fraction that increase fitness (Figure 16.1). Typically, deleterious gene duplicates are lost soon after they arise, neutral duplicates only rarely fix, and beneficial gene duplicates have a much larger chance of becoming established within a population. Consequently, gene duplicates that remain within a genome for long periods of time are a very biased subset of the ones that have occurred, a phenomenon known as the selective sieve. This helps explains, for example, why the number of copies of each transfer RNA in C. *elegans* correlates with codon usage in highly expressed genes (Duret, 2000). The simplest explanation for this is not that the tRNAs most in demand were duplicated at a higher rate, but that they were more likely to rise to fixation when they did occur.

Although it may be rare that a new gene duplicate confers a fitness advantage, beneficial duplicates do occasionally arise. The most likely source of a

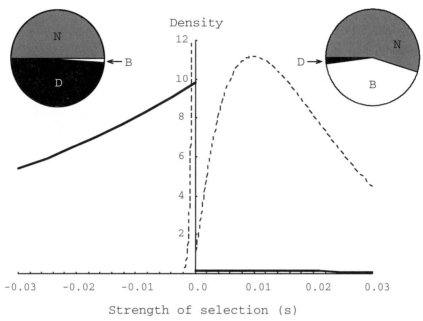

Figure 16.1. The selective sieve. Because selection affects the probability that a mutation will establish within a population, the effects of new gene duplicates on fitness will be very different from the effects of those gene duplicates that fix within a population. In the central figure, a probability density function for the selection coefficient (s) is illustrated, with the thick curves showing a hypothetical probability distribution among new duplicates, and the dashed curves showing the distribution among fixed duplicates (i.e., weighted by equation 1). It is assumed that the strength of selection follows a negative exponential distribution for both deleterious and beneficial duplicates, with the average magnitude of selection equal to 0.01. Fifty percent of new duplicates are assumed to be neutral (N), 49% deleterious (D), and only 1% beneficial (B), as illustrated in the left pie chart. Among fixed duplicates, however, these proportions change radically (right pie chart), with a full 43% of duplicates having a selective advantage. A population size of 1000 is assumed. The selective sieve would be even more effective in larger populations or with a larger average strength of selection.

fitness advantage is that, prior to duplication, there was insufficient gene product relative to the demands of the organism in its current environment. Of course, point mutations that increase expression would also have been advantageous under these circumstances, but gene duplications may be a relatively frequent mutation affecting expression levels. Such a scenario is thought to explain the extensive duplication of both transfer and ribosomal RNAs (Ohno, 1970). It is also thought that selection to increase enzyme production favored the spread of duplicates of *glutathione S-transferase* genes in houseflies (Wang *et al.*, 1991) and *esterase* genes in the mosquito, *Culex* (Mouchès *et al.*, 1986; Guillemaud *et al.*, 1999), both of which confer resistance to insecticides. Similarly, selection for metal tolerance may be related to a high incidence of *metallothionein* gene duplicates (Maroni *et al.*, 1987). If the benefit of gene duplicates often stems from selection for increased gene product, one would predict that the duplication of highly expressed genes is more likely to be advantageous, assuming that expression levels are correlated with demand for gene product. This prediction was confirmed in yeast, where genes with high mRNA levels, including heat shock, glucose metabolic, and cytosolic ribosomal genes, were more likely to be found in duplicate than genes with low levels of expression (Seoighe and Wolfe, 1999).

Another advantage to gene duplication may occur when carrying two different alleles of a gene is beneficial, i.e., when there is heterozygote advantage (Spofford, 1969; Ohno, 1970; see also Appendix). In this case, a gene duplicate carrying one allele can spread along with an alternative allele at the original gene, generating permanent heterozygosity. A striking example has been studied in *Culex pipiens* and again involves selection for insecticide resistance (Lenormand *et al.*, 1998). Organophosphate insecticides (OPs) target acetylcholinesterase (AChE), an enzyme that hydrolyzes acetylcholine and plays an important role in the normal functioning of the central nervous system. Mutations modifying the target of OPs led to resistant alleles ($Ace.1^R$), which have a higher fitness in the presence of OPs but which lower fitness in their absence due to the reduced activity of the $Ace.1^R$ allele. A gene duplication appeared that combined both resistant and sensitive alleles ($Ace.1^{RS}$). $Ace.1^{RS}$ individuals were found to be almost as resistant to insecticides, but they suffered from a much lower fitness cost because of the normal functioning of the sensitive allele. The duplication spread rapidly in the south of France, with an estimated selective advantage over $Ace.1^R$ of 3–6% (Lenormand *et al.*, 1998). In this case, the gene copies are tightly linked and the haplotype contains both alleles, a combination that we show in the Appendix is especially conducive to the spread of the gene duplicate when there is heterozygote advantage.

An often touted, but much less straightforward advantage, of gene duplication is that it can provide backup copies that protect an individual from either heritable (mutations) or nonheritable (developmental) errors. A critical point to remember, however, is that the benefit of such masking is roughly equal to the chance that an error occurs times the fitness consequence of the mutation or developmental error (Pál and Hurst, 2000). Thus a mutant allele or a developmental error that occurs in only one in a million individuals can account for only a miniscule benefit to gene duplication (s is proportional to 10^{-6}). Another problem pointed out by Pál and Hurst (2000) is that the majority of nonheritable errors should have minor fitness consequences, because they would affect only a fraction of the expected gene products within the organism (e.g., a translational error would affect only one protein within one cell). They argue that the most damaging errors are those involving gene silencing, which can be propagated to daughter cells, but they note that such errors are less common. A more substantial problem exists with the hypothesis that gene duplicates protect an individual from deleterious mutations: they also double the mutational target and reduce the efficiency by which selection can eliminate deleterious mutations by masking their effects. The result is that the frequency of mutant alleles rises in the population, especially among those individuals carrying the gene duplication. Clark (1994) showed that if deleterious mutations are fully redundant and if the two gene copies are completely linked, then the advantage of masking and the disadvantage of accumulating mutations exactly balance, so that the duplicate is no more likely to spread when rare than if it were neutral. In the Appendix, we generalize this model to allow for partial redundancy and for arbitrary linkage between the original and duplicate gene. We find that, unless there is perfect redundancy (i.e., unless mutations are completely recessive), tandem duplications are actually selected against because deleterious alleles remain associated with the gene duplicate for longer periods of time. In contrast, loosely linked duplications can gain a positive selective advantage from masking as long as the fitness of individuals carrying a deleterious mutation is sufficiently higher when a functioning duplicate gene is present than when it is not (see equation A7). Nevertheless, the magnitude of this selective advantage is never large (always being less than twice the mutation rate). Thus, masking deleterious mutations will have only a minor effect on the probability of fixation of a gene duplication (see equation 1).

C. Maintenance and divergence of gene duplicates

Once a gene duplicate has fixed within a population, its evolution is by no means over. In general, three eventual fates exist for duplicated genes: (1) maintenance of both copies performing the exact same functions (Figure 16.2B), (2) inactivation or loss of one copy (Figure 16.2C), or (3) divergence in function through neo-functionalization (Figure 16.2D), specialization (Figure 16.2E), or subfunctionalization (Figure 16.2F). Most empirical data on the likelihood of each of these fates comes from studies of ancient polyploids. A surprisingly high proportion of genes that duplicated via polyploidization have been retained over long periods of time: ~8% in yeast over ~100 MY (Seoighe and Wolfe, 1999), ~72% in maize over ~11 MY (Ahn and Tanksley, 1993; Gaut and Doebley, 1997), ~77% in *Xenopus* over ~30 MY (Hughes and Hughes, 1993), ~70% in salmonids over 25–100 MY (Bailey *et al.*, 1978), ~47% in catastomids over ~50 MY (Ferris and Whitt, 1979), and ~33% in vertebrates over ~ 500 MY (Nadeau and Sankoff, 1997). Many of these duplicates have diversified in function and, more typically, in expression patterns (e.g., Ferris and Whitt, 1979). Additional data on the fate of duplicated genes was also provided by the genome-wide analyses by Lynch and Connery (2000). They noted that there is a nearly exponential decline in the number of functional duplicates over time, where the time of each duplication

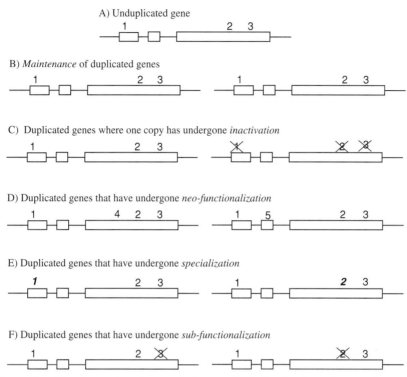

Figure 16.2. The ultimate fate of a gene duplication (for further details, see Force *et al.*, 1999). (A) A single-copy gene with three exons (boxes) is shown. For illustration purposes, we imagine that there are sites (1, 2, 3) that are critical for the performance of three different functions. These may be sites where regulatory proteins bind or where there is an active site in an enzyme. We consider the ultimate fate of a gene duplication that preserves the structure and function of the gene. (B) Both copies may be maintained indefinitely. (C) One copy may be inactivated by the accumulation of mutations that disrupt its functions (X). (D) The two copies may diverge and take on novel functions ("neo-functionalization"; 4, 5). (E) The two copies may specialize in function, each becoming better at performing one of the original functions of the single-copy gene (*1, 2*). (F) The two copies may accumulate deleterious mutations disrupting different functions ("subfunctionalization"), causing both to be maintained in order to perform all the functions of the original single-copy gene.

event is estimated by the number of silent site differences between the duplicate copies. Assuming a molecular clock, they infer an average half-life of duplicate genes of about 4 million years, which is a much higher inactivation rate than observed following polyploidization.

A large body of theory has been developed to examine which of the three ultimate fates is most likely and under what conditions (Bailey *et al.*, 1978; Takahata and Maruyama, 1979; Watterson, 1983; Walsh, 1987; Ohta, 1988a;

Hughes, 1994; Walsh, 1995; Nowak *et al.*, 1997; Force *et al.*, 1999; Lynch and Force, 2000). We turn now to a summary of these ideas.

1. Maintenance of function

Whether two gene copies can be maintained over long periods of evolutionary time depends strongly on whether the genes are redundant in function. If the two gene copies are perfectly redundant, i.e., if a single functional copy can completely mask the deleterious effects of mutations in other copies, they are very unlikely to be maintained (Fisher, 1935; Cooke *et al.*, 1997). With complete redundancy and total overlap in function, Fisher (1935) showed that the maintenance of both copies occurs only if one gene performs its function less efficiently and is less mutable than the other gene. Fisher also noted that both copies could be maintained if they had the exact same efficiency and number of mutations, but this would only work in an infinitely large population. Continual gene conversion, if frequent enough, could also ensure that both copies are maintained indefinitely (Walsh, 1987). On the other hand, the long-term maintenance of both gene copies is much more likely whenever redundancy is incomplete, such that mutations are deleterious even when they occur in only one gene copy (Takahata and Maruyama, 1979). In this case, selection acts against changes at critical sites in either copy, although the copies would diverge in sequence at silent sites. Such purifying selection has been observed in 17 duplicates pairs of genes in the anciently tetraploid *Xenopus laevis*, where Hughes and Hughes (1993) found that amino acid change was rare relative to the rate of silent substitution in both copies, suggesting that neither copy of the gene has been free to accumulate amino-acid altering mutations. Given that favorable duplicates comprise a much higher fraction of duplicates that fix within a population than their representation among new duplicates (as a result of the selective sieve; Figure 16.1), it is more likely that purifying selection will act to preserve the duplicates that have fixed. Thus, for example, the maintenance of 579 tRNAs within the *C. elegans* genome (Duret, 2000) may be explained by the need for a large pool of tRNAs within the cell, such that loss of a tRNA is deleterious even though functioning duplicates of the same tRNA exist within the genome. Another issue to consider is the source of the duplicated genes. Following a polyploidization event, the relative ratios of most (but not all; Galitski *et al.*, 1999) gene products remain roughly the same. Hence, the loss of a gene that has been duplicated by polyploidization may actually cause an underproduction of its gene product relative to those with which it interacts. Thus, selection to maintain gene copies may be stronger when the copies are produced by polyploidy than by gene duplication, which may help explain the apparent paradox that gene duplicates that arise following polyploidization appear to be retained for longer than gene duplicates that arise individually (Lynch and Conery, 2000).

2. Inactivation

If the integrity of the two gene copies is not actively maintained, the most likely fate of a gene duplicate is extinction, i.e., inactivation or loss (Haldane, 1933; Bailey et al., 1978; Takahata and Maruyama, 1979; Watterson, 1983; Walsh, 1995). When there is complete redundancy, Takahata and Maruyama (1979) showed that extinction is extremely rapid; typically, the half-life of a duplicate was less than 10 times the population size in generations. They concluded that their results are inconsistent with the percentage of gene duplicates that have been retained following polyploidization in salmonids and catostomids. Their model, however, only allowed deleterious mutations to accumulate in each gene. Walsh (1995) included both mutations that inactivate a gene and mutations that cause a gene duplicate to take on a novel function. Nevertheless, Walsh found that inactivation remains a very likely outcome, unless mutations causing novel functions are extremely common and the population size is large. For example, if mutations are 20,000 times more likely to inactivate a duplicate than they are to provide a novel function, then even if this novel function confers a 1% fitness advantage, inactivation occurs before neo-functionalization 99% of the time in a population with an effective size of 5000. Inactivated gene duplicates (pseudogenes and/or poorly expressed gene copies) have been observed in many gene families (e.g., the MHC class Ib genes; Hughes and Nei, 1989). Such examples of inactivation are characterized by a high rate of amino acid substitution, suggesting that purifying selection is weak or absent.

3. Divergence in function

Nevertheless, many gene duplicates do diversify in function over time, taking on novel functions, specializing in functions originally shared by the single-copy gene, or differentiating in the timing and/or location of expression (Figure 16.2D–F). Hughes (1994) reviewed several examples in which evidence has been found for positive selection having acted to diversify gene duplicates. Hughes also suggested that the chance of diversification might be substantially higher whenever the ancestral gene performed multiple functions, which he termed "gene sharing." In this case, selection might actively favor the specialization of each gene copy on different subsets of these functions. As an example, he cites δ-crystallin, a gene that encodes both an enzyme (argininosuccinate lyase) and an eye-lens crystallin in ducks; in other vertebrates, however, this gene appears to have duplicated such that each gene encodes only one of these products. Recently, it has been noted that positive selection for specialization is not necessary for this diversification process to occur. Force and colleagues (Force et al., 1999; Lynch and Force, 2000) noted that inactivation may itself lead to specialization. That is, if the original gene had several functional or regulatory domains, deleterious mutations disrupting different subfunctions could accumulate in each of the gene copies (Figure 16.2F). Subsequently, both gene copies would be essential and could not be further inactivated (see also Werth and Windham, 1991). This process, which they call "subfunctionalization," preserves the duplicates from loss or full inactivation over longer periods of evolutionary time, which increases the opportunity for adaptive mutations to occur. The fact that duplicate genes often differ in the timing and/or pattern of expression is consistent with this model (Ferris and Whitt, 1979; Hughes and Hughes, 1993; Force et al., 1999). In addition, Force (1999) noted that, with both the engrailed genes, eng1 and eng1b, in zebrafish and the ZAG1 and ZMM2 genes in maize, the shared expression pattern of the duplicated genes matches the total expression pattern of single-copy genes in related organisms lacking the duplication. This observation could, however, be a result of either positive selection favoring specialization (Hughes, 1994) or loss of subfunctions through deleterious mutation accumulation (Force et al., 1999; Lynch and Force, 2000).

III. DISCUSSION

Arguably, the duplication of genes, regulatory regions, and whole genomes has provided the most important raw material for evolution, for, without such duplications, it is likely that genomes would be limited to a handful of genes and organisms to a handful of components. Nevertheless, evolution has no foresight, and populations cannot evolve so as to maximize their future rate of adaptive evolution. In fact, the rate of gene and genome duplication is unlikely to be set at an optimal level for adaptive evolution. As with mutations in general (Sniegowski et al., 2000), because gene duplications and genome duplications are more likely to reduce fitness than they are to increase fitness, individuals carrying genes that increase the rate of duplication ("duplicators") will typically have a lower fitness, being burdened by the disruptive effects of gene duplication. Even if a beneficial duplication does arise in an individual carrying a "duplicator," recombination will ensure that the beneficial duplicate spreads from the "duplicator" genotype to "nonduplicator" genotypes within the population. Thus there is an asymmetry in the effectiveness of selection. Deleterious duplications immediately decrease the average fitness of duplicator genotypes, whereas beneficial duplications only increase the average fitness of duplicator genotypes to the extent that they remain associated with the genes causing the increased rate of duplication. In sexual populations, this beneficial association decays rapidly and provides only a weak and ineffective advantage to the duplicator genes. (See Sniegowski et al., 2000, for a review of the relevant theory on mutators, most of which may be applied directly to the case of duplicators.) Consequently, the most likely explanation for why gene and genome duplications occur is that it is too costly, in terms of time and energy, to eliminate these errors entirely.

The fact that gene and genome duplications are unlikely to have arisen to promote the evolvability of a population does not mean that they have not subsequently done so. Indeed, one could imagine that an organism early in evolution with only a handful of genes would have been more limited in evolutionary potential compared with any competitors that had undergone a spate of gene duplication. Consequently, it may be that organisms lacking duplications in their past have been more likely to go extinct or less likely to speciate over evolutionary time (although there is no clear evidence to this effect). This is, of course, a group selection argument. Although group selection is notoriously ineffective relative to individual selective forces acting on duplicators such as the ones described above, group selection can play an important role in determining which species have survived to comprise the biodiversity surrounding us. It remains a tantalizing possibility that the diversity and evolutionary success of ray-finned fishes and vertebrates, for example, were driven initially be genome-wide duplications. Yet the fact that there is no clear relationship between complexity (measured, for example, by the number of cell morphotypes; Valentine, 2000) and genome size, gene number, or proteome size suggests that we must be cautious in estimating the evolutionary importance of sheer numbers of genes. Complexity can and has evolved by other means, including the evolution of multidomain proteins, complex patterns of gene regulation, and alternative splicing.

Once a gene duplication appears, selection, drift, and mutation may each play a role in determining its ultimate fate. Although many theoretical analyses have focussed on the fate of a duplicate gene once it has already fixed within a population (with notable exceptions, e.g., Spofford, 1969; Ohta, 1987, 1988b, 1988c; Clark, 1994), evolutionary forces can have a major effect on the rate of spread and the probability of fixation of a new duplicate. Selection, for example, favoring a new gene duplicate improves the chance that the duplicate survives and rises in frequency. Thus, as we point out, duplicate genes that have fixed within a population will be relatively enriched for duplicates with a positive selective effect (Figure 16.1). Such fitness advantages may arise in a number of ways, including increasing the expression of underexpressed genes, allowing permanent heterozygosity when selection favors individuals with multiple alleles of a gene, and masking the deleterious effects of mutations. In an Appendix, we have constructed a model that examines the spread of a new gene duplication under various types of selection, with arbitrary linkage between the gene and its duplicate. We find that heterozygote advantage does indeed provide a large fitness advantage to a gene duplication. In fact, the strength of selection favoring a duplicated gene is comparable to the amount of selection favoring heterozygotes over homozygotes. Indeed, one reason why heterozygote advantage may be uncommon is that whenever this form of selection arises it leads, over evolutionary time, to the spread of gene duplicates such that every individual carries both alleles (Haldane, 1954).

In contrast, masking of deleterious mutations has only a very minor effect on the fitness of individuals bearing a duplicate gene. Even though it might be optimal to have a backup copy of a gene in case it mutates, selection is simply not powerful enough to cause the spread of gene duplicates to accomplish this benefit. With heterozygote advantage, a lower recombination rate (i.e., tighter linkage) between the gene and its duplicate promotes the spread of the duplication. With masking of deleterious mutations, a higher recombination rate (i.e., looser linkage) between the gene and its duplicate promotes the spread of the duplication. Given that duplications are much more strongly selected in the case of heterozygote advantage, these results suggest that tandem duplications are likely to have a stronger average benefit, making them more likely to establish when they occur compared to dispersed gene duplications. (See the Appendix for further details and discussion.)

Through selection or genetic drift, some gene duplications do fix within a population. However, they may still be lost or inactivated, especially if deleterious mutations accumulate as a result of the redundancy provided by gene copies. Although pseudogenes provide ample evidence for such inactivation, many duplicate genes have maintained their function over long periods of time, and still others have gained new functions. Maintenance of gene function is likely to occur in those cases where the effects of the duplicate gene are immediately advantageous. Divergence in gene function is likely to occur for genes that have multiple domains or multiple regulatory regions, both because selection might favor specialization of different gene copies on different functions and because the two gene copies might suffer deleterious mutations in different components, making both copies essential. Analyses of genomic sequences have clarified the role that different evolutionary processes have played in the origin, spread, maintenance, and diversification of duplicated genes (Lundin, 1993; Wolfe and Shields, 1997; Postlethwait et al., 1998; Lynch and Conery, 2000; Vision et al., 2000; Arabidopsis Genome Initiative, 2000). Undoubtedly, as more genomic sequence data become available, especially for closely related species, we will gain a clearer appreciation for the role that gene duplication has played in the evolution of life.

IV. APPENDIX: MODELING THE INVASION OF A DUPLICATE GENE

Most models of gene duplication assume that a gene duplicate has already fixed within the population or that duplicates are continuously produced (see Walsh, 1995). The two notable exceptions are the studies by Spofford (1969) and Clark (1994). Spofford focused on the spread of a chromosome carrying two different alleles (A and a) when there is a fitness advantage to individuals carrying both alleles, finding graphically that spread of the duplicate always occurred. Clark, on the other hand, modeled the situation in which deleterious mutations occur at both the original and duplicate genes. Assuming that these genes are completely linked and that only individuals lacking an A allele suffer a fitness loss, he found that the duplicate gene would spread only if it had a direct advantage. Here, we generalize these results by examining the initial rate of spread of a duplicate gene under a broader array of conditions. Specifically, we calculate the magnitude of selection for or against the gene duplicate under more general fitness regimes and with an arbitrary linkage relationship between the duplicate genes. To simplify matters, we ignore the ongoing production of gene duplicates (as was examined by Clark, 1994, and by Ohta, 1987, 1988b, 1988c) and gene conversion, both of which may play an important role in the establishment of a gene duplicate.

In our model, we consider an original locus (i) and its duplicate (j), separated by distance r. We denote a chromosome lacking a duplication by i- and one carrying the duplication by ij. Duplications on different chromosomes may be examined by setting r to $\frac{1}{2}$. At each gene there are two alleles A and a. Thus, with the duplication, there are four haplotypes to consider (AA, Aa, aA, and aa), whose frequencies are x_1, x_2, x_3, and x_4. Without the duplication, there are two additional haplotypes to consider (denoted as A- and a-), whose frequencies are x_5 and x_6. The fitness of a diploid individual carrying haplotypes k and l is denoted W_{kl}. At each gene, mutations are allowed to occur between alleles A and a at rate μ. To leading order in the mutation rate, mutations affect the invasion criterion only when there is no direct selection on the gene duplication and when the population is initially near a mutation-selection balance (Yong, 1998). Therefore we only discuss the effects of mutations in this case.

A. General diploid model

We model a diploid population that is at equilibrium when a gene duplication occurs. At this equilibrium, x_1, x_2, x_3, and x_4 equal zero, $x_5 = \hat{x}_5$, $x_6 = \hat{x}_6$, $\hat{x}_5 + \hat{x}_6 = 1$, and the mean fitness of the population is $\hat{W}$. A gene duplication then arises at low frequency. While the duplicate is rare, haplotypes bearing the duplication will change in frequency from one generation (x) to the next (x') according to the recursions:

$$
\begin{aligned}
x_1' &= (1 - \mu)^2 x_1^* \\
x_2' &= (1 - \mu)x_2^* + \mu(1 - \mu)x_1^* \\
x_3' &= (1 - \mu)x_3^* + \mu(1 - \mu)x_1^* \\
x_4' &= x_4^* + \mu^2 x_1^* + \mu x_2^* + \mu x_3^*
\end{aligned}
\tag{A1}
$$

where

$$x_1^* \approx \frac{x_1\hat{x}_5 W_{15} + r x_3\hat{x}_5 W_{35} + (1-r)x_1\hat{x}_6 W_{16}}{\hat{W}}$$

$$x_2^* \approx \frac{x_2\hat{x}_5 W_{25} + r x_4\hat{x}_5 W_{45} + (1-r)x_2\hat{x}_6 W_{26}}{\hat{W}}$$

$$x_3^* \approx \frac{x_3\hat{x}_6 W_{36} + r x_1\hat{x}_6 W_{16} + (1-r)x_3\hat{x}_5 W_{35}}{\hat{W}} \qquad (A2)$$

$$x_4^* \approx \frac{x_4\hat{x}_6 W_{46} + r x_2\hat{x}_6 W_{26} + (1-r)x_4\hat{x}_5 W_{45}}{\hat{W}}$$

Technically, equation (A1) is written assuming that mutations generate only a alleles from A alleles. However, if mutations occur rarely and are deleterious, the a allele will exist at low frequency within a population, and reverse mutations to it will be extremely rare. Therefore, back-mutations have little effect on the dynamics and can be safely ignored. Equations (A1) are linear functions of the frequencies of the duplicate gene because genotypes involving two duplication haplotypes are assumed to be exceedingly rare when the duplicate first appears and are ignored. To determine whether the duplicate gene spreads, we perform a local stability analysis of these equations (Bretscher, 1997). This involves finding the leading eigenvalue, λ_L, of the matrix form of equation (A1), which is known as a local stability matrix. $(\lambda_L - 1)$ corresponds, approximately, to the selection coefficient that acts on the gene duplicate while it is rare (Otto and Bourguet, 1999). Consequently, the duplicate will initially increase in frequency at an exponential rate whenever $\lambda_L > 1$ (or decrease if $\lambda_L < 1$). We shall first consider the rate of spread of the duplicate in a model with heterozygous advantage and then consider its spread when there are recurrent deleterious mutations.

B. Heterozygote advantage

The first case that we shall consider is heterozygote advantage. We assume that alleles A and a perform different functions such that the fitness of an individual is higher when both functions are performed. To simplify matters, we assume that all individuals carrying both A and a alleles (regardless of the number of them) have a fitness of $1(= W_{56} = W_{25} = W_{35} = W_{45} = W_{16} = W_{26} = W_{36})$, that individuals with only A alleles have a reduced fitness of $1 - s(= W_{55} = W_{15})$, and that individuals with only a alleles have a reduced fitness of $1 - t(= W_{66} = W_{46})$. This model differs from that examined by Spofford (1969), who focused on the case of a dimeric enzyme taking into account the probability that each type of dimer would be produced. In addition, the following derivation estimates the effective selection coefficient acting on a new duplicate, whereas Spofford focused on a numerical analysis of the dynamics of the duplicate gene.

Before the appearance of the duplication, the frequency of allele A approaches $\hat{x}_5 = t/(s+t)$, at which point the mean fitness of the population is $\hat{W} = 1 - st/(s+t)$ (Crow and Kimura, 1970). Without loss of generality, we label the alleles such that $s \leq t$, so that the frequency of allele A is greater than or equal to $\frac{1}{2}$. Under these assumptions, it can be shown that the leading eigenvalue describing the initial rate of spread of a gene duplicate is equal to:

$$\lambda_L = \frac{1}{\hat{W}} - \frac{1 - \hat{W} + r}{2\hat{W}} \left[1 - \sqrt{1 - \frac{4r\hat{x}_6(1 - \hat{W})}{(1 - \hat{W} + r)^2}} \right] \qquad (A3)$$

Under our assumptions that s, t, $\hat{x}_5$, $\hat{x}_6 > 0$, this leading eigenvalue is strictly greater than one. Thus, there will always be selection favoring the spread of a duplication. When the duplicate genes are tightly linked ($r \approx 0$), the leading eigenvalue becomes $1/\hat{W}$, which is greater than one under our assumptions. In this case, there is a slight caveat: if the first haplotype to appear is either AA or aa, the duplication will not be positively selected until either the Aa or aA haplotype is produced by mutation, conversion, or recombination. Increasing the recombination rate above zero always reduces the leading eigenvalue and hence slows the initial spread of the duplicate gene. When selection is weak ($s, t \ll 1$) and when the genes are not very tightly linked ($r > 0$), the eigenvalue is approximately:

$$\lambda_L \approx 1 + \hat{x}_5(1 - \hat{W}) = 1 + \frac{st^2}{(s+t)^2} \qquad (A4)$$

Equation (A4) indicates that the indirect or effective selection acting on the duplicate has the same order of magnitude as selection acting directly on the A and a alleles. Furthermore, selection for the duplicate is strongest when the two functions are approximately of equal benefit ($s \approx t$). Comparing the strength of selection on the duplicate ($\lambda_L - 1$), the duplicate experiences $(s+t)/t$ times the amount of selection when linkage is tight than when linkage is loose, assuming weak selection. This reaches a maximum of a twofold difference when $s \approx t$, indicating that a tandem duplication carrying both A and a alleles is twice as likely to fix within a population (replacing $\lambda_L - 1$ for the selection coefficient in equation 1) and will spread twice as quickly compared to an unlinked gene duplication. The advantage of tandem duplications is that the most fit chromosomes, i.e., ones bearing both A and a alleles, are unlikely to be broken apart by recombination. This may partially explain why tandem gene duplications are common in many gene families, although another obvious explanation is that the frequency with which duplications appear in tandem is higher as a result of unequal crossing over.

C. Mutation-selection balance

We next examine the fate of the gene duplication when the population is at mutation-selection balance. In the absence of the duplication, let the fitness of *AA, Aa,* and *aa* genotypes equal $W_{55} = 1$, $W_{56} = 1 - hs$, and $W_{66} = 1 - s$, respectively, where *s* measures the fitness cost of the mutation and *h* measures the degree of dominance of the *a* allele. We first assume that mutations are not fully recessive ($h > 0$). In large populations, the mutant allele (a) will reach a steady-state frequency of $\hat{x}_6 \approx \mu/(hs)$, at which point the population mean fitness is $\hat{W} \approx 1 - 2\mu$ (Crow and Kimura, 1970). These approximations are to leading order in the mutation rate and ignore terms such as μ^2. When the duplication first appears, we assume that it experiences no direct selection, such that the fitness of *AA/A*-individuals is also $1(= W_{15})$. Individuals carrying one mutant allele (*AA/a*- or *Aa/A*- or *aA/A*-) are assumed to have a fitness of $1 - ks (= W_{25} = W_{35})$, where *k* is the dominance coefficient of the *a* allele when paired with two A alleles. It can be shown that the fitness of genotypes bearing more than one deleterious mutation does not affect the spread of the gene duplication since these genotypes are much less common. Making these substitutions as well as $\hat{x}_5 = 1 - \hat{x}_6$ in equation (A1), we obtain a local stability matrix that is solely a function of the mutation rate and selection parameters. The leading eigenvalue of this matrix can be shown to equal

$$\lambda_L = 1 - \mu \left\{ \frac{k}{h} - \frac{(h-k)\,r\,(1-ks)}{h\,[r + ks\,(1-r)]} \right\} + O(\mu^2) \qquad \text{(A5)}$$

Assuming that the *a* allele is always deleterious, the term k/h is positive. Hence, for tandemly duplicated genes ($r \approx 0$), the leading eigenvalue is always less than one. Counterintuitively, this means that the duplication will spread at a slower rate than a purely neutral allele even if the duplication provides extra protection against the deleterious effects of mutations (i.e., $k < h$). The reason that tandem duplications are selected against is that mutations occur twice as often on chromosomes with two genes and are not strongly selected against when $k < h$. Consequently, deleterious mutant alleles accumulate on the chromosome bearing the duplication, generating indirect selection against the duplication. Note, however, that the strength of selection against the duplicate is proportional to the mutation rate, μ. Therefore, this effect is so slight that it will not completely prevent the fixation of any particular gene duplicate, but it will reduce the rate at which duplicate genes fix within large populations.

It can be shown that the leading eigenvalue, (A5), is always greater for gene duplicates that arise at a distance from the original gene (i.e., $\partial\lambda_L/\partial r > 0$).

Nevertheless, only when recombination is high enough,

$$r > \frac{k^2 s}{(h - 2k)(1 - ks)} \tag{A6}$$

and the masking advantage of the new duplicate is strong enough,

$$k < \frac{h}{2 + hs} \tag{A7}$$

will the gene duplication be positively selected ($\lambda_L - 1 > 0$). For weak selection and unlinked gene duplicates, these conditions require that the fitness reduction observed in AAa individuals be less than half that observed in Aa individuals ($k < h/2$). The reason that distant gene duplicates fare better is that recombination shuffles some of the deleterious mutant alleles that accumulate faster in individuals bearing duplicated genes to chromosomes lacking the gene duplication. This reduces the frequency of deleterious mutant alleles associated with the gene duplication. Even when the gene duplicate is favored, the strength of this indirect selective advantage is always less than twice the mutation rate, and hence masking deleterious mutations will only slightly increase the chance that loosely linked duplicates spread to fixation.

The results obtained in this section are similar to those obtained in models of ploidy evolution, where it has been shown that selection only favors the expansion of the diploid phase of a sexual organism (with "duplicates" of every gene) when recombination rates are sufficiently high and masking is sufficiently strong (Otto and Goldstein, 1992; Jenkins, 1995). The main qualitative difference is that duplication of the entire genome allows the effects to be greatly amplified, relative to the case of a single gene duplicate (Jenkins, 1995).

D. Mutation-selection balance with completely recessive mutations

In the above, we assumed that deleterious mutations have some deleterious effects even in heterozygous individuals. Here, we briefly review the results obtained when masking is complete ($h = k = 0$), such that only aa and aa/a- individuals have a reduced fitness of $1 - s$. In this case, before the appearance of the duplicate, $\hat{x}_6 \approx \sqrt{\mu/s}$ and $\hat{W} \approx 1 - \mu$ at the equilibrium between mutation and selection (Crow and Kimura, 1970). Repeating the local stability analysis, we find that the leading eigenvalue is always one, regardless of the recombination rate. This is the same result obtained by Clark (1994) for tandem duplications with lethal homozygous effects ($r = 0, s = 1$). Thus, the additional protection against deleterious mutations provided by a gene duplication makes no difference to the spread of the duplication when deleterious mutations are already fully masked in Aa heterozygotes lacking the duplicate.

E. Further results

Although the above analyses assume that the duplication is not directly selected for or against, it is straightforward to include such selection. Let s_D equal the direct selection coefficient (i.e., the change in fitness due to having a gene duplicate regardless of which alleles are carried). Similarly, let s_I equal the indirect selection that arises because carrying a gene duplicate changes the alleles that an individual is likely to carry, as measured in the above sections [i.e., $s_I = \lambda_L - 1$ from equation (A3) or (A5)]. Assuming that selection is relatively weak (S_D, $S_I \ll 1$), the total strength of selection acting on the gene duplicate while rare is simply the sum of these two effects (Yong, 1998; Otto and Bourguet, 1999). Thus, any amount of direct selection is likely to overwhelm the indirect selection that arises from deleterious mutations alone (from A5). In contrast, when there is heterozygote advantage, the strength of indirect selection (from A3) can be quite strong and may control the fate of a new gene duplicate.

Finally, although we have focused on the dynamics of gene duplications in diploid organisms, in his thesis, Yong (1998) also examined the dynamics of gene duplication in haploid organisms. Each generation, the haploid organisms were assumed to mate randomly to form diploid individuals, followed by meiosis to form daughter haploid organisms. One surprising difference in the results of the haploid and diploid models was that gene duplications in haploids never gain an advantage from masking deleterious mutations. In other words, in a haploid mutation-selection balance model, the leading eigenvalue (equivalent to A5) was always less than one, even though individuals carrying a gene duplication and a mutant allele (i.e., *Aa* individuals) were more fit than individuals carrying only a mutant allele (i.e., *a-* individuals). We speculate that this difference is caused by the fact that mutations accumulate more rapidly in gene duplicates in haploids than in diploids. Consider individuals carrying a new gene duplicate. If they are diploid, a new mutation occurs in the gene duplicate only one-third of the time; two-thirds of the time it will occur in the original gene. Because the gene duplication provides more opportunity for masking the effects of this new mutation, this mutation is sheltered from selection and has a higher chance of being passed to the offspring generation. Nevertheless, the gene duplicate is mutation free two-thirds of the time and, with loose linkage, has a good chance of segregating away from the mutation. In haploids, however, a new mutation that is sheltered by a gene duplication has a 50:50 chance of having occurred in the duplicate gene itself, in which case the duplicate and the new mutation are inextricably bound. Thus, even with free recombination, there is less opportunity for new gene duplicates to rid themselves of the higher frequency of mutant alleles that they shelter from selection. Hence, in haploids, the masking advantage of a gene duplication is always outweighed by the disadvantage that is incurred because mutations are sheltered from selection and reach a higher frequency at the duplicated gene.

Acknowledgments

The models described in the Appendix were inspired by the work of Andrew Clark and benefited greatly from discussions with him. Further mathematical details may be found in the Honour's thesis of PY, available from the Department of Zoology, University of British Columbia. We are grateful to Jay Dunlap, Thomas Lenormand, Toby Johnson, and Michael Whitlock for helpful discussions and comments on the manuscript. Funding to SPO was provided by a grant from the Natural Sciences and Engineering Research Council (Canada) and by a poste-rouge from the Centre National de la Recherche Scientifique (France).

References

Adams, M. D., Celniker, S. E., Holt, R. A., Evans, C. A., Gocayne, J. D., Amanatides, P. G., Scherer, S. E., Li, P. W., Hoskins, R. A., and Galle, R. F. (2000). The genome sequence of *Drosophila melanogaster*. *Science* **287**, 2185–2195.

Ahn, S., and Tanksley, S. D. (1993). Comparative linkage maps of the rice and maize genomes. *Proc. Natl. Acad. Sci. USA* **90**, 7980–7984.

Al-Kaff, N. S., Covey, S. N., Kreike, M. M., Page, A. M., Pinder, R., and Dale, P. J. (1998). Transcriptional and posttranscriptional plant gene silencing in response to a pathogen. *Science* **279**, 2113–2115.

Amores, A., Force, A., Yan, Y. L., Joly, J. S., Amemiya, C. T., Fritz, A., Ho, R. K., Langeland, J., Prince, V., and Wang, Y. L. (1998). Zebrafish *hox* clusters and vertebrate genome evolution. *Science* **282**, 1711–1714.

Arabidopsis Genome Initiative. (2000). Analysis of the genome sequence of the flowering plant *Arabidopsis thaliana*. *Nature* **408**, 796–815.

Bailey, G. S., Poulter, R. T., and Stockwell, P. A. (1978). Gene duplication in tetraploid fish: Model for gene silencing at unlinked duplicated loci. *Proc. Natl. Acad. Sci. USA* **75**, 5575–5579.

Baldo, A. M., and McClure, M. A. (1999). Evolution and horizontal transfer of dUTPase-encoding genes in viruses and their hosts. *J. Virol.* **73**, 7710–7721.

Blattner, F. R., Plunkett, G. 3rd, Bloch, C. A., Perna, N. T., Burland, V., Riley, M., Collado-Vides, J., Glasner, J. D., Rode, C. K., and Mayhew, G. F. (1997). The complete genome sequence of *Escherichia coli* K-12. *Science* **277**, 1453–1474.

Bretscher, O. (1997). "Linear Algebra with Applications." Prentice Hall, Upper Saddle River, NJ.

C. elegans Sequencing Consortium (1998). Genome sequence of the nematode *C. elegans*: A platform for investigating biology. *Science* **282**, 2012–2018.

Charlesworth, B., and Langley, C. H. (1991). Population genetics of transposable elements in *Drosophila*. In "Evolution at the Molecular Level" (R. K. Selander, A. G. Clark, and T. S. Whittam, eds.), pp. 222–247. Sinauer, Sunderland, MA.

Chervitz, S. A., Aravind, L., Sherlock, G., Ball, C. A., Koonin, E. V., Dwight, S. S., Harris, M. A., Dolinski, K., Mohr, S., and Smith, T. (1998). Comparison of the complete protein sets of worm and yeast: orthology and divergence. *Science* **282**, 2022–2028.

Clark, A. G. (1994). Invasion and maintenance of a gene duplication. *Proc. Natl. Acad. Sci. USA* **91**, 2950–2954.

Cooke, J., Nowak, M. A., Boerlijst, M., and Maynard-Smith, J. (1997). Evolutionary origins and maintenance of redundant gene expression during metazoan development. *Trends Genet.* **13**, 360–364.

Crollius, H. R., Jaillon, O., Bernot, A., Dasilva, C., Bouneau, L., Fischer, C., Fizames, C., Wincker, P., Brottier, P., and Quetier, F. (2000). Estimate of human gene number provided by genome-wide analysis using *Tetraodon nigroviridis* DNA sequence. *Nature Genet.* **25**, 235–238.

Crow, J. F., and Kimura, M. (1970). "An Introduction to Population Genetics Theory," pp. xiv, 591. Harper & Row, New York.

Deng, H. W., and Lynch, M. (1997). Inbreeding depression and inferred deleterious-mutation parameters in *Daphnia*. *Genetics* **147**, 147–155.

Duret, L. (2000). tRNA gene number and codon usage in the C. *elegans* genome are co-adapted for optimal translation of highly expressed genes. *Trends Genet.* **16**, 287–289.

Elena, S. F., and Lenski, R. E. (1997). Test of synergistic interactions among deleterious mutations in bacteria. *Nature* **390**, 395–398.

Ewing, B., and Green, P. (2000). Analysis of expressed sequence tags indicates 35,000 human genes. *Nature Genet.* **25**, 232–234.

Ferris, S. D., and Whitt, G. S. (1977). Loss of duplicate gene expression after polyploidisation. *Nature* **265**, 258–260.

Ferris, S. D., and Whitt, G. S. (1979). Evolution of the differential regulation of duplicate genes after polyploidization. *J. Mol. Evol.* **12**, 267–317.

Fisher, R. A. (1930). "The Genetical Theory of Natural Selection." Oxford University Press, Oxford, U.K.

Fisher, R. A. (1935). The sheltering of lethals. *Am. Naturalist* **69**, 446–455.

Fleischmann, R. D., Adams, M. D., White, O., Clayton, R. A, Kirkness, E. F., Kerlavage, A. R., Bult, C. J., Tomb, J. F., Dougherty, B. A., and Merrick, J. M. (1995). Whole-genome random sequencing and assembly of *Haemophilus influenzae* Rd. *Science* **269**, 496–512.

Force, A., Lynch, M., Pickett, F. B., Amores, A., Yan, Y. L., and Postlethwait, J. (1999). Preservation of duplicate genes by complementary, degenerate mutations. *Genetics* **151**, 1531–1545.

Galitski, T., Saldanha, A. J., Styles, C. A., Lander, E. S., and Fink, G. R. (1999). Ploidy regulation of gene expression. *Science* **285**, 251–254.

Gaut, B. S., and Doebley, J. F. (1997). DNA sequence evidence for the segmental allotetraploid origin of maize. *Proc. Natl. Acad. Sci. USA* **94**, 6809–6814.

Gelbart, W. M., and Chovnick, A. (1979). Spontaneous unequal exchange in the rosy region of *Drosophila melanogaster*. *Genetics* **92**, 849–859.

Gibson, T. J., and Spring, J. (2000). Evidence in favour of ancient octaploidy in the vertebrate genome. *Biochem. Soc. Trans.* **28**, 259–264.

Goldblatt, P. (1980). Polyploidy in angiosperms: Monocotyledons. *In* "Polyploidy: Biological Relevance" (W. H. Lewis, ed.), pp. 219–239. Plenum, New York.

Grant, V. (1963). "The Origin of Adaptations." Columbia University Press, New York.

Grant, V. (1981). "Plant Speciation." Columbia University Press, New York.

Guillemaud, T., Raymond, M., Tsagkarakou, A., Bernard, C., Rochard, P., and Pasteur, N. (1999). Quantitative variation and selection of esterase gene amplification in *Culex pipiens*. *Heredity* **83**, 87–99.

Haldane, J. B. S. (1933). The part played by recurrent mutation in evolution. *Am. Naturalist* **67**, 5–19.

Haldane, J. B. S. (1954). "The Biochemistry of Genetics." George Allen & Unwin, London.

Hughes, A. L. (1994). The evolution of functionally novel proteins after gene duplication. *Proc. R. Soc. Lond. B. Biol. Sci.* **256**, 119–124.

Hughes, A. L., and Nei, M. (1989). Evolution of the major histocompatibility complex: Independent origin of nonclassical class I genes in different groups of mammals. *Mol. Biol. Evol.* **6**, 559–579.

Hughes, M. K., and Hughes, A. L. (1993). Evolution of duplicate genes in a tetraploid animal, *Xenopus laevis*. *Mol. Biol. Evol.* **10**, 1360–1369.

Hutchison, C. A., Peterson, S. N., Gill, S. R., Cline, R. T., White, O., Fraser, C. M., Smith, H. O., and Venter, J. C. (1999). Global transposon mutagenesis and a minimal *Mycoplasma* genome. *Science* **286**, 2165–2169.

Jenkins, C. D., and Kirkpatrick, M. (1995). Deleterious mutations and the evolution of genetic life cycles. *Evolution* **49**, 512–520.

Kimura, M. (1983). "The Neutral Theory of Molecular Evolution." Cambridge University Press, Cambridge, U.K.

Leipoldt, M. (1983). Towards an understanding of the molecular mechanisms regulating gene expression during diploidization in phylogenetically polyploid lower vertebrates. *Hum. Genet.* **65,** 11–18.

Lenormand, T., Guillemaud, T., Bourguet, D., and Raymond, M. (1998). Appearance and sweep of a gene duplication: Adaptive response and potential for new functions in the mosquito *Culex pipiens. Evolution* **52,** 1705–1712.

Lewin, B. (1994). "Genes V." Oxford University Press, Oxford, U.K.

Liang, F., Holt, I., Pertea, G., Karamycheva, S., Salzberg, S. L., and Quackenbush, J. (2000). Gene index analysis of the human genome estimates approximately 120,000 genes. *Nature Genet* **25,** 239–240.

Löve, A., Löve, D., and Pichi Sermolli, R. E. G. (1977). "Cytotaxonomical Atlas of the Pteridophyta." J. Cramer, Vaduz, Germany.

Lundin, L. G. (1993). Evolution of the vertebrate genome as reflected in paralogous chromosomal regions in man and the house mouse. *Genomics* **16,** 1–19.

Lynch, M., and Conery, J. S. (2000). The evolutionary fate and consequences of duplicate genes. *Science* **290,** 1151–1155.

Lynch, M., and Force, A. (2000). The probability of duplicate gene preservation by subfunctionalization. *Genetics* **154,** 459–473.

Maeda, N., and Smithies, O. (1986). The evolution of multigene families: Human haptoglobin genes. *Annu. Rev. Genet.* **20,** 81–108.

Maroni, G., Wise, J., Young, J. E., and Otto, E. (1987). Metallothionein gene duplications and metal tolerance in natural populations of *Drosophila melanogaster. Genetics* **117,** 739–744.

Masterson, J. (1994). Stomatal size in fossil plants: Evidence for polyploidy in majority of angiosperms. *Science* **264,** 421–423.

Maynard Smith, J. (1998). "Evolutionary Genetics." Oxford University Press, Oxford, U.K.

McFadden, D., Kwong, L., Yam, I., and Langlois, S. (1993). Parental origin of triploidy in human fetuses: Evidence for genomic imprinting. *Hum. Genet.* **92,** 465–469.

Mittelsten Scheid, O., Jakovleva, L., Afsar, K., Maluszynska, J., and Paszkowski, J. (1996). A change in ploidy can modify epigenetic silencing. *Proc. Natl. Acad. Sci. USA* **93,** 7114–7119.

Mittelsten Scheid, O., Paszkowski, J., and Potrykus, I. (1991). Reversible inactivation of a transgene in *Arabidopsis thaliana. Mol. Gen. Genet.* **228,** 104–112.

Mouchès, C., Pasteur, N., Bergé, J. B., Hyrien, O., Raymond, M., de Saint Vincent, B. R., de Silvestri, M., and Georghiou, G. P. (1986). Amplification of an esterase gene is responsible for insecticide resistance in a California *Culex* mosquito. *Science* **233,** 778–780.

Mukai, T., Chigusa, S. I., Mettler, L. E., and Crow, J. F. (1972). Mutation rate and dominance of genes affecting viability in *Drosophila melanogaster. Genetics* **72,** 335–355.

Nadeau, J. H., and Sankoff, D. (1997). Comparable rates of gene loss and functional divergence after genome duplications early in vertebrate evolution. *Genetics* **147,** 1259–1266.

Nowak, M. A., Boerlijst, M. C., Cooke, J., and Smith, J. M. (1997). Evolution of genetic redundancy. *Nature* **388,** 167–171.

Nur, U. (1980). Evolution of unusual chromosome systems in scale insects (Coccoidea: Homoptera). *In* "Insect Cytogenetics" (R. L. Blackman, G. M. Hewitt, and M. Ashburner, eds.), pp. 97–117. Blackwell Scientific, London.

Ohno, S. (1970). "Evolution by Gene Duplication." George Allen & Unwin, London.

Ohta, T. (1987). Simulating evolution by gene duplication. *Genetics* **115,** 207–213.

Ohta, T. (1988a). Evolution by gene duplication and compensatory advantageous mutations. *Genetics* **120,** 841–847.

Ohta, T. (1988b). Further simulation studies on evolution by gene duplication. *Evolution* **42,** 375–386.

Ohta, T. (1988c). Time for acquiring a new gene by duplication. *Proc. Natl. Acad. Sci. USA* **85,** 3509–3512.

Otto, S. P., and Bourguet, D. (1999). Balanced polymorphisms and the evolution of dominance. *Am. Naturalist* **153,** 561–574.

Otto, S. P., and Goldstein, D. B. (1992). Recombination and the evolution of diploidy. *Genetics* **131,** 745–751.

Otto, S. P., and Whitton, J. (2000). Polyploid incidence and evolution. *Annu. Rev. Genet.* **34,** 401–437.

Pál, C., and Hurst, L. D. (2000). The evolution of gene number: Are heritable and non-heritable errors equally important? *Heredity* **84,** 393–400.

Postlethwait, J. H., Yan, Y. L., Gates, M. A., Horne, S., Amores, A., Brownlie, A., Donovan, A., Egan, E. S., Force, A., and Gong, Z. (1998). Vertebrate genome evolution and the zebrafish gene map. *Nature Genet.* **18,** 345–349.

Ramsey, J., and Schemske, D. W. (1998). Pathways, mechanisms, and rates of polyploid formation in flowering plants. *Annu. Rev. Ecol. Syst.* **29,** 467–501.

Ratcliff, F., Harrison, B. D., and Baulcombe, D. C. (1997). A similarity between viral defense and gene silencing in plants. *Science* **276,** 1558–1560.

Rubin, G. M., Yandell, M. D., Wortman, J. R., Gabor Miklos, G. L., Nelson, C. R., Hariharan, I. K., Fortini, M. E., Li, P. W., Apweiler, R., and Fleischmann, W. (2000). Comparative genomics of the eukaryotes. *Science* **287,** 2204–2215.

Selker, E. U. (1997). Epigenetic phenomena in filamentous fungi: Useful paradigms or repeat-induced confusion? *Trends Genet.* **13,** 296–308.

Selker, E. U. (1999). Gene silencing: Repeats that count. *Cell* **97,** 157–160.

Seoighe, C., and Wolfe, K. H. (1999). Yeast genome evolution in the post-genome era. *Curr. Opin. Microbiol.* **2,** 548–554.

Shaw, R. G., Byers, D. L., and Darmo, E. (2000). Spontaneous mutational effects on reproductive traits of *Arabidopsis thaliana*. *Genetics* **155,** 369–378.

Shigenobu, S., Watanabe, H., Hattori, M., Sakaki, Y., and Ishikawa, H. (2000). Genome sequence of the endocellular bacterial symbiont of aphids *Buchnera* sp. APS. *Nature* **407,** 81–86.

Simmons, M. J., and Crow, J. F. (1977). Mutations affecting fitness in *Drosophila* populations. *Annu. Rev. Genet.* **11,** 49–78.

Skrabanek, L., and Wolfe, K. H. (1998). Eukaryote genome duplication—Where's the evidence? *Curr. Opin. Genet. Dev.* **8,** 694–700.

Sniegowski, P. D., Gerrish, P. J., Johnson, T., and Shaver, A. (2000). The evolution of mutation rates: Separating causes from consequences. *Bioessays* **22,** 1057–1066.

Soltis, D. E., and Soltis, P. S. (1993). Molecular data and the dynamic nature of polyploidy. *Crit. Rev. Plant Sci.* **12,** 243–273.

Soltis, D. E., and Soltis, P. S. (1999). Polyploidy: Recurrent formation and genome evolution. *Trends Ecol. Evol.* **14,** 348–352.

Spofford, J. B. (1969). Heterosis and the evolution of duplications. *Am. Naturalist* **103.**

Stebbins, G. L. (1938). Cytological characteristics associated with the different growth habits in the dicotyledons. *Am. J. Bot.* **25,** 189–198.

Stebbins, G. L. (1980). Polyploidy in plants: Unsolved problems and prospects. *In* "Polyploidy: Biological Relevance" (W. H. Lewis, ed.), pp. 495–520. Plenum, New York..

Takahata, N., and Maruyama, T. (1979). Polymorpism and loss of duplicate gene expression: A theoretical study with application to tetraploid fish. *Proc. Natl. Acad. Sci. USA* **76,** 4521–4525.

Tekaia, F., and Dujon, B. (1999). Pervasiveness of gene conservation and persistence of duplicates in cellular genomes. *J. Mol. Evol.* **49,** 591–600.

Uchida, I. A., and Freeman, V. C. (1985). Triploidy and chromosomes. *Am. J. Obstet. Gynecol.* **151,** 65–69.

Valentine, J. W. (2000). Two genomic paths to the evolution of complexity in bodyplans. *Paleobiology* **26,** 513–519.

Vassilieva, L. L., Hook, A. M., and Lynch, M. (2000). The fitness effects of spontaneous mutations in *Caenorhabditis elegans*. *Evolution* **54,** 1234–1246.

Veiko, N. N., Lyapunova, N. A., Bogush, A. I., Tsvetkova, T. G., and Gromova, E. V. (1996). Ribosomal gene number in individual human genomes—data from comparative molecular and cytogenetic analysis. *Mol. Biol.* **30,** 641–647.

Vision, T. J., Brown, D. G., and Tanksley, S. D. (2000). The origins of genomic duplications in *Arabidopsis*. *Science* **290,** 2114–2117.

Vrijenhoek, R., Dawley, R., Cole, C., and Bogart, J. (1989). A list of known unisexual vertebrates. *In* "Evolution and Cytology of Unisexual Vertebrates" (R. Dawley, and J. Bogart, eds.), pp. 19–23. The University of the State of New York, New York.

Walsh, J. B. (1987). Sequence-dependent gene conversion: Can duplicated genes diverge fast enough to escape conversion? *Genetics* **117,** 543–557.

Walsh, J. B. (1995). How often do duplicated genes evolve new functions? *Genetics* **139,** 421–428.

Wang, J. Y., McCommas, S., and Syvanen, M. (1991). Molecular cloning of a glutathione S-transferase overproduced in an insecticide-resistant strain of the housefly (*Musca domestica*). *Mol. Gen. Genet.* **227,** 260–266.

Watterson, G. A. (1983). On the time for gene silencing at duplicate loci. *Genetics* **105,** 745–766.

Werth, C., and Windham, M. (1991). A model for divergent, allopatric speciation of polyploidy pteridophytes resulting from silencing of duplicate-gene expression. *Am. Naturalist* **137,** 515–526.

Wolf, S., Sharpe, L. T., Schmidt, H. J. A., Knau, H., Weitz, S., Kioschis, P., Poustka, A., Zrenner, E., Lichter, P., and Wissinger, B. (1999). Direct visual resolution of gene copy number in the human photopigment gene array. *Invest. Ophthalmol. Visual Sci.* **40,** 1585–1589.

Wolfe, K. H., and Shields, D. C. (1997). Molecular evidence for an ancient duplication of the entire yeast genome. *Nature* **387,** 708–713.

Wolffe, A. P., and Matzke, M. A. (1999). Epigenetics: Regulation through repression. *Science* **286,** 481–486.

Wu, C. T., and Morris, J. R. (1999). Transvection and other homology effects. *Curr. Opin. Genet. Dev.* **9,** 237–246.

Yong, P. (1998). Theoretical population genetic model of the invasion of an initial duplication, Honour's thesis, Department of Zoology, University of British Colombia, Vancouver.

17 Prions of Yeast as Epigenetic Phenomena: High Protein "Copy Number" Inducing Protein "Silencing"

Reed B. Wickner, Herman K. Edskes, B. Tibor Roberts, Michael Pierce, and Ulrich Baxa
Laboratory of Biochemistry and Genetics
National Institute of Diabetes, Digestive and Kidney Diseases
National Institutes of Health
Bethesda, Maryland 20892

ABSTRACT

Yeast infectious protein (prion) forms of the Ure2 and Sup35 proteins determine the nonchromosomal genes [URE3] and [PSI], and these are, therefore,
the basis for a kind of epigenetic phenomena. In many systems, introduction
of multiple copies of a DNA gene, or dsRNA copies of its sequence, results in
the epigenetic silencing of that gene. In parallel with these homology effects,
which act at the level of DNA or RNA, elevated copy number of the Ure2 and

Sup35 *proteins* increases the frequency of their own "silencing" by prion formation. Both [URE3] and [PSI] appear to be due to self-propagating amyloid formation of Ure2p and Sup35p, respectively. Another prion, [Het-s] of the filamentous fungus, *Podospora anserina,* is necessary for a normal cellular function, heterokaryon incompatibility. Since these prions are nonchromosomal genes, they are proteins acting as genes, a parallel to the fact that nucleic acids can catalyze enzymatic reactions.

I. INTRODUCTION

Not all inherited information is encoded as sequences of chromosomal DNA. In epigenetic phenomena, different phenotypes are found in cells with the same chromosomal constitution. To qualify as epigenetic, the phenotypes must be mitotically or meiotically inherited, and not be merely a response to altered conditions that is reversed upon return to the original environment. Examples include methylation patterns of DNA that, once established, are propagated by DNA methylases specific for hemimethylated sites. Hypermethylated regions are usually less transcriptionally active than the comparable unmethylated sequence. Other epigenetic phenomena are believed to be related to as yet poorly defined heritable "chromatin states." Posttranscriptional gene silencing in plants, fungi, and invertebrates has been found in many cases to involve sequence-specific mRNA degradation in response to multiple copies of endogenous or exogenous genes. (Matzke *et al.,* 2001; Grishok *et al.,* 2001).

A. What is a prion?

The word "prion" means infectious protein. For a protein to be infectious, it must somehow generate more copies of itself and travel from cell to cell and from individual host to individual host. This implies that the infectious protein must (1) encode itself, or (2) affect its own synthesis by the cell, or (3) affect its own form or structure after synthesis. In coining the word "prion," the first alternative ("reverse translation" or "protein-dependent protein synthesis") was suggested as a likely explanation for the mechanism of scrapie (Prusiner, 1982). Griffith, considering the same problem 15 years earlier (Griffith, 1967), imposed the additional condition that the rules of molecular biology not be violated and suggested mechanisms that embrace alternatives (2) and (3) above. One possibility he suggested was that a protein might enhance its own gene's transcription, and thus, once expressed, lead to runaway synthesis of itself, possibly to the cell's detriment. Griffith also suggested essentially what is now the prevailing model for prion propagation. He proposed that an altered conformer of the protein could oligomerize with one subunit of the normal form and, in this oligomer, push the normal form into the abnormal conformation.

To generalize possibility (3), a prion is an altered form of a cellular protein which may have lost its normal function, but has acquired the ability to convert the normal form of the protein into the same abnormal form. This could happen by a self-propagating conformational change, as first suggested by Griffith, or by a covalent alteration. One can imagine that a protein transacetylase or protein kinase normally modifies other proteins in order to regulate some cellular activity, but might also accidentally modify itself, and that the modified protein no longer recognizes other proteins but becomes very efficient at recognizing itself. There are many protein-modifying enzymes, some of which normally modify themselves, but protein modification has not yet been found to be the mechanism of a prion. Note that this mechanism requires some rather special conditions, namely, that the covalent modification of the enzyme makes it prefer its own unmodified self as a substrate. We will argue below that the apparent mechanism of the known prions is more likely to be general.

Prions were first suggested to be the basis of the mammalian transmissable spongiform encephalopathies (TSEs), such as scrapie of sheep, Creutzfeldt-Jakob disease of humans, and much later, mad cow disease (Griffith, 1967; Prusiner, 1982). However, the considerable evidence that these diseases are due to infectious proteins remains less than conclusive. We discovered two prions of yeast, [URE3] and [PSI], based on three unusual genetic properties that they shared, each of which is predicted for a prion, and each of which is incompatible with a nucleic acid replicon, such as a virus or plasmid (Wickner, 1994). Further studies have amply confirmed that [URE3] and [PSI] are indeed prions, and some insights into the mechanisms of their generation and propagation have been obtained. None of these three genetic criteria is known to be true of the mammalian diseases.

B. Proteins can mediate inheritance

It was once thought that all inheritance was carried out by nucleic acids and all enzymatic reactions were carried out by proteins. Cech's discovery that the *Tetrahymena* rRNA intron could self-splice (Kruger *et al.*, 1982) and Altman's revelation that the active part of RNase P was the RNA (Guerrier-Takada *et al.*, 1983) showed that nucleic acids can also carry out enzymatic reactions. It is clear that proteins can carry out inheritance. We showed that the nonchromosomal gene [URE3] is a self-propagating, self-inactivating form of the Ure2 proteins of *Saccharomyces cerevisiae* (and proposed the same explanation for [PSI] as a form of Sup35p) (Wickner, 1994). In essence, this means that proteins can be genes. The "central dogma" remains the rule, but the exceptions are of more than passing interest.

II. THREE GENETIC CRITERIA FOR PRIONS

We proposed that [URE3] is a prion form of Ure2p and [PSI] is a prion of Sup35p based on three genetic properties of these nonchromosomal genetic elements. Each property is expected to be true for a prion, but none is expected for a nucleic acid replicon or for other classes of epigenetic phenomena (Wickner, 1994) (Figure 17.1).

Genetic Properties of a Yeast Prion

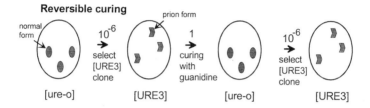

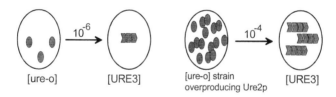

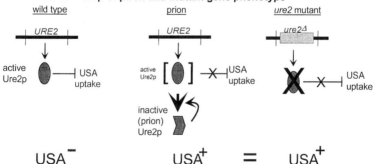

Figure 17.1. Three genetic criteria for a prion. A prion in yeast should, like yeast viruses and plasmids, behave as a nonchromosomal genetic element. Three unusual genetic properties were proposed to distinguish a prion from a nucleic acid replicon (Wickner, 1994). These genetic properties also distinguish a prion from other epigenetic phenomena. The figure outlines experiments showing that [URE3] satisfies all three genetic criteria as a prion of Ure2p. See text for details.

1. *Reversible curing.* Curing a cell of a nucleic acid replicon, such as the mitochondrial DNA or a virus or plasmid, results in a cell which can only reacquire that replicon by the introduction of that relicon from outside the cell. However, since a prion can arise by a spontaneous change of a protein encoded by the cellular genome, a strain cured of a prion can (at some low frequency) give rise to offspring which have again spontaneously developed the prion without the prion having been reintroduced from outside. The fact that prions can be cured without residual mutations in the gene for the protein also distinguishes them from repeat-induced mutation.

2. *De novo prion formation induced by overproduction of the protein.* A nucleic acid replicon will not arise *de novo* at a detectable frequency. Furthermore, overproduction of a cellular component on which such a nucleic acid replicon may depend will not change this fact. In contrast, overproduction of a protein that can undergo a prion change will increase the frequency with which that change occurs, simply because there will be more of the protein to undergo the change, and once the change has occurred, it should spread through the population of molecules. This relationship between the frequency appearance of the prion and the overproduction of the *protein,* as opposed to the mRNA, describes a critical distinction between prion biology and posttranscriptional gene silencing, where the production of excess RNA is the cause of the gene silencing. The key role of protein here also distinguishes prions from the RIP (repeat-induced point mutation) of *Neurospora,* in which high gene copy number is the inducer of silencing (reviewed in Hagemann and Selker, 1996).

3. *Phenotype relationship of mutants in the chromosomal gene for the protein and the presence of the prion.* Inactivation of a chromosomal gene for a protein needed for a nucleic acid replicon (such as a DNA polymerase needed for the replication of mitochondrial DNA) produces a phenotype that is similar to that caused by the loss of that replicon (inability to grow on glycerol as a carbon source in this case) and opposite to the phenotype produced by the presence of that replicon. Deletion of the chromosomal gene encoding a protein capable of undergoing a prion change should result in an inability to propagate that prion, because propagation of the prion depends on the continued conversion of the normal form to the abnormal form. Abrogation of the supply of the normal form will result in the abnormal form being gradually diluted out and eventually lost. The phenotype of a deletion in the gene for the protein is expected to be similar or identical to that caused by the presence of the prion. This similarity arises because in each case the normal form of the protein, and therefore its function, is absent or deficient; deletion of the gene precludes synthesis of the protein and the presence of the prion form results in the conversion of all or most of the protein to the altered (nonfunctional) form.

III. [URE3] AS A PRION OF THE URE2 PROTEIN

[URE3] was discovered by Francois Lacroute in 1971 as a non-Mendelian genetic element that enabled yeast cells to take up ureidosuccinate (an intermediate in uracil biosynthesis) in spite of the presence of a rich nitrogen source ((Lacroute, 1971; Figure 17.2). To understand this rather arcane phenotype, it is best first to explain the role of Ure2p in the regulation of nitrogen metabolism as elucidated by the studies of Lacroute, Magasanik, Cooper, and others. The relationship of ureidosuccinate and nitrogen metabolism arises from the coincident chemical

The Role of Ure2p in Nitrogen Catabolite Repression

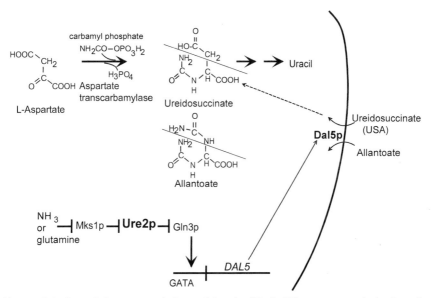

Figure 17.2. Control of nitrogen catabolism and the role of Ure2p. When yeast is supplied with a rich source of nitrogen, such as ammonia or glutamine, it turns off the transcription of genes encoding enzymes and transporters needed for the utilization of poor nitrogen sources (such as allantoate). This phenomenon is called nitrogen catabolite repression, and is executed by a signal transduction cascade involving Mks1p, Ure2p, and Gln3p (reviewed by Magasanik, 1992). Dal5p is the allantoate permease, and is tightly regulated by this system (Turoscy and Cooper, 1987). The chance chemical resemblance of allantoate to ureidosuccinate, an intermediate in uracil biosynthesis, makes uptake of ureidosuccinate subject to the same control system. Thus, inactivation of Ure2p by mutation of the URE2 gene or by a prion change of Ure2p results in uptake of ureidosuccinate even in the presence of ammonia.

resemblance of ureidosuccinate, an intermediate in the uracil biosynthesis pathway, to allantoate, a poor but usable nitrogen source for yeast. This resemblance results in the nitrogen source regulating the uptake of ureidosuccinate.

In the presence of a rich nitrogen source, yeast turns off the syntheseis of enzymes and transporters that are involved in the utilization of poor nitrogen sources (reviewed by Cooper, 1982; Magasanik, 1992). This response to a rich nitrogen source is called nitrogen catabolite repression, and involves a cascade of proteins recognizing the nitrogen state of the cell (Figure 17.2). Ure2p is a key transmitter of the state of the nitrogen supply and it acts by inhibiting Gln3p, a positive transcription factor for a wide array of genes encoding nitrogen utilization factors. Among these factors is Dal5p, the allantoate transporter, which can transport ureidosuccinate as well as its intended substrate. Thus, inactivation of Ure2p results in the constitutive expression of Dal5p and thus the constitutive ability of the cell to take up ureidosuccinate (Rai *et al.*, 1987; Turoscy and Cooper, 1987). Ure2p is also modified and regulated by the TOR (= target of rapamycin) pathway (Cardenas *et al.*, 1999; Hardwick *et al.*, 1999; Beck and Hall, 1999), with phosphorylation and dephosphorylation of Ure2p controlled by this system. However, the effects of TOR control on the [URE3] prion or on the nitrogen catabolite repression pathway, as classically defined, are not yet clear.

Aigle and Lacroute found that [URE3] showed irregular segregation in meiosis and was efficiently transmitted by cytoplasmic transfer (cytoduction) (Aigle and Lacroute, 1975). They found that *ure2* mutants were unable to maintain and replicate [URE3], and yet the phenotype of *ure2* mutants was nearly identical to that of [URE3] strains. This later proved to be the first clue that [URE3] was a prion (Wickner, 1994).

A. [URE3] satisfies the three genetic criteria for a prion

[URE3] can be efficiently cured by growth of cells in the presence of 3 mM guanidine (Aigle, 1979; Wickner, 1994). However, it is possible to isolate from the cured strains subclones that have again developed [URE3] *de novo* (Wickner, 1994). The frequency with which [URE3] arises in the cured strains is similar to the 10^{-6} frequency with which [URE3] arises in the original parent strain. Thus [URE3] is reversibly curable. It is important to note that it is not the fact that guanidine cures [URE3] that indicates it is a prion. Poliovirus replication was found to be inhibited by similar low concentrations of guanidine, an effect that proved to be due to an action on the virus-encoded RNA-dependent RNA polymerase (Tamm and Eggers, 1963). Rather, it is the fact that from the cured cells could be isolated rare subclones in which [URE3] had spontaneously reappeared.

Overproduction of the Ure2 protein increases the frequency with which [URE3] arises (Wickner, 1994) (from a spontaneous rate of $1–3 \times 10^{-6}$). The rate increases to $100–200 \times 10^{-6}$ when Ure2p is overproduced. Even higher rates are

Table 17.1. Comparison of Prions with Other "Homology-Dependent Inactivation" Systems

Phenomenon	Inducer	Mechanism	Organisms
Repeat-induced methylation	High **gene copy** inactivates **gene**	Methylation of repeated sequences	*Neurospora crassa*, flowering plants
Posttranscriptional gene silencing	High **transcript copy** inactivates (degrades) **transcripts**	dsRNA or overproduced transcripts is degraded, combines with protein to make sequence-specific RNAse	*Caenorhabditis*, *Drosophila*, *Arabidopsis*
Prions	High **protein copy** inactivates **protein**	Amyloid formation inactivates normal protein function	*Saccharomyces cerevisiae*, *Podospora anserina*

found upon overexpression of some fragments of Ure2p (see below). It was critical to show that it was overproduction of the Ure2 *protein* and not the presence of the gene in high copy number or the overproduction of the *URE2* mRNA that resulted in the appearance of [URE3]. This was done by showing that the reading frame of the prion domain, and not any particular part of the RNA, was critical for prion generation and propagation (Masison *et al.*, 1997). Moreover, the presence of the gene in high copy number does not itself affect the frequency with which [URE3] arises. Note that this result identifies [URE3] as a prion and distinguishes it from many other possible epigenetic mechanisms (Table 17.1).

The third criterion for a prion is also satisfied by [URE3]. [URE3] requires *URE2* for its propagation (Aigle and Lacroute, 1975; Wickner, 1994), and yet the phenotypes of *ure2* mutants are indistinguishable from those of [URE3] strains.

B. Ure2p is protease-resistant in [URE3] strains

The first biochemical clues of the mechanism of [URE3] was the finding that Ure2p, in extracts of [URE3] strains, was partially protease resistant compared to extracts of normal strains (Figure 17.3). Ure2p in normal extracts was completely digested at the earliest time point. However, Ure2p in [URE3] extracts was digested to form 30-kDa and 32-kDa transient species and then a stable 7–10-kDa species (Masison and Wickner, 1995). This pattern of digestion indicates that the prion form of Ure2p has a protease-resistant core and a protease-sensitive domain. In fact, the specificity of the antibody used in these experiments indicates that the prion domain (see below) is included in the protease-resistant part of Ure2p (Masison and Wickner, 1995).

C. Definition of the Ure2p prion domain

Deletion analysis showed that the N-terminal 65-amino acid residues of Ure2p were sufficient, when overexpressed, to induce the appearance of the [URE3]

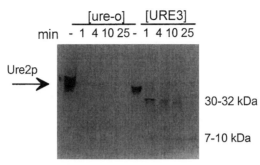

Figure 17.3. Ure2p is protease-resistant in [URE3] strains. Extracts of wild type ([ure-o]) and [URE3]
strains were treated for the indicated times with proteinase K, and aliquots were removed,
treated with protease inhibitor and quick frozen. Antibody to Ure2p was used to detect
the protein and its fragments on a western blot (Masison and Wickner, 1995).

prion at high efficiency (Masison and Wickner, 1995). In fact, Ure2p$^{1-65}$ induces
de novo [URE3] formation at a frequency approximately 100 times higher than
that obtained by overexpression of the full-length Ure2p. Further analysis has
shown that Ure2p$^{1-80}$ is somewhat better at [URE3] induction than is Ure2p$^{1-65}$
(Masison *et al.*, 1997; Maddelein and Wickner, 1999). Neither Ure2p$^{1-65}$ nor
Ure2p$^{1-80}$ can complement a *ure2*Δ mutant for the nitrogen regulation function
of the Ure2 protein. In contrast, deletion of residues 3–65 or residues 2–80 results
in a C-terminal domain that can fully complement *ure2*Δ, but that cannot induce
the appearance of [URE3], even when overexpressed (Masison and Wickner, 1995;
Maddelein and Wickner, 1999).

Indeed, Ure2p$^{1-65}$ is capable of propagating [URE3] in the complete absence of the C-terminal domain (Masison *et al.*, 1997). In contrast, [URE3] is
not propagated in a strain expressing only the C-terminal part of the molecule.
These results justify considering the N-terminal 65 or 80 residues as the prion
domain of Ure2p and the remaining C-terminal part as the nitrogen-regulation
domain.

As can be seen in Figure 17.4, the Ure2 prion domain is composed of 40%
asparagine residues and 20% serine + threonine residues. As we will see below, this
is amino acid composition similar to that of the prion domain of Sup35p, which is
rich in asparagine and glutamine, but differs from PrP (the putative mammalian
prion protein), which has no Asn or Gln-rich regions.

Further deletion analysis of *URE2* showed that residues 66–80 could
promote prion formation in the absence of residues 1–65, if other deletions of
prion-inhibiting regions were incorporated into the molecule (Maddelein and
Wickner, 1999). Moreover, a region in the middle of the C terminus, that has
neither asparagine nor glutamine residues, is important for the prion-inducing
activity of the otherwise full-length Ure2p (Maddelein and Wickner, 1999).

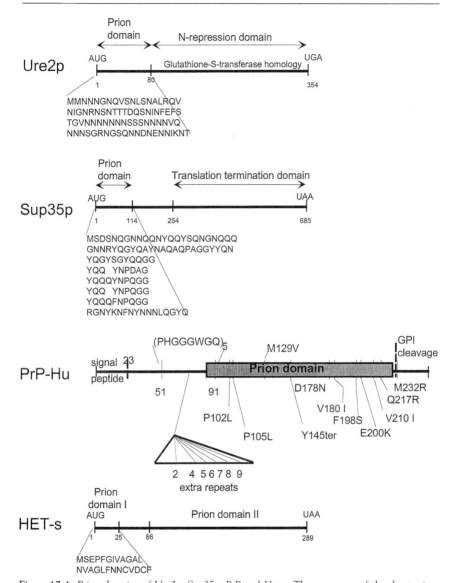

Figure 17.4. Prion domains of Ure2p, Sup35p, PrP and Het-s. The sequences of the short prion domains of Ure2p, Sup35p, and Het-s are shown. Het-s has two prion domains and the ure2p prion domain can be divided in two with each part active. The minimum extent of the PrP prion domain is not clearly defined, but it is within the indicated region. The mutations shown for PrP are those that produce inherited Creutzfeldt-Jakob disease (and its variants) in people. PrP-Hu is the human PrP protein.

D. Further genetic evidence for [URE3] as a prion of Ure2p

Since Ure2p is known to be a regulator of gene expression, a self-reinforcing
positive-feedback loop was among the alternative hypotheses to explain the
properties of [URE3]. Such a cycle has been demonstrated for *Escherichia coli*
growing on limiting lactose (Novick and Weiner, 1957) and for sex determination in *Drosophila* (Bell *et al.*, 1991). Several lines of evidence argue against this
possibility: (1) the nitrogen regulation domain and the prion domain of Ure2p
are nonoverlapping (Masison and Wickner, 1995); (2) Ure2p does not become
protease-resistant when cells are grown on a poor nitrogen source, so this change
is not one that normally accompanies derepressed nitrogen catabolism (Masison
et al., 1997); (3) the nitrogen source does not affect the frequency with which
[URE3] arises (Masison *et al.*, 1997).

In a further experiment to examine the nature of the prion change, the
prion domain and the nitrogen-regulation domain were expressed within the same
cell (Masison *et al.*, 1997). The method used involved passing [URE3]-containing
cytoplasm into the strain to be tested, growing the strain to be tested and examining its phenotype, and then passing the cytoplasm to a strain not carrying the
prion (designated [ure-o]) as a test of whether [URE3] had been propagated. It was
found that if only the C-terminal nitrogen-regulation domain was expressed, then
cells could not propagate [URE3], and even introducing [URE3] into such cells
by cytoplasmic transfer did not result in inactivation of the C-terminal domain.
Cells expressing only the N-terminal prion domain were derepressed for nitrogen regulation because they lacked the C-terminal domain of Ure2p. However,
such strains did not have [URE3]-donating ability. When [URE3] was introduced
into such a strain by cytoplasmic mixing, the phenotype did not change, but
now these cells could donate [URE3] to the initially [ure-o] cells. When both the
N-terminal and C-terminal fragments were expressed as separate molecules in the
same cell, they appeared to act independently. These cells initially were repressed
for nitrogen catabolism, indicating that the C-terminal fragment was active.
When [URE3] was introduced by cytoplasmic transfer, cells remained repressed
for nitrogen catabolism, but they were able to donate [URE3] to the [ure-o] cells.
This showed that [URE3] is not necessarily connected with the state of nitrogen
regulation, further proving that [URE3] is not a regulatory loop. It also showed
that the inactivation of the Ure2p C-terminal domain requires the covalent attachment of the prion domain (Masison *et al.*, 1997).

E. [URE3] really arises *de novo*

It was particularly important to show that [URE3] arises *de novo* and was not
present as a normal inapparent plasmid before the selection for [URE3] cells.
For example, the mitochondrial genome can give rise to dominant ρ^- mutants

(called suppressive petites) that are apparent because they eliminate the normal mitochondrial DNA genome. One could suppose that [URE3] is analogous to these ρ^- mutants and there is a normal nucleic acid genome (plasmid) of which [URE3] is a mutant. In this model, *ure2* mutants lose this normal plasmid and so produce their phenotype, while [URE3] strains eliminate this plasmid by competition and produce the same phenotype. This model would predict that replacing the *URE2* gene in a *ure2* mutant would not restore its phenotype, contrary to the observations of Coschigano and Magasanik in cloning the gene (Coschigano and Magasanik, 1991). Moreover, it should also be impossible to select [URE3] derivatives in such a *ure2* mutant complemented by the normal *URE2* gene. It was shown, however, that [URE3] derivatives of such a strain could be readily obtained, and that they had all the genetic properties of other [URE3] strains (Masison *et al.*, 1997).

IV. SELF-PROPAGATING Ure2p AMYLOID IS THE [URE3] PRION

A. Amyloid formation *in vitro* by Ure2p promoted by the prion domain

The synthetic Ure2 prion domain peptide (residues 1–65) spontaneously forms amyloid filaments when diluted from a denaturant into neutral buffered solution (Taylor *et al.*, 1999; Figure 17.5). These filaments are about 45 Å in diameter, have over 60% β-sheet structure, are resistant to digestion with proteinase K, and give yellow-green birefringence upon staining with Congo Red—all the properties of amyloid (Taylor *et al.*, 1999). Ure2p is found as a soluble dimer in extracts of normal yeast and is quite stable in buffers of pH 8 and with 0.2 M or higher salt, but forms nonspecific (nonfilamentous) aggregates at lower pH or lower ionic strength (Taylor *et al.*, 1999). Addition of the synthetic prion domain peptide Ure2p$^{1-65}$ to the soluble native Ure2p induces formation of a co-filament consisting of a 1:1 mixture of Ure2p$^{1-65}$ and full-length Ure2p. These co-filaments average 200 Å in diameter and likewise have all of the properties of amyloid (Taylor *et al.*, 1999). Small amounts of the co-filaments can seed amyloid formation by an excess of the native Ure2p, showing that the amyloid formation is self-propagating (Figure 17.5). The co-filaments and seeded filaments of Ure2p show high β-sheet content, yellow-green birefringence upon staining with Congo Red, and a pattern of protease-resistant fragments (Taylor *et al.*, 1999) similar to that seen for Ure2p in crude extracts of cells carrying the [URE3] prion (Masison and Wickner, 1995). In each case the most resistant fragment is a 7–10-kDa species that is detected specifically by antibody to the N-terminal prion domain.

That this *in vitro* amyloid formation is the explanation for the *in vivo* prion phenomenon is indicated by several lines of evidence. First, it is the same part of Ure2p that induces prion formation *in vivo* that induces amyloid formation *in vitro*. Other fragments of Ure2p that are inactive in prion induction *in vivo*

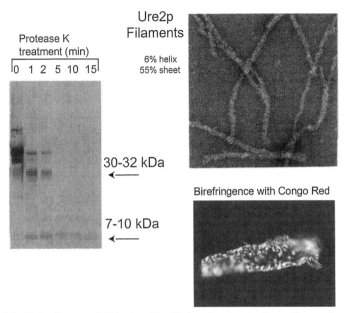

Figure 17.5. Ure2p forms amyloid *in vitro*. The Ure2p prion domain directs filament formation by full-length native Ure2p (upper right) to produce partially protease-resistant material (left) that produces yellow-green birefringence on staining with Congo Red (lower right) (Taylor *et al.*, 1999).

are inactive in inducing amyloid formation *in vitro*. The C-terminal nitrogen-regulation domain of Ure2p (Ure2cp) is not converted to amyloid by Ure2p$^{1-65}$, just as Ure2cp is not converted to the prion form *in vivo* by overexpression of the same prion domain. Another indication of the specificity of amyloid formation is that amyloid formed *in vitro* by the Aβ protein does not incorporate Ure2p into co-filaments. The second line of evidence connecting the amyloid with the prion is that Ure2p is aggregated in cells carrying [URE3], as shown by the Ure2p-GFP fusion studies, and it is the prion domain that directs this aggregation (Edskes *et al.*, 1999a). Finally, the pattern of protease-resistant fragments of Ure2p that is seen in extracts of [URE3] cells (Masison and Wickner, 1995) is the same as that produced from Ure2p amyloid made *in vitro* (Taylor *et al.*, 1999).

B. Direct evidence for Ure2p amyloid filaments in [URE3] cells

Recently, it has been possible to actually detect filaments—apparently amyloid filaments—in [URE3] cells (Speransky *et al.*, 2001). Speransky *et al.* examined thin sections of [URE3] cells fixed with glutaraldehyde and stained with osmium

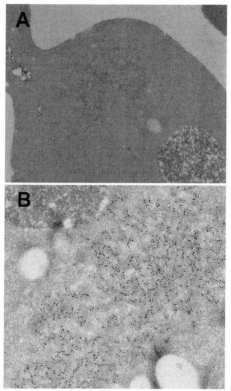

Figure 17.6. Networks of Ure2p *in vivo* in [URE3] strains. (A) Transmission electron microscopy of thin sections of [URE3] strains show filamentous networks not seen in [ure-o] cells (Speransky *et al.*, 2001). (B) Immunogold labeling shows that these filaments contain Ure2p (Speransky *et al.*, 2001).

tetroxide and uranyl acetate by electron microscopy. They observed large networks of filaments in [URE3] cells but not in [ure-o] cells (Figure 17.6). They showed that these filaments were composed of the Ure2 protein by immunogold labeling, using an antibody specific to the C-terminal domain of Ure2p. Interestingly, an antibody specific for the N-terminal domain only very poorly detected these filaments (Speransky *et al.*, 2001).

When the same [URE3] cells were lysed, a highly aggregated form of Ure2p was detected which, as in the electron microscopic studies, could be detected with antibody to the C-terminal domain of Ure2p, but not with antibody specific for the N-terminal prion domain. These aggregates were not solubilized by boiling in 2% sodium dodecyl sulfate and 3 M urea (Speransky *et al.*, 2001), a property consistent with their being amyloid (Sipe, 1992).

C. The Ure2 prion domain can induce β-sheet structure in another protein

A fusion of the Ure2p prion domain to glutathione-S-transferase (GST of *Schistosoma japonicum*) resulted in the formation of amyloid composed of the fusion protein (Schlumpberger *et al.*, 2000). The secondary structure of the amyloid formed by the fusion protein, measured by Fourier transform infrared spectroscopy, showed 44–49% β-sheet structure, indicating that 130–145 residues had this conformation. On its own, only 8% (19 residues) of GST forms a β-sheet structure. Thus, even assuming that all of the 65 residues of the Ure2p prion domain are in β-sheet conformation, the Ure2 prion domain has induced β-sheet structure in at least 61 residues of the GST part of the fusion protein (Schlumpberger *et al.*, 2000). The β-sheet structures may not be entirely due to amyloid formation, however, since up to 100 residues of the soluble form of the fusion protein was in β-sheet form. Further, the similarity sequence of the Ure2 nitrogen regulation domain sequence to that of GSTs (Coschigano and Magasanik, 1991), and the very close resemblance of the Ure2p nitrogen-regulation domain X-ray structure to that of several GSTs (Bousset *et al.*, 2001; Umland *et al.*, 2001), means that the GST of the fusion protein in the above experiments is not necessarily acting as a heterologous reporter.

D. Ure2p nitrogen-regulation domain structure resembles GSTs

The crystal structure of the Ure2p C-terminal fragment (residues 97–354) at 2.3 Å resolution has recently been determined (Bousset *et al.*, 2001; Umland *et al.*, 2001) (Figure 17.7; see color insert). Ure2p(97–354) has the fold of the glutatione S-transferase family, consistent with its sequence similarity to GSTs. However, the crystal structure demonstrates that Ure2p(97–354) lacks a properly positioned catalytic residue required for S-transferase activity, consistent with the failure of efforts to detect GST activity of Ure2p (Choi *et al.*, 1998; Coschigano and Magasanik, 1991). Replacement of residues His151–Ser158 with an alanine or deletion of the C-terminal residues Val347–Glu354 results in a dramatic increase in the efficiency with which the the remaining molecule induces the *de novo* generation of [URE3] (Masison and Wickner, 1995). The structure shows that each of these regions is close to the N terminus of the fragment and so may interact with the N-terminus, stabilizing it and preventing its conversion to the prion form (Umland *et al.*, 2001). Another region (Ser221–Ile227) eliminates the prion-inducing activity of the otherwise intact Ure2p, while not affecting the nitrogen regulating activity (Maddelein and Wickner, 1999). This segment is in the region of contact between the two monomers, and it is not yet clear how this segment may affect prion formation (Umland *et al.*, 2001). It will be of great importance to determine the sites of interaction of Ure2p with Gln3p and particularly with Mks1p (Edskes and Wickner, 2000; see Figure 17.2 and below). Knowing the complete structure of Ure2p would also be of great importance.

The structure of the prion domain is unknown, but unfolding studies suggest it has little structure (Perrett *et al.*, 1999). Ure2p is a dimer when purified from yeast (Taylor *et al.*, 1999) or from *E. coli* (Perrett *et al.*, 1999). Without the prion domain, Ure2p is still a dimer, and the concentration of guanidine needed to denature the deleted molecule is the same as that needed to denature the full length Ure2p (Perrett *et al.*, 1999). This indicates that the prion domain does not increase the overall stability of the protein. However, it is possible that structure within the prion domain or interaction between the prion domain and the C terminus is such that it is melted by guanidine concentrations that are similar to or lower than that which melts the structure of the C terminus. Indeed, two hybrid studies suggest interaction between C termini as well as between the C terminus and N terminus (Fernandez-Bellot *et al.*, 1999).

V. *MKS1* IS NECESSARY FOR *DE NOVO* GENERATION OF THE [URE3] PRION

A. Mks1p is involved in nitrogen regulation and is controlled by Ras-cAMP

Overexpression of the *MKS1* gene was found to cause derepression of the *DAL5* gene, an indication that Mks1p was involved in nitrogen regulation (Edskes *et al.*, 1999b). This was confirmed by the finding that *mks1Δ* strains were unable to derepress *DAL5* when they should do so, namely, when grown on a poor nitrogen source (proline) (Edskes *et al.*, 1999b). Since *mks1Δ-ure2Δ* double mutants had the phenotypes of *ure2Δ* strains, it was concluded that Ure2p is downstream of Mks1p in the nitrogen-regulation cascade (Edskes *et al.*, 1999b; Figure 17.2). In support of this model, overexpression of Ure2p suppresses the nitrogen derepression effect of overexpression of Mks1p. Mks1p must be an inhibitor of Ure2p, and Mks1p is itself inhibited by a good nitrogen source such as ammonia or glutamine.

The mechanism by which Mks1p inhibits Ure2p is as yet unclear. Mks1p overexpression does not lower the levels of Ure2p, nor does Mks1p visibly affect the pattern of Ure2p degradation products seen in extracts. It is suggested that Mks1p may interact directly with Ure2p, but this has not been demonstrated. Deletion analysis showed that Mks1p residues Met245–Asn340 are necessary to show the effect on nitrogen regulation (Edskes *et al.*, 1999b).

MKS1 was first discovered by Matsuura and Anraku as a gene whose strong overexpression slows cell growth, but this growth inhibitory effect was itself inhibited by the Ras-cAMP pathway (Matsuura and Anraku, 1993). The sequence of the 52-kDa Mks1p was unrevealing, and the mechanism of growth inhibition is not clear. The part of Mks1p necessary for growth inhibition is the same as the part needed for the nitrogen regulation effect (Edskes *et al.*, 1999b). Moreover, the growth inhibition was only seen on rich nitrogen sources, not on poor nitrogen sources (Edskes *et al.*, 1999b), suggesting again that the two effects are related.

MKS1 was also independently discovered as a gene repressing lysine biosynthesis, LYS80 (Feller *et al.*, 1997). The lys80/mks1 mutants have increased pools of lysine, α-ketoglutarate, and glutamate. However, the relation of this phenotype to the nitrogen regulation phenotype is not yet clear.

B. Mks1p is necessary for *de novo* generation of the [URE3] prion

Because Mks1p is apparently adjacent to Ure2p in the nitrogen-regulation cascade, its effects on [URE3] generation and propagation were examined (Edskes and Wickner, 2000). In an *mks1*Δ strain, the frequency of [URE3] arising was reduced to a nearly undetectable level. Even overexpression of Ure2p did not induce the appearance of [URE3] in an *mks1*Δ strain. Modestly increasing the level of Mks1p by introducing the gene on a single-copy plasmid resulted in 10-fold or greater increases in the frequency of [URE3] arising (Edskes and Wickner, 2000).

Although the generation frequency of [URE3] is reduced more than 1000-fold in *mks1*Δ strains, its propagation is unaltered. [URE3] introduced into MKS1 and *mks1*Δ strains shows no differences in stability or the concentration of guanidine necessary for its curing (Edskes and Wickner, 2000). Mks1p is thus the first protein shown to be necessary specifically for the generation of a prion.

C. The Ras-cAMP pathway controls [URE3] prion generation

The results of Matsuura and Anraku (1993) indicate that Mks1p activity is in some way inhibited by the Ras-cAMP pathway, although it is not clear whether this is by a direct effect, such as phosphorylation of Mks1p by the A-kinase, or by regulation of Mks1p expression. To determine whether, through Mks1p, the Ras-cAMP pathway affects prion generation, an "oncogenic" allele of RAS2, encoding Ras[val19], was introduced into normal cells. It was found that this reduced the frequency of [URE3] arising almost as dramatically as did a deletion of the MKS1 gene itself (Edskes and Wickner, 2000). This result shows that regulatory pathways can have a dramatic impact on prion generation. However, the mechanism by which Mks1p, and through it the Ras-cAMP pathway, can affect prion generation remains to be determined.

VI. CHAPERONES AFFECTING [URE3] PROPAGATION

A. Hsp104 is necessary for propagation of [URE3]

The involvement of chaperones in prion propagation was first demonstrated for [PSI] by Chernoff *et al.* (1995), and will be discussed below in detail. A search for genes involved in [URE3] propagation led to the finding that mutation of *HSP104* or deletion of the gene resulted in complete loss of [URE3] (Moriyama *et al.*, 2000). Overproduction of Hsp104 had no effect on the expression or propagation of [URE3] (Moriyama *et al.*, 2000); in contrast, overproduction of Hsp104 cures [PSI] (Chernoff *et al.*, 1995). The mechanism of the requirement for Hsp104 is not yet clear, and will be further discussed below.

B. Overproduction of Ydj1p cures [URE3]

Chaperones act together to prevent denaturation or to renature proteins. Among the Hsp40 group is Ydj1, a homolog of the *DnaJ* protein of *E. coli* (Lu and Cyr, 1998; Meacham *et al.*, 1999). Ydj1p can act with Hsp70s to block aggregation and to restore activity to chemically denatured proteins (Lu and Cyr, 1998). Phenotypes of *ydj1* mutants include effects on import of proteins into mitochondria, secretion of mating pheromones, and regulation of the activity of the cytoplasmic Hsp70s (Caplan and Douglas, 1991; Atencio and Yaffe, 1992; Becker *et al.*, 1996; Meacham *et al.*, 1999; Lu and Cyr, 1998).

Although overexpression of Hsp104 does not cure [URE3], it was found that overexpression of Ydj1p does result in the complete loss of this prion (Moriyama *et al.*, 2000). Because *ydj1*Δ mutants grow very slowly and the presence of [URE3] slows the growth of any strain, it was impractical to examine whether Ydj1p is necessary for [URE3] propagation.

As discussed below and shown in Table 17.2, and Fig. 17.8, the effects of different chaperones on different prions vary, and there is at present little insight into the nature of this specificity. Certainly the amyloid form of Ure2p found in [URE3] cells is more protease-resistant, and probably more stable to denaturants (Masison and Wickner, 1995; Speransky *et al.*, 2001) than are the aggregates

Table 17.2. Prion Systems in Mammals, Yeasts, and Fungi

Prion	Hosts	Manifestations	Protein
Transmissible spongiform encephalopathies	Mammals	Neurodegeration, death	PrP
[URE3]	*Saccharomyces cereivisae*	Derepressed nitrogen catabolism	Ure2p
[PSI]	*Saccharomyces cereivisae*	Translation termination read-through	Sup35p
[Het-s]	*Podospora anserina*	Loss of heterokaryon incompatibility	HET-s

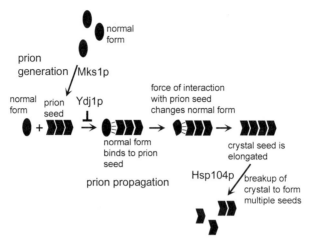

Figure 17.8. Role of chaperones and Mks1p in [URE3] generation and propagation.

of Sup35p in [PSI+] cells (Paushkin *et al.*, 1996). This difference in the self-propagating aggregates may be the basis for their different responses to various chaperones.

VII. [PSI] AS A PRION OF THE SUP35 PROTEIN

[PSI] was discovered in 1965 by Brian Cox as a nonchromosomal genetic element that increases the efficiency of a weak nonsense suppressor, SUQ5 (Cox, 1965; Cox *et al.*, 1988). SUQ5 is a serine-inserting ochre-suppressor tRNA (Liebman *et al.*, 1975; Waldron *et al.*, 1981), but [PSI] is capable of enhancing the efficiency of amber and opal suppression as well (Palmer *et al.*, 1979). While [PSI] makes weak suppressors strong, it can make strong suppressors lethal (Cox, 1971), and has some suppressor activity in strains with no identifiable suppressor mutation (Liebman and Sherman, 1979). [PSI] may be cured by growing cells in high-osmotic-strength medium (Singh *et al.*, 1979), but the cured strains can give rise to isolated rare subclones which have again acquired [PSI] (Lund and Cox, 1981). This result was interpreted to mean that [PSI] had not really been cured by high osmotic strength. However, attempts to associate [PSI] with any of the known nucleic acid replicons were unsuccessful (Leibowitz and Wickner, 1978; Tuite *et al.*, 1982).

The *sup35* and *sup45* mutations were identified as "omnipotent suppressors," capable of bypassing mutations in any nonsense codon. The products of the *SUP35* and *SUP45* genes were identified as the subunits of the translation termination factor, whose function is to recognize the termination codon, and release the completed peptide from the final tRNA (Zhouravleva *et al.*, 1995; Stansfield *et al.*, 1995).

A. [PSI] satisfies the genetic criteria for a prion of Sup35p

As noted above, although [PSI] is cured by growth in high-osmotic-strength medium, the cured strains can again become [PSI+]. Thus [PSI] is reversibly curable. The *SUP35* gene on a multicopy plasmid was found to induce the *de novo* appearance of [PSI+] (Chernoff *et al.*, 1993), and this effect was shown to be due to overproduction of the Sup35 protein, not the presence of the gene in high copy number or overexpression of the *SUP35* mRNA (Derkatch *et al.*, 1996; Sparrer *et al.*, 2000). Finally, the phenotype of [PSI+] strains is essentially the same as that of recessive *sup35* mutants, and the *SUP35* gene is necessary for the propagation of [PSI+] (Doel *et al.*, 1994; TerAvanesyan *et al.*, 1994). Thus, the relation of [PSI+] to Sup35p is the same as that of [URE3] to Ure2p, and [PSI] satisfies the genetic criteria for a prion (Wickner, 1994).

B. The prion domain of Sup35p is glutamine-rich

Deletion analysis of the *SUP35* gene showed that the amino-terminal 114 residues are essential for propagation of [PSI] (TerAvanesyan *et al.*, 1994; Figure 17.4). The same region is sufficient, when overproduced, to induce the *de novo* appearance of the [PSI] prion (Derkatch *et al.*, 1996). This part of the molecule is very rich in glutamine residues and has repeat sequences reminiscent of those found in the N terminus of PrP (see Figure 17.4). The glutamine residues were found to be particularly critical for prion induction and prion propagation (DePace *et al.*, 1998). An important role for the repeats was also demonstrated by showing that deletion of repeats decreased activity while adding extra repeats increased prion activity (TerAvanesyan *et al.*, 1994; Liu and Lindquist, 1999).

C. Self-propagating aggregation of Sup35p primed by extracts of [PSI+] cells

In extracts of normal cells, Sup35p is largely soluble and is associated with Sup45p as a heterodimer (Stansfield *et al.*, 1995; Paushkin *et al.*, 1996). However, in extracts of [PSI+] strains, Sup35p is easily pelleted by centrifugation, indicating that it is in an aggregated form (Paushkin *et al.*, 1996). Consistent with these findings, examination of the cellular distribution of Sup35–GFP fusion proteins revealed an even distribution in normal cells, but a punctate distribution in [PSI+] cells (Patino *et al.*, 1996).

The aggregation of Sup35p was self-propagating *in vitro*, since adding a small amount of aggregated Sup35p from a [PSI+] strain to an extract of a [psi−] strain resulted in aggregation of the Sup35p from the latter (Paushkin *et al.*, 1997).

This strongly indicated that the basis of the [PSI] prion was the self-propagating aggregation of Sup35p.

D. Amyloid formation by Sup35p prion domain *in vitro*

A peptide comprising residues 2-114 of Sup35p, essentially all of the prion domain, was found to form filaments spontaneously *in vitro* at pH 2 in 0.1% trifluoroacetic acid and 40% acetonitrile (King *et al.*, 1997). These filaments were high in β-sheet content as judged by circular dicroism. The filaments were also more protease-resistant than amorphous aggregates formed from the same peptide. Finally, the filaments showed the green-yellow birefringence on staining with Congo Red that is characteristic of amyloid (King *et al.*, 1997). The formation of these filaments could be promoted by addition of preformed filaments to the solution, showing that this is a self-seeding process. These filaments thus satisfy all of the criteria to be amyloid.

In another study, it was shown that full-length Sup35p could form filaments, as could any fragment of Sup35p that included the prion domain, but not fragments lacking the prion domain (Glover *et al.*, 1997), thus supporting the connection between the formation of filaments and prion propagation. In this study, too, the filament formation was found to be self-seeding, but it was of particular interest that extracts of [PSI+] strains could seed filament formation, but extracts of [psi−] cells could not, again suggesting that the filament formation reflects the *in vivo* prion propagation process (Glover *et al.*, 1997). The filaments were shown to be high in β-sheet content, consistent with being amyloid filaments.

E. Sup45p overexpression blocks induction of [PSI+]

Sup35p is normally present as a heterodimer with Sup45p (Zhouravleva *et al.*, 1995; Stansfield *et al.*, 1995). When overproduced, Sup35p induces the *de novo* formation of [PSI+] at rates 100-fold or more the spontaneous rate (Chernoff *et al.*, 1993; Derkatch *et al.*, 1996). However, when Sup45p is simultaneously overproduced, Sup35p loses most of its ability to induce the appearance of [PSI+] (Derkatch *et al.*, 1998). The authors suggest that the complex of Sup35p with its normal partner, Sup45p, stabilizes it in the normal form and thus lowers the probability that it will convert to the prion form.

VIII. CHAPERONES AND [PSI+]

A. Hsp104 levels are critical for propagation of [PSI+]

Chernoff *et al.* made the critical discovery that plasmids overproducing Hsp104 can diminish the [PSI+] phenotype and can cure the [PSI+] genetic element (Chernoff and Ono, 1992; Chernoff *et al.*, 1995). In addition, the deletion of the

HSP104 gene has the same effect as overproduction of the Hsp104 protein, namely, complete loss of [PSI+] (Chernoff *et al.*, 1995). These findings were of great importance in three ways: (1) they supported the prion model for [PSI] because chaperones are not generally known to be involved in the propagation of plasmids or viruses; (2) they suggested that the mechanism of prion propagation was altered conformation of Sup35p, since chaperones could be expected to affect protein conformation but not covalent changes; and (3) they suggested an approach for the treatment of prion diseases. Hsp104 has since been found to be essential for other prions, including [PIN+] (Derkatch *et al.*, 1997, 2000), [RNQ1] (Sondheimer and Lindquist, 2000), and [URE3] (Moriyama *et al.*, 2000), suggesting that Hsp104-mediated prion propagation may be a general phenomenon. There are no known mammalian Hsp104 homologs.

The mechanisms by which deficiency or excess Hsp104 cures strains of a prion have not yet been established. Because [PSI] and [URE3] involve aggregation of Sup35p and Ure2p, respectively, and Hsp104 is known to be able to disaggregate proteins (Parsell *et al.*, 1994), it is presumed that overproduction of Hsp104 cure strains by disaggregation, which then allows renaturation of the proteins. It has been suggested that Hsp104 can catalyze the partial unfolding of Sup35p, a process that may be necessary for the conversion of Sup35p to the prion form.

B. Hsp70s are necessary for [PSI] stability

Although overexpression of Hsp104 from an artificial promoter results in the curing of [PSI], exposing cells to heat shock does not (Cox *et al.*, 1988). One explanation for this apparent discrepancy is the finding that overexpression of Ssa1p, a member of the Hsp70 family of chaperones, decreases the efficiency of curing of [PSI] by overexpression of Hsp104 (Newnam *et al.*, 1999). *SSA1* is a member of the *SSA* family of genes encoding Hsp70s (*SSA1, 2, 3, 4*), which are very closely related by in sequence but differ substantially from each other in their regulation (Stone and Craig, 1990). Deletion of all of these genes is lethal (Craig, 1992), and deletion of the most active genes, *SSA1* and *SSA2*, does not result in loss of [PSI+] (Jung *et al.*, 2000). However, a missense mutation in *SSA1* resulting in loss of [PSI+] has been described, indicating an essential role for these Hsp70s in [PSI] prion propagation (Jung *et al.*, 2000). This mutant has normal levels of Hsp104 and appears to maintain the normal regulation of heat-shock proteins, suggesting that the instability of [PSI+] may be due to a direct effect of this Hsp70 on Sup35p.

The *SSB* gene family also encodes Hsp70 chaperones, in this case nonessential proteins thought to be involved in helping the cotranslational folding of nascent proteins (Nelson *et al.*, 1992; Pfund *et al.*, 1998). The modest effects of Ssb proteins is opposite to that of the Ssa proteins, with deletion decreasing the effectiveness of Hsp104 overproduction in curing [PSI] and overproduction of Ssb proteins helping Hsp104 to cure (Chernoff *et al.*, 1999). Overproduction of Ssb proteins alone does not cure [PSI].

Table 17.3. Chaperone Involvement in [URE3] and [PSI] Prions

	[URE3]	[PSI+]
Hsp104p	Excess → no effect	Excess → curing
	Δ → curing	Δ → curing
Ydj1p Hsp40	Excess → curing	Excess → no effect
Ssa Hsp70s	??	Required
Ssb Hsp70s	??	Excess → destabilize
		Δ → no effect

C. Ydj1p overproduction does not cure [PSI+]

Although overproduction of Ydj1p completely cures [URE3] (Moriyama *et al.*, 2000), [PSI+] is not detectably cured by the same treatment (D. Masison, personal communication). This suggests differences in the structures of the [URE3] and [PSI+] prion forms and in the way the cell treats these abnormal materials.

In summary, the role of chaperones in [PSI] and [URE3] propagation differs, but the reasons for the differences remain obscure (Table 17.3).

IX. PORTABLE PRION DOMAINS AND SYNTHETIC PRIONS

The Ure2p and Sup35p prion domains can act independently of their respective C-terminal functional domains (Masison and Wickner, 1995; Masison *et al.*, 1997; TerAvanesyan *et al.*, 1994). In addition, the C-terminal domain is unaffected without a covalently attached N-terminal prion domain (Masison *et al.*, 1997). Recently, it has been shown that the C-terminal functional domain of Sup35p can be used as a reporter for other prion domains, and that the prion domains can be transferred to other molecules to make them into prions. The green fluorescent protein was the first protein to be used as a heterologous reporter for prion change, when it was found that first Sup35p-GFP (Patino *et al.*, 1996) and later Ure2p-GFP (Edskes *et al.*, 1999a) fusion proteins were aggregated in [PSI+]- and [URE3]-carrying cells, respectively. Later, fusions of Sup35p to the glucocorticoid receptor protein were found to show glucocorticoid receptor activity in [psi−] cells, but not in [PSI+] cells, suggesting that the fusion protein, when aggregated due to its N terminus, cannot enter the nucleus to activate transcription (Li and Lindquist, 2000).

The inverse of the above approach is to keep the C-terminal domain constant, using it as a reporter of prion formation promoted by new prion domains. In one application of this approach, the *SUP35* homologs from a number of yeast species were cloned. It was found that in each species, the functional C-terminal domain was highly homologous (Kushnirov *et al.*, 2000; Santoso *et al.*,

2000; Chernoff *et al.*, 2000). The N-terminal domains had little or no homology with the *S. cerevisiae* sequence, but conserved the abundance of glutamine and asparagine residues and the presence of the octapeptide repeats. A fusion of the *Pichia methanolica* N terminus to the *S. cerevisiae* C terminus was shown to have all the properties expected of a prion (Kushnirov *et al.*, 2000; Chernoff *et al.*, 2000). Its overexpression resulted in colonies with a [PSI+]-like phenotype, and this phenotype was transmissible by cytoplasmic mixing to other strains was reversibly cured by guanidine or by deficiency of Hsp104 (Kushnirov *et al.*, 2000; Chernoff *et al.*, 2000). The hybrid protein was aggregated in prion-containing strains, was partially protease-resistant in extracts of these cells and could promote the co-aggregation of the normal form of the protein *in vitro* (Kushnirov *et al.*, 2000). Hybrid proteins with amino termini from *Candida albicans* and *Kluyveromyces lactis* were also shown to have some of the properties expected of [PSI+]-like prions (Santoso *et al.*, 2000).

Cells with the [PSI] of the *Pichia–Saccharomyces* hybrid Sup35p did not inactivate the purely *Saccharomyces* Sup35p, and vice versa (Kushnirov *et al.*, 2000). This work shows that many of the properties of the species barrier described in the mammalian TSEs (reviewed by Prusiner, 1999) can be seen in the yeast prion [PSI].

A. New prion domains among glutamine- and asparagine-rich regions

Because the prion domains of both Ure2p and Sup35p are rich in asparagine and glutamine residues, this feature has been used to search for potential new prion domains. Indeed, it was found that the N-terminal part (residues 8–24) of the Sup35 prion domain could be replaced by a run of glutamine residues (DePace *et al.*, 1998), and polyglutamine alone can promote protein aggregation in yeast (Krobitsch and Lindquist, 2000). Screening asparagine-glutamine-rich regions for the ability to promote aggregation of a fused GFP reporter domain, one gene, named *RNQ1* for "rich in N and Q residues," was found (Sondheimer and Lindquist, 2000). The Rnq1 protein was found to be evenly distributed in some strains and aggregated in others. Transfer of cytoplasm from a strain with the aggregated form to one with the even distribution made the latter change to the aggregated state (Sondheimer and Lindquist, 2000). This indicates that Rnq1p can be a prion. In this case, there is no (known) phenotype associated with either deletion of *RNQ1* or with the presence of the prion form of Rnq1p. Another asparagine–glutamine-rich region, found in the *S. cerevisiae* gene *YPL226w* and named *NEW1*, gave the Sup35 C-terminal region prion character, and so represents a potential prion domain (Santoso *et al.*, 2000).

It has long been noted that there are many natural sequences with runs of asparagine and glutamine, particularly in transcription factors. Which of these can be prion domains? Not all protein fragments found to be prion domains will be capable of becoming prions. It is known that the C-terminal functional domains of Ure2p (Masison and Wickner, 1995) and Sup35p (Kochneva-Pervukhova *et al.*, 1998) each stabilize their respective prion domains, by 100-fold and over 1000-fold, respectively. This suggests that there may well be protein segments capable of acting as prion domains in the absence of the stabilizing influence of their natural environment, but which are incapable of doing so otherwise.

X. CURING OF [URE3] AND [PSI+]

While the mammalian transmissible spongiform encephalopathies are invariably fatal and no cures effective in animals are known, the yeast prions are curable by several means. High osmotic strength eliminates [PSI+], even under conditions which preclude cell growth (Singh *et al.*, 1979). This is perhaps the only true curing method for yeast prions, as other methods may involve blocking prion replication with subsequent dilution of the prion form of the protein as the cells grow.

Growth in the presence of 3–5 mM guanidine HCl results in efficient curing of both [PSI+] and [URE3] (Tuite *et al.*, 1981; M. Aigle, cited in Cox *et al.*, 1988; Wickner, 1994). Although the detailed mechanism of guanidine curing is not yet clear, the kinetics of [PSI+] loss upon treatment with guanidine suggests that prion replication is arrested, and that loss is due to dilution of a limited number of [PSI+] particles as the cells divide (Eaglestone *et al.*, 2000).

The absence of Hsp104 results in curing of both [PSI+] (Chernoff *et al.*, 1995) and [URE3] (Moriyama *et al.*, 2000). This may be related to guanidine curing, since guanidine concentrations sufficient to cure [PSI+] or [URE3] strongly inhibit *in vivo* activity of Hsp104 (Jung and Masison, 2001). Overexpression of Hsp104 cures [PSI+] (Chernoff *et al.*, 1995), but does not cure [URE3] (Moriyama *et al.*, 2000). Overexpression of Ydj1p cures [URE3] (Moriyama *et al.*, 2000), but not [PSI+] (D. Masison, personal communication). Certain Hsp70 mutants (Ssa group) result in loss of [PSI+] (Jung *et al.*, 2000). The basis for the differences between the chaperone requirements between prions is as yet unclear.

Although fusions of Sup35p or Ure2p with GFP have been used to examine the state of aggregation of these proteins in prion-containing cells (Patino *et al.*, 1996; Edskes *et al.*, 1999a), the Ure2-GFP fusion proteins have a tendency, particularly when overproduced, to cure the [URE3] prion (Edskes *et al.*, 1999a). Likewise, fragments of Ure2p (but not the full-length protein), when overproduced, cure [URE3] at high frequency (Edskes *et al.*, 1999a). This is similar to the curing of scrapie-infected tissue culture cells by peptides derived from PrP

(Chabry *et al.*, 1998). It is likely that these fusion proteins and peptides cure by incorporation into the growing filaments, poisoning their growth.

XI. [Het-S], A PRION WITH A NORMAL CELLULAR FUNCTION

When one colony of a filamentous fungus meets another colony of the same strain, cellular processes of the two colonies fuse in order to share nutrients. However, this process is apparently carefully controlled, because there are risks in fusing with a strain which may be infected with a virus that can spread through the fusion process (Begueret *et al.*, 1994). To ensure relatedness between fusing colonies, the fusing cells somehow test the identity of alleles at 8–10 chromosomal loci called *het* loci (reviewed by (Begueret *et al.*, 1994; Saupe, 2000). The first cellular processes that meet carry out a trial fusion and if there is a difference at any one locus, the fused cells degenerate and a barrier to further fusions is established. This fusion phenomenon is called hyphal anastomosis, and the reaction of incompatibility is called vegetative incompatibility.

One such *het* locus is *het-s*, with alleles *het-s* and *het-S* encoding a protein of 289 amino acids that differs at only 13 residues between the alleles (Turcq *et al.*, 1991). In fact, the presence of histidine (*het-S*) or proline (*het-s*) at residue 33 is sufficient to determine the compatibility behavior of a strain (Deleu *et al.*, 1993).

It was found that *het-s* strains could have either of two phenotypes, depending on the presence or absence of a nonchromosomal genetic element called [Het-s] (Rizet, 1952). In the presence of [Het-s], the *het-s* cells could fuse with other *het-s* cells, but not with *het-S* cells. In the absence of [Het-s], *het-s* cells were neutral, fusing equally well with *het-s* and *het-S* cells. [Het-s] was often lost in the products of meiosis, but would arise again frequently in cells lacking it (Beisson-Schecroun, 1962).

Recently, Coustou *et al.* have found evidence that [Het-s] is a prion form of the HET-s protein (the product of the *het-s* allele) (Coustou *et al.*, 1997). The *het-s* gene is necessary for the propagation of the [Het-s] non-Mendelian genetic element, and overproduction of the HET-s protein, increases the frequency with which the [Het-s] element arises *de novo* (Coustou *et al.*, 1997). Moreover, the HET-s protein is more protease-resistant in [Het-s] cells than in cells lacking the non-Mendelian element (Coustou *et al.*, 1997).

Recently, an analysis of the prion and functional domains of HET-s has been carried out (Coustou *et al.*, 1999). As with Ure2p, two nonoverlapping parts of HET-s can serve as prion-inducing domains when overproduced. The N-terminal 25 amino acid residues constitute the shortest prion domain yet. Residues 86–289 can both induce the appearance of [Het-s] and can support the incompatibility reaction (Coustou *et al.*, 1999). Neither of these domains (or any other part of HET-s) has runs of asparagine or glutamine, so the chemical basis for the [Het-s] prion must differ substantially from that of [URE3] and [PSI+].

XII. ARE YEAST PRIONS ADAPTIVE, A CELLULAR STRESS, OR BOTH?

The [Het-s] prion of *Podospora anserina* is plainly necessary in this organism for what is in many fungi a widespread and apparently adaptive phenomenon (Begueret *et al.*, 1994; Coustou *et al.*, 1997; Saupe, 2000). Vegetative incompatibility limits hyphal anastomosis to genetically identical individuals, which are likely already to have the same infectious (and potentially harmful) entities, and therefore pose little new risk. Lack of identity at the *het-s/het-S* locus is insufficient to trigger the incompatibility reaction unless the Het-s protein is in the prion form.

This clear demonstration of a useful prion is a remarkable achievement, possibly setting the stage for other discoveries of prions of selective value. Yeast and other fungi are particularly good systems in which to investigate this possibility because, in addition to their facile genetic systems allowing ready distinction between chromosomal and nonchromosomal traits, they generally have mating or cell–cell fusion modes in which all of the cellular contents of each individual are contributed. In contrast, bacteria generally mate by the exchange of a single strand of DNA through a plasmid-encoded structure called a pilus. Phage infection provides another mode for the transfer of genetic information between individuals. Both of these modes prevent transfer of most of the cellular proteins, so that even if a self-propagating amyloid or other prion structure were to form, it would not be transferred to other cells, would therefore not be infectious, would not likely be detected, and in any case would not be a prion (which must, by definition, be infectious). In animals, the sperm transfer nuclear proteins and a limited number of cytoplasmic proteins, but the fact that animals eat has provided the infectivity route for the TSEs. Most plants do not eat, and pollen is largely limited in content to the donor's genome, but insects provide a possible route for the spread of plant prions as they do for many plant viruses.

The "cortical inheritance" phenomenon of *Paramecium* and other cilliated protozoa (Beisson and Sonneborn, 1965; Grimes and Aufderheide, 1991) is a phenomenon of nonchromosomal inheritance of cellular structure that closely parallels to the prion phenomenon and may not be unique. The surface of *Paramecium* is a structure that includes multiple cillia, asymmetrically shaped and arrayed in a specific asymmetric pattern. Either accidents of mating or purposeful microsurgery can change this pattern in a given individual. Sonneborn showed that this changed pattern was faithfully reproduced in all of the offspring of the original altered cell (Beisson and Sonneborn, 1965). This Lamarkian inheritance is in many ways similar to the yeast prions. The *Paramecia* inherit a self-propagating altered cellular structure, while the *Saccharomyces* inherit a self-propagating altered protein structure. In each case, the alteration can be purposefully induced, in *Paramecium* by microsurgery of the cell cortex, and in *Saccharomyces* by overexpression of Ure2p or Sup35p or fragments thereof. What other cellular structures might be self-propagating?

The conservation among yeast species of the ability of Sup35p to assume a prion form has been interpreted as evidence for the usefulness of the [PSI+] state. However, few would argue that the conservation of the ability of PrP to become a prion suggests that the transmissible spongiform encephalopathies are useful to any of the many mammalian species in which this lethal disease has been observed. Rather, we would suggest that evolution is not a perfect thing, and that the prion domains probably are at least marginally useful under some conditions, and that this provides sufficient selective pressure to maintain them in the population in spite of the occurence of these rare diseases. In the absence of the Ure2p prion domain, the protein is less efficient than normal in carrying out nitrogen regulation, and can only confer the wild-type phenotype when overexpressed. [URE3] cells are slow-growing under most conditions (as are *ure2* mutant cells).

Direct evidence for an advantage of carrying [PSI+] has been sought (Eaglestone *et al.*, 1999; True and Lindquist, 2000). Although one study found an advantage for cells growing at high concentrations of ethanol, this was not consistently seen in another study. In fact, the second study found that there were almost no consistent differences between strain pairs assumed to differ only in the presence or absence of [PSI+]. However, since [psi−] strains were produced by curing of [PSI+] strains by guanidine, which also cures [URE3], [PIN+] (see below), and [RNQ+], the variable results may have been due to the presence in some strains of one or more of these elements (or others as yet undefined).

There is some reluctance to view yeast as having diseases. Yeast retroviruses induce chromosomal rearrangements and gene inactivation and inappropriate activation, and these effects have been interpreted as selected because they allow variability in the offspring. Whether variability can ever be a selected trait is questionable. The myriad of DNA repair and recombination systems certainly do their best to correct DNA mutations, and every geneticist knows that most mutagen-induced mutations are deleterious. The L-A dsRNA virus does allow cells to propagate the M dsRNAs encoding the killer toxins (Wickner, 1996). This is clearly advantageous to individuals carrying the virus and the killer segment, but not to those that do not. Most wild-type yeast strains carry L-A and do not carry a killer toxin-encoding M dsRNA segment. There is no such potential benefit to cells carrying the L-BC dsRNA virus or the 20S or 23S RNA replicons. Each appears to be simply "selfish RNA." As yeast can be infested by these RNA parasites, it can also be parasitized by "selfish proteins."

Indeed, recently Jung *et al.* have found that the presence of [PSI+] leads to an increase in the basal levels of Hsp104 (Jung *et al.*, 2000). This suggests that the cell perceives [PSI+] as a stress. In an *ssal* Δ strain, the presence of [PSI+] results in an increase of the stress-inducible Ssa3/4p, again consistent with the cell's detecting the presence of the [PSI+] prion as a stress (Jung *et al.*, 2000). If the cell views [PSI+] as a stress, who are we to argue?

XIII. [PIN+] IS REQUIRED FOR INDUCIBILITY OF [PSI+]

Overproduction of Sup35p or fragments of Sup35p including the N-terminal prion domain induces the *de novo* appearance of [PSI+] in a wild-type strain (Chernoff *et al.*, 1993; Derkatch *et al.*, 1996; Kochneva-Pervukhova *et al.*, 1998). However, in some strains, the frequency with which [PSI+] arises upon overproduction of Sup35p is very low (Derkatch *et al.*, 1997). Crosses of high-yielding and low-yielding strains showed that the high-yielding strains carry a nonchromosomal genetic element, named [PIN+] (for [PSI] inducibility). Without [PIN+], most Sup35p fragments, or the full-length Sup35p, are unable to induce [PSI+] (Derkatch *et al.*, 1997).

The properties of [PIN+] suggested that it may itself be a prion. Like [PSI+] and [URE3], [PIN+] is cured by growth of cells on media containing 3–5 mM guanidine, but from the cured strains, [PIN+] derivatives may again be obtained (Derkatch *et al.*, 2000). This is the reversible curability property discussed above. In addition, [PIN+] requires Hsp104 for its stable propagation (Derkatch *et al.*, 2000), another property shared with [PSI+] (Chernoff *et al.*, 1995) and [URE3] (Moriyama *et al.*, 2000). However, [PIN+] is not cured by overexpression of Hsp104 (Derkatch *et al.*, 1997), like [URE3] and unlike [PSI+]. This fact also allows the elimination of [PSI+] from a strain to examine whether it has [PIN+]. These several lines of evidence suggested that [PIN+] is a prion, but definitve proof requires knowledge of the gene encoding the protein that changes into the putative prion.

Cells can have any combination of [PSI+] and [PIN+]. In particular, [PSI+] [pin−] strains may be made by partially curing [PSI+] [PIN+] strains with low levels of guanidine. Although there is a statistical tendency for cells to lose both [PSI] and [PIN] or to lose neither, it is not difficult to find, among the treated cells, ones which have lost [PIN+] but remain [PSI+] (Derkatch *et al.*, 2000). Several lines of evidence suggest that [PIN+] is not a property of Sup35p. [PIN] affects neither the level of suppression by [PSI+], nor the stability of the [PSI+] prion (Derkatch *et al.*, 2000). Strains completely lacking the Sup35p N-terminal prion domain can propagate [PIN+] (Derkatch *et al.*, 1997). Moreover, [PIN+] is not lost even when the *Pichia methanolica SUP35* is substituted for the *S. cerevisiae* gene (Chernoff *et al.*, 2000).

There is one curious exception to the requirement for [PIN+] for induction of the appearance of [PSI+]. Constructs made with restriction enzymes accidentally having the C-terminal sequence RVDLQACKLMIQYQRK fused to the Sup35p N-terminal prion domain fragment do not require [PIN+], although even these constructs are more efficient in inducing [PSI+] if they are [PIN+] (Derkatch et al., 2000). Other fortuitous extensions of the Sup35P N-terminal domain are not [PIN+]-independent in induction of [PSI+]. The basis of this phenomenon is not yet clear.

XIV. [KIL-d] IS AN EPIGENETIC PHENOMENON: IS IT A PRION?

The [KIL-d] nonchromosomal genetic element was found in a general screen for mutations affecting the expression and replication of the killer toxin-encoding M_1 dsRNA virus segment of S. cerevisiae (Wickner, 1976). Most yeast strains, including nearly all laboratory strains, carry a double-stranded RNA virus called L-A. L-A's single 4.5-kb segment encodes the viral major coat protein (Gag), and a Gag-Pol fusion protein formed by a -1 ribosomal frameshift. The structure of Gag-Pol ensures the packaging of both the viral RNA-dependent RNA polymerase and a single viral (+) strand, which is then converted within the particle into the dsRNA form (reviewed in Wickner, 1996). This single-segment virus does not substantially affect cell growth or, by itself, confer on the cell any phenotype. However, some strains carry a "satellite dsRNA" segment called M_1, which is entirely dependent on the L-A virus for its virus coat and replication functions. M_1 encodes the preprokiller toxin, and a protein conferring immunity to the toxin. This immunity protein is probably identical to the protoxin. The toxin is a secreted heterodimer, and ability to secrete an active toxin is denoted by the phenotype designation K^+, while resistance to toxin action is indicated by R^+. Subscripts are used if necessary to differentiate the different toxin and immunity specificities determined by different M dsRNAs. A wild-type strain carrying L-A and M_1 dsRNAs is $K_1^+R_1^+$ in phenotype, while a wild-type lacking M_1 dsRNA is K^-R^-.

The original [KIL-d] (for diploid-dependent) mutants had the $K^-R_1^+$ phenotype. Mating this strain with any other strain produced $K_1^+R_1^+$ diploids. Remarkably, nearly all of the meiotic segregants of these diploids were defective, with a rainbow of abnormal phenotypes including $K^-R_1^+$, $K_1^+R^-$, K^-R^-, and $K_1^+R_1^+$, even though all were derived from a mutant that was K^-R^+ (Wickner, 1976). Mating any of these defective segregants with a wild-type strain lacking the killer trait again produced diploids with the K^+R^+ phenotype (Wickner, 1976). This unusual behavior indicated that [KIL-d] is a nonchromosomal genetic element, since nearly all of the meiotic segregants were abnormal in phenotype. This suspicion was later confirmed by cytoduction experiments (Talloczy et al., 1998, 2000).

The effects of [KIL-d] plainly affect the dsRNA virus system, and so [KIL-d] was proposed to be a mutant of M_1 dsRNA (Wickner, 1976). However, [KIL-d] itself has recently been shown not to be located on the L-A or M_1 dsRNAs, although it is the expression of their information that is affected (Talloczy *et al.*, 1998, 2000). Eliminating M_1 and L-A by growth at high temperature or using the *mak3-1* mutant defective in an N-acetyltransferase required by the virus for its propagation does not result in loss of [KIL-d]. Mating the cured strain with a normal $K_1^+ R_1^+$ produces meiotic segregants with the rainbow of phenotypes indicating the presence of [KIL-d].

The restoration of wild-type phenotype by mating was suspected to be due to acquisition of heterozygosity at the mating-type locus which accompanies mating. However, it was shown that introduction of the the opposite mating-type locus on a plasmid did not restore the phenotype to normal (Talloczy *et al.*, 1998). The defect in the phenotype is not seen when a cDNA clone encoding the killer preprotoxin is introduced into the cells (Talloczy *et al.*, 2000). This shows that the defect does not involve the Kex1 or Kex2 proteins that are needed for processing and secretion of the preprotoxin.

A. The cytoduction

Transfer of cytoplasm (cytoduction) from a killer strain into a [KIL-d] strain does not change its phenotype; if it was $K^- R^+$ it remains $K^- R^+$. Transfer of cytoplasm from a $K^- R^+$ [KIL-d] strain to a normal killer, likewise leaves the phenotype of the recipient that of a normal killer. However, crossing these recipients with another wild-type strain and inducing meiosis produces the rainbow of phenotypes again. This revealing experiment shows that while the nucleus of the cell determines the current phenotype ($K^- R^+$, for example), the [KIL-d] genetic element is carried in the cytoplasm. Again, this confirms that [KIL-d] cells do not have defective dsRNAs.

The [KIL-d] phenomenon is still a genetic puzzle; it is particularly intriguing because its explanation is so obscure. The explanation will surely teach us important lessons about epigenetic phenomena, non-Mendelian genetic elements, and the dsRNA viruses.

XV. ARE TRANSMISSIBLE SPONGIFORM ENCEPHALOPATHIES OF MAMMALS DUE TO PRIONS?

It seems pedantic to challenge the prion hypothesis for the TSEs at this late date, but in reality there remains uncertainty and disagreement in the field on this central point (contrast reviews by Chesebro [1998] and Farquhar *et al.* [1998] with that by Prusiner [1998]). The prion concept arose in the late 1960s in essentially its modern form (Griffith, 1967) as an attempt to explain the extraordinary radiation resistance of the scrapie agent (Alper *et al.*, 1966, 1967). However, Griffith's precient formulation was to have limited impact over the next 15 years, as Gajdusek's "slow virus" view held sway in the field (Gajdusek, 1977).

Ironically, the first identification of PrP and its central role in scrapie was the discovery of the *Sinc* gene (for scrapie incubation period) of mice by Dickinson and co-workers in 1968 (Dickinson *et al.*, 1968). Of course, there was no suggestion that this gene encoded the infectious element itself. Prusiner's partially purified scrapie agent revealed a single main protein band, which he named PrP, and in suggesting that this protein was central to the infectious process, he coined the term "prion" (Bolton *et al.*, 1982; Prusiner, 1982). At that time, the word "prion" did not imply that the protein was the only component of the infectious material. However, cloning of the PrP gene showed that it was a chromosomal gene (Oesch *et al.*, 1985; Chesebro *et al.*, 1985). This ruled out the "slow virus" (Gajdusek, 1977), "reverse translation," and "protein-dependent protein synthesis" (Prusiner, 1982) models and led back to the Griffith model. Efforts to identify a scrapie genome were unsuccessful, in spite of many attempts, as were attempts to find a covalent modification of PrP that was unique to the scrapie form of the protein. Identification of Dickinson's *Sinc* gene as the gene for PrP showed that PrP is important for scrapie propagation (Carlson *et al.*, 1986). This was further demonstrated by showing that making mice transgenic for hamster PrP could overcome the species barrier to transmission from hamsters to mice (Prusiner *et al.*, 1990).

It was demonstrated that PrP is necessary for propagation of scrapie by showing that mice deleted for the PrP gene were immune to the disease and did not even replicate it in their tissues (Bueler *et al.*, 1993).

Why do these experiments not add up to a demonstration that scrapie is a prion disease, due to an abnormal form of PrP? All of the above results say that PrP is necessary or controls the development of scrapie. But none of them show that PrP is sufficient. Two experiments may have been interpreted in that way initially. Mutations in PrP in humans are associated with the development of Creutzfeldt-Jakob disease, CJD (and other clinical variants, Gerstmann-Straussler-Scheinker disease and fatal familial insomnia). Introducing one of these mutations, P101L, into the mouse PrP gene resulted in mice that all spontaneously develop a fatal scrapie-like syndrome with pathology indistinguishable from that of scrapie (Hsiao *et al.*, 1990). However, the brains of these mice are not infectious for normal mice. In contrast, human brain material from patients who have died of CJD due to the same mutation is infectious for normal mice (though at low efficiency). Moreover, single-copy P101L mice are healthy, but are efficiently infected with the same human material and their brains are highly infectious for normal mice (Manson *et al.*, 1999). Overproduction of PrP as a result of a high-copy transgene leads to another fatal syndrome, but again, the brains of these animals are not infectious for normal mice (Westaway *et al.*, 1994).

PrP from scrapie-infected animals is protease-resistant, while that from normal animals is protease-sensitive. The normal form has a largely α-helical conformation, while the scrapie form is higher in β-sheet. Numerous attempts to make PrP synthesized in bacteria or yeast convert *in vitro* into an infectious form have failed. Whether a critical component is missing from these preparations, or the correct conformation has simply not yet been prepared, remains to be determined.

Caughey and co-workers have developed perhaps the best evidence that PrP is sufficient for scrapie (Kocisko *et al.*, 1994, 1995; Bessen *et al.*, 1995). Incubating PrPSc isolated from infected animals with small amounts of radiolabeled PrPC (the normal form) results in conversion of much of the PrPC to protease-resistant form. This reaction has all of the specificity of the infectious process with a species barrier and scrapie strain specificity. So far, this reaction has been self-limited, but if conditions could be found in which it was autocatalytic, one could potentially demonstrate new infectious material.

XVI. CONCLUSIONS

Prions occupy a unique position in the panoply of unusual biological phenomena. Because they are genes composed entirely of protein, their discovery completes the symmetry with nucleic acids initially placed in imbalance by the discoveries by Cech and by Altman that nucleic acids could carry out enzymatic reactions. In comparing prions with other epigenetic phenomena, the demonstration that it is the Ure2p protein (for example), whose overproduction induces the *de novo* appearance of the nonchromosomal gene, is in striking contrast to other forms of homology-dependent inactivation (Table 17.1). In the other systems, it is generally the RNA or the gene in high copy which is inactivating.

How widespread is the prion phenomenon? Are all amyloidoses essentially in the prion family? Do more prions have normal functions? Are there other inherited structures such as the cortical inheritance of *Paramecium*? The expression "inheritance of organelles" is often used with a different meaning to mean simply the passage of structures from mother to daughter cell, without any implication of the "inherited" structure specifying the nature of such structure in the daughter cell or its offspring. We argue that this practice is confusing.

The utility of yeast and fungi in the study of prions has been amply demonstrated in the last few years. While the prion idea originated over 30 years ago from studies of sheep and mice, conclusive proof has only come from the study of yeast prions. While there is substantial evidence for the protein-only model for the transmissible spongiform encephalopathies, definitive proof in these systems has been elusive. The discoveries of many other cellular components affecting prion generation, propagation, and curing has opened new areas that will doubtless be of interest in studies of mammalian systems as well. Finally, the yeast systems are already finding use in discovery of new prions.

NOTE ADDED IN PROOF

Recently, Liebman and coworkers (Derkatch *et al.*, 2001) have found that [PIN+], the non-chromosomal gene necessary for *de novo* induction of [PSI+] (see p. 514), is a prion of the Rnq1 protein. Rnq1p had been shown capable of converting to an aggregated infectious form, but there was no associated phenotype (Sondheimer and Lindquist, 2000). Liebman also showed that the [URE3] prion can promote generation of new [PSI+] elements (Derkatch *et al.*, 2001). Thus, one prion can promote the generation of another. Since all of these prions are based on runs of asparagine or glutamine, it is possible that this phenomenon may be based on direct cross-seeding of one prion by another prions. It has also been found that overexpression of various proteins with asn or gln-rich domains can promote new prion generation without necessarily producing prions themselves (Derkatch *et al.*, 2001; Osherovich and Weissman, 2001).

References

Aigle, M. (1979). Contribution a l'etude de l'heredite non-chromosomique de *Saccharomyces cerevisiae* facteur [URE3] et plasmides hybrides. L'Universite Louis Pasteur de Strasbourg, Strasbourg, France, p. 95.

Aigle, M., and Lacroute, F. (1975). Genetical aspects of [URE3], a non-Mendelian, cytoplasmically inherited mutation in yeast. *Mol. Gen. Genet.* **136**, 327–335.

Alper, T., Cramp, W. A., Haig, D. A., and Clarke, M. C. (1967). Does the agent of scrapie replicate without nucleic acid?. *Nature* **214**, 764–766.

Alper, T., Haig, D. A., and Clarke, M. C. (1966). The exceptionally small size of the scrapie agent. *Biochem. Biophys. Res. Commun.* **22**, 278–284.

Atencio, D., and Yaffe, M. (1992). MAS5, a yeast homolog of DnaJ involved in mitochondrial import. *Mol. Cell. Biol.* **12**, 283–291.

Beck, T., and Hall, M. N. (1999). The TOR signalling pathway controls nuclear localization of nutrient-regulated transcription factors. *Nature* **402**, 689–692.

Becker, J., Walter, W., Yan, W., and Craig, E. A. (1996). Functional interaction of cytosolic hsp70 and a DnaJ-related protein, Ydj1p, in protein translocation in vivo. *Mol. Cell. Biol.* **16**, 4378–4386.

Begueret, J., Turq, B., and Clave, C. (1994). Vegetative incompatibility in filamentous fungi: *het* genes begin to talk. *Trends Genet.* **10**, 441–446.

Beisson, J., and Sonneborn, T. M. (1965). Cytoplasmic inheritance of the organization of the cell cortex in *Paramecium aurelia*. *Proc. Natl. Acad. Sci. USA* **53**, 275–282.

Beisson-Schecroun, J. (1962). Incompatibilte cellulaire et interactions nucleo-cytoplasmiques dans les phenomenes de barrage chez *Podospora anserina*. *Ann. Genet.* **4**, 3–50.

Bell, L. R., Horabin, J. L., Schedl, P., and Cline, T. W. (1991). Positive autoregulation of sex-lethal by alternative splicing maintains the female determined state in *Drosophila*. *Cell* **65**, 229–239.

Bessen, R. A., Kocisko, D. A., Raymond, G. J., Nandan, S., Landsbury, P. T., and Caughey, B. (1995). Non-genetic propagation of strain-specific properties of scrapie prion protein. *Nature* **375**, 698–700.

Bolton, D. C., McKinley, M. P., and Prusiner, S. B. (1982). Identification of a protein that purifies with the scrapie prion. *Science* **218**, 1309–1311.

Bousset, L., Beirhali, H., Janin, J., Melki, R., and Morera, S. (2001). Structure of the globular region of the prion protein Ure2 from the yeast *Saccharomyces cerevisiae*. *Structure* **9**, 39–46.

Bueler, H., Aguzzi, A., Sailer, A., Greiner, R.-A., Autenried, P., Aguet, M., and Weissmann, C. (1993). Mice devoid of PrP are resistant to Scrapie. *Cell* **73**, 1339–1347.

Caplan, A., and Douglas, M. G. (1991). Characterization of YDJ1: A yeast homolog of the bacterial *dnaJ* protein. *J. Cell Biol.* **114**, 609–621.

Cardenas, M. E., Cutler, N. S., Lorenz, M. C., Di Como, C. J., and Heitman, J. (1999). The TOR signaling cascade regulates gene expression in response to nutrients. *Genes Dev.* **13**, 3271–3279.

Carlson, G. A., Kingsbury, D. T., Goodman, P. A., Coleman, S., Marshall, S. T., DeArmond, S., Westaway, D., and Prusiner, S. B. (1986). Linkagae of prion protein and scrapie incubation time genes. *Cell* **46**, 503–511.

Chabry, J., Caughey, B., and Chesebro, B. (1998). Specific inhibition of *in vitro* formation of protease-resistant prion protein by synthetic peptides. *J. Biol. Chem.* **273**, 13203–13207.

Chernoff, Y. O., Derkach, I. L., and Inge-Vechtomov, S. G. (1993). Multicopy SUP35 gene induces de-novo appearance of psi-like factors in the yeast *Saccharomyces cerevisiae*. *Curr. Genet.* **24**, 268–270.

Chernoff, Y. O., Galkin, A. P., Lewitin, E., Chernova, T. A., Newnam, G. P., and Belenkly, S. M. (2000). Evolutionary conservation of prion-forming abilities of the yeast Sup35 protein. *Mol. Microbiol.* **35**, 865–876.

Chernoff, Y. O., Lindquist, S. L., Ono, B.-I., Inge-Vechtomov, S. G., and Liebman, S. W. (1995). Role of the chaperone protein Hsp104 in propagation of the yeast prion-like factor [psi$^+$]. *Science* **268**, 880–884.

Chernoff, Y. O., Newnam, G. P., Kumar, J., Allen, K., and Zink, A. D. (1999). Evidence for a protein mutator in yeast: Role of the Hsp70-related chaperone Ssb in formation, stability and toxicity of the [PSI+] prion. *Mol. Cell. Biol.* **19**, 8103–8112.

Chernoff, Y. O., and Ono, B.-I. (1992). Dosage-dependent modifiers of PSI-dependent omnipotent suppression in yeast. *In* "Protein Synthesis and Targeting in Yeast" (A. J. P. Brown, M. F. Tuite, and J. E. G. McCarthy, eds.), pp. 101–107. Springer-Verlag, Berlin.

Chesebro, B. (1998). BSE and prions: Uncertainties about the agent. *Science* **279**, 42–43.

Chesebro, B., Race, R., Wehrly, K., Nishio, J., Bloom, M., Lechner, D., Bergstrom, S., Robbins, K., Mayer, L., Keith, J. M., Garon, C., and Hasse, A. (1985). Identification of scrapie prion protein-specific mRNA in scrapie-infected brain. *Nature* **315**, 331–333.

Choi, J. H., Lou, W., and Vancura, A. (1998). A novel membrane-bound glutathione S-transferase functions in the statioinary phase of the yeast *Saccharomyces cerevisiae*. *J. Biol. Chem.* **273**, 29915–19922.

Cooper, T. G. (1982). Nitrogen metabolism in *Saccharomyces cerevisiae*. *In* "The Molecular Biology of the Yeast *Saccharomyces*: Metabolism and Gene Expression" (J. N. Strathern, E. W. Jones, and J. R. Broach, eds.), vol. 2, pp. 39–99. Cold Spring Harbor Laboratory Press, Cold Spring Harbor, NY.

Coschigano, P. W., and Magasanik, B. (1991). The URE2 gene product of *Saccharomyces cerevisiae* plays an important role in the cellular response to the nitrogen source and has homology to glutathione S-transferases. *Mol. Cell. Biol.* **11**, 822–832.

Coustou, V., Deleu, C., Saupe, S., and Begueret, J. (1997). The protein product of the *het-s* heterokaryon incompatibility gene of the fungus *Podospora anserina* behaves as a prion analog. *Proc. Natl. Acad. Sci. USA* **94**, 9773–9778.

Coustou, V., Deleu, C., Saupe, S. J., and Begueret, J. (1999). Mutational analysis of the [Het-s] prion analog of *Podospora anserina*: A short N-terminal peptide allows prion propagation. *Genetics* **153**, 1629–1640.

Cox, B. S. (1965). PSI, a cytoplasmic suppressor of super-suppressor in yeast. *Heredity* **20,** 505–521.

Cox, B. S. (1971). A recessive lethal super-suppressor mutation in yeast and other PSI phenomena. *Heredity* **26,** 211–232.

Cox, B. S., Tuite, M. F., and McLaughlin, C. S. (1988). The Psi factor of yeast: A problem in inheritance. *Yeast* **4,** 159–179.

Craig, E. A. (1992). The heat-shock response of *Saccharomyces cerevisiae*. *In* "The Molecular and Cellular Biology of the Yeast *Saccharomyces*" (E. W. Jones, J. R. Pringle, and J. R. Broach, eds.), vol. 2, pp. 501–573. Cold Spring Harbor Laboratory Press, Cold Spring Harbor, NY.

Deleu, C., Clave, C., and Begueret, J. (1993). A single amino acid difference is sufficient to elicit vegetative incompatibility in the fungus *Podospora anserina*. *Genetics* **135,** 45–52.

DePace, A. H., Santoso, A., Hillner, P., and Weissman, J. S. (1998). A critical role for amino-terminal glutamine/asparagine repeats in the formation and propagation of a yeast prion. *Cell* **93,** 1241–1252.

Derkatch, I. L., Bradley, M. E., Hong, J. Y., and Liebman, S. W. (2001). Prions affect the appearance of other prions: the story of *[PIN]*. *Cell* **106,** 171–182.

Derkatch, I. L., Bradley, M. E., and Liebman, S. W. (1998). Overexpression of the SUP45 gene encoding a Sup35p-binding protein inhibits the induction of the de novo appearance of the [PSI+] prion. *Proc. Natl. Acad. Sci. USA* **95,** 2400–2405.

Derkatch, I. L., Bradley, M. E., Masse, S. V., Zadorsky, S. P., Polozkov, G. V., Inge-Vechtomov, S. G., and Liebman, S. W. (2000). Dependence and independence of [PSI(+)] and [PIN(+)]: A two-prion system in yeast? *EMBO J.* **19,** 1942–1952.

Derkatch, I. L., Bradley, M. E., Zhou, P., Chernoff, Y. O., and Liebman, S. W. (1997). Genetic and environmental factors affecting the *de novo* appearance of the *[PSI+]* prion in *Saccharomyces cerevisiae*. *Genetics* **147,** 507–519.

Derkatch, I. L., Chernoff, Y. O., Kushnirov, V. V., Inge-Vechtomov, S. G., and Liebman, S. W. (1996). Genesis and variability of [PSI] prion factors in *Saccharomyces cerevisiae*. *Genetics* **144,** 1375–1386.

Dickinson, A. G., Meikle, V. M. H., and Fraser, H. (1968). Identification of a gene which controls the incubation period of some strains of scrapie in mice. *J. Comp. Pathol* **78,** 293–299.

Doel, S. M., McCready, S. J., Nierras, C. R., and Cox, B. S. (1994). The dominant *PNM2⁻* mutation which eliminates the [PSI] factor of *Saccharomyces cerevisiae* is the result of a missense mutation in the *SUP35* gene. *Genetics* **137,** 659–670.

Eaglestone, S. S., Cox, B. S., and Tuite, M. F. (1999). Translation termination efficiency can be regulated in *Saccharomyces cerevisiae* by environmental stress through a prion-mediated mechanism. *EMBO J.* **18,** 1974–1981.

Eaglestone, S. S., Ruddock, L. W., Cox, B. S., and Tuite, M. F. (2000). Guanidine hydrochloride blocks a critical step in the propagation of the prion-like determinant $[PSI^+]$ of *Saccharomyces cerevisiae*. *Proc. Natl. Acad. Sci. USA* **97,** 240–244.

Edskes, H. K., Gray, V. T., and Wickner, R. B. (1999a). The [URE3] prion is an aggregated form of Ure2p that can be cured by overexpression of Ure2p fragments. *Proc. Natl. Acad. Sci. USA* **96,** 1498–1503.

Edskes, H. K., Hanover, J. A., and Wickner, R. B. (1999b). Mks1p is a regulator of nitrogen catabolism upstream of Ure2p in *Saccharomyces cerevisiae*. *Genetics* **153,** 585–594.

Edskes, H. K., and Wickner, R. B. (2000). A protein required for prion generation: [URE3] induction requires the Ras-regulated Mks1 protein. *Proc. Natl. Acad. Sci. USA* **97,** 6625–6629.

Farquhar, C. F., Somerville, R. A., and Bruce, M. E. (1998). Straining the prion hypothesis. *Nature* **391,** 345–346.

Feller, A., Ramos, F., Peirard, A., and Dubois, E. (1997). Lys80p of *Saccharomyces cerevisiae*, previously proposed as a specific repressor of *LYS* genes, is a pleiotropic regulatory factor identical to Mks1p. *Yeast* **13,** 1337–1346.

Fernandez-Bellot, E., Guillemet, E., Baudin-Baillieu, A., Gaumer, S., Komar, A. A., and Cullin, C. (1999). Characterization of the interaction domains of Ure2p, a prion-like protein of yeast. *Biochem. J.* **338,** 403–407.

Gajdusek, D. C. (1977). Unconventional viruses and the origin and disappearance of Kuru. *Science* **197,** 943–960.

Glover, J. R., Kowal, A. S., Shirmer, E. C., Patino, M. M., Liu, J.-J., and Lindquist, S. (1997). Self-seeded fibers formed by Sup35, the protein determinant of [PSI+], a heritable prion-like factor of S. cerevisiae. *Cell* **89,** 811–819.

Griffith, J. S. (1967). Self-replication and scrapie. *Nature* **215,** 1043–1044.

Grimes, G. W., and Aufderheide, K. J. (1991). "Cellular Aspects of Pattern Formation: The Problem of Assembly." Karger, Basel.

Grishok, A., Pasquinelli, A. E., Conte, D., Li, N., Parrish, S., Ha, I., Baillie, D. L., Fire, A., Ruvkun, D., and Mello, C. C. (2001). Genes and mechanisms related to RNA interference requlate expression of the small temporal RNAs that control C. *elegans* developmental timing. *Cell* **106,** 23–34.

Guerrier-Takada, C., Gardiner, K., Marsh, T., Pace, N., and Altman, S. (1983). The RNA moiety of ribonuclease P is the catalytic subunit of the enzyme. *Cell* **35,** 849–857.

Hagemann, A. T., and Selker, E. U. (1996). Control and function of DNA methylation in *Neurospora crassa*. *In* "Epigenetic Mechanisms of Gene Regulation" (V. E. A. Russo, R. A. Martienssen, and A. D. Riggs, eds.), pp. 335–344. Cold Spring Harbor Laboratory Press, Cold Spring Harbor, NY.

Hardwick, J. S., Kuruvilla, F. G., Tong, J. K., Shamji, A. F., and Schreiber, S. L. (1999). Rapamycin-modulated transcription defines the subset of nutrient-sensitive signaling pathways directly controlled by the tor proteins [In Process Citation]. *Proc. Natl. Acad. Sci. USA* **96,** 14866–14870.

Hsiao, K. K., Scott, M., Foster, D., Groth, D. F., DeArmond, S. J., and Prusiner, S. B. (1990). Spontaneous neurodegeneration in transgenic mice with mutant prion protein of Gerstmann-Straussler syndrome. *Science* **250,** 1587–1590.

Jung, G., Jones, G., Wegrzyn, R. D., and Masison, D. C. (2000). A role for cytosolic Hsp70 in yeast [PSI+] prion propagation and [PSI+] as a cellular stress. *Genetics* **156,** 559–570.

Jung, G., and Masison, D. C. (2001). Guanidine hydrochloride inhibits Hsp104 activity *in vivo*: A possible explanation for its effect in curing yeast prions. *Curr. Microbiol.* **43,** 7–10.

King, C.-Y., Tittmann, P., Gross, H., Gebert, R., Aebi, M., and Wuthrich, K. (1997). Prion-inducing domain 2-114 of yeast Sup35 protein transforms in vitro into amyloid-like filaments. *Proc. Natl. Acad. Sci. USA* **94,** 6618–6622.

Kochneva-Pervukhova, N. V., Poznyakovski, A. I., Smirnov, V. N., and Ter-Avanesyan, M. D. (1998). C-terminal truncation of the Sup35 protein increases the frequency of de novo gneration of a prion-based [PSI+] determinant in Saccharmyces cerevisiae. *Curr. Genet.* **34,** 146–151.

Kocisko, D. A., Come, J. H., Priola, S. A., Chesebro, B., Raymond, G. J., Lansbury, P. T., and Caughey, B. (1994). Cell-free formation of protease-resistant prion protein. *Nature* **370,** 471–474.

Kocisko, D. A., Priola, S. A., Raymond, G. J., Chesebro, B., Landsbury, P. T., and Caughey, B. (1995). Species specificity in the cell-free conversion of prion protein to protease-resistant forms: a model for the scrapie species barrier. *Proc. Natl. Acad. Sci. USA* **92,** 3923–3927.

Krobitsch, S., and Lindquist, S. (2000). Aggregation of huntingtin in yeast varies with the length of the polyglutamine expansion and the expression of chaperone proteins. *Proc. Natl. Acad. Sci. USA* **97,** 1589–1594.

Kruger, K., Grabowski, P. J., Zaug, A. J., Sands, J., Gottschling, D. E., and Cech, T. R. (1982). Self-splicing RNA: Autoexcision and autocyclization of the ribosomal RNA intervening sequence of *Tetrahymena*. *Cell* **31,** 147–157.

Kushnirov, V. V., Kochneva-Pervukhova, N. V., Cechenova, M. B., Frolova, N. S., and Ter-Avanesyan, M. D. (2000). Prion properties of the Sup35 protein of yeast *Pichia methanolica*. *EMBO J.* **19,** 324–331.

Lacroute, F. (1971). Non-Mendelian mutation allowing ureidosuccinic acid uptake in yeast. *J. Bacteriol.* **106,** 519–522.

Leibowitz, M. J., and Wickner, R. B. (1978). Pet18: A chromosomal gene required for cell growth and for the maintenance of mitochondrial DNA and the killer plasmid of yeast. *Mol. Gen. Genet.* **165,** 115–121.

Li, L., and Lindquist, S. (2000). Creating a protein-based element of inheritance. *Science* **287,** 661–664.

Liebman, S. W., and Sherman, F. (1979). Extrachromosomal [PSI$^+$] determinant suppresses nonsense mutations in yeast. *J. Bacteriol.* **139,** 1068–1071.

Liebman, S. W., Stewart, J. W., and Sherman, F. (1975). Serine substitutions caused by an ochre suppressor in yeast. *J. Mol. Biol.* **94,** 595–610.

Liu, J. J., and Lindquist, S. (1999). Oligopeptide-repeat expansions modulate "protein-only" inheritance in yeast. *Nature* **400,** 573–576.

Lu, Z., and Cyr, D. M. (1998). Protein folding activity of Hsp70 is modified differentially by the Hsp40 co-chaperones Sis1 and Ydj1. *J. Biol. Chem.* **273,** 27824–27830.

Lund, P. M., and Cox, B. S. (1981). Reversion analysis of [psi−] mutations in *Saccharomyces cerevisiae*. *Genet. Res.* **37,** 173–182.

Maddelein, M.-L., and Wickner, R. B. (1999). Two Prion-Inducing Regions of Ure2p are Non-Overlapping. *Mol. Cell. Biol.* **19,** 4516–4524.

Magasanik, B. (1992). Regulation of nitrogen utilization. *In* "The Molecular and Cellular Biology of the Yeast *Saccharomyces*" (E. W. Jones, J. R. Pringle, and J. R. Broach, eds.), vol. 2, pp. 283–317. Cold Spring Harbor Laboratory Press, Cold Spring Harbor, NY.

Manson, J. C., Jamieson, E., Baybutt, H., Tuzi, N. L., Barron, R., McConnell, I., Somerville, R., Ironside, J., Will, R., Sy, M.-S., Melton, D. W., Hope, J., and Bostock, C. (1999). A single amino acid alteration (101L) introduced into murine PrP dramatically alters incubation time of transmissible spongiform encephalopathy. *EMBO J.* **18,** 6855–6864.

Masison, D. C., Maddelein, M.-L., and Wickner, R. B. (1997). The prion model for [URE3] of yeast: Spontaneous generation and requirements for propagation. *Proc. Natl. Acad. Sci. USA* **94,** 12503–12508.

Masison, D. C., and Wickner, R. B. (1995). Prion-inducing domain of yeast Ure2p and protease resistance of Ure2p in prion-containing cells. *Science* **270,** 93–95.

Matsuura, A., and Anraku, Y. (1993). Characterization of the *MKS1* gene, a new negative regulator of the ras-cyclic AMP pathway in *Saccharomyces cerevisiae*. *Mol. Gen. Genet.* **238,** 6–16.

Matzke, M., Matzke, A. J., and Kooter, J. M. (2001). RNA guiding gene silencing. *Science* **293,** 1080–1083.

Meacham, G. C., Browne, B. L., Zhang, W., Kellermayer, R., Bedwell, D. M., and Cyr, D. M. (1999). Mutations in the yeast Hsp40 chaperone protein Ydj1 cause defects in Axl1 biogenesis and pro-a-factor processing. *J. Biol. Chem.* **274,** 34396–34402.

Moriyama, H., Edskes, H. K., and Wickner, R. B. (2000). [URE3] prion propagation in *Saccharomyces cerevisiae*: Requirement for chaperone Hsp104 and curing by overexpressed chaperone Ydj1p. *Mol. Cell. Biol.* **20,** 8916–8922.

Nelson, R. J., Ziegelhoffer, T., Nicolet, C., Werner-Washburne, M., and Craig, E. A. (1992). The translation machinery and 70 kd heat shock protein cooperate in protein synthesis. *Cell* **71,** 97–105.

Newnam, G. P. R. D., Wegrzyn, S. L. L., and Chernoff, Y. O. (1999). Antagonistic interactions between yeast chaperones Hsp104 and Hsp70 in prion curing. *Mol. Cell. Biol.* **19,** 1325–1333.

Novick, A., and Weiner, M. (1957). Enzyme induction as an all-or-none phenomenon. *Proc. Natl. Acad. Sci. USA* **43,** 553–566.

Oesch, B., Westaway, D., Walchli, M., McKinley, M. P., Kent, S. B., Aebersold, R., Barry, R. A., Tempst, P., Templow, D. B., Hood, L. E., Prusiner, S. B., and Weissmann, C. (1985). A cellular gene encodes scrapie PrP 27-30 protein. *Cell* **40,** 735–746.

Osherovich, L. Z., and Weissman, J. S. (2001). Multiple Gln/Asn-rich prion domains confer susceptibility to induction of the yeast *[PSI+]* prion. *Cell* **106,** 183–194.

Palmer, E., Wilhelm, J., and Sherman, F. (1979). Phenotypic suppression of nonsense mutants in yeast by aminoglycoside antibiotics. *Nature* **277,** 148–150.

Parsell, D. A., Kowal, A. S., Singer, M. A., and Lindquist, S. (1994). Protein disaggregation mediated by heat-shock protein Hsp104. *Nature* **372,** 475–478.

Patino, M. M., Liu, J.-J., Glover, J. R., and Lindquist, S. (1996). Support for the prion hypothesis for inheritance of a phenotypic trait in yeast. *Science* **273,** 622–626.

Paushkin, S. V., Kushnirov, V. V., Smirnov, V. N., and Ter-Avanesyan, M. D. (1996). Propagation of the yeast prion-like [*psi*$^+$] determinant is mediated by oligomerization of the *SUP35*–encoded polypeptide chain release factor. *EMBO J.* **15,** 3127–3134.

Paushkin, S. V., Kushnirov, V. V., Smirnov, V. N., and Ter-Avanesyan, M. D. (1997). *In vitro* propagation of the prion-like state of yeast Sup35 protein. *Science* **277,** 381–383.

Perrett, S., Freeman, S. J., Butler, P. J. G., and Fersht, A. R. (1999). Equilibrium folding properties of the yeast prion protein determinant Ure2. *J. Mol. Biol.* **290,** 331–345.

Pfund, C., Lopez-Hoyo, N., Ziegelhoffer, T., Schilke, B. A., Lopez-Buesa, P., Walter, W. A., Wiedmann, M., and Craig, E. A. (1998). The molecular chaperone Ssb from *Saccharomyces cerevisiae* is a component of the ribosome-nascent chain complex. *EMBO J.* **17,** 3981–3989.

Prusiner, S. B. (1982). Novel proteinaceous infectious particles cause scrapie. *Science* **216,** 136–144.

Prusiner, S. B. (1998). Prions. *Proc. Natl. Acad. Sci. USA* **95,** 13363–13383.

Prusiner, S. B. (ed.) (1999). "Prion Biology and Diseases." Cold Spring Harbor Laboratory Press, Cold Spring Harbor, NY.

Prusiner, S. B., Scott, M., Foster, D., Pan, K.-M., Groth, D., Mirenda, C., Torchia, M., Yang, S.-L., Serban, D., Carlson, G. A., Hoppe, P. C., Westaway, D., and DeArmond, S. J. (1990). Transgenic studies implicate interactions between homologous PrP isoforms in scrapie prion replication. *Cell* **63,** 673–686.

Rai, R., Genbauffe, F., Lea, H. Z., and Cooper, T. G. (1987). Transcriptional regulation of the *DAL5* gene in *Saccharomyces cerevisiae*. *J. Bacteriol.* **169,** 3521–3524.

Rizet, G. (1952). Les phenomenes de barrage chez *Podospora anserina*: Analyse genetique des barrages entre les souches s et S. *Rev. Cytol. Biol. Veg.* **13,** 51–92.

Santoso, A., Chien, P., Osherovich, L. Z., and Weissman, J. S. (2000). Molecular basis of a yeast prion species barrier. *Cell* **100,** 277–288.

Saupe, S. J. (2000). Molecular genetics of heterokaryon incompatibility in filamentous ascomycetes. *Microbiol. Mol. Biol. Rev.* **64,** 489–502.

Schlumpberger, M., Willie, H., Baldwin, M. A., Butler, D. A., Herskowitz, I., and Prusiner, S. B. (2000). The prion domain of yeast Ure2p induces autocatalytic formation of amyloid fibers by a recombinant fusion protein. *Protein. Sci.* **9,** 440–451.

Singh, A. C., Helms, C., and Sherman, F. (1979). Mutation of the non-Mendelian suppressor [PSI] in yeast by hypertonic media. *Proc. Natl. Acad. Sci. USA* **76,** 1952–1956.

Sipe, J. D. (1992). Amyloidosis. *Annu. Rev. Biochem.* **61,** 947–975.

Sondheimer, N., and Lindquist, S. (2000). Rnq1: An epigenetic modifier of protein function in yeast. *Mol. Cell* **5,** 163–172.

Sparrer, H. E., Santoso, A., Szoka, F. C., and Weissman, J. S. (2000). Evidence for the prion hypothesis: Induction of the yeast [PSI+] factor by in vitro-converted Sup35 protein. *Science* **289,** 595–599.

Speransky, V., Taylor, K. L., Edskes, H. K., Wickner, R. B., and Steven, A. (2001). Prion filament networks in [URE3] cells of *Saccharomyces cerevisiae*. *J. Cell Biol.* **153,** 1327–1335.

Stansfield, I., Jones, K. M., Kushnirov, V. V., Dagkesamanskaya, A. R., Poznyakovski, A. I., Paushkin, S. V., Nierras, C. R., Cox, B. S., Ter-Avanesyan, M. D., and Tuite, M. F. (1995). The products of the *SUP45* (eRF1) and *SUP35* genes interact to mediate translation termination in *Saccharomyces cerevisiae*. *EMBO J.* **14,** 4365–4373.

Stone, D. E., and Craig, E. A. (1990). Self-regulation of 70 kilodalton heat shock proteins in *Saccharomyces cerevisiae*. *Mol. Cell. Biol.* **10,** 1622–1632.

Talloczy, Z., Mazar, R., Georgopoulos, D. E., Ramos, F., and Leibowitz, M. J. (2000). The [KIL-d] element specifically regulates viral gene expression in yeast. *Genetics* **155,** 601–609.

Talloczy, Z., Menon, S., Neigeborn, L., and Leibowitz, M. J. (1998). The [*KIL-d*] cytoplasmid genetic element of yeast results in epigenetic regulation of viral M double-stranded RNA gene expression. *Genetics* **150,** 21–30.

Tamm, I., and Eggers, H. J. (1963). Specific inhibition of replication of animal viruses. *Science* **142,** 24–33.

Taylor, K. L., Cheng, N., Williams, R. W., Steven, A. C., and Wickner, R. B. (1999). Prion domain initiation of amyloid formation *in vitro* from native Ure2p. *Science* **283,** 1339–1343.

TerAvanesyan, A., Dagkesamanskaya, A. R., Kushnirov, V. V., and Smirnov, V. N. (1994). The *SUP35* omnipotent suppressor gene is involved in the maintenance of the non-Mendelian determinant [psi+] in the yeast *Saccharomyces cerevisiae*. *Genetics* **137,** 671–676.

True, H. L., and Lindquist, S. L. (2000). A yeast prion provides a mechanism for genetic variation and phenotypic diversity. *Nature* **407,** 477–483.

Tuite, M. F., Lund, P. M., Futcher, A. B., Dobson, M. J., Cox, B. S., and McLaughlin, C. S. (1982). Relationship of the [psi] factor with other plasmids of *Saccharomyces cerevisiae*. *Plasmid* **8,** 103–111.

Tuite, M. F., Mundy, C. R., and Cox, B. S. (1981). Agents that cause a high frequency of genetic change from [*psi+*] to [*psi−*] in *Saccharomyces cerevisiae*. *Genetics* **98,** 691–711.

Turcq, B., Deleu, C., Denayrolles, M., and Begueret, J. (1991). Two allelic genes responsible for vegetative incompatibility in the fungus *Podospora anserina* are not essential for cell viability. *Mol. Gen. Genet.* **288,** 265–269.

Turoscy, V., and Cooper, T. G. (1987). Ureidosuccinate is transported by the allantoate transport system in *Saccharomyces cerevisiae*. *J. Bacteriol.* **169,** 2598–2600.

Umland, T. C., Taylor, K. L., Rhee, S., Wickner, R. B., and Davies, D. R. (2001). The crystal structure of the nitrogen catabolite regulatory fragment of the yeast prion protein Ure2p. *Proc. Natl. Acad. Sci. USA* **98,** 1459–1464.

Waldron, C., Cox, B. S., Wills, N., Gesteland, R. F., Piper, P. W., Colby, D., and Guthrie, C. (1981). Yeast ochre suppressor SUQ5-ol is an altered tRNA[ser]UCA. *Nucleic Acids Res.* **9,** 3077–3088.

Westaway, D., DeArmond, S. J., Cayetano-Canlas, J., Groth, D., Foster, D., Yang, S.-L., Torchia, M., Carlson, G. A., and Prusiner, S. B. (1994). Degeneration of skeletal muscle, peripheral nerves, and the central nervous system in transgenic mice overexpressing wild-type prion proteins. *Cell* **76,** 117–129.

Wickner, R. B. (1976). Mutants of the killer plasmid of *Saccharomyces cerevisiae* dependent on chromosomal diploidy for expression and maintenance. *Genetics* **82,** 273–285.

Wickner, R. B. (1994). Evidence for a prion analog in *S. cerevisiae*: The [URE3] non-Mendelian genetic element as an altered *URE2* protein. *Science* **264,** 566–569.

Wickner, R. B. (1996). Viruses of yeasts, fungi and parasitic microorganisms. *In* "Fields Virology" (B. N. Fields, D. M. Knipe, and P. M. Howley, eds.), vol. 1., pp. 557–585. Raven, New York.

Zhouravleva, G., Frolova, L., LeGoff, X., LeGuellec, R., Inge-Vectomov, S., Kisselev, L., and Philippe, M. (1995). Termination of translation in eukaryotes is governed by two interacting polypeptide chain release factors, eRF1 and eRF3. *EMBO J.* **14,** 4065–4072.

Index

A

abdominal-A gene, transvection, 410–411
abdominal-B gene, transvection, 410–411
Aggregation, self-propagating, Sup35p, 505
albino gene, *Neurospora* quelling, 280
Allele-specific imprinting
  maize endosperm, 190–193
  seed development, 205–206
Amyloid
  Sup35p prion domain *in vitro*, 506
  Ure2p, [URE3]
    amyloid formation, 497–498
    β-sheet structure, 500
    filament evidence, 498–499
    GST resemblance of nitrogen-regulation
      domain, 500–501
Animals, TEPRV and ERV comparison, 266
Antisense RNA
  bacteriophages, 372–374
  binding
    kinetics, 380–381
    mutation effects, 379–380
    pathways and topology, 381–383
  biological roles
    experimental determinations, 386–387
    half-life, 384–385
    protein role, 386
    structure role, 384
    targets, 385–386
  chromosomally encoded systems
    DsrA, 376
    MicF, 375
    other systems, 377–378
    overview, 374–375
    OxyS, 375–376
    RNAIII, 376–377
  gene regulation, 362–363
  plasmids
    ColE1-related plasmids, 369–370
    conjugation, 371–372

IncB–IncIα-related plasmids, 368–369
    natural control systems, 363–366
    overview, 366–367
    pIP501-related plasmids, 370
    postsegregational killing systems,
      370–371
    pT181-related plasmids, 370
    R1-related plasmid replication control, 368
  target RNA structure, 378–379
  transposons, 372
  Tsix, 35–36
Apomixis, genomic imprinting, 207–208
Arabidopsis
  endosperm, gene dosage and imprinting,
    180–181
  gene-specific imprinting, MEDEA locus,
    193–197
  genome-wide paternal imprinting
    chromatin structure, 185–186
    compartmentalization, 187–188
    DNA methylation, 186–187
    imprint establishment, 188–189
    lack of paternal activity, 182–184
    paternal gene expression, 184
  qde-2 homolog, 289
Asparagine, regions in Ure2p and Sup35p,
  509–510
Autosomal genes, imprinting in mouse embryo,
  176–177

B

b1, paramutation
  models, 227
  overview, 218–219
  sequence features, 225–226
  sequence identification, 221–225
Bacterial plasmids, antisense RNA
  ColE1-related plasmids, 369–370
  conjugation, 371–372
  IncB–IncIα-related plasmids, 368–369

527